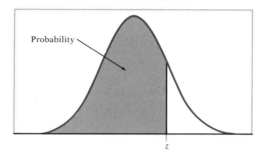

Probability

Table entry is probability at or below z.

Standard normal probabilities

z	.00	.01	.02	.03	.04	.05	.06	.07	.08	.09
0.0	.5000	.5040	.5080	.5120	.5160	.5199	.5239	.5279	.5319	.5359
0.1	.5398	.5438	.5478	.5517	.5557	.5596	.5636	.5675	.5714	.5753
0.2	.5793	.5832	.5871	.5910	.5948	.5987	.6026	.6064	.6103	.6141
0.3	.6179	.6217	.6255	.6293	.6331	.6368	.6406	.6443	.6480	.6517
0.4	.6554	.6591	.6628	.6664	.6700	.6736	.6772	.6808	.6844	.6879
0.5	.6915	.6950	.6985	.7019	.7054	.7088	.7123	.7157	.7190	.7224
0.6	.7257	.7291	.7324	.7357	.7389	.7422	.7454	.7486	.7517	.7549
0.7	.7580	.7611	.7642	.7673	.7704	.7734	.7764	.7794	.7823	.7852
0.8	.7881	.7910	.7939	.7967	.7995	.8023	.8051	.8078	.8106	.8133
0.9	.8159	.8186	.8212	.8238	.8264	.8289	.8315	.8340	.8365	.8389
1.0	.8413	.8438	.8461	.8485	.8508	.8531	.8554	.8577	.8599	.8621
1.1	.8643	.8665	.8686	.8708	.8729	.8749	.8770	.8790	.8810	.8830
1.2	.8849	.8869	.8888	.8907	.8925	.8944	.8962	.8980	.8997	.9015
1.3	.9032	.9049	.9066	.9082	.9099	.9115	.9131	.9147	.9162	.9177
1.4	.9192	.9207	.9222	.9236	.9251	.9265	.9279	.9292	.9306	.9319
1.5	.9332	.9345	.9357	.9370	.9382	.9394	.9406	.9418	.9429	.9441
1.6	.9452	.9463	.9474	.9484	.9495	.9505	.9515	.9525	.9535	.9545
1.7	.9554	.9564	.9573	.9582	.9591	.9599	.9608	.9616	.9625	.9633
1.8	.9641	.9649	.9656	.9664	.9671	.9678	.9686	.9693	.9699	.9706
1.9	.9713	.9719	.9726	.9732	.9738	.9744	.9750	.9756	.9761	.9767
2.0	.9772	.9778	.9783	.9788	.9793	.9798	.9803	.9808	.9812	.9817
2.1	.9821	.9826	.9830	.9834	.9838	.9842	.9846	.9850	.9854	.9857
2.2	.9861	.9864	.9868	.9871	.9875	.9878	.9881	.9884	.9887	.9890
2.3	.9893	.9896	.9898	.9901	.9904	.9906	.9909	.9911	.9913	.9916
2.4	.9918	.9920	.9922	.9925	.9927	.9929	.9931	.9932	.9934	.9936
2.5	.9938	.9940	.9941	.9943	.9945	.9946	.9948	.9949	.9951	.9952
2.6	.9953	.9955	.9956	.9957	.9959	.9960	.9961	.9962	.9963	.9964
2.7	.9965	.9966	.9967	.9968	.9969	.9970	.9971	.9972	.9973	.9974
2.8	.9974	.9975	.9976	.9977	.9977	.9978	.9979	.9979	.9980	.9981
2.9	.9981	.9982	.9982	.9983	.9984	.9984	.9985	.9985	.9986	.9986
3.0	.9987	.9987	.9987	.9988	.9988	.9989	.9989	.9989	.9990	.9990
3.1	.9990	.9991	.9991	.9991	.9992	.9992	.9992	.9992	.9993	.9993
3.2	.9993	.9993	.9994	.9994	.9994	.9994	.9994	.9995	.9995	.9995
3.3	.9995	.9995	.9995	.9996	.9996	.9996	.9996	.9996	.9996	.9997
3.4	.9997	.9997	.9997	.9997	.9997	.9997	.9997	.9997	.9997	.9998

Introduction
to the Practice
of Statistics

Introduction to the Practice of Statistics

David S. Moore
George P. McCabe

Purdue University

W. H. FREEMAN AND COMPANY
New York   Oxford

Cover art by Roy Wiemann

MINITAB is a registered trademark of Minitab, Inc.
SAS is a registered trademark of SAS Institute, Inc.

Library of Congress Cataloging in Publication Data

Moore, David S.
    Introduction to the practice of statistics.

    Includes bibliographies and index.
    1. Mathematical statistics.   I. McCabe, George P.
II. Title.
QA276.12.M65   1989   519.5   88-33566
ISBN 0-7167-1989-4

Printed in the United States of America

1 2 3 4 5 6 7 8 9 0 RRD 7 6 5 4 3 2 1 0 8 9

Contents

Starred Sections are optional.

Preface

Introduction to the Practice of Satistics is an elementary but serious introduction to modern statistics for general college audiences. This book is serious because our aim is to help readers think about data and use statistical methods with understanding. It is elementary in the level of mathematics required and in the statistical procedures presented. Students need only a working knowledge of algebra; that is, they must be able to read and use formulas without a detailed explanation of each step.

Statistics is interesting and useful because it is a means of using data to gain insight into real problems. As the continuing revolution in computing relieves the burden of calculating and graphing, an emphasis on statistical concepts and on insight from data becomes both more important and more practical. We have seen many statistical mistakes, but few that involved simply getting a calculation wrong. We therefore ask students to think about the background of the data, the design of the study that produced the data, the possible effect of outlying observations on their conclusions, and the reasoning that lies behind standard methods of inference. Users of statistics who form these habits from the beginning are well prepared to learn and use more advanced methods.

The title of the book expresses our intent to introduce readers to statistics as it is used in practice. Statistics in practice is concerned with gaining understanding from data; it is focused on problem-solving rather than on methods that may be useful in specific settings. A text cannot fully imitate practice, because it must teach specific methods in a logical order and must use data that are not the reader's own. Nonetheless, our interest and experience in applying statistics have influenced the nature of this book in several ways.

Focus on Statistical Reasoning and Data   We share the emerging consensus among statisticians that statistical education should focus on data and on statistical reasoning rather than on either the presentation of as many methods as possible or the mathematical theory of inference. Understanding statistical reasoning should be the most important objective of any reader.

We attempt to present statistics as a coherent discipline with important modes of thought that recur in many specific settings. The first three chapters, for example, concern the art of orgainizing and exploring data. We hope that readers will draw from these chapters some strategies for understanding data and not just a kit of useful tools. Later chapters similarly attempt to make clear the fundamental modes of thought in designs for data production and in the probabilistic reasoning of formal inference.

An emphasis on data accompanies the emphasis on conceptual understanding. Not all numbers are data. The number 10.3 alone is meaningless; it acquires meaning when we are told that it is the birth weight of a child in pounds or the percent of teenagers who are unemployed. Because context makes numbers meaningful, our examples and exercises are presented in the context of real-world problems. We often comment on issues of statistical practice raised by particular examples. The data presented in examples and in exercises are mostly real and, even when not, are based on real problems. Many of these problems are drawn from those brought to the statistical consulting service at Purdue University by students and faculty from many disciplines. We hope that the presence of background information, even in exercises intended for routine drill, will encourage readers to consider the meaning of their calculations as well as the calculations themselves.

Computers and Statistical Calculations   Statistical calculations are in practice performed by software packages on a computer. Many instructors make use of statistical software in the first course. We use either MINITAB or SAS in our own teaching, the choice depending on the needs of the students; many other packages are equally satisfactory. We have included some topics that reflect the dominance of software in practice, such as the interpretation of normal quantile plots and an explanation of the two-sample t statistic with approximate degrees of freedom. Some exercises require graphs and calculations that are tedious without a computer, while others present computer output as a basis for further work. But we have been careful to make the book easily usable by students without access to computing facilities. A scientific calculator that calculates means and standard deviations from keyed-in data is very helpful, but even a basic calculator is adequate.

Judgment and Statistics   Statistics in practice requires judgment. It is easy to list the mathematical assumptions that justify use of a particular procedure, but not always easy to decide when the procedure can be used in practice. Because judgment is developed by experience, an introductory course should present firm guidelines and not make unreasonable demands on the judgment of students. We have given guidelines—for example, on using the t procedures for means and avoiding the F procedures for variances—that we follow ourselves. Although we have cited the literature on which our recommendations rest, we recognize that other statisticians may follow different guidelines. We prefer to give imperfect rules rather than avoid the issue of when procedures are in fact useful in analyzing real problems. Similarly, some exercises require the use of judgment in addition to right-or-wrong calculations and conclusions. Both students and teachers should recognize that not every part of every exercise has a single correct answer. We enjoy and encourage classroom discussion of questions of interpretation. But, as is appropriate in a first course, most exercises are straightforward even at the cost of some oversimplification.

Teaching Experiences with Student Groups. In writing this book, we drew on our experiences with two groups of students. In teaching general undergraduates from a variety of disciplines, we cover the first 10 chapters omitting all but a few of the starred sections. This results in a modern introduction to basic statistics that is quite standard in content, though with some reordering of material and with a strong emphasis on data and reasoning. The Annenberg/Corporation for Public Broadcasting telecourse *Against All Odds: Inside Statistics* uses the text in this way. Our second group of students consists of advanced undergraduates and beginning graduates students in such fields as education, social sciences, and health sciences. These students are more scientifically sophisticated than general undergraduates, though not more mathematically advanced. In teaching this second group, we cover nearly the entire text except for Sections 5.4, 10.2, and 11.2. The starred sections therefore help distinguish between two somewhat different courses that can be taught from this book. Of course, any starred section can be omitted without impeding an understanding of later chapters.

Supplements Several supplements are available to adopters of *Introduction to the Practice of Statistics*. A combined Instructor's Manual and solutions Manual, free to all adopters, contains the following: overviews and teaching suggestions for each chapter, sample examinations with answers, tips for using the video programs from *Against All Odds: Inside Statistics* in a nontelecourse environment, and answers and worked-out solutions to all the exercises. A MINITAB computer supplement is available for sale to students using our text and any version of the MINITAB statistical software. This guide links the textbook and MINITAB by providing a selection of worked-out examples, data sets, problem-solving strategies, and exercises from the text in MINITAB format. Individuals or institutions may purchase some or all of the 26 television programs from *Against All Odds* on video tape from Annenberg/CPB.

Acknowledgments We are grateful to colleagues who commented on the manuscript and to students who studied from it. In particular, we would like to thank the following:

Rudolf J. Freund, Texas A & M University
Bernard Harris, University of Wisconsin, Madison
David Hildebrand, University of Pennsylvania
Lambert Koopmans, University of New Mexico

David R. Lund, University of Wisconsin, Eau Claire
Donald E. Ramirez, University of Virginia
Charles W. Sinclair, Portland State University
Jessica Utts, University of California, Davis
Robert Wardrop, University of Wisconsin, Madison
Anne Watkins, Los Angeles Pierce College
Jeffrey Witmer, Oberlin College
Mary Sue Younger, University of Tennessee, Knoxville
Douglas Zahn, Florida State University

Most of all we are grateful to the many people in varied disciplines and occupations with whom we have worked to gain understanding from data. They provided both material for this book and the experience that enabled us to write it. Perhaps even more important, working with people from many fields has constantly reminded us of the importance of statistical fundamentals in an age when computer routines and professional advice quickly handle detailed questions. If the publisher would allow it, we would call this book ''What You Should Know Before You Talk to a Statistician.'' We hope that users and potential users of statistics will find it helpful.

David S. Moore

George P. McCabe

The Telecourse
Against All Odds

Against All Odds: Inside Statistics is a 26-program telecourse on statistics and its applications sponsored by the Annenberg/Corporation for Public Broadcasting (CPB) Project. This telecourse, developed by David S. Moore, offers a visual introduction to modern statistics that closely parallels *Introduction to the Practice of Statistics*. In addition to broadcast of the series by public television stations, the programs are inexpensively available on videotape to individuals and institutions. Instructors who do not follow the telecourse format may find excerpts from this unique video series valuable as supplements to classroom instruction. More information and demonstration videotapes can be obtained by telephoning 1-800-LEARNER.

Two supplements are available that link the television programs to this text. A Telecourse Instructor's Manual contains an overview of the material in each video program and provides detailed advice on implementing a telecourse. The Telecourse Study Guide, available for sale to students, guides students in their study of the course material. Each unit in the study guide corresponds to a video program, and provides an overview of the content, learning objectives, assigned reading and exercises from this text, and self-test questions with fully worked solutions. The Study Guide also contains sample examinations with worked solutions. Instructors who are not using the video material may find the large number of additional exercises in the Study Guide useful.

What Is Statistics?

Statistics is the science of collecting, organizing, and interpreting numerical facts, which we call *data*. We are bombarded by data in our everyday life. Most of us associate "statistics" with the bits of data that appear in news reports: baseball batting averages, imported car sales, the latest poll of the president's popularity , and the average high temperature for today's date. Advertisements often claim that data show the superiority of the advertiser's product. All sides in public debates about economics, education, and social policy argue from data. Yet the usefulness of statistics goes far beyond these everyday examples.

The study and collection of data are important in the work of many professions, so that training in the science of statistics is valuable preparation for a variety of careers. Each month, for example, government statistical offices release the latest numerical information on unemployment and inflation. Economists and financial advisors as well as policy makers in government and business study these data in order to make informed decisions. Doctors must understand the origin and trustworthiness of the data that appear in medical journals if they are to offer their patients the most effective treatment. Politicians rely on data from polls of public opinion. Business decisions are based on market research data that reveal consumer tastes. Farmers study data from field trials of new crop varieties. Engineers gather data on the quality and reliability of manufactured products. Most areas of academic study make use of numbers, and therefore also make use of the methods of statistics.

We can no more escape data than we can avoid the use of words. Just as words on a page are meaningless to the illiterate or confusing to the partially educated, so data do not interpret themselves but must be read with understanding. Just as a writer can arrange words into convincing arguments or incoherent nonsense, so data can be compelling, misleading, or simply irrelevant. Numerical literacy, the ability to follow and understand numerical arguments, is important for everyone. The ability to express yourself numerically, to be an author rather than just a reader, is a vital skill in many professions and areas of study. The study of statistics is therefore essential to a sound education. We must learn how to read data, critically and with comprehension; we must learn how to produce data that provide clear answers to important questions; and we must learn sound methods for drawing trustworthy conclusions based on data.

Historically, the ideas and methods of statistics developed gradually as society grew interested in collecting and using data for a variety of applications. The earliest origins of statistics lie in the desire of rulers to count the number of inhabitants or measure the value of taxable land in their domains. As the physical sciences developed in the seventeenth and eighteenth centuries, the importance of careful measurements of weights, distances, and other physical quantities grew. Astronomers and surveyors striving for exactness had to deal with variation in their measurements. Many measurements should be better than a single measurement, even though they vary among themselves. How can we best combine many varying observations? Statistical methods that are still important were invented in order to analyze scientific measurements.

By the nineteenth century, the agricultural, life, and behavioral sciences also began to rely on data to answer fundamental questions. How are the heights of parents and children related? Does a new variety of wheat produce higher yields than the old, and under what conditions of rainfall and fertilizer? Can a person's mental ability and behavior be measured just as we measure height and reaction time? Effective methods for dealing with such questions developed slowly and with much debate.[1]

As methods for producing and understanding data grew in number and sophistication, the new discipline of statistics took shape in the twentieth century. Ideas and techniques that originated in the collection of government data, in the study of astronomical or biological measurements, and in the attempt to understand heredity or intelligence came together to form a unified "science of data." That science of data—statistics—is the topic of this text.

The first three chapters deal with statistical methods for organizing and describing data. These chapters progress from simpler to more complex data. Chapter 1 examines data on a single variable, Chapter 2 is devoted to the change of a variable over time, and Chapter 3 considers relationships among two or more variables. You will learn both how to examine data produced by others and how to organize and summarize your own data. These summaries will be first graphical, then numerical, then when appropriate in the form of a mathematical model that gives a compact description of the overall pattern of the data. Chapter 4 outlines arrangements (called designs) for producing data that answer specific questions. The principles presented in this chapter will help you to design proper samples and experiments, and to evaluate such investigations in your field of study.

The remaining seven chapters discuss statistical inference—formal methods for drawing conclusions from properly produced data. Statistical inference uses the language of probability to describe how reliable its conclusions are, so some basic facts about probability are needed to understand inference. Probability is the subject of Chapters 5 and 6. Chapter 7 introduces the reasoning of statistical inference. We emphasize that effective inference is based on good procedures for producing data (Chapter 4), careful examination of the data (Chapters 1 to 3), and an understanding of the nature of statistical inference as discussed in Chapter 7. Chapters 8 through 11 describe some of the most common specific methods of inference: for drawing conclusions about means and proportions from one and two samples, about relations in categorical data, regression and correlation, and analysis of variance.

The practice of statistics involves the use of many recipes for numerical calculation, some quite simple and some very complex. As you are learning how to use these recipes, remember that the goal of statistics is not calculation for its own sake, but gaining understanding from numbers. Many of the calculations can be automated by a calculator or computer, but you must supply the understanding. Chapters 8 to 11 present only a few of the many specific procedures for inference. The more complex procedures are always carried out by computers using specialized software. A thorough grasp of the principles of statistics will enable you to quickly learn more advanced methods as needed. On the other hand, a fancy computer analysis carried out without attention to basic principles will often produce elaborate nonsense. As you read, seek to understand the principles as well as the necessary details of methods and recipes.

NOTES

1. The rise of statistics from the physical, life, and behavioral sciences is described in detail by S.M. Stigler, *The History of Statistics: The Measurement of Uncertainty Before 1900*, Harvard-Belknap, Cambridge MA, 1986. Much of the information in the brief historical notes appearing throughout the text is drawn from this book.

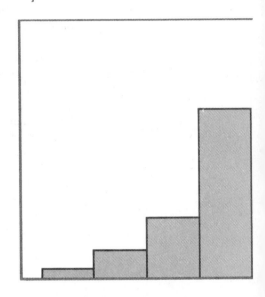

Prelude

Statistics begins with skills and principles for examining data. The first chapter concentrates on observations on a single quantity, or variable. We learn to look for a regular overall pattern in the data and then for important deviations from the pattern. We see how to present single-variable data graphically and how to use numerical measures to describe specific aspects of the data. Finally we discuss normal distributions, which describe the overall pattern in many sets of data.

- *Why do measurements of the speed of light made by an outstanding scientist vary and what is the pattern of variation?*

- *How can numerical calculations and graphs help us describe the differences among beef, meat, and poultry hot dogs in the number of calories each contains?*

- *How do normal distributions give us a compact description of the reading scores of seventh graders in Gary, Indiana?*

1

Looking at Data: Distributions

variable

value

Any characteristic of a person or thing that can be expressed as a number is called a *variable*. A *value* of the variable is the actual number that describes a particular person or thing. Height, sex, and annual income are variables that describe people; your specific height, sex, and income are the values of these variables that apply to you. An important initial step in organizing the way you think about data is to observe *how many variables* are present. This chapter looks at data that consist of values of a single variable, such as heights of people, incomes of families, or scores of school children on a reading test. Chapter 2 presents statistical methods for studying the growth or change of a single variable over time, such as the height of a child from birth to maturity or the average income of American families through the ups and downs of the economy. Chapter 3 examines relationships between two variables and the complexities of relationships among several variables. In this way we will advance from simpler to more complex data, gaining experience and learning new principles at each stage.

measured variable

categorical variable

In addition to the number of variables, you should note what *types of variables* are present in a set of data. Variables measured in a scale of equal units, such as height in centimeters or income in dollars, are called *measured variables*. The most common statistical methods are appropriate for measured variables. Other common variables are *categorical variables*. A categorical variable simply records into which of several categories a person or thing falls. Examples of categorical variables include the sex or political party affiliation of a person, the type of insecticide applied to a field, and the make of a car. We can assign numerical codes to categories, such as

$$\text{Female} = 1$$
$$\text{Male} = 0$$

Coded in this fashion, categorical variables fit our definition of variables as numerical characteristics. However, it is usually more convenient to use labels such as F for female and M for male rather than numerical codes. Notice that whether we record a person's sex as a number or a label, it does not make sense to do arithmetic with this categorical variable. We cannot meaningfully compute the "average sex" from a set of 0s and 1s.

counts

When categorical variables are of interest, we often work with *counts* or *percents* of the individuals in each category. For many statistical purposes, counts and percents can be treated like measured variables. In particular, arithmetic operations such as differences and averages make sense for both measured variables and counts, but not for categorical variables. We will call any variable that takes numerical values for which arithmetic makes

quantitative variable

sense a *quantitative variable*. This chapter and the next deal primarily with quantitative variables. The distinction between quantitative and categorical variables becomes important in Chapter 3.

1.1 DISPLAYING DISTRIBUTIONS

How do we begin to examine intelligently a set of values of a single measured variable? As a case study in the preliminary examination of data, we will look at the results of an important scientific study.

EXAMPLE 1.1

Light travels fast, but it is not transmitted instantaneously. Light takes over a second to reach us from the moon and over 10 billion years to reach us from the most distant objects yet observed in the expanding universe. Because radio and radar also travel at the speed of light, an accurate value for that speed is important to communicate with astronauts and orbiting satellites. An accurate value for the speed of light is also important to designers of computers, because electrical signals travel only at light speed.

The first reasonably accurate measurements of the speed of light were made a little over 100 years ago by A. A. Michelson and Simon Newcomb. Table 1.1 contains 66 measurements made by Newcomb between July and September 1882.[1]

■

Measurement

A set of numbers such as those in Table 1.1 is meaningless without some background information. We must ask several preliminary questions about any set of data. First, *What variable is being measured?* Newcomb measured how long light took to travel from his laboratory on the Potomac River to a mirror at the base of the Washington Monument and back, a total distance of about 7400 meters. Just as you can compute the speed of a car from the time required to drive a mile, Newcomb computed the speed of light from the travel time.

Table 1.1 Newcomb's measurements of the passage time of light

28	22	36	26	28	28
26	24	32	30	27	24
33	21	36	32	31	25
24	25	28	36	27	32
34	30	25	26	26	25
−44	23	21	30	33	29
27	29	28	22	26	27
16	31	29	36	32	28
40	19	37	23	32	29
−2	24	25	27	24	16
29	20	28	27	39	23

Answering the question "What variable is being measured?" requires a description of the instrument used to make the measurement. Then, we must judge whether the variable measured is appropriate for our purpose. This judgment often requires expert knowledge of the particular field of study. For example, Newcomb invented a novel and complicated apparatus to measure the passage time of light. We accept the judgment of physicists that this instrument is appropriate for its intended task and more accurate than earlier instruments.

Newcomb's study of the speed of light measured a clearly defined and easily understood variable, the time light takes to travel a fixed distance. Questions about measurement are often harder to answer in the social and behavioral sciences than in the physical sciences. We can agree that a tape measure is an appropriate way to measure a person's height. But how shall we measure her intelligence? Questionnaires and interview forms as well as tape measures are measuring instruments. A psychologist wishing to measure "general intelligence" might use the Wechsler Adult Intelligence Scale (WAIS). The WAIS is a standard "IQ test" that asks subjects to solve a large number of problems. The appropriateness of this instrument as a measure of intelligence is not accepted by all psychologists, many of whom disagree about what "intelligence" is and how to measure it. Because questions about measurement usually require knowledge of the particular field of study, we will say little about them.

Users of data should nonetheless be aware that taking numbers at face value, without thinking about the variable measured and the process used to measure it, can produce misleading results. The following examples illustrate the dangers.

EXAMPLE 1.2

A teenager argues that young people are safer drivers than the elderly. He cites government data showing that in 1985, 4472 drivers 65 years of age and over were involved in fatal accidents. In contrast, only 2790 drivers aged 16 and 17 had fatal accidents. The teenager's argument is incorrect because the variable (the count of drivers who had fatal accidents) is not appropriate. There are more than four times as many licensed drivers over age 65 as there are drivers aged 16 and 17. The larger group of drivers naturally had a larger number of accidents. A more appropriate variable is the *rate* of fatal accidents, that is, the fraction of drivers who were involved in fatal accidents in 1985. In fact, 66 of every 100,000 teenage drivers but only 24 of every 100,000 elderly drivers had fatal accidents in 1985. Fatal accidents are much more common among younger drivers. The use of *counts* where *rates* are more appropriate is a common misuse of statistics. ■

rate

EXAMPLE 1.3

One important measure of a nation's economic success is its unemployment rate, the percent of the active labor force that cannot find jobs. The unemployment rate in Japan has long been much lower than that in the United States. In 1986, for example, the U.S. unemployment rate was about 7%, while Japan's was about 2.5%. But examination shows that the measurement systems are not the same in the two countries. In the United States, any person without a job who is actively seeking work is counted as unemployed. But school-leavers seeking their first job are not considered unemployed in Japan, and self-employed workers are also omitted from the data. Because of these and other differences, the Japanese unemployment rate would almost double if measured by the American method. Conclusions about the two economies drawn without regard for such measurement differences may be incorrect.[2]   ■

The two remaining questions you should ask about any set of data are more straightforward: *What are the units of measurement?* and *How are the data recorded?* Newcomb's first measurement of the passsage time of light was 0.000024828 second. So his unit of measurement was seconds. But the entries in Table 1.1 do not look at all like 0.000024828. Such numbers are awkward to write and to do arithmetic with. We therefore move the decimal point nine places to the right, giving 24828, and then record only the deviation from 24800. Thus, 28 is short for the original 0.000024828 and −2 stands for 24798, or 0.000024798. This procedure is called *coding* the data. You should code whenever the actual data values contain many digits of which only a few digits vary from observation to observation. The data in coded form are easier for people to read. In addition, when you are using a calculator or a computer, reducing many digits to a few helps keep the results of arithmetic within the fixed number of digits that the machine can work with and store.

coding

We now have some understanding of what the numbers in Table 1.1 mean and where they came from. It is easier to understand data in our own area of study, but it is still important to ask the same three preliminary questions in every case. Then we can begin to look more closely at the data themselves.

Variation

The first feature to notice about the entries in Table 1.1 is that they vary. Newcomb's observations of the passage time of light are not all the same. Why should this be? After all, each observation records the travel time of light over the same path, measured by a skilled observer using the same apparatus each time. Newcomb knew that careful measurements almost always vary. The environment of every measurement is slightly different. The apparatus changes a bit with the temperature, the density of the atmosphere changes from day to day, and so on. Newcomb did his best to eliminate the sources of variation that he could anticipate, and his experimental skill gave his measurements less variation than those taken by less able scientists. But even the best experiments produce variable results. That is why Newcomb took many measurements rather than just one. The average of 66 passage times should be less variable than the result of a single measurement, because the average does not depend on the temperature and atmospheric density at a single time. (Section 1.2 shows how to compute several "averages," and

Chapter 6 shows how to express numerically the advantage of averaging 66 observations.)

Putting ourselves in Newcomb's place, we are tempted to compute the average passage time, convert this time to a new and better estimate of the speed of light, and rush off to make our reputation by publishing the result. The temptation to do a routine calculation and announce an answer arises whenever data have been collected to answer a specific question. Computers are very good at routine calculations, so that succumbing to temptation is the easy road. The first step toward statistical sophistication is to resist the temptation to calculate without thinking. Since variation is surely present in any set of data, we must first examine the nature of the variation—we may be very surprised at what we see.

Variation is even more obviously a fact of life when different people or things are measured. The reading ability of seventh graders varies from child to child, so many children should be tested to get a picture of reading skills in the seventh grade. Likewise, since the concentration of a drug may vary from lot to lot, the manufacturer should measure many lots to keep the concentration within the desired range. In such cases, the nature and magnitude of the variation may be as important as the average outcome.

Distribution

> The pattern of variation of a variable is called its distribution. The distribution records the numerical values of the variable and how often each value occurs.

The distribution of a variable is best displayed graphically. Some graphical tools for displaying distributions will be examined in the next few pages. With a better stocked tool kit, we will return to Newcomb's measurements and be able to understand them better.

Stemplots

Stemplots (also called stem-and-leaf plots) offer a quick way to picture the shape of a distribution while including the actual numerical values in the graph. A stemplot works best for small numbers of observations that are all greater than 0.

EXAMPLE 1.4   Here are the number of home runs that Babe Ruth hit in each of his 15 years with the New York Yankees (1920 to 1934).[3]

54   59   35   41   46   25   47   60   54   46   49   46   41   34   22

stemplot

stem

leaf

To make a *stemplot* for these data:

1 Separate each observation into a *stem* (the first digit in this case) and a *leaf* (the second digit); in general, stems may have as many digits as needed, but each leaf should contain only a single digit.

2 List the stems vertically in increasing order from top to bottom, draw a vertical line to the right of the stems, and add the leaves to the right of the line.

3 Arrange the leaves in increasing order from left to right.

Here is the resulting stemplot.

```
2 | 25
3 | 45
4 | 1166679
5 | 449
6 | 0
```

Once a distribution has been displayed by a stemplot, we can see its important features as follows:

1 *Locate the center of the distribution,* either by eye or by counting in from either end of the stemplot until half the observations are counted. Because Ruth hit 46 homers three times, less than 46 six times, and more than 46 six times, the distribution of his home runs is centered at 46. (This simple measure of the center of a distribution is called the *median.* We will use the term "median" informally until we study measures of center in detail in Section 1.2.)

median

2 *Examine the overall shape of the distribution.* Does it have one peak or several peaks? Is it approximately *symmetric* or is it *skewed* in one direction? A distribution is symmetric if the portions above and below its center are mirror images of each other. It is skewed to the right if the right tail (higher values) is much longer than the left tail (lower values). We do not insist on exact symmetry in giving a general description of the shape of a distribution. Babe Ruth's home runs have an approximately symmetric distribution.

symmetric

skewed

3 *Finally, look for marked deviations from the overall shape.* These may be *gaps* in the distribution, or they may be *outliers,* individual observations that fall well outside the overall pattern of the data. There are no gaps or outliers in the distribution of Ruth's home runs. In particular, his famous 60 home runs in 1927 was not "abnormal" in the context of his career distribution of home runs per year.

gaps

outliers

Ruth's record for a single year was broken by another Yankee, Roger Maris, who hit 61 home runs in 1961. Here is a stemplot of Maris' home run production during his 10 years in the American League.

```
0 | 8
1 | 346
2 | 368
3 | 39
4 |
5 |
6 | 1
```

Maris' record year is an outlier, an individual value that stands apart from his usual pattern. You should search for an explanation for any outlier. Often, outliers point to errors made in recording the data. In other cases, the unusual observation may be caused by equipment failure or other unusual circumstances. The outlier in Roger Maris' home run data marks an exceptional individual achievement.

back-to-back stemplot

When you wish to compare two related distributions, a *back-to-back stemplot* with common stems is useful. Here is a back-to-back stemplot comparing the home run data for Ruth and Maris.

```
         Ruth        Maris
                 | 0 | 8
                 | 1 | 346
              52 | 2 | 368
              54 | 3 | 39
          9766611 | 4 |
              944 | 5 |
                0 | 6 | 1
```

Stemplots do not work well for large data sets, where each stem must hold a large number of leaves. Fortunately, there are several modifications of the basic stemplot that are helpful when plotting the distribution of moderate amounts of data. You can increase the number of stems in a plot by *splitting each stem* in two, one with leaves 0 through 4 and the other with leaves 5 through 9. When the observed values have many digits, it is often best to *truncate* the numbers by dropping all but a few digits before making a stemplot. Since the purpose of a stemplot is to display a distribution effectively, it is better to be flexible rather than rule-bound in making the plot. Here is an example making use of both of the modifications mentioned above.

truncate

EXAMPLE 1.5

A marketing consultant observed 50 consecutive shoppers at a grocery store. One variable of interest was how much each shopper spent in the store. Here are the data (in dollars), arranged in increasing order.

2.32	6.61	6.90	8.04	9.45
10.26	11.34	11.63	12.66	12.95
13.67	13.72	14.35	14.52	14.55
15.01	15.33	16.55	17.15	18.22
18.30	18.71	19.54	19.55	20.58
20.89	20.91	21.13	23.85	26.04
27.07	28.76	29.15	30.54	31.99
32.82	33.26	33.80	34.76	36.22
37.52	39.28	40.80	43.97	45.58
52.36	61.57	63.85	64.30	69.49

To make a stemplot of this distribution, we truncate the purchases by dropping the "cents" so that the stems are tens of dollars and the leaves are dollars. This gives us the single-digit leaves that a stemplot requires. Then we split each stem.

```
0 | 2
0 | 6689
1 | 0112233444
1 | 556788899
2 | 00013
2 | 6789
3 | 012334
3 | 679
4 | 03
4 | 5
5 | 2
5 |
6 | 134
6 | 9
```

Turning the stemplot on its side so that the larger values fall to the right, we see that the distribution of amount spent is strongly skewed to the right. The center of the distribution is at $20 (found by counting 25 observations up from the lowest value). But 10% of the customers spent over $50. The shape revealed by the stemplot will interest the store management. If the big spenders can be studied in more detail, steps to attract more of them may be profitable. ∎

These examples illustrate the varied shapes of distributions. They also illustrate how simple calculations (of center, for example) can be helpful in clarifying the nature of a distribution. Making a graphical display is the first step toward understanding data; performing well-chosen calculations is the second. Section 1.2 will give more systematic suggestions for doing such calculations.

Most statistical software packages will produce a stemplot for you. If the observations have been entered into the MINITAB package as column

C1, for example, then the command

```
MTB> STEM C1
```

will make a stemplot. The program will make a decision about splitting stems. You can change that decision by a second command if you wish.

Histograms

A virtue of stemplots is that they display the actual values of the observations for detailed examination. But this feature makes stemplots awkward for large data sets. Moreover, the picture presented by a stemplot divides the observations into groups (stems) determined by the number system rather than by judgment. Histograms do not have these limitations. A histogram breaks the range of values of a variable into intervals of equal width, and displays only the count (or percent) of the observations that fall into each interval. We can choose any convenient number of intervals. Histograms are slower to construct by hand than are stemplots, and do not retain the actual values observed. For these reasons we prefer stemplots for small data sets. The construction of a histogram is best shown by example.

EXAMPLE 1.6

A study of the teaching of reading in the public schools of Gary, Indiana, began by examining the performance of seventh graders on the reading portion of the Iowa Test of Basic Skills.[4] In all, 947 Gary seventh graders took this test in 1984. Their vocabulary scores, expressed as an equivalent grade level, ranged from 2.0 (that is, equivalent to a beginning second grader) to 12.1. The scores of the first few of the 947 students are

| 5.4 | 6.8 | 2.0 | 7.6 | 6.6 | 7.8 | 8.1 | 5.6 | 6.0 | 7.9 ... |

histogram

To make a *histogram* of the distribution of scores, proceed as follows:

1 *Divide the range of the data into classes of equal width.* In this case, it is natural to use grade levels, so the classes are

$$2.0 \le \text{score} < 3.0$$
$$3.0 \le \text{score} < 4.0$$
$$\vdots$$
$$12.0 \le \text{score} < 13.0$$

We have specified the classes precisely so that each score falls in exactly one class. A score of 2.9 falls in the first class, a score of 3.0 falls in the second, and so on.

2 *Count the number of observations in each class.* These counts are called *frequencies,* and a table of frequencies for all classes is a *frequency table.* Table 1.2 is a frequency table of the Gary vocabulary scores. Frequency tables are a convenient summary of large sets of data.

frequency

frequency table

Table 1.2 Reading achievement scores
for seventh graders

Class	Number of students	Percent
2.0–2.9	9	0.95
3.0–3.9	28	2.96
4.0–4.9	59	6.23
5.0–5.9	165	17.42
6.0–6.9	244	25.77
7.0–7.9	206	21.75
8.0–8.9	146	15.42
9.0–9.9	60	6.34
10.0–10.9	24	2.53
11.0–11.9	5	0.53
12.0–12.9	1	0.11
Total	947	100.01

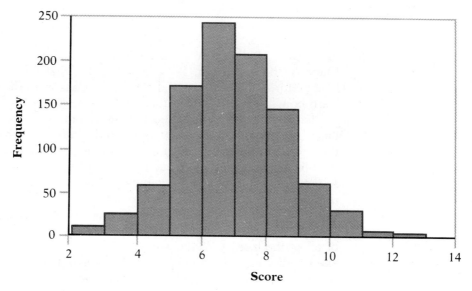

Figure 1.1 Histogram of the vocabulary scores of 947 seventh graders in Gary, Indiana.

3 *Draw the histogram.* In the histogram in Figure 1.1, the vocabulary
score scale is horizontal and the frequency scale vertical. Each bar
represents a class. The base of the bar covers the class, and the bar height
is the class frequency. The graph is drawn with no horizontal space
between the bars (unless a class is empty, so that its bar has 0 height). ■

The distribution in Figure 1.1 is centered in the 6.0 to 6.9 class. It is roughly symmetric in shape. The choice of classes can slightly affect the appearance of a histogram, as can the automatic choice of classes forced by a stemplot. This is another reason why exact symmetry is not necessary in order to describe a distribution as "symmetric." The distribution in Figure 1.1 is quite regular. The histogram shows no large gaps or obvious outliers, and both tails fall off quite smoothly from a single center peak. This is a bit surprising. Since adolescent girls as a group read better than boys of the same age, we might have expected a "double-peaked" distribution with this general shape:

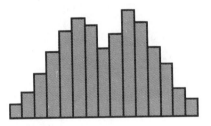

More elaborate calculations show that there is indeed an effect based on sex in the Gary reading scores, but the effect is not large enough to change the shape of the histogram. We are still at the first step of examining data, looking for major patterns and clear deviations from them. In time we will learn how to discern finer detail.

Large sets of data are almost always reported in the form of frequency tables, since it is not practical to publish the individual observations. In addition to the frequency (count) for each class, the fraction or percentage of the observations that fall in each class can be reported. These fractions are sometimes called *relative frequencies*. Table 1.2 gives both frequencies and relative frequencies for the Gary reading scores. When you are reporting statistical results, however, it is clearer to use the nontechnical terms "number" and "percent." Expressed as percents, the relative frequencies of all of the classes should add to 100%. The percents in Table 1.2 total 100.01% because of *roundoff errors*. Each is rounded to the nearest 0.01%, so that the rounded numbers do not have a sum exactly equal to 100%.

relative frequency

roundoff error

A histogram of relative frequencies has the same appearance as a frequency histogram such as Figure 1.1. Simply relabel the vertical scale to read in percents. Histograms of relative frequencies are preferable for comparing two distributions with different numbers of observations. The bars of a frequency histogram for reading scores in Los Angeles would tower over a similar histogram for Gary, since there are many more seventh graders in Los Angeles. All relative frequency histograms have the same vertical scale (0% to 100%), so that the shapes of the two distributions can more easily be compared.

out? Are the data intended to answer specific questions? Are they appropriate for that purpose?

2 Always examine your data. An informative picture comes first and is usually supplemented by some numerical calculations.

3 Look first for an overall pattern, then for deviations from that pattern, such as outliers.

What of Simon Newcomb and the speed of light? A laboratory scientist knows that even the most careful measurements will vary but still hopes for a symmetric single-peaked distribution like that of Figure 1.1. Then the best estimate of the true value of the measured quantity is the center of the distribution. (This is not true if the measurement process is biased so as to produce systematically high or low readings—that is a fatal flaw that statistics cannot remedy.) The histogram in Figure 1.2 shows that Newcomb was not so fortunate—his data are afflicted with outliers. What can be done?

The handling of outliers is another matter requiring judgment. Sometimes outliers are of special interest as evidence of an extraordinary event. An outlier from the background distribution of brightness seen from a surveillance satellite may represent a missile launch. An outlier from the background measurement of electrical activity in a detector used by high-energy physicists may be evidence of a new elementary particle. In such cases the overall distribution simply provides the background against which extraor-

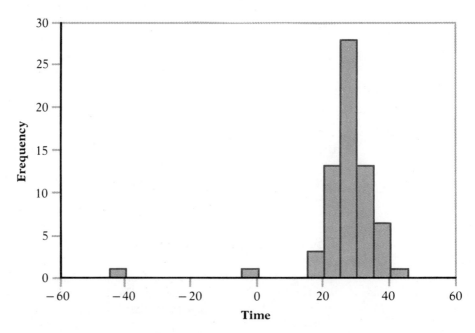

Figure 1.2 Histogram of Simon Newcomb's 66 measurements of the passage time of light, from Table 1.1.

dinary events stand out. But Newcomb had hoped for a well-behaved distribution with a clear center, and the two outliers (-44 and -2 in Table 1.1) were disturbing.

When outliers are surprising and unwanted, you should first search for a clear cause for each outlier, such as equipment failure during the experiment or an error in typing the data. Almost all large data sets have errors, often caused by mistakes in entering the data into a computer. Outliers often betray these errors and allow us to correct them by looking at the original data records. If equipment failure or some other abnormal condition caused the outlier, we can delete it from the data with a clear conscience. When no cause is found, the decision is difficult. The outlier may be evidence of an extraordinary occurrence or of unexpected variability in the data. Newcomb finally dropped the worst outlier (-44), but retained the other. He based his estimate of the speed of light on the average (the mean, in the language of Section 1.2) of his observations. The mean of all 66 observations is 26.21; the mean of the 65 retained observations is 27.29. The strong effect of the single value -44 on the mean is one motivation for discarding it when our interest is in the center of the distribution as a whole.

We can gain more insight into Newcomb's data from a different kind of graph. When data represent similar observations made over time, it is wise to plot them either against time or against the order in which the observations were taken. Figure 1.3 plots Newcomb's passage times for light against the

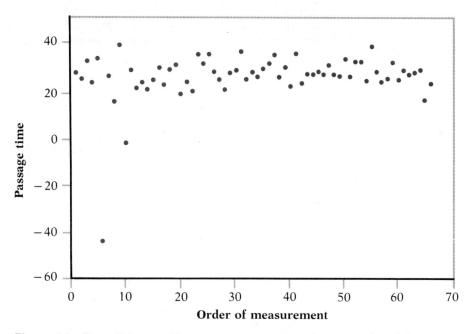

Figure 1.3 Plot of Newcomb's measurements against the time order of the observations.

time order. There is some suggestion in this plot that the variability (the vertical spread in the plot) is decreasing over time. In particular, both outlying observations were made early on. Perhaps, as he gained experience, Newcomb became more adept at using his equipment. Learning effects like this are quite common; the experienced data analyst will check for them routinely. If we allow Newcomb 20 observations for learning, the mean of the remaining 46 measurements is 28.15. The best modern measurements suggest that the "true value" for the passage time in Newcomb's experiment is 33.02. Eliminating the low outliers or allowing for learning does move the average results closer to the true value. But adjustments based on judgment alone, as these were, are somewhat suspect. If it is possible, always find the reason for an outlier.

EXAMPLE 1.7

In 1985 British scientists reported a hole in the ozone layer of the earth's atmosphere over the South Pole. This is disturbing, since ozone protects us from cancer-causing ultraviolet radiation. The British report was at first disregarded, since it was based on ground instruments looking up. More comprehensive observations from satellite instruments looking down had shown nothing unusual. Then, examination of the satellite data revealed that the South Pole ozone readings were so low that the computer software used to analyze the data had automatically suppressed these values as erroneous outliers! Readings dating back to 1979 were reanalyzed and showed a large and growing hole in the ozone layer that is unexplained and possibly dangerous.[5] Computers analyzing large volumes of data are often programmed to suppress outliers as protection against errors in the data. As the example of the hole in the ozone layer illustrates, suppressing an outlier without investigating it can keep valuable information out of the sight. ■

SUMMARY

Measurement is the process of representing a characteristic by numbers.

A **variable** is any numerical characteristic. The values of a variable vary when measurements are made on different people or things, or at different times.

A large number of observations on a single variable can be summarized in a table of **frequencies** (counts) or **relative frequencies** (percents or fractions).

The **distribution** of a variable is its pattern of variation, as described by the values of the variable and their frequencies or relative frequencies.

A distribution is displayed by a **stemplot** or by a **histogram**. Stemplots separate each observation into a **stem** and a **leaf**, while histograms are based on the frequency or relative frequency of classes of values.

When examining a distribution, first locate its **center**. Then look at the **overall shape** and at clear **deviations** from that shape.

The shape of a distribution can be approximately **symmetric** (each side of the center is a mirror image of the other) or **skewed** (one tail extends farther from the center than the other). The number of peaks is another aspect of overall shape.

Deviations from the overall shape of a distribution include gaps and **outliers** (individual observations that appear not to be in accord with the remaining data).

SECTION 1.1 EXERCISES

1.1   You want to compare the "size" of several statistics textbooks. Give at least three possible numerical variables that make precise the vague idea of the "size" of a book. In what *unit* is each variable measured? What *measuring instrument* does each require? Which variable is most appropriate for estimating how long it would take you to read the book? Which is most appropriate for deciding whether the book will fit easily into your book bag?

1.2   You are studying the relationship between political attitudes and length of hair among male students. You will measure political attitudes with a standard questionnaire. How will you measure length of hair? Give precise instructions that an assistant could follow. Include a statement of the *unit* and the *measuring instrument* that your assistant is to use.

1.3   Various studies have attempted to rank cities in terms of how desirable it is to live and work in each city. Describe five variables that you would measure for each city if you were designing such a study. Give reasons for each of your choices.

1.4   The number of deaths from cancer in the United States has risen steadily over time. In 1985, for example, about 462,000 people died of cancer, up from 331,000 deaths in 1970. A member of Congress says that these numbers show that no progress has been made in treating cancer. Explain how the number of people dying of cancer could increase even if treatment of the disease were improving. Then describe at least one variable that would be a more appropriate measure of the effectiveness of medical treatment for a potentially fatal disease.

1.5   Scientists who study human growth use different measures of the size of an individual. Weight, height, and weight divided by height are three of the most common measures. If you were interested in studying the short-term effects of a digestive illness, which of these three variables would you study? Why?

1.6 You are writing an article for a consumer magazine based on a survey of the magazine's readers on the reliability of their household appliances. Of 13,376 readers who reported owning Brand A dishwashers, 2942 required a service call during the past year. Only 192 service calls were reported by the 480 readers who owned Brand B dishwashers. Describe an appropriate variable to measure the reliability of a make of dishwasher, and compute the values of this variable for Brand A and for Brand B.

1.7 The usual method of determining heart rate is to take the pulse and count the number of beats in a given time period. The results are generally reported as beats per minute; for instance, if the time period is 15 seconds, the count is multiplied by four. Take your pulse for two 15-second periods, two 30-second periods, and two 1-minute periods. Convert the counts to beats per minute and report the results. Which procedure do you think gives the best results? Why?

1.8 In the previous exercise, some beats probably occurred just before or after the end of your time interval. In what way would this problem affect your results? Does the seriousness of the problem depend upon the length of the time interval? Consider the following alternative procedure. Count the time that it takes for the heart to beat a certain number of times, say 80. Describe how you would convert this measure to beats per minute. Make two measurements on yourself and report the results. Do you think the alternative procedure is better? Give reasons.

1.9 Each year *Fortune* magazine lists the top 500 companies in the United States, ranked according to their total annual sales in dollars. Describe three other variables that could reasonably be used to measure the "size" of a company.

1.10 Cost is often an important factor in deciding which variable to measure in a study. Sometimes the cost can be determined directly in dollars, while in other cases it may be expressed in terms of the time required to take the measurement. A crude but inexpensive measure is sometimes preferable to a more precise measure that is too expensive. A blood test, for example, is a faster and cheaper way to diagnose many diseases than exploratory surgery.

Foresters need quick measurements of the size of a tree, since they must measure many trees in a woodlot in order to assess its overall condition. List three variables that measure the size of a tree, and arrange them in order from most costly to least costly.

1.11 A quality engineer in an automobile engine plant measures a critical dimension on each of a sample of crankshafts at regular intervals.

Since the purpose of a histogram is to display the shape of a distribution, we must be attentive to the visual aspects of the display. Our eyes respond to the *area* of the bars, so an accurate picture requires that the area of each bar be exactly proportional to the frequency of its class. When the classes are of equal width, we need only make the bar heights proportional to the frequencies. We recommend equal class widths whenever the raw data are available to you. However, at times you may need to draw a histogram with unequal class widths, particularly when examining data published in the form of a frequency table with unequal classes. Much government economic and social data, for example, are published in this form. In such cases, you must vary the bar heights in such a way that the areas (height × width) are proportional to the frequencies. See Exercise 1.30 for an example.

Before drawing a histogram you must decide how many classes to use. This is a matter of judgment. Too few classes will lump most observations together, while too many will place only a few observations in each class. Neither extreme will show clearly the shape of the distribution. Narrow classes preserve more detail but the heights of the bars often vary irregularly. Wide classes often give a more regular picture of the overall shape, but they lose detail by lumping a wider interval of values into each class. Use natural classes, such as the grade levels in Example 1.6, whenever possible. Fortunately, a broad range of choices will usually give a similar impression of the distribution.

Statistical software on a computer can ease the task of drawing a histogram. The program will choose the number of classes and produce a crude histogram for you to inspect. You can then instruct the program to change the choice of classes if you wish. If the data have been entered into the MINITAB statistical package as column C1, for example, then the command

```
MTB> HISTOGRAM C1
```

will make a histogram. If you are using an ordinary terminal and printer, the bars of the histogram will simply be horizontal rows of asterisks. You can use the program to explore different choices of classes and count the frequencies, and then draw a proper histogram if you need an attractive graph.

Looking at Data

Some principles have emerged from our initial experiences with data that will remain valid as we advance. These are guidelines based on experience rather than hard and fast rules that must always be followed.

    1 To interpret data, you must first learn something of their context: What exactly was measured? How was the measurement carried

The dimension is supposed to be 224 millimeters (mm), but some variation will occur in production. Here are the latest measurements.

224.120	224.001	224.017	223.982	223.989	223.961
223.960	224.089	223.987	223.976	223.902	223.980
224.098	224.057	223.913	223.999		

The engineer *codes* these measurements to make them easier to work with. The coded value is the number of thousandths of a millimeter above 223 mm. (For example, 224.120 mm is coded as 1120.) Give the coded value for each of the measurements in the sample.

1.12 Another quality measurement in the same plant concerns the eccentricity of a valve assembly. (Eccentricity is a measure of how off-center two supposedly concentric circles are.) Here are the data, measured in inches.

.001084	.001131	.000887	.000639	.001216	.000903
.000977	.001088	.000940	.001069	.000667	.000536

Explain how you would code these data for easier calculation. (A good coding results in positive whole numbers.) Then give the coded values for each of the eccentricities listed.

1.13 The histograms in Figure 1.4 display the distributions of two weather-related variables. Figure 1.4(a) shows the number of days of

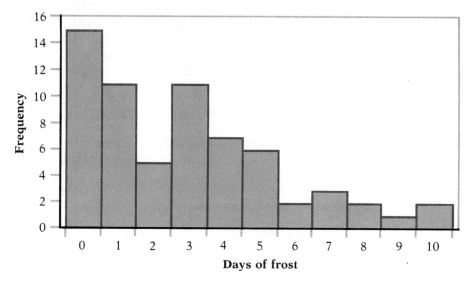

Figure I.4(a) Histogram of the number of frost days during April in Greenwich, England, over a 65-year period, for Exercise 1.13.

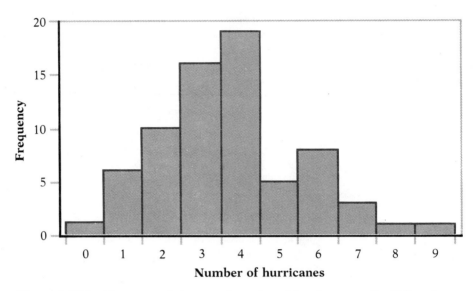

Figure 1.4(b)   Histogram of the annual number of hurricanes on the U.S. east coast over a 70-year period, for Exercise 1.13.

frost (minimum temperature below freezing) in Greenwich, England, in the month of April over a 65-year period. Figure 1.4(b) shows the number of hurricanes reaching the east coast of the United States each year over a 70-year period. Give a brief description of the overall shape of each of these distributions. What is the most important difference between the two shapes? About where does the center of each distribution lie? (Data on frosts from C. E. Brooks and N. Carruthers, *Handbook of Statistical Methods in Meteorology*, H. M. Stationery Office, 1953. Hurricane data from H. C. S. Thom, *Some Methods of Climatological Analysis*, World Meteorological Organization, Geneva, Switzerland, 1966.)

1.14   Figure 1.5 displays the distribution of the lengths of words used in articles in *Popular Science* magazine. These data were collected by students who opened the magazine arbitrarily and recorded the lengths of all words in the first complete paragraph on the page. Describe the main features of this distribution: Is it symmetric, right skewed, or left skewed? Single or double peaked? Are there gaps or outliers? (Statistical methods have been effectively used to distinguish between works by different authors. Although the distribution of word lengths is a rather crude measure, *Popular Science* uses notably more long words than do most authors of serious fiction.)

1.15   Figure 1.6 is a histogram of the mean scores on the verbal part of the Scholastic Aptitude Test (SAT) in each of the 50 states in the 1982– 1983 school year. In some states most college-bound students take this

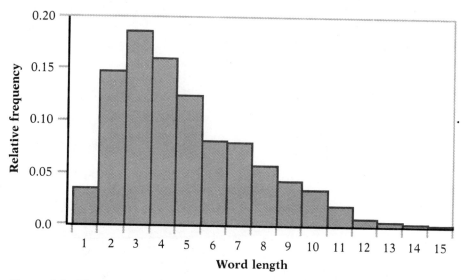

Figure 1.5 The distribution of the lengths of words appearing in articles in *Popular Science*, for Exercise 1.14.

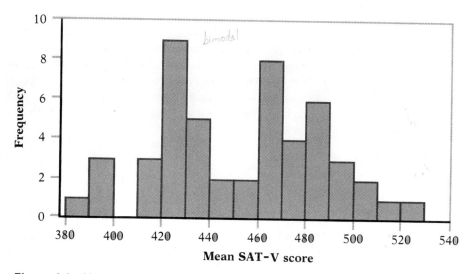

Figure 1.6 Histogram of the mean scores on the verbal part of the SAT examination for the 50 states in the 1982–1983 school year, for Exercise 1.15.

test, while in others most students take the American College Testing (ACT) examination instead. In the ACT states, only students who hope to enter major universities take the SAT. Describe the overall shape of this distribution. How are the two groups of states just described visible in the histogram?

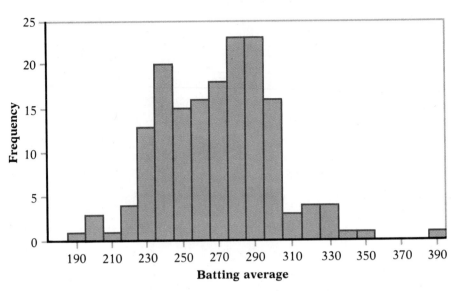

Figure I.7 Batting averages of American League players in 1980, for Exercise 1.16.

1.16 Figure 1.7 displays the distribution of batting averages for all 167 American League baseball players who batted at least 200 times in the 1980 season. Is the overall shape (ignoring the outlier) roughly symmetric or clearly skewed? What is the approximate batting average of a typical American League regular player? (The outlier is the .390 batting average of George Brett, the highest batting average in the major leagues since Ted Williams hit .406 in 1941. See Exercise 1.72 for a comparison of Brett and Williams.)

1.17 The Survey of Study Habits and Attitudes (SSHA) is a psychological test that evaluates the motivation, study habits, and attitude toward school of college students. A selective private college gives the SSHA to a sample of 18 of its incoming freshman women. Their scores are

| 154 | 109 | 137 | 115 | 152 | 140 | 154 | 178 | 101 |
| 103 | 126 | 126 | 137 | 165 | 165 | 129 | 200 | 148 |

Make a stemplot of these data. The overall shape of the distribution is indistinct, as often happens when only a few observations are available. Are there any outliers? What is the approximate value of the median score (the score such that half are higher and half lower)?

1.18 Here are data on the percent of residents 65 years of age and over in each of the 50 states. Make a stemplot of these data using percent as stems and tenths of a percent as leaves. Then make a second stemplot splitting each stem in two (for leaves 0 to 4 and leaves 5 to

9). Which display do you prefer? Describe the shape of the distribution. Is it roughly symmetric or distinctly skewed? Where is the center of the distribution? Are there any clear outliers?

Ala.	11.5	Hawaii	8.2	Mass.	12.8	N.Mex.	9.1	S.Dak.	13.5
Alaska	3.0	Idaho	10.2	Mich.	10.2	N.Y.	12.4	Tenn.	11.5
Ariz.	11.7	Ill.	11.2	Minn.	12.0	N.C.	10.6	Tex.	9.6
Ark.	13.9	Ind.	11.1	Miss.	11.7	N.Dak	12.4	Utah	7.5
Calif.	10.3	Iowa	12.0	Mo.	13.3	Ohio	11.1	Vt.	11.5
Colo.	8.6	Kan.	13.1	Mont.	11.0	Okla	12.3	Va.	9.6
Conn.	12.0	Ky.	11.4	Neb.	13.2	Oreg.	11.9	Wash.	10.6
Del.	10.3	La.	9.6	Nev.	8.5	Pa.	13.2	W.Va.	12.4
Fla.	17.3	Maine	12.7	N.H.	12.7	R.I.	13.6	Wis.	12.2
Ga.	9.6	Md.	9.6	N.J.	11.9	S.C.	9.5	Wyo.	7.8

1.19 Thomas the cat is helping researchers study the fleas that cause discomfort to him and his fellow cats. One part of the research concerns the egg production of the flea *Ctenocephalides felis*. The researchers deposit 25 female and 10 male fleas in Thomas' fur and count the number of flea eggs produced each day. The number of eggs produced in 27 consecutive days is as follows:

436	495	575	444	754	915	945	655	782	704
590	411	547	584	550	487	585	549	475	435
523	390	425	415	450	395	405			

Truncate each count by dropping the last digit, and present the distribution in a stemplot. Describe the general shape and the center of the distribution. Are there any unusual values? (Data provided by Sayed Gaafar and Michael Dryden, Purdue University School of Veterinary Medicine.)

1.20 Climatologists interested in flooding gather statistics on the daily rainfall in various cities. The following data set gives the *maximum* daily rainfall (in inches) for each of the years 1941 to 1970 in South Bend, Indiana. (Successive years follow each other across the rows in the table.)

1.88	2.23	2.58	2.07	2.94	2.29	3.14	2.15	1.95	2.51
2.86	1.48	1.12	2.76	3.10	2.05	2.23	1.70	1.57	2.81
1.24	3.29	1.87	1.50	2.99	3.48	2.12	4.69	2.29	2.12

Make a stemplot for these data. Describe the general shape of the distribution and any prominent deviations from the overall pattern. (You are expected to truncate and split stems as needed in making stemplots.)

1.21 Make a stemplot for the data of Exercise 1.11 after you have coded the original measurements. Would a stemplot of the original data have the same appearance? Describe the shape of the distribution.

1.22 The *1987 Old Farmer's Almanac* gives the growing season for selected U.S. cities as reported by the National Climatic Center. The growing season is defined as the average number of days between the last frost in the spring and the first frost in the fall. The values are

279	244	318	262	335	321	165	180	201	252
145	192	217	179	182	210	271	302	169	192
156	181	156	125	166	248	198	220	134	189
141	142	211	196	169	· 237	136	203	184	224
178	279	201	173	252	149	229	300	217	203
148	220	175	188	160	176	128			

(a) Make a stemplot for these data. Describe the general shape of the distribution. Are there any unusual values?

(b) The values for Los Angeles, San Diego, San Francisco, and Miami are represented by an * in the almanac rather than a number. There is a footnote explaining why. Given the above definition of growing season, why do you think there were difficulties with measuring this variable for these cities?

1.23 There are many ways to measure the reading ability of children. Research designed to improve reading performance is dependent on good measures of the outcome. One frequently used test is the DRP or Degree of Reading Power. In a research study on third grade students, the DRP was administered to 44 students. Their scores were

40	26	39	14	42	18	25	43	46	27	19
47	19	26	35	34	15	44	40	38	31	46
52	25	35	35	33	29	34	41	49	28	52
47	35	48	22	33	41	51	27	14	54	45

Make a stemplot of these data and then make a histogram. Which display do you prefer and why? Describe the main features of the distribution. (Data provided by Maribeth Cassidy Schmitt, from her Ph.D. dissertation, *"The effects of an elaborated directed reading activity on the metacomprehension skills of third graders,"* Purdue University 1987.)

1.24 Make a histogram of the distribution of growing season from the data of Exercise 1.22. You must first choose classes and make a frequency table.

1.25 In 1798 the English scientist Henry Cavendish measured the density of the earth by careful work with a torsion balance. The variable recorded was the density of the earth as a multiple of the density of water. Here are Cavendish's 29 measurements.

5.50	5.61	4.88	4.07	5.26	5.55	5.36	5.29	5.58	5.65
5.57	5.53	5.62	5.29	5.44	5.34	5.79	5.10	5.27	5.39
5.42	5.47	5.63	5.34	5.46	5.30	5.75	5.86	5.85	

Present these measurements graphically by either a stemplot or a histogram and explain the reason for your choice. Then briefly discuss the main features of the distribution. In particular, what is your estimate of the density of the earth based on these measurements? (Cavendish's data are reported by S. M. Stigler in the article cited in Note 1.)

1.26 The distribution of the ages of a nation's population has a strong influence on economic and social conditions. The following table shows the age distribution of U.S. residents in 1950 and in 2075, in millions of persons. The 1950 data come from that year's census, while the 2075 data are projections made by the Census Bureau.

Age group	1950	2075
Under 10 years	29.3	34.9
10–19 years	21.8	35.7
20–29 years	24.0	36.8
30–39 years	22.8	38.1
40–49 years	19.3	37.8
50–59 years	15.5	37.5
60–69 years	11.0	34.5
70–79 years	5.5	27.2
80–89 years	1.6	18.8
90–99 years	0.1	7.7
100–109 years	—	1.7
Total	151.1	310.6

(a) Make a histogram of the 1950 age distribution and describe the main features of the distribution. In particular, look at the number of children relative to the rest of the population.

(b) Make a histogram of the projected age distribution for the year 2075. What are the most important changes in the U.S. age distribution projected for the 125-year period between 1950 and 2075?

1.27 There is some evidence that increasing the amount of calcium in the diet can lower blood pressure. In a medical experiment one group of men was given a daily calcium supplement, while a control group received a placebo (a dummy pill). The seated systolic blood pressure of all the men was measured before the treatments began and again after 12 weeks. The blood pressure distributions in the two groups should have been similar at the beginning of the experiment. Here are the initial blood pressure readings for the two groups.

Calcium group

| 107 | 110 | 123 | 129 | 112 | 111 | 107 | 112 | 136 | 102 |

Placebo group

| 123 | 109 | 112 | 102 | 98 | 114 | 119 | 112 | 110 | 117 | 130 |

Make a back-to-back stemplot of these data. Does your plot show any major differences in the two groups before treatments began? In particular, are the centers of the two blood pressure distributions close together? (See Example 8.12 for a discussion of the medical experiment from which these data are taken.)

1.28 Plant scientists have developed varieties of corn that have increased amounts of the essential amino acid lysine. In a test of the protein quality of this corn, an experimental group of 20 one-day-old male chicks was fed a ration containing the new corn. A control group of another 20 chicks was fed a ration that was identical except that it contained normal corn. Here are the weight gains (in grams) after 21 days.

Control				Experimental			
380	321	366	356	361	447	401	375
283	349	402	462	434	403	393	426
356	410	329	399	406	318	467	407
350	384	316	272	427	420	477	392
345	455	360	431	430	339	410	326

Make a back-to-back stemplot of these data. Does it appear that the the chicks fed high-lysine corn grew faster? Are there any outliers or other problems? (Based on G. L. Cromwell et al., "A comparison of the nutritive value of *opaque-2, floury-2* and normal corn for the chick," *Poultry Science*, 47 (1968), pp. 840–847.)

1.29 A manufacturing company is reviewing the salaries of its full-time employees below the executive level at a large plant. The clerical staff is almost entirely female, while a majority of the production workers and technical staff are male. As a result, the distributions of salaries for male and female employees may be quite different. Table 1.3 gives the frequencies and relative frequencies of the salaries for women and men. Make histograms from these data, choosing the type that is

Table 1.3 The salary distributions of female and male workers in a large manufacturing plant

Salary	Women		Men	
($1000)	Number	%	Number	%
10–15	89	11.8	26	1.1
15–20	192	25.4	221	9.0
20–25	236	31.2	677	27.6
25–30	111	14.7	823	33.6
30–35	86	11.4	365	14.9
35–40	25	3.3	182	7.4
40–45	11	1.5	91	3.7
45–50	3	0.4	33	1.4
55–60	2	0.3	19	0.8
60–65	0	0	11	0.4
65–70	1	0.1	3	0.1
Total	756	100.1	2451	100.0

most appropriate for comparing the two distributions. Then describe the overall shape of the two salary distributions and the chief differences between them.

1.30 A report on the recent graduates of a large state university includes the following relative frequency table of the first-year salaries of last year's graduates. Salaries are in $1000 units, and it is understood that each class includes its left endpoint but not its right endpoint—for example, a salary of exactly $15,000 belongs in the second class.

Salary	10–15	15–20	20–25	25–30	30–35	35–45	45–55	55–75
Percent	4	10	21	26	18	13	5	3

The last three classes are wider than the others. An accurate histogram must take this into account.

(a) Make a histogram in which the base of each bar covers a class and the height is the percent of graduates with salaries in that class. The areas of the three rightmost bars overstate the percent who have salaries in those classes.

(b) To make a correct histogram, the area of each bar must be proportional to the percent of graduates in that class. Most classes are $5000 wide. A class *twice* as wide ($10,000) should have a bar *half* as tall as the percent in that class. This keeps the area proportional to the percent. How should you treat the height of the bar for a class $20,000 wide? Make a correct histogram with the heights of the bars for the last three classes adjusted so that the areas of the bars reflect the percent in each class.

1.31 In the National Health Survey, heights of 52,744 males between the ages of 18 and 79 were measured. The following is a frequency table of the data as given in a government publication.

Height	Count	Height	Count	Height	Count
Under 62 in	675	66 in	7021	71 in	3216
62 in	874	67 in	6249	72 in	2817
63 in	1720	68 in	9379	73 in	1103
64 in	3691	69 in	5421	74 in	581
65 in	3488	70 in	6239	75 in and over	270

This table has *open classes* on both ends. For example, we know that 675 men were less than 62 inches tall but we do not know the lower endpoint of this interval. In making a histogram we cannot draw bars for open classes. Instead, put a sign such as * at each end and write a note stating how many observations fell below (or above) the classes drawn. Make a histogram for the National Height Survey data, and discuss the center and shape of the distribution of heights of American men.

1.2   DESCRIBING DISTRIBUTIONS

"Face it. A hot dog isn't a carrot stick." So said *Consumer Reports*, commenting on the low nutritional quality of the all-American frank. Table 1.4 shows the magazine's laboratory test results for calories and milligrams of sodium (mostly due to salt) in a number of major brands of hot dogs. There are three types: all beef, "meat" (mainly pork and beef, but government regulations allow up to 15% poultry meat), and poultry. Since people concerned about health may prefer low-calorie, low-salt hot dogs, we ask: Are there any systematic differences among the three types in these two variables? We can begin to answer this question by comparing stemplots of the three distributions of calorie content for the three types, and by making a similar comparison for sodium content. The comparisons can be made more explicit by using some numerical tools for describing distributions. This section will develop these tools.

Measuring Center

We saw in Section 1.1 that a numerical measure of the center of a distribution is helpful in inspecting a stemplot or a histogram. The most common measure of center is the ordinary arithmetic average, or mean.

Table 1.4 Calories and sodium in hot dogs by type

Beef		Meat		Poultry	
Calories	Sodium	Calories	Sodium	Calories	Sodium
186	495	173	458	129	430
181	477	191	506	132	375
176	425	182	473	102	396
149	322	190	545	106	383
184	482	172	496	94	387
190	587	147	360	102	542
158	370	146	387	87	359
139	322	139	386	99	357
175	479	175	507	170	528
148	375	136	393	113	513
152	330	179	405	135	426
111	300	153	372	142	513
141	386	107	144	86	358
153	401	195	511	143	581
190	645	135	405	152	588
157	440	140	428	146	522
131	317	138	339	144	545
149	319				
135	298				
132	253				

Source: *Consumer Reports,* June 1986, pp. 366–367.

Mean

If n observations are denoted by $x_1, x_2, \ldots, x_n$, their mean is

$$\bar{x} = \frac{1}{n}(x_1 + x_2 + \cdots + x_n)$$

or in more compact notation

$$\bar{x} = \frac{1}{n}\sum x_i \qquad (1.1)$$

The $\sum$ in the recipe for the mean is short for "add them all up." The bar over the x indicates the mean of all the x-values. This notation is so common that writers in many fields use $\bar{x}$, $\bar{y}$, etc. without additional explanation. The subscripts on the observations x_i are just a way of keeping the n observations distinct. They do not necessarily indicate order or any other special facts about the data.

EXAMPLE 1.8

From the data in Example 1.4, the mean number of home runs hit in a year by Babe Ruth is

$$\bar{x} = \frac{1}{15}(54 + 59 + \cdots + 22) = \frac{659}{15} = 43.9$$

Roger Maris' mean, on the other hand, is

$$\bar{y} = \frac{1}{10}(8 + 13 + \cdots + 61) = \frac{261}{10} = 26.1$$

Ruth's superiority is evident from these averages.   ■

The distribution of Maris' yearly home run production shows a clear outlier, his record 61 homers in 1961. If we exclude this as atypical, the mean number of home runs in the other 9 years of his American League career is 22.2. The single outlier has increased the 10-year average by about four homers per year. This illustrates an important weakness of the mean as a measure of center: It is sensitive to the influence of a few extreme observations. These may be outliers, but a skewed distribution that has no outliers can also pull the mean toward its long tail. Because the mean can-not resist the influence of extreme observations, we say that it is not a *resistant measure* of center. A measure that is resistant does more than limit the influence of outliers; its value does not respond strongly to changes in a few observations, no matter how large those changes may be. The mean fails this requirement, because we can make the mean as large as we wish by making a large enough increase in just one observation.

resistant measure

A resistant measure of center, the median, was already used informally in Section 1.1: The median is the midpoint of a distribution, the point such that half of the observations fall above it, and half below. This definition will now be made more explicit.

Median

To compute the median of a distribution:

  1 Arrange all observations in order of size, from smallest to largest.

  2 If the number n of observations is odd, the median M is the center observation in the ordered list. The location of the median is found by counting $(n + 1)/2$ observations up from the the bottom of the list.

  3 If the number n of observations is even, the median M is the average of the two center observations in the ordered list. The location of the median is again found by counting $(n + 1)/2$ observations up from the bottom of the list.

Note that the formula $(n + 1)/2$ does *not* give the median, only the location of the median in the ordered list. Medians require little arithmetic, so they are easy to calculate by hand for small sets of data. However, arranging even a moderate number of observations in order is very tedious, so if a calculator is used, it is usually quicker to calculate the mean than it is to locate the median.

EXAMPLE 1.9

To find the median number of home runs hit in a year by Babe Ruth, we arrange the data in increasing order

| 22 | 25 | 34 | 35 | 41 | 41 | 46 | **46** | 46 | 47 | 49 | 54 | 54 | 59 | 60 |

The median is the bold 46, the eighth observation in the ordered list. You can locate the center by eye—there are seven observations to the left and seven to the right—or use the recipe $(n + 1)/2 = 16/2 = 8$ to locate the median in the list.

Roger Maris' median home run production is found in a similar way. His yearly totals in increasing order are

| 8 | 13 | 14 | 16 | **23** | **26** | 28 | 33 | 39 | 61 |

Since the number of observations $n = 10$ is even, there is a center pair of numbers rather than a single center point. This pair of numbers is shown in bold in the list. The median M is therefore

$$M = \frac{23 + 26}{2} = \frac{49}{2} = 24.5$$

The recipe $(n + 1)/2 = 11/2 = 5.5$ for the position of the median in the list means that the median is "halfway between the fifth and sixth observations." ∎

Notice that since a stemplot arranges the observations in increasing order you can compute a median from a stemplot without rewriting the data. Notice also that the median, unlike the mean, is resistant. Maris' record 61 home runs is simply one observation falling above the median. Even if this observation had been erroneously entered as 610, the median would have the same value. The mean and median will agree for symmetric distributions. But because the mean is sensitive to extreme observations, it will move away from the median toward the long tail in skewed distributions. This effect can be substantial. For example, the Insurance Information Institute pointed out in 1984 that the average award in product liability cases exceeded $1 million. This, said the Institute, is why liability insurance coverage is so expensive. Opponents of increases in insurance premiums replied that the median award was only $271,000. The two measures differ dramatically because the average (the mean) allows a few huge awards to overwhelm the effects of many smaller cases. However, it is not true that using the median is always preferable to using the mean simply because it is resistant. The mean is computed from the actual values of the observations, and contains more information than the median. Insurance companies, for example, are concerned about the total claims payments they must make. This total can be recovered from the mean award, but not from the median award.

We now have two general strategies for dealing with outliers and other irregularities in data. The first strategy requires that outliers be detected and their causes investigated. The offending observations can then often be corrected, deleted for good reason, or otherwise given individual attention. The second strategy is to use resistant methods, so that outliers have little influence over our conclusions. In general, we prefer to examine data carefully and consider any irregularity in the light of the individual situation. This will sometimes lead to a decision to employ resistant methods.

In the case of Roger Maris, for example, the median is more typical of his performance than is the mean, but it obscures the great event of his career. A better summary might be: "Maris hit 61 home runs in 1961, and averaged 22.2 homers in his other 9 years in the league." Newcomb's data, on the other hand, are supposed to be repeated measurements of the same quantity, but they contain unexplained outliers. Even though the outlying values may not be explained, they should not be given much influence over our estimate of the velocity of light. We can delete the outliers from a computation of the mean, or we can report the median. If the two outliers are omitted, the mean of the remaining 64 measurements is 27.75. The median of 66 observations has position $(66 + 1)/2 = 33.5$ in the ordered list of passage times from Table 1.1, and can be found to be 27. Either of these values is preferable to the overall mean (26.21) as a statement of Newcomb's value for the travel time of light.

Resistant Measures of Spread

Even the briefest summary of a distribution requires an indication of how spread out, or variable, the data are in addition to a measure of their center. Two nations with identical median personal incomes will look very different if in one there are extremes of wealth and poverty, while in the other a general equality of incomes is the rule. A drug whose mean potency is just what the doctor ordered is defective if some lots have dangerously high potency and others are too weak to be effective.

*p*th percentile   The spread or variability of a distribution can be indicated by giving several percentiles. *The pth percentile of the distribution is the value such that p percent of the observations fall at or below it.* The median is just the 50th percentile, so the use of percentiles to report spread is particularly appropriate when the median is used to measure center. The most commonly used percentiles other than the median are the *quartiles.* The first quartile is the 25th percentile, and the third quartile is the 75th percentile. (The second quartile is the median itself.) To calculate a percentile, arrange the observations in increasing order and count up the required percent from the bottom of the list. Our definition of percentiles is a bit inexact, because there is not always a value with exactly p percent of the data at or below it. We will be content to take the nearest observation for most percentiles, but the quartiles are important enough to require an exact recipe.

Quartiles

> To calculate the quartiles, first locate the median in the ordered list of observations. The first quartile is the median of the observations below the location of the median, and the third quartile is the median of the observations above that location.

EXAMPLE 1.10

In Example 1.5 we arranged the shopping data in order in a stemplot, after truncating to eliminate the cents. The truncated data are sufficient to give us a quick numerical summary of the distribution. Here they are, rewritten for convenience from the stemplot.

2	6	6	8	9	10	11	11	12	12	13	13	14	14	14	15	15
16	17	18	18	18	19	19	20	‖		20	20	21	23	26	27	28	29
30	31	32	33	33	34	36	37	39	40	43	45	52	61	63	64	69

There are $n = 50$ observations, so the position of the median is $(50 + 1)/2 = 25.5$, or midway between the 25th and 26th observations. This location is marked by ‖ in the list. Both center observations are $20, so the median is $20. To find the first quartile, consider the 25 observations falling below the location ‖ of the median. Note carefully that the lower of the two center observations is included in this group, since the location ‖ falls between two data points. The median of these 25 observations is the thirteenth in order, or $14. This is the first quartile. You can find the third quartile similarly as the thirteenth value above ‖, or $33. We can summarize these results in compact form as

$$Q_1 = \$14 \qquad M = \$20 \qquad Q_3 = \$33$$

Other percentiles are found more informally. For example, we compute the 95th percentile for the 50 observations to be the 48th in the ordered list, because $0.95 \times 50 = 47.5$, which we round to 48. The 95th percentile is therefore $63. ∎

When there are an odd number of observations, the median is the unique center observation, and the rule for finding the quartiles excludes this center value. For example, the median of Ruth's 15 home run totals

22	25	34	35	41	41	46	**46**	46	47	49	54	54	59	60

is the bold 46 (the eighth entry). The first quartile is the median of the seven observations falling below this point in the list, $Q_1 = 35$. Similarly, $Q_3 = 54$. Some statistical computing systems use slightly different rules to compute the median and the quartiles, so the results given by a computer may not agree exactly with the results found by using our rules. However, any differences will be small.

The quartiles together with the median give some indication of the center, spread, and shape of a distribution. In Example 1.10, the fact that Q_3 lies farther from M than does Q_1 reflects the right skewness of the data. The distance between the quartiles is a simple measure of spread that gives the range covered by the middle half of the data. This distance is called the *interquartile range*, or in symbols

interquartile range

$$IQR = Q_3 - Q_1$$

In Example 1.10, $IQR = \$33 - \$14 = \$19$. The quartiles and the IQR are unaffected by changes in either tail of the distribution. They are therefore resistant, because changes in a few data points have no further effect once these points move outside the quartiles. You should be aware that a single numerical measure of spread, such as IQR, is useful primarily as a description of symmetric distributions. The individual quartiles Q_1 and Q_3 (given with the median M) are more informative in the case of a skewed distribution because they portray the unequal spread of the two sides of the distribution.

Since Q_1, M, and Q_3 contain no information about the tails, a fuller summary of the shape of a distribution can be obtained by giving the highest and lowest values as well.

Five-number summary

> The five-number summary of a distribution consists of the median M, the quartiles Q_1 and Q_3, and the smallest and largest individual observations, written in the order
>
> Minimum, Q_1, M, Q_3, Maximum

The five-number summary for the grocery spending data of Example 1.10 is

$$\$2 \quad \$14 \quad \$20 \quad \$33 \quad \$69$$

The IQR can be computed from the five-number summary, as can other helpful differences such as the distances of the quartiles from the median.

One common rule of thumb for identifying suspected outliers singles out values falling at least $1.5 \times IQR$ above the third quartile or below the first quartile. In Example 1.10, $1.5 \times IQR = 1.5 \times \$19 = \$28.5$, so any values at or below $\$14 - \$28.5 = -\$14.5$ and at or above $\$33 + \$28.5 = \$61.5$ are tagged. In the case of the strongly skewed distribution of grocery purchases, the tagged values ($\$63$, $\$64$, and $\$69$) do not appear to be outliers in the sense of deviating from the overall pattern of the distribution. Using the $1.5 \times IQR$ rule to spot outliers does not replace using judgment, but it is helpful whenever the process must be automated.

boxplot

The five-number summary leads to another effective visual representation of a distribution, the *boxplot*. Figure 1.8(a) is a boxplot of the distribution of grocery store purchases from Example 1.10. In a boxplot,

1 The ends of the box are at the quartiles, so that the length of the box is the *IQR*.

2 The median is marked by a line within the box.

3 The two colored vertical lines (called whiskers) outside the box extend to the smallest and largest observations.

The center, spread, and overall range of the distribution are immediately apparent from a boxplot. There are many variations on the boxplot idea. Because the advantages of a boxplot over a stemplot are greatest for large data sets, modifications that are appropriate for large amounts of data are particularly important. Since the smallest and largest observations in a large data set are often extreme observations that tell us little about the shape of the distribution, we can replace them in both the five-number summary and the boxplot by the 10th and 90th percentiles. This would be appropriate, for example, in a summary of the incomes of all 660,000 lawyers in the United States.

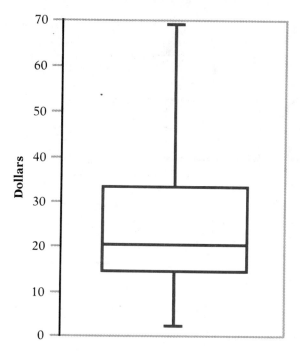

Figure 1.8(a) Boxplot of the amount spent by 50 shoppers at a grocery store, from Example 1.10.

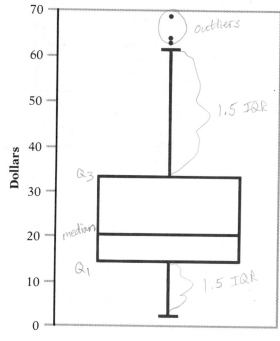

Figure 1.8(b) Modified boxplot of the grocery spending data.

For moderate amounts of data, it is valuable to plot potential outliers individually. To do this in a boxplot, extend the whiskers to the extreme high and low observations *only if* these values are less than $1.5 \times IQR$ beyond the quartiles. Otherwise, end the whiskers at the most extreme observations still within $1.5 \times IQR$ of the quartiles and plot the remaining cases individually. This plot is a *modified boxplot.* Figure 1.8(b) is a modified boxplot of the grocery purchase distribution. The upper whisker extends to 61, which is the smallest observation within the $1.5 \times IQR$ limit, and the suspected outliers $63, $64, and $69 are marked separately. We recommend modified boxplots except for very large numbers of observations.

modified boxplot

Boxplots are particularly suited for comparing distributions. Let us therefore return to the calorie content of hot dogs.

EXAMPLE 1.11

The five-number summaries of the distributions of calories for the three types of hot dogs in Table 1.4 are

Beef	111	140	152.5	178.5	190
Meat	107	139	153	179	195
Poultry	86	102	129	143	170

Calculate the *IQR* for each type, and check that no observations fall more than $1.5 \times IQR$ outside the quartiles. Figure 1.9 presents the boxplots based on these calculations. We see at once that poultry franks as a group contain fewer calories

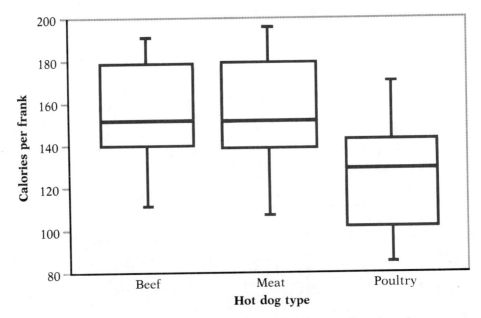

Figure 1.9 Boxplots of the calorie content of three types of hot dogs, from Table 1.4.

than beef or meat hot dogs. The median number of calories in a poultry hot dog is less than the first quartile of either of the other distributions. But each type shows quite a large spread among brands, so simply buying a poultry frank does not guarantee a low-calorie food. ■

Figure 1.9 is also testimony to the fact that boxplots are generally inferior to stemplots as displays of a single distribution. Here is a stemplot of the calorie content of the 17 brands of meat hot dogs.

```
10 | 7
11 |
12 |
13 | 5689
14 | 067
15 | 3
16 |
17 | 2359
18 | 2
19 | 015
```

There are two distinct clusters of brands (the distribution has two peaks with a gap between them) and one outlier in the lower tail. The boxplot hid the clusters. The quartiles $Q_1 = 139$ and $Q_3 = 179$ are roughly at the centers of the clusters, so much of the *IQR* is the spread between the two clusters. Because of this, the $1.5 \times IQR$ rule used in drawing the boxplot did not call attention to the outlier. Returning to the *Consumer Reports* article from which the data were taken supplies an explanation for the outlier. It is the calorie content of *Eat Slim Veal Hot Dogs*, the only frank in the group that is not composed of some mix of pork, beef, and poultry. We do not have enough information to explain why the other brands fall into two distinct clusters.

As this example illustrates, we recommend boxplots primarily for comparing several distributions. A stemplot (or, for large data sets, a histogram) is usually a clearer display for a single distribution, especially when accompanied by the median and quartiles as numerical signposts.

The distribution of the calorie content of meat hot dogs also illustrates an important limitation of our entire program to find statistical measures for specific aspects of distributions. Neither the mean nor the median conveys much useful information about this distribution, because a clear "center" is not present. The median falls near the gap between the two clusters and is not a typical calorie value. The main message given by a double-peaked distribution is simply its shape, and shape is best portrayed visually. The routine methods of statistics compute numerical measures and draw conclusions based on their values. These methods are extremely useful, and they will be studied carefully in later chapters. But they cannot be applied blindly, by feeding data to a computer program, because *statistical measures and methods based on them are generally meaningful only for distributions of sufficiently regular shape.* This principle will become clearer as we progress, but it is good to be aware at the beginning that quickly resorting to fancy calculations is the mark of a statistical amateur. The trained practitioner looks, thinks, and chooses calculations selectively.

The Standard Deviation

The most common measure of the spread of a distribution is not the *IQR* or any other resistant measure, but the standard deviation. This is a measure of spread about the mean, and should be used only when the mean is employed as your measure of center.

Variance and standard deviation

The variance of n observations $x_1, x_2, \ldots, x_n$ is

$$s^2 = \frac{1}{n-1}\left[(x_1 - \bar{x})^2 + (x_2 - \bar{x})^2 + \cdots + (x_n - \bar{x})^2\right]$$

or more compactly

$$s^2 = \frac{1}{n-1}\sum(x_i - \bar{x})^2 \tag{1.2}$$

The standard deviation s is the square root of the variance s^2.

The idea behind the variance and the standard deviation as measures of spread is as follows: The deviations $x_i - \bar{x}$ display the spread of the values x_i about their mean $\bar{x}$. Some of these deviations will be positive and some negative because some of the observations fall on each side of the mean. So we cannot simply add the deviations to get an overall measure of spread—in fact, the sum of the deviations $x_i - \bar{x}$ will always be zero. Squaring the deviations makes them all positive, and observations far from the mean in either direction will have large positive squared deviations. The variance is roughly the average squared deviation. So s^2 will be large if the observations are widely spread about their mean, and small if the observations are all close to the mean. Because the variance involves squaring the deviations, it does not have the same unit of measurement as the original observations. Lengths measured in centimeters have a variance measured in squared centimeters. Taking the square root remedies this, so that the standard deviation s measures dispersion about the mean in the original scale.

If the variance is the average of the squares of the deviations of the observations from their mean, why do we average by dividing by $n - 1$ rather than n? Because the sum of the deviations $x_i - \bar{x}$ is always zero, the last deviation can be found once we know the first $n - 1$. So we are not averaging n unrelated numbers. Because only $n - 1$ of the squared deviations can vary freely, the average is found by dividing by $n - 1$. The number $n - 1$ is called the *degrees of freedom* of the variance or standard deviation.

EXAMPLE 1.12

Babe Ruth's 15 yearly home run totals were

54   59   35   41   46   25   47   60   54   46   49   46   41   34   22

As we saw in Example 1.8, the mean of this distribution is

$$\bar{x} = \frac{1}{n} \sum x_i = \frac{659}{15} = 43.93$$

To compute the variance from Equation 1.2 we subtract 43.93 from each observation, square the resulting deviations, add, and divide by 14 as follows:

$$s^2 = \frac{1}{14} [(54 - 43.93)^2 + (59 - 43.93)^2 + \cdots + (22 - 43.93)^2]$$

$$= \frac{101.40 + 227.10 + \cdots + 480.92}{14}$$

$$= \frac{1770.93}{14} = 126.495$$

The standard deviation is therefore $s = \sqrt{126.495} = 11.25$. ∎

Computing the variance is our first substantial piece of arithmetic, so some comments on calculation are in order. You will find the task of calculating s^2 and other statistical measures more or less tedious depending on the resources at hand. Many calculators will compute s or s^2 directly from the keyed-in data, with no intermediate work on your part. Interactive computer systems are even more desirable because their display and editing capabilities help avoid mistakes in entering the data. Interactive statistical systems will also do graphical displays and the more elaborate statistical procedures that we will meet later. Calculating descriptive measures like s step by step as in Example 1.12 is rapidly becoming an anachronism. We nonetheless recommend that you do a few examples in detail to be certain that you understand what the recipe means.

You will also find that the exact numerical value of your answer will depend on how you did the calculation. In general, whenever you write down an intermediate result without saving all the digits produced by your calculator, you introduce *roundoff error* into the following steps in the calculation. In Example 1.12, the mean $\bar{x}$ is not exactly 43.93, so our calculation of s will suffer from roundoff error. Since all calculators and computers can handle only a fixed number of digits, roundoff error is always with us. Usually these errors are small, and it is the business of program designers to minimize them in constructing the programs for a calculator or computer system. Roundoff errors will therefore not be of great concern to us. If you must use intermediate results that will later be keyed back into your calculator, save at least one more digit than the data you started with. Most of the calculations in this book are done on a computer, so if you check them you may get slightly difference results.

Computing formula for the variance* If you do statistical calculations with a basic calculator, you will need to know alternative formulas that are designed for easier calculation of such quantities as s^2. Equation 1.2 is the *defining formula* for the variance. That is, this equation shows how s^2 measures spread about the mean. But Equation 1.2 is awkward to use because you must first subtract the mean $\bar{x}$ from each individual observation. A bit of algebra shows that an equivalent formula is

$$s^2 = \frac{1}{n-1}\left[\sum x_i^2 - \frac{1}{n}\left(\sum x_i\right)^2\right] \qquad (1.3)$$

This is a *computing formula* for the variance; it obscures the meaning of s^2 but leads to much shorter calculations. Equation 1.3 uses the basic quantities $\sum x_i$ and $\sum x_i^2$, which can be obtained quickly on a calculator with a memory and a square button without the need to write down intermediate results.

EXAMPLE 1.13

Returning to the calculation of Example 1.12, we first find the building block sums

$$\sum x_i = 54 + 59 + \cdots + 22 = 659$$
$$\sum x_i^2 = 54^2 + 59^2 + \cdots + 22^2 = 30{,}723$$

Since $n = 15$, the variance calculated from Equation 1.3 is

$$s^2 = \frac{1}{14}\left[30{,}723 - \frac{659^2}{15}\right]$$
$$= \frac{1}{14}\left[30{,}723 - \frac{434{,}281}{15}\right]$$
$$= \frac{1}{14}\left[30{,}723 - 28{,}952.07\right] = \frac{1770.93}{14} = 126.495$$

It is considerably faster to do this arithmetic on your calculator than to write it down. The standard deviation is then $s = \sqrt{126.495} = 11.25$ as before. ∎

Attention to the detail of getting calculations right, whether through computing formulas as in Example 1.13 or through instruction sets for a computer system, is a necessary part of statistical work. But do not let this detail distract you from a clear view of what you are doing and why. What

---

* This explanation is optional if you have a calculator or computing system that will compute s directly from the data.

we are now doing is measuring spread. Here are the basic properties of the standard deviation s as a measure of spread.

- s measures spread about the mean and should be used only when the mean is chosen as the measure of center.
- $s = 0$ only when there is no spread, that is, when all observations have the same value. Otherwise $s > 0$.

The standard deviation is not a resistant measure of the spread of a distribution. In fact, the use of *squared* deviations renders s even more sensitive to a few extreme observations than is the mean. For example, Roger Maris' 61-home-run year raises his mean homers per season from 22.2 for the remaining 9 years to 26.1, and it increases the standard deviation from 10.2 for 9 years to 15.6 for all 10 years. A skewed distribution with a few observations in the single long tail will have a large standard deviation, and the number s does not give much helpful information in such a case.

You may rightly feel that the importance of the standard deviation is not yet clear. If we must measure spread about the mean, why not just take the average distance of observations from the mean? Or the median distance, which is more resistant to outlying values? Squaring the distances seems to reduce the resistance for no good reason. These doubts are well founded. The standard deviation is not a very helpful measure of spread for distributions in general. We commented earlier that the usefulness of many statistical procedures is tied to distributions of particular shapes. This is distinctly true of the standard deviation. The omnipresence of s in statistical work is due to its intimate connection with the normal distributions, a specific class of distributions that we shall meet shortly. For the preliminary work of describing a data set, however, measures based on percentiles are superior.

Changing the Unit of Measurement

The same variable can be recorded in different units of measurement. Americans commonly record distances in miles and temperatures in degrees Fahrenheit, while Europeans measure distances in kilometers and temperatures in degrees Celsius. Fortunately, it is easy to convert all of our numerical descriptions of a distribution from one unit to another. This is true **linear** because a change in the measurement unit is a *linear transformation* in the **transformation** measurements.

A linear transformation changes the original variable x into the new variable x^* given by

$$x^* = a + bx$$

Adding the constant a shifts all values of x upward or downward by the same amount. In particular, such a shift changes the origin (zero point) of the variable. Multiplying by the positive constant b changes the size of the unit of measurement.

EXAMPLE 1.14

If a distance x is measured in miles, the same distance in kilometers is

$$x^* = 1.609x$$

That is, 10 miles is the same as 16.09 kilometers. Here $a = 0$ and $b = 1.609$. This transformation changes the unit without changing the origin—a distance of 0 miles is the same as a distance of 0 kilometers.

A temperature x measured in degrees Fahrenheit must be re-expressed in degrees Celsius to be easily understood by Europeans. The transformation is

$$x^* = \frac{5}{9}(x - 32) = -\frac{160}{9} + \frac{5}{9}x$$

Thus, the high of 95°F on a hot American summer day translates into 35°C. In this case

$$a = -\frac{160}{9} \quad \text{and} \quad b = \frac{5}{9}$$

This linear transformation changes both the unit size and the origin of the measurements. The origin in the Celsius scale (0°C, the temperature at which water freezes) is 32° in the Fahrenheit scale. ■

Linear transformations do not affect the shape of a distribution. If measurements on a variable x have a right-skewed distribution, any new variable x^* obtained by a linear transformation $x^* = a + bx$ (for $b > 0$) will also have a right-skewed distribution. Similarly, if the distribution of x is symmetric, the distribution of x^* remains symmetric.

Although a linear transformation preserves the basic shape of a distribution, the center and spread will generally change. Because linear changes of measurement scale are common, we must be aware of their effect on numerical descriptive measures of center and spread.

Effect of linear transformation

Suppose that $x_1, x_2, \ldots, x_n$ are observations with mean $\bar{x}$, median M, quartiles Q_1 and Q_3, interquartile range IQR, and standard deviation s. The linear transformation $x_i^* = a + bx_i$ with $b > 0$ produces observations $x_1^*, x_2^*, \ldots, x_n^*$ for which

- The mean, median, and quartiles are all transformed by multiplying by b and then adding a. For example, $\overline{x^*} = a + b\bar{x}$.
- The interquartile range and standard deviation are multiplied by b. For example, $s_{x^*} = bs$.

Notice that the two measures of spread, *IQR* and *s*, are not affected by adding the constant *a* to all of the observations. Adding a constant changes the location of the distribution but leaves the spread unaltered. Multiplying all observations by *b* does change the spread. If the extremes of the *x* values are −3 and 10, for example, the extremes of 5*x* are −15 and 50, so any measure of spread should be five times as large.

In discussing Newcomb's measurements of the speed of light, we noticed that the entries in Table 1.1 had been *coded* to produce simpler numbers. Coding is also a linear transformation; that is, the coded value *x** is obtained from the original measurement *x* by $x^* = a + bx$. The facts we have given about linear transformations relate the mean, for example, of the coded values to the mean of the original data.

SUMMARY

A **resistant measure** of any aspect of a distribution is relatively unaffected by changes in the numerical value of a small proportion of the total number of observations, no matter how large these changes are.

The **center** of a distribution can be measured by the **mean** $\bar{x}$ or by the **median** *M*. The mean is the arithmetic average, while the median is the midpoint. The median is a resistant measure of center but the mean is not.

Resistant measures of the **spread** or **variability** of a distribution are provided by the **quartiles** Q_1 and Q_3 or by the **interquartile range** $IQR = Q_3 - Q_1$. The **five-number summary** consisting of the median, quartiles, and high and low extremes provides a quick overall description of a distribution.

Boxplots based on the five-number summary are useful for comparing several distributions. The identification of potential outliers as observations falling at least $1.5 \times IQR$ outside the quartiles is helpful in drawing **modified boxplots** and in other contexts. These measures of spread are appropriate when the median is used to indicate the center of the distribution.

The **variance** s^2 and especially its square root the **standard deviation** *s* are common but nonresistant measures of spread about the mean as center. They are most useful for the normal distributions introduced in the next section.

Linear transformations have the form $x^* = a + bx$. They change the origin if $a \neq 0$, and the size of the unit of measurement if $b \neq 1$. Linear transformations do not alter the overall shape of a distribution. A linear transformation multiplies a measure of spread by *b* when $b > 0$, and changes a percentile or measure of center *m* into $a + bm$.

Numerical measures of particular aspects of a distribution, such as center and spread, do not report the entire shape of general distributions. In some cases, particularly distributions with multiple peaks and gaps, these measures may not be very informative.

SECTION 1.2 EXERCISES

1.32 A college rowing coach tests the 10 members of the women's varsity rowing team on a Stanford Rowing Ergometer (a kind of stationary rowing machine). The variable measured is revolutions of the ergometer's flywheel in a 1-minute session. The data are

446	552	527	504	450	583	501	545	549	506

Make a stemplot of these data by dropping the last digit and using the first digit as the stem. Then compute the mean and the median of the untruncated ergometer scores. Explain the similarity or difference in these two measures in terms of the symmetry or skewness of the distribution.

1.33 Exercise 1.17 gives the scores of 18 freshman women on the Survey of Study Habits and Attitudes. Find the mean score and the median score for this group. Explain the relationship between these two measures in terms of the main features of the distribution of scores.

1.34 Last year a small accounting firm paid each of its five clerks $22,000, two junior accountants $50,000 each, and the firm's owner $270,000. What is the mean salary paid at this firm? How many of the employees earn less than the mean? What is the median salary?

1.35 Exercise 1.11 gives measurements on 16 automobile engine crankshafts. Compute the mean and the median of these measurements as they appear in that exercise. Explain the relationship between the two measures of center in terms of the symmetry or skewness of the distribution.

1.36 This year, the firm in Exercise 1.34 pays the same salaries to the clerks and junior accountants, while the owner's take increases to $430,000. How does this change affect the mean? How does it affect the median?

1.37 In Exercise 1.11, you were asked to code the crankshaft measurements given. Write your coding procedure in the form $x^* = a + bx$, where x is the original measurement and x^* is the coded value. Then use your results from Exercise 1.35 to give the mean and median for the coded data.

1.38 According to the Department of Commerce, the mean and median price of new houses sold in the United States in mid-1988 were $141,200 and $117,800. Which of these numbers is the mean and which is the median? Explain your answer.

1.39 In computing the median income of any group, some federal agencies omit all members of the group who had no income. Give an example to show that the reported median income of a group can go down even though the group becomes economically better off. Is this also true of the mean income?

1.40 Exercise 1.17 gives the scores of 18 freshman women on the Survey of Study Habits and Attitudes (SSHA). The college also administered the test to a sample of 20 freshman men. Their scores were

| 108 | 140 | 114 | 91 | 180 | 115 | 126 | 92 | 169 | 146 |
| 109 | 132 | 75 | 88 | 113 | 151 | 70 | 115 | 187 | 104 |

(a) Make a back-to-back stemplot of the men's and women's scores. What is the most noticeable contrast between the two distributions?
(b) Find the five-number summaries for both distributions. Are there any outliers according to the $1.5 \times IQR$ criterion?
(c) Make side-by-side boxplots of the distributions. Use modified boxplots that plot outliers separately.

1.41 (a) Compute the five-number summary for the maximum rainfall data of Exercise 1.20.
(b) Compute the IQR for these data and draw a modified boxplot that plots individually any suspected outliers by the $1.5 \times IQR$ criterion.
(c) Based on the shape of the distribution, do you expect the mean to fall distinctly above the median, close to the median, or distinctly below the median?

1.42 In the previous exercise you computed a five-number summary in inches. Now you are writing an article for a journal that requires use of the metric system. Give the five-number summary in centimeters. (One inch is 2.54 cm.)

1.43 The following are the golf scores of 12 members of a women's golf team in tournament play:

| 89 | 90 | 87 | 95 | 86 | 81 | 102 | 105 | 83 | 88 | 91 | 79 |

(a) Present the distribution by a stemplot and describe its main features.
(b) Compute the mean, variance, and standard deviation of these golf scores.

(c) Then compute the median, the quartiles, and the *IQR*. Are there any suspected outliers by the 1.5 × *IQR* criterion?

(d) Based on the shape of the distribution, would you report the standard deviation or the quartiles as a measure of spread?

1.44 The rowing ergometer scores given in Exercise 1.32 appear to have a symmetric distribution. The researcher who administered the test described the results by giving the mean and the standard deviation. You found the mean in Exercise 1.32. Now find the standard deviation of the scores.

1.45 A health-conscious medical student takes her resting pulse each night after eating. The past week's measurements (in beats per minute) were

62   57   53   69   60   61   58

Compute the mean and the standard deviation of these data. (There are too few observations to give reliable information about the shape of the distribution. The medical student chooses $\bar{x}$ and s because she knows that pulse rate under stable conditions tends to have a symmetric and roughly normal distribution.)

1.46 This is a standard deviation contest. You must choose four numbers from the whole numbers 0 to 10, with repeats allowed.

(a) Choose four numbers that have the smallest possible standard deviation.

(b) Choose four numbers that have the largest possible standard deviation.

(c) Is more than one choice possible in either (a) or (b)? Explain.

1.47 This exercise requires a calculator that is preprogrammed to find the standard deviation, or statistical software on a computer. The observations

10,001   10,002   10,003

have mean $\bar{x} = 10,002$ and standard deviation $s = 1$. Adding more 0s in the center of these numbers, the next set becomes

100,001   100,002   100,003

The standard deviation remains $s = 1$ as 0s are added. Use your calculator or computer to calculate the standard deviation of these numbers, adding extra 0s until you get an incorrect answer. How soon did you go wrong? This example demonstrates that calculators and computers cannot handle an arbitrary number of digits correctly. Coding the data is the remedy for this problem.

1.48 Compute the mean and standard deviation of Cavendish's observations given in Exercise 1.25. Based on the general shape of this distribution, would you be willing to report the mean and standard deviation as helpful descriptive measures? If so, the mean would be used to estimate the actual density of the earth.

1.49 Compute the standard deviation of the crankshaft measurements given in Exercise 1.11 after coding. (You found the mean in Exercise 1.37.) Then report the standard deviation in the original units (mm). Based on the general shape of this distribution, would you be willing to report the mean and standard deviation as helpful descriptive measures?

1.50 From the data given in Exercise 1.18, give a brief numerical summary of the distribution of percent of residents over 65 by state. Defend your choice of numerical measures.

1.51 In Exercise 1.19 you displayed the distribution of counts of flea eggs on Thomas the cat. Based on the shape of the distribution and the presence or absence of outliers, would you prefer the median and quartiles or the mean and standard deviation as a helpful numerical summary? Compute the summary that you have chosen.

1.52 In each of the following cases, give the values of *a* and *b* for the linear transformation $x^* = a + bx$.

$y = mx + b$
$x^* = a + bx$
$x^* = 0 + \frac{1}{2}(x)$
$66\ in = 66(\frac{1}{12}) = 5.5\ ft.$

(a) *x* is measured in feet and transformed to yards. 0 $\frac{1}{3}$
(b) *x* is measured in hours and transformed to minutes. 0 60
(c) *x* is measured in degrees Fahrenheit and is transformed to the difference between *x* and the "normal" body temperature of 98.6 degrees. $y = -98.6$ $x = 0$ $x^* = -98.6 + (1)x$ $x - 98.6$

1.53 Most statistical software will calculate descriptive measures in response to simple commands, once the data are entered. Suppose, for example, that the growing season data given in Exercise 1.22 have been entered into the MINITAB statistical system as a data set named GROW. We then have the following command and output:

```
MTB> DESCRIBE 'GROW'
```

	N	MEAN	MEDIAN	TRMEAN	STDEV	SEMEAN
GROW	57	203.65	192.00	200.92	51.87	6.87

	MIN	MAX	Q1	Q3
GROW	125.00	335.00	167.50	233.00

In this MINITAB output, TRMEAN is the 5% trimmed mean (see Exercise 1.63), and SEMEAN is the standard error of the mean, a quantity that will be presented in Chapter 6. The other quantities are familiar to us. The exact recipe that MINITAB uses to find the quartiles is not the same as ours, so the quartiles in the output may differ slightly from those obtained by our method.

(a) Use the output to find the *IQR*. Are there any suspected outliers by the $1.5 \times IQR$ criterion?

(b) Draw a modified boxplot of the distribution. (In fact, MINITAB can be instructed to do both (a) and (b) for us, although these are easy to carry out by hand once the computer has done the lengthier calculations.)

1.54 Assume that a suitably defined average growing season for the four cities mentioned in Exercise 1.22(b) would be above 300 days. Find the median and quartiles of the 61 observations that result when these cities are included, and compare your results with the values given in Exercise 1.22. Explain why you cannot compute the mean and standard deviation of the expanded data set.

1.55 Exercise 1.22 gives the growing season x in days for selected U.S. cities.

(a) Find the linear transformation $x^* = a + bx$ that will express the growing season in weeks.

(b) Using what you have learned about the effect of a linear transformation and the results for the original data set from Exercise 1.53, calculate the mean, the standard deviation, and the median of the growing season in weeks.

1.56 Find the 20th and the 80th percentiles of the distribution of residents over age 65 from Exercise 1.18.

1.57 Find the *quintiles* (the 20th, 40th, 60th, and 80th percentiles) of the growing season data in Exercise 1.22. For quite large sets of data, the quintiles or the *deciles* (10th, 20th, 30th, etc. percentiles) give a more detailed summary than the quartiles.

1.58 We have already compared the distribution of calories in the three types of hot dogs represented in Table 1.4. The following computer output gives the basic descriptive measures for the three distributions of sodium content. Using this information, draw side-by-side boxplots and comment on any differences between the distributions. (The notation of the output is that of PROC UNIVARIATE in the SAS software system, which also reports other measures that are not needed at this point.)

```
VARIABLE=BEEF

    N          20        100% MAX    645.00
    MEAN     401.15       75% Q3     478.00
    STD DEV  102.43       50% MED    380.50
                          25% Q1     320.50
                           0% MIN    253.00
```

```
VARIABLE=MEAT

        N               17          100% MAX        545.00
        MEAN         418.53          75% Q3         496.00
        STD DEV       93.87          50% MED        405.00
                                     25% Q1         386.00
                                      0% MIN        144.00

VARIABLE=POULTRY

        N               17          100% MAX        588.00
        MEAN         459.00          75% Q3         528.00
        STD DEV       84.74          50% MED        430.00
                                     25% Q1         383.00
                                      0% MIN        357.00
```

1.59 Compute five-number summaries of the blood pressure distributions for the calcium and placebo groups in the experiment in Exercise 1.27. Draw side-by-side boxplots to compare the two distributions. Are any important differences between the groups apparent? (Remember that the boxplots will be less informative than the back-to-back stemplot of Exercise 1.27.)

1.60 Compute five-number summaries of the weight gains in the experimental and control groups of chicks in Exercise 1.28. Draw side-by-side boxplots to compare the distributions. Write a short statement about the observed differences, backing your verbal claims by citing appropriate numbers. (We will see in Exercise 1.108 that both distributions are approximately normal. Therefore, analysis of this experiment would, in practice, be based on means and standard deviations. Exercise 8.45 will ask you to carry out a formal analysis.)

1.61 Refer again to the data in Exercise 1.28, which give weight gains in grams (g) for two groups of chicks.
  (a) One ounce equals 28.35 g. What is the linear transformation $x^* = a + bx$ that restates the weight gains in ounces?
  (b) Using the results of the previous exercise and what you know about the effect of linear transformations, calculate the five-number summary for each group in ounces.
  (c) Make side-by-side boxplots to compare the two groups using the transformed weight scale. Describe the similarities and differences between these side-by-side boxplots and the ones you made in the previous exercise.

1.62 Give an approximate five-number summary of the salary data in Exercise 1.30 by pretending that all salaries fall at the midpoint of their class. (Replace the extremes by the 10th and 90th percentiles.)

1.63 The *trimmed mean* is a measure of center that is more resistant than the mean but that uses more of the available information than the median. To compute the 10% trimmed mean, discard the highest 10% and the lowest 10% of the observations and compute the mean of the remaining 80%. Trimming eliminates the effect of a small number of outliers.

Compute the 10% trimmed mean of the maximum rainfall data given in Exercise 1.20. Then compute the 20% trimmed mean. Compare the values of these measures with the median and the ordinary untrimmed mean.

1.3 THE NORMAL DISTRIBUTIONS

We now have command of a kit of graphical and numerical tools for describing distributions. But these tools alone do not give us a full picture of a substantial data set such as the 947 vocabulary scores for Gary, Indiana seventh graders summarized in Table 1.2. A stemplot is impractical, a frequency table and histogram suppress much detail and depend upon a choice of classes, and numerical measures such as the median and percentiles report only selected aspects of the data. Is there a way to describe an entire distribution in a single expression? If we are willing to accept a description of the overall shape of the distribution, omitting outliers and other deviations from a regular pattern, the answer is yes. Such a description is provided by a *mathematical model* for the distribution.

The use of idealized mathematical descriptions for complex objects and phenomena is a common and powerful tool in science, not least in statistics. Suppose, for example, that we are watching a caterpillar crawl forward. Once each minute we measure the distance it has moved, plot the observed values against time, and see that the points fall close to a straight line. It would be natural to draw a straight line through our graph as a compact description of these data. The straight line, with its equation of the form $y = a + bt$ (where y is distance and t is time), is a mathematical model of the caterpillar's progress. It is an idealized description because the points on the graph do not fall *exactly* on the line. We want to give a similarly compact description of the distribution of a variable, to replace stemplots and histograms. Just as a straight line is one of many types of curves that can be used to describe a plot of distance traveled against time, there are many types of mathematical distributions that can be used to describe a set of single-variable data. This section will concentrate on one type, the normal distributions.

Density Curves

The rationale for using mathematical models to describe distributions lies close at hand. We will approximate a histogram by a smooth curve that displays the shape of the distribution without the lumpiness of the histogram.

Because a smooth curve can be described by a single equation, this gives us a compact description of the entire distribution. But because the curve smooths out some of the detail presented in the histogram, the description is idealized and not completely accurate. Figure 1.10 shows a histogram of the Gary vocabulary test scores from Example 1.6. We used twice as many classes as in Figure 1.1 in order to display more detail. The corresponding normal curve is drawn over the histogram. You can see that the curve describes the overall shape in idealized form. The curve is actually a better description of the distribution of test scores than Figure 1.10 suggests, since some of the unevenness in the histogram is due to the accident of the specific choice of classes.

The vertical scale for a frequency histogram depends on the total number of observations. The scale for a relative frequency histogram is less arbitrary, because only the percentage or fraction of data in each class is recorded. The area of the bars in a histogram is proportional to the relative frequencies. The simplest choice of a curve to describe a distribution makes area not only proportional to relative frequency but exactly equal to it. Since the relative frequency of all the observations together is 1, we require that our curve have *total area 1* underneath it. The area under the curve and above any range of values of the variable is then the proportion of observations falling in this range, thought of as a fraction of the whole. The curve describes the shape of the distribution and has the advantage that *area = relative frequency*. It **density curve** is called the *density curve* of the distribution.

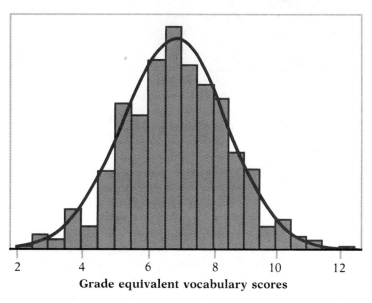

2   4   6   8   10   12
Grade equivalent vocabulary scores

Figure 1.10 Histogram of the vocabulary scores from Example 1.6, showing the approximation by a normal curve.

Density curves, like distributions, come in many shapes. Figure 1.11 shows two density curves, a symmetric normal density curve and a right-skewed curve. A density curve of the appropriate shape is often an adequate description of the overall pattern of a distribution. Outliers, which are deviations from the overall pattern, are not described by the curve.

The measures of center and spread described above apply to density curves as well as to actual sets of observations. Think first about percentiles. Since a proportion p (expressed as a percent) of all observations falls below the pth percentile, *the pth percentile on a density curve is the point with p percent of the area lying to the left* and the remaining $100-p$ percent to the right. In particular, the median is the "equal-area point," the point with half the total area on each side. As Figure 1.11(a) illustrates, the median of a symmetric density curve falls at the center of symmetry. On an asymmetric curve, the equal-area point is not so easy to locate by eye. There are mathematical ways of finding the median for any density curve, which were used to mark the median on the skewed curve in Figure 1.11(b). You can judge that the areas on either side of this point are equal. You can also roughly locate the quartiles by dividing the area into quarters as accurately as possible by eye. The *IQR* is then the distance between the first and third quartiles.

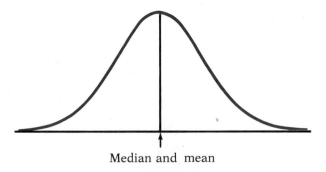

Median and mean

Figure 1.11(a)   The median and mean of a symmetric density curve.

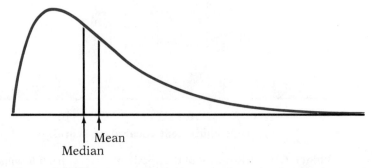

Mean

Median

Figure 1.11(b)   The median and mean of a right-skewed density curve.

What about the mean? The mean of a set of observations is just their arithmetic average. If we think of the observations as weights strung out along a thin rod, the mean is the point at which the rod would just balance on a fulcrum beneath. This interpretation extends to density curves: *The mean is the point at which the curve would balance if made of solid material.* Figure 1.12 illustrates this interpretation of the mean. The mean of a symmetric curve is therefore at the center of symmetry, and the mean and median are equal for symmetric distributions. As part of the idealization inherent in a mathematical model, symmetric density curves are exactly

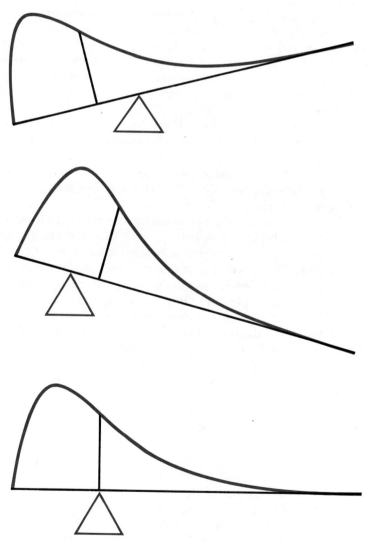

Figure 1.12 The mean is the balance point of a density curve.

symmetric even though real data will rarely show perfect symmetry. The mean (or balance point) of a skewed distribution is pulled toward the long tail of a skewed density curve more than is the median. This is because a small area (a few observations) far out in the tail can tip the curve more than the same area near the center. Locating the mean of a skewed distribution by eye is a bit tricky. There are mathematical ways of calculating the mean for any density curve, so that we are able to mark the mean as well as the median on the skewed curve in Figure 1.11(b).

The mean and percentiles can be located at least approximately by eye on any density curve. This is not true of the standard deviation, in keeping with our earlier remark that the standard deviation is not a natural measure for most distributions. However, when necessary, we can call on the results of more advanced mathematics to learn the value of the standard deviation. The study of mathematical methods for doing calculations with density curves belongs to theoretical statistics. Though we are concentrating on statistical practice, we often make use of the results of mathematical study. The ability to apply mathematical methods is a major advantage of using mathematical models to describe data.

Because a density curve is an idealized description of the distribution of data, we should distinguish between the mean and standard deviation of the density curve and the numbers $\bar{x}$ and s computed from the actual observations. The usual notation for the mean of an idealized distribution is μ, the Greek letter "mu." The standard deviation is denoted by σ, the Greek letter "sigma."

Our usual strategy in examining data is to look first for an overall pattern and then for deviations from that pattern. The overall pattern can be presented in ways that are based entirely on the data, by describing the major features of a stemplot or histogram. Sometimes no briefer description is adequate, as in the discussion of the calorie content of hot dogs that followed Example 1.11. But statisticians usually try to find a mathematical model that gives a compact description of the overall pattern of the data. When the data are observations on a single variable, the models employed are density curves for the distribution of data. We will meet many other examples of the process of giving models for data (and then inspecting the deviations of the actual data from the model) as we progress.

Normal Distributions

normal distribution

One particularly important class of density curves has already appeared in Figures 1.10 and 1.11(a). The *normal distributions* are symmetric, single-peaked, bell-shaped density curves. All normal distributions have the same overall shape; the exact density curve for a particular normal distribution is described by giving its mean μ and its standard deviation σ. The mean is, of course, located at the center of the curve, and is the same as the median. Changing μ without changing σ moves the normal curve to a new location without altering its spread. The standard deviation controls the spread of a

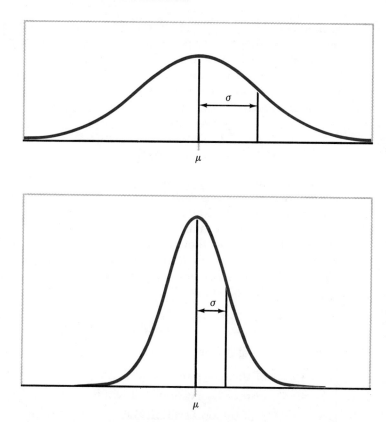

Figure I.I3   Two normal curves, showing the mean μ and standard deviation σ.

normal curve. Figure 1.13 shows normal curves with the same mean but different values of σ. We can even locate σ by eye on a normal density curve. As we move out in either direction from the center μ, the curve changes from falling ever more steeply

to falling ever less steeply

The points at which this change of curvature takes place are located at distance σ on either side of the mean μ. You can feel the change as you run a pencil along a normal curve, and so find the standard deviation.

Remember that μ and σ alone do not specify the shape of general distributions, and that the shape of density curves in general does not reveal σ. These are special properties of normal distributions.

There are other symmetric bell-shaped density curves that are not normal. The normal density curves are specified by a particular equation: The height of the density curve at the point x is given by

$$\frac{1}{\sigma\sqrt{2\pi}} e^{-\frac{1}{2}(\frac{x-\mu}{\sigma})^2} \qquad (1.4)$$

We will not make direct use of this fact, although it is the basis of mathematical work with normal distributions. Note that the equation of the curve is completely specified by μ and σ.

Though there are many normal curves, as Figure 1.13 illustrates, they are tightly bound together. In fact, *all normal distributions are the same if measurements are made in units of σ about the mean μ.* In particular, all normal distributions have the properties illustrated in Figure 1.14 and specified in the following rule.

The 68–95–99.7 rule

In any normal distribution:

- 68% of the observations fall within σ of the mean μ.
- 95% of the observations fall within 2σ of μ.
- 99.7% of the observations fall within 3σ of μ.

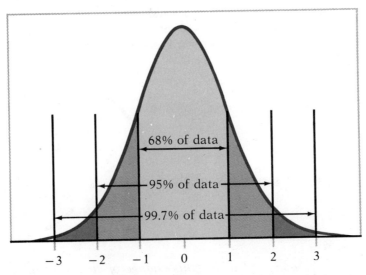

Figure 1.14 The 68–95–99.7 rule for normal distributions.

The 68–95–99.7 rule enables us to think about normal distributions without constantly making detailed calculations.

EXAMPLE 1.15

The distribution of heights of American women aged 18 to 24 is approximately normal with mean $\mu = 65.5$ inches and standard deviation $\sigma = 2.5$ inches. From the 68–95–99.7 rule we can estimate that the middle 95% of the distribution lies between 60.5 inches and 70.5 inches tall. Therefore, the tallest 2.5% of these women are taller than 70.5 inches. (The extreme 5% fall more than 2σ, or 5 inches, from the mean. Because the normal distributions are symmetric, half of these women are on the tall side.) Almost all young women are between 58 inches and 73 inches in height, according to the 99.7 part of the rule. Figure 1.15 shows the application of the 68–95–99.7 rule in this example. ■

The normal distributions provide good models for some distributions of real data. Distributions that are often close to normal include scores on tests taken by a broad population (e.g., tests of scholastic ability and many psychological tests); repeated careful measurements of the same quantity (e.g., Newcomb's data in Table 1.1 without the outliers); and characteristics of homogeneous biological populations (e.g., lengths of cockroaches, yields of corn, and moisture loss in packaged chicken meat). The distributions of some other common variables are usually skewed, and therefore distinctly non-normal. Examples include economic variables such as personal income and gross sales of business firms, the survival times of cancer patients after treatment, and the service lifetime of mechanical or electronic components.

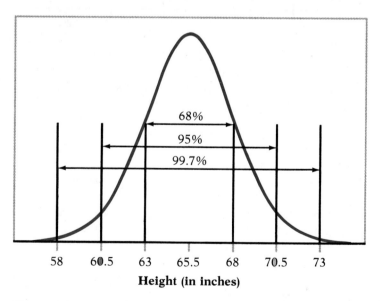

Figure 1.15   The 68–95–99.7 rule applied to the heights of young women.

While experience can suggest whether or not a normal model is plausible in a particular case, it is risky to assume that a distribution is normal without actually inspecting the data.

EXAMPLE 1.16

We expect a large set of Iowa Test scores to be approximately normal, and in fact Figure 1.10 suggests that the normal model is adequate for the 947 Gary vocabulary scores. Since it is difficult to assess the fit of a normal curve to a histogram by eye, let us use the 68–95–99.7 rule to check more closely. Having entered the scores into a statistical computing system as the values of a variable GARY, we obtain the mean and standard deviation as follows. (We used the MINITAB system; most other systems do these calculations in response to similar commands.)

```
MTB> MEAN 'GARY'
    MEAN = 6.8585

MTB> STDEV 'GARY'
    STDEV = 1.5952
```

Now that we know that $\bar{x} = 6.8585$ and $s = 1.5952$, we imitate the description of the normal distribution in Figure 1.14 by finding the actual counts of Gary vocabulary scores in intervals of length s about the mean $\bar{x}$. This requires more elaborate computing. Here are the counts.

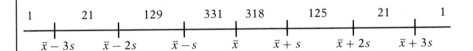

The symmetry of this distribution is evident, and its adherence to the 68–95–99.7 rule is striking: There are 649 scores (68.5%) within one standard deviation of the mean, 903 (95.4%) within two standard deviations, and 945 (99.8%) of the 947 scores within three. These counts confirm that the normal model with $\mu = 6.86$ and $\sigma = 1.595$ fits these data well. ∎

Adopting a density curve as a model for a distribution should not be done lightly, because the curve simultaneously specifies every aspect of the distribution. Tests for the adequacy of particular models are therefore important. Comparing actual counts with the 68–95–99.7 rule, as in Example 1.16, provides a quick check for normality.

What effect does changing the units of measurement or coding the data have on a normal distribution? We know that a linear transformation does not change the general shape of a distribution, and that the mean and standard deviation change in a simple manner. In fact, *any variable obtained* *from a normal variable by a linear transformation remains normal.* Of course, the mean and standard deviation will change in the usual way, so if x has the normal distribution with mean μ and standard deviation σ, then $x^* = a + bx$ for a positive b has the normal distribution with mean $a + b\mu$ and standard deviation $b\sigma$.

EXAMPLE 1.17

A food company bakes cakes in a large oven at a temperature of 350°F. The temperature x is not exactly constant, but varies according to a normal distribution with mean 350°F and standard deviation 2°F. If instruments that measure temperature on the Celsius scale are used, what is now the distribution of the oven temperature? The temperature in the Celsius scale is

$$x^* = \frac{5}{9}(x - 32) = -\frac{160}{9} + \frac{5}{9}x$$

Therefore, the Celsius temperature x^* has the normal distribution with mean

$$\mu_{x^*} = -\frac{160}{9} + \frac{5}{9}\,350 = 176.7°$$

and standard deviation

$$\sigma_{x^*} = \frac{5}{9}\,2 = 1.1°$$

The change of scale did not change the normal shape of the distribution, but it did change the numerical value of the center and the spread. ■

Normal Distribution Calculations

The normal distributions are commonly used models for the distributions of observed variables, so much so that a compact notation is helpful. If X is a measured variable having the normal distribution with mean μ and standard deviation σ, we say that X is $N(\mu, \sigma)$.

Because area under a density curve is relative frequency, any question about relative frequencies in a distribution described by a density curve can be answered by calculating areas under the curve. In the case of normal distributions, these calculations are made easier by the fact that, except for changes in location and scale, there is only one normal curve. The 68–95–99.7 rule is one consequence of this fact. A more sweeping consequence is the opportunity to find relative frequencies for any normal distribution from a single table of areas.

All normal distributions are the same when measurements are made in standard deviation units about the mean as origin. Here is a more formal statement of this fact.

Standard normal distribution

If a variable X has a normal distribution with mean μ and standard deviation σ, then the standardized variable

$$Z = \frac{X - \mu}{\sigma} \qquad (1.5)$$

has the normal distribution $N(0, 1)$ with mean 0 and standard deviation 1. This is called the standard normal distribution.

---

standardized

When the distribution of a variable is approximately normal, observations are often *standardized* by subtracting the mean and dividing by the standard deviation. A standardized observation states how many standard deviations the original observation falls away from the mean—and in which direction. Observations larger than the mean are positive when standardized, while observations smaller than the mean are negative. Standardization is a form of coding that changes the mean to 0 and the standard deviation to 1.

EXAMPLE 1.18

The heights of American young women are (approximately) normally distributed with $\mu = 65.5$ inches and $\sigma = 2.5$ inches. The standardized height

$$Z = \frac{\text{height} - 65.5}{2.5}$$

then follows the standard normal distribution. A woman's standardized height is the number of standard deviations by which her height differs from the mean height of all American young women. A woman 69 inches tall, for example, has a standardized height of

$$Z = \frac{69 - 65.5}{2.5} = 1.4$$

or 1.4 standard deviations above the mean. Similarly, a woman 5 feet (60 inches) tall has a standardized height of

$$Z = \frac{60 - 65.5}{2.5} = -2.2$$

or 2.2 standardized deviations less than the mean height. ∎

What fraction of all American young women are less than 69 inches tall? This relative frequency is the area under the $N(65.5, 2.5)$ curve to the left of the point 69. But because the standardized height corresponding to 69 inches is 1.4, this area is the same as the area under the *standard* normal curve below the point 1.4. *Table A in the back of the book gives areas under the standard normal curve*, calculated using Equation 1.4. The table entry for each value z is the area to the left of z.

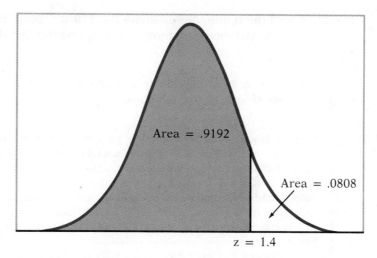

Area = .9192

Area = .0808

z = 1.4

Figure 1.16 The area under a standard normal curve below the point z = 1.4 is 0.9192.

EXAMPLE 1.19

To find the area to the left of 1.4, we enter Table A under $z = 1.40$. Locate 1.4 in the left-hand column, then the remaining digit in the top row. The entry opposite 1.4 and under 0.00 is 0.9192, which is the area we seek. Figure 1.16 illustrates the relationship between the value $z = 1.40$ and the area 0.9192. This area is the relative frequency of the event that $Z < 1.4$. Since the total area under the curve is 1, the area lying *above* 1.4 is $1 - 0.9192 = 0.0808$. Therefore, the fraction of young women who are less than 69 inches tall is 0.9192 (or about 92%) and the fraction who are taller than 69 inches is 0.0808 (or about 8%).

Sometimes we encounter a value of z more extreme than those appearing in Table A. For example, the area to the left of $z = -4$ is not given directly in the table. But since -4 is less than -3.4, this area is smaller than the entry for $z = -3.40$, which is 0.0003. ∎

Notice that since there is *no* area under a smooth curve and exactly over the point 69, the area below 69 (the relative frequency of $X < 69$) is the same as the area at or below this point (the relative frequency of $X \le 69$). This is not true of actual data sets, which may contain a woman just 69 inches tall. It is a consequence of the idealized smoothing of normal models for data.

Any normal relative frequency can be found by a procedure similar to the one used in Examples 1.18 and 1.19:

- Formulate the problem in terms of an observed variable X.
- Restate the problem in terms of a standard normal variable Z by standardizing X, using Equation 1.5.

- Find the required area under the $N(0, 1)$ curve, making use of Table A, the symmetry of normal curves, and the fact that the total area is 1.

As the following examples illustrate, drawing a picture will help you do normal relative frequency calculations.

EXAMPLE 1.20

A video display tube for computer graphics terminals has a fine mesh screen behind the viewing surface. During assembly the mesh is stretched and welded onto a metal frame. Too little tension at this stage will cause wrinkles, while too much will tear the mesh. The tension is measured by an electrical device with output readings in millivolts (mV). At the present time, the tension readings for successive tubes follow the $N(275, 43)$ distribution. The minimum acceptable tension corresponds to a reading of 200 mV. What proportion of tubes exceed this limit?

Following the procedure above, with X denoting the tension measurement:

- We want the relative frequency that $X > 200$, where X is $N(275, 43)$.
- This is the same as the relative frequency that

$$\frac{X - 275}{43} > \frac{200 - 275}{43} = -1.74$$

or $Z > -1.74$ for a standard normal Z.

- From Table A, we see that the relative frequency that $Z < -1.74$ is 0.0409. Since the total area under the curve is 1, the area *to the right of* -1.74 is $1 - 0.0409 = 0.9591$. That is, about 96% of the tubes exceed the minimum. See Figure 1.17(a) for an illustration of the areas. ∎

Notice that because of the symmetry of the normal curves, the area above -1.74 is equal to the area below 1.74, so we could have found the relative frequency in Example 1.20 directly as the table entry for 1.74. There are usually several correct ways of doing normal distribution calculations from Table A.

EXAMPLE 1.21

In the production setting of Example 1.20, tension above 375 mV will usually tear the mesh. The acceptable range of tension readings is therefore 200 mV to 375 mV. What proportion of tubes have acceptable tension values?

- We must find the relative frequency that $200 < X < 375$.
- Standardizing, this is the same as

$$\frac{200 - 275}{43} < \frac{X - 275}{43} < \frac{375 - 275}{43}$$

or $-1.74 < Z < 2.33$.

- The area between -1.74 and 2.33 is the area below 2.33 *minus* the area below -1.74, or from Table A, $0.9901 - 0.0409 = 0.9492$. A picture like

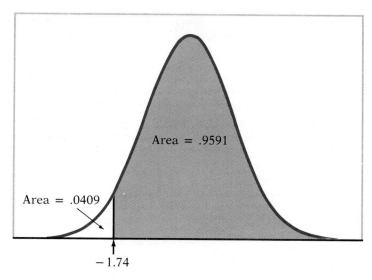

Figure 1.17(a) The area under the standard normal curve for Example 1.20.

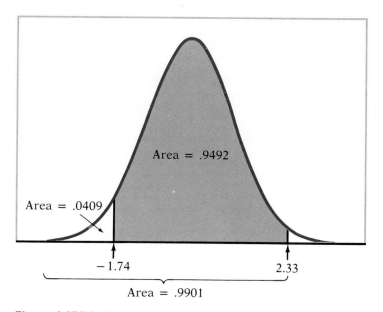

Figure 1.17(b) The area under the standard normal curve for Example 1.21.

Figure 1.17(b) helps us see the areas in question. Thus, 95% of the tubes have acceptable tension.

Can this procedure be improved upon? A moment's thought suggests that adjusting the operation so that the mean tension is exactly centered between the limits 200 mV and 375 mV will make the proportion of acceptable outcomes as high as possible for $\sigma = 43$ mV. This adjustment may be quite easy. Further improvement requires reducing the standard deviation. Since σ measures the inherent variability in the process, reducing σ will require actual improvement of the tensioning operation to make it less variable. ∎

Examples 1.20 and 1.21 illustrate the use of Table A to find the relative frequency of a given event, such as "tension reading between 200 mV and 375 mV." If we instead wish to know the observed values corresponding to a given relative frequency, Table A can help here also.

EXAMPLE 1.22

Scores on the Scholastic Aptitude Test (SAT) for verbal ability of high school seniors follow the $N(430, 100)$ distribution. (This is approximately true for current scores. The range of possible scores is 200 to 800.) How high must a student score in order to place in the top 10% of all students taking the SAT? Figure 1.18 poses this question in graphical form.

- Let X stand for a student's SAT score. We want to find the value x such that the event $X \geq x$ has relative frequency 0.1. That is, the event $X < x$ has relative frequency 0.9.
- In terms of standardized scores, this event is the same as

$$\frac{X - 430}{100} < \frac{x - 430}{100}$$

or

$$Z < \frac{x - 430}{100}$$

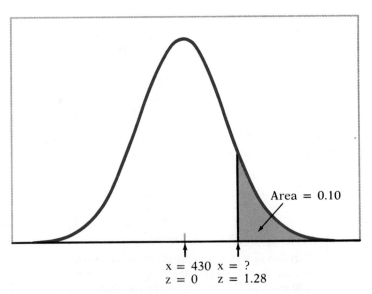

Figure 1.18 Locating the point on a normal curve with area 0.10 above it.

■ Look in the body of Table A for the entry closest to 0.90. It is 0.8997 and is the area below 1.28. So $Z < 1.28$ has relative frequency 0.1; therefore

$$\frac{x - 430}{100} = 1.28$$

Solving this equation for x gives

$$x = 430 + (1.28)(100) = 558$$

A student must score at least 558 to place in the highest 10%. ■

Remember that these calculations are valid only for normal distributions. Nonnormal distributions, like nonnormal people, are common and interesting.

Assessing Normality

The decision to describe a distribution by a normal model may determine the later steps in analyzing the data. Both relative frequency calculations and statistical inference based on such calculations follow from the choice of a model. Such a momentous choice should not be made blindly.

A histogram or stemplot can reveal distinctly nonnormal features of a distribution, such as outliers (Newcomb's histogram, Figure 1.2), pronounced skewness (the grocery purchase stemplot, Example 1.5), or gaps and clusters (the hot dog stemplot, Example 1.11). The effectiveness of these plots for assessing whether the distribution is normal can be improved by marking the points $\bar{x}$, $\bar{x} \pm s$, and $\bar{x} \pm 2s$ on the x axis. This gives the scale natural to normal distributions. We can then compare the count of observations in each interval with the 68–95–99.7 rule, as in Example 1.16.

quantile-quantile plot

A more sensitive assessment of the adequacy of the normal model for a set of data is provided by a *quantile-quantile plot*. "Quantile" is an alternative name for percentile when we speak in fractions rather than in percents. The general idea of a quantile-quantile plot is to compare two distributions by plotting their quantiles (or percentiles) against each other. If the distributions are nearly the same, their quantiles will be nearly the same. The quantile-quantile plot will then be close to the line $x = y$. If not, the deviation from this line will reveal exactly how the distributions differ. We are interested in only one application of this general idea: To compare the observed distribution of a variable to the theoretical normal distribution.

normal quantile plot

The idea of a *normal quantile plot* for a set of observations is to regard each observation as a quantile of the observed distribution, and plot it against the corresponding quantile of the standard normal distribution. The smallest of 20 observations, for example, is the 0.05 quantile of the data because 1/20, or 0.05, of the observations lie at or below it. So we plot the smallest observation on the horizontal axis, and on the vertical axis the z from Table A with area 0.05 below it. In this way, each observation is plotted against the z value that has the same fraction of the distribution below it. If the observations follow a standard normal distribution, the points will fall close to the line $x = y$ because the actual quantiles of the data will be close to the ideal quantiles from Table A. Because any normal variable X is related to a standard normal variable Z by a linear equation $X = \mu + \sigma Z$, the points will fall close to *some* straight line if the observations follow *any* normal distribution.

Normal quantile plots (often called normal probability plots) are the recommended tool for assessing normality. It is important to know how to interpret these plots rather than how to construct them, because it is not practical to draw normal quantile plots by hand. Fortunately, almost all statistical computing systems draw normal quantile plots, so users of statistics can easily check the normality or nonnormality of their data. Suppose that your observations have been entered into the MINITAB statistical package as column C1, for example. The following commands will draw a normal quantile plot:

```
MTB> NSCORES C1 C2
MTB> PLOT C2 C1
```

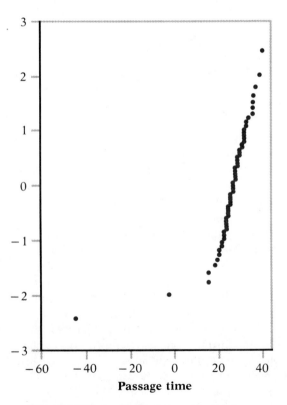

Figure 1.19(a) Normal quantile plot for Newcomb's passage time data, Table 1.1.

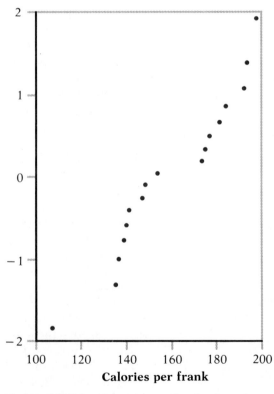

Figure 1.19(b) Normal quantile plot for calories in brands of meat hot dogs, Table 1.4.

There are several variations on the normal quantile plot, so you must ascertain which variety your computer system favors. We will draw the plot with the observed values on the horizontal scale and the standard normal quantiles on the vertical scale. Four such plots appear in Figure 1.19. The vertical scale in the normal quantile plots of Figure 1.19 extends from -3 to 3 because almost all of a standard normal curve lies between these values. If the data were perfectly normal, the plotted points would form a straight line. Approximately normal data will produce plotted points that lie close to a line. Let us examine these four plots in turn.

Figure 1.19(a) is a normal quantile plot of Newcomb's passage time data. Most of the points lie close to a straight line, indicating that a normal model fits well. The two outliers deviate from the line and show how the plot responds to long left tails or to low outliers: *In a left-skewed distribution the smallest observations fall distinctly to the left of a line drawn through the main body of points.* Similarly, *right skewness or high outliers cause the largest observations to fall distinctly to the right of a line drawn through the smaller observations.*

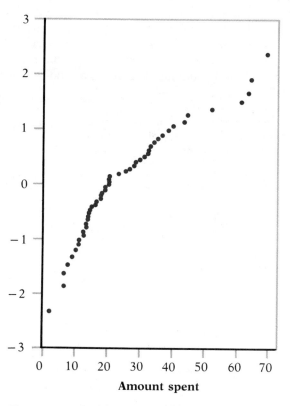

Figure 1.19(c) Normal quantile plot for amount spent by grocery shoppers, Example 1.10.

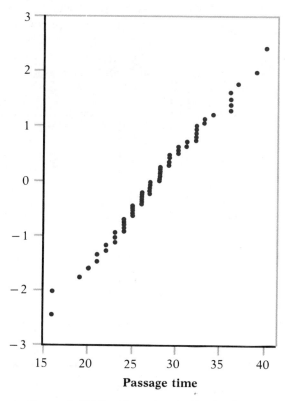

Figure 1.19(d) Normal quantile plot for Newcomb's data with the outliers omitted.

The plot of the data on the calorie content of meat hot dogs in Figure 1.19(b) displays clearly the two clusters and the low outlier that were apparent in the stemplot following Example 1.11. The right skewness of the lower cluster is also visible in the curvature of those points. The stemplot gives a clearer picture of this small and definitely nonnormal data set, but comparing Figure 1.19(b) with the stemplot helps us understand how normal quantile plots behave.

The data on amounts spent for groceries are plotted in Figure 1.19(c). The strong right skewness of this distribution is apparent if you lay a straightedge along the leftmost points, which correspond to the smaller observations. The larger observations fall systematically to the right of this line, indicating right skewness. Unlike Figure 1.19(a), there are no individual outliers, just a right tail that extends farther than in a normal distribution.

Figure 1.19(d) is another normal quantile plot of Newcomb's measurements of the passage time of light, this time with the two outliers omitted. The effect of omitting the outliers is to magnify the plot of the remaining data. As we saw from Figure 1.19(a), a normal distribution fits quite well. The only important deviations from normality are the numerous short vertical stacks of points in the plot. Each stack represents repeated observations having the same value—there are six measurements at 27 and seven at 28, for example. This phenomenon is called *granularity*, since the data come in lumps rather than being smoothly distributed as the normal model specifies. Minor granularity, as in this case due to the limited precision of the measurements, does not hinder adopting a normal distribution as a model. The

granularity

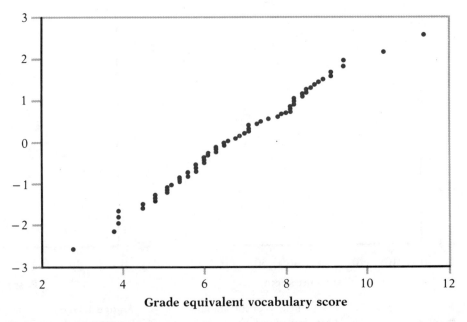

Figure 1.20   Normal quantile plot of the Gary vocabulary scores.

data would presumably be even closer to the normal model if Newcomb had been able to measure one more decimal place, thus spreading out the stacks.

What of the Gary vocabulary scores, whose normality we have assessed by cruder means in Figure 1.10 and Example 1.16? A plot of 947 individual points, many taking the same values, will wear holes in your paper. Figure 1.20 therefore presents a normal quantile plot based on the score of every tenth student in the list. It is remarkably similar to Figure 1.19(d), showing excellent fit to a normal model except for some granularity due to the reporting of scores only to the nearest tenth of a grade level. A normal distribution does provide an accurate model for this large data set.

As Figures 1.19(d) and 1.20 illustrate, real data will almost always show some departure from the theoretical normal model. It is important to confine your examination of a normal quantile plot to searching for shapes that show clear departures from normality. Don't overreact to minor wiggles in the plot. When we discuss statistical methods that are based on the normal model, we will pay attention to the sensitivity of each method to departures from normality. Many common methods work well when the data are approximately normal. It is this fact, along with the accuracy of the normal model for some common types of data, that explains the frequent use of the normal distributions in statistics.

SUMMARY

The overall pattern of a distribution can often be described compactly by a **density curve**. A density curve is a curve that always remains on or above the horizontal axis and has total area 1 underneath it. Area under a density curve is **relative frequency**.

The mean μ (balance point), the median (equal-area point), and other percentiles can be located on a density curve. The standard deviation σ cannot be located by eye on most density curves. The mean and median are equal for symmetric density curves, but the mean of a skewed curve is located farther toward the long tail than the median.

The **normal distributions** are specified by bell-shaped symmetric density curves. The mean μ and standard deviation σ completely describe the normal distribution $N(\mu, \sigma)$. The mean is the center of symmetry and σ is the distance from μ to the change of curvature points on either side. All normal distributions are the same when measurements are made in units of σ about the mean. These are called **standardized measurements**. In particular, all normal distributions conform to the **68–95–99.7 rule**.

If X has the $N(\mu, \sigma)$ distribution, then the **standardized variable** $Z = (X - \mu)/\sigma$ has the **standard normal** distribution $N(0, 1)$. Relative frequencies for any normal distribution can be calculated from Table A, which gives relative frequencies for the events $Z < z$ for many values of z.

The adequacy of a normal model for a set of data is best assessed by a **normal quantile plot**, which is available in most statistical software packages. Deviations of the points on such a plot from a straight line show deviations of the data from normality.

SECTION 1.3 EXERCISES

1.64 Figure 1.21 displays three density curves, with several points marked on each. At which of these points on each curve do the mean and the median fall?

1.65 Figure 1.22 displays the density curve of a *uniform* distribution. The curve takes the constant value 1 over the interval from 0 to 1, and is zero elsewhere. This means that data described by this distribution take values that are uniformly spread between 0 and 1. Use the fact that areas under a density curve are relative frequencies to answer the following questions.

  (a) What fraction of the observations lie above 0.8?
  (b) What fraction of the observations lie below 0.6?
  (c) What fraction of the observations lie between 0.25 and 0.75?

1.66 What are the mean and the median of the distribution in Figure 1.22? What are the quartiles?

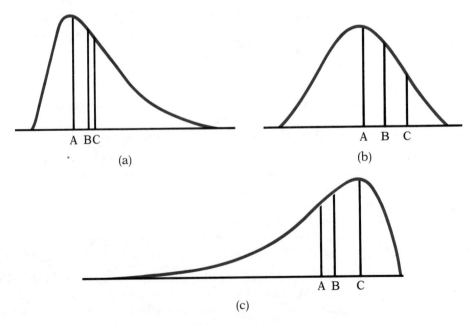

Figure 1.21 Density curves for Exercise 1.64.

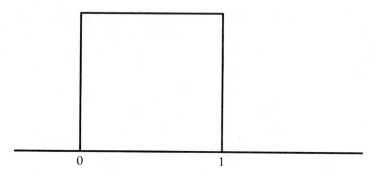

Figure 1.22 A uniform density curve for Exercise 1.65.

1.67 The Environmental Protection Agency requires that the exhaust of each model of motor vehicle be tested for the level of several pollutants. The level of oxides of nitrogen (NOX) in the exhaust of one light truck model was found to vary among individual trucks according to a normal distribution with mean $\mu = 1.45$ grams (g) per mile driven and standard deviation $\sigma = 0.40$ g per mile. Sketch the density curve of this normal distribution, with the scale of g per mile marked on the horizontal axis.

1.68 In a study of elite distance runners, the mean weight was reported to be 63.1 kilograms (kg), with a standard deviation of 4.8 kg. Assuming that the distribution of weights is normal, sketch the density curve of the weight distribution with the horizontal axis marked in kilograms. (Based on M. L. Pollock, et al., ''Body composition of elite class distance runners,'' in P. Milvy, (ed.) *The Marathon: Physiological, Medical, Epidemiological, and Psychological Studies*, New York Academy of Sciences, 1977.)

1.69 Give an interval that contains the middle 95% of NOX levels in the exhaust of trucks using the model described in Exercise 1.67.

1.70 Use the 68–95–99.7 rule to find intervals centered at the mean that will include 68%, 95%, and 99.7% of the weights of the elite runners described in Exercise 1.68.

1.71 Eleanor scores 680 on the mathematics part of the Scholastic Aptitude Test. The distribution of SAT scores in a reference population is normal with mean 500 and standard deviation 100. Gerald takes the American College Testing mathematics test and scores 27. ACT scores are normally distributed with mean 18 and standard deviation 6. Find the standardized scores for both students. Assuming that both tests measure the same kind of ability, who has the higher score?

1.72 Three landmarks of baseball achievement are Ty Cobb's batting average of .420 in 1911, Ted Williams' .406 in 1941, and George

Brett's .390 in 1980. These batting averages cannot be compared directly because the distribution of major league batting averages has changed over the decades. The distributions are quite symmetric and (except for outliers such as Cobb, Williams, and Brett) reasonably normal. While the mean batting average has been held roughly constant by rule changes and by the balance between batting and pitching, the standard deviation has dropped over time. Here are the facts:

Decade	Mean	Std. dev.
1910s	.266	.0371
1940s	.267	.0326
1970s	.261	.0317

Compute the standardized batting averages for Cobb, Williams, and Brett to compare how far each stood above his peers. (Data from Stephen Jay Gould, "Entropic homogeneity isn't why no one hits .400 any more," *Discover*, August 1986, pp. 60–66. Gould does not standardize, but gives a speculative discussion instead.)

1.73 Using Table A, find the proportion of observations from a standard normal distribution that satisfy each of the following statements. In each case, sketch a standard normal curve and shade the area under the curve that is the answer to the question.

(a) $Z < 2.85$
(b) $Z > 2.85$
(c) $Z > -1.66$
(d) $-1.66 < Z < 2.85$

1.74 Using Table A, find the relative frequency of each of the following events in a standard normal distribution. In each case, sketch a standard normal curve with the area representing the relative frequency shaded.

(a) $Z \leq -2.25$
(b) $Z \geq -2.25$
(c) $Z > 1.77$
(d) $-2.25 < Z < 1.77$

1.75 Use Table A to find the value z of a standard normal variable Z that satisfies each of the following conditions. (Use the value of z from Table A that comes closest to satisfying the condition.) In each case, sketch a standard normal curve with your value of z marked on the axis.

(a) The point z with 25% of the observations falling below it
(b) The point z with 40% of the observations falling above it

1.76 The variable Z has a standard normal distribution.

(a) Find the value z of Z such that the event $Z < z$ has relative frequency 0.8.

(b) Find the value z of Z such that the event $Z > z$ has relative frequency 0.35.

1.77 The height X of American young women has approximately the normal distribution with mean $\mu = 65.5$ inches and standard deviation $\sigma = 2.5$ inches. Use Table A to find the relative frequencies of each of the following events. In each case, sketch a normal curve and shade the area that represents the relative frequency.

(a) $X < 67$

(b) $64 < X < 67$

1.78 The Graduate Record Examinations (GRE) are widely used to help predict the performance of applicants to graduate schools. The range of possible scores on a GRE is 200 to 900. The psychology department at a university finds that the scores of its applicants on the quantitative GRE are approximately normal with mean $\mu = 544$ and standard deviation $\sigma = 103$. Use Table A to find the relative frequency of applicants whose score X satisfies each of the following conditions:

(a) $X > 700$

(b) $X < 500$

(c) $500 < X < 800$

1.79 A patient is said to be hypokalemic (low potassium in the blood) if the measured level of potassium is 3.5 or less. An individual's potassium level is not a constant, however, but varies from day to day. In addition, measurement errors may occur in the procedure. Suppose that the variation follows a normal distribution. Judy has a mean potassium level of 3.8 with a standard deviation of 0.2. If she were measured on many days, what would be the proportion of days on which the measurement would indicate hypokalemia?

1.80 The Acculturation Rating Scale for Mexican Americans (ARSMA) is a psychological test that evaluates the degree to which Mexican Americans are adapted to Mexican/Spanish versus Anglo/English culture. The distribution of ARSMA scores in a population used to develop the test is approximately normal with mean 3.0 and standard deviation 0.8. The range of possible scores is 1.0 to 5.0, with higher scores showing more Anglo/English acculturation. A researcher believes that Mexicans will have an average score near 1.7, and that first-generation Mexican Americans will average about 2.1 on the ARSMA scale. What proportion of the original population has scores below 1.7? Between 1.7 and 2.1?

1.81 Refer to Exercise 1.68. If 90% of elite distance runners weigh less than Peter, what is Peter's weight?

1.82 How high a score on the ARSMA test of Exercise 1.80 must a Mexican American obtain in order to place in the 30% of the original population who are most oriented toward Anglo/English culture? What scores fall in the lowest 30%, that is, the 30% who are most Mexican/Spanish in their acculturation?

1.83 In a study of London bus drivers, measurements of daily food intake in calories were taken. The drivers had a mean intake of 2821 calories with a standard deviation of 436 calories. The distribution was approximately normal. Calculate the proportion of drivers whose daily consumption falls in each of the following ranges: less than 2000, 2000 to 2500, 2500 to 3000, 3000 to 3500, and above 3500.

1.84 The scores of a reference population on the Wechsler Intelligence Scale for Children (WISC) are normally distributed with $\mu = 100$ and $\sigma = 15$.

(a) What percent of this population have WISC scores below 100?
(b) Below 80?
(c) Above 140?
(d) Between 100 and 120?

1.85 The distribution of scores on the WISC is described in the previous exercise. What score must a child achieve on the WISC in order to fall in the top 5% of the population? In the top 1%?

1.86 The distribution of a dimension on auto engine crankshafts (see the sample data in Exercise 1.11) is approximately normal with $\mu = 224$ mm and $\sigma = 0.03$ mm. Crankshafts with dimensions between 223.92 mm and 224.08 mm are acceptable. What percent of all crankshafts produced are acceptable? What percent of the 16 specific crankshafts measured in Exercise 1.11 were acceptable?

1.87 Compare the percents of the 16 measurements in Exercise 1.11 that fall within one, two, and three standard deviations of the mean of the $N(224, 0.03)$ distribution with the percents in the 68–95–99.7 rule. Is there any strong indication that the data do not follow this distribution?

1.88 The median of any normal distribution is the same as its mean. We can also use Table A to evaluate the quartiles and related descriptive measures for normal distributions.

(a) What are the first and third quartiles of the standard normal distribution? (These numbers give the location of the quartiles for any normal distribution, in standard deviation units away from the mean.)
(b) What is the value of the *IQR* for the standard normal distribution?

(c) What percent of the observations in the standard normal distribution are suspected outliers according to the 1.5 × *IQR* criterion? (This percent is the same for any normal distribution.)

1.89 Figure 1.23 is a normal quantile plot of the DRP scores from Exercise 1.23. Are these scores approximately normally distributed? Discuss any major deviations from normality that appear in the plot.

1.90 In a medical experiment, 72 guinea pigs are injected with tubercle bacilli to study their resistance to infection. The survival time (in days) after infection is measured. The actual data appear in Exercise 8.33. Figure 1.24 is a normal quantile plot of the survival times. In what way does the distribution of survival times depart from normality? (Based on data in T. Bjerkedal, "Acquisition of resistance in guinea pigs infected with different doses of virulent tubercle bacilli," *American Journal of Hygiene*, 72 (1960), pp. 130–148.)

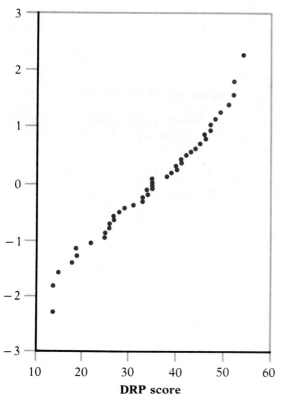

Figure 1.23 Normal quantile plot of the DRP scores for Exercise 1.89.

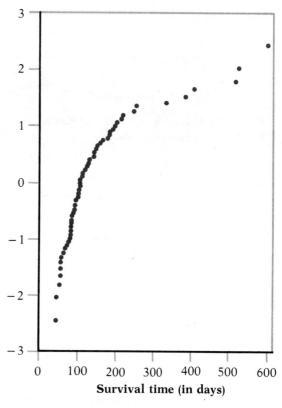

Figure 1.24 Normal quantile plot of guinea pig survival times for Exercise 1.90.

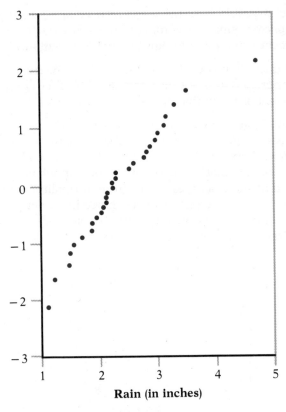

Figure 1.25 Normal quantile plot of maximum daily rainfall for Exercise 1.91.

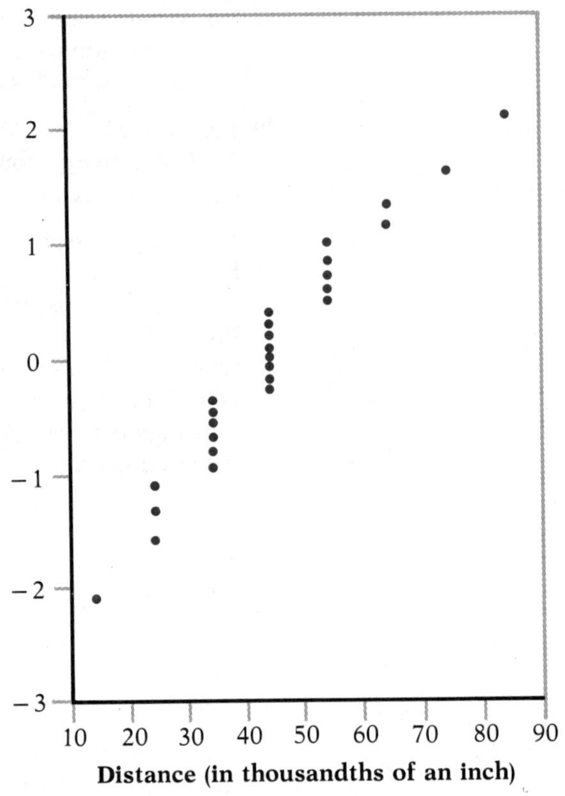

Figure 1.26 Normal quantile plot of distance measurements for Exercise 1.92.

1.91 Figure 1.25 is a normal quantile plot of the maximum rainfall data from Exercise 1.20. Is the distribution approximately normal? Are there any clear deviations from the normal shape?

1.92 The distance between two mounting holes is important to the performance of an electrical meter. The manufacturer measures this distance regularly for quality control purposes, recording the data in coded form, as thousandths of an inch above 0.600 inch. For example, 0.644 is recorded as 44. Figure 1.26 is a normal quantile plot of the distances for the last 27 meters measured. Is the overall shape of the distribution approximately normal? Why do you think that many of the points appear in vertical stacks? (Data provided by Charles Hicks.)

1.93 Assess the normality of Cavendish's measurements in Exercise 1.25 by finding the percent that fall within one, two, and three standard deviations of the mean. (The mean and standard deviation were computed in Exercise 1.48.) Is there a clear indication of nonnormality?

The remaining exercises for this section require the use of computer software that will make normal quantile plots.

1.94 Make a normal quantile plot for Cavendish's measurements in Exercise 1.25. Are the data approximately normal or is there a clear deviation from normality?

1.95 Make a normal quantile plot of the crankshaft measurements in Exercise 1.11. Describe the nature of any clear departure from normality that you see.

1.96 Table 1.4 gives the calorie content of a number of brands of beef, meat, and poultry hot dogs. We have already compared these distributions to some extent in Example 1.11 and Figure 1.9. Figure 1.19(b) is a normal quantile plot for meat hot dogs. Make normal quantile plots for the calorie content of beef and poultry hot dogs. Then briefly compare the three distributions on the basis of their normal quantile plots.

NOTES

1 Newcomb's data and the background information about his work appear in S. M. Stigler, "Do robust estimators work with real data?" *Annals of Statistics*, 5 (1977), pp. 1055–1078.

2 *The Economist*, Aug. 30, 1986.

3 Data from *The Baseball Encyclopedia*, 3d ed., Macmillan, New York, 1976. Maris' home run data are from the same source.

4 Data from Gary Community School Corporation, courtesy of Celeste Foster, Department of Education, Purdue University, West Lafayette, Ind.

5 James Gleick, "Hole in ozone over South Pole worries scientists," *The New York Times*, July 29, 1986.

CHAPTER I EXERCISES

1.97 Voting patterns were affected by many social changes in the years between 1940 and 1980. Prior to the civil rights movement in the 1960s, blacks were effectively prevented from voting in some southern states. And until 1971 citizens between the ages of 18 and 21 could not vote in national elections. Figure 1.27(a) displays the distribution of the percent of the voting age population who actually voted in the 1940 presidential election in each of the 48 states that existed at that time. Figure 1.27(b) shows the distribution of the percent of the voting age population who voted in the 1980

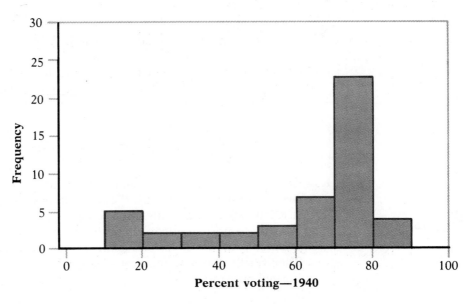

Figure 1.27(a)   Distribution of the percent of the voting age population who voted in each of the states in the 1940 presidential election, for Exercise 1.97.

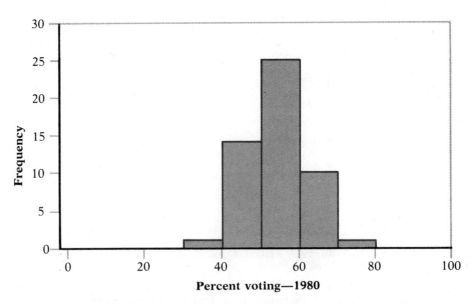

Figure 1.27(b)   Distribution of the percent of the voting age population who voted in each of the states in the 1980 presidential election, for Exercise 1.97.

presidential election in the 50 states and the District of Columbia. To allow a direct comparison, the horizontal and vertical scales and the class widths are the same for the two histograms. Describe the most important changes in the shape of the distribution between 1940 and 1980. In particular, how is the addition of black voters reflected in the shapes of the distributions?

1.98 Discuss in detail the distribution of the eccentricity measurements in Exercise 1.12. Include at least one graphical display of the distribution, and numerical measures that are appropriate for a distribution of this shape. Is the distribution approximately normal? (You may use either the original measurements or the coded values.)

1.99 You are planning a sample survey of households in Indiana. You decide to select households separately within counties. (Chapter 4 will present more information on planning sample surveys.) To aid in the planning, here are the populations of the state's 92 counties (in thousands of persons) according to the 1980 Census.

30	294	65	10	16	36	12	20	41	89
25	32	10	28	34	24	34	129	34	137
28	61	19	20	19	33	81	30	82	44
27	70	53	87	36	37	26	23	30	23
77	42	60	26	523	109	42	139	765	39
11	40	99	36	52	15	35	5	19	16
16	19	13	120	26	13	29	30	24	20
242	20	40	19	22	25	21	7	122	17
7	168	18	112	37	9	41	22	76	25
24	26								

(a) Make either a stemplot or a histogram for these data. If suitable software is available, make a normal quantile plot. Briefly describe the overall shape of the distribution of county population counts.

(b) Compute the mean and the standard deviation.

(c) Compute the five-number summary and the *IQR*.

(d) Do the mean and the median differ substantially? If so, why?

(e) The 68–95–99.7 rule states that about 68% of the counties should have populations within one standard deviation of the mean. What percentage of the counties do in fact lie within one standard deviation of the mean? What do you conclude about the applicability of the rule to this distribution?

(f) Suppose that you were given only the summary measures and not the full data. Can you recover the approximate total population of Indiana from the median? From the mean? What is the state's population?

(g) Give a brief summary of the distribution, suitable for a report that will not contain the original data. Make use of appropriate numerical measures from those you have computed.

(h) Suggest some groupings of counties based on population size that would be useful in conducting the sample survey.

1.100 The following data are the founding dates of wineries in New Zealand:

1975	1980	1973	1974	1943	1902	1947	1942
1937	1983	1934	1936	1935	1902	1937	1961
1979	1969	1976	1925	1922	1937	1973	1933
1948	1944	1904	1981	1872	1918	1860	1960
1876	1980	1967	1974	1977	1979	1981	1978

(a) Make either a stemplot or a histogram for these data. If suitable software is available, make a normal quantile plot.

(b) Compute the mean and the standard deviation.

(c) Compute the five-number summary and the *IQR*.

(d) Give a careful description of the main features of the distribution. How useful are the numerical measures you have computed?

(e) There is another winery for which a precise founding date is not available. We do know, however, that it was founded after World War II. Assuming that the date is after 1945, recompute the median. Do you think that the median is a good descriptive measure for these data? Why or why not?

1.101 The following table gives the weights in pounds of the players on a Big Ten football team as listed in the team's preseason publicity brochure. We will investigate the shape of this distribution.

160	165	202	245	150	235	241	195	181	256
200	265	168	233	215	215	185	185	263	214
230	260	190	185	294	180	179	295	175	157
218	208	145	205	210	171	276	220	189	222
175	170	157	195	260	186	225	188	260	192
214	266	265	188	196	228	270	163	183	160
260	178	260	155	210	200	276	290	165	220
210	200	185	195	196	260	165	175	224	235
225	228	215	225	229	214	193	162	294	280
183	185	263	276	290	218	184	190	180	225
220	250	253	220	155	160	176	200	230	218
158	215	225	184	195	282	252	218	235	182
180	180	180	224	214	210	255	270	200	234
175									

(a) Make either a stemplot or a histogram, and explain your choice of graphical display. If you have suitable software, also make a normal quantile plot.

(b) Describe the main features of the distribution. Is it symmetric? Approximately normal? Does it appear that there are groups of players with different weight distributions, such as linemen and defensive backs? Are there any outliers?

(c) Give a brief numerical summary in light of your findings in (b).

(d) For publication you must restate your summary in kilograms rather than pounds; do this. (One kilogram equals 2.2 pounds, so 160 pounds is 72.73 kg.)

(e) There is a special kind of granularity in this data set; describe it. (Hint: Look at the last digit.)

1.102 The following table gives the 1987 baseball income (in thousands of dollars) for the members of the St. Louis Cardinals, National League champions that year. Examine this distribution, present it graphically and numerically in the way you think appropriate, and describe the notable features of the distribution. (Data from *The New York Times*, Nov. 1, 1987.)

Smith	1940	Dayley	325	Lake	85
Clark	1300	Pendleton	210	Ford	70
Pena	1200	Dawley	175	Morris	65
Tudor	1050	Coleman	160	Johnson	62.5
Herr	925	Worrell	160	Lindeman	62.5
McGee	765	Lawless	115	Magrane	62.5
Forsch	750	Conroy	110	Pagnozzi	62.5
Cox	600	Driessen	100	Peters	62.5
Lahti	450	Mathews	100	Tunnell	62.5
Horton	410	Oquendo	100		

1.103 Exercise 8.33 reports the survival time (in days) of 72 guinea pigs after they were injected with disease organisms in a medical experiment. Make a stemplot of these data. Briefly describe the overall shape of the distribution. Do you prefer the five-number summary or the mean and standard deviation as a brief numerical summary of the center and spread of these data? Calculate the measures that you prefer.

1.104 Explain why the shape of the distribution of DRP reading scores as displayed in the normal quantile plot in Figure 1.23 suggests that the mean and standard deviation are adequate measures of center and spread for these data. Then calculate the mean and standard deviation from the scores given in Exercise 1.23.

1.105 Refer again to the DRP reading scores whose mean and standard deviation you found in the previous exercise. Find the linear transformation—i. e., the values of a and b in $x^* = a + bx$—that preserves the order of the scores and produces transformed scores having mean 100 and standard deviation 10. (Hint: State the relations between the means of x and x^*, and between the standard deviations of x and x^*. Then put in the known values for x and the desired values for x^* and solve for a and b.)

1.106 A test designed to measure reading ability of children was administered to a large group of third and sixth grade children. The mean score for the third graders was 75 and the standard deviation was 10. For the sixth graders the mean was 82 and the standard deviation was 11. We want to transform the scores so that a meaningful comparison can be made between the score of a third grader and the score of a sixth grader relative to the scores of other children in the same grade.

   (a) Find the linear transformation—i. e., the values of a and b—that will transform the third grade scores so that they have a mean of 100 and a standard deviation of 20.
   (b) Do the same for the sixth grade scores.
   (c) Suppose that a third grade student scores 78 on the test. Find the transformed score. Do the same for a sixth grade student who scores 78.
   (d) Suppose that the distribution of scores in each grade is normal. Then both transformed scores have the $N(100, 20)$ distribution. What percent of third graders have scores less than 78? What percent of sixth graders have scores less than 78?

1.107 The Chapin Social Insight Test evaluates how accurately the subject appraises other people. In the reference population used to develop the test, scores are approximately normally distributed with mean 25 and standard deviation 5. The range of possible scores is 0 to 41.

   (a) What proportion of the population has scores below 20 on the Chapin test?
   (b) What proportion has scores below 10?
   (c) How high a score must you have in order to be in the top quarter of the population in social insight?

1.108 Consider the weight gain data for chicks fed normal and high-lysine corn in Exercise 1.28.

   (a) Make a normal quantile plot for each group if your software allows. Both distributions are approximately normal.
   (b) We can therefore summarize the distributions by computing the mean and standard deviation for each group; do this. Did

the high-lysine chicks gain more weight? Is the variabililty in weight gains substantially different in the two groups?

1.109 Most statistical software packages have routines for generating values of variables having specified distributions. The following MINITAB commands, for example, produce 100 observations from the $N(20, 5)$ distribution and then compute several descriptive measures for these observations.

```
MTB> RANDOM 100 Cl;
   SUBC> NORMAL 20, 5.
MTB> DESCRIBE Cl
```

Use your statistical software to generate 100 observations from the $N(20, 5)$ distribution. Compute the mean and standard deviation $\bar{x}$ and s of the 100 values you obtain. How close are $\bar{x}$ and s to the μ and σ of the distribution from which they are observations?

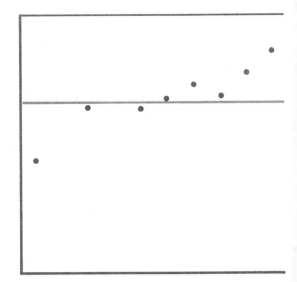

Prelude

In this chapter we move from examining the distribution of a single variable to exploring the change in a variable over time. We first see what it means for a variable to remain stable over time. Then we discuss two important regular patterns of change over time: straight-line growth and exponential growth. Finally we turn to more complex patterns of change over time and see how statistical methods can make these patterns clearer to us.

- *How can you distinguish systematic change from chance variation in a record of your commuting time, blood pressure, or spending on food?*
- *How can fitting a straight line help us study the growth of children?*
- *How does the idea of exponential growth enable us to understand the increase of world oil production in the twentieth century and to spot interruptions and changes in the pattern of increase?*
- *What kinds of variation may be present in a time series and how does statistical smoothing help us see regular patterns of change?*

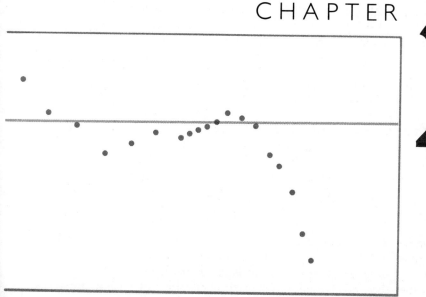

Looking at Data: Change and Growth

A statistical investigation only rarely focuses on the distribution of a single variable. We are more often interested in comparisons among several distributions, in changes in a variable over time, or in relationships among several variables. Chapter 1 provided graphical and numerical tools for comparing several distributions, as well as for examining the distribution of a single variable. In this chapter we will concentrate on methods for studying the change in a variable over time. Our approach will be familiar:

- Combine graphical display with numerical summaries.
- Seek overall patterns and deviations from those patterns.
- Seek compact mathematical models for the data in addition to descriptive measures of specific aspects of the data.

time series Many interesting data sets are *time series*, or measurements of a variable taken at regular intervals over time. Many government economic and social statistics are time series. Examples include the monthly unemployment rate, the quarterly gross national product, and the annual survey of victimization by crime. Manufacturers use statistical process control, based on regular measurements over time, to check that their production processes are turning out goods of the desired quality. The changes over time in variables such as tumor size or concentration of a drug in the bloodstream are vital in medicine. Personal data, such as your weight, commuting time, or bank balance, vary over time in ways that are important to you.

We begin the study of change and growth in Section 2.1 with a statistical description of stability over time. Sections 2.2 and 2.3 present simple mathematical models for regular growth, a particularly important kind of change. In Section 2.4 we look at the complex patterns of change that are typical of general time series.

2.1 STATISTICAL CONTROL

There are many situations in which our goal is to hold a variable steady over time. You may monitor your weight or blood pressure, and plan to change your behavior if either changes. Manufacturers watch the results of regular measurements made during production and plan to take action if the quality deteriorates. In these cases, the statistical study of change over time is intended to anticipate and prevent change. We will use the following case study to introduce statistical tools for this purpose.

EXAMPLE 2.1 How long does it take you to travel from home to work or school? Professor Simon, who lives a few miles outside of a college town, kept a record of his commuting time. The data (in minutes) appear in Table 2.1. They cover most of the fall semester, although on some dates the professor was out of town or forgot to set his stopwatch.

Table 2.1 Commuting time (minutes), September 10 to December 19

Date	Time	Date	Time	Date	Time
Sept. 10	8.25	Oct. 10	7.75	Nov. 22	7.75
Sept. 11	7.83	Oct. 13	7.92	Nov. 24	7.42
Sept. 12	8.30	Oct. 16	8.00	Nov. 28	6.75
Sept. 15	8.42	Oct. 17	8.08	Nov. 29	7.42
Sept. 17	8.50	Oct. 22	8.42	Dec. 2	8.50
Sept. 22	8.67	Oct. 23	8.75	Dec. 4	8.67
Sept. 23	8.17	Oct. 24	8.08	Dec. 5	10.17
Sept. 24	9.00	Oct. 27	9.75	Dec. 8	8.75
Sept. 25	9.00	Oct. 28	8.33	Dec. 10	8.58
Sept. 29	8.17	Nov. 3	7.83	Dec. 11	8.67
Sept. 30	7.92	Nov. 5	7.92	Dec. 12	9.17
Oct. 3	9.00	Nov. 19	8.58	Dec. 15	9.08
Oct. 8	8.50	Nov. 20	7.83	Dec. 17	8.83
Oct. 9	9.00	Nov. 21	8.42	Dec. 19	8.67

trend

outliers

The presence of variation in the data raises several questions. Is there a *trend*, a systematic long-term change in travel time, hidden in the day to day variation? A systematic increase, for example, might be due to new housing bringing heavier traffic. Are there any *outliers*? These unusually high or low measurements may be responses to unusual conditions, such as an accident or a foot of snow on the road. ■

The Idea of Statistical Control

The questions just asked in Example 2.1 can be rephrased as one: Is the distribution of the length of the drive to work stable or are changes occurring over time? Put in other terms, are there signs of unusual variation in the data in addition to the regular pattern of variation? The same question is asked when a manufacturer measures a key characteristic of a product during production, such as the thickness of a coat of primer sprayed onto new cars by a robot painter. There will always be variation from car to car, and some variation is acceptable. Yet the priming process should be stable enough that the quality of the painting is predictable during continued production.

Statistical control

> A variable that continues to be described by the same distribution when observed over time is said to be in statistical control, or simply in control.

We begin our study of growth and change by examining control versus lack of control. We wish to distinguish between the variation inherent in the variable observed (whether commuting time or primer thickness on new cars) and the extraordinary variation that signals a real change (bad weather that slows travel, or a new batch of primer that produces too thin a layer).

We are already quite advanced in the art of thinking statistically when we describe a variable as stable or in control if its distribution does not change with time. We do not insist that the variable itself remain unchanged. Measured variables—whether commuting time, Newcomb's records of the passage time of light, or the thickness of primer on new cars—do vary. Statistical control is a sophisticated concept because it recognizes that variability will be present and requires only that the pattern of variability remain the same.

The first step in examining data that are produced over time is to *plot the data against time* or at least against the order in which the measurements were made. Figure 2.1 displays the commuting time data plotted against time. The date on the horizontal axis is given in days, counting from the date of the first measurement. Time should always be marked on the horizontal axis so that changes in the vertical direction reflect changes in the variable being followed. There is considerable spread (vertical scatter) even among dates that are close together. This large day to day variation makes it difficult to see longer-term changes. A few points (October 27, November 28, and December 5) seem to fall outside that general range of

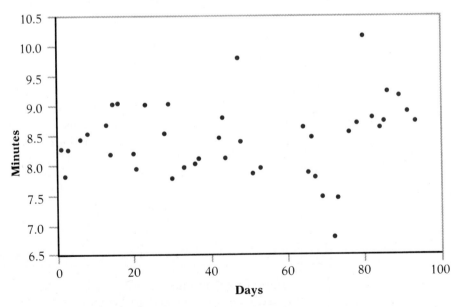

Figure 2.1 Commuting times from Table 2.1 plotted in time order.

variation. There is also some suggestion in Figure 2.1 that Professor Simon's travel time was increasing toward the end of the period. Can we verify these perceptions more objectively?

Control Charts

Some calculations can help us examine the time plot more closely. Experience suggests that if we travel the same route repeatedly under the same conditions, the distribution of travel time would be approximately normal. The actual measurements in Table 2.1 have mean $\bar{x} = 8.40$ minutes and standard deviation $s = 0.62$ minute. If the distribution is stable over time, the observations should continue to be approximately described by the $N(8.40, 0.62)$ distribution. Figure 2.2 uses this idea by drawing a solid center line at the level $\bar{x}$ and dashed *control limits* at $\bar{x} + 2s$ and $\bar{x} - 2s$. If the professor's travel time is in control, only about 5% of the data points should fall outside the control limits. In fact, 3 of the 42 points (7.1%) fall outside these limits and none fall outside the wider limits $\bar{x} \pm 3s$ within which almost all values from a normal distribution should be contained. In particular, the apparent trend has not carried the travel time outside the range that would be expected if the distribution was normal and stable over time. Figure 2.2 is called a control chart for the commuting times.

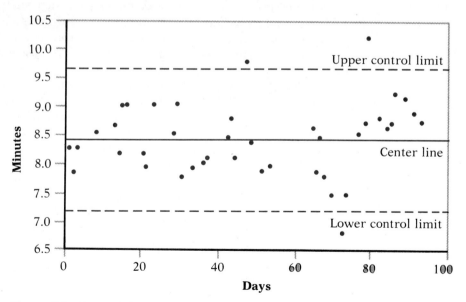

Figure 2.2 Control chart for the commuting times from Table 2.1.

Control chart

> A control chart for a variable x is a plot of the observations on x in time order with a solid center line at height $\bar{x}$ and dashed control limits at $\bar{x} \pm 2s$ or $\bar{x} \pm 3s$.

The control limits help us to distinguish between the variation that is inherent in the variable's distribution and extraordinary variation due to some special cause that changes the distribution. Any point outside the control limits should be investigated to see if the distribution has been disturbed by an unusual occurrence. In fact, all three out-of-control points in Figure 2.2 can be explained. On October 27, the drive to work was delayed by a truck backing into a loading dock. November 28 was the day after Thanksgiving, a holiday when the absence of both students and staff speeded the drive to campus. On December 5, ice on the windshield forced Professor Simon to stop and clear the glass.

The reasoning behind Figure 2.2 does, however, have some weaknesses. The first is that control limits based on normal distributions are not always appropriate, because not all variables have a normal distribution. The second weakness is more subtle. We would like the control limits to reflect the usual variation in the length of the professor's commute over a short period of time. Data taken over longer periods are said to be in control if they remain within these limits. But the limits in Figure 2.2 were based on the standard deviation s of all 42 observations over a period of more than three months. The variation measured by this s includes long-term change such as the apparent upward trend as well as the expected short-term variation in travel time. So s is too large, the control limits are too far from the mean, and the chart is not very sensitive to systematic changes in the distribution of commuting time.

These two weaknesses can be remedied by using more sophisticated control charts. Control charts of many kinds are used heavily in manufacturing process control. We are more concerned with the idea of statistical control than with the details of specific control charts. There is one more aspect of Figure 2.2 that is revealing, however, and that suggests the kind of additional tools we need to make full use of control charts. If the distribution of the professor's drive to work is stable over time and roughly normal, about half of the measurements should be above the mean and half below. The last 10 measurements in Figure 2.2 are all above the mean. This is like tossing 10 heads in a row with a coin—you then suspect that the coin is loaded, since 10 straight heads are unlikely with a balanced coin. Similarly, a *run* of 10 values above the calculated mean $\bar{x}$ suggests that the true mean travel time has shifted upward. Looked at in this light, the trend in Figure 2.2 does show lack of statistical control. In fact, a spell of ice and snow in December lengthened the drive to work and brought about the run in the chart.

Looking at runs in a control chart requires a new kind of calculation, evaluating the chance or *probability* that an event may occur. Probability calculations are at the heart of more sophisticated statistical methods for studying data. We will develop an understanding of probability before returning to control charts in Chapter 6. We are again reminded that pictures must often be supplemented by more mathematical approaches to the study of data.

SUMMARY

The pattern of growth or change in a variable over time is often of great interest. A **plot against time** displays the nature of change or growth. Since repeated measurements on a variable usually show some variation, we call a variable stable over time if its distribution does not change. Such a variable is said to be **in statistical control**.

A **control chart** helps us both assess the control of past data and monitor new data for changes. A control chart for measurements x has a center line at $\bar{x}$ and **control limits** at $\bar{x} \pm 2s$ or sometimes at $\bar{x} \pm 3s$. Points outside these limits, or long **runs** on one side of the center line, suggest that x is no longer in control.

SECTION 2.1 EXERCISES

2.1 Joe has regularly recorded his weight, measured at the gym after a workout, for several years. The mean is 162 pounds and the standard deviation 1.5 pounds, with no signs of lack of control. An injury keeps Joe away from the gym for several months. The data below give his weight, measured once each week for the first 16 weeks after he returns from the injury. Use the mean and standard deviation based on Joe's previous measurements to make a control chart for these measurements. Mark any points that fall outside the control limits. What is the longest run of points above or below the center line? How is Joe's weight changing over time during this period?

Week	1	2	3	4	5	6	7	8
Weight	168.7	167.6	165.8	167.5	165.3	163.4	163.0	165.5

Week	9	10	11	12	13	14	15	16
Weight	162.6	160.8	162.3	162.7	160.9	161.3	162.1	161.0

2.2 Refer to the crankshaft data given in Exercise 1.11 (see also Exercises 1.35 and 1.49). The measurements are in time order when read across the rows; that is, the first six crankshafts produced are on the first row and so forth. Draw a control chart with center line and control limits. Are there any values outside of the control limits? Is there any indication that the process is changing over time?

2.3   Blood pressure is a basic indicator of cardiovascular fitness. It can be influenced by a variety of factors such as age, stress, and exercise. Nancy measures her blood pressure every Sunday morning. The following table gives values of Nancy's systolic blood pressure (SBP) for a period of 30 weeks.

Week	SBP	Week	SBP	Week	SBP
1	132	11	125	21	128
2	133	12	126	22	125
3	138	13	140	23	126
4	133	14	133	24	126
5	141	15	124	25	124
6	131	16	135	26	130
7	131	17	141	27	123
8	129	18	133	28	130
9	139	19	124	29	124
10	136	20	139	30	122

After entering the data as the variable SBP into the MINITAB statistical computing system, we obtain their mean and standard deviation as follows:

```
MTB > MEAN 'SBP'
MEAN = 130.7

MTB > STDEV 'SBP'
STDEV = 5.9
```

Plot the data on a control chart with a center line and control limits. Are there any values outside of the control limits? What is the length of the longest run of points on the same side of the center line? Can you see any pattern or trend? If so, describe it.

2.4   The table below gives a family of four's monthly expenditures for food for the calendar years 1982 through 1985. The amounts recorded were spent for food to be prepared at home, and do not include the costs of eating out or of nonfood items bought at the same time as groceries.

(a) To examine the stability of this family's spending on food, draw a control chart with center line and control limits. The mean monthly spending computed from these values is $\bar{x} = \$322$, and the standard deviation is $s = \$71$.

(b) Circle all points on your graph that are outside of the control limits. Beside the circles write the name of the month corresponding to each point that is out of control.

(c) Many families take vacations in the summer. Use this information and the definition of food expenditures given above to propose an explanation for the values that are out of control.

(d) How might you expect the results to change if the costs of eating out were included in the monthly totals?

Month	1982	1983	1984	1985
Jan.	268	304	345	373
Feb.	320	227	322	272
Mar.	400	320	367	409
Apr.	333	367	367	410
May	286	322	379	380
June	179	121	162	145
July	337	362	314	281
Aug.	355	271	402	346
Sept.	355	300	312	400
Oct.	243	324	338	360
Nov.	344	420	405	386
Dec.	396	321	246	263

2.5 Publication of the national average score on the Scholastic Aptitude Tests (SATs) each year is often the occasion for commentators to claim that American high school students are learning more (if the scores are up) or less (if the scores are down). Here are the nationwide mean SAT verbal scores for the 12 years 1976 to 1987.

1976	1977	1978	1979	1980	1981	1982	1983	1984	1985	1986	1987
431	429	429	427	424	424	426	425	426	431	431	430

Computation gives $\bar{x} = 427.75$ and $s = 2.73$.

(a) Make a control chart for these scores. Is there any sign of a systematic change over time?

(b) When the 1987 scores were released, a news article began "In a disappointing sign for efforts to improve American high schools, SAT verbal scores fell last year." Write a short letter to the editor explaining in plain language why the change from 1986 to 1987 is of no importance.

2.6 In training for a marathon, runners often keep detailed records of the miles they run. The following table gives Ingrid's totals for the 13 weeks prior to the race.

Week	Miles	Week	Miles	Week	Miles
1	60	6	45	11	55
2	49	7	53	12	65
3	63	8	49	13	18
4	54	9	43		
5	45	10	62		

(a) Draw a control chart with center line and control limits.

(b) The 13th week appears unusual, but runners usually reduce their training just before a race. Compute the standardized value of the mileage for the 13th week.

(c) Because we understand the reason for the large deviation in the 13th week, omit that observation and compute $\bar{x}$ and s for the data from the first 12 weeks only. Draw the new lines on your control chart and comment on the statistical control of Ingrid's training pattern.

2.7 In industrial settings, it sometimes happens that a target value for the mean μ is given, and that the desired standard deviation σ is known because the precision of the process is known. In this case, a center line at μ and control limits at $\mu \pm 2\sigma$ can be based on the target values rather than estimated from the data. The purpose of the control chart in this case is not simply to maintain statistical control, but to maintain control at the given values of μ and σ.

   In the manufacture of integrated circuit chips, an important variable is the width of the microscopic lines that are etched into the chip to serve as electrical conductors. Each wafer of chips contains a line-width test pattern used for quality control measurements. For one type of chip, the target width is $\mu = 3.00$ (the units are millionths of a meter) and experience shows that the etching process produces lines with $\sigma = 0.15$ when functioning properly. Below are the line widths for the last 25 wafers produced, in time order reading across the rows. Make a control chart and mark any points that are out of control. Then examine the chart for runs and briefly describe your findings about the state of the process.

2.75	2.88	3.01	3.09	2.98	2.88	2.97	2.98	3.47
2.98	3.08	3.17	3.06	3.39	3.12	3.17	3.36	3.21
3.37	2.93	3.32	3.39	3.18	3.36	3.22		

2.8 The data in Exercise 1.25 are repeated laboratory measurements of the same quantity, made by Cavendish in the eighteenth century.

When read across the rows, the data are in time order. Many laboratories now keep control charts to ensure that the measuring process remains stable. The mean and standard deviation of Cavendish's data were calculated in Exercise 1.48. Prepare a control chart. Is there any sign of systematic change over time or of individual points out of control?

2.9 We often hear that "crime is out of control." Let's see if this is true in the statistical sense. We must first face the measurement problem. Many crimes are never reported, so changes in government data can reflect changes in reporting rather than changes in crime. Since almost all murders are reported, we will look at data on murders. The actual number of murders tends to grow with the size of the population, so we will study the *rate* of murders per 100,000 resident population. Here are the data for every other year from 1960 through 1986:

Year	Rate	Year	Rate
1960	4.7	1974	9.8
1962	4.8	1976	8.8
1964	5.1	1978	9.0
1966	5.9	1980	10.2
1968	7.3	1982	9.1
1970	8.3	1984	7.9
1972	9.0	1986	8.6

Investigate the control of this time series. Describe any systematic pattern that is visible in the data in addition to any signs of lack of control.

2.10 A homeowner notices unusually high electricity use. After reading the meter each day for a week, he calls an electrician on June 22. The electrician finds and fixes a current leak in the circuit breaker box. The homeowner then records his power consumption for an additional 13 days to confirm that the problem has been resolved. Here are the data for electricity use in kilowatt-hours (kWh):

Date	kWh	Date	kWh	Date	kWh
June 15	57.0	June 22	49.1	June 29	29.9
June 16	54.2	June 23	28.9	June 30	27.6
June 17	51.8	June 24	40.7	July 1	31.8
June 18	41.6	June 25	28.1	July 2	29.2
June 19	40.3	June 26	31.4	July 3	43.7
June 20	48.9	June 27	34.0	July 4	30.5
June 21	49.3	June 28	35.6	July 5	30.4

(a) Make a plot of electricity consumption against time for all 21 days. How does the effect of the repair of the current leak show up on your graph?

(b) The mean power consumption over this period is $\bar{x} = 38.76$ kWh, and the standard deviation is $s = 9.69$ kWh. Draw a center line and control limits on your graph. Are any points out of control? What are the longest runs of points on one side of the center line, and on what dates do they occur?

(c) Calculate the standard deviation for the 13 days after the leak was repaired, beginning with June 23. Why is the overall standard deviation $s = 9.69$ so much larger than your result? Because the control chart used the large overall s, the control limits are far from the center line and it is difficult for a point to be out of control.

2.11 Choose a player on your school's basketball team who normally takes at least 10 shots in a game. Obtain records of how many shots he or she took and how many were made in each game during a season. Make a control chart for the percent of shots made per game. Are there points out of control? Can you discover a reason for these points? In many cases, you will find that "good games" and "bad games" do not result in points falling outside the control limits. Sportswriters and fans tend to overlook chance variation in athletic performance, preferring to invent a reason for each above or below average performance.

2.12 Record your pulse rate once each waking hour for several days. (See Exercises 1.7 and 1.8 for suggestions on making accurate measurements.) Make a note of the circumstances of each measurement to help in your later interpretation of the data. Compute the mean and the standard deviation and prepare a control chart. Comment on the overall state of control and explain any unusual points.

2.2 LINEAR GROWTH

How do children grow? The height of a child is certainly not stable but increases over time. The pattern of growth varies from child to child, so we can best understand the general pattern by following the average height of a number of children. Table 2.2 presents the mean heights of a group of children in Kalama, an Egyptian village that is the site of a study of nutrition in developing countries.[1] The data were obtained by measuring the heights of all 161 children in the village each month over several years.

The first step in examining data that are produced over time is to plot the data against time. Accordingly, Figure 2.3 plots mean height against age. Following our usual strategy for examining data, we look first for an overall

Table 2.2 Mean height of
Kalama children

Age (months)	Height (cm)
18	76.1
19	77.0
20	78.1
21	78.2
22	78.8
23	79.7
24	79.9
25	81.1
26	81.2
27	81.8
28	82.8
29	83.5

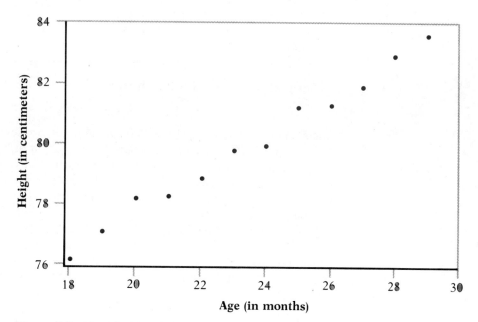

Figure 2.3 Mean height of children in Kalama, Egypt plotted against age from 18 to 29 months, from Table 2.2.

linear growth

pattern and then for deviations from that pattern. The overall pattern of growth is clear: The points follow a straight line. That is, the children display *linear growth* over time. The deviation from this pattern is the wobble that prevents the points from forming a perfectly straight line. There are no outliers—no points that fall clearly outside the straight-line pattern. The overall pattern of linear growth describes these data very well, and the deviations are of little importance.

Fitting a Line to Data

Linear (straight-line) growth says that the mean height increases by a fixed amount each month. How much do these children grow in a month? Can we predict the mean height of Kalama children at 32 months of age? To answer such questions efficiently, we write a mathematical model for the straight-line pattern.

Straight lines

Suppose that y is a measured variable (plotted on the vertical axis) and t is time (plotted on the horizontal axis). The equation of a straight line has the form

$$y = a + bt$$

In this equation, b is the slope, the amount by which y changes when t increases by one unit. The number a is the intercept, the value of y when $t = 0$.

In our case, a straight line has the form

$$\text{height} = a + (b \times \text{age})$$

The slope b is the amount by which the height increases when age increases by one month. The intercept a would be the height at birth (age = 0) if the straight-line pattern of growth were true starting at birth. If we can write the equation for a straight line that passes through the points in Figure 2.3, we will have a compact description of the data.

Because no straight line passes exactly through all the points in Figure 2.3, we must find a line that fits the points as well as possible. You might fit a line by eye by holding a transparent straightedge over the graph and moving it about until you are satisfied. This is not a bad way to draw a line on the graph, but it does not produce an equation without more work on your part. There are many mathematical ways of fitting a line to a set of data. The oldest and most common is the *method of least squares.** This approach looks at the deviations of the plotted points from the line in the vertical direction, on the grounds that the purpose of the line is to predict height for any given age. When a line fits reasonably well, it will pass below some points and above others. That is, some of the deviations will be positive

---

* The method of least squares was first published by the French mathematician Adrien Marie Legendre (1752–1833) in 1805. It at once became the standard method for combining observations in astronomy and surveying, and remains one of the most widely used statistical tools.

and some negative. The *squares* of the deviations are all positive, however. The sum of the squares of the deviations measures the size of the entire set of deviations. The least squares method fits to the data the line that makes this sum of squares smallest.

The least squares line

> The least squares line is the line that makes the sum of the squares of the deviations of the data points from the line in the vertical direction as small as possible.

The equation of the least squares line can be calculated from data like those in Table 2.2. We will learn how to do this in the next chapter when we study the dependence of one variable on another in more generality. For now, we are interested in how to interpret and use a line fitted to the data once we know its equation. It is common to learn the equation of the least squares line without doing the calculations, because some calculators and all statistical computing systems will calculate the line from data that you enter. The command to do this usually involves the term "regression," which is used somewhat vaguely to describe the dependence of one variable on another. In the MINITAB system, for example, we find the least squares line for the data in Table 2.2 as follows.

```
MTB > REGRESS 'HEIGHT' ON 1, 'AGE'
```

The output generated in response to this command begins as follows.

```
THE REGRESSION EQUATION IS
HEIGHT = 64.9 + 0.635 AGE


PREDICTOR        COEF       STDEV      T-RATIO
CONSTANT       64.9283     0.5084      127.71
AGE            0.63496     0.02140      29.66
```

The REGRESSION EQUATION in the output is the equation of the least squares line. We will not need most of the entries in the table below the regression equation until Chapter 10. Notice that the COEF column in the table gives the intercept and the slope more accurately than the summary equation. This added accuracy is often needed in problems. Rather than print this table for each example, we will often abbreviate computer results by writing the regression equation with extra decimal places obtained from the table. If you use MINITAB yourself, you can duplicate our results by looking in the table that follows the regression equation.

Figure 2.4 shows the least squares line $y = 64.93 + 0.635t$ drawn over the plot of height versus age. This equation shows that these children gain about 0.6 cm in height each month.

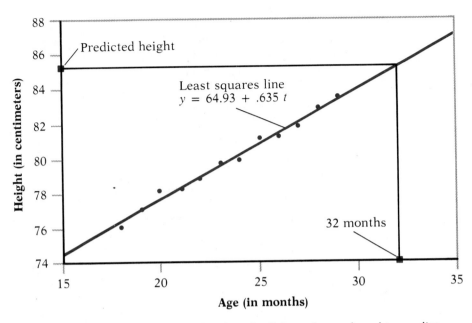

Figure 2.4 The least squares line fitted to the Kalama data and used to predict height at a given age.

EXAMPLE 2.2

prediction

A line fitted to data is a compact mathematical model for linear growth over time. In particular, it can be used to *predict* values of the variable whose growth the line describes. What do we predict to be the mean height of Kalama children at 32 months of age? Substituting 32 for the age in the equation of the fitted line gives

$$\text{height} = 64.93 + (.635 \times 32) = 85.25$$

If you have fitted a line by eye on graph paper, you can use it for prediction by simply finding the height (marked on the vertical scale) of the point on the line above age 32 on the horizontal scale. You do not need to know the equation of the line. Figure 2.4 illustrates the process of prediction.

Can we predict the height of Kalama residents at age 20 years? Since 20 years is 240 months, we substitute 240 for the age. We obtain a predicted height of about 217 cm, which is over 7 feet. Blind calculation has produced an unreasonable result. Our data covered only ages from 18 to 29 months. As people grow older, they gain height more slowly, so our fitted line is not useful at ages far removed from the data that produced it. ∎

extrapolation

Making use of a fitted line or other model outside the range of the data to which the model was actually fitted is called *extrapolation*. As Example 2.2 illustrates, extrapolation is dangerous. Notice that age 0 (birth) lies well outside the range of the data in Table 2.2. Taking the intercept 64.93 cm (over 25 inches) as an estimate of height at birth is therefore another example of extrapolation. The linear growth pattern does not begin at birth, and this estimate of height at birth is very inaccurate. In addition, we are well aware that the pattern we discovered in Figure 2.3 does not hold for all children everywhere. We cannot use data from Kalama to predict the height of children in Cairo, much less Los Angeles. Statistical methods such as least squares cannot go beyond the limitations of the data that feed them. Do not let calculation override common sense.

Residuals

A mathematical model, such as a straight line, gives a compact statement of the overall pattern of a set of data. Such models have the added advantage of being completely explicit. We can use a fitted line but not a verbal description of the pattern to predict future values, for example. When a model expresses the overall pattern, deviations from that pattern can also be stated explicitly.

Residual

> A **residual** is the difference between an observed value of a variable and the value predicted by the model. That is,
>
> residual = observed value − predicted value

EXAMPLE 2.3

According to Table 2.2, the observed mean height of Kalama children at 24 months was 79.9 cm. The fitted line predicts that the height should be

$$\text{height} = 64.93 + (.635 \times 24) = 80.17$$

The residual at 24 months is therefore

$$\text{residual} = \text{observed value} - \text{predicted value}$$
$$= 79.9 - 80.17$$
$$= -.27$$

The 12 data points used in fitting the least squares line in Figure 2.4 produce 12 residuals. They are

−.26	.01	.47	−.06	−.10	.17
−.27	.30	−.24	−.27	.09	.16

Although residuals can be calculated from any fitted line, the residuals from the least squares line have some special properties.

- The sum (and therefore the mean) of the residuals is zero.
- The variance of the residuals (the spread about the line) is the smallest that is possible for any line through the given data.

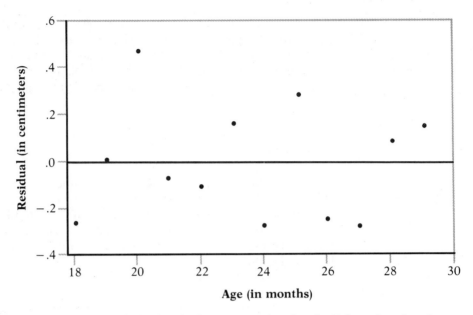

Figure 2.5 Residual plot for the least squares line for the Kalama data, based on the residuals in Example 2.3.

The residuals are the deviations from the line—adding the residual to the predicted height at any age gives the actual observed height. The principle, that in examining data we seek an overall pattern and deviations from it, can be expressed as an equation that is easy to remember.

$$DATA = FIT + RESIDUAL$$

residual plot

A plot of the residuals against time helps us examine the fit of our model by magnifying the deviations of the actual data points from the fitted line. Figure 2.5 is a *residual plot* for the Kalama data. Since the mean of the residuals is always zero, the horizontal line at zero helps orient us. This line (residual = 0) corresponds to the fitted line in Figure 2.4. The residuals in Figure 2.5 show the irregular horizontal pattern that is typical of data that do not deviate from the model in any systematic way. Residual plots will be discussed in more detail as part of our more general study of straight-line relations in the next chapter.

EXAMPLE 2.4

We are aware that we cannot extrapolate the linear growth pattern of Kalama children between 18 and 29 months to ages far outside that range. In particular, the pattern cannot be extrapolated backwards to birth. Figure 2.6 shows the growth in mean height of Egyptian children who were followed from birth in another village, Nahya.[2] In all, 170 infants were measured each month during their first year of life. The pattern is not linear. Children grow more rapidly during the first few months of life than at later times. We have nonetheless drawn the least squares line on the plot. You can fit a least squares line to any set of growth measure-

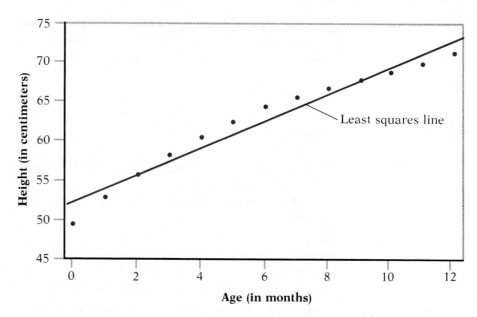

Figure 2.6 Mean height of children in Nahya, Egypt plotted against age from birth to 12 months. See Example 2.4.

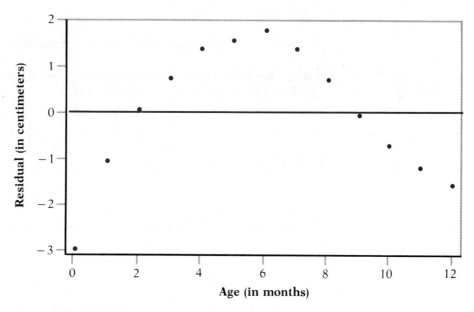

Figure 2.7 Residual plot for the least squares line for the Nahya data.

ments, whether or not a linear pattern is evident: The least squares method does not know whether it is being applied wisely or foolishly—that is your responsibility.

The fitted line in Figure 2.6 does help us see the nonlinear pattern more clearly. The residual plot (Figure 2.7) magnifies the deviations of the points from the line. The systematic lack of fit in Figure 2.7 contrasts clearly with the unsystematic scatter in Figure 2.5. The linear growth model has extracted the systematic pattern in the growth of 18 to 29-month-old children, but fails for 0 to 12-month olds.[3]   ∎

SUMMARY

The simplest form of change over time is **linear growth**, when a variable increases by a fixed amount in each equal time period. A straight line fitted to the plot against time then describes the overall pattern of the data.

The most common method of fitting a line to a plot against time is **least squares**. The least squares line for a variable y against time t is the straight line $y = a + bt$ that minimizes the sum of the squares of the vertical deviations of the observed y values from the line.

A fitted line can be used to predict the value of the variable at any time, but **extrapolation** beyond the time period spanned by the data is risky.

The **residuals** found as

$$\text{residual} = \text{observed value} - \text{predicted value}$$

show the deviations from the linear pattern. The deviations are clearest in a **residual plot** of the residuals against time.

SECTION 2.2 EXERCISES

2.13 (Review of straight lines) Fred keeps his savings in his mattress. He began with $500 from his mother, and adds $100 each year. His total savings y after t years are then given by the linear equation

$$y = 500 + 100t$$

    (a) Draw a graph of this equation. (Hint: Choose two values of t, such as 0 and 10. Compute the corresponding values of y from the equation. Plot these two points on graph paper and draw the straight line joining them.)

    (b) After 20 years, how much will Fred have in his mattress?

    (c) If Fred had added $200 instead of $100 each year to his initial $500, what is the equation that describes his savings after t years?

2.14 (Review of straight lines) Sound travels at a speed of 1500 meters per second in sea water. You dive into the sea from your yacht. Give an equation for the distance y at which a shark can hear your splash in terms of the number of seconds t since you hit the water.

2.15 (Review of straight lines) During the period after birth, a male white rat gains exactly 40 grams (g) per week. (This rat is unusually regular in his growth, but 40 g per week is a realistic rate.)

    (a) If the rat weighed 100 g at birth, give an equation for his weight after t weeks. What is the slope of this line?

    (b) Draw a graph of this line between birth and 10 weeks of age.

    (c) Would you be willing to use this line to predict the rat's weight at age 2 years? Do the prediction and think about the reasonableness of the result. (There are 454 g in a pound. To help you assess the result, note that a large cat weighs about 10 pounds.)

2.16 (Review of straight lines) Cellular mobile telephone service costs $30 per month plus $33 for each hour of use. Give an equation for the amount y of the monthly bill in terms of the number t of hours of use that month. You use about 15 hours of time each month. What will your monthly bill be? Another mobile telephone company offers air time at $50 per month plus $25 per hour of use. Would you save money by switching to this company?

2.17 Researchers studying acid rain measured the acidity of precipitation in an isolated wilderness area in Colorado for 150 consecutive weeks in the years 1975 to 1978. The acidity of a solution is measured by pH, with lower pH values indicating that the solution is more acid. The acid rain researchers observed a linear pattern over time. They reported that the least squares line

$$\text{pH} = 5.43 - (.0053 \times \text{weeks})$$

fit the data well. (See William M. Lewis and Michael C. Grant, "Acid precipitation in the western United States," *Science*, 207 (1980), pp.176–177.)

    (a) Draw a graph of this line. Note that the linear change is decreasing rather than increasing.

    (b) According to the fitted line, what was the pH at the beginning of the study (weeks = 0)? At the end (weeks = 150)?

    (c) What is the slope of the fitted line? Explain clearly what this slope says about the change in the pH of the precipitation in this wilderness area.

2.18 Sarah's parents are concerned that she seems short for her age. Their pediatrician has the following record of Sarah's height:

Age (months)	Height (cm)
36	86
48	90
51	91
54	93
57	94
60	95

(a) Plot these points on graph paper. Then use a transparent straightedge to fit a line through the points. What is the approximate slope of your line? (Hint: Find the predicted height at ages 48 and 60 months from the line on your graph. The slope is the difference between these heights divided by 12, the difference in age.)

(b) Normally growing girls gain about 6 cm in height between ages 4 (48 months) and 5 (60 months). What slope does this correspond to? Is Sarah's growth more rapid (larger slope) than normal, or less rapid (smaller slope)? Sarah's pediatrician noticed the unusual slope of her growth curve and decided to do some additional tests. She was found to have a growth hormone deficiency that can be treated with doses of a synthetic hormone.

(c) Girls grow nearly linearly between ages 4 and 9. Use your line from (a) to predict Sarah's height at age 9 years if she continues to grow at the same rate.

2.19 An experiment in a physics course requires students to drop a ball from a height and measure its downward speed in meters per second (m/sec) at several points as it falls. The speed is supposed to increase linearly over time because the ball is subject to the steady pull of gravity. It is not easy to measure the speed after a fixed time, so the student measurements will usually not fall exactly on a straight line. Bob's team got these results:

Time (seconds)	0	.2	.4	.6	.8
Speed (m/sec)	0	1.82	3.58	6.01	7.88

(a) Plot these data on graph paper. Using a transparent straightedge, fit a line through the points.

(b) What is the approximate slope of your line? (Finding the slope is the goal of the experiment, because the slope estimates the acceleration due to gravity. More accurate measurements would give the slope as 9.8.)

(c) Extend your line, and predict the speed of the ball after 2 seconds. Do you think that extrapolation is trustworthy in this case?

2.20 Some industrial processes drift steadily when in statistical control. For example, a cutting tool that is machining the outer diameter of a rod will wear gradually, so that the diameter of the finished parts will increase over time. The following table gives the diameter of the last rod produced each half hour. The data are coded so that the units are thousandths of an inch above 0.2450 inch. The cutting tool must be replaced when the diameter reaches 0.2500 inch, which is 50 on the coded scale.

Time	1	2	3	4	5	6	7
Diameter	39.2	38.4	39.9	40.7	40.1	40.5	42.7

Time	8	9	10	11	12	13	14
Diameter	42.3	42.5	44.5	43.0	44.2	46.3	45.4

(a) Make a graph of the rod diameter against time. Then use a transparent straightedge to fit a straight line to the data, and draw the line on your graph.
(b) Extend your line until it crosses the horizontal line at $y = 50$. This is the point at which the tool must be changed. Use your line to predict when the diameter will reach 50.

2.21 A group of 60 female Americans is followed from birth to age 21. Their median heights during childhood are as follows. Age is recorded in years and height in centimeters.

Age	4	4.5	5	5.5	6	6.5	7	7.5	8	8.5	9
Height	103.6	107.6	109.8	111.8	115.6	117.4	121.6	125.7	127.3	130.8	133.8

(a) Plot the heights against time. The overall pattern is linear.
(b) The following computer work in MINITAB gives the least squares line:

```
MTB> REGRESS 'HEIGHT' ON 1, 'AGE'

THE REGRESSION EQUATION IS
HEIGHT = 79.6245 + 6.0018 AGE
```

Draw this line on your plot.
(c) Use this line to predict the median height at age 14. The study found the median height at age 14 to be 163.1 cm. Was your prediction accurate?
(d) Now use the fitted line to predict the height at age 20. The study found (many years later) that the median height of these women at age 20 was 167.3 cm. Why was your prediction so inaccurate?

2.22 We will now examine the residuals from the line fitted to the heights of female Americans in the previous exercise.

(a) What is the predicted height at age 8? Calculate the residual at age 8.

(b) The 11 residuals for the 11 points to which the line was fitted are

$$-.03 \qquad .97 \qquad .17 \qquad -.83 \qquad -.04 \qquad -1.24$$
$$-.04 \qquad 1.06 \qquad -.34 \qquad .16 \qquad .16$$

Check that the sum of these residuals is 0 (at least up to roundoff error).

(c) Make a plot of the residuals against age. Does the plot show any systematic deviations from the line?

2.23 The nutrition project in Kalama, Egypt measured the weights as well as the heights of the 161 children in the village. Here are the mean weights in kilograms (kg) for ages from 18 to 29 months:

Age (months)	18	19	20	21	22	23	24	25	26	27	28	29
Weight (kg)	10.0	10.1	10.3	10.3	10.7	10.9	11.2	11.3	11.5	11.7	11.9	12.3

(a) Plot the mean weight against time.

(b) Is the overall pattern linear? Are there any clear deviations from the pattern?

(c) The least squares line appears on MINITAB output as

```
MTB> REGRESS 'WEIGHT' ON 1, 'AGE'

THE REGRESSION EQUATION IS
WEIGHT = 6.14 + 0.208 AGE
```

Draw this line on your plot.

(d) Based on the fitted line, what would you predict to be the weight of a 21-month-old Kalama child?

(e) Find the value of the residual at 21 months.

2.24 The weights of the Nahya children (Example 2.4) were also recorded. Here are the mean weights of the 170 children in that village:

Age (months)	1	2	3	4	5	6	7.	8	9	10	11	12
Weight (kg)	4.3	5.1	5.7	6.3	6.8	7.1	7.2	7.2	7.2	7.2	7.5	7.8

(a) Plot the mean weight against time.

(b) A hasty user of statistics enters the data into the MINITAB computing system and computes the least squares line without plotting the data. The result is

MTB > REGRESS 'WEIGHT' ON 1, 'AGE'

THE REGRESSION LINE IS
WEIGHT = 4.88 + 0.267 AGE

Plot this line on your graph. Is it an acceptable summary of the overall pattern of growth?

2.25 Fortunately for the hasty user in the previous exercise, the computing system prints out the residuals from the least squares line. They are

−.85	−.31	.02	.35	.58	.62
.45	.18	−.08	−.35	−.32	−.28

Check that these residuals do have sum 0 (up to roundoff error). Then plot the residuals against age, and describe carefully the pattern that you see.

2.3 EXPONENTIAL GROWTH*

Petroleum has become the most important single source of energy for the developed nations in this century, and more recently the cause of economic dislocation and threats of war. Table 2.3 and Figure 2.8 show the trend of

Table 2.3 Annual world crude oil production, 1880–1984

Year	Mbbl	Year	Mbbl	Year	Mbbl
1880	30	1935	1655	1968	14,104
1890	77	1940	2150	1970	16,690
1900	149	1945	2595	1972	18,584
1905	215	1950	3803	1974	20,389
1910	328	1955	5626	1976	20,188
1915	432	1960	7674	1978	21,922
1920	689	1962	8882	1980	21,732
1925	1069	1964	10,310	1982	19,403
1930	1412	1966	12,016	1984	19,608

* This material is important in biology, business, and some other areas of application, but can be omitted without loss of continuity.

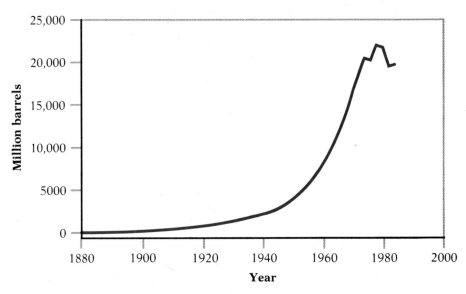

Figure 2.8   Annual world crude oil production, 1880 to 1984, from Table 2.3.

annual world crude oil production, measured in millions of barrels (Mbbl) per year.[4] Oil production has increased much faster than linear growth, but the pattern of growth follows a smooth curve until 1973, when a mideast war touched off a vast price increase and a change in the previous pattern of production. Can we describe the growth of oil production from 1880 to 1973 by a simple mathematical model? We can indeed, and the resulting model for exponential growth is second only to the linear growth model in its usefulness.

The Nature of Exponential Growth

exponential growth

A variable grows linearly over time if it *adds* a fixed increment in each equal period. *Exponential growth* occurs when a variable is *multiplied* by a fixed number in each time period. To grasp the effect of multiplicative growth, consider a population of bacteria in which each bacterium splits in two each hour. Beginning with a single bacterium, we have 2 after one hour, 4 at the end of two hours, 8 after three hours, then 16, 32, 64, 128, and so on. These first few numbers are deceiving. After 1 day of doubling each hour, there are 2^{24} or 16,777,216 bacteria in the population. That number then doubles the next hour! Try successive multiplications by 2 on your calculator to see for yourself the very rapid increase after a slow start. Figure 2.9 shows the growth of the bacteria population over 24 hours. For the first 15 hours, the population is too small to rise visibly above the zero level on the graph.

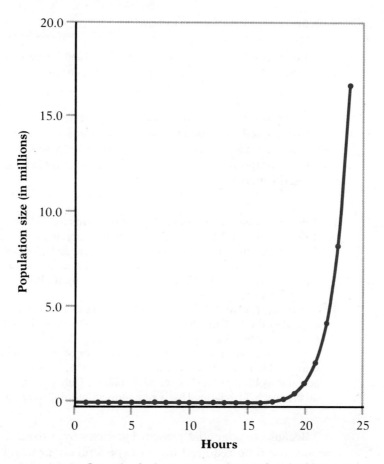

Figure 2.9 Growth of a bacteria population that doubles each hour.

It is characteristic of exponential growth that the increase appears slow for a long period, then seems to explode. Both Figures 2.8 and 2.9 display explosive growth after a long period of gradual increase. Our first reaction to the apparently sudden surge is to search for some explanation, something new that distinguishes the high-growth present from the low-growth past. But the nature of the increase has not changed at all. Only our understanding of the long-term consequences of exponential growth needs correcting. An old story tells how a king of Persia learned this lesson. There was once, the story goes, a courtier who asked the king for a simple reward: a grain of rice on the first square of a chessboard, 2 grains on the second square, then 4, 8, 16, and so on through all 64 squares. The king foolishly granted the request. We now know better. Similarly, we should know better than to risk our prosperity on the assumption that oil production will grow exponentially forever.

Linear versus exponential growth

> Linear growth increases by a fixed *amount* in each time period; exponential growth increases by a fixed *percentage* of the previous total.

Populations of living things—like our bacteria—tend to grow exponentially if not restrained by outside limits such as lack of food or space. More pleasantly, money also displays exponential growth when returns to an investment are compounded. Compounding means that last period's income earns income this period.

EXAMPLE 2.5

A dollar invested at an annual rate of 6% turns into $1.06 in a year. The original dollar remains and has earned $0.06 in interest. That is, 6% annual interest means that any amount on deposit for the entire year is multiplied by 1.06. If the $1.06 remains invested for a second year, the new amount is therefore 1.06×1.06, or 1.06^2. That is only $1.12, but this in turn is multiplied by 1.06 during the third year, and so on.

If the Native Americans who sold Manhattan Island for $24 in 1626 had deposited the $24 in a savings account at 6% annual interest, they would now have over $36 billion. Our savings accounts don't make us billionaires because we don't stay around long enough. A century of growth at 6% per year turns $24 into $8143. That's 1.06^{100} times $24. By 1826, two centuries after the sale, the account would hold a bit over $2.7 million. Only after a patient 302 years do we finally reach $1 billion. That's real money, but 302 years is a long time. ■

Because exponential growth increases by a fixed percent in each time period, the time required for an exponentially growing quantity to double remains fixed, no matter what the current level of the quantity may be. If you invest money at 6% interest compounded annually, your investment **doubling time** will double in 12 years. This *doubling time* is a convenient and easily understood way to state the rate of exponential growth. The doubling time is approximately given by a simple rule.

The rule of 70

> If a quantity grows exponentially at a rate of R% per period, the doubling time is approximately $70/R$ periods.

An investment yielding 5% per year compounded will double your money in about $70/5 = 14$ years; a return of 6% has a doubling time of approximately $70/6 = 11.7$ years; and an investment with a 10% return doubles about every 7 years.

The Logarithm Transformation

We can see multiplication at work in the growth of a biological population or of money deposited at compound interest. It is not surprising that the exponential model fits these cases. But there is no obvious multiplication going on in the case of world oil production. Is this really exponential growth? The shape of the growth curve for oil production in Figure 2.8 does indeed generally resemble the exponential curve in Figure 2.9. But our eyes are not very good at comparing curves of roughly similar shape. We need a better way to check whether growth is exponential. Our eyes are quite good at judging whether or not points lie along a straight line. So we will apply a mathematical transformation that changes exponential growth into linear growth—and patterns of growth that are not exponential into something other than linear.

logarithm   The necessary transformation is carried out by taking the *logarithm* of the data points. The logarithm of any positive number y is the power of 10 that equals y. That is, $\log y$ is defined by the fact that

$$10^{\log y} = y$$

For example, $\log 100 = 2$ because $10^2 = 100$. Use a calculator with a **LOG** button to compute logarithms. Better yet, most interactive statistical software systems will calculate the logarithms of an entire data set in response to a single command. The essential property of the logarithm for our purposes is that it straightens out an exponential growth curve. *If a variable grows exponentially, its logarithm grows linearly.*

EXAMPLE 2.6   Figure 2.9 shows the exponential growth of a population of bacteria that doubles each hour. The logarithms of the number of bacteria should show straight-line growth. Use your calculator to find some of the logarithms. The population counts for the first few hours are 2, 4, 8, ... The logarithms are approximately

$$\log 2 = .30$$
$$\log 4 = .60$$
$$\log 8 = .90$$

Figure 2.10 graphs the logarithms of the bacteria population counts from Figure 2.9 against time. The promised straight line has indeed appeared. After 15 hours, for example, the population contains 2^{15} or 32,768 bacteria. The logarithm of 32,768 is 4.52, and this point appears above the 15-hour mark in Figure 2.10. (We have calculated and plotted common logarithms, or logarithms with base 10. There are other types of logarithms as well. Since all transform exponential growth into a straight line, you need not be concerned about which logarithm to use.) ∎

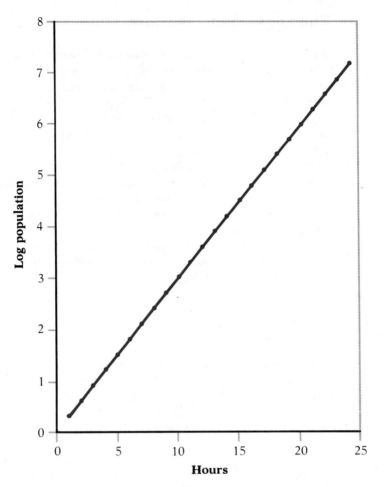

Figure 2.10   The logarithms of the bacteria population size from Figure 2.9. See
Example 2.6.

Our artificial example of bacterial growth was exactly exponential. Real
data will not fit the exponential model so perfectly, so applying logarithms
will not produce a perfect straight line. Figure 2.11 is a graph of the logarithm
of the annual world crude oil production from Table 2.3. The individual
points have been connected by lines to make the pattern clear. The graph
is quite close to a straight line from 1880 to 1973. A clear deviation from
the line begins in 1973. Closer examination shows a smaller deviation during
the 1930s, when growth slowed during the Great Depression. The increase
in oil production has indeed been close to exponential over most of the past
century, but with important deviations.

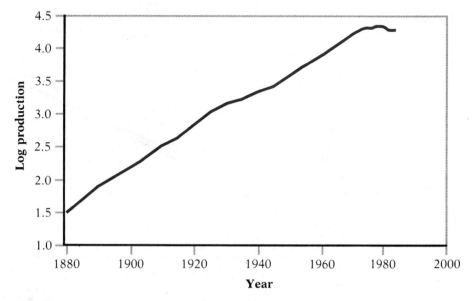

Figure 2.11 The logarithms of annual world crude oil production, from Table 2.3.

The logarithmic transformation enables us to judge the appropriateness of a somewhat complicated mathematical model by reducing the question to judging whether points lie on a straight line. Comparison of Figures 2.8 and 2.11 reveals another advantage: Logarithms compress the vertical scale of the plot, making it easier to see deviations from fit in the earlier stages of exponential growth. The slowdown in oil production during the Depression is visible in Figure 2.11, but not in Figure 2.8. On the other hand, the post-1973 disturbance, occurring at the top of the curve, is clearer in Figure 2.8.

Residuals Again

Just as in the case of linear growth, calculating and plotting residuals allow a closer examination of the fit of the exponential growth model to data. Before we can find the residuals, however, we must fit to the data an explicit model described by an equation. Since taking logarithms of data that display exponential growth produces a straight line, we can fit a straight line to the logarithms and obtain residuals as the deviations of the actual logarithms from this line. Figure 2.12 shows the result of fitting the least squares line to the logarithms of annual oil production from 1880 to 1972. We did not allow data for years after 1972 to enter into the calculation because we know that the pattern of exponential growth ended in 1973.

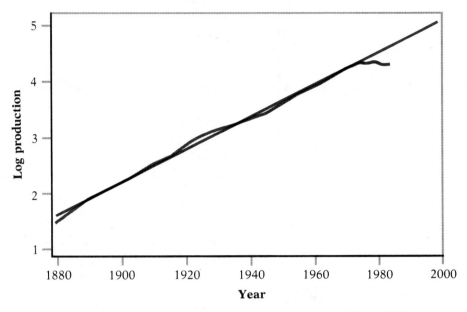

Figure 2.12 The least squares line for world oil production, 1880 to 1972.

EXAMPLE 2.7

We enter the annual oil production for the years 1880 to 1972 into the MINITAB computing system as values of the variable C1, and the corresponding years as the variable C2. MINITAB then computes the logarithms and finds the least squares line for the logarithms as follows:

```
MTB > LET C3 = LOGT(C1)
MTB > REGRESS C3 ON 1, C2

THE REGRESSION LINE IS
C3 = -52.686 + 0.02888 C2
```

That is, the computer finds the least squares line to be

$$\text{log production} = -52.686 + (.02888 \times \text{year})$$

The residuals of the data from the fitted line are computed in the usual way, as

$$\text{residual} = \text{observed value} - \text{predicted value}$$

For the year 1945, the value of the logarithm predicted by the least squares line is

$$\text{log production} = -52.686 + (.02888 \times 1945)$$
$$= 3.4856$$

Since 2595 million barrels of oil were actually produced in 1945, the residual for that year is

$$\text{residual} = \text{observed value} - \text{predicted value}$$
$$= \text{log } 2595 - 3.4856$$
$$= 3.4141 - 3.4856$$
$$= -.0715$$

Most computer least squares routines will compute and print a table of the residuals, and many will also produce residual plots. For example, a table of the residuals is part of the output from the MINITAB command REGRESS. ∎

The deviations from the overall pattern of exponential growth in oil production are easier to see in Figure 2.12 than in the previous Figure 2.11, which lacked the fitted line. The residual plot in Figure 2.13 is clearer yet. Recall that DATA = FIT + RESIDUAL. The horizontal line at zero represents the fit, the fitted line in Figure 2.12. The pattern of residuals about the horizontal line shows systematic departures from the overall fit. Because we are plotting logarithms, straight-line patterns of residuals have a specific interpretation. The line of fit (zero residual) represents exponential growth at the average rate observed in the entire period 1880 to 1972. A straight-line rising pattern of residuals shows a period of growth at a faster rate, while a declining pattern shows a slower rate of growth.

The most dramatic departure of the residuals from zero marks the end of exponential growth after 1972. But we can now see other systematic deviations. Oil production was increasing more rapidly than the long-term rate in the years between 1900 and the beginning of the Depression in 1929. Production increased more slowly not only during the Depression but also during World War II. That is, the turning points in the residual plot occur in 1925 (the last point before 1929) and 1945 (the end of World War II). Only after the war did oil production return to an above-average growth rate, which lasted until 1973. (The production of oil was of course increasing during the entire period up to 1973—the runs of declining residuals show periods of slower growth, not an actual drop in oil production.)

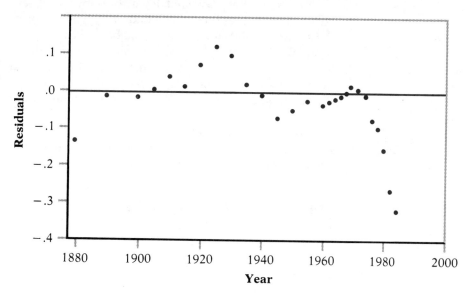

Figure 2.13 Residual plot for the least squares line for world oil production.

Our tools for data analysis have given us quite a detailed understanding of the history of world oil production. The long-term picture is one of exponential growth from the beginnings of commercial petroleum use to the Arab oil boycott of 1973. But the long-term rate of growth is an average over several periods of somewhat slower or somewhat faster growth, periods that coincide with major events in the economic history of the twentieth century.

The use of logarithms to obtain a linear pattern from exponential growth does complicate prediction a bit. Because the least squares line is fitted to the logarithms of the observations, it also predicts the logarithm. To express the prediction in easily understood terms, we must restate it in terms of the original units of measurements. To do this, use the fact that any number y can be obtained from its common (base 10) logarithm $\log y$ by

$$y = 10^{\log y}$$

EXAMPLE 2.8

The least squares line obtained in Example 2.7 predicts that in 1977

$$\log \text{production} = -52.686 + (.02888 \times 1977)$$
$$= 4.4098$$

The predicted production in the original units (millions of barrels of oil) is therefore

$$\text{production} = 10^{4.4098} = 25{,}692$$

The actual production in 1977 was 21,787 million barrels. The prediction is much too high because the exponential growth model fails to fit post-1973 oil production. ■

Could we have avoided logarithms by fitting an exponential growth curve to the actual production data displayed in Figure 2.8? Yes. But there are several reasons for transforming before fitting the model and calculating residuals. First, straight lines are simple, while the explicit mathematical form of the exponential growth model is more involved. We will discuss the mathematics of exponential models in the next section. You will then have the equation of a curve to describe exponential growth without logarithms. A second reason for working with the logarithms is that it is easy to calculate the least squares line. Fitting an exponential curve to a set of data is more difficult. Many statistical computing systems (not all) will carry out such nonlinear fits, but the calculations are not feasible without a computer.

More About Exponential Models*

If a variable y changes linearly with time, its value at a time t is described by the equation

$$y = a + bt$$

---

* This section is more mathematical than the rest of the text. It can be omitted without loss of continuity.

Exponential change can be described by a similar equation if we work not with the original variable y but with its logarithm. That is, if y grows exponentially, then

$$\log y = a + bt$$

This fact is very convenient because straight lines are easy to graph and assess visually. However, it is sometimes useful to express exponential growth by an equation that describes the variable directly. Some such equations have exponential curves like Figure 2.9 as their graphs. There are other types of exponential change as well.

We can discover the form of exponential models from the examples of bacteria growth and compound interest. The bacteria population that doubles each hour has size y after t hours of growth given by

$$y = 2^t$$

If you deposit $24 in a bank account that earns 6% compound interest per year, the value of your account after t years is

$$y = 24(1.06)^t$$

These examples illustrate the general equation for exponential change.

Exponential curves

The equation for an exponential curve is

$$y = Ac^t \qquad (2.1)$$

In this equation, the constants c and A are positive numbers and the time t takes only positive values. The constant c is the rate of exponential increase or decrease, and A is the starting value of the variable y.

The term "exponential" reflects the fact that the time t appears as an exponent in Equation 2.1. The equation also makes clear the multiplicative nature of exponential growth: When the time t increases by 1, the variable y is multiplied by c. So y increases over time when c is greater than 1, and decreases when c is less than 1. That is why the constant number c in Equation 2.1 is called the rate of increase or decrease. If $c = 2$, then y doubles in each time period, as the population of bacteria did. If $c = 1/2$, y is halved in each time period. The value of y when $t = 0$ is A because when $t = 0$, $c^t = c^0 = 1$. The remaining constant A in Equation 2.1 is therefore the starting value of y.

Figure 2.14 shows three exponential growth curves. In all three cases, the rate c is greater than 1, so that y increases as time increases. Curves 1 and 2 have the same A, with Curve 2 having a larger rate c. Curve 3 has

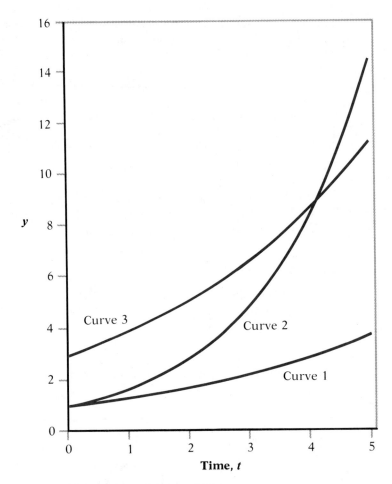

Figure 2.14 Exponential growth curves.

the same value of c as Curve 1, but starts from a larger A. Notice that Curve 2, with its faster growth rate, eventually surpasses Curve 3 despite its lower starting point.

If you are acquainted with the rules for manipulating logarithms, you can verify that taking the logarithm of both sides in Equation 2.1 gives

$$\log y = \log A + (t \times \log c)$$

As expected, $\log y$ follows a straight line when plotted against t. The slope of this line is $\log c$, which is positive when c exceeds 1.

Figure 2.15 shows another type of exponential curve. These curves represent *exponential decay*. They are described by Equation 2.1 with $c < 1$. As the time t increases, the value of y decreases. As the figure illustrates, the decrease in y is more rapid when the rate c is farther from 1. Although

exponential decay

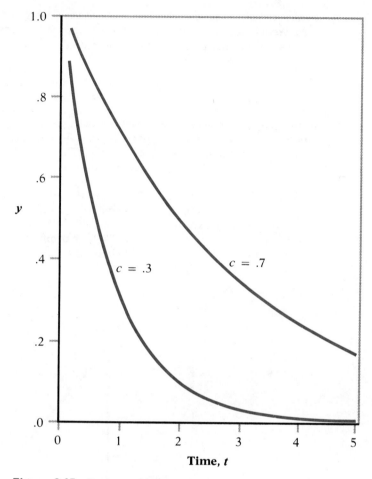

Figure 2.15 Exponential decay curves.

y increases without bound for $c > 1$, *y* never becomes negative when $c < 1$, but approaches ever closer to zero. The logarithm of *y* follows a straight line when plotted against *t*, but the line slopes downward because its slope log *c* is negative. An important example of exponential decay is radioactivity. Radioactive elements emit radiation with an intensity that decays exponentially with time. That is, the same percentage decrease takes place in every equal time period. Thus the danger from such radiation decreases if we wait. The rate of decrease (our constant *c*) depends on the type of radioactive material. Unfortunately, this rate is very slow in some cases.

The model of exponential decay, like that of exponential growth, is most trustworthy when we understand the mechanism that leads to multiplicative change. When exponential curves describe the overall pattern of data, but we do not understand why exponential change is occurring, we should be particularly wary of extrapolation.

EXAMPLE 2.9

Infant mortality is an important measure of the quality and availability of health care. We measure infant mortality by the number of children who die within the first year of life out of every 1000 who are born alive. Infant mortality in the United States has declined as follows.

1920	1930	1940	1950	1960	1970	1980
85.8	64.6	47.0	29.2	26.0	20.0	12.6

Figure 2.16(a) is a plot of U.S. infant mortality against time. While the pattern is a bit irregular, it suggests exponential decay. To check this pattern, we plot the logarithms against time in Figure 2.16(b). The least squares line on the plot helps us assess linearity. The overall pattern is linear. We can see that infant mortality dropped more rapidly than the long-term rate during the decade of the 1940s and the decade of the 1970s. These deviations from the pattern of exponential decay might form the starting point for a study of policies to lower infant mortality. Although exponential decay provides a helpful description of 60 years' change in infant death rates, it is unlikely that infant mortality will in fact decline to zero over time. We do not believe that the model will hold good for the indefinite future. ■

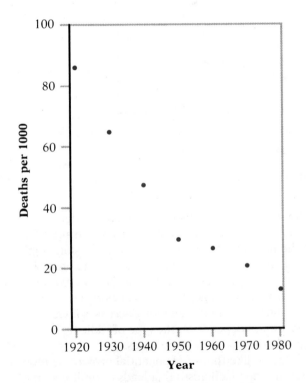

Figure 2.16(a)   U.S. infant mortality, 1920 to 1980, for Example 2.9.

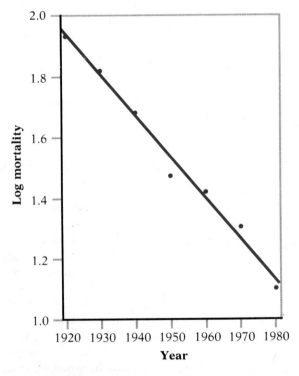

Figure 2.16(b)   Logarithms of U.S. infant mortality, 1920 to 1980, with the least squares line.

SUMMARY

When a variable is multiplied by a fixed amount greater than 1 in each equal time period, **exponential growth** results. Multiplication by a fixed amount less than 1 in each equal time period results in **exponential decay**.

A plot against time of exponential growth or decay becomes linear (a straight line) if we take the **logarithm** before plotting. This fact makes it possible to detect exponential change easily.

Deviations from the overall pattern of exponential growth or decay are most easily examined by fitting a line to the logarithms of the data and plotting the residuals from this line.

SECTION 2.3 EXERCISES

2.26 (Exact exponential growth) The common intestinal bacterium *E. coli* is one of the fastest growing bacteria. Under ideal conditions, the number of *E. coli* in a colony doubles about every 15 minutes until restrained by lack of resources. Starting from a single bacterium, how many *E. coli* will there be in 1 hour? In 5 hours?

2.27 (Exact exponential growth) A clever courtier, offered a reward by an ancient king of Persia, asked for a grain of rice on the first square of a chess board, 2 grains on the second square, then 4, 8, 16, and so on.

  (a) Make a table of the number of grains on each of the first 10 squares of the board.
  (b) Plot the number of grains on each square against the number of the square for squares 1 to 10, and connect the points with a smooth curve. This is an exponential curve.
  (c) How many grains of rice should the king deliver for the 64th (and final) square?
  (d) Take the logarithm of each of your numbers of grains from (a). Plot these logarithms against the number of squares from 1 to 10. You should get a straight line.
  (e) From your graph in (d) find the approximate values of the slope b and the intercept a for the line. Use the equation $y = a + bx$ to predict the logarithm of the amount for the 64th square. Check your result by taking the logarithm of the amount you found in (c).

2.28 (Exact exponential growth) Alice is given a savings bond at birth. The bond is initially worth $500 and earns interest at 7.5% each year. This means that the value is multiplied by 1.075 each year.

(a) Find the value of the bond at the end of 1 year, 2 years, and so on up to 10 years.

(b) Plot the value y against years t on graph paper. Connect them with a smooth curve. This is an exponential curve.

(c) Take the logarithm of each of the values y that you found in (a). Plot the logarithm log y against years t on graph paper. You should obtain a straight line.

(d) According to the rule of 70, how many years are needed to double Alice's $500 at 7.5% interest per year? The answer is not a whole number of years—check it by comparing the value of the bond at the end of the year before and the year after the time given by the rule of 70.

2.29 Recall Fred from Exercise 2.13. He and Alice were born the same year, and each began life with $500. Fred added $100 each year, but earned no interest. Alice added nothing, but earned interest at 7.5% annually. After 25 years, Fred and Alice are getting married. Who has more money?

2.30 Biological populations can grow exponentially if not restrained by predators or lack of food. The gypsy moth outbreaks that occasionally devastate the forests of the northeast illustrate approximate exponential growth. It is easier to count the number of acres defoliated by the moths than to count the moths themselves. Here are data on an outbreak in Massachusetts. (Data provided by Chuck Schwalbe, U.S. Department of Agriculture.)

Year	Acres
1978	63,042
1979	226,260
1980	907,075
1981	2,826,095

(a) Plot the number y of acres defoliated against the year t. The pattern of growth appears exponential.

(b) Verify that y is being multiplied by about 4 each year by calculating the ratio of acres defoliated each year to the previous year. (Start with 1979 to 1978, when the ratio is $226,260/63,042 = 3.6$.)

(c) Take the logarithm of each number y and plot the logarithms against the year t. The linear pattern confirms that the growth is exponential.

(d) The least squares line fitted to the four points is

$$\log y = -1094.51 + (.55577 \times \text{year})$$

Draw this line on your graph in (c).

(e) Use the line to predict the number of acres defoliated in 1982. (Predict log y by substituting $t = 1982$ in the equation. Then use the fact that $y = 10^{\log y}$ to predict y.) The actual number for 1982 was 1,383,265—far less than the prediction. The exponential growth of the gypsy moth population was cut off by a viral disease, and the population quickly collapsed back to a lower level.

2.31 Federal expenditures on social insurance (chiefly social security and Medicare) increased rapidly after 1960. Here are the amounts spent, in millions of dollars:

1960	1965	1970	1975	1980
14,307	21,807	45,246	99,715	191,162

(a) Plot social insurance expenditures against time. Does the pattern appear closer to linear growth or to exponential growth?
(b) Take the logarithm of the amounts spent. Plot these logarithms against time. Do you think that the exponential growth model fits well?
(c) After entering the data into the MINITAB statistical system, with year as C1 and expenditures as C2, we obtain the least squares line for the logarithms as follows:

```
MTB > LET C3 = LOGT(C2)
MTB > REGRESS C3 ON 1, C1

THE REGRESSION EQUATION IS
C3 = -110.04 + 0.05824 C1
```

That is, the least squares line is

$$\log y = -110.04 + (.05824 \times \text{year})$$

Draw this line on your graph from (b).
(d) Use this line to predict the logarithm of social insurance outlays for 1985. Then compute

$$y = 10^{\log y}$$

to predict the amount y spent in 1985.
(e) The actual amount (in millions) spent in 1985 was $313,107. Take the logarithm of this amount and add the 1985 point to your graph in (b). Does it fall close to the line? Did the trend of exponential growth in spending for social insurance change in a major way after President Reagan took office in 1981?

2.32 The following table shows the growth of the population of Europe (millions of persons) between 400 B.C. and 1950:

Date	Population	Date	Population	Date	Population
400 B.C.	23	1200	61	1600	90
1 A.D.	37	1250	69	1650	103
200	67	1300	73	1700	115
700	27	1350	51	1750	125
1000	42	1400	45	1800	187
1050	46	1450	60	1850	274
1100	48	1500	69	1900	423
1150	50	1550	78	1950	594

(a) Plot population against time.

(b) The graph shows that the population of Europe dropped at the collapse of the Roman Empire (200–500 A.D.) and at the time of the Black Death (1348 A.D.). Growth has been uninterrupted since 1400. Plot the logarithm of population against time, beginning in 1400.

(c) Was growth exponential between 1400 and 1950? If not, what overall pattern do you see?

2.33 The following table gives the resident population of the United States from 1790 to 1980, in millions of persons:

Date	Population	Date	Population	Date	Population
1790	3.9	1860	31.4	1930	122.8
1800	5.3	1870	39.8	1940	131.7
1810	7.2	1880	50.2	1950	151.3
1820	9.6	1890	62.9	1960	179.3
1830	12.9	1900	76.0	1970	203.3
1840	17.1	1910	92.0	1980	226.5
1850	23.2	1920	105.7		

(a) Plot population against time. The growth of the American population appears exponential.

(b) Plot the logarithms of population against time. The pattern of growth is now clear. An expert says that "the population of the United States increased exponentially from 1790 to about 1880. After 1880 growth was still approximately exponential, but at a slower rate." Explain how this description is obtained from the graph.

(c) Look closely at your graphs. One point deviates more than the others from the overall pattern. What year does this point correspond to? Can you suggest an historical explanation for the deviation?

(d) Lay a transparent straightedge over the graph of the logarithm of population from 1900 to 1980. Use this line to predict the population of the United States in 1990.

2.34 The number of motor vehicles (cars, trucks, and buses) registered in the United States grew as follows:

Year	Vehicles (millions)	Year	Vehicles (millions)
1940	32.4	1965	90.4
1945	31.0	1970	108.4
1950	49.2	1975	132.9
1955	62.7	1980	155.8
1960	73.9	1985	170.2

(a) Plot the number of vehicles against time.
(b) Compute the logarithm of the number of vehicles and plot the logarithms against time.
(c) Look at the years 1950 to 1980. Was the growth in motor vehicle registrations more nearly linear or more nearly exponential during this period? (Use a straightedge to assess the two graphs.)
(d) The year 1940 fits the pattern well, but 1945 and 1985 do not. Can you suggest explanations for the negative residuals in those years?

2.35 The least squares line fit to the 1950 to 1980 data from the previous exercise is approximately
$$\log y = -30.62 + (.01657 \times \text{year})$$
The computing system gives the residuals from the exact least squares line for the years 1950 to 1980 as

1950	1955	1960	1965	1970	1975	1980
−.0116	.0109	−.0006	.0040	.0000	.0056	−.0082

(a) Use this line to predict motor vehicle registrations in 1940, 1945, and 1985.
(b) Make a residual plot for 1950 to 1980. Are there any signs of systematic deviation from exponential growth?
(c) Add the residuals for 1940, 1945, and 1985 to your plot. As usual, a plot of residuals magnifies deviations.

2.36 The Department of Agriculture publishes an index of agricultural productivity. This index is a measure of the output per hour of labor on American farms. Farm productivity increased as follows:

Year	1940	1950	1960	1970	1980
Productivity	12	20	37	66	113

(a) Did agricultural productivity grow exponentially during these decades?

(b) One reason for the growth of farm productivity is the growth of inputs such as fertilizer. Here are data on the amount of fertilizer consumed in the United States (in thousands of tons).

Year	1920	1930	1940	1950	1960	1970
Fertilizer	7176	8171	9360	18,343	24,877	39,591

(c) Make appropriate graphs of these data. Describe in words the pattern of growth in fertilizer usage. Is the exponential model a helpful overall description?

2.4 TIMES SERIES*

Patterns in Time Series

We have thus far encountered models for stability over time (statistical control) and for systematic growth of a rather simple kind (linear or exponential). Many time series, including those of greatest interest in economics and finance, show much more complicated behavior. These time series cannot be adequately described by a simple systematic model (such as an exponential curve) and residuals that show deviations from the model. We therefore fall back on a verbal description of the overall pattern and deviations from it.

The overall pattern of a time series can be hidden by irregular and erratic movements. To see the overall patterns more clearly, it is helpful to plot the *average* of observations on many individuals rather than measurements taken on a single individual. For example, a stock market index (an average of the price of many stocks) plotted against time is less irregular than the price of a single stock.

EXAMPLE 2.10

index number

Figure 2.17 is a graph of the average price of fresh oranges over the 7-year period 1979 to 1985.[5] This information is collected each month by the Census Bureau as part of the government's reporting of retail prices. The monthly Consumer Price Index is the most publicized product of this effort. The price is presented as an *index number*. That is, the price scale gives the price as a percent of the average price of oranges in 1967. This is indicated by the legend "1967 = 100." The first data point is 250.4 for January 1979, so at that time oranges cost about 250% of their 1967 price. The index number is based on the retail price of oranges at many stores in all parts of the country. The price of oranges at a single store

---

* This material is important in economics and other areas of application, but can be omitted without loss of continuity.

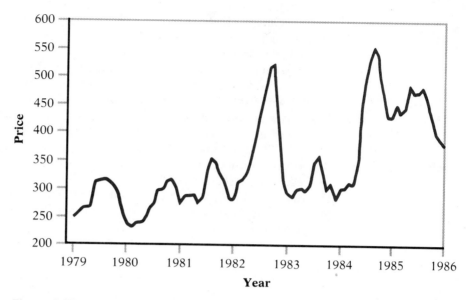

Figure 2.17 Index number (1967 = 100) for monthly prices of fresh oranges, 1979 to 1985, for Example 2.10.

will drop when the manager decides to advertise a sale on oranges, and it will rise when the sale ends. The index number is a kind of nationwide average price that is less variable than the price at one store. ■

trend

seasonal variation

irregular fluctuations

seasonally adjusted

Figure 2.17 shows several features that are common in time series. There is a *trend*, a long-term change in the level of the variable. In this case there is a trend of increasing price. Superimposed on this trend is a strong *seasonal variation*, a regular rise and fall that recurs each year. Orange prices are usually highest in August or September when the supply is lowest. Prices then fall in anticipation of the harvest, and are lowest in January or February when the harvest is complete and oranges are plentiful. But perhaps the most notable features of this time series are the dramatic increases in price in 1982 and again in 1984. The seasonal low in the winter of 1984–1985 is higher than any previous seasonal high except those of 1982 and 1984. These *irregular fluctuations* are price shocks due to unusual events—a freeze in Florida and a failure of the orange crop in Brazil.

Trends and seasonal variation can be described by mathematical models, though we will not give the details. Since many economic time series show strong seasonal variation, government agencies often adjust for this variation before releasing economic data. The data are then said to be *seasonally adjusted*. Seasonal adjustment helps avoid misinterpretation. A rise in the unemployment rate from December to January, for example, does not mean that the economy is slipping. Unemployment almost always rises in January as temporary Christmas help is laid off and outdoor employment in the north drops because of bad weather. The seasonally adjusted unemployment rate reports an increase only if unemployment rises more than normal from December to January.

The orange price shocks of 1982 and 1984 could not have been predicted by such a model. Such large fluctuations are much more important than the small deviations from the models we used in Sections 2.2 and 2.3. Economic time series are often subject to outside shocks due to strikes, bad weather, and other unpredictable events.

cycles

In addition to trends, seasonal variation, and irregular fluctuations, time series sometimes show *cycles*, distinct up and down movements that are less regular than seasonal variation and are not explained by seasonal effects. Figure 2.18 is a time series of great interest to many people—common stock prices. It may surprise you to learn that some people suspect regular cycles in stock prices.

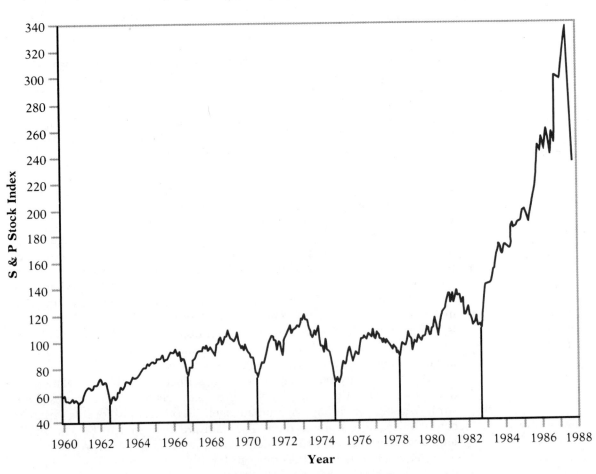

Figure 2.18 The S&P 500 Stock Index, 1960 to 1987, with market bottoms marked. See Example 2.11.

EXAMPLE 2.11

The variable graphed in Figure 2.18 is the Standard & Poor's 500 Composite Stock Price Index, commonly known as the S&P 500. The S&P 500 is an average of the stock prices for 500 companies, but it is not simply the mean of the prices. It is instead a weighted average in which each company's stock influences the index in proportion to its total market value (stock price × number of shares.) A change in the price of IBM stock, for example, moves the index more than the same change in the price of any other stock, because IBM leads the pack in the total market value of its shares. The S&P 500 is a widely used measure of the overall level of U.S. common stock prices. It is one of the Commerce Department's official economic indicators, for example. ■

Figure 2.18 shows an increasing trend, though with a long trendless period during the 1970s. For more than 20 years there also appeared to be a 4-year cycle. Stock prices bottomed out in 1962, 1966, 1970, 1974, 1978, and 1982. The peaks are less regular but occurred roughly midway between the bottoms.

Why should stock prices show a four-year cycle? The cynics suggest that the 4-year cycle of American politics may have something to do with it. Stock market tops occur roughly at presidential election time, with the bottoms midway between major elections. Is it possible that the party in power attempts to pump up the economy, and with it the stock market, as the election approaches? Necessary but unpleasant economic medicine could then be administered during the period between elections. This hypothesis remains unproven. What is more, planning your stock purchases and sales around the 4-year cycle is no way to make money. Selling your stocks in anticipation of the next market bottom in 1986 would have been unprofitable: The S&P 500 rose 14.6% in 1986. Buying stocks in mid-1987 in anticipation of the 1988 election would have been disastrous, since the crash of October 19, 1987 intervened. Extrapolation, remember, is risky. You should be particularly wary of betting on a pattern based solely on visual inspection of a time series when the reasons for the observed variation are not well understood.

Smoothing Time Series

Although concise models for general time series are beyond our reach, we can fall back on descriptive measures to help summarize the overall pattern of a time series. The longer-term patterns—trends, cycles, and seasonal variation—in many time series are obscured by short-term irregular fluctuations. When we have no clear model in mind, we emphasize the larger patterns by smoothing out the fluctuations. A freehand curve drawn on the time plot will do this, but we prefer methods that are less subjective and that can be turned over to a computer.

The idea behind the most common smoothing methods is to replace each value of the time series by the center of several adjacent values. The centers fluctuate more slowly than the individual observations but still respond to long-term changes. Because of the resistance of the median, medians of adjacent values are particularly good smoothers.

Running median of three

> The running median of three smooths a time series by replacing each data point by the median of the point itself and the observations immediately before and after it in time order.

Other running medians (of five, for example) work similarly. The name describes the way in which the operation of finding medians "runs" along the sequence of data.

EXAMPLE 2.12

Return to Professor Simon's commuting times from Table 2.1. The first 10 observations in this series, expressed in minutes, are

8.25   7.83   8.30   8.42   8.50   8.67   8.17   9.00   9.00   8.17

We will use running medians of three to smooth these data to search for any regular change over time. Since the first observation is not the center of three, we leave it unchanged. (The final observation is also left unchanged. There are methods for smoothing endpoints, but we will not discuss them.) The second data point, 7.83, is replaced by the median of the three values 8.25, 7.83, 8.30, which is 8.25. The third value, 8.30, is replaced by the median of 7.83, 8.30, 8.42; this median is 8.30, so the third observation is not changed. Running medians of three or five are very fast to calculate by eye, since the median is just the middle-sized value of three or five adjacent observations. Proceeding along the data, the first 10 values in the smoothed time series are

8.25   8.25   8.30   8.42   8.50   8.50   8.67   9.00   9.00   8.17

Figure 2.19 shows the original time series as points and the smoothed version as a line.   ∎

Notice that a running median of three is completely resistant to a single outlier in any run of three consecutive data points. The three outliers that appear in Figure 2.19 were effectively smoothed away. The line in Figure 2.19 shows no obvious pattern, though there may be an upward trend at the end of the series. This trend, however, is based on relatively few observations. The weakness of graphical methods not tied to any model is that our eyes may see patterns that are simply accidental. The control chart of Figure 2.2, with its run of 10 observations above the mean, gives firmer evidence that the apparent trend reflects a real change in the distribution of commuting times. The weakness of more formal methods such as control charts is that the models behind them (such as normal distributions for control charts) may not be correct. Neither type of statistical method removes the need for good judgment.

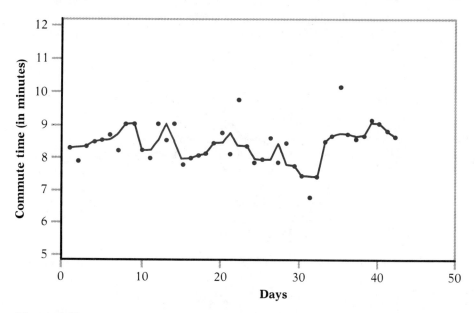

Figure 2.19 The commuting times from Table 2.1 smoothed by running medians of three. See Example 2.12.

A running median of three will be pulled out by two consecutive outliers, but running medians of five will resist two-point spikes in the data. In general, medians of five are smoother than medians of three but also less like the original data. The wider the window over which we average at each point, the smoother the result will be but less detail will be retained from the original time series.

repeated running median
Another way to further smooth a time series is the *repeated running median*, which applies running medians repeatedly until no further change in the smoothed time series occurs. In the case of the commuting time data from Table 2.1, two applications of running medians of three produce a time series that is not changed if we again take running medians of three. Figure 2.20 shows the result of repeated running medians in this example. Repeated smoothing has smoothed out the three sharp upward spikes in the center of the once-smoothed series in Figure 2.19. Some statistical computing systems offer smoothing programs that eliminate the work involved in doing repeated smoothing by hand.

A smoothed time series represents the overall pattern of the original series. As always, we should not focus exclusively on the overall pattern, because the residuals or deviations from the smoothed series are equally important. We can compute and plot residuals from a smoothed time series just as from a linear or exponential model. The distinction is simply that the FIT is no longer the product of a mathematical model for the overall pattern but rather a description based entirely on the data.

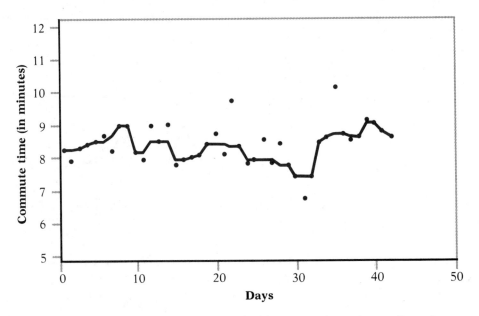

Figure 2.20   The commuting times smoothed by repeated running medians of three.

SUMMARY

General time series may show **trends**, **seasonal variation**, and **cycles**, often combined with large **irregular fluctuations**.

Irregular fluctuations can hide the longer-term pattern of a time series. One way to obtain a smoother time series is to plot an average of observations on many individuals.

Averaging over time by means of smoothing operations, such as **running medians**, can also help display the main features of a time series.

SECTION 2.4 EXERCISES

2.37 The years around 1970 brought unrest to many U.S. cities. Here are data on the number of civil disturbances (a polite name for riots) in each 3-month period during the years 1968 to 1972.

(a) Make a plot of these data against time. Connect the points in your plot by straight line segments to make the pattern clearer.

(b) Describe the trend and seasonal variation that you see in your graph. Can you suggest an explanation for the seasonal variation in civil disorders?

Period	Count	Period	Count
1968, Jan.–Mar.	6	1970, July–Sept.	20
Apr.–June	46	Oct.–Dec.	6
July–Sept.	25	1971, Jan.–Mar.	12
Oct.–Dec.	3	Apr.–June	21
1969, Jan–Mar.	5	July–Sept.	5
Apr.–June	27	Oct.–Dec.	1
July–Sept.	19	1972, Jan.–Mar.	3
Oct.–Dec.	6	Apr.–June	8
1970, Jan.–Mar.	26	July–Sept.	5
Apr.–June	24	Oct.–Dec.	5

(c) If we number the 3-month periods in the table from 1 to 20 and call the resulting variable TIME, the least squares line for these data is given by MINITAB as follows:

```
MTB > REGRESS 'RIOT' ON 1, 'TIME'

THE REGRESSION EQUATION IS
RIOT = 23.38 - 0.927 TIME
```

Draw this line on your graph from part (a). A linear trend can be described by a least squares line even when other patterns such as seasonal variation are present.

2.38 We often look at time series data to see the effect of a social change or new policy. Here are data on motor vehicle deaths in the United States. Because the *number* of deaths will tend to rise as motorists drive more miles, we look instead at the *rate* of deaths, which is the number of deaths per 100 million miles driven.

Year	Rate	Year	Rate
1960	5.1	1974	3.5
1962	5.1	1976	3.3
1964	5.4	1978	3.3
1966	5.5	1980	3.3
1968	5.2	1982	2.8
1970	4.7	1984	2.6
1972	4.3	1987	2.4

(a) Make a plot of these death rates against time. Describe any patterns that are apparent in the data.
(b) In 1974 the national speed limit was lowered to 55 miles per hour in the wake of the increase in oil prices after the 1973

mideast war. The *number* of highway deaths dropped when the speed limit was lowered. Some said that the lower speed limit saved lives. Others said that deaths were down only because higher gasoline prices led to fewer miles driven. Explain why these data refute the second explanation.

2.39 College administrators concerned about resource planning are very interested in the numbers of students enrolled in different degree programs. The following table gives the numbers of students enrolled in undergraduate degree programs in the School of Science at Purdue University for the years 1963 to 1983:

Year	Students	Year	Students	Year	Students
1963	1567	1970	2461	1977	2565
1964	1596	1971	2550	1978	2538
1965	1755	1972	2639	1979	2453
1966	1895	1973	2689	1980	2463
1967	2099	1974	2628	1981	2766
1968	2237	1975	2727	1982	2809
1969	2355	1976	2586	1983	2921

(a) Plot the number of students versus year.

(b) Give a short summary of the changes in enrollment during the years 1963 to 1983. Since the same pattern does not describe the entire two decades, call attention to shorter periods that show a clear pattern.

2.40 Here are the numbers of international airline passengers (in thousands) for each month of the years 1954 to 1956. Plot this time series. Identify the major patterns present (trend, cycles,

Month	1954	1955	1956
Jan.	204	242	284
Feb.	188	233	277
Mar.	235	267	317
Apr.	227	269	313
May	234	270	318
June	264	315	374
July	302	364	413
Aug.	293	347	405
Sept.	259	312	355
Oct.	229	274	306
Nov.	203	237	271
Dec.	229	278	306

seasonal variation). Suggest an explanation for the patterns that you see. (Part of a larger data set given by G. E. P. Box and G. M. Jenkins, *Time Series Analysis*, Holden-Day, Oakland, Calif., 1976, p. 531.)

2.41 As the earth revolves around the sun, the number of daylight hours varies in a regular pattern. The following table gives the daylight hours (in hours and minutes) for Boston on the first day of each month in 1987:

Month	Hours	Month	Hours	Month	Hours
Jan.	9:08	May	14:04	Sept.	13:10
Feb.	10:00	June	15:04	Oct.	11:45
Mar.	11:14	July	15:14	Nov.	10:21
Apr.	12:43	Aug.	14:28	Dec.	9:19

(a) Plot the number of daylight hours versus month. (Hint: First convert all numbers to minutes.) Connect the points with a smooth curve.

(b) Explain why the point for January at the beginning of the plot should have a value close to that for December at the end of the plot.

(c) Recall that the values are given for the first of each month. Using your smooth curve, give an estimate for the length of the day on December 15. The actual value is 9:06. Compute the difference between the actual value and your guess.

(d) Again using your smooth curve, estimate the length of the day for April 15. The actual value is 13:22. Compute the difference between the actual value and your guess.

(e) Using your smooth curve you could give an estimate for any day of the year. From what you learned in parts (c) and (d) of this exercise, give an estimate of the worst error (difference between the actual value and your guess) that you would expect if you guessed the number of daylight hours for each day of the year in Boston using your smooth curve. Justify your answer by a careful consideration of your plot.

2.42 Smooth the civil disturbances data in Exercise 2.37 by computing running medians of three. Draw the smoothed time series on the same graph as the original observations. Does the smoothed series show the trend of the original series? The seasonal variation?

2.43 Smooth the annual motor vehicle death counts in Exercise 2.38 by running medians of three. Plot the smoothed series and compare it with the original time series. Does the smoothing emphasize or hide the most interesting feature of these data?

2.44 Continue the smoothing of the civil disturbance data (Exercises 2.37 and 2.42) by repeated running medians of three. Compare the visibility of the trend and seasonal variation in a plot of the smoothed data with your earlier results.

2.45 In 1957 John J. Kelly won the Boston Marathon in a time of 2 hours, 20 minutes, and 5 seconds; in 1988 Ibrahim Hussein won with a time of 2 hours, 8 minutes, and 43 seconds. The following table gives the winning times (in minutes, rounded to the nearest minute) for the years 1957 to 1988:

Year	Time	Year	Time	Year	Time
1957	140	1968	142	1979	129
1958	146	1969	134	1980	132
1959	143	1970	131	1981	129
1960	141	1971	139	1982	129
1961	144	1972	136	1983	129
1962	144	1973	136	1984	131
1963	139	1974	134	1985	134
1964	148	1975	130	1986	128
1965	137	1976	140	1987	132
1966	137	1977	135	1988	129
1967	136	1978	130		

(a) Plot the winning times against year.
(b) Smooth the winning times by running medians of three. Plot the running medians on your graph from (a) and connect them by line segments to display the smoothed time series.
(c) Give a brief verbal description of the pattern of Boston Marathon winning times over these three decades.

2.46 We will compare two further smoothings of the Boston Marathon winning times given in the previous exercise.

(a) Smooth this time series by running medians of five, and plot the smoothed series on a graph with the original data. (Note that the first two and the last two points are unchanged by running medians of five.)
(b) Starting with the result of the previous exercise, smooth the marathon data by repeated running medians of three. Plot this alternative smoothed series on your graph from (a) in another color. Are there any important differences between the two smoothed series?

2.47 It is known that there are seasonal patterns in the growth of children. We can examine the pattern for some children in Egypt using data from Kalama. The following table gives the average weights of

Egyptian toddlers who reach the age of 24 months in each month of the year. (Since there are only a few children who are exactly 24 months old in any one month, a fairly complicated statistical procedure has been used to obtain the table entries from the weights of all children between 18 and 30 months of age.) Changes in this standardized weight indicate slower or faster average growth in the entire group of children.

Month	Weight (kg)	Month	Weight (kg)	Month	Weight (kg)
Jan.	11.26	May	10.87	Sept.	10.87
Feb.	11.39	June	10.80	Oct.	10.74
Mar.	11.37	July	10.77	Nov.	10.89
Apr.	11.10	Aug.	10.79	Dec.	11.25

Plot the data. Describe any seasonal patterns that appear in the plot. In particular, which months correspond to the highest weights? Which correspond to the lowest?

2.48 Refer to the previous exercise. The corresponding data for height are given in the following table:

Month	Height (cm)	Month	Height (cm)	Month	Height (cm)
Jan.	79.80	May	80.10	Sept.	79.94
Feb.	79.66	June	80.23	Oct.	79.80
Mar.	79.91	July	80.07	Nov.	79.39
Apr.	80.00	Aug.	79.79	Dec.	79.51

Plot the data. Describe any seasonal patterns that appear in the plot. Which months correspond to the largest heights? Which correspond to the smallest? Describe the similarities and differences between this plot and the weight plot from the previous exercise.

2.49 Refer to the previous two exercises. In villages like Kalama, the prevalence of certain types of illness may slow the growth of children. In particular, diarrhea is a serious problem and a major cause of death for small children. The following table gives the average percentage of days they are ill with diarrhea per month for the children whose weight and height information is given in the previous exercises.

Month	Percentage	Month	Percentage	Month	Percentage
Jan.	2.54	May	8.42	Sept.	1.20
Feb.	1.82	June	6.87	Oct.	2.51
Mar.	2.28	July	4.51	Nov.	2.34
Apr.	6.86	Aug.	3.42	Dec.	1.27

(a) Plot the data. Describe any seasonal patterns that appear in the plot. Which months correspond to the highest average percentage of days ill with diarrhea? Which months correspond to the lowest?

(b) According to Egyptian physicians who work with these children, there are two peaks in the frequency of this disease—one that is very high and a secondary smaller peak. Is there evidence in these data to support these observations?

(c) Draw a smooth curve through the points of your plot, and also through your plots in the two previous exercises. Comparing these plots, can you formulate any theories concerning the relationship between diarrhea and the growth of these children? Note that illness in one month may not have an effect on growth until subsequent months. On the other hand, a child who is not growing well may be more susceptible to illness.

NOTES

1 Data courtesy of Linda D. McCabe and the Agency for International Development.

2 Zeinab E. M. Afifi, "Principal components analysis of growth of Nahya infants: Size, velocity and two physique factors," *Human Biology*, 57 (1985), pp. 659–669.

3 There are as yet no fully satisfactory mathematical models for the growth in height of humans from birth to adulthood. The models that have been suggested are necessarily much more complicated than those we consider in this chapter. You can see examples in M. A. Preece and M. J. Baines, "A new family of mathematical models describing the human growth curve," *Annals of Human Biology*, 5 (1978), pp. 1–24.

4 Data from the Energy Information Administration, recorded in Robert H. Romer, *Energy: An Introduction to Physics*, W. H. Freeman, San Francisco, 1976 for 1880 to 1972, and in *The World Almanac and Book of Facts 1986*, Newspaper Enterprise Association, New York, 1985 for more recent years.

5 The data upon which Figure 2.17 is based appear in the monthly issues of the Bureau of Labor Statistics' *CPI Detailed Report*.

CHAPTER 2 EXERCISES

2.50 The number of people living on American farms has declined steadily during this century. Here are data on the farm population (millions of persons) from 1935 to 1980:

Year	1935	1040	1945	1950	1955	1960	1965	1970	1975	1980
Population	32.1	30.5	24.4	23.0	19.1	15.6	12.4	9.7	8.9	7.2

Figure 2.21(a) is a plot of these data, with the least squares line added. The equation of the least squares line is

$$\text{population} = 1166.93 - .5869t$$

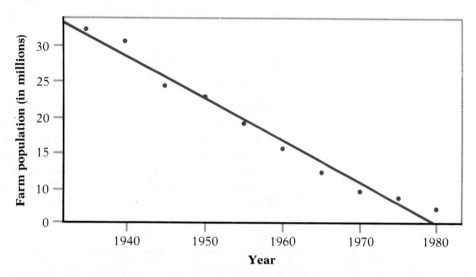

Figure 2.21(a)   U.S. farm population, 1935 to 1980, with the least squares line. See Exercise 2.50.

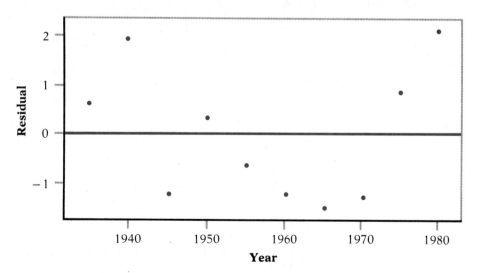

Figure 2.21(b)   Residual plot for the least squares line in Figure 2.21(a).

Figure 2.21(b) is a plot of the residuals from this least squares fit against time.

(a) According to the fitted line, how much did the farm population decline each year during this period?

(b) Use the least squares line to predict the farm population in the year 1945. Then compute the residual for 1945 and check your work by looking at Figure 2.21(b). (Your value will differ slightly from the value plotted because of roundoff error.)

(c) Circle the residual for 1945 in Figure 2.21(b). Because that year marks the end of World War II, it is exceptional. Describe the pattern of residuals that emerges in Figure 2.21(b) when the 1945 point is ignored. During which periods was the decline in farm population faster or slower than the long-term linear trend?

2.51 Do mold colonies grow exponentially? In an investigation of the growth of molds, flasks containing a growth medium are inoculated with equal amounts of spores of the mold *Aspergillus nidulans*. The size of a colony is measured by analyzing how much remains of a radioactive tracer substance that is consumed by the mold as it grows. Each size measurement requires destroying that colony, so the data below refer to 30 separate colonies. To smooth the pattern, we take the mean size of the three colonies measured at each time.

Hours	Colony size			Mean
0	1.25	1.60	.85	1.23
3	1.18	1.05	1.32	1.18
6	.80	1.01	1.02	.94
9	1.28	1.46	2.37	1.70
12	2.12	2.09	2.17	2.13
15	4.18	3.94	3.85	3.99
18	9.95	7.42	9.68	9.02
21	16.36	13.66	12.78	14.27
24	25.01	36.82	39.83	33.89
36	138.34	116.84	111.60	122.26

Figure 2.22(a) graphs the mean colony size against time, while Figure 2.22(b) is a graph of the logarithm of the mean colony size against time. (Early experiments are described by A. P. J. Trinci, "A kinetic study of the growth of *Aspergillus nidulans* and other fungi," *Journal of General Microbiology*, 57 (1969), pp. 11–24. These data were provided by Dr. Thomas Kuczek, Department of Statistics, Purdue University.)

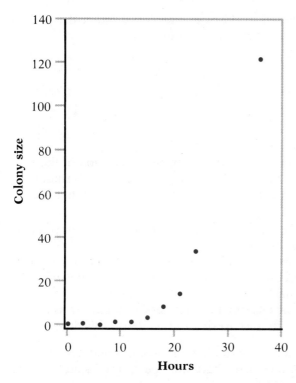

Figure 2.22(a) Mean size of mold colonies. See Exercise 2.51.

Figure 2.22(b) The logarithms of mold colony size, with the least squares line.

(a) On the basis of data like these, microbiologists divide the growth of mold colonies into three phases that follow each other in time. Exponential growth occurs during only one of these phases. Briefly describe the three phases, making specific reference to the graphs to support your description.

(b) The line in Figure 2.22(b) is the least squares line fitted only to the data from the exponential growth phase. The equation of the line is

$$\text{log size} = -.213 + .0636t$$

Use this line to predict the size of a colony 10 hours after inoculation. (Note that the line predicts the logarithm; you must then obtain the size from its logarithm.)

(c) By how much is the size of a colony multiplied each hour during the exponential growth phase?

2.52 Refer to Exercise 2.4 on the family food expenditures. Suppose that when you entered the data into the computer, you got a little tired

toward the end and entered the last value as 26300 rather than
263.00. The computer then gives these results.

```
MTB > MEAN 'FOOD'
MEAN = 869.542

MTB > STDEV 'FOOD'
STDEV = 3749.23
```

(It is good practice to ask for a more complete description of the
data, such as that given by DESCRIBE in MINITAB. A description
that prints the minimum and maximum values would call that 26300
to your attention.)

(a) Construct a control chart for this data set using $\bar{x}$ and s as printed
out above. Explain the resulting pattern.

(b) How many points are within one standard deviation of the mean?
How does this compare to the number that you would expect if
the data were normal?

(c) Using all that you have learned in Chapters 1 and 2, list as many
methods as possible for discovering the outlier in this data set.

2.53 Refer to the previous exercise. Assume that the bad value has escaped
detection, as it might if a control chart were being produced by a
computer without human intervention.

(a) Construct a control chart using the median as a center line and
1.5 times the interquartile range in place of two standard
deviations.

(b) How does this compare with the original control chart?

(c) What are the advantages of using resistant measures to establish
control limits?

2.54 Here are the death rates per 100,000 women age 45 to 54 from the
two types of cancer that kill the most women: breast cancer and
respiratory/intrathoracic (mainly lung) cancer. Plot the two time
series together and describe the pattern. Now extrapolate your curves
(this is risky) to predict when lung cancer will pass breast cancer as
the leading killer of women.

Year	1940	1950	1960	1970	1975	1980	1985
Breast	47.5	49.9	51.4	52.6	50.4	48.9	46.7
Lung	6.2	6.7	10.1	22.2	28.0	34.8	35.9

2.55 Organ transplants are becoming more and more common. The
following table gives the numbers of transplants of four organs for
the years 1982 to 1985:

Year	1982	1983	1984	1985
Heart	103	172	346	719
Liver	62	164	308	602
Kidney	5358	6112	6968	7695

(a) Plot each variable and its logarithm over time, and draw a smooth curve through each series.

(b) Which types of organ transplant were increasing approximately linearly during this period, which were increasing approximately exponentially, and which followed neither pattern of growth?

2.56 Recall Thomas the cat and his fleas from Exercise 1.19. The table given in that exercise lists the number of eggs that Thomas' fleas produced on 27 consecutive days. Read the table across the rows; i.e., the first row gives the first 10 days.

(a) Plot Thomas' flea data versus day.

(b) Draw a center line and control limits.

(c) Circle any points that are beyond the control limits and label these points with the number of the day.

(d) Describe any patterns that you see. Are there any periods of relatively consistent increase? Of consistent decrease?

2.57 Refer to the previous exercise. Thomas has a colleague named Creole who is also helping the experimenters study flea eggs. Creole's fleas were bred in a California laboratory and are of a different species than Thomas'. The data for Creole are given in the following table:

Day	Eggs	Day	Eggs	Day	Eggs	Day	Eggs	Day	Eggs
1	444	6	552	11	486	16	465	21	502
2	524	7	459	12	446	17	483	22	445
3	444	8	504	13	451	18	472	23	513
4	492	9	505	14	505	19	430	24	401
5	555	10	465	15	478	20	438	25	490

(a) Plot Creole's flea data versus day.

(b) Draw a center line and control limits.

(c) Circle any points that are beyond the control limits and label these points with the number of the day.

(d) Describe any patterns that you see. Are there any periods of relatively consistent increase? Of consistent decrease?

(e) Summarize the ways in which Thomas' flea data differ from Creole's.

2.58 Electricity consumption in a given household can be expected to vary over time. In warm months air conditioning can increase electricity consumption. Sometimes people take vacations and their electricity usage drops to very low levels. The table below gives the monthly electricity usage (in kilowatt-hours) for one household over a period of 4 years. Computation from these data gives $\bar{x} = 1211.479$ and $s = 222.3497$.

Month	Year 1	Year 2	Year 3	Year 4
Jan.	1146	1311	1369	1528
Feb.	1126	1325	1133	1224
Mar.	1243	1022	1105	1033
Apr.	1062	1043	1065	1224
May	1125	1259	1216	1202
June	854	898	890	767
July	1645	1716	1154	1272
Aug.	1776	1807	1273	1599
Sept.	1213	1089	1023	1190
Oct.	1280	1119	1030	1359
Nov.	1326	1140	1162	1214
Dec.	1193	1028	1094	1289

(a) Plot the data in the form of a control chart with center line and control limits. Circle all values outside of the control limits. Beside the circles write the name of the month corresponding to the value that is out of control.

(b) Do the points that are out of control have any similarities? Offer an explanation for these points and for any other pattern that you notice.

(c) For each month compute the mean for the four values corresponding to the 4 years. Plot these means versus the month. Describe the pattern you see.

(d) The plots you constructed in (a) and (c) give different types of information. Summarize the information that you get from each. What are the advantages of each type of plot?

2.59 Aquaculture is fish farming, the growing of fish for food in large tanks. Because the fish deplete the oxygen dissolved in the water, a system to recharge the oxygen is an important part of an aquaculture system. In an experiment to compare several recharge systems, water with no dissolved oxygen was recharged and the growth of dissolved oxygen over time was studied. Here are the data for one system. Time is measured in minutes, and oxygen in milligrams per liter of water. (From Gary E. Miller, "Evaluation of a trickling biofilter in a recirculating aquaculture system containing channel catfish, *Ictalurus punctatis*," M.S. Thesis, Purdue University, 1982.)

Time	0.0	1.0	2.0	3.0	4.0	5.0	6.0	7.0
Oxygen	.00	.42	.58	.55	.88	1.13	1.38	1.58

Time	8.0	9.0	10.0	12.5	15.0	17.5	20.0	22.5
Oxygen	1.78	2.04	2.17	2.72	3.13	3.46	3.88	4.25

Time	25.0	27.5	30.0	35.0	40.0	45.0	50.0	
Oxygen	4.63	4.88	5.17	5.71	6.21	6.55	6.88	

Plot the oxygen content against time. Describe the pattern of growth in words. Is this pattern linear? Is it exponential? As this example suggests, there are many types of regular growth that are not described by our two models.

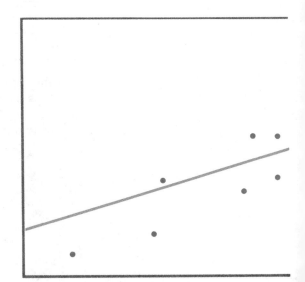

Prelude

With this chapter we reach the important topic of relations between
two variables and learn how an additional variable can influence
such a relation. Once again graphs and numerical calculations
supplement each other in describing the data. Relations that have
a simple pattern, especially straight-line relations, are easiest to
describe. The study of relationships raises the question of causation,
a subtle but important issue in statistics and in science.

- *How does regression help us describe the relation between winter
  temperatures and how much heating fuel a home consumes?*

- *How does correlation measure the strength and direction of the
  straight-line relation between the age at which a child first speaks
  and the child's later score on an aptitude test, and what are the
  drawbacks of this measure?*

- *What kind of evidence links cigarette smoking to lung cancer? Can
  the evidence convince us that smoking actually causes lung cancer?*

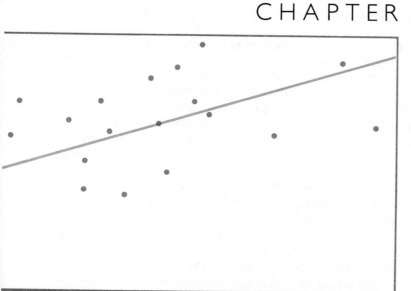

CHAPTER

3

Looking at Data: Relationships

The focus of many statistical studies is on relationships between two or more variables. Did the Democratic party's share of presidential votes in each state change in any important way between the 1980 and 1984 elections? Does planting more corn on an acre of farm land change the yield at harvest, and if so, what is the best planting rate? Do groups of people who smoke more have a higher death rate from lung cancer? Such questions are our topic in this chapter, and these examples are among those that we will consider.

When you first examine the relationship between two variables, you should ask some preliminary questions. Most of these questions have already been raised in Chapter 1: What exactly are the variables? How were they measured? Are both variables quantitative or is at least one a categorical variable? *Quantitative variables* take numerical values for which numerical descriptions such as means and standard deviations are meaningful. Variables measured on a scale of equal units—such as length in centimeters or income in dollars—are quantitative variables. So are counts of individuals, and percents or fractions based on counts. *Categorical variables*, on the other hand, are essentially labels that tell us into which class an individual falls. The sex or occupation of a person, the make of a car, and the species of an animal are all categorical variables.

We have concentrated on quantitative variables to this point. But when data on several variables are being examined, categorical variables are usually present and are essential aids in organizing the data. Here is a small portion of a typical data set, as printed out by the statistical computing system used to analyze it.

OBS	ID	AGE	SEX	JOB	WT	SBP
1	1083	39	M	T	183	132
2	1381	27	F	E	116	117
3	1502	57	M	E	172	144
4	1481	26	M	T	139	110
5	1666	48	F	T	132	150

These data record medical measurements of the employees of a large company that offers regular free physical examinations. Each row gives data for one employee, or in statistical language, one case.

Case

> A case is an individual person, animal, or thing for which values of the variables are recorded.

The employees are given an identification number (ID) so that their names will not be included on the printout. The computing system has numbered the cases consecutively under the heading ❶BS. Each column after the case number and ID contains the values of a specific variable. Five variables are recorded for each case. Three of these—the employee's age, weight, and systolic blood pressure—are quantitative. The others are the employee's sex (male or female) and job category (executive or technical), which are categorical variables.

In studying relationships among variables, we must pay more attention to categorical variables than in earlier chapters. We are interested in relations between two quantitative variables (such as a person's weight and blood pressure), between a quantitative and a categorical variable (such as sex and blood pressure), and between two categorical variables (such as sex and job category). Some parts of this chapter (Sections 3.2 and 3.3) will focus on relations between quantitative variables. Categorical variables are considered in Section 3.4.

When we examine more than one variable, a new question becomes important. Is your purpose simply to explore the nature of the relationship, or do you hope to show that one of the variables can explain changes in the other? In looking at the Democrats' share of the popular vote for president in each state in 1980 and in 1984 (Table 3.1), we do not wish to explain the 1984 data by the 1980 data but rather to see a pattern that may reflect changing political conditions.[1] But in another case, the agronomists who carefully planted corn at different rates per acre and recorded the yield (Table 3.2) are indeed interested in a cause and effect relationship. They believe that the planting rate will affect the yield and their purpose is to recommend the best planting rate to farmers. In such cases, we distinguish the explanatory variable (plants per acre) from the response variable (yield of corn).

Response variable, explanatory variable

A response variable measures an outcome of a study. An explanatory variable attempts to explain the observed outcomes.

In many studies, the goal is to show that changes in one or more explanatory variables actually cause changes in a response variable. For example, medical researchers studying a new drug to treat high blood pressure give different doses of the drug to each of several groups of patients and measure the change in blood pressure after several weeks of treatment. The explanatory variable is the dosage and the response variable is the change in blood pressure. The researchers hope to show that different doses of the drug cause changes in blood pressure. Not all explanatory–response relationships involve direct causation, however. Some of the statistical techniques in this chapter require us to distinguish explanatory from response variables; others make no use of this distinction. Explanatory variables are often called *independent variables*, and response variables are often referred to as *dependent variables*. The idea behind this language is that the response variable depends on the explanatory variable. However, since the words "independent" and "dependent" have other meanings in statistics that are unrelated to the explanatory–response distinction, we prefer to avoid those words here.

The techniques used to study relations among variables are more complex than the methods we developed in Chapter 1 to examine the distribution of a single variable or those developed in Chapter 2 to examine the variation of a single variable over time. Fortunately, statistical analysis of several-variable data builds on the tools mastered in those chapters for examining individual variables. And the principles that guide our work remain the same:

- Combine graphical display with numerical summaries.
- Seek overall patterns and deviations from those patterns.
- Seek compact mathematical models for the data in addition to descriptive measures of specific aspects of the data.

Table 3.1 Percent of the presidential votes won by the Democratic candidate, 1980 and 1984

State	1980	1984	State	1980	1984
Ala.	48.7	38.7	Mont.	33.3	38.7
Alaska	30.1	30.9	Neb.	26.4	29.0
Ariz.	28.9	32.9	Nev.	27.9	32.7
Ark.	48.3	38.8	N.H.	28.6	31.1
Calif.	36.9	41.8	N.J.	39.2	39.5
Colo.	32.0	35.6	N.Mex.	37.5	39.7
Conn.	39.0	39.0	N.Y.	44.8	46.0
Del.	45.3	40.0	N.C.	47.5	38.0
Fla.	38.8	34.7	N.Dak.	26.7	34.3
Ga.	56.3	39.8	Ohio	35.5	40.5
Hawaii	45.6	44.3	Okla.	35.4	31.0
Idaho	25.7	26.7	Oreg.	40.1	43.9
Ill.	42.3	43.5	Pa.	43.1	46.3
Ind.	38.2	37.9	R.I.	48.1	48.2
Iowa	39.1	46.3	S.C.	48.6	35.9
Kan.	33.9	33.0	S.Dak.	32.1	36.7
Ky.	48.1	39.7	Tenn.	48.7	41.8
La.	46.4	38.6	Tex.	41.8	36.2
Maine	43.1	38.9	Utah	20.9	24.9
Md.	47.6	47.2	Vt.	39.3	41.3
Mass.	42.3	48.6	Va.	40.9	37.3
Mich.	43.1	40.4	Wash.	38.2	43.2
Minn.	47.7	50.1	W.Va.	50.1	44.7
Miss.	48.6	37.7	Wis.	44.0	45.4
Mo.	44.7	40.0	Wyo.	28.7	28.6

3.1 SCATTERPLOTS

Relationships between two quantitative variables are best displayed graphically. The most useful graph for this purpose is a *scatterplot*. Here is an example of the effective use of scatterplots.

EXAMPLE 3.1

Table 3.1 shows the percent of the popular vote that was won by the Democratic presidential candidates in the 1980 and 1984 elections. Both candidates, Jimmy Carter in 1980 and Walter Mondale in 1984, were defeated by the Republican Ronald Reagan. (In 1980 an independent candidate, John Anderson, captured 6.7% of the national vote.) We know that many states have persisting political traditions, so we expect similar behavior in two successive elections. It is possible to see this relationship in the columns of numbers in the table, but it is very difficult to assess the strength of the relationship or to see any significant changes from 1980 to 1984. A picture is needed.

Figure 3.1 is a scatterplot of the data in Table 3.1. Each point on the plot represents a single case—that is, a single state. The horizontal coordinate x_i is the percent who voted Democrat in that state's 1980 presidential vote. The vertical coordinate y_i is the percent who voted Democrat in 1984. Thus, Alabama appears as the point (48.7, 38.7), for example. Since both variables have the same units (percent), we use the same scale on both axes. The resulting plot outline is square. ■

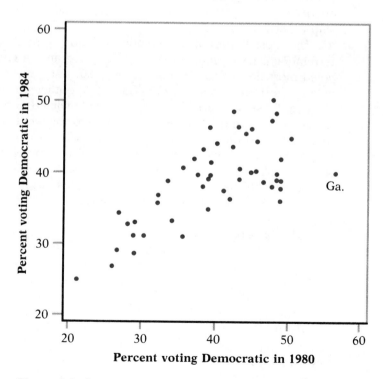

Figure 3.1 Percent of votes for Democrats in the 1980 and 1984 presidential elections, by state. See Example 3.1.

Interpreting Scatterplots

To interpret a scatterplot, look first for an overall pattern. This pattern should reveal the direction, form, and strength of the relationship between 1980 and 1984 voting in the states. The direction is clear from Figure 3.1: States with a high percent who voted Democrat in 1980 tended to also vote Democrat in 1984. That is, the percent voting Democrat in 1980 and in 1984 are positively associated.

Positive association, negative association

> Two variables are positively associated when above-average values of one tend to accompany above-average values of the other, and below-average values tend similarly to occur together. Two variables are negatively associated when above-average values of one accompany below-average values of the other, and vice versa.

In addition to the positive association, the scatterplot in Figure 3.1 shows the form of the relationship: It is roughly linear, though with much scatter about the linear pattern. The large scatter indicates that the linear relationship is not very strong.

The most important systematic deviation from the overall linear pattern in Figure 3.1 occurs at the right of the graph. A cluster of states there voted most heavily for Democrats in 1980, but were markedly less favorable to the Democrats in 1984. The single outlier on the extreme right is Georgia, President Carter's home state. Following that hint, we suspect that the south as a whole was more receptive to the southerner Carter in 1980 than to the northern liberal Mondale in 1984. To show this effect on the graph, Figure 3.2 uses a different symbol (o) to represent the 10 states south of Washington, D.C. and east of the Mississippi River; Louisiana, through which the river flows, is also included.

Figure 3.2 generally sustains our surmise about the south. In fact, we can refine our crude geographical definition of "the south" by examining the political behavior shown in Figure 3.2. The unemphasized point (●) in the middle of the southern cluster is Arkansas, which appears to be southern in voting pattern even though it lies west of the Mississippi. The two emphasized points (o) lying in the political mainstream (to the left of the cluster) are Florida and Virginia. Neither is fully southern in its behavior.

In dividing the states into "southern" and "nonsouthern," we introduced a third variable into the scatterplot. This is a categorical variable that has only two values. The two values are displayed by the two different plotting symbols. Using different symbols to plot points is a good way to incorporate a categorical variable into a scatterplot.[2]

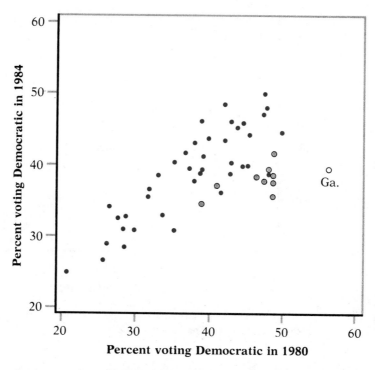

Figure 3.2 The 1980 and 1984 percent of votes for Democrats, with the south emphasized (⊙).

EXAMPLE 3.2

News reports sometimes rate state educational systems by comparing the mean scores of seniors in each state on college entrance examinations. This method is misleading, because the percent of high school seniors who take any particular college entrance test varies greatly from state to state. Figure 3.3 is a scatterplot of the mean score on the Scholastic Aptitude Test (SAT) mathematics examination for high school seniors in each state versus the percent of graduates in each state who took the test.

The *negative association* between these variables is evident: SAT scores tend to be lower in states where the percent of students who take the test is higher. Only 3% of the seniors in the three highest-scoring states took the SAT. The overall pattern shows two *clusters* of points. At the upper left are states where only students seeking admission to colleges that require the SAT take that test. Most students in these states take a different college entrance examination, the American College Testing (ACT) examination. The students who elect to take the SAT tend to be above average academically, so the mean SAT scores for states in this cluster are high. The other cluster, at the lower right of the scatterplot, contains states in which a high percent of college-bound seniors take the SAT. The mean scores are lower here because a less selective group of students take the test. There is little difference in mean SAT scores among these states, even though the percent of seniors who take the test varies from 30% to nearly 70%.

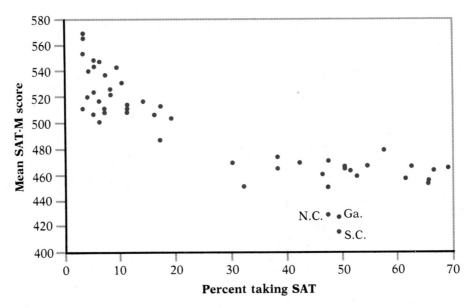

Figure 3.3 Mean SAT mathematics score and percent of high school seniors who took the test, by state. See Example 3.2.

> The points that are individually labeled seem to lie a bit outside the lower cluster. The mean SAT scores in these states are lower than in other states in which a similar percent of students take the test. It is possible that the test scores point to educational deficiencies in these states. ∎

Scatterplots showing relationships between other quantitative variables may appear quite different from the clouds of points in Figures 3.1 and 3.3. This is particularly true in experiments in which measurements of a response variable are taken at only a few selected levels of the explanatory variable. The following example illustrates the use of scatterplots in such a setting:

EXAMPLE 3.3

How much corn per acre should a farmer plant to obtain the highest yield? Too few plants will clearly result in a low yield. On the other hand, if there are too many plants they will compete with each other for moisture and nutrients, and yields will again fall. The amount of moisture is critical: In dry seasons the best planting rate is lower than in wet seasons. Rather than try to forecast the weather, a farmer can ensure even moisture by irrigating his fields. Table 3.2 shows the results of several years of field experiments in Nebraska.[3] Each entry is the mean yield of four small plots planted at the indicated rate per acre. All plots were irrigated, and all were fertilized and cultivated identically. The yield of each plot should therefore depend only on the planting rate—and of course on the uncontrolled aspects of each growing season such as temperature and wind. The experiment lasted several years in order to avoid misleading conclusions due to the peculiarities of a single growing season.

Table 3.2 Average irrigated corn yields (bushels per acre)

Plants per acre	1956	1958	1959	1960	Mean
12,000	150.1	113.0	118.4	142.6	131.0
16,000	166.9	120.7	135.2	149.8	143.2
20,000	165.3	130.1	139.6	149.9	146.2
24,000		134.7	138.4	156.1	143.1
28,000			119.0	150.5	134.8
Mean	160.8	124.6	130.1	149.8	

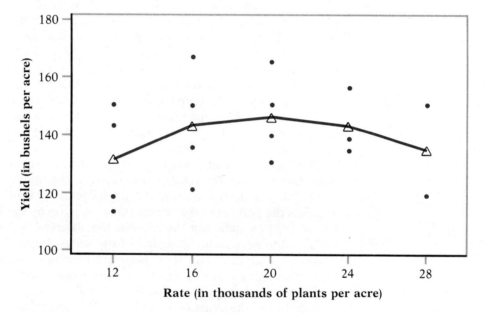

Figure 3.4 Yield of corn versus planting rate for irrigated cornfields in Nebraska. See Example 3.3.

The scatterplot in Figure 3.4 displays the results of this experiment. Since planting rate is the explanatory variable, it is plotted horizontally as the x variable. The yield is the vertical variable y. The vertical spread of points over each planting rate shows the year-to-year variation. The overall pattern is revealed by plotting the mean yield for each planting rate (averaged over all years). These means are marked by triangles and joined by line segments. As expected, the form of the relationship is not linear. Yields first increase with the planting rate, then decrease when too many plants are crowded in. Since there is no consistent direction, we cannot describe the association as either positive or negative. It appears that the best choice is about 20,000 plants per acre. ∎

Plotting the means as was done in Figure 3.4 is an excellent way to summarize an experiment in which repeated observations are made at a few fixed levels of a variable such as planting rate. The scatterplot also indicates how much variability lies behind each mean. In this case, however, we must regard the means with great caution. As the gaps in Table 3.2 show, the agronomists used only the three lowest planting rates in all 4 years, adding the higher rates only as the need to study them became apparent. The means therefore cover different spans of time. In particular, 1956 was a year of very high yields, so that the means that include 1956 are biased upward relative to those that do not.

A closer look at Table 3.2 shows that 24,000 plants per acre was superior to 20,000 in 2 of the 3 years in which both rates were planted. The agronomists concluded that maximum yield is already approached at 16,000 plants per acre, but yields continue to increase somewhat up to 24,000 plants per acre. *Incomplete data*, such as those in Example 3.3, are common in practice. The agronomists' data would have been more convincing if all five planting rates had been used in all 4 years—but it was only after the first results were in that the researchers realized that higher planting rates might give better results. Incomplete data often complicate a statistical analysis, whether it is an informal analysis such as that of Example 3.3 or a formal analysis of the type that we will learn about in Chapter 11. You should therefore be alert to missing data.

Examples 3.1 and 3.3, though different in most respects, share a most important feature: *The relationship between the two variables plotted cannot be fully understood without knowledge about a third variable.* Figure 3.1 plots the percent of the presidential vote won by the Democrats in 1980 versus 1984 by state, but the relationship observed is partly explained by a political and geographic grouping of the states. Figure 3.4 plots corn yield versus planting rate, but the experiment spanned several growing seasons that differed from each other. And if the agronomists had not carefully controlled many other variables (moisture, fertilizer, etc.), these variables would have confused the situation completely. You should be cautious in drawing conclusions from a strong relationship appearing in a scatterplot until you understand what other variables may be lurking in the background.

Smoothing Scatterplots

The strength of a scatterplot is that it provides a complete picture of the relationship between two variables—at least as far as that relationship is reflected in the available data. A complete picture is often too detailed for easy interpretation, so we seek to describe the plot in terms of an overall pattern and deviations from that pattern. Though we can often do this by eye, more systematic methods of extracting the overall pattern are desirable. This is called *smoothing* a scatterplot. When we are plotting a response variable y against an explanatory variable x, Example 3.3 gives us a hint as to

smoothing

how to proceed. We smoothed Figure 3.4 by averaging the y values separately for each x value. Though not all scatterplots have many y values at the same value of x, as did Figure 3.4, we can smooth a scatterplot by slicing it into vertical strips and computing the mean or median y in each strip. For initial analysis, the median is preferred to the mean because it is more resistant.

Median trace

> To construct the median trace of a scatterplot, first slice the plot into vertical strips of equal width. Compute the median of the y values in each strip and plot the median vertically above the horizontal midpoint of the strip. Connect the medians by straight line segments to form the median trace.

The median trace displays the overall pattern of the dependence of y on x. The scatter of the observations above and below the median trace displays deviations from the pattern.

EXAMPLE 3.4

Prior to 1970, American young men were drafted into military service by local draft boards who followed a complex system of preferences and exemptions. Congress decided that a random selection process—a draft lottery—would be fairer. The first draft lottery was held in 1970. The 366 possible birth dates were placed into identical plastic capsules, poured into a rotating drum, and picked out one by one. The first birth date drawn won draft number 1, the next 2, and so on. Eligible men were then drafted in order of their draft numbers, those with the lowest numbers first.

The outcome of the 1970 draft lottery appears in the scatterplot in Figure 3.5. Birth dates, numbered 1 to 366 beginning with January 1, are plotted horizontally. The draft number assigned each date by the lottery is plotted on the vertical scale. A properly conducted lottery should produce *no* systematic relationship between these variables. Figure 3.5 certainly shows no clear overall pattern. Yet it was charged that the lottery was biased against men born late in the year, that these men received systematically lower draft numbers than men born earlier. An investigation showed that birth dates had indeed been inserted into capsules and poured into a box 1 month at a time before being placed into the drum.

With this hint that a month effect might be present, we will smooth Figure 3.5 by a median trace. Imagine that the scatterplot is sliced vertically like a loaf of bread. Each slice contains 1 month's birth dates. We calculate the median draft number for each month, and plot it vertically above the horizontal midpoint of each slice. Figure 3.6 shows the median trace superimposed on the scatterplot. The downward trend late in the year is now apparent. In 1971, the Department of Defense reassigned the officers who had conducted the 1970 lottery and asked statisticians from the National Bureau of Standards to design a truly random selection procedure.[4]   ■

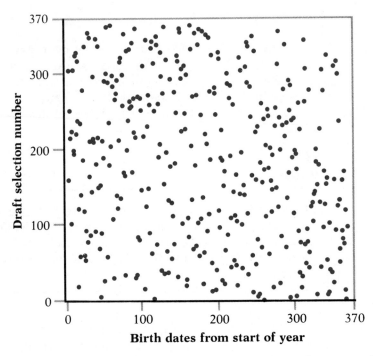

Figure 3.5 The 1970 draft lottery: draft selection numbers versus birth dates. See Example 3.4.

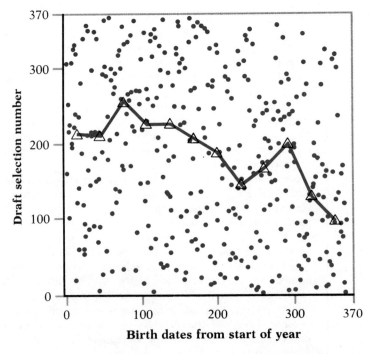

Figure 3.6 A median trace by month of birth for the 1970 draft lottery.

Example 3.4 demonstrates that smoothing can reveal relationships that are not obvious from a scatterplot alone. In this case there is a *negative association* between birth date and draft number: Later birth dates tend to have lower draft numbers. Once the median trace shows us what to look for, we can see that the point cloud in Figure 3.5 is a bit thin in the upper right region, indicating that men born late in the year won few high draft numbers. The combination of graphing and calculating once again proves its effectiveness.

To draw a median trace, first slice the scatterplot. The number of slices chosen determines the degree of smoothing provided by the median trace. Fewer slices smooth the data more, while more slices allow the median trace to follow the ups and downs of y more closely. In general, the number of slices should increase with the number of cases; beyond this, choosing the slices is a matter of judgment. It is often best to take advantage of naturally occurring slices such as the months in Example 3.4.

Slicing a scatterplot into vertical strips is the basis for other graphical displays as well. Figure 3.7 shows *boxplots* for each of the monthly slices of the draft lottery data. The sequence of medians shows the smoothed pattern; the quartiles and extremes show the scatter about the pattern portrayed by the individual points in Figure 3.6. Smoothing a scatterplot either by the median trace or by side-by-side boxplots thus displays both an overall pattern and deviations from it.

The 1970 draft lottery raises one final question. How can we be confident that the lower draft numbers assigned to men born later in the year were not merely the play of chance? After all, repeated random drawings of birth

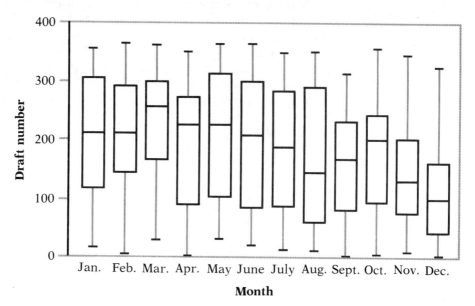

Figure 3.7 Boxplots by month for selection numbers in the 1970 draft lottery.

dates would give a different order each time. Some would appear—after the fact—to favor January, and others to favor December. So the pattern that Figure 3.5 seemed to show may be an accident. Quite true. We must judge whether this pattern is stronger than could reasonably arise from the play of chance in a truly random drawing. This judgment requires a calculation of probabilities. Such a calculation shows that an association between birth date and draft number as strong as the one observed in 1970 would occur less than once in 1000 random drawings. There is in fact good evidence that the 1970 lottery was unfair.

Categorical Explanatory Variables

Variations on scatterplots are also the preferred means for showing relations between a categorical explanatory variable and a quantitative response. These displays are very similar to those already discussed. Suppose that the agronomists of Example 3.3 had compared the yields of five varieties of corn rather than five planting rates. The plot in Figure 3.4 remains helpful if the varieties A, B, C, D, and E are marked at equal intervals on the horizontal axis in place of the planting rate. In particular, a graph of the mean (or median) responses for each category will show the overall nature of the relationship. If there are too many observations in each category to plot individually, as in Figure 3.4, side-by-side boxplots or stemplots can replace the scatterplot of response values above each category label in the graph. Figure 1.9 is such a graph. There the categorical explanatory variable is hot dog type (beef, meat, or poultry), and the response is the number of calories in each hot dog.

Many categorical variables, like corn variety or hot dog type, have no natural order from smallest to largest. In this case we cannot speak of a positive or negative association with the response variable. If the mean responses increase as we go from right to left in the plot, we could make them decrease by writing the categories in the opposite order. In such cases the plot simply presents a side-by-side comparison of several distributions. The categorical variable labels the distributions. Some categorical variables do have a least-to-most order. We can then speak of the direction of the association between the categorical explanatory variable and the quantitative response. Here is an example.

EXAMPLE 3.5

What is the relationship between family income and the educational level of the householder? (In government records, the householder is the adult who owns or rents a dwelling.) Educational level is the explanatory variable. It appears in government data as a categorical variable with these values:

A Less than 8 years of elementary school

B 8 years of elementary school

C 1 to 3 years of high school

D 4 years of high school

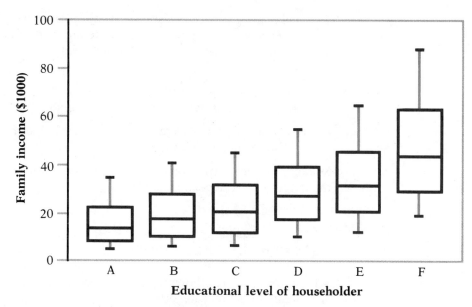

Figure 3.8 Boxplots by educational level for family incomes in 1985. See Example 3.5.

E 1 to 3 years of college

F 4 or more years of college

The categories A to F are ordered from least education to most education. The side-by-side boxplots of Figure 3.8 show the distributions of family income in 1985 for white families in these educational categories. The plot shows a positive association between educational level and income. That is, family income tends to increase as the educational level of the householder increases. What is more, the variability of income also increases with education. Some college-educated households have quite low incomes. Additional education offers the opportunity to earn more but does not guarantee a high income. ■

SUMMARY

A **categorical variable** records into which of two or more groups an observation falls, while a **quantitative variable** takes numerical values for which arithmetic operations make sense.

When changes in a variable x are thought to explain or even cause changes in a second variable y, x is called an **explanatory variable** and y is called a **response variable**.

A **scatterplot** is a plot of observations x_i and y_i as points in the plane, where x_i and y_i are the values of quantitative variables x and y for the same **case**, that is, the same person, animal, or object.

The explanatory variable, if any, is always plotted on the horizontal scale of a scatterplot. Plotting points with different symbols allows us to see the effect of a categorical variable in a scatterplot.

In examining a scatterplot, look for an overall pattern showing the form, direction, and strength of the relationship, and then for outliers or other deviations from that pattern. **Linear relationships** are an important form.

If the relationship has a clear direction, we speak of either **positive association** or **negative association**.

Smoothing a scatterplot by a **median trace** or other method helps reveal the nature of the dependence of y on x.

SECTION 3.1 EXERCISES

3.1 In each of the following cases, tell whether the variable is quantitative or categorical:

(a) The name of the manufacturer of a TV set

(b) The number of insects on a corn plant

(c) The score on a test of math anxiety for a student taking a statistics course

(d) The major area of study for the student in (c)

(e) The number of pages in a book

(f) Your height in inches

3.2 In each of the following cases, tell whether you would be interested simply in exploring the relationship between the two variables or whether you would want to view one of the variables as an explanatory variable and the other as a response variable. In the latter case, state which is the explanatory variable and which is the response variable.

(a) The amount of time spent studying for a statistics exam and the grade on the exam

(b) The height and weight of a person

(c) The amount of yearly rainfall and the yield of a crop

(d) A student's scores on the SAT math exam and on the SAT verbal exam

(e) The occupational class of a father and of his son

3.3 Vehicle manufacturers are required to test their vehicles for the amount of each of several pollutants in the exhaust. Even among identical vehicles the amount of a pollutant varies, so several vehicles must be tested. Figure 3.9 plots the amounts of two pollutants, carbon monoxide and nitrogen oxides, for 46 identical vehicles. Both variables are measured in grams of the pollutant per

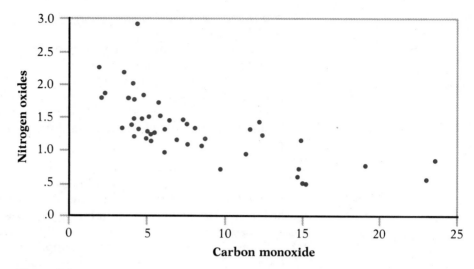

Figure 3.9 Nitrogen oxides versus carbon monoxide in the exhaust of 46 vehicles (Exercise 3.3).

mile driven. (Data from Thomas J. Lorenzen, "Determining statistical characteristics of a vehicle emissions audit procedure," *Technometrics*, 22 (1980), pp. 483–493.)

(a) Describe the nature of the relationship. Is the association positive or negative? Is it nearly linear or clearly curved? Are there any outliers?

(b) A writer on automobiles says, "When an engine is properly built and properly tuned, it emits few pollutants. If the engine is out of tune, it emits more of all the important pollutants. You can find out how badly a vehicle is polluting the air by measuring any one pollutant. If that value is acceptable, the other emissions will also be OK." Do the data in Figure 3.9 support this claim?

3.4 The following are the golf scores of 12 members of a women's golf team in two rounds of tournament play. (A golf score is the number of strokes required to complete the course, so low scores are better.)

Player	1	2	3	4	5	6	7	8	9	10	11	12
Round 1	89	90	87	95	86	81	102	105	83	88	91	79
Round 2	94	85	89	89	81	76	107	89	87	91	88	80

(a) Plot the scores from round 2 versus the scores from round 1.

(b) Is there an association between the two scores? If so, is it positive or negative? Explain why you would expect scores in two rounds of a tournament to have an association like the one you observed.

(c) A good golfer can have an unusually bad round and a weaker golfer can have an unusually good round. Either of these situations could produce an outlier. Circle the outlier in your scatterplot. Can you tell from the data given whether the unusual value is produced by a good player or a poor player? What other data would you need to distinguish between the two possibilities?

3.5 When water flows across farm land, some of the soil is washed away, resulting in erosion. An experiment was conducted to investigate the effect of the rate of water flow on the amount of soil washed away. Flow is measured in liters per second and the eroded soil is measured in kilograms. The data are given in the following table. (From G. R. Foster, W. R. Ostercamp, and L. J. Lane, *Effect of Discharge Rate on Rill Erosion*, presented at the 1982 Winter Meeting of the American Society of Agricultural Engineers.)

Flow rate	.31	.85	1.26	2.47	3.75
Eroded soil	.82	1.95	2.18	3.01	6.07

(a) Plot the data.
(b) Describe the pattern that you see. Would it be reasonable to describe the overall pattern by a straight line? Is the association positive or negative?

3.6 McDonald's "Big Mac" hamburger is sold in many countries around the world. By comparing the cost of a Big Mac in the local currency to its cost in the United States, we can find the exchange rate between the American dollar and that currency that would make the cost of a Big Mac the same in both countries. *The Economist* did this, and compared the result with the actual exchange rates between the dollar and foreign currencies. The following table gives some of the data. The entries are the value of 1 dollar in foreign currency; for example, at the official exchange rate 1 U.S. dollar was worth 1.64 Australian dollars. (From *The Economist*, Sept. 6, 1986, p. 77.)

(a) Make a scatterplot of the official exchange rate *y* versus the Big Mac exchange rate *x*.
(b) Describe the overall pattern of your scatterplot. Is the association positive or negative? Is there a clearly linear pattern? Are there any distinct outliers? How well does comparing the prices of a Big Mac predict the official value of the dollar in foreign currencies?

Country	Big Mac dollar value	Official dollar value
Australia	1.09	1.64
Brazil	7.80	13.80
Britain	.69	.67
Canada	1.18	1.39
France	10.30	6.65
Hong Kong	4.75	7.80
Ireland	.74	.74
Holland	2.72	2.28
Singapore	1.75	2.15
Sweden	10.30	6.87
W. Germany	2.66	2.02

3.7 Do heavier cars cost more than lighter cars? Figure 3.10 is a plot of the base price in dollars and the weight in pounds for all 1986 model four-door sedans listed in an auto guide. Cars made by American manufacturers are plotted with a "●" and cars of foreign make are plotted with an "○."

(a) Describe the overall relationship between the weight of a car and its price. Is the association strong (weight and price closely connected) or weak? Is it generally positive or negative?

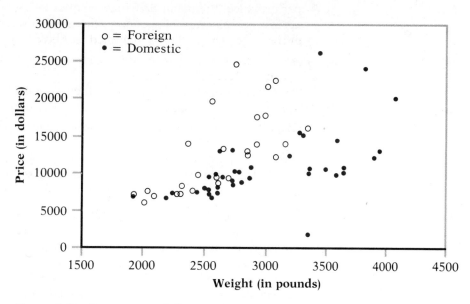

Figure 3.10 Base price in dollars versus weight in pounds for 1986 four-door sedans (Exercise 3.7).

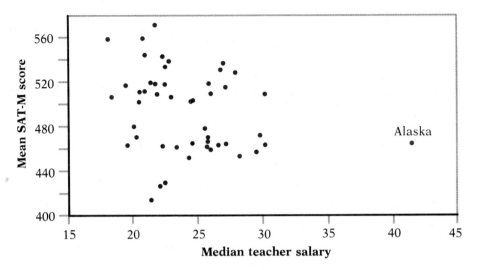

Figure 3.11　Mean SAT mathematics score versus median teacher salary by state (Exercise 3.8).

(b) Describe the major differences between domestic and foreign cars as they appear in the plot.

3.8　Does increasing public spending on education improve student performance? Figure 3.11 plots one measure of academic performance, the mean SAT mathematics test score in each state, against one measure of spending, the median salary paid to teachers in the state. The outlier is identified as Alaska.

(a) Is Alaska an outlier in both variables or only in one? Why do you think Alaska is an outlier?

(b) Describe the overall pattern of the relationship between teachers' salaries and students' SAT mathematics scores.

(Experts in education say there is not a clear association between spending on public education and SAT scores. As Example 3.2 shows, the percent of high school students who take the SAT varies greatly from state to state, so the mean SAT score is based on different kinds of students in different states.)

3.9　The following data refer to an outbreak of botulism, a form of food poisoning that may be fatal. Each case is a person who contracted botulism in the outbreak. The variables recorded are the subject's age in years, the incubation period (the time in hours between eating the infected food and the first signs of illness), and whether the subject survived (S) or died (D). (Modified from data provided by Professor Dana Quade, University of North Carolina, Chapel Hill, N.C.)

Case	1	2	3	4	5	6	7	8	9
Age	29	39	44	37	42	17	38	43	51
Incubation	13	46	43	34	20	20	18	72	19
Outcome	D	S	S	D	D	S	D	S	D

Case	10	11	12	13	14	15	16	17	18
Age	30	32	59	33	31	32	32	36	50
Incubation	36	48	44	21	32	86	48	28	16
Outcome	D	D	S	D	D	S	D	S	D

(a) Make a scatterplot of incubation period against age, using different symbols for cases that were fatal and cases where the victim survived.

(b) Is there an overall relationship between age and incubation period? If so, describe it.

(c) More important, is there a relationship between either age or incubation period and whether the victim survived? Describe any relations that seem important here.

(d) Are there any unusual cases that may require individual investigation?

3.10 We wish to predict the level of nitrogen oxide (NOX) emissions from the level of carbon monoxide (CO) emissions for a vehicle of the type considered in Exercise 3.3. Figure 3.9 displays a moderately strong relationship between the two variables. To construct a median trace, we compute the median NOX level for the vehicles in each 5-g/mile-wide slice of CO levels. Here is the result of these calculations:

CO level	Count	Median NOX
$0 \leq CO < 5$	16	1.785
$5 \leq CO < 10$	18	1.270
$10 \leq CO < 15$	8	1.055
$15 \leq CO < 20$	2	.635
$20 \leq CO < 25$	2	.715

Draw the median trace on a graph with the same scales as those used in Figure 3.9. Describe the overall relation displayed by the median trace.

3.11 Refer back to the flea data given in Exercise 1.19 for Thomas the cat.

(a) Divide the data into consecutive 5-day periods and find the median for each period (discard the data for the final two days).

Plot all of the data and superimpose the median trace. Describe
the pattern displayed by the trace.

(b) Repeat (a) using 3-day periods. For this problem do you prefer
5-day periods or 3-day periods? Why?

3.12 Compute the median trace for 5-day periods from the flea data given
in Exercise 2.57 for Creole the cat. Compare the overall patterns of
flea reproduction for Creole and Thomas as displayed by the median
traces.

3.13 To demonstrate the effect of nematodes (microscopic worms) on
plant growth, a botanist prepares 16 identical planting pots and then
introduces different numbers of nematodes into the pots. A tomato
seedling is transplanted into each plot. Here are data on the increase
in height of the seedlings (in centimeters) 16 days after planting.
(Data provided by Matthew Moore.)

Nematodes	Seedling growth			
0	10.8	9.1	13.5	9.2
1000	11.1	11.1	8.2	11.3
5000	5.4	4.6	7.4	5.0
10,000	5.8	5.3	3.2	7.5

(a) Make a scatterplot of the response variable (growth) against the
explanatory variable (nematode count). Then compute the mean
growth for each group of seedlings, plot the means against the
nematode counts, and connect these four points with line segments.

(b) Briefly describe the conclusions about the effects of nematodes
on plant growth that these data suggest.

3.14 The presence of harmful insects in farm fields is detected by erecting
boards covered with a sticky material and then examining the insects
trapped on the board. Some colors are more attractive to insects
than others. In an experiment aimed at determining the best color
for attracting cereal leaf beetles, six boards in each of four colors
were placed in a field of oats in July. The following table gives
data on the number of cereal leaf beetles trapped. (Modified from
M. C. Wilson and R. E. Shade, "Relative attractiveness of various
luminescent colors to the cereal leaf beetle and the meadow

Board color	Insects trapped					
Lemon yellow	45	59	48	46	38	47
White	21	12	14	17	13	17
Green	37	32	15	25	39	41
Blue	16	11	20	21	14	7

spittlebug," *Journal of Economic Entomology*, 60 (1967), pp. 578–580.)

(a) Make a plot of the counts of insects trapped against board color (space the four colors equally on the horizontal axis). Compute the mean count for each color, add the means to your plot, and connect the means with line segments.

(b) Based on the data, state your conclusions about the attractiveness of these colors to the beetles.

(c) Does it make sense to speak of a positive or negative association between board color and insect count?

3.15 When animals of the same species live together, they often establish a clear "pecking order." Lower ranking individuals defer to higher ranking animals in many ways, usually avoiding open conflict. A researcher on animal behavior wants to study the relationship between pecking order and physical characteristics such as weight. He confines four chickens in each of seven pens and observes the pecking order that emerges in each pen. Here is a table of the weights (in grams) of the chickens, arranged by pecking order. That is, the first row gives the weights of the dominant chicken in the seven pens, the second row the weights of the number 2 chicken in each pen, and so on. (Data collected by D. L. Cunningham, Cornell University, Ithaca, N.Y.)

Pecking order	Weight (g)						
	Pen 1	Pen 2	Pen 3	Pen 4	Pen 5	Pen 6	Pen 7
1	1880	1300	1600	1380	1800	1000	1680
2	1920	1700	1830	1520	1780	1740	1460
3	1600	1500	1520	1520	1360	1520	1760
4	1830	1880	1820	1380	2000	2000	1800

(a) Make a plot of these data that is appropriate to study the effect of weight on pecking order. Include in your plot any means that might be helpful.

(b) We might expect that heavier chickens would tend to stand higher in the pecking order. Do the data give clear evidence for or against this expectation?

3.16 The tensile modulus of elasticity is a measure of strength for wood products. In an experiment 50 strips of yellow poplar wood were each measured twice. The units of measurement are millions of pounds per square inch, the two measures are denoted by T1 and T2, and the number of the strip is denoted by S. The data are given in the following table. Analysis of this larger data set is best done using statistical software. (Data provided by Michael Triche and

Professor Michael Hunt, Forestry Department, Purdue University, West Lafayette, Ind.)

S	T1	T2	S	T1	T2	S	T1	T2	S	T1	T2
1	1.58	1.55	2	1.53	1.54	3	1.38	1.37	4	1.24	1.25
5	1.59	1.57	6	1.52	1.51	7	1.47	1.45	8	1.45	1.44
9	1.54	1.53	10	1.49	1.49	11	1.51	1.51	12	1.60	1.58
13	1.34	1.32	14	1.33	1.33	15	1.72	1.70	16	1.63	1.62
17	1.51	1.50	18	1.42	1.42	19	1.90	1.87	20	1.81	1.79
21	1.88	1.86	22	1.68	1.65	23	1.56	1.54	24	1.68	1.67
25	1.75	1.74	26	1.63	1.62	27	1.65	1.65	28	1.76	1.75
29	1.58	1.56	30	1.69	1.67	31	1.60	1.58	32	1.57	1.56
33	1.70	1.69	34	1.41	1.40	35	1.33	1.33	36	1.59	1.59
37	1.64	1.66	38	1.74	1.74	39	1.89	1.91	40	1.73	1.75
41	1.80	1.81	42	1.67	1.68	43	1.63	1.64	44	1.87	1.88
45	2.02	2.04	46	1.79	1.81	47	1.86	1.88	48	1.72	1.73
49	1.83	1.83	50	1.68	1.70						

(a) Plot T2 versus T1.

(b) Is the association between T1 and T2 positive or negative? Is this the direction of association that you would expect before looking at your plot? Why?

(c) Plot T1 versus S and T2 versus S. Do you see any pattern? If so, describe it. What questions would you ask the experimenters in seeking an explanation for the pattern observed?

3.2 LEAST SQUARES REGRESSION

When we smooth a scatterplot by using a median trace or some other tool, we are attempting to summarize the dependence of the response y on the explanatory variable x without specifying what form that dependence should take. A single equation that describes the dependence of y on x provides a more compact summary. The simplest such equation is a linear equation of the form

$$y = a + bx$$

whose graph is a straight line. A strong linear pattern in a scatterplot can be summarized by a linear equation.

EXAMPLE 3.6

Warren heats his home with natural gas. The amount of gas required to heat the home depends on the outdoor temperature—the colder the weather, the more gas will be consumed. As long as the family's habits, the insulation of the house, and other such factors do not change, Warren should be able to predict gas consumption from the outdoor temperature. He therefore measures his household's natural gas consumption each month during one heating season, from October to the following June. Because months differ in length, he divides gas consumption by the number of days in the month to obtain the daily gas consumption.

Outdoor temperature influences gas consumption only when it is cold enough to require heating. The usual measure of the need for heating is *heating degree days*. To find the number of heating degree days for a given day, first record the high and low temperature for the day. The average temperature is taken to be the mean of the high and low temperatures. One heating degree day is accumulated for each degree this average falls below 65°F. An average temperature of 20°F, for example, corresponds to 45 degree days. Warren obtains the average number of degree days per day for each month from weather records for his community.

Here are the data.[5] The explanatory variable x is heating degree days per day for the month, and the response variable y is gas consumption per day in hundreds of cubic feet.

Month	Oct.	Nov.	Dec.	Jan.	Feb.	Mar.	Apr.	May	June
x	15.6	26.8	37.8	36.4	35.5	18.6	15.3	7.9	.0
y	5.2	6.1	8.7	8.5	8.8	4.9	4.5	2.5	1.1

A scatterplot (Figure 3.12) shows that the relationship is strongly linear. The deviations from the straight-line pattern reflect the use of gas for cooking, windows left open, and other influences on gas consumption. ■

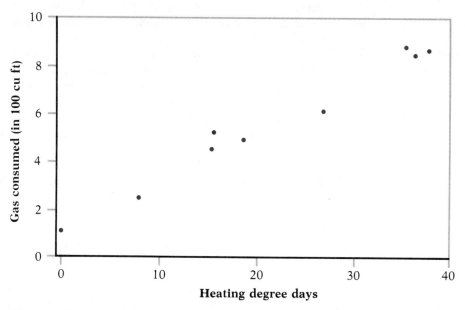

Figure 3.12 Residential natural gas consumption versus heating degree days. See Example 3.6.

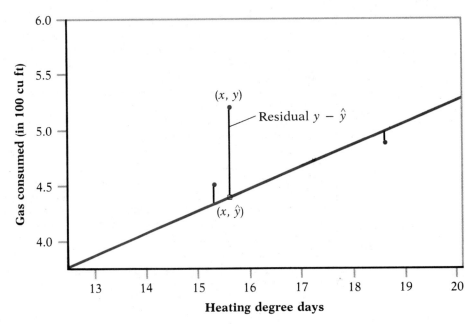

Figure 3.13 The least squares method: fitted line and residuals.

Computing the Regression Line

regression line

Warren's goal is to use his data to predict gas consumption at different temperatures. To do this, he will draw a straight line through the scatterplot in Figure 3.12. A line that shows the dependence of gas consumption on degree days is called a *regression line* of the response variable (consumption) on the explanatory variable (degree days).* Warren could find a satisfactory regression line by moving a transparent straightedge through the points until the fit seems good. We prefer a more objective method. When a scatterplot shows a less strong linear pattern than the one in Figure 3.12, it is difficult to fit by eye a line that is appropriate for predicting y from x. Chapter 2 introduced the most common technique for fitting a line to data, the *method of least squares*. That chapter dealt only with linear growth over time; now we will use least squares regression lines to describe the overall pattern of any linear relationship.

To develop the least squares idea in detail, we magnify the center portion of Figure 3.12 and draw a line through the points. See Figure 3.13. How well does this line fit? Since our goal is to predict y when we are given a

---

* The term "regression" and the general methods for studying relationships now included under this term were introduced by the English gentleman scientist Francis Galton (1822–1911). Galton was engaged in the study of heredity. One of his observations was that children of tall parents tended to be taller than average but not as tall as their parents. This "regression toward mediocrity" gave these statistical methods their name.

value of x, the error in our prediction is measured in the y, or vertical direction. We therefore concentrate on the deviations of the data points from the line in the vertical direction. These deviations are drawn in Figure 3.13. They are the residuals when this particular line is used to predict y from x.

We represent n observations on two variables x and y as

$$(x_1, y_1), (x_2, y_2), \ldots, (x_n, y_n)$$

If a line $y = a + bx$ is drawn through the scatterplot of these observations, the line gives the value of y corresponding to x_i as $\hat{y}_i = a + bx_i$. The notation $\hat{y}$ (read "y hat") will be used to distinguish the y-value predicted by the line from the actually observed y. The ith residual is the vertical deviation of the ith data point from the line, which is

$$\text{residual} = \text{observed } y - \text{predicted } y$$
$$= y_i - \hat{y}_i$$
$$= y_i - a - bx_i$$

Some of these residuals are positive and some are negative. *The method of least squares chooses the line that makes the sum of the squares of the residuals as small as possible.* Finding this line amounts to finding the values of the intercept a and the slope b that minimize the quantity

$$\sum(y_i - a - bx_i)^2$$

for the given observations x_i and y_i. In the case of the heating data given in Example 3.6, we must choose the a and b that minimize

$$(5.2 - a - 15.6b)^2 + (6.1 - a - 26.8b)^2 + \cdots + (1.1 - a - .0b)^2$$

Here is the solution to this mathematical problem.

Least squares regression line

> The least squares regression line of y on x calculated from n observations on these two variables is given by $\hat{y} = a + bx$, where
>
> $$b = \frac{\sum(x - \bar{x})(y - \bar{y})}{\sum(x - \bar{x})^2} \qquad (3.1)$$
>
> $$a = \bar{y} - b\bar{x}$$

The sums in Equation 3.1 run over all of the observed values x_i or y_i; the subscripts were omitted for easier reading. There are many alternative expressions for a and b. Many calculators and all statistical software systems will compute a and b for least squares regression lines. Use these resources if they are available to you rather than calculate by hand.

If you do statistical calculations with a basic calculator, you will want to use an alternative *computing formula* for the slope b of the least squares line. Like the similar Equation 1.3 for the variance, this formula avoids subtracting the mean from each observation and makes efficient use of basic sums and sums of squares. The computing formula is

computing formula for b

$$b = \frac{n \sum xy - (\sum x)(\sum y)}{n \sum x^2 - (\sum x)^2} \tag{3.2}$$

EXAMPLE 3.7

To calculate the least squares regression line of gas consumption on degree days from the data in Example 3.6, proceed as follows. First compute the building block sums

$$\sum x = 15.6 + \cdots + .0 = 193.9$$
$$\sum x^2 = 15.6^2 + \cdots + .0^2 = 5618.11$$
$$\sum y = 5.2 + \cdots + 1.1 = 50.3$$
$$\sum xy = (15.6)(5.2) + \cdots + (.0)(1.1) = 1375.0$$

Then from Equation 3.2

$$b = \frac{n \sum xy - (\sum x)(\sum y)}{n \sum x^2 - (\sum x)^2}$$
$$= \frac{(9)(1375.0) - (193.9)(50.3)}{(9)(5618.11) - (193.9)^2}$$
$$= \frac{2621.83}{12,965.78} = .202$$
$$a = \bar{y} - b\bar{x}$$
$$= \frac{50.3}{9} - .202 \frac{193.9}{9} = 1.23$$

The resulting line is

$$\hat{y} = 1.23 + .202x$$

It is shown in Figure 3.14 superimposed on the original scatterplot. ∎

slope

In the regression line $\hat{y} = 1.23 + 0.202x$, the *slope* of the line is $b = 0.202$. The slope is the increase in y corresponding to an increase of one unit in x. Warren estimates that each additional degree day increases his gas consumption by 0.202 hundred cubic feet (or 20.2 cubic feet) per day.

intercept

The number $a = 1.23$ is the *intercept*, the value of $\hat{y}$ when $x = 0$. In this case, $x = 0$ indicates an average temperature of 65° or higher, so there is no demand for heating. Warren expects to use 123 cubic feet of gas per day when $x = 0$. This represents gas used for heating water and for cooking.

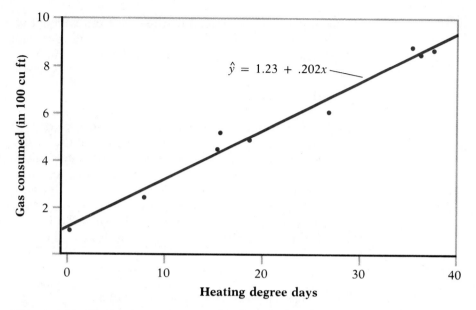

Figure 3.14 The least squares regression line of natural gas consumption on heating degree days.

Warren adds insulation in his attic during the summer, hoping to reduce his gas consumption for heating his home. The next February, gas consumption is 870 cubic feet per day. Was the added insulation effective? We cannot simply compare this year's gas consumption with last February's rate (880 cubic feet per day) because the average temperatures for the two months were not the same. In fact, this February had an average of 40 degree days per day. We therefore *predict* from the regression equation how much gas the house would have used at 40 degree days per day last winter. Our prediction is

$$\hat{y} = 1.23 + (.202)(40) = 9.31$$

or 931 cubic feet. Warren estimates that he saved about 61 cubic feet per day by adding insulation.

The accuracy of predictions from a regression line depends on how much scatter about the line the data show. When, as in this example, the data points are all very close to the line, we are confident that our prediction is reliable. If, however, the data show a linear pattern with considerable spread (as in Figure 3.1), we may agree that a regression line summarizes the pattern but we will also put less confidence in a prediction based on the line. We will learn in Chapter 10 how to give a numerical statement of our level of confidence in predictions. Of course, the usefulness of any regression line is limited by the data from which it was computed. Warren must collect new data and compute a new regression line if he wishes to predict his gas consumption with the new insulation in place.

Fit and Residuals

The dependence of gas consumption on degree days (Examples 3.6 and 3.7) was so strong that we can be confident that the least squares regression line tells almost the whole story about these data. This happy circumstance is not always the case. The deviations from the fitted regression line often give important additional information. These deviations are best studied by calculating and plotting the n residuals $y_i - \hat{y}_i$. As we saw in Chapter 2, the residuals from a least squares regression line have the property that their sum (and therefore their mean) is always 0.

EXAMPLE 3.8

In Example 3.6, the observed consumption of gas for October, when $x = 15.6$ degree days, was $y = 5.2$. The predicted consumption for this x from the least squares regression line of Example 3.7 is

$$\hat{y} = 1.23 + (.202)(15.6) = 4.38$$

The October residual is therefore

$$e_1 = 5.2 - 4.38 = .82$$

This residual appears in Figure 3.13 as the length of the vertical line joining y and $\hat{y}$. ■

The distribution of the residuals can be examined using the tools presented in Chapter 1. A stemplot helps to check the symmetry of the distribution and to spot outliers. A normal probability plot can tell whether the residuals are approximately normally distributed. Since formal statistical methods for predicting y from x using a regression line depend on the shape of the distribution of residuals, questions of symmetry and normality will eventually interest us. Our present concerns, however, are descriptive. Is the relationship between y and x close to linear? Are there outlying observations that require special attention? Is there evidence that y is influenced by variables other than x? Scatterplots of the residuals against other variables are used to investigate questions like these.

You should always plot the residuals against the corresponding values of the explanatory variable x. If all is well, the pattern of this plot will be an unstructured horizontal band centered at 0 (the mean of the residuals) and symmetric about 0. Figure 3.15(a) displays an idealized version of this pattern. A curved pattern like the one in Figure 3.15(b) shows a nonlinear dependence of the response y on x. A fan-shaped pattern like the one in Figure 3.15(c) shows that the variation of y about the line increases as x increases; predictions of y will therefore be more precise for smaller values of x, where y shows less spread about the line.

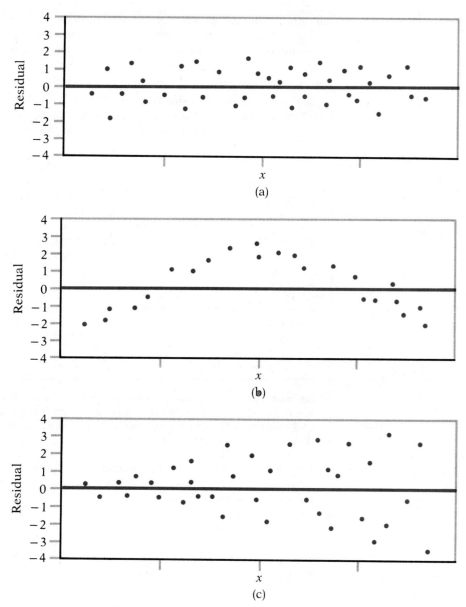

Figure 3.15 Idealized patterns in plots of least squares residuals.

Other residual plots can give other kinds of information. A plot of the residuals against the time order in which the data were collected may reveal a dependence of y on some variable that changes over time. Since we have seen that background variables can strongly affect the appearance of a scatterplot, residuals should be plotted against time and against any other variables. In all such plots, the residuals are plotted vertically and time or any other potential explanatory variable horizontally. A sloping linear band or other strong deviation from the pattern shown in Figure 3.15(a) indicates a dependence of y on the new variable. A sloping linear band cannot appear when the residuals are plotted against x because the residuals have no remaining linear dependence on x. Of course, in practice, patterns will not be as neat as the schematic drawings in Figure 3.15. Any set of data will show some irregularity; don't overreact to vague patterns in a residual plot. Here are two examples of plotting and interpreting residuals.

EXAMPLE 3.9

The mathematics department of a large state university must plan in advance the number of sections and instructors required for elementary courses. The department hopes that the number of students in these courses can be predicted from the number of entering freshmen, which is known before the new students actually choose courses. The table below contains the data for recent years.[6] The explanatory variable x is the number of students in the freshman class, and the response variable y is the number of students who enroll in mathematics courses at the 100 level.

Year	1980	1981	1982	1983	1984	1985	1986	1987
x	4595	4827	4427	4258	3995	4330	4265	4351
y	7364	7547	7099	6894	6572	7156	7232	7450

Equation 3.2 could be used to compute the regression line from these data. Instead, we enter the data into the MINITAB statistical computing system, calling x FRESH and y MATH for easy recall. The regression command and the first part of the printout appear as follows:

```
MTB > REGRESS 'MATH' ON 1, 'FRESH'
```

The regression equation is

```
MATH = 2492.69 + 1.0663 FRESH
```

Other information, including a table of the residuals, follows on the printout. We see from the output that the least squares regression line is

$$\hat{y} = 2492.69 + 1.0663x$$ ∎

A scatterplot (Figure 3.16) of the data from Example 3.9 with the regression line shows a reasonably linear fit. There is a cluster of points with similar values near the center. A plot of the residuals against x (Figure 3.17) magnifies the vertical deviations of the points from the line. It is apparent from either graph that a slightly different line would fit the five lower points very well, so that the three points above the line represent a somewhat different relation between the number of freshmen x and mathematics enrollments y.

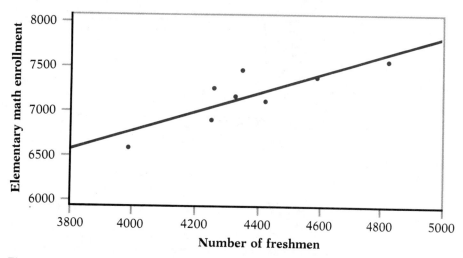

Figure 3.16 The regression of elementary mathematics enrollment on number of freshmen at a large university. See Example 3.9.

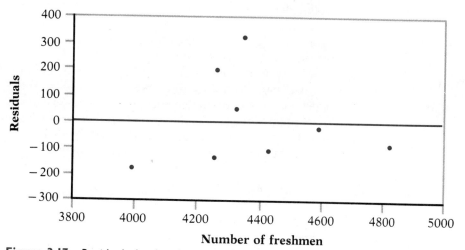

Figure 3.17 Residual plot for the regression of mathematics enrollment on number of freshmen.

A second plot of the residuals clarifies the situation. Figure 3.18 is a plot of the residuals against time. We now see that the five negative residuals are for the years 1980 to 1984, and the three positive residuals represent the years 1985 to 1987. This pattern suggests that a change took place between 1984 and 1985 causing a higher proportion of students to take mathematics courses beginning in 1985. In fact, one of the schools in the university

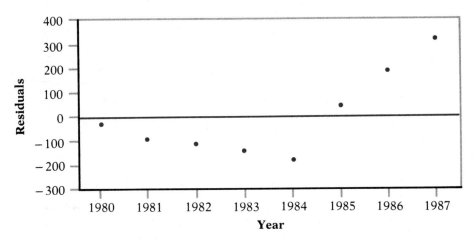

Figure 3.18 Plot of the residuals versus time for the regression of mathematics enrollment on number of freshmen.

changed its program to require that entering students take another mathematics course. Notice that Figure 3.18 does *not* say that mathematics enrollments changed linearly over time up to 1984—a glance at the table in Example 3.9 shows that they did not. Figure 3.18 is a plot of the residuals from the regression of the enrollments on the count of freshmen, not a plot of the enrollments themselves.

The contrast in Figure 3.18 between the years up to 1984 and the following years is another reminder that an observed relationship may not be trustworthy when changes occur in the underlying situation. Because the least squares regression line from Example 3.9 makes use of data from 1984 and earlier, the mathematics department should not use it in future years to predict elementary mathematics enrollments from the count of freshmen.

EXAMPLE 3.10

A study of cognitive development in young children recorded the age (in months) at which each of 21 children spoke their first word and their Gesell score, the result of an aptitude test taken much later. The data appear in Table 3.3 and in the scatterplot of Figure 3.19.[7] Notice that cases 3 and 13, and cases 16 and 21, have identical values for both variables. When drawing the scatterplot, we used a different symbol to show that one point stands for two cases.

The purpose of the study was to ascertain if age at first word (x) could predict the later test score y. It is also plausible, however, to try to guess the age when a child first spoke from the test score. Both the regression line of y on x and the regression line of x on y are therefore displayed in Figure 3.19. *These regression lines are very different*: One minimizes vertical deviations and the other minimizes horizontal deviations. The lines intersect at the point $(\bar{x}, \bar{y})$, through which any least squares regression line always passes. It is essential in a regression setting to clearly distinguish the explanatory from the response variable. This example also indicates why it is hard to fit an appropriate line by eye—we cannot easily concentrate on lack of fit in one direction only. ■

Table 3.3 Age (in months) at first word and Gesell adaptive score

Case	Age	Score	Case	Age	Score
1	15	95	12	9	96
2	26	71	13	10	83
3	10	83	14	11	84
4	9	91	15	11	102
5	15	102	16	10	100
6	20	87	17	12	105
7	18	93	18	42	57
8	11	100	19	17	121
9	8	104	20	11	86
10	20	94	21	10	100
11	7	113			

Figure 3.20 shows the regression of Gesell score on age at first word (the colored line) and highlights two cases, the children numbered 18 and 19 in Table 3.3. Case 19 is an outlier that lies well away from the general straight-line pattern of the other points. Case 18 is not an outlier if we take the overall pattern to be roughly linear, for this point falls in that pattern. The residual plot (Figure 3.21) reinforces these conclusions. Case 19 produces

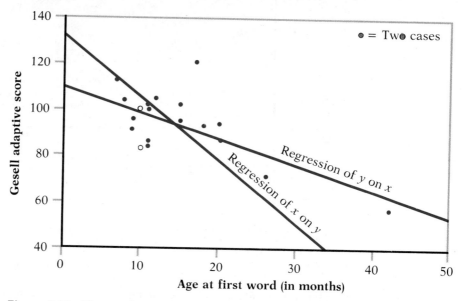

Figure 3.19 The two least squares regression lines of Gesell score on age at first word and of age at first word on Gesell score. See Example 3.10.

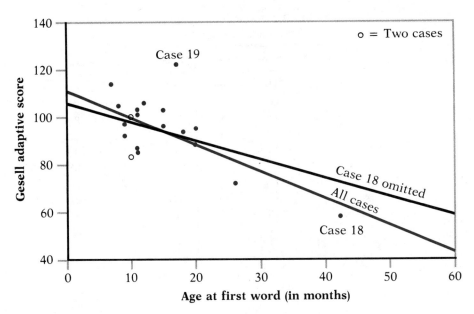

Figure 3.20 The regression of Gesell score on age at first word, with and without an influential case.

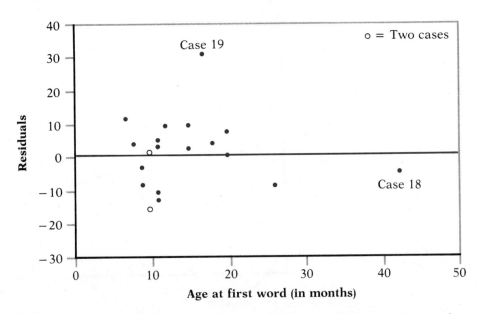

Figure 3.21 Residual plot for the regression of Gesell score on age at first word; case 19 is an outlier and case 18 is an influential case that does not have a large residual.

the largest residual and case 18 a rather small residual. Yet case 18 is very important. Because of its extreme position on the age scale, it has a strong influence on the position of the regression line. The black line in Figure 3.20 is the least squares regression line computed from all 20 cases except case 18. Notice how this single case pulls the colored regression line toward itself. Case 19, on the other hand, has less influence on the position of the regression line because this child's age at first word (17 months) lies close to the mean age $\bar{x} = 14.4$ months. The regression line must pass through the point $(\bar{x}, \bar{y})$, so it is difficult for an observation at an age close to $\bar{x}$ to move the line to any great degree.

Outliers and Influential Observations

Example 3.9 illustrates the power of residuals to detect deviations from linear dependence of y on x and even to suggest explanations for departures from linearity. Example 3.10, however, points out a weakness of regression residuals. Residuals call attention to outliers but not to other influential cases.

Outlier, influential observation

> An outlier in a regression is a data point that produces a large residual. An observation is influential if removing it would markedly change the position of the regression line.

In other words, an outlier lies outside the linear pattern, and so has a large residual. But you must also be on the alert for influential observations that are not outliers. Least squares regression is not resistant. The position of the least squares regression line is heavily influenced by observations that are extreme in x. The influence of these points often guarantees that they are not outliers, however, because they draw the regression line toward themselves. Influential observations are easy to detect (as long as there is only a single explanatory variable), but merely looking for large residuals doesn't do the job. Many statistical computing systems produce tables and plots of regression residuals, but if you rely on these alone, you may miss the most influential cases. The basic scatterplot of y versus x will alert you to observations that are extreme in x and may therefore be influential.

An influential observation should be investigated to ensure that it is correct. Even if no error is found, you should ask whether this observation belongs to the population you are studying. A statistician would ask the child development researcher in Example 3.10 whether the child of case 18 is so slow to speak that this case should not be allowed to influence the analysis. If this case is excluded, much of the evidence for a connection between the age at which a child begins to talk and later aptitude scores vanishes. If the case is retained, we need data on other children who were slow to begin talking, so that the analysis is no longer so heavily dependent on a single child.

In addition to showing outliers and influential observations, examination of residuals may also disclose the influence of lurking variables.

Lurking variable

> A lurking variable is a variable that has an important effect on the response but that is not included among the explanatory variables studied.

Since lurking variables are often unrecognized and unmeasured, detecting their effect is a challenge. The most effective tool is an understanding of the background of the data that allows you to guess what lurking variables might be present. One useful method for detecting lurking variables is to plot both the response variable and the regression residuals against the time order of the observations, since many variables change with time. This was the case in Figure 3.18, where the change in course requirements over time was a lurking variable. Here is another example.

EXAMPLE 3.11   A study of the manufacture of molded plastic parts examined the effect of time spent in the mold (x) on the strength (y) of the part. Several batches of hot plastic were pressed for 10 seconds, then several more batches for 20 seconds, and so on. Regression showed a strong dependence of strength on molding time. A plot of strength against the order in which the batches were molded, however, showed an even stronger dependence. This led the engineers to realize that the mold grew constantly warmer as more batches were processed, and that this lurking variable explained the changing strength.[8]

This experiment was poorly designed; the effect of time in the mold cannot be separated from the effect of any lurking variable that was changing over time because all 10-second batches were molded first, then all 20-second batches, and so on. As we shall see in Chapter 4, properly designed experiments can prevent the effects of lurking variables from being confused with the effects of the explanatory variable or variables.   ■

A regression line is a compact description of the overall pattern of dependence of y on x. Since the straight line is fitted to the data, it can be described briefly as a FIT. The definition of residuals then says that

$$DATA = FIT + RESIDUAL$$

There are many ways of obtaining an overall FIT for two-variable data. The median trace produces a FIT that is not forced to be a straight line. Some statistical software systems provide alternative methods of fitting a straight line to data that are more resistant than least squares regression. Examining residuals, especially through plots, can help assess the success of a FIT produced by any method. Residuals from least squares regression have several special properties (such as mean 0) that make them simpler to work with.

If your software allows resistant fitting of a line to data, you can check the results of the least squares fit by also doing a resistant fit. If the two lines do not agree closely, search for influential observations or other explanations for the lack of agreement. Despite its lack of resistance, the least squares method continues to dominate statistical practice. To use this—and other—statistical methods wisely, you must never neglect the detailed examination of your data.

SUMMARY

When a scatterplot suggests that the dependence of y on x can be summarized by a straight line, the **least squares regression line** of y on x can be calculated. This fitted line can be used to predict y for a given value of x.

The fit of a regression line is examined by plotting the **residuals**, or differences between the observed and predicted values of y. Be on the lookout for **outliers**, which are points with unusually large residuals, and also for nonlinear patterns and uneven variation about the line.

Influential observations, individual points that substantially change the regression line, must also be spotted and examined. Influential observations are often outliers in the x variable, but they may not have large residuals.

Evidence of the effects of **lurking variables** on y may be provided by plots of y and the residuals against the time order of the observations.

SECTION 3.2 EXERCISES

3.17 Exercise 2.18 presented the following data on Sarah's growth between the ages of 36 months and 60 months:

Age (months)	36	48	51	54	57	60
Height (cm)	86	90	91	93	94	95

(a) Use Equation 3.2 to compute the slope b of the least squares regression line of height on age; then find the intercept a from $a = \bar{y} - b\bar{x}$. According to the regression line, how much does Sarah grow each month?

(b) Make a scatterplot of the data and draw the least squares line on the plot.

(c) Use your regression line to estimate Sarah's height at 42 months of age.

3.18 A student who waits on tables at a Chinese restaurant in a college neighborhood records the cost of meals and the tip left by single diners. Here are some of the data.

Meal	$3.50	$4.79	$5.24	$3.62	$5.35
Tip	$ .25	$ .50	$ .60	$ .40	$ .75

(a) Compute the least squares regression for these data starting with Equation 3.2.
(b) Make a scatterplot of the data and draw the regression line on your plot.
(c) The next diner orders a meal costing $3.89. Use your regression line to predict the tip.

3.19 Use your regression line from Exercise 3.17 to predict Sarah's height at each of 36, 48, 51, 54, 57, and 60 months. Then compute the residuals for the regression line by subtracting these predicted heights from the actually observed heights at these ages. Verify that the residuals have sum 0 (up to roundoff error). Make a plot of the residuals against age and briefly describe the pattern of residuals.

3.20 Compute the five residuals for the data in Exercise 3.18, using the regression line that you have already computed. Verify that the residuals have sum 0 (up to roundoff error). Plot the residuals against the cost of the meals and comment on the pattern of the residuals.

3.21 Runners are concerned about their form when racing. One measure of form is the stride rate, defined as the number of steps taken per second. A runner is inefficient when the rate is either too high or too low. Of course, as the speed increases, the stride rate should also increase. In a study of 21 of the best American female runners, the stride rate was measured for different speeds. The following table gives the speeds (in feet per second) and the average stride rates for these women. (Data from R. C. Nelson, C. M. Brooks, and N. L. Pike, "Biomechanical comparison of male and female distance runners," in P. Milvy (ed.), *The Marathon: Physiological, Medical, Epidemiological, and Psychological Studies*, New York Academy of Sciences, 1977, pp. 793–807.)

Speed	15.86	16.88	17.50	18.62	19.97	21.06	22.11
Stride rate	3.05	3.12	3.17	3.25	3.36	3.46	3.55

(a) Plot the data with speed on the x axis and stride rate on the y axis. Does a straight line adequately fit these data?
(b) Compute the slope and intercept for the least squares line, using

the computing formula 3.2 or a statistical calculator or software. Graph the least squares line on your plot from (a).

(c) For each of the speeds given, compute the predicted value using the least squares line.

(d) Using the results of (c), compute the residuals. Verify that the residuals add to zero.

(e) Plot the residuals versus speed. Describe the pattern. What does the plot indicate about the adequacy of the linear fit?

3.22 Refer to the previous exercise. The corresponding data for a group of 24 elite American male runners are given in the following table:

Speed	15.86	16.88	17.50	18.62	19.97	21.06	22.11
Stride rate	2.92	2.98	3.03	3.11	3.22	3.31	3.41

Explain why the stride rate at each speed is lower for men than for women. Then answer the questions given in the previous exercise using the data for males.

3.23 Research on digestion requires accurate measurements of blood flow through the lining of the stomach. A promising way to make such measurements easily is to inject mildly radioactive microscopic spheres into the bloodstream. The spheres lodge in tiny blood vessels at a rate that is in proportion to blood flow; their radioactivity allows blood flow to be measured from outside the body. Medical researchers compared blood flow in the stomachs of dogs, measured by use of microspheres, with simultaneous measurements taken using a catheter inserted into a vein. The data, in milliliters of blood per minute (ml/minute), appear below. (Based on L. H. Archibald, F. G. Moody, and M. Simons, "Measurement of gastric blood flow with radioactive microspheres," *Journal of Applied Physiology*, 38 (1975), pp. 1051–1056.)

Spheres	4.0	4.7	6.3	8.2	12.0	15.9	17.4	18.1	20.2	23.9
Vein	3.3	8.3	4.5	9.3	10.7	16.4	15.4	17.6	21.0	21.7

(a) Make a scatterplot of these data, with the microsphere measurement as the explanatory variable. There is a strongly linear pattern.

(b) Calculate the least squares regression line of venous flow on microsphere flow. Draw your regression line on the scatterplot.

(c) Predict the venous measurement for microsphere measurements of 5, 10, 15, and 20 ml/minute. If the microsphere measurements are within about 10% to 15% of the predicted venous measurements, the researchers will simply use the microsphere measurements in future studies. Is this condition satisfied over this range of blood flow?

3.24 Suppose that the last customer in Exercise 3.18 had ordered a large meal and left no tip. The data are now as follows:

Meal	$3.50	$4.70	$5.24	$3.62	$15.69
Tip	$ .25	$ .50	$ .60	$ .40	0

(a) Make a scatterplot of these data. Which observation will be most influential? Why?

(b) Fit a line by eye to the first four observations and draw this line on your plot. The least squares regression line fitted to all five observations is $\hat{y} = 0.57 - 0.034x$. Draw this line on your graph as well. A comparison of the two lines shows the influence of the final observation.

(c) Does it appear from the graph that the influential observation has a larger residual from the least squares line than the other observations? (You need not actually compute the residuals.)

3.25 One component of air pollution is airborne particulate matter such as dust and smoke. Particulate pollution is measured by using a vacuum motor to draw air through a filter for 24 hours. The filter is weighed at the beginning and end of the period, and the weight gained is a measure of the concentration of particles in the air. In a study of pollution, measurements were taken every 6 days with identical instruments in the center of a small city and at a rural location 10 miles southwest of the city. Because the prevailing winds blow from the west, it was suspected that the rural readings would be generally lower than the city readings, but that the city readings could be predicted from the rural readings. The following table gives readings taken every 6 days between May 2 and November 26, 1986. The entry NA means that the reading for that date is not available, usually because of equipment failure. (Data provided by Matthew Moore.)

Rural	NA	67	42	33	46	NA	43	54	NA	NA	NA	NA
City	39	68	42	34	48	82	45	NA	NA	60	57	NA

Rural	38	88	108	57	70	42	43	39	NA	52	48	56
City	39	NA	123	59	71	41	42	38	NA	57	50	58

Rural	44	51	21	74	48	84	51	43	45	41	47	35
City	45	69	23	72	49	86	51	42	46	NA	44	42

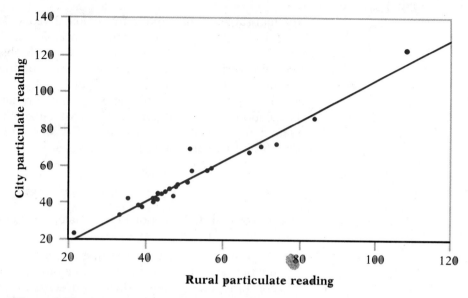

Figure 3.22 The regression of city particulate concentration on rural particulate concentration for the same days (Exercise 3.25).

To assess the success of predicting the city particulate reading from the rural reading, the 26 complete cases (with both readings present) are analyzed. Computer work finds the least squares regression line of the city reading y on the rural reading x to be

$$\hat{y} = -2.580 + 1.0935x$$

Figure 3.22 is a scatterplot for the 26 complete cases with the regression line drawn.

(a) Which observation in Figure 3.22 appears to be the most influential? Is this the observation with the largest residual (vertical distance from the line)?

(b) Locate in the table the observation you chose from the graph in (a) and compute its residual.

(c) Do the data suggest that using the least squares line for prediction will give approximately correct results over the range of values appearing in the data? (The incompleteness of the data does not seriously weaken this conclusion if equipment failures are independent of the variables being studied.)

(d) On the fourteenth date in the series, the rural reading was 88 and the city reading was not available. What do you estimate the city reading to be for that date?

3.26 To study the energy savings resulting from adding solar heating panels to a house, researchers measured the natural gas consumption of the house for more than a year, then installed solar panels and observed the natural gas consumption for almost 2 years. The variables are as in Example 3.6: The explanatory variable x is degree days per day during the several weeks covered by each observation, and the response variable y is gas consumption (in hundreds of cubic feet) per day during the same period. Figure 3.23 plots y against x, with separate symbols for observations taken before and after the installation of the solar panels. The least squares regression lines were computed separately for the before and after data, and are drawn on the plot. The regression lines are

$$\text{Before:} \quad \hat{y} = 1.089 + .189x$$
$$\text{After:} \quad \hat{y} = 0.853 + .157x$$

(Data provided by Professor Robert Dale, Purdue University, West Lafayette, Ind.)

(a) Does the scatterplot suggest that a linear model is appropriate for the relationship between degree days and natural gas consumption? Do any individual observations appear to be outliers (large residuals) or highly influential?

(b) About how much additional natural gas was consumed per day for each additional degree day before the panels were added? After the panels were added?

(c) The average daily temperature during January in this location is about 30°, which corresponds to 35 degree days per day. Use the

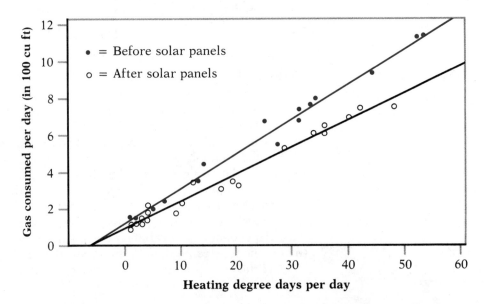

Figure 3.23 The regression of residential natural gas consumption on heating degree days before and after installation of solar heating panels (Exercise 3.26).

regression lines to predict daily gas usage for a day with 35 degree days before and after installation of the panels. If the price of natural gas is $0.60 per hundred cubic feet, how much money do the solar panels save in the 31 days of a typical January?

3.27 Table 1.4 gives the calories and sodium content for each of 17 brands of meat hot dogs. The distribution of calories was examined in the discussion following Example 1.11, and the distribution of sodium in Exercise 1.58. Now we examine the relation between calories and sodium. Figure 3.24 is a scatterplot of the data from Table 1.4.

(a) Describe the main features of the relationship. (The discussion following Example 1.11 may help.)

(b) The plot shows two least squares regression lines. One was calculated using all of the observations, while the other omitted the brand of veal hot dogs that is an outlier in both variables measured. Which line (colored or black) was calculated from all of the data? Explain your answer.

(c) The regression line that ignores the outlier is

$$\hat{y} = 46.90 + 2.401x$$

A new brand of meat hot dog (not made with veal) has 150 calories per frank. How many milligrams of sodium do you estimate that one of these hot dogs contains?

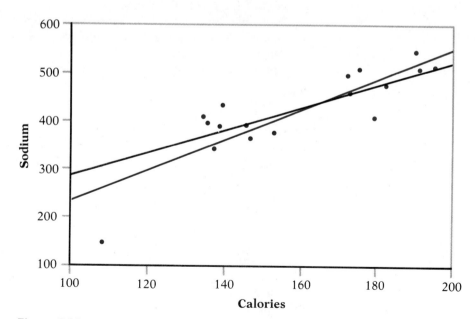

Figure 3.24 The regression of sodium content on calories for brands of hot dogs, calculated with and without an outlier (Exercise 3.27).

3.28 Are baseball players paid according to their performance? To study this question, a statistician analyzed the salaries of over 260 major league hitters along with such explanatory variables as career batting average, career home runs per time at bat, and years in the major leagues. This is a *multiple regression* with several explanatory variables. More detail on multiple regression appears in Chapter 10, but the fit of the model is assessed just as we have done in this chapter, by calculating and plotting the residuals

$$\text{residual} = \text{observed } y - \text{predicted } y$$

(This analysis was done by Crystal Richard.)

(a) Figure 3.25(a) is a plot of the residuals versus the predicted salary. This plot was produced by the SAS statistical software system that was used to analyze the data. Notice that when points are too close together to plot separately, SAS uses letters of the alphabet to show how many points there are at each position. Describe the pattern that appears on this residual plot. Will the regression model predict high or low salaries more precisely?

(b) After studying the residuals in more detail, the statistician decided to predict the logarithm of salary rather than the salary itself. One reason was that while salaries are not normally distributed (the distribution is skewed to the right), their logarithms are nearly normal. When the response variable is the logarithm of salary, a plot of the residuals against the predicted value is satisfactory—it looks like Figure 3.15(a). Figure 3.25(b) is a plot of the residuals against the number of years the player has been in the major leagues. Describe the pattern that you see. Will the model overestimate or underestimate the salaries of players who are new to the majors? Of players who have been in the major leagues about 8 years? Of players with more than 15 years in the majors?

3.29 Refer to the golf data given in Exercise 3.4.

(a) Plot the data with the round 1 scores on the x axis and the round 2 scores on the y axis. Does it appear reasonable to fit a straight line to these data? Are there any apparent outliers or influential observations?

(b) Compute the slope and intercept for the least squares line. Graph the least squares line on your plot from (a).

(c) For each of the round 1 scores, compute the predicted value for the round 2 score using the least squares line.

(d) Using the result of (c), compute the residuals. Verify that the residuals add to zero.

(e) Plot the residuals versus the round 1 scores. Describe the pattern. What does the plot indicate about the adequacy of the linear fit?

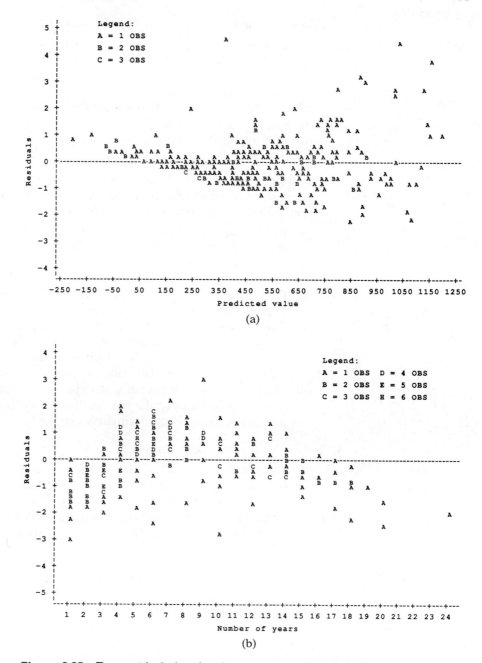

Figure 3.25 Two residual plots for the regression of baseball players' salaries on their performance (Exercise 3.28).

3.30 Refer to the erosion data given in Exercise 3.5.

(a) Compute the slope and intercept for the least squares line. Graph the least squares line on your plot from (a) of that exercise.

(b) For each of the flow rates, compute the predicted value using the least squares line.

(c) Using the results of (b), compute the residuals. Verify that the residuals add to zero.

(d) Plot the residuals versus flow rate. Describe the pattern. What does the plot indicate about the adequacy of the linear fit?

3.31 Refer to the data in Exercise 3.16 giving two measurements of the strength of wood strips. As before, you should use a statistical computing system to analyze this larger data set.

(a) Plot the data with T1 on the x axis and T2 on the y axis. Does it appear reasonable to fit a straight line to these data?

(b) Compute the least squares regression line. Graph the least squares line on your plot from (a).

(c) For each case compute the predicted value using the least squares line and find the residuals.

(d) Plot the residuals versus T1. Describe the pattern. What does the plot indicate about the adequacy of the linear fit?

(e) Plot the residuals versus S, which gives the time order of the measurements taken. Describe the plot.

(f) Make a stemplot (or, if your system permits, a normal quantile plot) of the residuals. Describe the distribution of the residuals. Is it nearly symmetric? Approximately normal?

3.32 Refer to the previous exercise.

(a) Plot the data with T2 on the x axis and T1 on the y axis.

(b) Compute the slope and intercept for the least squares line which views T1 as the response variable and T2 as the explanatory variable. Graph the least squares line on your plot from (a).

(c) On a plot of the data with T1 on the x axis and T2 on the y axis, graph the least squares line from the previous exercise and also the one that you just found in part (b) of this exercise. (Hint: For each line simply find two points on it, plot the points and connect them with a line.)

(d) Find the mean of each variable and verify the fact that the two lines intersect at the point corresponding to these means.

3.33 If a statistical computing system is available, a more thorough analysis of the particulate pollution data in Exercise 3.25 is possible. Enter the data into the computer.

(a) Do a regression of the city readings y on the rural readings x; verify the regression equation given in Exercise 3.25 and obtain the residuals.

(b) Plot the residuals against both x and the time order of the observations, and comment on the results.

(c) Make a stemplot (or, if your system allows, a normal quantile plot) of the residuals. Is the distribution of the residuals nearly symmetric? Does it appear to be approximately normal?

3.34 The following table gives the results of a study of a sensitive chemical technique called gas chromatography which is used to detect very small amounts of a substance. Five measurements were taken for each of four amounts of the substance being investigated. The explanatory variable x is the amount of substance in the specimen, measured in nanograms (ng), or units of 10^{-9} gram. The response variable y is the output reading from the gas chromatograph. The purpose of the study is to calibrate the apparatus by relating y to x. (Data from D. A. Kurtz (ed.), *Trace Residue Analysis*, American Chemical Society Symposium Series No. 284, 1985, Appendix.)

Amount (ng)	Response				
.25	6.55	7.98	6.54	6.37	7.96
1.00	29.7	30.0	30.1	29.5	29.1
5.00	211	204	212	213	205
20.00	929	905	922	928	919

(a) Make a scatterplot of these data. The relationship appears to be approximately linear, but the wide variation in the response values makes it hard to see detail in this graph.

(b) Compute the least squares regression line of y on x, and plot this line on your graph.

(c) Now compute the residuals and make a plot of the residuals against x. It is much easier to see deviations from linearity in the residual plot. Describe carefully the pattern displayed by the residuals.

3.3 CORRELATION

We have to this point concentrated on analyzing data having a clear explanatory–response structure. In order to fit a regression line, we must know which is the explanatory variable and which is the response. What tools are available when we are interested in the relation or association between two variables but do not wish to claim that one explains the other?

The basic scatterplot, of course, continues to portray the direction, form, and strength of any relationship between two quantitative variables. But the interpretation of a scatterplot by eye is surprisingly subjective. Changing the horizontal or vertical scale, for example, greatly affects our perception of the strength of a linear or other pattern. Even the amount of white space around the point cloud in a scatterplot can fool us. As Figure 3.26 illustrates, a scatterplot appears to show a stronger relationship when the point cloud is reduced in size relative to its surroundings. The two scatterplots in this figure

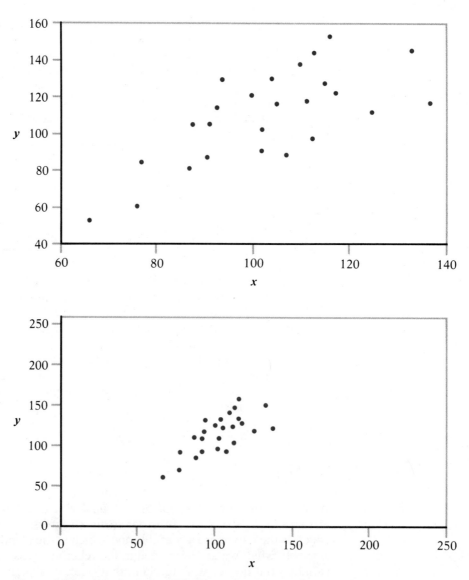

Figure 3.26 Two identical scatterplots; the linear pattern in the lower plot appears stronger because of the surrounding white space.

are identical in every respect except that the lower plot is drawn smaller in a large field and therefore appears to show a stronger linear pattern.[9]

When we are concerned with linear patterns in a scatterplot, we can use an important numerical measure to aid our visual perception. The *correlation coefficient* measures the strength of the linear association between two quantitative variables. It does not distinguish explanatory from response variables, and it is not affected by changes in the unit of measurement of either or both variables.* The word "correlation" is often used as a vague synonym for "association." Because correlation is a specific numerical measure that applies only to linear association and only to quantitative variables, we will use the word only in this sense. There is a positive association between educational level and income in Example 3.5, for example, but correlation is not meaningful because educational level is a categorical variable.

Computing the Correlation

We again have n observations on two variables x and y, denoted by

$$(x_1, y_1), (x_2, y_2), \ldots, (x_n, y_n)$$

Unlike the regression setting, x and y are not necessarily explanatory and response variables, although they may be. Here is the definition of the correlation coefficient.

Correlation coefficient

> The correlation coefficient for variables x and y computed from n cases is
>
> $$r = \frac{1}{n-1} \sum \left(\frac{x - \bar{x}}{s_x} \right) \left(\frac{y - \bar{y}}{s_y} \right) \qquad (3.3)$$

Here $\bar{x}$ and s_x are the mean and standard deviation of the x observations alone, and similarly $\bar{y}$ and s_y refer to the y observations. As usual, the sum runs over all of the cases for which the variables x and y have been measured.

We can also give a *computing formula* for the correlation coefficient r that eliminates the need to calculate the deviations $x - \bar{x}$ and $y - \bar{y}$ from the means. Like the earlier computing formulas, Equation 1.3 for the variance and Equation 3.2 for the slope of the least squares line, this formula is built up from basic sums. Because it is good practice to compute the means and

---

* Correlation and its relation to regression is another of Francis Galton's contributions, made in 1888.

standard deviations of each variable as part of the overall description, we give a form of the computing equation that uses the standard deviations.

$$r = \frac{\sum xy - \frac{1}{n}(\sum x)(\sum y)}{(n-1)s_x s_y} \tag{3.4}$$

In statistical practice, r is usually computed by a statistical calculator or by computer software rather than from a formula such as Equation 3.4.

The defining formula Equation 3.3 suggests why r is a measure of association between x and y. Suppose that x and y are the height and the weight of a person. The n cases represent measurements on n people. Height and weight are positively associated, so that larger than average values of x tend to occur together with larger than average values of y. Therefore, the deviations $x - \bar{x}$ and $y - \bar{y}$ from the means will tend to either both be positive (for larger people) or both be negative (for smaller people). In either case, the product $(x - \bar{x})(y - \bar{y})$ will be positive. Hence, r will be positive and will be larger as the positive association grows stronger. In the case of negative association, on the other hand, the deviations $x - \bar{x}$ and $y - \bar{y}$ will tend to have opposite signs, so the sign of r will be negative.

standardized deviations
The use in Equation 3.3 of the *standardized deviations* $(x - \bar{x})/s_x$ and $(y - \bar{y})/s_y$ implies that r measures association between x and y when both variables are measured in standard deviation units about the mean as origin. Changing the unit of measurement of either variable—for example, recording weight in kilograms rather than pounds—does not affect the value of r because both variables are in effect reduced to a standard scale before r is calculated. The properties of correlation will be explored in detail after an example that illustrates the calculation of r.

EXAMPLE 3.12

We will compute the correlation between gas consumption y and heating degree days x for the data below, which first appeared in Example 3.6.

Month	Oct.	Nov.	Dec.	Jan.	Feb.	Mar.	Apr.	May	June
x	15.6	26.8	37.8	36.4	35.5	18.6	15.3	7.9	.0
y	5.2	6.1	8.7	8.5	8.8	4.9	4.5	2.5	1.1

There are $n = 9$ cases. Figure 3.12 shows strong positive linear association in this example. As in the regression calculation of Example 3.7, first calculate the building block sums.

$$\sum x = 15.6 + \cdots + .0 = 193.9$$
$$\sum x^2 = (15.6)^2 + \cdots + (.0)^2 = 5618.11$$
$$\sum y = 5.2 + \cdots + 1.1 = 50.3$$
$$\sum y^2 = (5.2)^2 + \cdots + (1.1)^2 = 341.35$$
$$\sum xy = (15.6)(5.2) + \cdots + (.0)(1.1) = 1375.0$$

Second, calculate the means and standard deviations of both variables. The means and standard deviations are useful in their own right as descriptions of the distributions of x and y.

$$\bar{x} = \frac{1}{n}\sum x = \frac{193.9}{9} = 21.54$$

$$\bar{y} = \frac{1}{n}\sum y = \frac{50.3}{9} = 5.59$$

$$s_x^2 = \frac{1}{n-1}\left[\sum x^2 - \frac{1}{n}(\sum x)^2\right]$$

$$= \frac{1}{8}\left[5618.11 - \frac{(193.9)^2}{9}\right]$$

$$= \frac{1}{8}[5618.11 - 4177.468] = 180.080$$

$$s_x = \sqrt{180.080} = 13.419$$

$$s_y^2 = \frac{1}{n-1}\left[\sum y^2 - \frac{1}{n}(\sum y)^2\right]$$

$$= \frac{1}{8}\left[341.35 - \frac{(50.3)^2}{9}\right]$$

$$= \frac{1}{8}[341.35 - 281.121] = 7.529$$

$$s_y = \sqrt{7.529} = 2.744$$

Finally, substitute into the computing formula Equation 3.4 for the correlation coefficient.

$$r = \frac{\sum xy - \frac{1}{n}(\sum x)(\sum y)}{(n-1)s_x s_y}$$

$$= \frac{1375.0 - \frac{1}{9}(193.9)(50.3)}{(8)(13.419)(2.744)}$$

$$= \frac{291.31}{294.57} = .989$$

As in other multistep calculations, the exact answer depends on how many decimal places are carried in the intervening steps. A computer or statistical calculator will generally give more accurate answers. After a few practice runs to ensure that you understand what the formula for r says, you should automate your arithmetic if possible. ■

To interpret the numerical value of the correlation coefficient r, you must understand its behavior. Here are the basic properties of r.

1 The value of r always falls between -1 and 1. Positive r indicates positive association between the variables and negative r indicates negative association.

2 The extreme values $r = -1$ and $r = 1$ occur only in the case of perfect linear association, when the points in a scatterplot lie exactly along a straight line. Values of r close to 1 or -1 indicate that the points lie close to a straight line.

3 The value of r is not changed when the unit of measurement of x, y, or both changes. The correlation r has no unit of measurement; it is a dimensionless number between -1 and 1.

4 Correlation measures only the strength of *linear* association between two variables. Curved relationships between variables, no matter how strong, need not be reflected in the correlation.

The standardization of x and y in Equation 3.3 serves to constrain r to the range -1 to 1. The linear association increases in strength as r moves

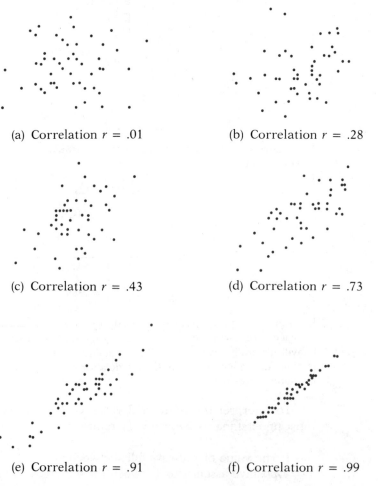

(a) Correlation $r = .01$ (b) Correlation $r = .28$

(c) Correlation $r = .43$ (d) Correlation $r = .73$

(e) Correlation $r = .91$ (f) Correlation $r = .99$

Figure 3.27   How the correlation coefficient measures the strength of linear association.

away from 0 toward either -1 or 1. The sign of r indicates only the direction of the association, so that $r = -0.7$ and $r = 0.7$ indicate linear association of the same strength but opposite directions. Understanding that r measures association in a standard scale helps avoid misinterpretation of scatterplots. Stretching or compressing the x or y scale can dramatically alter the appearance of a scatterplot, but does not change the correlation. It is therefore not always easy to guess the value of r from visual inspection of a scatterplot. The scatterplots in Figure 3.27 illustrate how values of r closer to 1 or -1 correspond to stronger linear association. To make the essential meaning of r clear, the standard deviations of both variables in these plots are equal and the horizontal and vertical scales are the same. In general, it is not so easy to guess the value of r from the appearance of a scatterplot.

The nature of the correlation as a measure of linear association is also illustrated by the real data that we have examined. The very linear home heating data in Example 3.12 (Figure 3.12) have a correlation close to 1 ($r = 0.989$). The linear relationship between the percent of votes for Democrats in 1980 and in 1984 (Figure 3.1) is positive but less strong; the correlation is $r = 0.703$. Figure 3.3 shows a quite strong negative association ($r = -0.849$) between mean SAT score and the percent of each state's high school seniors who take the SAT. The association between birth date and draft lottery number in Figure 3.5 is weak and slightly negative; the correlation is $r = -0.226$.

Finally, the correlation coefficient measures the strength of *linear* association only. It is possible to create examples of strong nonlinear association in which the correlation coefficient is small, or even 0 (see Exercise 3.40). For example, the strong nonlinear dependence of corn yield on planting rate was noted in Example 3.3. The correlation between these variables is $r = 0.135$, showing a very small linear association. Correlation is therefore not a general measure of all the types of association that may be visible in a scatterplot.

Correlation and Regression

Although correlation does not presuppose an explanatory-response relationship as regression does, the correlation coefficient r is meaningful for regression as well. In fact, the numerical value of r is most clearly interpreted from the following fact about regression.

r^2 in regression

> The square of the correlation coefficient, r^2, is the fraction of the variation in the values of y that is explained by the least squares regression of y on x. Moreover, the roles of x and y in this interpretation can be interchanged.

To understand this fact intuitively, consider again the scatterplot of household natural gas consumption y versus heating degree days x which appears in Figure 3.28(a). The horizontal dashed line marks the mean gas consumption $\bar{y}$. Gas consumption shows considerable variation from month

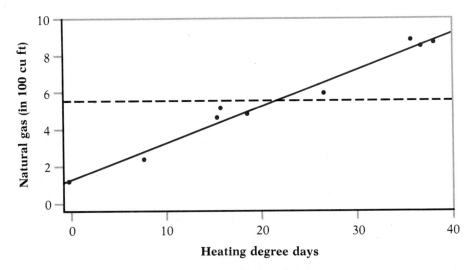

Figure 3.28(a)   High r^2: The variation in y about the regression line is much less than the variation in y about the mean $\bar{y}$; most of the variation in y is explained by the linear relationship of y and x.

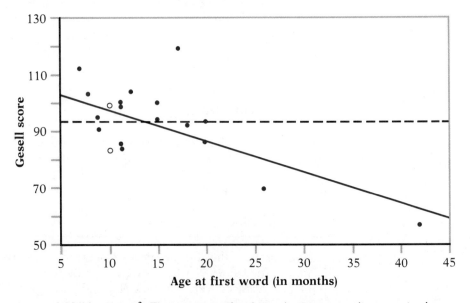

Figure 3.28(b)   Low r^2: The variation of y about the regression line remains large; less of the variation in y is explained by the linear relationship between y and x.

to month, as indicated by the vertical deviations of the points from the dashed line. The colored line is the least squares regression line. The vertical deviations of the points from this line are small. That is, when x changes, y changes with it, and this linear relationship accounts for almost all of the variation in y. Since $r = 0.989$ in this example, $r^2 = 0.978$ and we can say that the linear regression explains 97.8% of the observed variation in gas consumption.

On the other hand, the scatterplot of Gesell score y versus age at first word x in Figure 3.28(b) shows a weaker linear relationship. The vertical variation about the colored regression line, although smaller than the variation about the mean $\bar{y}$ (dashed line), is still large. The linear tie between x and y explains a smaller fraction of the observed variation in y. In fact, since $r = -0.640$ and $r^2 = 0.410$, we can say that 41% of the variation *in either variable* is explained by linear regression on the other variable. As Figure 3.19 illustrates, the regression lines of y on x and of x on y are quite different. But there is only a single correlation r between x and y (in either order), and r^2 helps interpret both regressions. The correlation between Gesell score and age at first word is negative, since late talkers tend to have lower aptitude scores. The interpretation of r through r^2 makes it clear that the magnitude of r, not its sign, measures the strength of a linear association.

We have gained not only more insight into interpreting correlation, but also a valuable numerical measure of the usefulness of a least squares regression line. Since the goal of regression is to explain y by linear dependence on x, r^2 is a direct measure of the success of the regression and is almost always reported along with the regression results. This close connection with correlation is specifically a property of least squares regression and is not shared by more resistant methods of fitting a line to a scatterplot.

The use of r^2 to measure how successfully a regression line explains the observed variation in y is not the only connection between correlation and regression. There is a close relationship between r and the slope b of the least squares regression line $\hat{y} = a + bx$. Some algebra based on Equations 3.1 and 3.3 establishes the following fact:

Regression slope

If s_x and s_y are the standard deviations of the observed x_i and y_i, and r is the correlation coefficient, then the slope of the least squares regression line of y on x is

$$b = r\frac{s_y}{s_x} \tag{3.5}$$

That is, the least squares regression line of y on x is the line with slope rs_y/s_x that passes through the point $(\bar{x}, \bar{y})$. Regression can therefore be described entirely in terms of the basic descriptive measures $\bar{x}$, s_x, $\bar{y}$, s_y, and r. If both x and y are standardized variables, so that their means are 0 and their standard deviations are 1, then the regression line has slope r and passes through the origin.

Equation 3.5 is the formula for the slope b that is easiest to understand and remember. This equation says that along the regression line, *a change of one standard deviation in x corresponds to a change of r standard deviations in y*. When the variables are perfectly correlated ($r = 1$ or $r = -1$), the change in the response y is the same (in standard deviation units) as the change in x. Otherwise, since $-1 \leq r \leq 1$, the change in y is less than the change in x. As the correlation grows less strong, y moves less in response to changes in x.

EXAMPLE 3.13

In Example 3.12 we found that for heating degree days x and natural gas consumption y,

$$\bar{x} = 21.54 \quad \text{and} \quad s_x = 13.42$$
$$\bar{y} = 5.59 \quad \text{and} \quad s_y = 2.74$$
$$r = .989$$

The slope of the regression line of gas usage on degree days is therefore

$$b = .989\, \frac{2.74}{13.42} = .202$$

in agreement with the result of Example 3.7.

The regression line passes through the point $(\bar{x}, \bar{y})$, which is (21.54, 5.59). Along the line, y increases by $b = 0.202$ when x increases by 1. In terms of correlation, y increases by $r = 0.989$ standard deviations when x increases by one standard deviation. ∎

Interpreting Correlation and Regression

Limitations of correlation and regression   Correlation and regression are powerful tools for measuring the association between two variables and for expressing the dependence of one variable on the other. These tools must be used with an awareness of their limitations, beginning with the fact that they apply to *only linear* association or dependence. Also remember that *neither r nor the least squares regression line is resistant*. One influential observation or incorrectly entered data point can greatly change these measures.

EXAMPLE 3.14

We saw in Example 3.10 that case 18 is an influential observation in the regression of the Gesell score y on age at first word x for young children. In fact, the correlation based on all 21 children is $r = -0.640$. Since $(-0.64)^2 = 0.41$, age at first word appears to explain 41% of the variation in Gesell score among children. But if case 18 is omitted, the correlation for the remaining 20 children is only $r = -0.335$. Only 11% of the variation in aptitude score among these 20 children is explained by the age at which they first spoke. Excluding case 19, which is an outlier, we have $r = -0.756$ and $r^2 = 0.572$ or 57%. Case 19 is also influential, though not as much as case 18. The least squares line of y on x for the complete data set is

$$\hat{y} = 109.87 - 1.127x$$

but if case 18 is excluded the regression line becomes

$$\hat{y} = 105.63 - .779x$$

These regression lines are graphed in Figure 3.20. This single observation of case 18 dramatically changes both the correlation and the fitted line. A decision to exclude this child as not belonging to the same population as the other children will weaken the study's conclusion that Gesell score can be partially predicted from age at first word. Just as calculation often adds to the information provided by a scatterplot, a plot is essential if calculation is not to be blind. Without a plot to help spot the influential observation, numerical calculations for these data can be seriously misleading. ■

Lurking variables We have seen repeatedly that the effect of variables not included in a study can render a correlation or regression misleading. Examples 3.9 and 3.11 both illustrate the effect of such lurking variables. To give another example, there is a strong positive correlation over time between teachers' salaries and sales of liquor. Both increase with rising price levels and general prosperity, creating a strong association. Such correlations are sometimes called "nonsense correlations," but the correlation is perfectly real. What *is* nonsense is the conclusion that because the correlation exists, teachers must be spending their salary increases on liquor. *Even a strong correlation does not imply any cause and effect relationship.* The question of causation is important enough to merit separate treatment in Section 3.5. For now, just remember that a correlation between two variables x and y can reflect many types of relationship between x, y, and other variables not explicitly recorded.

The effect of lurking variables can hide a true relationship between x and y as well as create an apparent relationship, as the following example illustrates.

EXAMPLE 3.15

A study of housing conditions and health in the city of Hull, England measured a large number of variables for each of the wards in the city. (A ward is a small, relatively homogeneous geographic area.) Two of the variables were an index x of overcrowding and an index y of the lack of indoor toilets. Since x and y are both measures of inadequate housing, we expect a high correlation. In fact, the correlation was only $r = 0.08$. How can this be? Investigation disclosed that some poor wards were dominated by public housing (called council housing in England). These wards had high values of x but low values of y because council housing always includes indoor toilets. Other deprived wards lacked council housing, and in these wards high values of x were accompanied by high values of y. Because the relationship between x and y differed in council and noncouncil wards, analyzing all wards together obscured the nature of the relationship.[10] ■

Figure 3.29 shows in simplified form how groups formed by a lurking (categorical) variable as in Example 3.15 can make the correlation r misleading. The groups appear as clusters of points in the scatterplot. There is a strong relationship between x and y within each of the clusters; in fact, $r = 0.85$ and $r = 0.91$ in the two clusters. However, because similar values of x correspond to quite different values of y in the two clusters, x alone is of little value in predicting y. The correlation for all points displayed is therefore low; in fact, $r = 0.14$. This example is another reminder to plot the data rather than to simply calculate numerical measures such as the correlation.

Prediction   A regression line is often used to *predict* the response y to a given value x of the explanatory variable. This is clearly valid when the regression reflects a cause and effect relationship and when r^2 is high enough to give us confidence that changes in x explain most of the variation in y. For example, we can predict household natural gas consumption at different outside temperatures (degree days), using the regression line calculated in Example 3.7. However, *successful prediction does not require a causal relationship.* If both x and y respond to the same underlying unmeasured variables, it may be possible to predict y from x even though x has no direct influence on y.

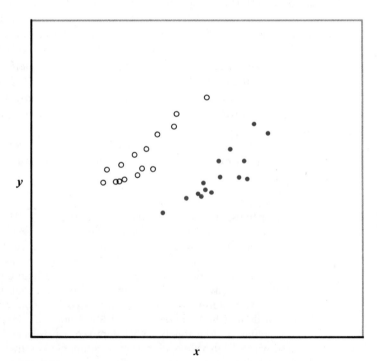

Figure 3.29   This scatterplot has a low r^2 even though there is a strong correlation within each of the two clusters.

EXAMPLE 3.16

Decisions on which applicants to admit to graduate school are based on a number of criteria, such as the applicant's undergraduate grades. Another criterion is scores on the Graduate Record Examinations (GRE), national tests of both aptitude and knowledge administered by the Educational Testing Service. There is no causal relationship between the score x a college senior achieves on the GRE and the student's grade point average (GPA) y as a first-year graduate student the following year. But because both x and y respond to the student's level of ability and knowledge, it is plausible to predict y from x.

The success of this prediction depends on the strength of the association between GRE scores and later GPA in graduate school. A study[11] of a large number of first-year students at many graduate schools in many disciplines showed that for students of economics

- The correlation between the GRE verbal aptitude score and graduate GPA was $r = 0.09$.

- The correlation between the GRE quantitative aptitude score and graduate GPA was $r = 0.34$.

- The correlation between the GRE economics advanced test score and graduate GPA was $r = 0.45$.

- The correlation between undergraduate GPA and graduate GPA was $r = 0.27$.

These results show that the verbal aptitude test bears little relation to success as a graduate student of economics; but they also show that a student's score on the GRE economics advanced test is a better predictor of success than the undergraduate GPA. However, even the GRE economics test accounts for only 20% of the variation in first-year graduate GPA [because $r^2 = (0.45)^2 = 0.20$]. Although prediction from a regression line makes sense in this setting, prediction based on the GRE advanced score alone will be quite unreliable. ∎

Notice that for the purpose of assessing whether GRE scores are helpful in making admissions decisions, the data in Example 3.16 are incomplete in a systematic way. We have GPA information only for students who were admitted to graduate school. Because many students with low GRE scores were not admitted, the data refer primarily to students who did well on the GRE. The reported correlations describe the relation between GRE scores and graduate GPAs for students who make it to graduate school. They do not tell us whether students with GRE scores too low to allow them admission would in fact have earned low grades if they had entered graduate school.

Even when prediction is logically justified and r^2 is high, several additional cautions are in order. Chapter 2 noted the danger of *extrapolation*, the use of a regression line for prediction at values of x removed from the range of x-values used to fit the line. Most relationships remain linear only over a restricted range of x, so extrapolation can yield silly results. An investigation of the firing time y of blasting caps used in mining showed a strong linear response to the voltage x applied to the detonator. This is true, however, only over the range of voltages used in practice. The overall relationship between voltage and firing time is similar to that shown in Figure 3.30.

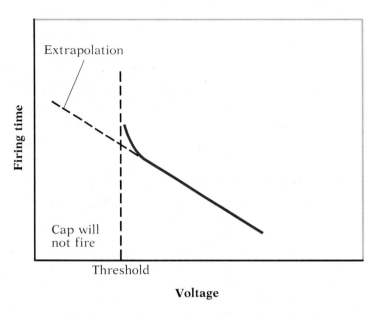

Figure 3.30 Beware of extrapolation: The relationship between the voltage applied and the firing time of a blasting cap is linear only in a restricted range of voltages.

Below a threshold voltage, the blasting cap will not detonate at all. Extrapolation of the regression line to low voltages is meaningless.

Using averaged data   Many regression or correlation studies work with averages or other measures that combine information from many individuals. You should note this carefully and resist the temptation to apply the results of such studies to individuals. The regression of natural gas consumption on heating degree days, for example, was based on daily data averaged over each month. The very high correlation observed does not apply to individual days. Prediction of gas usage from outside temperature on a single day would be much less reliable than is suggested by the high r^2 in the regression of Examples 3.7 and 3.12. The reason for this is that averaging over an entire month smooths out much of the day-to-day variation due to doors left open, house guests using more gas to heat water, and so on. Figure 3.31 shows the gas consumption and degree days for several individual days in April (A), May (M), and November (N). There is considerable variation within each month and the correlation for these individual observations would be moderate. But when only the three monthly averages (marked as ●) are recorded, the result is three of the points in Figure 3.12, with a correlation near 1. *Correlations based on averages are usually too high when applied to individuals.* This is another reminder that it is important to note exactly what variables were measured in a statistical study.

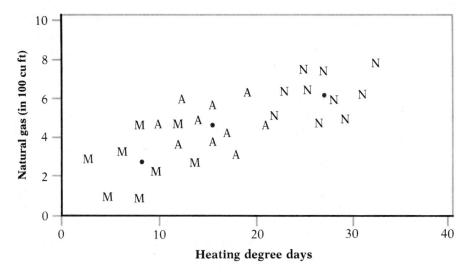

Figure 3.31 Natural gas consumption plotted against degree days for individual days in April (A), May (M), and November (N), and the averages of both variables in the three months (•). The correlation is much higher for the averaged data.

Correlation is not everything   Finally, remember that *correlation is not a complete description of two-variable data.* The means and standard deviations of both x and y should be given along with the correlation. (Because correlation and regression make use of means and standard deviations, these nonresistant measures are appropriate to accompany a correlation.) Conclusions based on correlations alone may require rethinking in the light of a more complete description of the data.

EXAMPLE 3.17

Competitive divers are scored on their form by a panel of judges, who use a scale from 1 to 10. The subjective nature of this scoring often results in controversy. We have the scores awarded by two judges, Ivan and George, on a large number of dives. How should we assess their agreement? Some computation shows that the correlation between their scores is $r = 0.9$. But the mean of Ivan's scores is 3 points lower than George's mean.

These facts do not contradict each other. They are simply different kinds of information. Ivan awards much lower scores than George, as the mean scores reveal. But because Ivan gives *every* dive a score about 3 points lower than George, the correlation remains high. Remember that adding or subtracting the same number to all values of either x or y does not change the correlation. If Ivan and George both rate several divers, the contest is consistently scored because, as the high correlation shows, Ivan and George agree on which dives are better than others. But if Ivan scores one diver and George another, we must add 3 points to Ivan's scores to arrive at a fair comparison.   ■

Many statistical studies of important issues rely on correlation and regression to describe complex relationships among many variables. The examples in this section demonstrate that even when only two variables are involved, the numerical results of correlation or regression must be interpreted in the light of an understanding both of the behavior of these statistical procedures and of the full background of the issue under study. The numerical work is easily automated, and you should let a computer do it. Interpretation requires an informed and skeptical human mind.

SUMMARY

The **correlation coefficient** r measures the strength and direction of the linear association between two quantitative variables x and y. Correlation always satisfies $-1 \leq r \leq 1$, and $r = \pm 1$ only in the case of perfect linear association. The value of r is not affected by changes in the unit of measurement of either variable.

Correlation and regression are closely connected. The squared correlation coefficient r^2 is the fraction of the variation of one variable that is explained by least squares regression on the other variable. The regression line of y on x is the line with slope $b = rs_y/s_x$ that passes through the point $(\bar{x}, \bar{y})$.

A correlation or regression should be interpreted with due attention to the following: the possible effects of lurking variables, the lack of resistance of these procedures, the danger of extrapolation, the fact that correlations based on averages are usually too high for individuals, and an understanding that correlation and regression measure only linear relationships to the exclusion of other important aspects of the data.

SECTION 3.3 EXERCISES

3.35 A student wonders if people of similar heights tend to date each other. She measures herself, her dormitory roommate, and the women in the adjoining rooms; then she measures the next man each woman dates. Here are the data (heights in inches).

Women	66	64	66	65	70	65
Men	72	68	70	68	71	65

(a) Make a scatterplot of these data. Based on the scatterplot, do you expect the correlation to be positive or negative? Near ± 1 or not?

(b) Use Equation 3.4 to compute the correlation r between the heights of the men and women.

(c) How would r change if all the men were 12 inches shorter than the heights given in the table? Does the correlation help answer the question of whether women tend to date men taller than themselves?

(d) If every woman dated a man exactly 3 inches taller than she, what would be the correlation between male and female heights?

3.36 The growth of young children is nearly linear. Here again are data on Sarah's height at several ages. (See Exercise 3.17 for the regression line.)

Age (months)	36	48	51	54	57	60
Height (cm)	86	90	91	93	94	95

(a) From a scatterplot of height versus age, explain why you would expect the correlation to be close to 1.

(b) Compute the correlation coefficient r between height and age, using Equation 3.4.

(c) If Sarah were 4 centimeters taller at every age, how would the value of r change? *doesn't*

3.37 Compute the mean and standard deviation for the heights of the men and women in Exercise 3.35. Use your results and the correlation found in Exercise 3.35 to compute the slope of the regression line of male height on female height. What is the slope of the regression of female height on male height (when male height is on the x axis and female height is on the y axis)? If both lines were drawn on the same graph, with female height on the x axis, at what point would they intersect?

3.38 Compute the mean and the standard deviation of both height and age in Exercise 3.36. Use these values and the correlation from Exercise 3.36 to find the slope of the regression line of height on age. (Compare your result with the slope you found in Exercise 3.17.) What is the slope of the regression line of age on height?

3.39 Exercise 3.6 compares two methods of computing the value of a dollar in foreign currencies: by comparing the cost of a McDonald's Big Mac and the official exchange rate.

(a) Calculate the correlation r between the Big Mac value and the official value of the dollar.

(b) The Big Mac value of the dollar in Japan was 231 Japanese yen, while the official exchange rate was 154 yen to the dollar. Add this observation to the data set and calculate the correlation again. Explain carefully why adding this one observation causes such a large increase in r.

3.40 The gas mileage of an automobile first increases and then decreases as speed increases. Suppose that this relationship is very regular, as shown by the following data on speed (miles per hour) and mileage (miles per gallon):

Speed	20	30	40	50	60
Mileage	24	28	30	28	24

Make a scatterplot of mileage versus speed. Show that the correlation between speed and mileage is $r = 0$. (Note that to show that $r = 0$, you need only compute the numerator in Equation 3.4.) Explain why the correlation is 0 even though there is a strong association between speed and mileage.

3.41 A college newspaper interviews a psychologist about a proposed system for rating the teaching ability of faculty members. The psychologist says, ''The evidence indicates that the correlation between a faculty member's research productivity and teaching rating is close to zero.'' The paper reports this as ''Professor McDaniel said that good researchers tend to be poor teachers, and vice versa.'' Explain why the paper's report is wrong. Write a statement in plain language (don't use the word ''correlation'') explaining the psychologist's meaning.

3.42 Each of the following statements contains a blunder. Explain in each case what is wrong.

  (a) ''There is a high correlation between the sex of American workers and their income.''
  (b) ''We found a high correlation ($r = 1.09$) between students' ratings of faculty teaching and ratings made by other faculty members.''
  (c) ''The correlation between planting rate and yield of corn was found to be $r = 0.23$ bushel.''

3.43 A study of class attendance and grades among freshmen at a state university showed that in general students who attended a higher percent of their classes earned higher grades. Class attendance explained 16% of the variation in grade index among the freshmen studied. What is the numerical value of the correlation between percent of classes attended and grade index?

3.44 Suppose that the heights of the men in Exercise 3.35 were measured in centimeters (cm) rather than in inches, but that the heights of the women remained in inches. (There are 2.54 cm to an inch.)

(a) What would now be the correlation between male and female height? (Use information from Exercise 3.35—don't compute the new r directly.)

(b) What would be the slope of the regression line of male height on female height? (Use your calculations from Exercise 3.37—don't compute a new regression line. Hint: Use Equation 3.5 for the slope b.)

3.45 Changing the units of measurement can dramatically alter the appearance of a scatterplot. Consider the following data:

x	-4	-4	-3	3	4	4
y	.5	$-.6$	$-.5$	.5	.5	$-.6$

(a) Draw x and y axes each extending from -6 to 6. Plot the data on these axes. Then plot $x^* = x/10$ against $y^* = 10y$ on the same axes using a different plotting symbol. The two plots are very different in appearance.

(b) The correlation between x and y is about $r = 0.25$. What must be the correlation between x^* and y^*?

(c) Will the regression line of y^* on x^* have the same slope as the regression line of y on x? Explain your answer. (Hint: Look at Equation 3.5 for the slope.)

3.46 Return to the scatterplot in Figure 3.2 showing the percent of presidential votes cast for Democrats in 1980 and 1984, with the south emphasized. The correlation for all 50 states is $r = 0.703$. Would r be higher or lower if the 10 southern states were omitted? Why?

3.47 The full MINITAB computer output for Example 3.9 contains the entry R-sq = 69.4%. Explain what this means in this specific example, in language that can be understood by someone who knows no statistics.

3.48 Figure 3.10 shows how the price of four-door sedans varies with their weight.

(a) If only foreign models are considered, the correlation between weight and price is $r = 0.707$. What percent of the variation in the prices of these foreign cars can be explained by the fact that price increases linearly as weight increases?

(b) From examination of Figure 3.10, do you think that the linear relation of price to weight explains a higher or a lower percent of the price variation of the domestic models in the study? Why?

3.49 A study of erosion (see Exercises 3.5 and 3.30) produced the following data on the rate at which water flows across land and the resulting amount of erosion:

Flow rate	.31	.85	1.26	2.47	3.75
Eroded soil	.82	1.95	2.18	3.01	6.07

What percent of the variation in the amount of erosion can be explained by the fact that as the flow rate increases, erosion increases with it in a linear manner?

3.50 The mean height of American women in their early twenties is about 65.5 inches and the standard deviation is about 2.5 inches. The mean height of men the same age is about 68.5 inches, with standard deviation about 2.7 inches. If the correlation between the heights of husbands and wives is about $r = 0.5$, what is the slope of the regression line of the husband's height on the wife's height in young couples? Draw a graph of this regression line. Predict the height of the husband of a woman who is 67 inches tall.

3.51 In a large economics class, the correlation between a student's total score prior to the final examination and the final examination score is $r = 0.6$. The pre-exam totals for all students in the course have mean 280 and standard deviation 30. The final exam scores have mean 75 and standard deviation 8. The professor has lost Julie's final exam but knows that her total before the exam was 300. He decides to predict her final exam score from her pre-exam total.

(a) What is the slope of the regression of final exam scores on pre-exam total scores in this course?

(b) Draw a graph of this regression line and use it to predict Julie's final examination score.

3.52 The British government conducts regular surveys of household spending. The following table shows the average weekly household spending on tobacco products and alcoholic beverages for each of the

Region	Alcohol	Tobacco
North	£6.47	£4.03
Yorkshire	6.13	3.76
Northeast	6.19	3.77
East Midlands	4.89	3.34
West Midlands	5.63	3.47
East Anglia	4.52	2.92
Southeast	5.89	3.20
Southwest	4.79	2.71
Wales	5.27	3.53
Scotland	6.08	4.51
Northern Ireland	4.02	4.56

11 regions of Great Britain. (Data from British official statistics, *Family Expenditure Survey,* Department of Employment, 1981.)

(a) Make a scatterplot of spending on tobacco against spending on alcohol.

(b) Describe the pattern of the plot. Circle the most influential observation.

(c) The correlation is only $r = 0.224$. Compute the correlation for the 10 regions with Northern Ireland omitted. Explain why this r differs so greatly from the r for all 11 cases.

3.53 Example 3.12 illustrates the computation of the correlation coefficient r. Redo that computation, rounding off all intermediate steps to the nearest whole number. Compare your answer with the result in the example. (Use the building block sums found in the example as your starting point, but round them to the nearest whole number before proceeding.) Remember that your numerical answers for b, r, and other descriptive measures that require long calculations will vary as you carry more or fewer significant digits in the intermediate steps.

3.54 There is a strong positive correlation between years of education and income for economists employed by business firms. (In particular, economists with doctorates earn more than economists with only a bachelor's degree.) There is also a strong positive correlation between years of education and income for economists employed by colleges and universities. But when all economists are considered, there is a *negative* correlation between education and income. The explanation for this is that business pays high salaries and employs mostly economists with bachelor's degrees, while colleges pay lower salaries and employ mostly economists with doctorates. Sketch a scatterplot with two groups of cases (business and academic) which illustrates how a strong positive correlation within each group and a negative overall correlation can occur at the same time. (Hint: Begin by studying Figure 3.29.)

3.55 If you have a statistical computing system, enter the data on the relation between Gesell score and age at first word (Table 3.3). Compute the correlation and the least squares regression line for the data with *both* cases 18 and 19 omitted. How do the results, particularly r^2, compare with those given in Example 3.14?

3.56 Refer to the data on wood strength used in Exercises 3.16, 3.31, and 3.32.

(a) Compute the correlation between T1 and T2.

(b) Using the slope of the least squares line, the standard deviations of x and y, and the correlation found in (a), verify that Equation 3.5 holds for this set of data.

3.57 Refer to the data on women's stride rates given in Exercise 3.21.

(a) Compute the correlation between speed and stride rate.

(b) What proportion of the variation in stride rate is explained by speed for this set of data?

(c) Repeat (a) and (b) using only the data for speeds 15.86, 16.88, 17.50, and 18.62. Do the results change appreciably? From the plotting that you have done, would you expect the least squares line to change appreciably? Explain what you have learned from this part of the exercise.

3.58 The data in Exercise 3.21 relating stride rate in strides per second to running speed give the average stride rate of 21 elite female runners at each speed. Suppose that you had data on many individual time periods for all 21 runners. If you plotted each individual stride rate at each speed and computed the correlation for these individual data, would you expect the correlation between stride rate and speed to be lower than, about the same as, or higher than the correlation for the published data? Sketch a scatterplot of stride rate versus speed for individual runners to illustrate your answer.

3.59 The price of seafood varies with species and time. The following table gives the prices in cents per pound received in 1970 and 1980 (PR70 and PR80) by fishermen and vessel owners for several species:

Species	PR70	PR80
Cod	13.1	27.3
Flounder	15.3	42.4
Haddock	25.8	38.7
Menhaden	1.8	4.5
Ocean perch	4.9	23.0
Salmon, chinook	55.4	166.3
Salmon, coho	39.3	109.7
Tuna, albacore	26.7	80.1
Clams, soft-shelled	47.5	150.7
Clams, blue hard-shelled	6.6	20.3
Lobsters, American	94.7	189.7
Oysters, eastern	61.1	131.3
Sea scallops	135.6	404.2
Shrimp	47.6	149.0

(a) Plot the data with PR70 on the x axis and PR80 on the y axis.

(b) Describe the overall pattern. Are there any outliers or points that may be highly influential? If so, label them.

(c) Compute the correlation for the entire set of data.

(d) What proportion of the variation in 1980 prices is explained by the 1970 prices?

(e) Recompute the correlation discarding the cases that you labeled in (b).

(f) To what extent do you think the correlation provides a good measure of the relationship between the 1970 and 1980 prices for this set of data? Explain your answer.

3.4 RELATIONS IN CATEGORICAL DATA

Up to this point our focus has been on relationships between quantitative variables, although categorical variables played an important role in Section 3.1. Now our focus will shift to describing relationships between two or more categorical variables. Some variables—such as sex, race, and occupation—are inherently categorical. In other cases, categorical variables are created by grouping values of a quantitative variable into classes. Published data are often reported in this form to save space. Analysis of categorical data is based on the counts or percents of the cases that fall into various categories.

EXAMPLE 3.18

two-way table

Table 3.4 presents Census Bureau data on the educational attainment of Americans of different ages.[12] Because many persons under 25 years of age have not completed their education, they are not included in the table. Both variables, age and education, have been grouped into categories. The entries in this *two-way table* are the frequencies, or counts, of persons in each age by education class. Although both age and education as presented in this table are categorical variables, both have a natural order from least to most. The order of the rows and the columns in Table 3.4 reflects the order of the categories. ∎

Table 3.4 Educational attainment by age, 1984 (thousands of persons)

Education	Age group					Total
	25–34	35–44	45–54	55–64	≥65	
Did not complete high school	5416	5030	5777	7606	13,746	37,577
Completed high school	16,431	11,855	9435	8795	7558	54,073
College, 1–3 years	8555	5576	3124	2524	2503	22,282
College, 4 or more years	9771	7596	3904	3109	2483	26,862
Total	40,173	30,058	22,240	22,033	26,291	140,794

Analyzing Two-Way Tables

How can we best grasp the information contained in Table 3.4 about the educational attainment of Americans? Notice first the abundance of information in a two-way table. The "Total" column at the right of the table gives the distribution of all Americans over 25 years of age by education. The "Total" row at the bottom gives their distribution by age. The bottom row contains the column totals, and the rightmost column contains the row totals. The distributions of both variables can therefore be calculated from the body of the table. These are often called the *marginal distributions* because they appear at the bottom and right margins of the table. If the row and column totals are missing, the first thing to do in studying a two-way table is to calculate the marginal distributions by summing each row and each column.

marginal distribution

The row and column sums in Table 3.4 show some discrepancies. The sum of the "35–44" years column, for example, is 30,057. But the total is given in the table as 30,058. The explanation is *roundoff error*: The table entries are in *thousands* of persons, and each is rounded to the nearest thousand. The Census Bureau obtained the column total of persons aged 35 to 44 by rounding the exact number of such persons to the nearest thousand: The result was 30,058,000. Adding the column entries, each of which is already rounded, gives a slightly different result.

The body of Table 3.4 contains much more information than the two marginal distributions of age alone and education alone. The nature of the relationship between age and education cannot be deduced from the separate distributions but requires the full table. *Relationships among categorical variables are described by calculating appropriate percentages from the counts given.*

EXAMPLE 3.19

We are interested in the association between age and college education among adult Americans. Table 3.4 shows that 9,771,000 persons aged 25 to 34 have completed college, while only 3,109,000 persons in the 55 to 64 age group have done so. These counts do not display the association, because there are many more people in the younger age group. We must instead compare the percent in each age group who have finished college.

For persons aged 25 to 34, the percent who have completed college is the number of persons in this age group who completed college as a percent of the total number of 25 to 34 year olds (the column total). This percent is

$$\frac{9771}{40,173} = .2432 = 24.3\%$$

Similar calculations for each age group give

Age group	25–34	35–44	45–54	55–64	≥65
Percent with 4 years of college	24.3	25.3	17.6	14.1	9.4

These percentages make it clear that a college education is much more common among younger Americans.

For the over-25 population as a whole, the percent who completed college is

$$\frac{26{,}862}{140{,}794} = .1908 = 19.1\%$$

This is the row total who completed college as a percent of the table total, which appears at the lower right corner of the table. ∎

bar graph

If you are presenting the information in Example 3.19 to a general audience, a *bar graph* like Figure 3.32 may be helpful. Each bar represents one age group. The height of the bar is the percent of that age group with at least 4 years of college. Although bar graphs are somewhat similar to histograms, their details and uses are distinct. A histogram shows the distribution of frequencies or relative frequencies among the values of a single variable, while a bar graph compares the size of different items. The horizontal axis of a bar graph need not have any measurement scale but may simply identify the items being compared. The items compared in Figure 3.32 are the five age groups. Because each bar in a bar graph describes a different item, the bars are usually drawn with space between them.

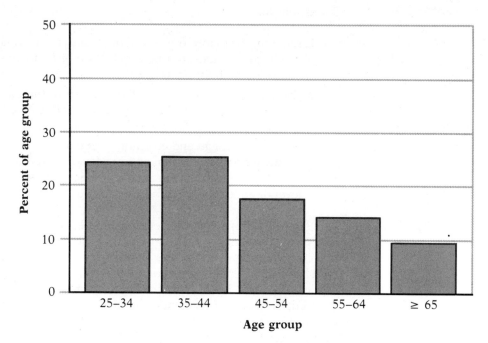

Figure 3.32 A bar graph comparing the percent of people in five age groups who have completed college. See Example 3.19.

If we wish to compare the complete distributions of education in the five age groups, we must go farther. Consider the 25 to 34 age group, represented by the first column in Table 3.4. The percent distribution of educational level within this age group is found by computing each count as a percent of the column total. The results are

Education	1–3 years of high school	4 years of high school	1–3 years of college	≥4 years of college
Percent	13.5	40.9	21.3	24.3

conditional distribution

These percents add to 100%, since all 25 to 34 year-olds fall into one of the educational categories. The four percents together make up the *conditional distribution* of education, given that a person is 25 to 34 years of age. The term "conditional" is used because the distribution refers only to people who satisfy the condition that they be 25 to 34 years old.

The conditional distributions of education in the different age groups can be expected to differ from one another. They will also differ from the marginal (overall) distribution of education computed from the right-most column in Table 3.4. The influence of age on education is seen by comparing the five conditional distributions. Figure 3.33 does this graphically by means **segmented bar** of a *segmented bar graph*, which compares all of the age groups in Table **graph** 3.4 as follows:

1 Each bar describes one age group, that is, one column in Table 3.4. The segments indicate the breakdown of that age group by educational attainment.

2 Each bar has height 100%. The division of the bar into segments shows what percent of that age group falls into each educational category.

3 Because both age and education have a clear order from least to most, the bars are arranged in order of increasing age from left to right; the segments in each bar are arranged in order of increasing level of education from bottom to top.

Figure 3.33 presents the relationship between age and education in visual form. We can see, for example, that the percent of Americans who did not complete high school (the bottom segment in each bar) rises rapidly with age. The percent who completed college (the top segment) remains stable until over 44 years old, and then declines. These two comparisons are clear-cut because the segments compared all have a common starting point, either at the bottom or at the top of the bars. In order to compare the percent in each age group who completed high school, we must compare the lengths of segments that both start and end at different points, a task that our eyes do less well. Segmented bar graphs do not present information as

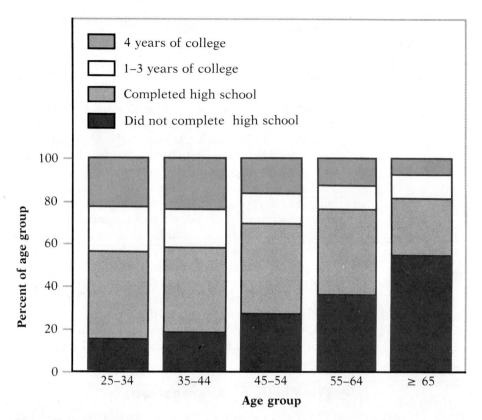

Figure 3.33 A segmented bar graph comparing the distribution of educational levels within five age groups.

vividly as most of the other graphs used in statistics, primarily because of the large amount of information given in a two-way table.[13] The graph does not present all of the information in the table. For example, lengths in Figure 3.33 represent percents, not counts. This choice results in a clearer comparison of the five distributions of education within each age group. Table 3.4 shows the actual number of people in each age group by educational category, but Figure 3.33 does not present this information.

In calculating the percentages and constructing the segmented bar graph, we concentrated on the distribution of educational attainment in each age group and the way this distribution changes with age. That is, we regarded age as the explanatory variable and education as the response variable. To do this, we considered each table entry as a part of its column total. This produces the conditional distribution of education given age. We might also be interested in the distribution of age among persons having a certain level of education. This is the conditional distribution of age given education. Each table entry would then be considered as part of the corresponding *row total*. For example, the percentage of persons 65 years or older among those who did not complete high school is

$$\frac{13{,}746}{37{,}577} = .3658 = 36.6\%$$

Figure 3.33 summarizes the conditional distributions of education within each age group. It does not present information about the conditional distributions of age within each educational group.

A two-way table contains, in compact form, two marginal distributions and two sets of conditional distributions. Interpretation of this information requires that you have a clear awareness of what facts are relevant to your purpose. If you are studying trends in the training of the American work force, conditioning on age reveals the more extensive education of younger people. If, on the other hand, you are planning a program to improve the skills of people who did not finish high school, the age distribution within this educational group is essential information. There is no single graph (such as a scatterplot) that portrays the form of the relationship between categorical variables, and no single numerical measure (such as the correlation coefficient) that summarizes the strength of simple relations in an easily understood way. We must rely on well-chosen percentages, perhaps summarized graphically, or on more advanced statistical methods.[14]

The Perils of Aggregation

As is the case with quantitative variables, the effects of lurking variables that are not recorded can change or even reverse relationships between two categorical variables. Here is a hypothetical example that demonstrates what surprises can await the unsuspecting consumer of data.

EXAMPLE 3.20

In an attempt to help consumers make informed decisions about health care, the government releases data about patient outcomes in a large number of hospitals. You are interested in comparing Hospital A and Hospital B, which serve your community. Here are the data on the survival of patients after surgery in these two hospitals. All patients undergoing surgery in a recent time period are included; "survived" means that the patient lived at least 6 weeks following surgery.

	Hospital A	Hospital B
Died	63	16
Survived	2037	784
Total	2100	800

The evidence seems clear: Hospital A loses 3% (63/2100) of its surgery patients, whereas Hospital B loses only 2% (16/800). You should choose Hospital B if you need surgery.

Not all surgery cases are equally serious, however. Later in the government report you find data on the outcome of surgery broken down by the condition of the patient before the operation. Patients are classified as being in either "poor" or "good" condition. Here are the more detailed data.

	Good condition			Poor condition	
	Hospital A	Hospital B		Hospital A	Hospital B
Died	6	8	Died	57	8
Survived	594	592	Survived	1443	192
Total	600	600	Total	1500	200

Aha! Hospital A beats Hospital B for patients in good condition: Only 1% (6/600) died in Hospital A, compared with 1.3% (8/600) in Hospital B. And Hospital A wins again for patients in poor condition, losing 3.8% (57/1500) to Hospital B's 4% (8/200). So Hospital A is safer for both patients in good condition and patients in poor condition. If you are facing surgery you should choose Hospital A. ■

The result of Example 3.20 is surprising and perhaps disturbing: Hospital A has a higher overall rate of patient deaths from surgery even though its rate is lower than that of Hospital B for each class of patients considered separately. This phenomenon is called Simpson's paradox.

Simpson's paradox

> Simpson's paradox refers to the reversal of the direction of a comparison or an association when data from several groups are combined to form a single group.

Although the apparent paradox is unexpected, the explanation is clear when the data are examined. Hospital A is a university medical center that attracts seriously ill patients from a wide region, while Hospital B is a local hospital that handles relatively few major surgery cases. Hospital A operated on 1500 patients in poor condition, while Hospital B saw only 200 such cases. Because patients in poor condition are more likely to die, this difference in the patients increases Hospital A's death rate despite its superior performance on each class of patients. The original two-way table, which did not take account of the condition of the patients, was misleading. The death rates compiled and released by the Federal Health Care Financing Agency do allow for the patient's condition and diagnosis. Hospitals continue to complain that there are differences among patients that can make the government's mortality data misleading.

The table of outcome by hospital by patient condition in Example 3.20 **three-way table** is a *three-way table* that reports the frequencies of each combination of levels of three categorical variables. As the example illustrates, a three-way table is presented as several two-way tables, one for each level of the third variable. In Example 3.20, there is a separate table of patient outcome by hospital for each patient condition. The original two-way table is obtained by adding the corresponding entries in these tables, that is, by *aggregating* the data over patient conditions. Aggregating has the effect of ignoring patient condition, which then becomes a lurking variable. The peril of drawing incorrect conclusions from rates based on aggregated categorical data is so great—and so little understood—that special caution is required. Here are some real-life examples of this phenomenon.

- In 1979 the National Science Foundation conducted a study of persons who received a degree in science or engineering in 1977–78. The study found that at the bachelor's degree level the average salary of women with full-time jobs was only 77% of the average male salary. However, women's salaries in every individual field of science and engineering were at least 92% of male salaries. The aggregate data suggested a wide sex difference in salaries because women were concentrated in the life and social sciences, where salaries are lower than in the physical sciences and engineering.[15]
- Comparisons of the actual rate of federal income tax paid by individuals in different years often show that the average tax rate has increased with time, even though the rate in every income category has decreased. Why? Because inflation has moved more taxpayers into higher income brackets which are taxed at higher rates.[16]
- At one point during World War II, the percentage of women among U.S. industrial workers was rising in every individual industry, but falling overall. Total employment was rising most rapidly in heavy industry, where few women were employed, so male employment rose faster even though women were gaining in every industry.[17]

The effects of lurking variables are often profound; sometimes they may even reverse the direction of an association based on aggregated data. How far to go in examining unreported variables is a matter of judgment, and even a matter for controversy. A charge of employment discrimination based on data showing adverse treatment of women or minorities, for example, will often be met by a defense claiming that other variables such as education and work experience explain the apparent discrimination. As Example 3.20 illustrates, the question is certainly not settled by first appearances. Careful examination of all relevant variables is needed to uncover the truth.

SUMMARY

A **two-way table** of counts or percents describes the relationship between two categorical variables. Two-way tables are often used to summarize large amounts of data by grouping outcomes into categories.

A two-way table allows us to calculate the **marginal distribution** of each variable alone from the row sums and column sums. It also allows us to obtain the **conditional distribution** of one variable given a specific level of the other by considering table entries as proportions of their row or column sums. A **segmented bar graph** provides a visual comparison for a set of conditional distributions.

Relationships among three categorical variables can be described by a **three-way table**, which is printed as separate two-way tables for each level of the third variable. A comparison between two variables that holds for each level of a third variable can be changed or even reversed when the data are aggregated by summing over all levels of the third variable. **Simpson's paradox** refers to the reversal of a comparison by aggregation.

SECTION 3.4 EXERCISES

3.60 A survey was designed to study how the operations of a group of businesses vary with their size. Companies were classified as small, medium, and large. Questionnaires were sent to 200 randomly selected businesses of each size, for a total of 600 questionnaires. Since not all questionnaires in a survey of this type are returned, it was decided to examine whether or not the response rate varied with the size of the business. The data are given in the following two-way table:

Size	Response	No response	Total
Small	125	75	200
Medium	81	119	200
Large	40	160	200

(a) Compute the response percentage for all 600 companies.

(b) Compute the response percentage for each of the three sizes of companies.

(c) Draw a bar graph that compares the response percentage for the three sizes of companies.

(d) Using the total number of responses as a base, compute the percent of responses that come from small businesses. Repeat the calculation for the medium and large businesses.

(e) The sampling plan was designed to obtain equal numbers of responses from small, medium, and large companies. In preparing an analysis of the survey results, do you think it would be reasonable to proceed as if the responses represented companies of each size equally?

3.61 A survey of commercial poultry raising operations was conducted to assess the extent to which rodents are a problem in poultry houses and to find out the means used to control the problem. The operations surveyed were classified into two types (egg and turkey), and the extent of the problem was classified as mild, moderate, or severe. The numbers of responses in each category are given in the following two-way table. (Data taken from R. M. Corrigan and R. E. Williams, "The house mouse in poultry operations: pest significance and a novel baiting strategy for its control," *Proceedings of the Twelfth Vertebrate Pest Conference*, 1986, pp. 120–126.)

Type	Mild	Moderate	Severe
Egg	34	33	7
Turkey	22	22	4

(a) Give the overall (marginal) percentages for mild, moderate, and severe problems.
(b) Give the percentages as in (a) for each type of operation. (These are the conditional distributions of the extent of the problem, given the type of operation.)
(c) Construct a segmented bar chart showing how the extent of the problem varies with the type of operation.
(d) Do you think it is reasonable to ignore the type of operation in discussing rodent problems in poultry houses? Explain your answer.

3.62 The following two-way table shows the relationship between the age of American civilians and their participation in the 1984 presidential election. Each column refers to an age group and contains the *percent* of that age group who voted in 1984, the percent who registered but did not vote, and the percent who did not register. That is, the percents in each column add to 100% of an age group. The data appear in this form in government publications.

	Age Group				
Voting status	18–20	21–24	25–44	45–64	≥65
Voted	36.7	43.5	58.4	69.8	67.7
Registered only	10.3	10.8	8.2	6.8	9.2
Did not register	53.0	45.7	33.4	23.4	23.1

(a) Display these data in a segmented bar graph that compares the voting behavior of the five age groups.
(b) Based on your graph, write a brief description of the relationship between voting behavior and age.
(c) If you were planning a political campaign, you would be interested in the percent of each age group among registered

voters. Can that percent be computed from the data in the table? If so, do it. If not, what information is lacking?

3.63 The following two-way table shows the relationship between the income of American households in 1985 and the educational level of the householder. The educational categories are the same as in Figure 3.33. Household income is recorded in thousands of dollars. The table entries are the counts of households (in thousands) in each education by income class.

| Education | Income class | | | | |
	<15	15–34	35–49	≥50	Total
1–3 years of high school	13,151	7564	1814	874	23,402
4 years of high school	9465	13,463	5217	3126	31,271
1–3 years of college	3427	6486	2983	2416	15,300
≥4 years of college	1885	5989	3937	6636	18,485

13052   78458

(a) Find the sum of the counts in the first row of the table. Why do you think that this sum differs slightly from the total given for that row?

(b) Compute (in percents) the conditional distribution of income among persons who did not complete high school.

(c) Compute the conditional distribution of income in each of the remaining three education classes.

(d) Make a segmented bar graph that shows how income varies with educational level.

(e) Write a brief description of the relationship between household income and education.

3.64 You are studying low-income households, defined as those with income less than $15,000. Use the data in the previous exercise to compute the distribution (in percent) of educational levels in low-income households. (This is the conditional distribution of education, given that the household has low income.) Describe your results in words.

3.65 Use the data in Exercise 3.63 to compute (in percents) the two marginal distributions, the distribution of householder education among all households, and the distribution of income among all households. Which income class included the greatest number of American households in 1985?

3.66 Many studies have suggested a link between high blood pressure and cardiovascular disease. A group of white males aged 35 to 64 were classified on systolic blood pressure as "low" (less than 140 mm Hg) or "high" (140 mm Hg or higher), and then followed for 5 years.

The following table gives the number of men in each blood pressure category and the number of deaths from cardiovascular disease during the 5-year period. (Data taken from J. Stamler, "The mass treatment of hypertensive disease: defining the problem," *Mild Hypertension: To Treat or Not to Treat*, New York Academy of Sciences, 1978, pp. 333–358.)

Blood pressure	Deaths	Total
Low	21	2676
High	55	3338

(a) Calculate the mortality rate (deaths as a fraction of the total) for each group of men.

(b) Do these data support the idea that there is a link between high blood pressure and death from cardiovascular disease? Explain your answer.

3.67 The following two-way table categorizes suicides committed in 1983 by the sex of the victim and the method used ("hanging" also includes strangulation and suffocation). Based on these data, write a brief account of differences in suicide between males and females. Be sure to cite appropriate counts or percents to justify your statements.

Method	Male	Female
Firearms	13,959	2641
Poison	3148	2469
Hanging	3222	709
Other	1457	690
Total	21,786	6509

3.68 Upper Wabash Tech has two professional schools, business and law. Here is a three-way table of applicants to these professional schools, categorized by sex, school, and admission decision. (Although these data are fictitious, similar though less simple situations occur in reality. See P. J. Bickel and J. W. O'Connell, "Is there a sex bias in graduate admissions?" *Science*, 187 (1975), pp. 398–404.)

	Business			Law	
	Admit	Deny		Admit	Deny
Male	480	120	Male	10	90
Female	180	20	Female	100	200

(a) Make a two-way table of sex by admission decision for the combined professional schools by summing entries in the three-way table.

(b) Compute separately the percents of male and female applicants admitted from your two-way table. Upper Wabash Tech's professional schools admit a higher percent of male applicants than of female applicants.

(c) Now compute separately the percents of male and female applicants admitted by the business school and by the law school. Each school admits a higher percent of female applicants.

(d) Explain carefully, as if speaking to a skeptical reporter, how it can happen that Upper Wabash appears to favor males when each school individually favors females.

3.69 In 1987 Reggie Jackson retired from baseball after a career of 21 years in the major leagues. He had a reputation for playing particularly well in crucial situations such as World Series games. For regular season play and World Series games the following table gives the numbers of Reggie's times at bat and hits:

	Hits	At bats
Regular season	2584	9864
World Series	35	98

(a) Compute Reggie's batting average for all games combined. (A player's batting average is the proportion of times that a player gets a hit, i.e., "Hits" divided by "At bats.")

(b) Compute Reggie's batting average for regular season and World Series games separately. Draw a bar graph to compare the two batting averages. Do the results support the idea that he played better than usual in World Series games?

3.70 A study of the salaries of full professors at Upper Wabash Tech showed that the median salary for female professors was considerably less than the median male salary. Further investigation showed that the median salaries for male and female full professors were about the same in every department (English, physics, etc.) of the university. Explain how equal salaries in every department can still result in a higher overall median salary for men.

3.71 Some baseball experts believe that the chance of a player getting a hit depends on whether the pitcher throws right-handed or left-handed. Consider two players, Joe and Moe, both of whom bat right-handed. Their performance against right-handed and left-handed pitchers is recorded in the following table.

Player	Pitcher	Hits	At bats
Joe	Right	40	100
Joe	Left	80	400
Moe	Right	120	400
Moe	Left	10	100

(a) Construct a two-way table of player (Joe or Moe) versus outcome (hit or no hit) summed over the different types of pitching.

(b) Compute the overall batting average (hits divided by total times at bat) for each player. On the basis of these calculations, who is the better hitter?

(c) For each of the two types of pitching, construct a two-way table of player by outcome (hit or no hit). When arranged side by side, these two two-way tables form a three-way table of player by pitcher type by outcome.

(d) Using the table in (c) compute batting averages to compare the two players for each type of pitching. Who is the better hitter against right-handed pitching? Who is the better hitter against left-handed pitching?

(e) Summarize the results of your calculations and explain them as if you were speaking to a skeptical manager. Which hitter would you prefer to have on your team?

3.72 The influence of race on imposition of the death penalty for murder has been much studied and contested in the courts. The following three-way table classifies 326 cases in which the defendant was convicted of murder. The three variables are the defendant's race, the victim's race, and whether or not the defendant was sentenced to death. (Data from M. Radelet, "Racial characteristics and imposition of the death penalty," *American Sociological Review*, 46 (1981), pp. 918–927.)

White defendant	Death penalty		Black defendant	Death penalty	
	Yes	No		Yes	No
White victim	19	132	White victim	11	52
Black victim	0	9	Black victim	6	97

(a) From these data make a two-way table of defendant's race by death penalty.

(b) Show that Simpson's paradox holds: A higher percent of white defendants are sentenced to death overall, but for both black and white victims a higher percent of black defendants are sentenced to death.

(c) Basing your reasoning on the data, explain why the paradox holds in language that a judge could understand.

3.5 THE QUESTION OF CAUSATION

In many studies of the relationship between two variables, the goal is to establish that changes in the explanatory variable *cause* changes in the response variable. Even when a strong association is present, the conclusion that this association is due to a causal link between the variables is often elusive. Traffic fatalities dropped after the United States adopted a 55-mph speed limit in 1974—but did the reduced speed limit *cause* the decline in road deaths? Nations with strict gun controls generally have much lower murder rates than does the United States—but do gun controls *cause* a drop in murders? Such questions are difficult to answer because many variables other than the proposed explanatory variable influence the response. The reduced highway speed limit coincided with a many-fold increase in gasoline prices which discouraged driving and so lessened our exposure to road accidents. Nations differ profoundly in so many ways that isolating the effect of one factor such as gun control is probably impossible. The issues involved in establishing causation will be raised in this section by a discussion of a single example, the connection between cigarette smoking and death from lung cancer.

Smoking and Lung Cancer

The relationship between smoking and lung cancer has been observed and contested for over 50 years. Here is an example of the connection that has been observed.

EXAMPLE 3.21

Table 3.5 gives a summary of a study carried out by government statisticians in England.[18] The data concern 25 occupational groups and are condensed from data on thousands of individual men. The explanatory variable x is a smoking ratio—that is, a measure of the number of cigarettes smoked per day by men in each occupation relative to the number smoked by all men of the same age. This smoking ratio is 100 if men in an occupation are exactly average in their smoking, it is below 100 if they smoke less than average, and above 100 if they smoke more than average. The response variable y is the standardized mortality ratio for deaths from lung cancer. It is also measured relative to the entire population of men of the same ages as those studied, and is greater or less than 100 when there are more or fewer deaths from lung cancer than would be expected based on the experience of all English men.

Table 3.5 Indices of cigarette smoking and lung cancer mortality for English occupational groups

Occupational group	Smoking	Mortality
Farmers, foresters, and fisherman	77	84
Miners and quarrymen	137	116
Gas, coke, and chemical makers	117	123
Glass and ceramics makers	94	128
Furnace, forge, foundry, and rolling mill workers	116	155
Electrical and electronics workers	102	101
Engineering and allied trades	111	118
Woodworkers	93	113
Leather workers	88	104
Textile workers	102	88
Clothing workers	91	104
Food, drink, and tobacco workers	104	129
Paper and printing workers	107	86
Makers of other products	112	96
Construction workers	113	144
Painters and decorators	110	139
Drivers of stationary engines, cranes, etc.	125	113
Laborers not included elsewhere	133	146
Transport and communications workers	115	128
Warehousemen, storekeepers, packers, and bottlers	105	115
Clerical workers	87	79
Sales workers	91	85
Service, sport, and recreation workers	100	120
Administrators and managers	76	60
Professionals, technical workers, and artists	66	51

A scatterplot (Figure 3.34) of the data in Table 3.5 shows a moderately strong linear association. The correlation is $r = 0.716$. Because $r^2 = 0.51$, differences in smoking behavior among occupations explain 51% of the variation in deaths from lung cancer. The least squares regression line also appears in Figure 3.34. The residual plot in Figure 3.35 may show a slight curved pattern but this is by no means strong enough to lead us to abandon the linear relationship. The residuals are satisfactory in other respects: There are no obvious outliers or influential observations. The data demonstrate a clear connection between smoking and death from lung cancer. How should this connection be interpreted? ■

Because the data of Example 3.21 do not refer to individuals but are averaged over occupational groups, the correlation may be higher than would be the case for individuals (recall Figure 3.31). Many other studies, however, have shown a strong tie between cigarette smoking and the risk

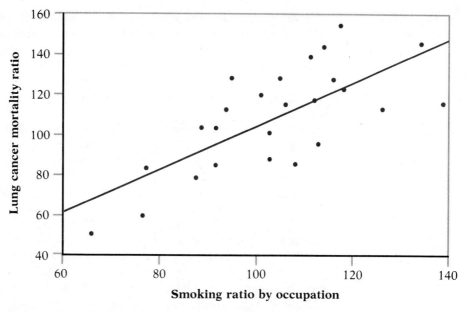

Figure 3.34   The regression of a measure of deaths from lung cancer on a measure of smoking for British occupational groups. See Example 3.21.

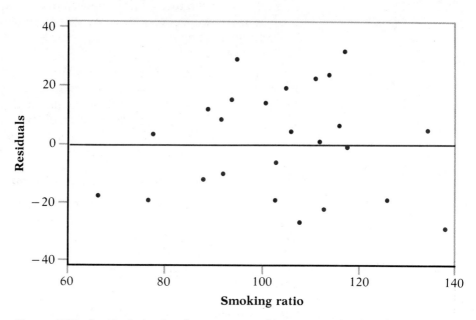

Figure 3.35   Residual plot for the regression of lung cancer deaths on smoking.

of death from lung cancer. The most convincing are *prospective studies* in which smokers and nonsmokers are identified and then followed for many years to observe their medical history and the eventual cause of death. Prospective studies have consistently found that heavy smokers are 20 or more times as likely to contract lung cancer as nonsmokers. Since the evidence linking smoking and lung cancer is so strong, we will not dwell on the peculiarities of Example 3.21, but instead seek explanations for a connection that has been clearly demonstrated to exist. The obvious explanation is that smoking causes lung cancer. However, we have seen enough of the effects of lurking variables to know that we must look further.

What unrecorded variables might explain the strong association between smoking and lung cancer? An old favorite is the "genetic hypothesis." According to this hypothesis, there may be a hereditary trait that predisposes some people both to nicotine addiction and to lung cancer. These people would have high rates of lung cancer even if they never smoked. Another possibility, less favored by the tobacco industry because of its unflattering picture of smokers, is the "sloppy lifestyle" hypothesis. Perhaps smoking is most prevalent among people who also drink too much, don't exercise, eat unhealthy food, and so on. If some other component of a sloppy lifestyle caused lung cancer, smoking would wrongly appear to be the culprit.

These alternative explanations remind us that a strong association between two variables x and y can reflect any of several underlying relationships.

- *Causation*: Changes in x cause changes in y—for example, a change in outdoor temperature causes changes in natural gas consumption for heating. If we can change x, we can bring about a change in y. Quitting smoking may reduce a person's chance of getting lung cancer if causation holds.

- *Common response*: Both x and y respond to changes in some unobserved variable or variables—for example, both GRE scores and graduate GPA respond to a student's ability and level of knowledge. In this case y can sometimes be predicted from x, but intervening to change x would not bring about a change in y. The genetic hypothesis claims that both smoking behavior and lung cancer are responses to a genetic predisposition; quitting smoking does not change heredity and will have no effect on future lung cancer.

- *Confounding*: The effect on y of the explanatory variable x is hopelessly mixed up with the effects on y of other variables, as when time in the mold was confounded with the mold temperature in Example 3.11. The "sloppy lifestyle" hypothesis claims that smoking is confounded with other types of behavior, so that we have no information about the effect of smoking alone on health.

Figure 3.36 illustrates these three kinds of relationship in schematic form. Both common response and confounding involve the influence of a lurking variable (or variables) z on the response variable y. The distinction

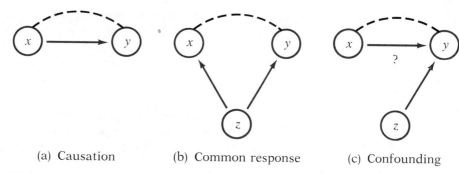

(a) Causation   (b) Common response   (c) Confounding

Figure 3.36   Variables x and y show a strong association (dashed line). This association may be the result of any of several causal relationships (solid arrows). (a) Causation: Changes in x cause changes in y. (b) Common response: Changes in both x and y are caused by changes in a lurking variable z. (c) Confounding: The effect (if any) of x on y is confounded with the effect on y of a lurking variable z.

between these two types of relationship is less important than the common element, a lurking variable. We reserve "common response" for the situation in which z clearly influences both x and y. Confounding simply means that the influences of x and z on y cannot be distinguished from each other.

EXAMPLE 3.22   An article in a women's magazine reported that women who nurse their babies feel more receptive toward their infants than mothers who bottle-feed. The author concluded that breast-feeding (x) leads to a more positive attitude (y) toward the child. But women choose whether to nurse or bottle feed and this choice may reflect already existing attitudes (z) toward their infants. That is, mothers who already feel more positive about the child may choose to nurse, while those to whom the baby is a nuisance more often choose the bottle. The effects of the explanatory variable (type of feeding) on the response (mother's attitude) are confounded with the effects of the mother's already established attitude. This confounding means that we do not know whether breast-feeding causes mothers to feel more receptive toward their infants.   ■

Establishing Causation

Common response and confounding are active competitors of causation in explaining observed associations between explanatory and response variables. How can a direct causal link between x and y be established? The best method—indeed, the only fully compelling method—of establishing causation is to conduct a carefully designed *experiment* in which the effects of possible lurking variables are controlled. To experiment means to actively change x and observe the response y. Much of Chapter 4 is devoted to the art of designing convincing experiments. We can imagine selecting people at birth, forcing some to smoke and others to abstain and observing the subjects until death. Such an experiment would settle the issue, since smoking or not smoking would be imposed on the subjects independent of

experiment

their heredity or lifestyle. It is because experiments such as this cannot be done—because people *choose* whether or not to smoke—that the question of whether cigarette smoking causes lung cancer remains arguable.

When experiments are not possible, good evidence for causation is less direct and requires a combination of several factors.

- The association between x and y must be observed in many studies of different types among different groups. This reduces the chance that the observed association is due to a defect in one type of study or a peculiarity in one group of subjects.

- The association must continue to hold when the effects of plausible third variables are taken into account—for example, by looking at three-way tables.

- There must be a plausible explanation for a direct influence of x on y, so that the causal link does not depend on the observed association alone. Equally plausible explanations linking y to other variables z that might be involved in confounding or common response should be absent.

The claim that cigarette smoking causes lung cancer meets these criteria. A strong association has been observed in numerous studies of many types in many countries. Consistent results are given by prospective studies that follow individuals over time and by cross-sectional studies of diverse groups at a fixed point in time (such as Example 3.21). Many possible sources of common response or confounding have been examined in these studies and have not been found to explain the association between smoking and lung cancer. In particular, the "sloppy lifestyle" hypothesis is not supported by the evidence. As far as the genetic hypothesis is concerned, attention has recently even been given to nonsmokers who are exposed to tobacco smoke from nearby smokers. It has been found that lifelong nonsmokers who live with smokers get lung cancer 35% more often than those who do not live with smokers.[19] The genetic hypothesis cannot explain this effect. Nor can it account for the fact that lung cancer among women, once rare, has increased in step with smoking among women. There is a lag of about 30 years between increased smoking and increased lung cancer. Long the leading cause of cancer deaths among men, lung cancer is about to surpass breast cancer as the cancer that kills the most women (see Exercise 2.54 for data).

Moreover, animal experiments have demonstrated conclusively that tobacco smoke contains substances that cause tumors. Some of these experiments have even attached animals to "smoking machines" so that they breathe cigarette smoke. Because the smoke contains substances that in high concentrations cause tumors in animals exposed to it for a short time, it is plausible that lower concentrations over many years cause tumors in human lungs. There is therefore a known pathway by which smoking could cause lung cancer. No similar pathway has been discovered for alternative explanations.

This evidence is quite conclusive. Most physicians are now convinced that smoking causes lung cancer and has other serious consequences as well. The evidence for causation here is about as strong as nonexperimental evidence can be. Though association does not always indicate causation, sometimes it does. Nonetheless, the evidence linking cigarette smoking to lung cancer is not as strong as the evidence of causation that can be provided by well-designed experiments.

Many important issues in which statistical evidence plays a central role concern relationships, like that between smoking and health, that cannot be clarified by experiments. It is easy to claim either too little or too much for the role of statistics in understanding these issues. This chapter has begun to equip you with both effective tools for understanding relationships and with cautions about the use of these tools.

SUMMARY

An observed association between two variables can be due to **causation**, **common response**, or **confounding**. Both common response and confounding involve the effect of other variables on the response.

That an association is due to causation is best established by an **experiment** in which the explanatory variable is directly changed and other influences on the response are controlled.

In the absence of experimental evidence, causation should be only cautiously accepted. Good evidence of causation requires that the association be observed in many varied studies, that examination of the effects of other variables not remove the association, and that a clear explanation for the alleged causation exist.

SECTION 3.5 EXERCISES

Each of Exercises 3.73 through 3.77 reports an observed association. In each case, suggest possible explanations for the association other than a causal relationship between x and y. Express your suggestion in a diagram like those in Figure 3.36. (Note that more than one explanation can contribute to a single association. Even if x does cause changes in y, a second explanation can also be present.)

3.73 A study of elementary school children, ages 6 to 11, finds a high positive correlation between shoe size x and score y on a test of reading comprehension.

3.74 There is a strong positive association between educational level x and income y of American adults. A 1987 Census Bureau study found that the median monthly income of high school dropouts was $693 per month, while high school graduates averaged $1045 a month, and holders of a bachelor's degree earned $1841. Is this entirely because more education opens the door to better jobs? (From an article in *The New York Times*, Oct. 3, 1987.)

3.75 The National Halothane Study was a major investigation of the safety of the anesthetics used in surgery. Records of over 850,000 operations performed in 34 major hospitals showed the following death rates y for four common anesthetics x. (See L. E. Moses and F. Mosteller, "Safety of anesthetics," in J. M. Tanur et al. (eds.), *Statistics: A Guide to the Unknown*, 2d ed., Holden-Day, San Francisco, 1978, pp. 16–25.)

Anesthetic	A	B	C	D
Death rate	1.7%	1.7%	3.4%	1.9%

It is possible that the anesthetics cause some deaths, and that anesthetic C is particularly dangerous.

3.76 The Bureau of Labor Statistics conducts extensive surveys of the earnings of workers in various job categories. One such survey found that in the South Bend, Indiana metropolitan area the median earnings of Class D secretaries (low-level clerical workers) were higher than the median for Class B secretaries, who have more skill and more responsible jobs. This negative association between job level x and pay y seems puzzling, since each individual employer pays Class B secretaries more. (From V. L. Ward, "Measuring wage relationships among selected occupations," *Monthly Labor Review*, May 1980.)

3.77 A study shows that there is a clear positive correlation between the size of a hospital (measured by number of beds x) and the median number of days y that patients remain in the hospital. Are the large hospitals padding their bills by keeping patients longer?

3.78 A group of college students believes that herb tea has remarkable restorative powers. To test this belief, they make weekly visits to a local nursing home, visiting with the residents and serving them herb tea. The nursing home staff reports that after several months many of the residents are more cheerful and healthy. A skeptical sociologist commends the students for their good deeds but scoffs at the idea that herb tea helped the residents. It's all a matter of confounding, says the sociologist. Identify the explanatory and response variables in this informal study. Then explain what other variables are confounded with the explanatory variable.

3.79 Members of a high school language club believe that study of a foreign language improves a student's command of English. From school records, they obtain the scores on an English achievement test given to all seniors. The average score of seniors who had studied a foreign language for at least two years is much higher than the average score of seniors who studied no foreign language. The club's advisor says that she also thinks language study helps English but that the data are not good evidence for this effect. Identify the explanatory and response variables in this study. Then explain why confounding prevents the conclusion that language study improves students' English scores.

3.80 There is an observed association between the level of cholesterol in the blood and the formation of deposits in the arteries, which in turn increases the risk of a heart attack. Blood cholesterol levels may be influenced by many factors, including heredity and diet. It is suspected that diets high in red meat, eggs, and dairy products may cause high cholesterol levels. Outline the kinds of information you would want in order to produce evidence for or against the claim that this kind of diet raises blood cholesterol.

3.81 Return to the study of the safety of anesthetics from Exercise 3.75. Suppose that, for ethical reasons, surgeons are unwilling to do an experiment in which similar patients are given different anesthetics. However, you have very detailed records for the 850,000 operations that were investigated in the study cited in Exercise 3.75. What kinds of information would you seek in these records in order to establish that anesthetic C does or does not cause extra deaths among patients? Is there any other information that might be helpful?

NOTES

1 The data are from the *Statistical Abstract of the United States*, 1987.

2 A sophisticated treatment of improvements and additions to scatterplots is W. S. Cleveland and R. McGill, "The many faces of a scatterplot," *Journal of the American Statistical Association*, 79 (1984), pp. 807–822.

3 The data are from W. L. Colville and D. P. McGill, "Effect of rate and method of planting on several plant characters and yield of irrigated corn," *Agronomy Journal*, 54 (1962), pp. 235–238.

4 The draft lottery is analyzed in detail by S. E. Fienberg, "Randomization and social affairs: the 1970 draft lottery," *Science*, 171 (1971), pp. 255–261. The design of the 1971 lottery is described in another article in that same issue.

5 These data, for a residence in West Lafayette, Indiana, were supplied by Professor Robert Dale, Purdue University, West Lafayette, Ind.

6 Data provided by Dr. Peter Cook, Department of Mathematics, Purdue University, West Lafayette, Ind.

7 These data were originally collected by L. M. Linde of UCLA, but were first published by M. R. Mickey, O. J. Dunn, and V. Clark, "Note on the use of stepwise regression in detecting outliers," *Computers and Biomedical Research*, 1 (1967), pp. 105–111. The data have been used by several authors. We found them in N. R. Draper and J. A. John, "Influential observations and outliers in regression," *Technometrics*, 23 (1981), pp. 21–26.

8 This example is modeled on one cited by B. L. Joiner, "Lurking variables: some examples," *The American Statistician*, 35 (1981), pp. 227–233. This article contains excellent advice on detecting lurking variables in regression studies with many explanatory variables.

9 A careful study of this phenomenon is W. S. Cleveland, P. Diaconis, and R. McGill, "Variables on scatterplots look more highly correlated when the scales are increased," *Science*, 216 (1982), pp. 1138–1141.

10 This example is drawn from M. Goldstein, "Preliminary inspection of multivariate data," *The American Statistician*, 36 (1982), pp. 358–362.

11 Kenneth M. Wilson, *The Validation of GRE Scores as Predictors of First-Year Performance in Graduate Study*, Educational Testing Service, Princeton, N.J., 1979, p. 21.

12 The data are from the U.S. Bureau of the Census as reported in *The World Almanac and Book of Facts*, 1986.

13 An alternative to segmented bar charts is presented in William S. Cleveland, *The Elements of Graphing Data*, Wadsworth, Monterey, Calif., 1985, pp. 259–262.

14 A clear and comprehensive discussion of numerical measures of association for categorical data appears in chapter 3 of Albert M. Liebetrau, *Measures of Association*, Sage Publications, Beverly Hills, Calif., 1983.

15 National Science Foundation, *Science Indicators 1980*, U.S. Government Printing Office, 1981, pp. 152–153.

16 A specific example comparing the years 1974 and 1978 appears in C. H. Wagner, "Simpson's paradox in real life," *The American Statistician*, 36 (1982), pp. 46–48.

17 This example is reported by Nathan Mantel in a letter to *The American Statistician*, 36 (1982), p. 395.

18 The original source for these data is *Occupational Mortality: The Registrar General's Decennial Supplement for England and Wales, 1970–1972*, Her Majesty's Stationery Office, London, 1978. We found the data in a course of The Open University, *Statistics in Society*, Unit C4, The Open University Press, Milton Keynes, England, 1983.

19 See "Involuntary smokers face health risks," *Science*, 234 (1986), p. 1066, for this and other evidence on the risks of passive smoking drawn from a report of the National Research Council.

CHAPTER 3 EXERCISES

Some of these exercises involve extensive computation and plotting. They are intended for students who have access to a statistical computing system.

3.82 Table 1.4 gives the sodium content (in milligrams) and the number of calories for each of 20 brands of beef hot dogs. We examined the distribution of calories in Example 1.11 and the distribution of sodium in Exercise 1.58, comparing these distributions with those found for meat hot dogs and poultry hot dogs. Now we wish to know whether sodium and calories are related. In particular, we suspect that hot dogs that are high in calories are also high in sodium. The data are entered into the MINITAB statistical system, calories as the variable BCAL and sodium as BSOD. We then compute descriptive measures as follows:

```
MTB > MEAN 'BCAL'
    MEAN = 156.85

MTB > STD 'BCAL'
    ST. DEV. = 22.642

MTB > MEAN 'BSOD'
    MEAN = 401.15

MTB > STD 'BSOD'
    ST. DEV. = 102.435

MTB > COR 'BCAL' 'BSOD'
    CORRELATION OF BCAL AND BSOD = 0.8871
```

(a) Plot sodium (y) against calories (x) for the 20 brands of beef hot dogs in Table 1.4. Describe the overall pattern of the relation and any major deviations from the pattern. In particular, do brands that are high in calories tend also to be high in sodium?

(b) Use the information in the MINITAB output to find the equation of the least squares regression line of sodium on calories. Add this line to your graph in (a). Are there any conspicuous outliers or influential observations? What percent of the variation in sodium level among brands can be explained by the linear relation between sodium and calories?

(c) A new brand of beef hot dogs has 180 calories per frank. Predict its sodium content.

3.83 With the aid of statistical software, we can explore in more detail the relation between calories and sodium in hot dogs. Enter the data for all three types of hot dogs into your computer.

(a) Meat and beef hot dogs are quite similar. Combine the data for these types of hot dogs and make a plot of sodium (y) against calories (x). In the discussion of Example 1.11 we noted that the meat hot dog data contain an outlier in x, a veal hot dog that has unusually few calories. Mark the point for this brand on your scatterplot. Does it appear that it will be influential when the regression line is fit?

(b) Compute the least squares regression line of sodium on calories for all cases, and then for all cases except the veal hot dog marked in (a). Draw both lines on your scatterplot. Was the case noted highly influential?

(c) Figure 1.9 suggests that poultry hot dogs are quite different from the other types, at least in the distribution of calories. Make a new plot of sodium against calories, with meat and beef hot dogs plotted using one symbol and poultry hot dogs plotted using a different symbol. Describe the major differences (if any) in the calorie-sodium relationship for the two groups plotted.

(d) Compute the least squares regression line of sodium on calories for poultry hot dogs. Draw this line and the regression line for meat and beef hot dogs on your graph from (c). Give the r^2 values for both regressions. If the regression lines are used to predict sodium from calories, for which group of hot dogs will the prediction be more reliable? Would you be willing to use a single regression line to predict sodium from calories for all types of hot dogs?

3.84 Here are the mean Scholastic Aptitude Test mathematics (SAT-M) and verbal (SAT-V) scores for each of seven ethnic or racial groups in 1985. (College Board data reported in the *Chronicle of Higher Education*, Sept. 3, 1986.)

Group	SAT-V	SAT-M
American Indians	392	428
Asian-Americans	404	518
Blacks	346	376
Mexican-Americans	382	426
Puerto Ricans	368	409
Whites	449	490
Others	391	448

(a) Make a scatterplot that is appropriate for examining how well SAT math scores can be predicted from SAT verbal scores.

(b) The plot shows a generally linear pattern with one outlier. Mark the outlier on your graph. Which group produced the outlier?

Explain why you think the math and verbal scores for this group fall outside the overall linear pattern.

(c) The correlation between SAT-M and SAT-V is $r = 0.821$. What percent of the variation in average SAT-M scores among ethnic groups can be explained by variation in their average SAT-V scores? Will the correlation between SAT-M and SAT-V for *individual students* probably be lower or higher than the r for groups? Why?

$r^2 = (.821)^2$ *more variation*

(d) Compute the correlation between SAT-M and SAT-V for the six groups that remain when the outlier noted in (b) is omitted. Explain why your new r is higher than the r from (c).

3.85 The same article that contains the data in the previous exercise also reports median SAT scores for different income classes. Here are the medians for SAT-V:

Family income	Percentage of students	SAT-V
Under $6000	3.6	350
$6000–11,999	7.9	376
$12,000–17,999	10.2	398
$18,000–23,999	12.1	413
$24,000–29,999	11.7	427
$30,000–39,999	18.6	434
$40,000–49,999	13.4	446
$50,000 and over	22.6	465

(a) Draw the median trace for a scatterplot of SAT-V against family income. (The individual points on the scatterplot cannot be plotted because we do not have individual scores and incomes, but there is enough information to draw a median trace using the given income classes. For convenience, assume that the top income class ends at $100,000.) Describe the nature of the relationship between SAT-V scores and family income.

(b) The data in the previous exercise show an association between ethnic group and SAT scores. A possible partial explanation is that the SAT exams reflect the culture of the largest group (whites), who therefore have an advantage on the tests. The data of this exercise suggest another possible explanation for the association. Explain carefully, using the language of Figure 3.36.

3.86 Return once more to the SAT data presented in Exercise 3.84.

(a) The least squares regression line of SAT-M on SAT-V is

$$SAT\text{-}M = -40.962 + 1.238 SAT\text{-}V$$

Draw this line on a scatterplot of the data. Mark the two observations that appear most unusual.

(b) The residuals for this regression are (in the same order that the data appear in Exercise 3.84)

$$-16.3 \quad 58.9 \quad -11.3 \quad -5.9 \quad -5.6 \quad -24.8 \quad 5.0$$

Make a residual plot, and mark the two unusual points from (a). Both are potentially influential.

(c) Compute the least squares regression lines for the SAT data with each of the two unusual points omitted in turn (each regression is therefore based on six cases). Draw both lines on your plot from (a) for comparison with the regression line from all cases. Which case was more influential in changing the regression line?

3.87 What is the relationship between the number of home runs that a major league baseball team hits and its team batting average? On one hand, we might expect better hitting teams to have both more home runs and higher batting averages. For individual players this is true: The correlation is about $r = 0.2$. On the other hand, teams have different styles of play and play their home games in different stadiums. Some may favor the home run, while other teams rely on singles and speed. The following table gives the data for American League teams in the 1987 season. Analyze the data and report your conclusions.

Team	Home runs	Batting average
Boston	174	.277
Milwaukee	163	.276
Seattle	161	.272
Detroit	230	.272
Toronto	214	.269
Texas	194	.266
Cleveland	187	.263
Kansas City	167	.262
New York	196	.262
Minnesota	196	.261
Oakland	199	.259
Chicago	173	.258
Baltimore	213	.257
California	172	.252

3.88 Exercises 3.21 and 3.22 give data on the stride rates for female and male runners at various speeds.

(a) Plot the data for both groups on one graph using different symbols to distinguish between the points for females and those for males.

(b) In earlier exercises, we computed separate regression lines for female and male runners. Suppose now that the data came to you without identification as to sex. Compute the least squares line from all of the data and plot it on your graph.

(c) Compute the residuals from this line for each observation. Make a plot of the residuals against speed. How does the fact that the data come from two distinct groups show up in the residual plot?

3.89 Environmentalists, government officials, and vehicle manufacturers are all interested in studying the emissions produced when gasoline is burned in internal combustion engines. Of particular concern are hydrocarbons (HC), carbon monoxide (CO), and nitrogen oxides (NOX). Table 3.6 gives data for these three emissions for a sample of

Table 3.6 Amounts of three pollutants (in grams per mile) emitted by light-duty engines

EN	HC	CO	NOX	EN	HC	CO	NOX
1	.50	5.01	1.28	2	.65	14.67	.72
3	.46	8.60	1.17	4	.41	4.42	1.31
5	.41	4.95	1.16	6	.39	7.24	1.45
7	.44	7.51	1.08	8	.55	12.30	1.22
9	.72	14.59	.60	10	.64	7.98	1.32
11	.83	11.53	1.32	12	.38	4.10	1.47
13	.38	5.21	1.24	14	.50	12.10	1.44
15	.60	9.62	.71	16	.73	14.97	.51
17	.83	15.13	.49	18	.57	5.04	1.49
19	.34	3.95	1.38	20	.41	3.38	1.33
21	.37	4.12	1.20	22	1.02	23.53	.86
23	.87	19.00	.78	24	1.10	22.92	.57
25	.65	11.20	.95	26	.43	3.81	1.79
27	.48	3.45	2.20	28	.41	1.85	2.27
29	.51	4.10	1.78	30	.41	2.26	1.87
31	.47	4.74	1.83	32	.52	4.29	2.94
33	.56	5.36	1.26	34	.70	14.83	1.16
35	.51	5.69	1.73	36	.52	6.35	1.45
37	.57	6.02	1.31	38	.51	5.79	1.51
39	.36	2.03	1.80	40	.48	4.62	1.47
41	.52	6.78	1.15	42	.61	8.43	1.06
43	.58	6.02	.97	44	.46	3.99	2.01
45	.47	5.22	1.12	46	.55	7.47	1.39

46 engines of the same type. In the table EN is an engine identifier. Figure 3.9 is a plot of NOX against CO for these engines. With the aid of statistical software, a detailed study is feasible. (Data taken from T. J. Lorenzen, "Determining statistical characteristics of a vehicle emissions audit procedure," *Technometrics*, 22 (1980), pp. 483–493.)

(a) Plot HC against CO, HC against NOX, and NOX against CO.

(b) For each plot, describe the nature of the association. In particular, which pollutants are positively associated and which are negatively associated?

(c) Inspect your plots carefully and circle any points that appear to be outliers or influential observations. Next to the circles, indicate the EN identifier. Does any consistent pattern of unusual points appear in all three plots?

(d) Compute correlations for each pair of pollutants. If you found unusual points, compute the correlations both with and without these points.

(e) Compute the least squares lines corresponding to the three plots—use CO to predict HC and NOX, and NOX to predict HC.

(f) Compute the least squares lines omitting the unusual points as a check on the influence of these points.

(g) Summarize the results of your analysis. In particular, is any one of these pollutants a good predictor of another pollutant?

3.90 There are different ways to measure the amount of money spent on education. Average salary paid to teachers and expenditures per pupil are two commonly used measures. Table 3.7, based on information compiled by the National Education Association and reported in *The New York Times* on Nov. 8, 1986, gives values for these variables by state. The states are classified according to region: NE (New England), MA (Middle Atlantic), ENC (East North Central), WNC (West North Central), SA (South Atlantic), ESC (East South Central), WSC (West South Central), MN (Mountain), and PA (Pacific).

(a) Make a modified box plot for pay. Label any outlying cases (those plotted individually in the modified boxplot) with the state identifier. Do the same for spending. Do the distributions appear symmetric or skewed?

(b) Plot pay versus spending with spending on the *x* axis and pay on the *y* axis.

(c) Describe the pattern. Is there any association? If so, is it positive or negative? Explain why you might expect to see an association of this kind before looking at the data.

(d) On your plot, circle any outlying points found in (a) and any additional points that appear to be outliers in the scatterplot. Label the circled points with the state identifier. Comment on

Table 3.7 Average teachers' pay and per-pupil educational spending in the 50 states (in thousands of dollars)

State	Region	Pay	Spend	State	Region	Pay	Spend
Maine	NE	19.6	3.35	N.H.	NE	20.3	3.11
Vt.	NE	20.3	3.55	Mass.	NE	26.8	4.64
R.I.	NE	29.5	4.67	Conn.	NE	26.6	4.89
N.Y.	MA	30.7	5.71	N.J.	MA	27.2	5.54
Pa.	MA	25.9	4.17	Ohio	ENC	24.5	3.55
Ind.	ENC	24.3	3.16	Ill.	ENC	27.2	3.62
Mich.	ENC	30.2	3.78	Wis.	ENC	26.5	4.25
Minn.	WNC	27.4	3.98	Iowa	WNC	21.7	3.57
Mo.	WNC	22.0	3.16	N.Dak.	WNC	20.8	3.06
S.Dak.	WNC	18.1	2.97	Neb.	WNC	20.9	3.29
Kan.	WNC	27.6	3.91	Del.	SA	24.6	4.52
Md.	SA	27.2	4.35	D.C.	SA	34.0	5.02
Va.	SA	23.4	3.59	W.Va.	SA	20.6	2.82
N.C.	SA	22.8	3.37	S.C.	SA	21.6	2.92
Ga.	SA	22.1	2.98	Fla.	SA	22.3	3.73
Ky.	ESC	20.9	2.85	Tenn.	ESC	21.8	2.53
Ala.	ESC	22.9	2.73	Miss.	ESC	18.4	2.31
Ark.	WSC	19.5	2.64	La.	WSC	20.5	3.12
Okla.	WSC	21.4	2.75	Tex.	WSC	25.2	3.43
Mont.	MN	22.5	3.95	Idaho	MN	21.0	2.51
Wyo.	MN	27.2	5.44	Colo.	MN	25.9	4.04
N.Mex.	MN	22.6	3.40	Ariz.	MN	26.6	2.83
Utah	MN	22.3	2.30	Nev.	MN	25.6	2.93
Wash.	PA	26.0	3.71	Oreg.	PA	25.8	4.12
Calif.	PA	29.1	3.61	Alaska	PA	41.5	8.35
Hawaii	PA	25.8	3.77				

whether or not the points found in (a) are outliers from the overall relationship.

(e) Compute the least squares regression line for predicting pay from spending. Graph the line on your plot. Give a numerical measure of the success of overall spending on education in explaining variations in teacher pay among states.

(f) Repeat (e) excluding Alaska. Do any of your conclusions change substantially?

3.91 Refer to the previous exercise.

(a) Construct side-by-side boxplots for education spending in the nine regions. Note that if there are fewer than five states in a region, you will need to make some compromises. State clearly how you deal with this situation. For each region, label the high and low values with the state identifier.

(b) Repeat part (a) for the variable pay.

(c) Do you see appreciable differences in spending and pay by region? Are the differences consistent for the two variables, i.e., are regions that tend to be high in spending also high in pay and vice versa?

(d) Construct a new data set giving the median spending and median pay for the nine regions (this data set will have nine points). Plot these nine points on a graph of the original data. Use a different symbol to distinguish these points from the data for individual states.

(e) Compute the least squares regression line for median pay against median spending in the nine regions. Graph this line on the plot you produced in (d).

(f) How does this line compare to the original least squares line? In particular, is r^2 higher or lower for the regional medians than for the individual states?

3.92 Refer to the previous two exercises.

(a) Using the least squares line for all states, compute the residuals.

(b) Display the residuals by region using side-by-side boxplots. Draw a center line at zero.

(c) Do the residuals appear to differ by region? Describe any patterns that are noteworthy.

3.93 Wood scientists are conducting a research project to investigate the possibility of making wood products out of flakes of aspen and pine to use in place of conventional products. The following data were collected to examine the relationship between the length and the strength of some composite beams. Length is measured in inches and strength is expressed as pounds per square inch. (Data provided by Jim Bateman and Michael Hunt, Forestry Department, Purdue University, West Lafayette, Ind.)

Length	5	6	7	8	9	10	11	12	13	14
Strength	446	371	334	296	249	254	244	246	239	234

(a) Plot the data with length on the x axis and strength on the y axis.

(b) Describe the patterns that you see in the plot. Would it be reasonable to fit a straight line to the entire set of data? Explain why or why not.

(c) Are there one or more ranges of lengths where a straight-line fit would be reasonable? Explain your answer.

(d) Fit a least squares line to the entire set of data. Graph the line on your plot.

(e) Compute the residuals and plot them versus length. Describe the pattern. What do you conclude about the adequacy of your line as a summary of these data?

(f) Refer to (c). If you answered yes, construct least squares lines for each subset of the data and graph the lines on a plot of the original data.

(g) Summarize your analysis of this data set.

3.94 Recent studies have shown that earlier reports seriously underestimated the health risks associated with being overweight. The error was caused by overlooking some important lurking variables. In particular, smoking tends to both reduce weight and lead to earlier death. Illustrate Simpson's paradox by stating a simplified version of this situation. That is, make up a three-way table of overweight (yes or no) by early death (yes or no) by smoker (yes or no) such that:

- Overweight smokers and overweight nonsmokers both tend to die earlier than those who are not overweight.
- But when smokers and nonsmokers are combined into a two-way table of overweight by early death, persons who are not overweight tend to die earlier.

3.95 The following table gives the resident population of voting age and the votes cast for president, both in thousands, for presidential elections between 1960 and 1984:

Year	Population	Votes
1960	109,672	68,838
1964	114,090	70,645
1968	120,285	73,212
1972	140,777	77,719
1976	152,308	81,556
1980	164,381	86,515
1984	174,447	92,653

(a) For each year compute the percent of people who voted. Make a graph of the percent who voted against year and describe the change over time of participation in presidential elections.

(b) Before proposing political explanations for this change, we should examine possible lurking variables. Exercise 3.62 showed that voting behavior varies greatly with age. Use this fact to propose a partial explanation for the trend you saw in (a).

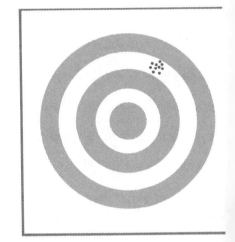

Prelude

Having learned some of the skills for exploring and describing data, we now turn to statistical methods used when producing our own data. Experiments produce many scientific data and sample surveys are popular among social scientists, government data collectors, and political pollsters. The deliberate use of chance in both experiments and sampling is one of the central ideas of statistics. We discuss not only the statistical techniques for designing reliable experiments and sample surveys, but some of the practical issues that must be dealt with in actual use.

- *How can a well-designed randomized comparative experiment give clear evidence of the effectiveness of a new medical treatment or discover better operating conditions for an industrial process?*

- *How can a randomly selected sample of only about 1500 persons give reliable information about, for example, the churchgoing habits of American adults as a group?*

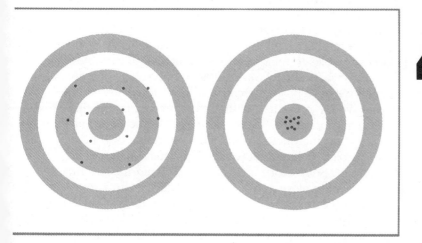

CHAPTER

4

Producing Data

Numerical data are the raw material for the growth of knowledge. Human efforts to understand the world have been most successful in areas of knowledge where we have learned what to measure and how to measure it. Statistics is a tool that helps measurement produce knowledge rather than confusion. As such, it must be concerned with producing data as well as with interpreting already available data.

Chapters 1 to 3 looked in some detail at the art of analyzing data. They showed how to uncover the nature of a set of data by applying numerical and graphical skills. The application of these skills is organized by strategies such as seeking an overall pattern and deviations from it, moving from graphs to numerical descriptions to mathematical models, and advancing from the distributions of individual variables to relationships among several variables.

It is helpful to distinguish between two purposes in analyzing data depending on whether or not we have posed specific questions to be answered from the data. Sometimes we are presented with data that seem interesting and that deserve careful exploration. We do not know what the data may show, and we may have no specific questions in mind before we begin our examination. Our inquiry into the behavior of world oil production over time (Section 2.3) was of this kind. Such work is often called *exploratory data analysis*, on the analogy of an explorer reporting the nature of unknown lands. Exploratory analysis relies heavily on graphical methods and seeks patterns in the data that suggest novel conclusions or questions for further study. However, exploratory analysis alone can rarely provide convincing evidence for its conclusions, since striking patterns in data can arise from many sources.

exploratory data analysis

The organized advance of human knowledge depends on people asking specific questions and producing data to answer these questions. Providing clear answers to questions posed in advance is a second purpose in analyzing data. "What is the speed of light?" (Section 1.1); "How many plants per acre give the highest yield of corn?" (Section 3.1); "Does smoking cigarettes cause lung cancer?" (Section 3.5). Data that answer such questions do not grow naturally. They are as much the product of intelligent effort as hybrid tomatoes and television sets. This chapter is devoted to developing the skills needed for producing trustworthy data and for judging the quality of data produced by others. The techniques of producing data discussed in this chapter are simple and involve no formulas or graphs, but they are among the most important ideas in statistics.

formal statistical inference

Statistical techniques for producing data open the door to a further advance in data analysis—*formal statistical inference*—which answers specific questions with a known degree of confidence. Statistical inference uses the descriptive tools already presented in combination with new kinds of reasoning. It is primarily numerical rather than graphical. The later chapters of this book are devoted to inference. Even in the setting of inference, however, the ideas of exploratory analysis retain their importance. Large data sets usually contain errors that exploration can uncover, and surprising results can lie hidden in the most routine study. The wise statistician always explores the data before proceeding to a formal analysis.

4.1 FIRST STEPS

The first questions you must answer when you plan the production of data are: ''What shall I measure?'' and ''How shall I measure it?'' That is, what variables (weight, number of tumors, aggressiveness) will be recorded and what instruments or methods (balance scale, pathologist's examination, written personality questionnaire) will be used to give numerical values to these variables? These are central questions in the development of any science but they are not questions to which statistics gives answers. Statistics is not completely a bystander in the discussion of measurement—it can help to describe the accuracy of measurements and often helps to clarify the meaning of a variable by showing its relation to other variables. But by and large, questions of measurement belong to the substantive fields of science, not to the methodological field of statistics. We will therefore take for granted that all the variables we work with have specific definitions and are satisfactorily measured.

designs

Statistics begins to play an essential role when the focus turns from measurements on individual cases (whether ball bearings, rats, or people) to arrangements for collecting data from many individuals. These arrangements or patterns for producing data are called *designs*. Some questions that a design must answer are: How many individuals shall we collect data from? How shall we select the individuals to be studied? If, as in many experiments, several groups of individuals are to receive different treatments, how shall we form the groups? If there is no systematic design for producing data, we may be misled by haphazard or incomplete data, or by confounding.

The Need for Design

How many homeless people are there in the United States? The homeless are visible in any large city, sleeping on heat vents and clutching the shopping bags that hold their few belongings. But is homelessness a massive problem or a matter affecting relatively few people? We need data. In the absence of data, it is tempting to draw conclusions based on observation of a few individual cases.

Anecdotal evidence

Anecdotal evidence is based on haphazardly selected individual cases, which often come to our attention because they are striking in some way. These cases need not be representative of any larger group of cases.

We can see and talk to individual street people. Basing their recommendation on the hardships of these individuals, advocates for the homeless urge large programs of subsidized housing to make homes available to the very poor. But quite different conclusions are also possible. Many individual street people are clearly mentally disturbed. Some experts therefore trace the problem of homelessness to the program of "deinstitutionalizing" the mentally ill, releasing them from large institutions into communities that were not prepared to offer care. What is needed, they say, is not a massive housing program but a mental health program on a much smaller scale. As these conflicting recommendations illustrate, anecdotal evidence is no guide to policy.

EXAMPLE 4.1

An effective response to the problem of homelessness requires data on the number of the homeless and their condition. Such data are not available from the usual sources. The U.S. Census, for example, counts only the population in dwelling units. In the absence of carefully designed data collection, estimates of the number of homeless people in the United States have ranged from about 250,000 to over 3 million. Guesses about the proportion of mentally disturbed people among the homeless have varied from 20% to 90%.

The first attempt to collect reliable data was made in Chicago. The Chicago Homeless Study searched a sample of several hundred blocks during the night, counting homeless people and interviewing them. The researchers also counted and interviewed people spending the night in shelters. Though the data apply only to Chicago, they support the view that the homeless are relatively few in number and often unable to function in society. The number of homeless in Chicago on any one night was estimated at 2000 to 3000 persons. About 80% of those interviewed had been in a mental institution, jail, or drug treatment center. The data reveal a small population with many severe problems that simple provision of housing will not address.[1]   ■

Example 4.1 illustrates the weakness of anecdotal evidence and also the fact that available data may be inadequate.

Available data

> Available data are data that were produced in the past for some other purpose but that may help answer a present question.

Available data from the library are the only data used in most student papers and reports. Since producing new data is expensive, available data are used whenever possible. Census data, though unable to reveal the size of the homeless population, are a staple source of information for social science, economics, and market research. However, the clearest answers to present questions are usually based on data produced to answer those specific questions. Statistical designs for producing data rely on either *sampling* or *experiments*.

Sampling

sampling

The Chicago Homeless Study set out to collect data to answer specific questions about the number and condition of homeless people. It was not possible to search the entire city of Chicago in a single night to count all homeless people. Instead, the researchers produced their data by using a statistical design to choose a part of the city to be searched. That is, they selected a *sample* of city blocks to represent the larger *population* of all blocks in the city. The idea of *sampling* is to study a part in order to gain information about the whole. Data are often produced by sampling a population of people or things. Opinion polls, for example, report the views of the entire country based on interviews with a sample of about 1500 people. Government reports on employment and unemployment are produced from a monthly sample of about 70,000 households. The quality of manufactured items is monitored by inspecting small samples on a regular basis.

census

In all of these cases, as in the Chicago Homeless Study, the expense of examining every item in the population makes sampling a practical necessity. Timeliness is another reason for preferring a sample to a *census*, which is an attempt to contact every individual in the entire population. We want information on current unemployment and public opinion next week, not next year. Finally, a carefully conducted sample is often more accurate than a census. Accountants, for example, sample a firm's inventory to verify the accuracy of the records. Attempting to count every last part in the warehouse would be not only expensive but also inaccurate. Bored people do not count carefully.

For conclusions based on a sample to be valid for the entire population, a sound design for selecting the sample is required. Sampling designs are discussed in Section 4.3.

Experiments

experiment

The purpose of sampling is to collect information about some population by selecting and measuring a sample from the population. The goal is a picture of the population disturbed as little as possible by the act of gathering information. An *experiment*, in contrast, deliberately imposes some treatment on the experimental units or subjects in order to observe the response. A merely observational study, even one based on a sound statistical sample, is a poor way to gauge the effect of an intervention. To see how nature responds to a change, we must actually impose the change.

EXAMPLE 4.2

A majority of adult recipients of welfare are mothers of young children. Observational studies of welfare mothers show that many of them are able to increase their earnings and leave the welfare system even though they have preschool children. Some of them take advantage of voluntary job training programs to improve their skills. Should participation in job-training and job-search programs be required of all able-bodied welfare mothers? Observational studies cannot tell us what the effects of such a policy would be. Even if the mothers studied are a properly chosen sample of all welfare recipients, those who undergo training and find jobs may differ in many ways from those who do not. They are observed to have more education, for example, but they may also differ in values and motivation, things that cannot be observed. To see if a required jobs program will help mothers escape welfare, such a program must actually be tried. Choose two similar groups of mothers when they apply for welfare. Require one group to participate in a job training program, while not offering the program to the other group. Comparing the income and work record of the two groups after several years will show whether requiring job programs has the desired effect.[2]   ■

In Example 4.2, the effect of voluntary training programs on success in finding work is *confounded* with the special character of mothers who seek out training. We have already seen that observational studies of the effect of one variable on another often fail because the explanatory variable is confounded with other variables. When our goal is to understand cause and effect, experiments are the only source of fully convincing data. Because experiments allow us to pin down the effects of specific variables of interest to us, they are the preferred method for gaining knowledge in science, medicine, and industry.

Experiments, like samples, require careful design. Because of the unique importance of experiments as a method for producing data, the discussion of statistical designs for data collection begins in Section 4.2 with the principles underlying the design of experiments.

SUMMARY

Data analysis is sometimes **exploratory** in nature. Exploratory analysis asks what the data tell us about the variables and their relations to each other. The conclusions of an exploratory analysis may not generalize beyond the specific data studied.

Statistical inference produces answers to specific questions, along with a statement of how confident we can be that the answer is correct. The conclusions of statistical inference are usually intended to apply beyond the specific cases studied. Successful statistical inference usually requires **production of data** intended to answer the exact questions posed.

Anecdotal evidence based on a few individual cases is rarely trustworthy. **Available data** collected for other purposes, such as census data, are sometimes helpful. Data intended to answer specific questions are usually produced by sampling or experimentation.

Sampling selects a part of a population of interest to represent the whole. **Experiments** are distinguished from observational studies by the active imposition of some treatment on the subjects of the experiments.

SECTION 4.1 EXERCISES

4.1 A letter to the editor of *Organic Gardening* magazine (August 1980) stated, "Today I noticed about eight stinkbugs on the sunflower stalks. Immediately I checked my okra, for I was sure that they'd be under attack. There wasn't one stinkbug on them. I'd never read that stinkbugs are attracted to sunflowers but I'll surely interplant them with my okra from now on."

Explain briefly why this anecdote does not provide good evidence that sunflowers attract stinkbugs away from okra. In your explanation, suggest some factors that might account for all of the bugs being on the sunflowers.

4.2 When the discussion turns to the pros and cons of wearing automobile seat belts, Herman always brings up the case of a friend who survived an accident because he was not wearing seat belts. The friend was thrown out of the car and landed on a grassy bank, suffering only minor injuries, while the car burst into flames and was destroyed. Explain briefly why this anecdote does not provide good evidence that it is safer not to wear seat belts.

4.3 Explain carefully why each of the following studies is *not* an experiment.

(a) The question of whether a radical mastectomy (removal of breast, chest muscles, and lymph nodes) is more effective than a simple mastectomy (removal of the breast only) in prolonging the life of women with breast cancer has been debated intensely. To study this question, a medical team examines the records of five large hospitals and compares the survival times after surgery of all women who have had either operation.

(b) It has been suggested that there is a "gender gap" in political party preference in the United States, with women more likely than men to prefer Democratic candidates. A political scientist selects a large sample of registered voters, both men and women, and asks them whether they voted for the Democratic or Republican candidate in the last Congressional election.

4.4 Even though the studies in Exercise 4.3 are not experiments, they have explanatory and response variables. What are these variables in each case?

4.5 A study of the effect of living in public housing on family stability and other variables in poverty-level households was carried out as follows. A list of all applicants for public housing during the previous year was obtained. Some applicants had been accepted, while others had been turned down by the housing authority. Both groups were interviewed and compared. Was this study an experiment? Why or why not? What are the explanatory and response variables in the study?

4.6 A study of the effect of abortions on the health of subsequent children was conducted as follows. The names of women who had had abortions were obtained from medical records in New York City hospitals. Birth records were then searched to locate all women in this group who bore a child within five years of the abortion. Then hospital records were examined again for information about the health of the newborn child. Was this study an experiment? Why or why not?

4.7 Some people believe that exercise raises the body's metabolic rate for as long as 12 to 24 hours, enabling it to continue to burn off fat after the workout has ended. In a study of this effect, subjects were asked to walk briskly on a treadmill for several hours. Their metabolic rates were measured before, immediately after, and 12 hours after the exercise. The study was criticized because eating raises the metabolic rate, and no record was kept of what the subjects ate after exercising. Was this study an experiment? Why or why not? What are the explanatory and response variables?

4.8 Before a new variety of frozen muffins is put on the market, it is subjected to extensive taste testing. People are asked to taste the new muffin and a competing brand, and to say which they prefer. (Both muffins are unidentified in the test.) Is this an experiment? Why or why not?

4.2 DESIGN OF EXPERIMENTS

A study is an experiment when we actually do something to people, animals, or objects in order to observe the response. Here is the basic vocabulary of experiments.

Experimental units, subjects, treatment

> The objects on which the experiment is done are the experimental units. When the units are human beings, they are called subjects. A specific experimental condition applied to the units is called a treatment.

Since the purpose of an experiment is to reveal the response of one variable to changes in other variables, the distinction between explanatory and

factor
level

response variables remains important. The explanatory variables in an experiment are often called *factors*. When more than one factor is present, each treatment is a combination of a specific value (often called a *level*) of each factor.

EXAMPLE 4.3

In a study of the absorption of a drug into the bloodstream, an injection of the drug (the treatment) is given to 25 persons (the experimental subjects). The response variable is the concentration of the drug in the subjects' blood, measured 30 minutes after the injection. This experiment has a single factor with only one level. If three different doses of the drug are injected, there is still a single factor (the dosage of the drug), but now with three levels. ∎

EXAMPLE 4.4

A mathematics education researcher is studying where in high school mathematics texts it is most effective to insert questions. She wants to know whether it is better to present questions as motivation before the text passage or as review after the passage. The result may depend on the type of question asked: simple fact, computation, or word problem. The researcher therefore prepares six versions of an instructional unit in elementary algebra. All versions have the same text except that questions are added to each version in a different fashion, according to the scheme of Figure 4.1.

Six groups of students are selected and each group studies a different version of the unit. The students are the subjects in the experiment. All of them are given the same homework assignment for the material and then all take the same test. The response variable is the test score. There are two factors, the position and the type of questions added to the reading material. Position has two levels (before, after) and type has three levels (simple fact, computation, word problem). Each of the six combinations of one level of each factor is a treatment. ∎

Both of these examples illustrate our basic vocabulary for talking about experiments. They also suggest the advantages of experiments over observational studies in which the researcher observes units or subjects and measures variables of interest but does not impose any treatment. Experimentation allows us to study the effects of the specific treatments that are of interest. Moreover, we can control the environment of the experimental units to hold constant factors that are of no interest to us, such as the text content and homework assignment in Example 4.4. In the ideal laboratory

	Factor B—Question type		
	Simple fact	Computation	Word problem
Before	1	2	3
After	4	5	6

Factor A—Question position

Figure 4.1 The treatments in a two-factor experimental design. (See Example 4.4.)

case, all outside factors are thus controlled. Like most ideals, such control is realized only rarely in practice and certainly not in the schoolroom. Nonetheless, a well-designed experiment makes it possible to draw conclusions about the effect of one variable on another.

A less obvious advantage of experiments is that we can effectively study the combined effects of several factors simultaneously. The interaction of several factors can produce effects that could not have been predicted from looking at the effects of each factor alone. In the absence of careful planning, the researcher of Example 4.4 might well have done a single-factor experiment, simply comparing the effectiveness of factual questions before versus after the reading. Later, if she realized that the type of question might be important, she might have repeated the experiment with word problems. These repeated single-factor experiments, done at different times and perhaps in different schools, are less reliable and require more effort than a single two-factor experiment.

Comparative Experiments

The design of an experiment first describes the layout of the treatments as combinations of levels of the factors, as illustrated by Figure 4.1. Laboratory experiments in the sciences and engineering often have a simple design with only a single treatment, which is applied to all of the experimental units. The design of such an experiment can be displayed as

$$\text{Treatment} \longrightarrow \text{Observation} \qquad (4.1)$$

or, if before-and-after measurements are made,

$$\text{Observation 1} \longrightarrow \text{Treatment} \longrightarrow \text{Observation 2} \qquad (4.2)$$

For example, we may subject a beam to a load (treatment) and measure its deflection (observation). Alas, when experiments are conducted in the field or with living subjects, simple designs such as (4.1) and (4.2) often yield invalid data. That is, it is impossible to tell whether or not the treatment had an effect on the units. An example will show what can go wrong.

EXAMPLE 4.5

Ulcers in the upper intestine are unfortunately common in modern society; they are the product of stress and of acids secreted in the stomach. Here is a clever treatment for ulcer patients: The patient swallows a deflated balloon with tubes attached and then a refrigerated solution is pumped through the balloon for an hour. This "gastric freezing" therapy promised to reduce acid secretion by cooling the stomach and so relieve ulcers. An experiment reported in the *Journal of the American Medical Association* showed that gastric freezing did reduce acid secretion and relieve ulcer pain. The treatment was safe and easy, and was widely used for several years.

placebo effect

Unfortunately, the design of the experiment was defective. The positive response of the patients may have been due to the *placebo effect*. A placebo is a dummy treatment such as a sugar pill. Many patients respond favorably to *any* treatment, even a placebo, presumably because of their trust in the physician and expectations of a cure. This is the placebo effect. The placebo effect is quite strong. It extends to measurable physical responses, such as acid secretion in the stomach, as well as to subjective responses, such as lessening of pain. A better designed experiment, done several years later, divided ulcer patients into two groups. One group was treated by gastric freezing as before, while the other received a placebo treatment in which the solution in the balloon was at body temperature rather than freezing. The results: 34% of the 82 patients in the treatment group improved, versus 38% of the 78 patients in the control (placebo) group. This and other properly designed experiments showed that gastric freezing had no effect, and its use was abandoned.[3]   ■

The original gastric freezing experiment had a design of the form (4.2):

$$\text{Observe pain} \longrightarrow \text{Gastric freezing} \longrightarrow \text{Observe pain}$$

The resulting data were misleading because of confounding with the placebo effect. The data also reflect any special features of that particular study, such as a physician with a soothing manner. Confounding of the treatment with these environmental variables occurs because the environment cannot be completely controlled, as it might in a chemistry laboratory experiment. Fortunately, the remedy is simple: Experiments should *compare* treatments rather than attempt to assess a single treatment in isolation. When the two groups of patients in the second experiment are compared, the placebo effect and environmental variables operate on both groups. The only difference between the groups is the actual effect of gastric freezing. If the freezing group had experienced more relief from ulcer pain, the explanation must be that gastric freezing is effective. The group of patients who were given **control group** a sham treatment is called a *control group*, because it enables us to control the effects of environmental variables on the outcome. *Control of the effects of outside variables is the first principle of statistical design of experiments.* Comparison of several treatments is the simplest form of control.

Without comparison of treatments, experimental results in medicine and the behavioral sciences can be dominated by such influences as the details of the experimental arrangement, the selection of subjects, and the placebo effect. The result is often bias.

Bias

The design of a study is biased if it systematically favors certain outcomes.

An uncontrolled study of a new medical therapy, for example, is biased in favor of finding the treatment effective because of the placebo effect. It should not surprise you to learn that uncontrolled studies in medicine give new therapies a markedly higher success rate than proper comparative experiments. Medical researchers who trust their professional judgment to assess the success of a treatment from a series of patients without control are overconfident.

EXAMPLE 4.6

Lack of adequate controls can also destroy the credibility of nonexperimental studies. The Love Canal neighborhood near Niagara Falls was the site of the first major controversy over the health effects of toxic chemicals dumped decades earlier. The emotional climax of the Love Canal affair came in May 1980 when the U.S. Environmental Protection Agency reported evidence of chromosome damage among Love Canal residents. Already worried, the residents demanded that they be relocated, and the federal government soon did so.

But when the chromosome study was reviewed by an expert panel, it appeared that the study's author had compared the chromosomes of Love Canal residents to those of a population he had studied earlier. This, the panel felt, is no more valid than comparing patients treated by gastric freezing to past ulcer patients who received no treatment. They pointed out that everyone has some chromosome damage, which varies with age, sex, geographic location, and other variables. Moreover, the way in which the cells are processed in the laboratory affects the count of damaged chromosomes. Finally, whether a chromosome is damaged or not is sometimes hard to judge. Adequate control of these factors requires that Love Canal chromosomes be compared with those of a control population similar in age, sex, and location; that cells from both populations be processed in the same way and at the same time; and that the person counting the damaged chromosomes do all the counts without knowing which cells are from the "suspect" population. Lack of such control made the Love Canal chromosome study meaningless.[4]   ■

Randomization

The first aspect of an experimental design is the layout of treatments, with comparison as the leading principle. The second aspect is the rule by which the experimental units are assigned to the treatments. Comparison of the effects of several treatments is valid only when all treatments are applied to similar groups of experimental units. If one variety of corn is planted on more fertile ground, or if one cancer drug is given to less seriously ill patients, comparisons among treatments are meaningless. Systematic differences among the groups of experimental units in a comparative experiment are a source of bias. Allocating the available units or subjects among the treatments must therefore be done with care.

Experimenters often attempt to match groups by elaborate balancing acts. Medical researchers, for example, try to match the patients in a "new drug" experimental group and a "standard drug" control group by age, sex, physical condition, smoker or not, and so on. Such attempts are helpful but not adequate—there are too many nonexperimental variables that might affect the outcome. The experimenter is unable to measure some of these variables and will not think of others until after the experiment. Some important variables, such as how advanced a cancer patient's disease is, are so subjective that an experimenter might bias the study by, for example, assigning less advanced cancer cases to a promising new treatment in the subconscious hope that it will help them.

The remedy is to rely on chance to make an assignment that does not depend on any characteristic of the experimental unit and that does not rely on the judgment of the experimenter in any way. The use of chance can be combined with matching, but the simplest design creates groups by chance alone. Here is an example.

EXAMPLE 4.7

A food company assesses the nutritional adequacy of a new "instant breakfast" product by feeding it to newly weaned male white rats and measuring their weight gain over a 28-day period. A control group of rats will be fed a standard diet for comparison. The researchers choose to use 30 rats for the experiment and will divide them into two groups of 15. To do this in a completely unbiased fashion, they put the names (or cage numbers) of the 30 rats in a hat, mix them up, and draw 15. These rats form the experimental group and the remaining 15 make up the control group. ∎

randomization

The use of chance to divide experimental units into groups is called randomization. *Randomization is the second major principle of statistical design of experiments.** Combining comparison and randomization, we ar-arrive at the simplest randomized comparative design:

$$\text{Random allocation} \begin{cases} \text{Group 1} \longrightarrow \text{Treatment 1} \longrightarrow \text{Observation} \\ \\ \text{Group 2} \longrightarrow \text{Treatment 2} \longrightarrow \text{Observation} \end{cases} \tag{4.3}$$

The logic behind a randomized comparative design is straightforward. Randomization produces groups of experimental units that should be similar in all respects before the treatments are applied. Comparative design ensures that influences other than the experimental treatments operate equally on all groups. Therefore, differences in the response variable must be due to the effects of the treatments—that is, the treatments are not just associated with the observed differences in the response, they must actually cause them.

* The use of randomization in experimental design was pioneered by R. A. Fisher (1890–1962). Fisher worked with agricultural field experiments and genetics at the Rothamsted Experiment Station in England. He invented statistical design of experiments and formal statistical methods for analyzing experimental data, and developed them to an advanced state.

Experimenters were slow to accept randomized comparative designs when they were first put forward in the 1920s. Many researchers are overly confident of their ability to assign units to groups without bias and even to judge the effectiveness of a treatment without controls. They dislike randomization because it makes the outcome of the study depend on chance. If the researchers in Example 4.7 drew names out of the hat a second time, they would get a different 15 rats in the experimental group and no doubt a somewhat different result when they compared the average weight gains of the two groups.

These objections can be effectively refuted. Many studies have demonstrated that researchers cannot in fact avoid bias in using their expert judgment. As for randomization, it does indeed render the outcomes dependent on chance but in a regular manner that can be described by the laws of probability. "Random" in statistics does not mean "haphazard." Feeding the new breakfast food to the first 15 rats that the technician could catch would be haphazard, and perhaps biased if these rats were smaller, slower, or friendlier than the others. Randomization, on the other hand, gives each rat an equal chance to be chosen.

The role of chance does require us to elaborate the logic of randomized comparative experiments a bit. We cannot say that *any* difference in average weight gain between the two groups of rats, however small, must be due to the diets. Some differences will appear even if the diets are identical because the rats are not exactly alike. Some rats will grow faster than others, and by chance more of the faster growing rats may end up in one group than in the other. So even though there are no systematic differences among the groups, there will still be random differences. It is the business of statistical inference, using the laws of probability, to say whether the observed difference is too large to occur plausibly as a result of chance alone. A difference too large to attribute plausibly to chance is called *statistically significant*. If we observe significant differences among the groups after a comparative randomized experiment, we can conclude that they are caused by the treatments.

statistically significant

This is a subtle idea that will be worked out in detail in Chapter 7. One important point should be made immediately, however. You would not trust the results of an experiment that fed each diet to only one rat—the role of chance is too large if we use two rats and toss a coin to decide which is fed the new diet. The more rats used, the more likely it is that randomization will create groups that are alike on the average. When differences among the rats are averaged out, only the effects of the different treatments will remain. So experiments with many subjects are better able to detect differences among the effects of the treatments than similar experiments with fewer subjects. Here is a third principle of statistical design of experiments, *replication: Repeat each treatment on a large enough number of experimental units or subjects to allow the systematic effects of the treatments to be seen.*

replication

How to Randomize

table of random digits

You can understand the idea of randomization by thinking of drawing names from a hat. In practice we randomize more quickly by using a *table of random digits*. Table B at the back of the book is such a table. A table of random digits is a long string of the digits 0, 1, 2, 3, 4, 5, 6, 7, 8, 9 with these two properties:

1 Each entry in the table is equally likely to be any of the 10 digits 0, 1, 2, 3, 4, 5, 6, 7, 8, 9.

2 The entries are independent of each other. That is, knowledge of one part of the table gives no information about any other part.

You can think of Table B as the result of asking an assistant (or a computer) to mix the digits 0 to 9 in a hat, draw one, then replace the digit drawn, mix again, draw a second digit, and so on. The assistant's mixing and drawing saves us the work of mixing and drawing when we need to randomize.

In Table B, each entry is equally likely to be any of the 10 possibilities. Each pair of entries is equally likely to be any of the 100 possible pairs 00, 01, . . . , 99. Each triple of entries is equally likely to be any of the 1000 possibilities 000, 001, . . . , 999, and so on. The division of the table entries into groups of five is merely a device to make the table easier to read, and the numbered rows merely make it easier to say where you started in the table. These conveniences do not affect the nature of the table as a long string of randomly chosen digits. How to use Table B for experimental randomization is best learned from an example.

EXAMPLE 4.8

In a nutrition experiment like that of Example 4.7, we must divide 30 rats at random into two groups of 15 rats each. First, give each rat a numerical label, using as few digits as possible. Two digits are needed to label 30 rats, so use labels

$$01, 02, 03, . . . , 29, 30$$

Next, enter Table B anywhere and read two-digit groups. Suppose we enter at line 130, which is

$$69051\ 64817\ 87174\ 09517\ 84534\ 06489\ 87201\ 97245$$

The first 10 two-digit groups in this line are

$$69\ 05\ 16\ 48\ 17\ 87\ 17\ 40\ 95\ 17$$

Each successive two-digit group is a label. The labels 00 and 31 to 99 are not used in this example and we ignore them. The first 15 labels between 01 and 30 that we encounter in the table choose rats for the experimental group. Of the first 10 labels in line 130, five are ignored because they are too high (over 30). The others are 05, 16, 17, 17, and 17. The rats labeled 05, 16, and 17 go into the experimental group. The second and third 17s are ignored because that rat is already in the group. Now run your finger across line 130 (and continue to lines 131,

132, and so on) until 15 rats are chosen. They are the rats labeled

05, 16, 17, 20, 19, 04, 25, 29, 18, 07, 13, 02, 23, 27, 21

These rats form the experimental group; the remaining 15 are the control group.

■

In practice, randomization consists of two steps: Assigning labels to the experimental units and then using Table B to select labels at random. You need not try to scramble the labels as you assign them. Table B will do the required randomizing, so assign labels in any convenient manner, such as in alphabetical order for human subjects. You can assign each subject several labels to speed use of the table. Just be certain that all subjects have the same number of labels and that all labels have the same number of digits. Only then will all subjects have the same chance of being chosen. You can read digits from Table B in any order—along a row, down a column, and so on—because the table has no order. As an easy standard practice, we recommend that you read along rows.

Most statistical software systems will do randomization for you, eliminating the need for Table B. For example, the MINITAB command

```
MTB > RANDOM 15 C1;
    INTEGERS 1 TO 30.
```

chooses 15 labels at random from 01, 02, . . . , 30 and places them in column C1. This is the randomization required for Example 4.8.

EXAMPLE 4.9   Many electronic devices contain circuit packs, which are small boards on which many hundreds of components are mounted. During production of a circuit pack, about 2000 connections must be soldered to electrically connect the components. These connections are soldered simultaneously by passing the packs through a standing wave of molten solder. If the conveyor carrying the circuit packs moves either too fast or too slowly, the quality of the soldered connections deteriorates. The engineer charged with setting up the production process for a new circuit pack conducts an experiment to determine the best speed for the conveyor in the wave soldering operation. The candidate speeds are 5, 6, and 7 feet per minute (ft/min). The engineer chooses to use 15 circuit packs for the experiment.

In this example, the experimental unit is a circuit pack, the explanatory variable is the conveyor speed, and the response variable is the count of imperfectly soldered connections in the pack. There are three treatments to be compared. The outline of the design is

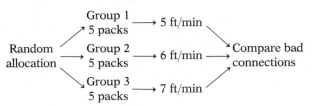

To carry out the random assignment of circuit packs to treatments, label the 15 packs 01 to 15. Starting at line 110 of Table B, the first five labels encountered select Group 1. This group consists of the packs labeled 06, 08, 12, 04, and 11. Continue through the table until five more packs are selected to make up Group 2. They are 02, 14, 03, 09, and 15. The remaining five packs form Group 3. Because there are only 15 packs, most labels are unused and the randomization requires searching through five lines in Table B. Fortunately, this can be done quickly.   ■

The diagram of the experiment in Example 4.9 is more specific than the generic diagram 4.3, shown earlier. The response variable (bad connections) is indicated in the diagram, and the number of units in each group (5 packs) is also shown. We will see later that for statistical inference there are some advantages to having equal numbers of units in all groups. Groups of equal size should usually be chosen unless there is some compelling reason not to do so.

completely randomized design

When all experimental units are allocated at random among all treatments, the experimental design is *completely randomized*. Design 4.3 is the simplest such design, comparing only two treatments. Completely randomized designs can compare more than two treatments; this is done by allocating the experimental units at random to as many groups as there are treatments, as was done in Example 4.9. The process is tedious but straightforward.

Cautions About Experimentation

Although experiments are the preferred means of gaining knowledge about the effects of explanatory variables on a response, even experimental evidence should be examined with a critical eye.

hidden bias

First, the way the experiment is conducted may produce a *hidden bias* despite the use of comparison and randomization. Great care must be taken to deal with all experimental units or subjects in the exact same way, so that the treatments are the only systematic differences present. Sometimes violations of this principle are obvious. A study of the roasting of meat in large commercial ovens once found a mysteriously large weight loss in the roasts cooked at the right front corner of the oven. The reason turned out to be that in every run of the experiment, a meat thermometer was thrust into the roast at that location, allowing juice to escape.

double-blind experiment

Less obvious violations of equal treatment can occur in medical and behavioral experiments. Suppose that in the comparative gastric freezing experiment of Example 4.5 the person interviewing the subjects knew which patients received the "real" therapy and which received "only" the placebo. This knowledge may subconsciously affect the interviewer's attitude toward the patient and recording of the patient's report of the degree of pain relief experienced. The experiment should therefore be *double-blind*—that is, neither the subjects nor the evaluators should know which treatment a subject received. The gastric freezing experiment was double-blind, with attention to such details as ensuring that the tube in the mouth of all subjects was cold, whether or not the fluid in the balloon was refrigerated. Careful planning and attention to detail are needed to avoid hidden bias.

lack of realism in experiments

The most serious potential weakness of experiments is *lack of realism*. Some examples of this are: A behavioral experiment on the effects of stress exposes student subjects to an artificial situation for a few hours—do the results apply to prolonged stress on the job? An industrial experiment uses a small-scale pilot production process to find the choices of catalyst concentration and temperature that maximize yield—will these be the best choices for the actual operation of a large plant? A new computer-based teaching method works well in a few classes taught by the designers of the computer program—will the method succeed in typical schools staffed by overworked teachers?

Most experimenters want to generalize their conclusions to some setting wider than that of the actual experiment. Statistical analysis of the original experiment cannot tell us how far the results will generalize. Rather, the experimenter must argue based on an understanding of psychology or chemical engineering or education that the experimental results do describe the wider world. Other psychologists or engineers or educators may disagree. This is one reason why a single experiment is rarely completely convincing, despite the compelling logic of experimental design. The true scope of a new finding must usually be explored by a number of experiments in various settings.

A convincing case that an experiment is sufficiently realistic to produce useful information is based not on statistics but on the experimenter's knowledge of the subject-matter of the experiment. The attention to detail required to avoid hidden bias also rests on subject-matter knowledge. Good experiments combine statistical principles with understanding of a specific field of study.

More About Design

To this point, we have met only completely randomized experimental designs for comparing levels of a single factor. Completely randomized designs can have more than a single factor. The education experiment of Example 4.4 has two factors—type and placement of questions in mathematics textbooks—and their combinations form six treatments. A completely randomized design assigns the students at random to these treatments. Once the layout of treatments is set, the randomization needed for a completely randomized design remains straightforward.

Completely randomized designs are the simplest statistical designs for experiments; they illustrate clearly the principles of control, randomization, and replication. However, completely randomized designs are often inferior to more elaborate statistical designs. In particular, matching the subjects in various ways can produce more precise results than simple randomization.

EXAMPLE 4.10

A psychologist is interested in the effect of room temperature on the performance of tasks requiring manual dexterity. She chooses temperatures of 70°F and 90°F as treatments. The response variable is the number of correct insertions, during a 30-minute period, in an elaborate peg and hole apparatus that requires the use of both hands simultaneously. Each subject is trained on the apparatus and then asked to make as many insertions as possible in 30 minutes of continuous effort.

A completely randomized design of the form 4.3 could be used, dividing the subjects into two groups. One group works at 70°F and the other at 90°F. Since individuals differ greatly in dexterity, the wide variation in individual scores will hide the systematic effect of room temperature unless there are many subjects in each group. It is more efficient to test each subject at both temperatures, so that a score at 90°F can be compared to the same person's score at 70°F. To avoid fatigue, each subject's two runs are performed on different days. Because learning or boredom may change the second score relative to the first, the order in which the treatments are given to each subject should be random. Toss a coin for each subject—if the coin falls heads, that subject is tested first at 70°F; if tails, at 90°F. ■

The psychologist in Example 4.10 used a *paired comparisons* design. This design uses the principles of comparison of treatments, randomization, and replication on several subjects. However, the randomization is not complete (all subjects randomly assigned to treatment groups) but restricted to assigning the order of the treatments for each subject. By allowing each subject to serve as his or her own control, the paired comparisons design reduces the effect of individual variation in dexterity and is able to detect the effect of temperature with fewer subjects. Paired comparisons are an example of block designs.

Block

> A block is a group of experimental units or subjects that are known before the experiment to be similar in some way that is expected to affect the response to the treatments.

In Example 4.10, each subject formed a separate block. In general, paired comparisons designs compare two treatments using blocks that contain either a single subject or a matched pair of subjects. For example, a block in a paired comparisons educational experiment might contain two students having the same age, race, sex, socioeconomic status, and achievement test scores. More general block designs can have blocks of any size.

block design   In a *block design*, the random assignment of units to treatments is carried out separately within each block. This allows the old idea of matching treatment groups to be combined with the newer idea of randomization.

Blocks are another form of *control*. They control the effects of outside variables by bringing those variables into the experiment to form the blocks. Here are some typical examples of blocks.

- The progress of a type of cancer differs in women and men. A clinical experiment to compare three therapies for this cancer therefore treats sex as a blocking variable. Two separate randomizations are done, the first assigning the male subjects to the treatments and the second assigning the female subjects. Figure 4.2 outlines the design of this experiment. Note that there is no randomization involved in making up the blocks. They are groups of subjects that differ in some way (sex in this case) that is apparent before the experiment begins.

- The soil type and fertility of farmland vary widely by location. Because of this, a test of the effect of tillage type (two types) and pesticide application (three application schedules) on soybean yields uses small fields as blocks. Each block is divided into six plots, and the six treatments are randomly assigned to plots separately within each block.

- A social policy experiment will assess the effect on family income of several proposed new welfare systems and compare them with the present welfare system. Because the income of a family under any welfare system is strongly related to its present income, the families who agree to participate are divided into blocks of similar income levels. The families in each block are then allocated at random among the welfare systems.

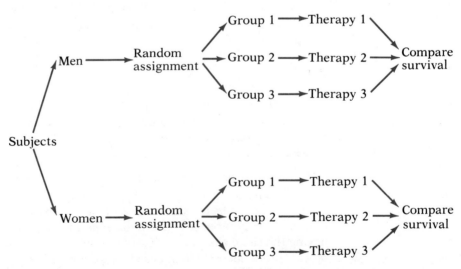

Figure 4.2   The outline of a randomized block design. The blocks consist of male and female subjects, while the treatments are three therapies for cancer.

Experiments with several factors and several blocking variables can be quite complicated to design and analyze. A detailed discussion would quickly take us beyond the basic practice of statistics. The design of complex experiments is work for professional statisticians. However, the idea of blocking is an important additional principle of statistical design of experiments. A wise experimenter will form blocks based on the most important unavoidable sources of variability among the experimental units. Randomization will then neutralize the effects of the remaining variation, and allow an unbiased comparison of the treatments.

SUMMARY

In an experiment, one or more **treatments** are imposed on the experimental **units** or **subjects**. Each treatment is a combination of levels of the explanatory variables, or **factors**.

The **design** of an experiment refers to the choice of treatments and the manner in which the experimental units or subjects are assigned to the treatments.

The basic principles of statistical design of experiments are **control**, **randomization**, and **replication**.

Experiments should be comparative in order to avoid **confounding** of the treatments with other influences, such as environmental variables.

Randomization creates treatment groups that are similar (except for chance variation) before the treatments are applied. Randomization and comparison together prevent **bias**, or systematic favoritism, in experiments.

Randomization can be carried out by giving numerical labels to the experimental units and using a **table of random digits** to choose treatment groups.

Replication of the treatments on many units reduces the role of chance variation and makes the experiment more sensitive to differences among the treatments.

The validity of experimental results may be threatened by **hidden bias** or **lack of realism**. Statistics alone does not protect against these threats.

In addition to comparison, a second form of control is to restrict randomization by forming **blocks** of experimental units that are similar in some way that is important to the response. Randomization is then carried out separately within each block.

SECTION 4.2 EXERCISES

For each of the experimental situations described in Exercises 4.9 to 4.11, identify the experimental units, the explanatory variable(s), and the response variable.

4.9 A manufacturer of food products uses package liners that are sealed at the top by applying heated jaws after the package is filled. The customer peels the sealed pieces apart to open the package. What effect does the temperature of the jaws have on the force required to peel the liner? To answer this question, the engineers obtain 20 pairs of pieces of package liner. Five pairs are sealed at each of 250°F, 275°F, 300°F, and 325°F. Then the peel strength of each seal is measured.

4.10 Can aspirin help prevent heart attacks? The Physicians' Health Study, a large experiment involving 22,000 male physicians, attempted to answer this question. One group of about 11,000 physicians took an aspirin every second day, while the rest took a placebo. After several years it was found that subjects in the aspirin group had significantly fewer heart attacks than subjects in the placebo group.

4.11 New varieties of corn with altered amino acid patterns may have higher nutritive value than standard corn, which is low in the amino acid lysine. An experiment is conducted to compare two new varieties, called opaque-2 and floury-2, with normal corn. Corn-soybean meal diets using each type of corn are formulated at three different protein levels: 12%, 16%, and 20%. There are thus nine diets in all. Researchers assign 10 one-day-old male chicks to each diet and record their weight gains after 21 days. The weight gain of the chicks is a measure of the nutritive value of their diet.

4.12 List the specific treatments studied in the Physicians' Health Study (Exercise 4.10). Then use a diagram to outline an appropriate completely randomized design for this study.

4.13 List the specific treatments studied in the package liner experiment of Exercise 4.9. Then use a diagram to describe a completely randomized experimental design that is appropriate for this study.

4.14 List the specific treatments compared in the high-lysine corn experiment of Exercise 4.11. Use a diagram to describe a completely randomized design for this experiment.

4.15 The following situations were not experiments. Can an experiment be done to answer the questions raised or not? If so, briefly outline its design.

(a) The comparison of two surgical procedures for breast cancer in Exercise 4.3(a)

(b) The "gender gap" issue of Exercise 4.3(b)

4.16 You are testing a new medication for relief of migraine headache pain. You intend to give the drug to migraine sufferers and ask them one hour later to estimate what percent of their pain has been relieved. You have 40 patients available to serve as subjects.

(a) Outline an appropriate design for the experiment, taking the placebo effect into account.

(b) The names of the subjects are given below. Use Table B beginning at line 131 to do the randomization required by your design, and list the subjects to whom you will give the drug.

Abrams	Danielson	Duttman	Lippman	Rosen
Adamson	Durr	Hipp	Marsden	Solomon
Afifi	Edwards	Hruska	McNeil	Thompson
Bikalis	Fluharty	Iselin	Morse	Travers
Burkett	Fratianna	Janle	O'Brian	Turing
Cefalo	Gerson	Kaplan	Obramowitz	Ullmann
Cranston	Green	Krushchev	Plochman	Williams
Curzakis	Gutierrez	Lattimore	Roberts	Wise

4.17 A horticulturist is comparing two methods (call them A and B) of growing potatoes. Standard potato cuttings will be planted in small plots of ground; the response variables are number of tubers per plant and fresh weight (weight when just harvested) of vegetable growth per plant. The horticulturist has 20 plots available for the experiment. Sketch the outline of a field divided into 20 small plots. Then outline the experimental design and use line 145 of Table B to do the required randomization. Mark on your sketch which growing method you will use in each plot.

4.18 Exercise 3.14 reports some of the data from an experiment to determine the best color for attracting cereal leaf beetles to boards on which they will be trapped. Four colors (treatments) are being compared. The response variable is the count of beetles trapped. One board will be attached to each of 16 poles evenly spaced throughout the center of a field of oats. Sketch a rectangular field with the locations of 16 poles. Outline the design of a completely randomized experiment to compare the colors. Then use Table B, starting at line 115, to randomly assign colors to the poles. Mark on your sketch the color assigned to each pole.

4.19 Use Table B, starting at line 120, to do the randomization required by your design for the package liner experiment in Exercise 4.13.

4.20 You decide to use a completely randomized design in the two-factor mathematics education experiment described in Example 4.4. Since it is too disruptive to assign school children at random to the treatment groups, you will use 12 classes of the same grade level instead. Outline the design of an experiment in which you assign these classes at random to the treatments. Then use Table B, beginning at line 101, to do the randomization required.

4.21 The horticulturist of Exercise 4.17 realizes that different genetic lines of potatoes may respond differently to the two growing methods. He therefore enlarges the experiment to include five genetic lines as a second factor. Use a diagram like Figure 4.1 to describe the treatments in this two-factor experiment. Outline a completely randomized design based on 20 plots of ground. Return to Table B at line 145 to do the randomization required to allocate the 20 plots among treatments. Make a sketch of the field divided into 20 plots and mark the treatment assigned to each plot.

4.22 Briefly discuss the scientific advantages of a well-designed experiment to compare breast cancer treatments over the observational study of Exercise 4.3(a).

4.23 An experiment that was publicized as proof that meditation lowers anxiety was conducted as follows. The experimenter interviewed the subjects and rated their levels of anxiety. Then the subjects were randomly assigned to two groups. The experimenter taught one group how to meditate and they meditated daily for a month. The other group was simply told to relax more. At the end of the month, the experimenter interviewed all the subjects again and rated their anxiety levels. The meditation group now had less anxiety. Psychologists said that the results were suspect because the ratings were not blind. Explain what this means and how lack of blindness could bias the reported results.

4.24 A chemist wants to compare the variability of a new and simpler assay method with the standard method. She prepares a batch of solution, divides it into 40 specimens, and then selects 20 at random. She asks her technician to analyze these 20 specimens using the new method and the remaining 20 using the standard method. Since all of the specimens come from the same solution, all assays should give the same answer except for variation caused by differences in carrying out the chemical analysis. After performing the analyses with the new method the technician falls ill, so the chemist performs the 20 assays using the old method herself. The results using the new method are much more variable than those using the old method, and the chemist therefore rejects the new assay technique. This experiment is fatally flawed despite a randomized comparative design. Explain why the results cannot be trusted.

4.25 Is the right hand generally stronger than the left in right-handed people? A crude test of hand strength can be conducted by placing a bathroom scale on a shelf with the end protruding, and then squeezing the scale with the thumb below and four fingers above it. The reading of the scale shows the force exerted. Describe the design of a paired comparisons experiment to compare the strength of the right and left hands, using 10 right-handed people as subjects. Use a coin to do the required randomization.

4.26 Twenty overweight females have agreed to participate in a study of the effectiveness of four reducing regimens, A, B, C, and D. The researcher first compares each subject's weight with her "ideal" weight and then calculates how many pounds overweight she is. The subjects and their excess weights are

Birnbaum	35	Hernandez	25	Moses	25	Smith	29
Brown	34	Jackson	33	Nevesky	39	Stall	33
Brunk	30	Kendall	28	Obrach	30	Suggs	35
Dixon	34	Loren	32	Rakov	30	Wilansky	42
Festinger	24	Mann	28	Siegel	27	Williams	22

The response variable is the weight lost after 8 weeks of treatment. Because the initial amount overweight will influence the response variable, a randomized block design is used.

(a) Arrange the subjects in order of increasing excess weight. Form five blocks by grouping the four least overweight, the next four, and so on.

(b) Use Table B to do the required random assignment of subjects to the four reducing regimens separately within each block. Be sure to explain exactly how you used the table.

4.27 Return to the mathematics education experiment of Exercise 4.20. Six of the 12 available classes are in one school district and the other six are in another district. Differences between the districts, in curriculum and otherwise, may have a strong effect on the response. You therefore decide to use a randomized block design with the two districts as blocks. The six treatments will be assigned at random to the six classes within each block separately. Outline the design with a diagram. Then use Table B, beginning at line 125, to do the randomization. Report your result in a table that lists the six classes in each district and the treatment you assigned to each class.

4.28 Return to the experiment of Exercises 3.14 and 4.18 on the attractiveness of several colors to cereal leaf beetles. The researchers decide to use two oat fields in different locations and to space eight poles equally within each field. Outline a randomized block design using the fields as blocks. Then use Table B, beginning at line 105, to carry out the random assignment of colors to poles. Report your results by means of a sketch of the two fields with the color at each pole noted.

4.3   SAMPLING DESIGN

A political scientist wants to know what fraction of the public consider themselves Democrats. A quality engineer must determine what fraction of the bearings rolling off an assembly line are defective. Government economists inquire about average household income. In all these cases, we want to gather information about a large group of people or things. But we will not, as in an experiment, impose a treatment in order to observe the response. Time, cost, and inconvenience usually forbid inspecting every bearing or contacting every household. In such cases, we gather information about only part of the group in order to draw conclusions about the whole.

Population, units, sample

> The entire group of objects or people about which information is wanted is called the population. Individual members of the population are called units.
>
> A sample is a part of the population that is actually examined in order to gather information.

Notice that population is defined in terms of our desire for knowledge. If we wish to draw conclusions about all U.S. college students, that group is our population even if only local students are available for questioning. The sample is the part from which we draw conclusions about the whole. A poorly designed sampling procedure can produce misleading conclusions.

EXAMPLE 4.11   A coil mill produces large coils of thin steel for use in manufacturing home appliances. The quality engineer wants to submit a sample of 5-cm squares to detailed laboratory examination. She asks a technician to cut a sample of 10 such squares. Wanting to provide "good" pieces of steel, the technician carefully avoids the visible defects in the coil material when cutting the sample. The laboratory results are wonderful but the customers complain about the material they are receiving. ■

EXAMPLE 4.12   The advice columnist Ann Landers once asked her readers, "If you had it to do over again, would you have children?" A few weeks later, her column was headlined "70% OF PARENTS SAY KIDS NOT WORTH IT." Indeed, 70% of the nearly 10,000 parents who wrote in said they would not have children if they could make

voluntary response samples

the choice again. The results of this sample are worthless as indicators of opinion in the population of all American parents. Because the sample was self-selected, it consisted of people who felt strongly enough to take the trouble to write to Ann Landers, Many of them, as their letters showed, were angry at their children and alienated from them. *Voluntary response samples* of this kind overrepresent people with strong opinions, most often negative opinions. It is not surprising that a statistically designed opinion poll on the same issue a few months later found that 91% of parents *would* have children again. Ann Landers' poorly designed sampling method produced a 70% ''No'' result when the truth about the population was close to 90% ''Yes.'' ∎

In both Examples 4.11 and 4.12, the sample was selected in a manner that guaranteed that it would not be representative of the entire population. These sampling schemes display *bias*, or systematic error, in favoring some parts of the population over others. The remedy is to give every unit in the population an equal chance to be chosen for the sample. This should be done in a mechanical way, so that there is neither favoritism by the sampler (as in Example 4.11) nor self-selection by respondents (as in Example 4.12). Such *random selection* will give all units an equal chance, just as randomization eliminated bias in assigning experimental subjects.

random selection

Simple Random Samples

Just as we spoke of the design of experiments, we can discuss the design of sampling schemes. The design is the pattern of randomization used to select the sample. The simplest design amounts to placing names in a hat (the population) and drawing out a handful (the sample). This is simple random sampling.

Simple random sample

> A simple random sample (SRS) of size n consists of n units from the population chosen in such a way that every set of n units has an equal chance to be the sample actually selected.

Each treatment group in a completely randomized experimental design is in effect an SRS drawn from the available experimental units. An SRS is selected by labeling all units and then using a table of random digits to select a sample of the desired size, just as in experimental randomization. In practice this is done by a computer if the population is large. Notice that an SRS not only gives each unit an equal chance to be chosen (thus avoiding bias in the choice), but also gives every possible sample an equal chance to be chosen. There are other random sampling designs that give each unit, but not each sample, an equal chance. One such design, systematic random sampling, is discussed in Exercise 4.37.

EXAMPLE 4.13

An academic department wishes to choose a three-member advisory committee at random from the members of the department. To choose an SRS of size 3 from the 28 faculty members listed below, first label the members of the population as shown.

00	Abbott	07	Goodwin	14	Pillotte	21	Theobald
01	Cicirelli	08	Haglund	15	Raman	22	Vader
02	Crane	09	Johnson	16	Riemann	23	Wang
03	Dunsmore	10	Keegan	17	Rodriguez	24	Wieczoreck
04	Engle	11	Lechtenberg	18	Rowe	25	Williams
05	Fitzpatrick	12	Murray	19	Sommers	26	Wilson
06	Gaidos	13	Nguyen	20	Stone	27	Zink

Now enter Table B, and read two-digit groups until you have chosen three committee members. If you enter at line 140, the committee consists of Murray (12), Nguyen (13), and Engle (04). ■

Other Sampling Designs

probability sampling design

The general framework for sampling is a *probability sampling design*. Any such design must give each member of the population a known nonzero chance to be selected. Some probability sampling designs (such as an SRS) give each member of the population an equal chance to be selected. This may not be true in more elaborate sampling designs. In every case, however, the use of chance to select the sample is the essential principle of statistical sampling.

Designs for sampling from large geographically dispersed populations are usually much more complex than an SRS. For example, it is common to restrict the random selection by dividing the population into groups of similar units, called *strata*, and then selecting a separate SRS in each stratum. This is a *stratified random sample*. The strata, like the blocks in a block design for an experiment, are based on facts known before the sample is taken. A stratified design can produce more exact information than an SRS of the same size by taking advantage of the fact that units in the same stratum are similar to each other. If the units in each stratum are all identical, for example, just one unit from each stratum is enough to completely describe the population. Once the strata are stated, you can obtain a stratified random sample by taking several SRSs, one from each stratum.

stratified random sample

EXAMPLE 4.14

A radio or television station that performs a piece of music owes a royalty to the composer. ASCAP (the American Society of Composers, Authors and Publishers) sells licenses permitting performance of the works of any of its members. ASCAP must then pay the proper royalties to the composers whose music was actually played. The major television networks keep program logs of all music played, but local radio and television stations do not. Since there are over a billion ASCAP-licensed performances each year, a detailed accounting is too expensive and cumbersome. Here is a case for sampling.

ASCAP allocates royalties among its members by taping a random sample of broadcasts. The sample of local commercial radio stations, for example, consists of 60,000 hours of broadcast time each year. Radio stations are stratified by type of community (metropolitan, rural), by geographic location (New England, Pacific, etc.), and by the size of the license fee paid to ASCAP (which reflects the size of the audience). In all, there are 432 strata. Tapes are made at random hours for randomly selected members of each stratum. They are reviewed by experts who can recognize almost every piece of music ever written, and the composers are then paid according to their popularity.[5] ■

Another common means of restricting random selection is to perform the selection in stages. This is often done for national samples of families, households, or individual people. For example, government data on employment and unemployment are gathered by the Current Population Survey, which conducts interviews in over 70,000 households each month. It is not practical to maintain a list of all U.S. households from which to select an SRS. Moreover, the cost of sending interviewers to the widely scattered households in an SRS would be excessive. The Current Population Survey

multistage sample design

therefore uses a *multistage sample design*. The final sample consists of clusters of nearby households. Most opinion polls and other national samples are also multistage, though interviewing in most national samples today is done by telephone rather than in person, eliminating the economic need for clustering.

A national multistage sample proceeds somewhat as follows:

Stage 1 Take a sample from the 3000 counties in the United States

Stage 2 Select a sample of townships within each of the counties chosen

Stage 3 Select a sample of blocks within the chosen townships

Stage 4 Take a sample of households within each block

Lists of counties and townships are easily available, and from that point the sampling can be based on a local map if necessary. No national list of households is required. Moreover, the sample households occur in clusters in the same block and so can be contacted by a single interviewer. The sample at each stage of a multistage design may be an SRS. Stratified samples are also common—for example, counties might be grouped into rural and metropolitan strata before sampling.

Analysis of data from sample designs more complex than an SRS takes us beyond basic statistics. But the SRS is the building block of more complex designs, and analysis of more complex data differs more in complexity of detail than in fundamental concepts.

Cautions About Sample Surveys

The statistical idea of random selection eliminates bias in the choice of a sample from a list of the population. When the population consists of human beings, however, accurate information from a sample requires much more than a good sampling design.[6] To begin, an accurate and complete list of the population is needed. Since such a list is rarely available, most samples suffer from some degree of *undercoverage*. A sample survey of households, for example, will miss not only the homeless but also prison inmates and students in dormitories. An opinion poll conducted by telephone will miss the 7% to 8% of the American population without residential phones. The results of national sample surveys therefore have some bias if the people not covered—who most often are poor people—differ from the rest of the population.

undercoverage

A more serious source of bias is *nonresponse*, which occurs when a selected individual either cannot be contacted or refuses to cooperate. Nonresponse to nongovernmental surveys often reaches 30%, even with careful planning and several callbacks. Since nonresponse is much higher in urban areas, most sample surveys substitute other accessible people in the same area to avoid favoring rural areas in the final sample. If accessible people differ from those who are rarely at home or who refuse to answer questions, some bias remains.

nonresponse

In addition, the behavior of the respondent or of the interviewer can cause *response bias* in sample results. Respondents may lie, especially if asked about illegal or antisocial behavior. The sample then underestimates the presence of such behavior in the population. An interviewer whose attitude suggests that some answers are more desirable than others will get these answers more often. The race or sex of the interviewer may influence responses to questions about race relations or attitudes toward feminism. Answers to questions that ask respondents to recall past events are often inaccurate because of faulty memory. In particular, many people will "telescope" events in the past, bringing them forward in memory to more recent time periods. "Have you visited a dentist in the last six months?" will often draw a "Yes" from someone who last visited a dentist eight months ago.[7] Because of these effects, careful training of interviewers and careful supervision to avoid variation among interviewers are important to good sample survey practice. Because it is so difficult to train large numbers of interviewers, a sample survey is often at least as accurate as a census.

response bias

The *wording of questions* is the most important influence on the answers given to a sample survey. Opinion polls on nuclear arms reduction, for example, have produced strongly differing conclusions depending on the questions asked. Suppose that you want to show by a statistically sound opinion poll that most people oppose nuclear disarmament. First, since more people favor arms reduction in general than agree with any one proposal, make your question as specific as possible. Second, word the question to suggest some unpleasant effect of arms reduction. Here is a question that drew a 58%

response opposing a nuclear freeze at a time when other questions showed much more favorable attitudes.[8]

> Do you agree or disagree with the following statement: A freeze in nuclear weapons should be opposed because it would do nothing to reduce the danger of thousands of nuclear weapons already in place and would leave the Soviet Union in a position of nuclear superiority.

Never trust the results of a sample survey until you have read the exact questions posed. The sampling design, the amount of nonresponse, and the date of the survey are also important. Good statistical design is a part, but only a part, of a trustworthy survey.

SUMMARY

In a sample survey, a **sample** is selected from the **population** of all people or things about which we desire information. Conclusions about the population are based on examination of the sample.

The **design** of a sample refers to the method used to select the sample from the population. **Probability sampling designs** use random selection to give each member of the population a known nonzero chance to be selected for the sample.

The basic probability sample is a **simple random sample (SRS)**, which gives every possible sample of a given size the same chance of being selected. Simple random samples are chosen by labeling the members of the population and using random digits to select the sample.

In **stratified random sampling** the population is divided into **strata**, groups of units that are similar in some way that is important to the response. A separate SRS is then selected from each stratum. **Multistage samples** select successively smaller regions within the population in stages. Each stage may employ an SRS, a stratified sample, or another type of sample.

Failure to use probability sampling often results in **bias**, or systematic errors in the way the sample represents the population. **Voluntary response** samples in which the respondents choose themselves are particularly prone to large bias.

In human populations, even probability samples can suffer from bias due to **undercoverage** or **nonresponse**, from **response bias** due to the behavior of the interviewer or the respondent, or from misleading results due to badly worded questions.

SECTION 4.3 EXERCISES

4.29 For each of the following situations, identify the population as exactly as possible—that is, identify the basic unit and specify which units fall in the population. If the information given is not complete, complete the description of the population in a reasonable way.

    (a) Each week the Gallup Poll questions a sample of about 1500 adult U.S. residents to discover the opinions of Americans on a wide variety of issues.

    (b) The decennial census tries to gather basic information from every household in the United States. But a "long form" requesting much additional information is sent to a sample of about 20% of U.S. households.

    (c) A machinery manufacturer purchases voltage regulators from a supplier. There are reports that variation in the output voltage of the regulators is affecting the performance of the finished products. To assess the quality of the supplier's production, the manufacturer subjects a sample of 5 regulators from the last shipment to careful laboratory analysis.

4.30 Follow the same instructions given in Exercise 4.29 for each of the following situations.

    (a) A sociologist wants to study the attitudes of American male college students toward marriage and husband-wife relations. She gives a questionnaire to an SRS of the men enrolled in Sociology 101 at her college.

    (b) A member of Congress wants to know what his constituents think of proposed legislation on health insurance. His staff reports that 228 letters have been received on the subject, of which 193 oppose the legislation.

    (c) An insurance company wants to monitor the quality of its procedures for handling loss claims from its auto insurance policyholders. Each month the company selects an SRS of all auto insurance claims filed that month to examine them for accuracy and promptness.

4.31 The author Shere Hite undertook a study of women's attitudes toward sex and love by distributing 100,000 questionnaires through women's groups. Only 4.5% of the questionnaires were returned. Based on this sample of women, Hite wrote *Women and Love*, a best-selling book claiming that women today are fed up with men. For example, 91% of the divorced women in the sample said that they had initiated the divorce, and 70% of the married women said that they had committed adultery.

    Explain briefly why Hite's sampling method is nearly certain to produce a strong bias. Are the sample results cited (91% and 70%)

much higher or much lower than the truth for the population of all adult American women?

4.32 Some television stations take quick polls of public opinion by announcing a question on the air and asking viewers to call one of two telephone numbers to register their opinions as "Yes" or "No." Telephone companies make available "900" numbers for this purpose; dialing such a number results in a small charge to your telephone bill. One such call-in poll finds that 73% of those who called in are opposed to a proposed local gun control ordinance. Explain why this sampling method is biased. Is the percent of the population who oppose gun control probably higher or lower than the 73% of the sample who are opposed?

4.33 A manufacturer of specialty chemicals chooses 3 from each lot of 25 containers of a reagent to test them for purity and potency. Below are the control numbers stamped on the bottles in the current lot. Use Table B at line 111 to choose an SRS of 3 of these bottles.

A1096	A1097	A1098	A1101	A1108
A1112	A1113	A1117	A2109	A2211
A2220	B0986	B1011	B1096	B1101
B1102	B1103	B1110	B1119	B1137
B1189	B1223	B1277	B1286	B1299

4.34 Figure 4.3 is a map of a census tract in Cedar Rapids, Iowa. Census tracts are small homogeneous areas with an average population of about 4000. Each block in the tract is marked with a Census Bureau identifying number. A sample of blocks from a census tract is often the next to last stage in a multistage sample. Select 5 blocks from the tract illustrated by using Table B at line 125 to choose an SRS.

4.35 The people listed below are enrolled in a statistics course taught by means of television. Use Table B at line 139 to choose 6 to be interviewed in detail about the quality of the course.

Agarwal	Dewald	Hixson	Puri
Anderson	Fernandez	Klassen	Rosenkranz
Baxer	Frank	Mihalko	Rubin
Bowman	Fuhrmann	Moser	Samuels
Bruvold	Goel	Naber	Shen
Casella	Gupta	Petrucelli	Shyr
Cote	Hicks	Pillai	Sundheim

4.36 In using Table B repeatedly to choose samples or to do experimental randomization, you should not always begin at the same place, such as line 101. Why not?

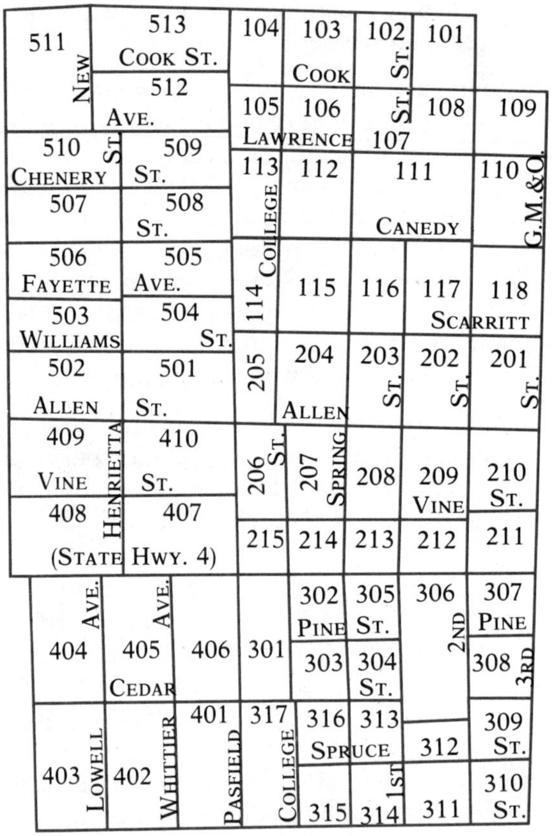

Figure 4.3 Map of a census tract in Cedar Rapids, Iowa (Exercise 4.33).

4.37 *Systematic random samples* are often used to choose a sample of apartments in a large building or dwelling units in a block at the last stage of a multistage sample. Choosing such a sample will be described by an example. Suppose that we must choose 4 addresses out of 100. Since $100/4 = 25$, we can think of the list as four separate lists of 25 addresses. Choose 1 of the first 25 at random, using Table B. The sample contains this address and the addresses 25, 50, and 75 places down the list from it. If 13 is chosen, for example, then the systematic random sample consists of the addresses numbered 13, 38, 63, and 88.

(a) Use Table B to choose a systematic random sample of 5 addresses from a list of 200. Enter the table at line 120.

(b) Like an SRS, a systematic sample gives all units the same chance to be chosen. Explain why this is true, and then explain carefully why a systematic sample is nonetheless *not* an SRS.

4.38 A club contains 30 student members and 10 faculty members. The students are

Abel	Fisher	Huber	Moran	Reinmann
Carson	Golomb	Jack	Moskowitz	Silvers
Cryer	Griswold	Jones	Neyman	Sobar
David	Hein	Kiefer	O'Brien	Thompson
Deming	Hernandez	Klotz	Pearl	Utts
Elashoff	Holland	Lamb	Potter	Vlasic

and the faculty members are

Andrews	DuMouchel	Kim	Moore	Rabinowitz
Besicovitch	Gupta	Lightman	Phillips	Vincent

The club can send four students and two faculty members to a convention and decides to choose those who will go by random selection. Use Table B to choose a stratified random sample of 4 students and 2 faculty members.

4.39 A university has 2000 male and 500 female faculty members. The equal opportunity employment officer wants to poll the opinions of a random sample of faculty members. In order to give adequate attention to female faculty opinion, he decides to choose a stratified random sample of 200 males and 200 females. He has alphabetized lists of female and male faculty members. Explain how you would assign labels and use Table B to choose the desired sample. Enter Table B at line 122 and give the labels of the first 5 females and the first 5 males in the sample.

4.40 If a stratified random sample of 50 female and 200 male faculty members had been selected in Exercise 4.39, each member of the faculty would have had one chance in 10 to be chosen. This sample design would have given every unit in the population the same chance to be chosen for the sample. Is it an SRS? Explain your answer.

4.41 The list of units from which a sample is actually selected is called the *sampling frame*. Ideally, the frame should list every unit in the population, but in practice this is often difficult. A frame that is not representative of the population is a common source of bias.

(a) Suppose that a sample of households in a community is selected at random from the telephone directory. What households are omitted from this frame? What types of people do you think are likely to live in them? These people will probably be underrepresented in the sample.

(b) It is more common in telephone surveys to use random digit dialing equipment that selects the last four digits of a telephone number at random after being given the exchange (the first three digits). Which of the households you mentioned in answer to (a) will be included in the sampling frame by random digit dialing?

4.42 Bias is present in each of the following cases. Identify the source of the bias and specify the direction of the bias (that is, whether the sample result will be systematically above or below the true population result).

(a) A flour company wants to know what fraction of Minneapolis households bake their own bread. An SRS of 500 residential addresses is drawn and interviewers are sent to these addresses. The interviewers are employed during regular working hours on weekdays and they interview only during those hours.

(b) The Miami Police Department wants to know if black residents of Miami are satisfied with police service in their neighborhood. A questionnaire is prepared. An SRS of 300 mailing addresses in predominantly black neighborhoods is chosen, and a uniformed police officer is sent to each address to interview an adult resident.

4.43 Comment on each of the following as a potential sample survey question. Is the question clear? Is it slanted toward a desired response?

(a) "Does your family use food stamps?"

(b) "Which of the following best represents your opinion on gun control?

1. The government should confiscate our guns.
2. We have the right to keep and bear arms.

(c) "A freeze in nuclear weapons should be favored because it would begin a much-needed process to stop everyone in the world from building nuclear weapons now and reduce the possibility of nuclear war in the future. Do you agree or disagree?"

(d) "In view of escalating environmental degradation and incipient resource depletion, would you favor economic incentives for recycling resource-intensive consumer goods?"

4.4 WHY RANDOMIZE?

Statistical designs for both sampling and experimentation are built upon the deliberate introduction of chance into the process of collecting data. In both cases, randomization is used to eliminate any favoritism, or bias, in allocating units to treatments or in choosing units to be in the sample. The justification for randomization and its consequences will be examined more carefully in this section.

The goal of both sampling and experimentation is to draw conclusions about an underlying population based on data from selected units. In an experiment, we tacitly assume that the experimental units are representative of a wider population to which our inferences will extend. An experimenter who compares the performance on a dexterity test of college students before and after they take a stimulant, for example, will argue that results for college students extend to a wider population of adults. Even though a population does not enter into the mechanics of experimental design, experimentation as well as sampling aims to make inferences about a population. Since statistics deals with numerical variables, most often a number that is computed from the data is used to make inferences about an unknown number that describes the population.

Parameter, statistic

> A parameter is a number describing the population.
>
> A statistic is a number that can be computed from the data without making use of any unknown parameters.

Some examples will illustrate the distinction between parameters and statistics.

EXAMPLE 4.15

The Gallup Poll has collected data about churchgoing in the United States over several decades. The population is all U.S. residents age 18 and over. A recent poll interviewed a sample containing 1785 randomly selected adults and asked them, "Did you, yourself, happen to attend church or synagogue in the last seven days?" Of the respondents, 1035 or 58% said "No." The number 58% is a *statistic*. The corresponding *parameter* is the percent of all adult U.S. residents who would have said "No" if asked the same question. ∎

EXAMPLE 4.16

A type of semiconductor logic chip is manufactured in circular wafers containing 120 chips. One stage in the process is to bake a chemical coating on the wafer; the next is to etch in the circuits. Will baking at 90°F or 105°F lead to fewer defective chips per wafer? A group of wafers is divided at random into one group to be baked at 90°F and another to be baked at 105°F. The wafers are baked individually in a random order. The first group averaged 57.3 defective chips per wafer, the second 63.8 defective chips.

The *statistic* of interest is the difference $63.8 - 57.3 = 6.5$. That is, baking at 90°F produced 6.5 fewer defective chips per wafer on the average. The *parameter* is the average difference in the counts of defective chips between the population of all wafers baked at 90°F and the population of all wafers baked at 105°F. These populations are hypothetical, since in production the wafers will be baked at one temperature only. It is common in experiments to seek information about a "What if?" population that cannot be randomly sampled because its units do not all exist in actual fact. ∎

For the sake of simplicity, the rest of this section will focus on the situation illustrated by Example 4.15: Estimation of a population proportion by a sample proportion. All of the ideas involved extend to estimation of other parameters by other statistics and to experimentation as well as sampling.

Sampling Distributions

The opinion poll of Example 4.15 offers information on the churchgoing habits of Americans. Let us examine it in more detail. We want to estimate the proportion p of the population that did not attend church or synagogue the week of the poll. The poll reports that of 1785 randomly selected adults, 1035 had not attended church or synagogue the previous week. The sample proportion

$$\hat{p} = \frac{1035}{1785} = 0.58$$

is a statistic that can be used to estimate the unknown parameter p.

How can $\hat{p}$, based on only 1785 of the more than 180 million American adults, be an accurate estimate of p? After all, a second random sample taken at the same time would select different respondents and no doubt produce a different value of $\hat{p}$. This basic fact is called *sampling variability*: the value of a statistic varies in repeated random sampling.

sampling variability

To understand why sampling variability is not fatal, we will produce some data ourselves. Suppose that an opinion poll takes an SRS of size 100 from the population of all adult U.S. residents. Suppose also that in fact 60% of the population did not attend church or synagogue last week (so $p = 0.6$). The population can be imitated by a huge table of random digits, with each digit standing for a person. Six of the digits (say, 0 to 5) stand for people who did not attend church. The remaining four digits, 6 to 9, stand for those who attended. Since all digits in a random number table are equally likely, this assignment produces a population proportion of nonchurchgoers equal to $p = 0.6$. An SRS of 100 people from the population can be imitated by taking 100 consecutive digits from Table B. The statistic $\hat{p}$ is the proportion of 0s to 5s in the sample. For example, the first 100 entries in Table B contain 63 digits between 0 and 5, so $\hat{p} = 63/100 = 0.63$. A second SRS based on the second 100 entries in Table B gives a different result, $\hat{p} = 0.56$. That's sampling variability.

simulation

We used the table of random digits to *simulate* drawing an SRS from a large population. Simulation is a powerful tool for studying random phe-

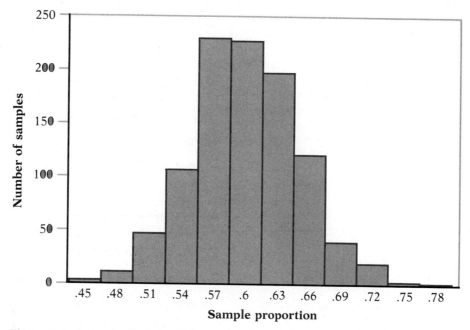

Figure 4.4 The sampling distribution of the sample proportion $\hat{p}$ in 1000 repetitions of an SRS of size 100 from a population with population proportion $p = .6$.

nomena. It is much faster to use Table B than to actually draw repeated SRSs, and much faster yet to use a computer programmed to produce random digits. Figure 4.4 displays the results of simulating 1000 separate SRSs of size 100 from a population with $p = 0.6$. This frequency histogram shows the sampling distribution of the sample proportion $\hat{p}$.

Sampling distribution

> The sampling distribution of a statistic is the distribution of values taken by the statistic in a large number of samples from the same population.

Although the values of $\hat{p}$ vary from sample to sample, the behavior of the statistic in repeated sampling is regular and predictable. The sampling distribution describes this regular pattern of behavior.

We can apply our familiar tools for describing distributions to Figure 4.4. There are no gaps or outliers. The distribution is symmetric and appears to be approximately normal. Figure 4.5 is a normal quantile plot of the 1000 observed $\hat{p}$'s, with only every twentieth point plotted because of the large number of observations. We see that the sampling distribution is very close to normal. The only important deviation is slight granularity caused by the

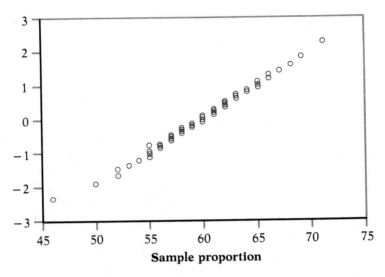

Figure 4.5 Normal quantile plot of the sample proportions in Figure 4.4.

fact that $\hat{p}$ from a sample of size 100 can take only values that are multiples of 0.01. It is a remarkable fact that the sampling distributions of several common statistics are approximately normal. This is another reason why the normal distributions are so important in statistics.

The appearance of the sampling distribution in Figures 4.4 and 4.5 is a consequence of the random selection of samples. Haphazard sampling does not give such regular and predictable results. When randomization is used in a design for producing data, statistics computed from the data have a definite pattern of behavior over many repetitions, even though the result of a single repetition is uncertain. The reasoning of formal statistical inference is based on the long-term regularity expressed in sampling distributions.

Bias

The fact that statistics from random samples have definite sampling distributions allows a more careful answer to the question of how trustworthy a statistic is as an estimate of a parameter. First, we can describe *bias* more exactly by speaking of the bias of a statistic rather than bias in a sampling method. Bias concerns the center of the sampling distribution. Notice that the center of the distribution in Figure 4.4 is very close to the true value $p = 0.6$ of the population from which the samples were drawn. In fact, the mean of the 1000 $\hat{p}$'s is 0.6025 and their median is exactly 0.6. Bias is measured numerically as the difference between the center of the sampling distribution and the true parameter value.

Unbiased statistic

A statistic used to estimate a parameter is called unbiased if the center of its sampling distribution is equal to the true value of the parameter being estimated.

Unbiasedness means that the center of the distribution reflects the truth about the population. If we draw an SRS from a population in which 50% did not attend church or synagogue, the center of the sampling distribution of $\hat{p}$ will be 0.5 instead of 0.6. An unbiased statistic computed from an individual sample sometimes falls above the true value of the parameter and sometimes below it. Since its sampling distribution is centered at the true value, however, there is no systematic tendency to overestimate or underestimate the parameter.

Figure 4.6 illustrates the idea of bias as systematic deviation from the truth. Think of the parameter as the bull's-eye on a target and the sample

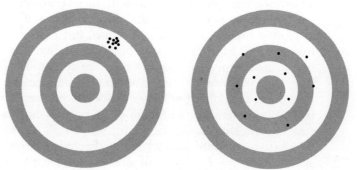

(a) High bias; low variability　　(b) Low bias; high variability

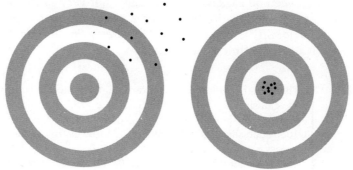

(c) High bias; high variability　　(d) Low bias; low variability

Figure 4.6 Bias and variability.

estimate as an arrow shot at the target. Unbiasedness means that our aim is properly aligned. Even though the arrows do not all go through the same hole in the target they are centered around the bull's-eye. Bias occurs when the arrows fall regularly away from the bull's-eye in a single direction. Avoiding bias requires both a proper sampling design and a proper choice of the statistic used to estimate the unknown parameter. Common statistics such as a sample proportion $\hat{p}$ will be biased if the sampling procedure is biased in the sense of favoring some part of the population. The proportion $\hat{p}$ of parents who would not have children again in the voluntary response sample of Example 4.12, for example, is strongly biased as an estimate of the population proportion p.

Variability

The point of archery is to shoot the arrows as close as possible to the bull's-eye. To achieve this the archer needs small bias, but also small variability so that the arrows fall close together. As Figure 4.6 shows, bias and variability are separate properties. Both large and small bias can combine with either large or small variability. Properly chosen statistics computed from random samples have little or no bias. But Figure 4.4 shows that estimates from an SRS can have quite a bit of variability. The variability of a statistic in repeated samples is measured by the spread of its sampling distribution. Our 1000 $\hat{p}$'s range from 0.46 to 0.77. Since the distribution is nearly normal, the variability is better described by the standard deviation, which is 0.049. Remember the 68–95–99.7 rule for normal distributions? In this case, it says that in the long run, 95% of all samples will give a $\hat{p}$ within 0.098 (two standard deviations) of the mean. Lack of bias puts the mean at the true $p = 0.6$, so 95% of all samples give an estimate $\hat{p}$ between 0.502 and 0.698. That is, if in fact 60% of the population does not attend church or synagogue, the estimates from repeated SRSs of size 100 will usually fall between 50.2% and 69.8%. That's not very satisfactory.

Randomization produces a sampling distribution and eliminates bias. *Variability is controlled by the size of the sample*: Larger samples produce statistics with less variability.

EXAMPLE 4.17   The actual opinion poll in Example 4.15 interviewed not 100 people, but 1785. We can simulate repeated SRSs of size 1785 from the same population ($p = 0.6$). Figure 4.7 shows the sampling distribution of the sample proportions $\hat{p}$ from 1000 such samples. The scale is the same as that of Figure 4.4, so the comparison is easy.

The center remains close to 0.6 (no bias). However, the sampling variability is much reduced. In fact, the smallest and largest among the $\hat{p}$'s are 0.569 and 0.623, so that all of the 1000 samples gave a result close to the truth. The sampling distribution is again very close to normal. Figure 4.8 displays the distribution on a different scale which makes the normal shape more visible. The standard

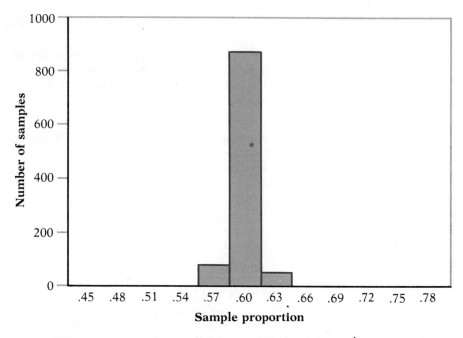

Figure 4.7 The sampling distribution of the sample proportion $\hat{p}$ in 1000 repetitions of an SRS of size 1785 from the same population as in Figure 4.4. The graph has the same scale as Figure 4.4, so the smaller variability in the larger sample is clearly visible. (See Example 4.17.)

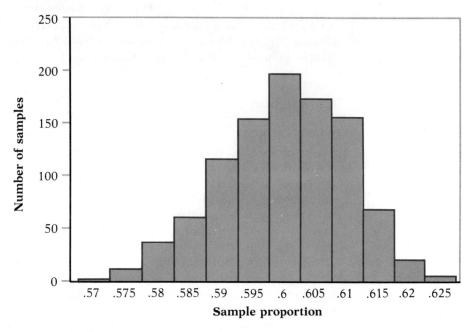

Figure 4.8 The same sampling distribution as in Figure 4.7, on a scale that shows more clearly the normal shape of the distribution.

deviation of the distribution is about 0.01. Therefore, 95% of all samples will give an estimate within about 0.02 of the mean, or between 0.58 and 0.62. *An SRS of size 1785 can in fact be trusted to give sample estimates that are very close to the truth about the entire population.* ■

One important and surprising aspect of this conclusion is that the spread of the sampling distribution does not depend very much on the size of the population.

As long as the population is much larger than the sample (say, 100 times as large), the spread of the sampling distribution is controlled entirely by the size of the sample.

This remarkable fact may not seem intuitively reasonable. To see why it is plausible, imagine sampling harvested corn by thrusting a scoop into a lot of corn kernels. The scoop doesn't know whether it is surrounded by a bag of corn or by an entire truckload. As long as the corn is well mixed (so that the scoop selects a random sample), the variability of the results depends only on the size of the scoop.

The fact that the variability of sample results is controlled by the size of the sample has important consequences for sampling design. A statistic from an SRS of size 1785 from the more than 240,000,000 residents of the United States is just as precise as an SRS of size 1785 from the 700,000 inhabitants of San Francisco. This is good news for designers of national samples but bad news for those who want accurate information about the citizens of San Francisco. If both use an SRS, both must use the same size sample to obtain equally trustworthy results.

Conclusion

Why randomize? Because the act of randomizing guarantees that the behavior of statistics is governed by a sampling distribution. The form of the distribution is known, and in many cases is approximately normal. Often the center of the distribution lies at the true parameter value, so that the notion that randomization eliminates bias is made more explicit. The spread of the distribution describes the variability of the statistic and can be made as small as we wish at the expense of a larger sample. In a randomized experiment, variability can be similarly reduced at the expense of larger groups of subjects for each treatment.

These facts lie at the heart of formal statistical inference. Later chapters will have much to say in more technical language about sampling distributions and the way statistical conclusions are based on them. What any user of statistics must understand is that all the technical talk has its basis in a simple question: What would happen if the sample or the experiment was repeated many times? The reasoning applies not only to an SRS but also to the complex sampling designs actually used by opinion polls and other national sample surveys. The same conclusions hold as well for randomized experimental designs. The details vary with the design but the basic facts are true whenever randomization is used to produce data.

Remember that proper statistical design is not the only aspect of a good sample or experiment. The sampling distribution shows only how a statistic varies due to the operation of chance in randomization. It shows nothing about possible bias due to undercoverage or nonresponse in a sample, or to lack of realism in an experiment. The true distance of a statistic from the parameter it is estimating can be much larger than might be suggested by a look at the sampling distribution. What is worse, there is no way to say how large the added error is. The real world is less orderly than statistics textbooks imply.

SUMMARY

A number that describes a population is called a **parameter**. A number that can be computed from the data is called a **statistic**. The purpose of sampling or experimentation is usually to use statistics to make statements about unknown parameters.

A statistic from a probability sample or randomized experiment has a **sampling distribution** that describes how the statistic varies in repeated data collection. Formal statistical inference is based on the sampling distributions of statistics.

A statistic as an estimator of a parameter may suffer from **bias** or from high **variability**. Bias means that the center of the sampling distribution is not equal to the true value of the parameter. The variability of the statistic is described by the spread of its sampling distribution.

Properly chosen statistics from randomized data production designs have no bias resulting from the way the sample is selected or the way the experimental units are assigned to treatments. The variability of the statistic is determined by the size of the sample or by the size of the experimental groups.

SECTION 4.4 EXERCISES

State whether each boldface number in Exercises 4.44 to 4.47 is a parameter or a statistic.

4.44 The Bureau of Labor Statistics last month interviewed 70,000 members of the U.S. labor force, of whom **7.2%** were unemployed.

4.45 A carload lot of ball bearings has a mean diameter of **2.5003** cm. This is within the specifications for acceptance of the lot by the purchaser. By chance, however, the acceptance sampling procedure inspects 100 bearings from the lot with a mean diameter of **2.5009** cm. Since this is outside the specified limits, the lot is mistakenly rejected.

4.46 A telephone sales company in Los Angeles uses a device that dials residential phone numbers in that city at random. Of the first 100 numbers dialed, **23** are unlisted. This is not surprising because **38%** of all Los Angeles residential phone numbers are unlisted.

4.47 A researcher investigating the effects of a toxic compound in food conducts a randomized comparative experiment with young male white rats. A control group is fed a normal diet, while the experimental group is fed a diet with 2500 parts per million of the toxic material. After 8 weeks, the mean weight gain is **335** grams for the control group and **289** grams for the experimental group.

4.48 An entomologist samples a field for egg masses of a harmful insect by placing a yard-square frame at random locations and examining the ground within the frame carefully. He wishes to estimate the proportion of square yards in which egg masses are present. Suppose that in a large field egg masses are present in 20% of all possible yard-square areas—that is, $p = 0.2$ in this population.

(a) Use Table B to simulate the presence or absence of egg masses in each square yard of an SRS of 10 square yards from the field. Be sure to explain clearly which digits you used to represent the presence and the absence of egg masses. What proportion of your 10 sample areas had egg masses?

(b) Repeat (a) with different lines from Table B until you have simulated the results of 20 SRSs of size 10. What proportion of the square yards in each of your 20 samples had egg masses? Make a stemplot from these 20 values to display the sampling distribution of $\hat{p}$ in this case. What is the median of this distribution? What is its shape?

4.49 An opinion poll asks, "Are you afraid to go outside at night within a mile of your home because of crime?" Suppose that the proportion of all adult U.S. residents who would say "Yes" to this question is $p = 0.4$.

(a) Use Table B to simulate the result of an SRS of 20 adults. Be sure to explain clearly which digits you used to represent each of "Yes" and "No." What proportion of your 20 responses were "Yes"?

(b) Repeat (a) using different lines in Table B until you have simulated the results of 10 SRSs of size 20 from the same population.

Compute the proportion of "Yes" responses in each sample. Find the median of these 10 proportions. Is it close to p?

4.50 The table below contains the results of simulating on a personal computer 100 repetitions of the drawing of an SRS of size 200 from a large lot of bearings, 10% of which do not conform to the specifications. The numbers in the table are the counts of nonconforming bearings in each sample of 200.

17	23	18	27	15	17	18	13	16	18	20	15	18	16	21
17	18	19	16	23	20	18	18	17	19	13	27	22	23	26
17	13	16	14	24	22	16	21	24	21	30	24	17	14	16
16	17	24	21	16	17	23	18	23	22	24	23	23	20	19
20	18	20	25	16	24	24	24	15	22	22	16	28	15	22
9	19	16	19	19	25	24	20	15	21	25	24	19	19	20
28	18	17	17	25	17	17	18	19	18					

(a) Make a frequency table for the counts, and list for each count of defectives the corresponding value of the sample proportion of defectives

$$\hat{p} = \frac{\text{count of defectives}}{200}$$

Then draw a frequency histogram for the values of $\hat{p}$.

(b) Where is the center of this sampling distribution? Does the shape appear to be approximately normal? Does the statistic $\hat{p}$ appear to have large or small bias as an estimate of the population proportion, which is $p = 0.10$ in this particular population?

(c) If we had simulated the repeated selection of SRSs of size 1000 instead of 200 from this same population, where would be the center of the sampling distribution of the sample proportion $\hat{p}$?

4.51 Figure 4.9 shows histograms of four sampling distributions of statistics intended to estimate the same parameter. Label each distribution relative to the others as large or small bias and as large or small variability.

4.52 A management student is planning to take a survey of student attitudes toward part-time work while attending college. He develops a questionnaire and plans to ask 25 randomly selected students to fill it out. His faculty advisor approves the questionnaire but urges that the sample size be increased to at least 100 students. Why is the larger sample helpful?

4.53 For a study of state and federal taxation, the Internal Revenue Service is planning to select an SRS of individual federal income tax returns from each state. One variable of interest is the proportion

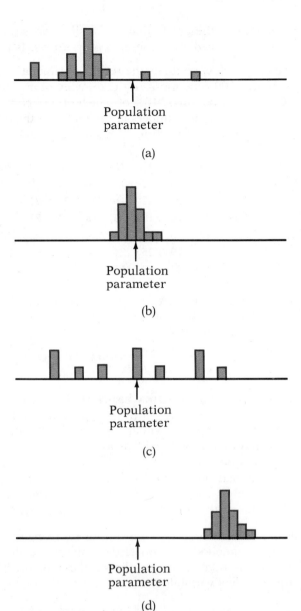

Population
parameter

(a)

Population
parameter

(b)

Population
parameter

(c)

Population
parameter

(d)

Figure 4.9 Which of these sampling distributions displays large or small bias and high or low variability (Exercise 4.51)?

of returns claiming itemized deductions. The total number of tax returns in a state varies from over 12 million in California to less than 210,000 in Wyoming.

(a) Will the sampling variability of the sample proportion change from state to state if an SRS of 2000 tax returns is selected in each state? Explain your answer.

(b) Will the sampling variability of the sample proportion change from state to state if an SRS of 1% of all tax returns is selected in each state? Explain your answer.

4.54 A national opinion poll recently estimated that 44% ($\hat{p} = 0.44$) of all American adults agree that parents of school-age children should be given vouchers good for education at any public or private school of their choice. The polling organization used a probability sampling method for which the sample proportion has a normal distribution with standard deviation about 0.015. The poll therefore announced a "margin of error" of 0.03 (two standard deviations) for its result. If a sample was drawn by the same method from the state of New Jersey (population 7.6 million) instead of from the entire United States (population 240 million), would this margin of error be larger or smaller? Explain your answer.

Most statistical software packages allow you to do simulations. This makes it practical to explore sampling variability more thoroughly than if you use Table B for simulation. Suppose, for example, that you wish to simulate 100 repetitions of an SRS of size 50 from a large population of which the proportion $p = 0.25$ would say "Yes" to a certain question. The following MINITAB commands carry out the simulation and store the 100 counts of "Yes" results as C1:

```
MTB> RANDOM 100 C1;
       BINOMIAL N = 50, P = .25.
```

You can then find the sample proportions $\hat{p}$ by dividing the counts by the sample size 50. Chapter 6 will explain why "binomial" is the magic word here. The next two exercises involve simulations to be carried out using a computer.

4.55 Draw 100 samples of size $n = 50$ from populations with $p = 0.1$, $p = 0.3$, and $p = 0.5$. Make a stemplot of the 100 values of $\hat{p}$ obtained in each simulation. Compare your three stemplots. Do they show about the same variability? How does changing the parameter p affect the sampling distribution? If your software permits, make a normal quantile plot of the $\hat{p}$ values from the population with $p = 0.5$. Is the sampling distribution approximately normal?

4.56 Draw 100 samples of each of the sizes $n = 50$, $n = 200$, and $n = 800$ from a population with $p = 0.6$. Prepare a frequency histogram of the $\hat{p}$ values for each simulation, using the same horizontal and vertical scales so that the three graphs can be compared easily. How does increasing the size of an SRS affect the sampling distribution of $\hat{p}$?

NOTES

1 P. H. Rossi, J. D. Wright, G. A. Fisher, and G. Wills, "The urban homeless: estimating composition and size," *Science*, 235 (1987), pp. 1336–1339.

2 Both observational studies and experiments on the issue of requiring job programs for welfare recipients are discussed in Constance Holden, "Is the time ripe for welfare reform?" *Science*, 238 (1987), pp. 607–609.

3 L. L. Miao, "Gastric freezing: an example of the evaluation of medical therapy by randomized clinical trials," in J. P. Bunker, B. A. Barnes, and F. Mosteller (eds.), *Costs, Risks and Benefits of Surgery*, Oxford University Press, New York, 1977, pp. 198–211.

4 G. B. Kolata, "Love Canal: false alarm caused by botched study," *Science*, 208 (1980), pp. 1239–1242.

5 The information in this example is taken from *The Ascap Survey and Your Royalties*, ASCAP, New York, undated.

6 For more detail on the material of this section and complete references, see P. E. Converse and M. W. Traugott, "Assessing the accuracy of polls and surveys," *Science*, 234 (1986), pp. 1094–1098.

7 For more detail on the limits of memory in surveys, see N. M. Bradburn, L. J. Rips, and S. K. Shevell, "Answering autobiographical questions: the impact of memory and inference on surveys," *Science*, 236 (1987), pp. 157–161.

8 This question and the response are reported in *The New York Times*, May 6, 1982. A different question, given in Exercise 4.43(c), asked at the same time drew a 56% response in favor of a nuclear freeze.

CHAPTER 4 EXERCISES

4.57 Do consumers prefer the taste of Pepsi or Coke in a blind test in which neither cola is identified? Describe briefly the design of a paired comparisons experiment to investigate this question.

4.58 Is the number of days a letter takes to reach another city affected by the time of day it is mailed and whether or not the zip code is used? Describe briefly the design of a two-factor experiment to investigate this question. Be sure to specify the treatments exactly and to tell how you will handle lurking variables such as the day of the week on which the letter is mailed.

4.59 The previous two exercises illustrate the use of statistically designed experiments to answer questions that arise in everyday life. Select a question of interest to you that an experiment might answer and carefully discuss how you would design an appropriate experiment.

4.60 There are several psychological tests available to measure the extent to which Mexican Americans are oriented toward Mexican/Spanish or Anglo/English culture. Two such tests are the Bicultural Inventory (BI) and the Acculturation Rating Scale for Mexican Americans (ARSMA). To study the correlation between the scores on these two tests, researchers will give both tests to a group of 22 Mexican Americans.

(a) Briefly describe a paired comparisons design for this study. In particular, how will you use randomization in your design?

(b) You have an alphabetical list of the subjects (numbered 1 to 22). Carry out the randomization required by your design and report the result.

4.61 A medical study of heart surgery investigates the effect of drugs called β-blockers on the pulse rate of the patient during surgery. The pulse rate will be measured at a specific point during the operation. The investigators decide to use as subjects 30 patients facing heart surgery who have consented to take part in the study. You have a list of these patients, numbered 1 to 30 in alphabetical order.

(a) Outline in graphical form a completely randomized experimental design for this study.

(b) Enter Table B at line 125 to carry out the randomization required by your design and report the result.

4.62 In a study of the relationship between physical fitness and personality, middle-aged college faculty members who have volunteered for an exercise program are divided into low-fitness and high-fitness groups on the basis of a physical examination. All subjects then take the Cattell Sixteen Personality Factor Questionnaire and the results for the two groups are compared. Is this study an experiment? Explain your answer.

4.63 A university's financial aid office wants to know how much it can expect students to earn from summer employment. This information will be used to set the level of financial aid. The population contains 3478 students who have completed at least one year of study but have not yet graduated. A questionnaire will be sent to an SRS of 100 of these students, drawn from an alphabetized list.

(a) Describe how you will label the students in order to select the sample.

(b) Use Table B, beginning at line 105, to select the *first five* students in the sample.

4.64 A labor organization wants to study the attitudes of college faculty members toward collective bargaining. These attitudes appear to differ depending on the type of college. The American Association of University Professors classifies colleges as follows:

Class I   Offer doctorate degrees and award at least
           15 per year

Class IIA Award degrees above the bachelor's but are
           not in Class I
Class IIB Award no degrees beyond the bachelor's
Class III Two-year colleges

Discuss the design of a sample of faculty from colleges in your state, with total sample size about 200.

4.65 You want to investigate the attitudes of students at your school about the faculty's commitment to teaching. The student government will pay the costs of a sample of about 500 students.

(a) Specify the exact population for your study; for example, will you include part-time students?

(b) Describe your sample design. Will you use a stratified sample?

(c) Briefly discuss the practical difficulties that you anticipate; for example, how will you contact the students in your sample?

4.66 You are participating in the design of a medical experiment to investigate whether a calcium supplement in the diet will reduce the blood pressure of middle-aged men. Preliminary work suggests that calcium may be effective and that the effect may be greater for black men than for white men.

(a) Outline in graphic form the design of an appropriate experiment.

(b) Choosing the sizes of the treatment groups requires more statistical expertise. We will learn more about this aspect of design in later chapters. Explain in plain language the advantage of using larger groups of subjects.

4.67 A chemical engineer is designing the production process for a new product. The chemical reaction that produces the product may have higher or lower yield, depending on the temperature and the stirring rate in the vessel in which the reaction takes place. The engineer decides to investigate the effects of combinations of two temperatures (50°C and 60°C) and three stirring rates (60 rpm, 90 rpm, and 120 rpm) on the yield of the process. Two batches of the feedstock will be processed at each combination of temperature and stirring rate.

(a) How many factors are there in this experiment? How many treatments? Identify each of the treatments. How many experimental units (batches of feedstock) does the experiment require?

(b) Outline in graphic form the design of an appropriate experiment.

(c) The randomization in this experiment determines the order in which batches of the feedstock will be processed according to each treatment. Use Table B starting at line 128 to carry out the randomization and state the result.

4.68 The requirement that human subjects must give their informed consent to participate in an experiment can greatly reduce the number of available subjects. In trials of treatments for cancer, for example, patients must agree to be randomly assigned to the standard therapy or to an experimental therapy. The patients who do not wish their treatment to be decided by a coin toss are dropped from the experiment and given the standard therapy. Why is it not correct to keep these patients in the experiment as part of the control group, since the control group receives the standard therapy?

4.69 You are on the staff of a member of Congress who is considering a controversial bill that would provide for government-sponsored insurance to cover care in nursing homes. You report that 1128 letters dealing with the issue have been received, of which 871 oppose the legislation. "I'm surprised that most of my constituents oppose the bill. I thought it would be quite popular," says the congresswoman. Are you convinced that a majority of the voters oppose the bill? State briefly how you would explain the statistical issue to the congresswoman.

4.70 A study of the effects of running on personality involved 231 male runners who each ran about 20 miles a week. The runners were given the Cattell Sixteen Personality Factors Questionnaire, a 187-item multiple-choice test often used by psychologists. A news report (*The New York Times*, Feb. 15, 1988) stated, "The researchers found statistically significant personality differences between the runners and the 30-year-old male population as a whole." A headline on the article said, "Research has shown that running can alter one's moods." Explain carefully, to someone who knows no statistics, why the headline is misleading.

4.71 To demonstrate how randomization reduces confounding, return to the nutrition experiment described in Example 4.7. Suppose that the 30 rats are labeled 01 to 30. Suppose also that, unknown to the experimenter, the 10 rats labeled 01 to 10 have a genetic defect that will cause them to grow more slowly than normal rats. If the experimenter simply put rats 01 to 15 in the experimental group and rats 16 to 30 in the control group, this lurking variable would bias the experiment against the new food product.

Use Table B to assign 15 rats at random to the experimental group as in Example 4.8. Record how many of the 10 rats with genetic defects are placed in the experimental group and how many are in the control group. Repeat the randomization using different lines in Table B until you have done five random assignments. What is the mean number of genetically defective rats in experimental and control groups in your five repetitions?

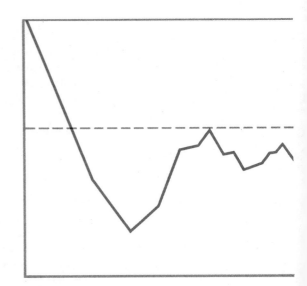

Prelude

In this chapter we prepare the way for the study of statistical inference by introducing the basic ideas of probability. Probability provides a mathematical description of randomness, such as the chance variation observed in the outcomes of randomized experiments and random samples. We will learn some fundamental rules of probability and discuss how we can misunderstand random behavior if we are ignorant of these rules.

- *How can simple rules of probability help us understand the inheritance of blood types and the reliability of undersea cables?*

- *How do our tools for describing distributions of data, such as means and standard deviations, extend to the task of describing the idealized probability distribution of a random variable?*

- *What does the law of large numbers tell us about gambling as a business for a casino, and how do gamblers and basketball players often misunderstand random behavior?*

5

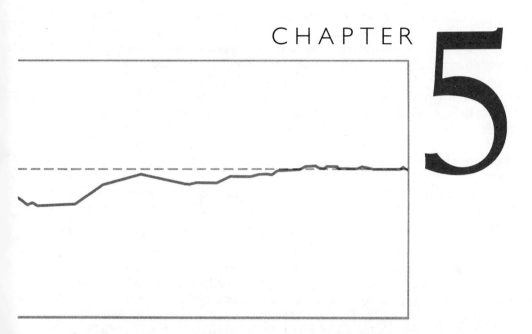

Probability: The Study of Randomness

Many things in the world can be predicted, given a sufficiently advanced science and a sufficiently large computer. Drop an apple from a known distance above the ground and elementary physics can tell you how long it will take to fall. Observe the place and motion of an asteroid and rather advanced physics can tell you where it will be a year from now. Other things in the world are unpredictable—the toss of a coin, the time between emissions of alpha particles by a radioactive source, the sexes of the next litter of lab rats. But there is a regularity to these particular phenomena, and to many others, that places them between strict predictability and unanalyzable haphazardness. Although an individual outcome is indeed uncertain, there is a regular *pattern* of outcomes that emerges in many repetitions. The next toss of the coin cannot be foretold but in the long run very close to half the tosses will be heads and half will be tails.

Random phenomenon

> A phenomenon is called random if the outcome of a single repetition is uncertain but there is nonetheless a regular distribution of relative frequencies in a large number of repetitions.

The value of a statistic computed from a probability sample or from a randomized experiment is random in this sense. Figures 4.4 and 4.8 are examples of the regular distribution that emerges when many outcomes are recorded. The long-run regularity of random phenomena can be described mathematically, just as the fall of an apple or the motion of an asteroid can be predicted. The mathematical study of randomness is called *probability theory*.

The Idea of Probability

The mathematics of probability begins with the observed fact that some phenomena are random—that is, the relative frequencies of their outcomes seem to settle down to fixed values in the long run. Consider tossing a single coin. The relative frequency of heads is quite erratic in two, five, or ten tosses. But after several thousand tosses it remains stable, changing very little over thousands of further tosses. Here are some examples of coin tossing.

- The French naturalist Buffon (1707–1788) tossed a coin 4040 times. Result: 2048 heads, or relative frequency $2048/4040 = 0.5069$ for heads.
- Around 1900, the English statistician Karl Pearson heroically tossed a coin 24,000 times. Result: 12,012 heads, a relative frequency of 0.5005.

- While imprisoned by the Germans during World War II, the English mathematician John Kerrich tossed a coin 10,000 times. Result: 5067 heads, a relative frequency of 0.5067.

If in many tosses of a coin the proportion of heads observed becomes close to 1/2, we say that the probability of a head on any single toss is 1/2. In intuitive terms, *probability is long-term relative frequency*. From this point of view, correct probabilities can only be determined empirically, by actually watching many tosses of the coin to see if about 1/2 the outcomes are heads. Of course, we can never observe a probability exactly, because we could always continue tossing the coin. Mathematical probability is an idealization based on imagining what would happen to the relative frequencies in an indefinitely long series of trials.

EXAMPLE 5.1

Here is a case slightly less obvious than tossing a single coin repeatedly to illustrate the fact that relative frequencies stabilize in the long run. We used a computer to toss *four* coins 1000 times and to record the number of heads among the four coins on each trial. We then found the relative frequency of the outcome "exactly 2 heads" after each trial. The results of the first few trials were

Trial	Outcome	Relative frequency of exactly 2 heads
1	3 heads	$\frac{0}{1} = 0$
2	0 heads	$\frac{0}{2} = 0$
3	0 heads	$\frac{0}{3} = 0$
4	2 heads	$\frac{1}{4} = .25$
5	4 heads	$\frac{1}{5} = .20$
6	2 heads	$\frac{2}{6} = .33$

Figure 5.1 plots the relative frequency of the outcome "2 heads" against the number of trials. This relative frequency is 0 for the first three trials, then jumps up to $1/4 = 0.25$ when the fourth toss yields 2 heads. The relative frequency is quite variable during the first 100 trials but gradually becomes more stable. Shortly, we will compute that the mathematical probability of 2 heads in four tosses of a balanced coin is 0.375. This probability is marked on the graph as a black horizontal line. The gradual approach of the relative frequency to the probability is clearly visible. (The horizontal scale in Figure 5.1 is not linear but logarithmic.

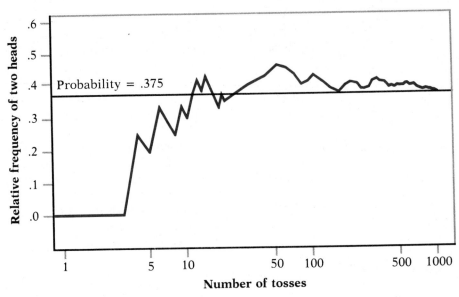

Figure 5.1 The long-term behavior of the relative frequency of two heads in many tosses of four coins. The relative frequency approaches the probability 0.375.

This choice makes it easier to see the erratic early fluctuations in the relative frequency while still allowing 1000 tosses to fit in the graph.) ■

Long-term relative frequency is not the only intuitive interpretation of probability. From another point of view, observing many outcomes is only one source of personal opinion about the chance that the next toss will produce a head. Probability describes my personal opinion as reflected, for example, in the odds I am willing to accept in betting on the outcome. If I will accept even odds on the toss of a coin, my personal probability of a head is 1/2. *Probability as personal opinion* frees us from having to imagine many repetitions. This freedom has the great advantage that a probability can even be assigned to one-time events (for example, "I think the probability that the Bears will win the Superbowl this year is 1/4"). On the other hand, the broad scope of personal probability tends to lead us away from long-term regularity of relative frequencies as an observed fact that we want to describe mathematically.

personal probability

The mathematics of probability is motivated by these common-sense notions of what probability means. The mathematics makes more sense if you keep in mind long-term relative frequency or personal assessment of chance. Statistical designs for sampling or experimentation deliberately introduce randomness so that there will be a regular long-term pattern. We will therefore concentrate on the relative frequency interpretation of probability.

The Uses of Probability

Probability theory originated in the study of games of chance. Tossing dice, dealing shuffled cards, and spinning a roulette wheel are examples of deliberate randomization that are similar to random sampling. Although games of chance are ancient, they were not studied by mathematicians until the sixteenth and seventeenth centuries. It is only a mild simplification to say that probability as a branch of mathematics arose when seventeenth-century French gamblers asked the mathematicians Blaise Pascal and Pierre de Fermat for help. Gambling is still with us, in casinos and state lotteries. We will discuss games of chance as simple examples that illustrate the principles of probability. Careful measurements in astronomy and surveying led to further advances in probability in the eighteenth and nineteenth centuries because the results of repeated measurements are random and can be described by relative frequency distributions much like those arising from random sampling. Similar distributions appear in data on human lifespans (mortality tables) and in data on lengths or weights in a population of skulls, leaves, or cockroaches.[1] In the twentieth century, probability has been used to describe the flow of traffic through a highway system, a telephone interchange, or a computer processor; the genetic makeup of individuals or populations; the energy states of subatomic particles; the spread of epidemics or rumors; and the rate of return on risky investments. Although we are interested in probability because of its usefulness in statistics, the mathematics of chance is important in many fields of study.

In applying probability calculations to the analysis of data, our intent is to replace the uncertainties of "eyeball appraisal" by the exactness and power of mathematics. You or I might guess that the beam above our heads is strong enough to keep the building from collapsing but we feel safer knowing that an engineer has calculated it. Similarly, we might guess that the observed mean difference between the response of the treatment and control groups indicates a real difference between treatments, but we will feel more confident if we can back our eyeball appraisal with calculations.

5.1 PROBABILITY MODELS

Earlier chapters gave mathematical models for linear and exponential growth (in the form of equations) and for some distributions of data (in the form of normal density curves). Now we must give a mathematical description of randomness. To see how to proceed, think first about a very simple random phenomenon, tossing a coin once. When we toss a coin, we cannot know the outcome in advance. What do we know? We are willing to say that the outcome will be either heads or tails. Because the coin appears to be balanced, we believe that each of these outcomes has probability 1/2. This description of coin tossing has two parts:

- A list of possible outcomes
- A probability for each outcome

Such a description is the basis of all probability models. We will begin by describing the outcomes of a random phenomenon and then learn how to assign probabilities to the outcomes.

Sample Spaces

The first element of a probability model is a statement of the outcomes that are possible.

Sample space

> The sample space S of a random phenomenon is the set of all possible outcomes.

The name "sample space" is natural in random sampling, where each possible outcome is a sample and the sample space contains all possible samples.

To specify S, we must state first what constitutes an individual outcome and then which outcomes can occur. We often have some freedom in defining the sample space, so the choice of S is a matter of convenience as well as correctness. The idea of a sample space, and the freedom we may have in specifying it, are best illustrated by some examples.

EXAMPLE 5.2 Toss a coin. There are only two possible outcomes, and the sample space is

$$S = \{\text{Heads, Tails}\}$$

or, abbreviated, $S = \{\text{H, T}\}$. ■

EXAMPLE 5.3 Let your pencil point fall blindly into Table B of random digits and record the value of the digit you land on. Plainly,

$$S = \{0, 1, 2, 3, 4, 5, 6, 7, 8, 9\}$$ ■

EXAMPLE 5.4 Toss a coin four times and record the results. That's a bit vague. To be exact, record the results of each of the four tosses in order. A typical outcome is then HTTH, and the sample space S is the set of all such strings of four H's and T's.

Suppose that your only interest is the number of heads in four tosses. Now you can be exact in a simpler fashion: The random phenomenon is to toss a coin four times and count the number of heads. The sample space is

$$S = \{0, 1, 2, 3, 4\}$$

This example illustrates the importance of carefully specifying what constitutes an individual outcome. ∎

Although these examples seem remote from the practice of statistics, the connection is surprisingly close. Suppose that in the course of conducting an opinion poll you select four people at random from a large population and ask each if he or she favors reducing federal spending on low-interest student loans. The answers in each case are "Yes" or "No." The possible outcomes—the sample space—are exactly as in Example 5.4 if heads is simply replaced by "Yes" and tails by "No." Similarly, the possible outcomes of an SRS of 1500 people are the same in principle as the possible outcomes of tossing a coin 1500 times. One of the great advantages of mathematics is that the essential features of quite different phenomena can be described by the same mathematical model.

EXAMPLE 5.5

Many computing systems have a function that generates a random number between 0 and 1. For example, the MINITAB command

```
MTB > RANDOM 1 C1;
   SUBC > UNIFORM 0 1.
```

puts such a number into C1. Since the computer routine generates a random number between 0 and 1, take

$$S = \{\text{all numbers between 0 and 1}\}$$

This S is a mathematical idealization. Any specific random number generator produces numbers with some limited number of decimal places so that, strictly speaking, not all numbers between 0 and 1 are possible outcomes. The entire interval from 0 to 1 is easier to think about. It also has the advantage of being a suitable sample space for different computers that produce random numbers with different numbers of significant digits. ∎

Random variables Sample spaces need not be sets of numbers. In Example 5.4, the first S for tossing four coins consisted of all possible strings of four H's or T's. When we were interested only in the number of heads, however, we could use a simpler sample space whose outcomes are numbers. Since our goal is to describe the random behavior of statistics, which are numbers computed from a sample, it is often convenient to concentrate on numerical outcomes only.

Random variable

> A random variable is a numerical outcome of a random phenomenon.

Random variables are denoted by capital letters near the end of the alphabet. Like sample spaces, they are most easily understood by example.

EXAMPLE 5.6

Let the random variable X be the number of heads in four tosses of a coin. The possible values of X are 0, 1, 2, 3, and 4. Tossing a coin four times will give X one of these possible values. Tossing four more times will give X another and probably different value. X is called a random *variable* because its values vary when the coin tossing is repeated. ∎

In Example 5.4, two choices of sample space were offered for four tosses of a coin. The second amounted to a list of the possible values of X, the number of heads. The first choice of S listed the outcomes of the four tosses in order, such as HTTH. From such a detailed outcome, the value of X can be obtained. If the outcome is HTTH, then $X = 2$; if the outcome is TTTT, then $X = 0$; if the outcome is THTH, $X = 2$ once more. Mathematically, the random variable X is a function that assigns a number to each member of the sample space S. This is to say again, in fancier language, that X is a numerical outcome of a random phenomenon.

Assigning Probabilities

Whether we choose to give a sample space S or to directly describe a random variable X, we are listing the possible outcomes of a random phenomenon. To complete a mathematical model for the random phenomenon, we must also give the probability with which these outcomes occur.

The true long-term relative frequency of any outcome—say, "2 heads in four tosses of a coin"—can be found only by experiment, and then only approximately. How then can we describe probability mathematically? Rather than immediately attempt to give "correct" probabilities, let's confront the easier task of laying down rules that any assignment of probabilities must satisfy. We need to assign probabilities not only to single outcomes but also to sets of outcomes.

Event

> An event is a set of outcomes of a random phenomenon, that is, a subset of the sample space.

EXAMPLE 5.7

Take the sample space S for four tosses of a coin to be the list of all possible outcomes in the form HTHH. Then, "exactly 2 heads" is an event. We will call this event A. The event A expressed as a set of outcomes is

$$A = \{\text{HHTT, HTHT, HTTH, THHT, THTH, TTHH}\}$$

Listing the outcomes in an event such as A is tedious; the verbal description "exactly 2 heads in four tosses" is shorter. If we let the random variable X be the number of heads, we have an even shorter description: A is the event that

$X = 2$, which we write as $\{X = 2\}$. The event "2 or more heads" can similarly be described as $\{X \geq 2\}$. We will most often describe random outcomes by random variables, and the events of most interest to us will be sets of values of a random variable.   ■

In a probability model, events have probabilities. We will write the probability of the event A as $P(A)$. What properties must any assignment of probabilities to events have? A relative frequency is always a number between 0 and 1. The sample space S contains all possible outcomes and therefore always has relative frequency 1. If we think of probability as long-run relative frequency, we are led to two rules that any assignment of probabilities must obey.

Basic probability rules

Rule 1. Any probability $P(A)$ is a number between 0 and 1—that is, $0 \leq P(A) \leq 1$.

Rule 2. The collection S of all possible outcomes has probability 1— that is, $P(S) = 1$.

There are also other rules that any legitimate assignment of probabilities to events must obey. Before we meet these rules, let's look at actual assignments of probability in Examples 5.2 and 5.3.

EXAMPLE 5.8

In tossing a coin (Example 5.2) there are only two outcomes in S. We simply assign probabilites to each of them. If we believe that the coin is balanced, then we assign

$$P(H) = \frac{1}{2}$$

$$P(T) = \frac{1}{2}$$

If observation convinces us that the coin is unbalanced so as to land tails more often, we might assign probabilities

$$P(H) = .4$$
$$P(T) = .6$$ ■

In Example 5.8, any assignment of two probabilities (numbers between 0 and 1) with sum 1 satisfies Rules 1 and 2. We can determine which assignment is correct for a particular coin only by observing many tosses or perhaps by carefully weighing and measuring the coin to determine its balance. This reinforces the lesson that our mathematics only says which assignments of probability make sense, not which one is actually correct.

EXAMPLE 5.9

The successive digits in Table B (Example 5.3) were produced by a careful randomization, so when we drop a pencil point blindly into the table the digit X selected is equally likely to be any of the 10 candidates. Since the total probability must be 1, the probabilities of the outcomes of the random variable X are

$$P(X = 0) = .1 \qquad P(X = 5) = .1$$
$$P(X = 1) = .1 \qquad P(X = 6) = .1$$
$$P(X = 2) = .1 \qquad P(X = 7) = .1$$
$$P(X = 3) = .1 \qquad P(X = 8) = .1$$
$$P(X = 4) = .1 \qquad P(X = 9) = .1$$

This assignment of probabilities to individual outcomes can be summarized in a table as follows:

Outcome	0	1	2	3	4	5	6	7	8	9
Probability	.1	.1	.1	.1	.1	.1	.1	.1	.1	.1

We must assign probability to all events, not just to individual outcomes. The probability of an event in this example is simply the sum of the probabilities of the outcomes making up the event. For example, the probability that an odd digit is chosen is

$$P(X \text{ is odd}) = P(1) + P(3) + P(5) + P(7) + P(9) = .5$$

This assignment of probability satisfies Rules 1 and 2.   ∎

Examples 5.8 and 5.9 illustrate one way of assigning probabilities to events when a random phenomenon has only a finite number of possible outcomes.

Probabilities in a finite sample space

> 1. Assign a probability to each individual outcome. These probabilities must be numbers between 0 and 1, and they must have sum 1.
> 2. The probability of any event is the sum of the probabilities of the outcomes making up the event.

In Example 5.9, all outcomes were equally likely. This need not always be the case, as the following example illustrates.

EXAMPLE 5.10

Suppose that we select one M&M candy at random from a large bag and record its color. The sample space is

$$S = \{\text{brown, green, orange, red, tan, yellow}\}$$

The probability of each outcome will be the proportion of all M&M's made that have that color. (As usual, this is an idealization that imagines the result of looking at one piece of candy after another forever.) From the manufacturer we can learn that the correct assignment of probabilities to the outcomes is

Color	Brown	Red	Yellow	Green	Orange	Tan
Probability	.3	.2	.2	.1	.1	.1

Because these probabilities have sum 1, we can assign probabilities to events by adding them up, just as we did in Example 5.9. For example,

$$P(\text{red or orange}) = P(\text{red}) + P(\text{orange})$$
$$= .2 + .1 = .3 \qquad \blacksquare$$

Probability Rules

Rules 1 and 2 concern the assignment of probabilities to single events. To complete a basic description of probability, we must learn a few additional rules that govern how probabilities of events combine. Three such rules are important for statistics. In each case, we first describe a particular relation between two events and then the probability rule that applies to this relation.

Disjoint events   Suppose that a large dormitory contains 40% freshmen, 20% seniors, and 40% belonging to other classes. The relative frequency of students who are *either* freshmen *or* seniors is 60%. We can add the relative frequencies of freshmen and seniors because no student can be both a freshman and a senior. This simple fact about relative frequencies leads to a third basic law of probability.

Disjoint events

> Events that have no outcomes in common are called disjoint events.

Select a student at random from the inhabitants of the dormitory and record the student's class. "Freshman" and "senior" are disjoint events. Figure 5.2 illustrates disjoint events A and B as nonoverlapping areas. Diagrams like Figure 5.2 that picture the sample space S as an area in the plane and events as areas within S are called *Venn diagrams*. Venn diagrams are help-
Venn diagrams   ful in showing relations such as disjointness.

Two disjoint events cannot both occur on the same trial of a random phenomenon. The long-term relative frequency that "one or the other" occurs is therefore the sum of their individual relative frequencies. So another rule that any legitimate assignment of probability must satisfy is the addition rule for disjoint events.

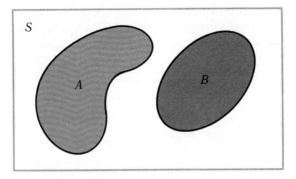

Figure 5.2   Venn diagram showing disjoint events *A* and *B*.

Addition rule for disjoint events

Rule 3. If events *A* and *B* are disjoint, then
$$P(A \text{ or } B) = P(A) + P(B)$$

EXAMPLE 5.11

The addition rule is satisfied when we assign probability by summing the probabilities of individual outcomes. Example 5.9 gives an assignment of probabilities for choosing a random digit *X* between 0 and 9. Consider the events

$$A = \{X \text{ is odd}\}$$
$$B = \{X \text{ is divisible by } 4\}$$

The outcomes in *A* are 1, 3, 5, 7, and 9, so $P(A) = 0.5$. The event *B* contains outcomes 4 and 8, with $P(B) = 0.2$. What is the probability of the event $\{A \text{ or } B\}$ that the random digit is *either* odd *or* divisible by 4? This event contains all of the outcomes in either *A* or *B*, namely 1, 3, 4, 5, 7, 8, and 9. Adding up the probabilities of these outcomes from the table in Example 5.9 gives probability 0.7. Because *A* and *B* have no outcomes in common, the same result is obtained by adding $P(A)$ and $P(B)$.

$$P(A \text{ or } B) = P(A) + P(B)$$
$$= .5 + .2 = .7 \qquad \blacksquare$$

Complements   The addition rule leads to another important law of probability, the complement rule. If 20% of the students in a dormitory are seniors, then 80% are not seniors. That is, the relative frequency that a student is a senior and the relative frequency that a student is *not* a senior must add to 100%. This is true because every student either is or is not a senior and no student is both.

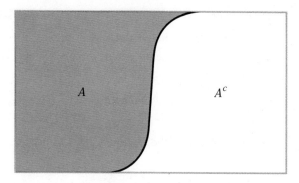

Figure 5.3 Venn diagram showing the complement A^c of an event A.

Complement of an event

If A is any event, the event that A does *not* occur is called the complement of A, written A^c.

Figure 5.3 is a Venn diagram that illustrates how an event and its complement partition the sample space between them. Notice that A and A^c are always disjoint events. The complement A^c contains exactly those outcomes that are not in A. The long-term relative frequency of A^c is therefore one minus the relative frequency of A because A^c occurs exactly when A does not. This fact is another of our basic rules for probability.

Complement rule

Rule 4. For any event A, the probability that A does not occur is

$$P(A^c) = 1 - P(A)$$

EXAMPLE 5.12

A sociologist studies social mobility in England by recording the social class of a large sample of fathers and their sons. Social class is determined by such factors as education and occupation. The social classes are ordered from Class 1 (lowest) to Class 5 (highest). Here are the probabilities that the son of a lower class (Class 1) father will end up in each social class.

Son's class	1	2	3	4	5
Probability	.48	.38	.08	.05	.01

Consider the events

$$A = \{\text{Son remains in Class 1}\}$$
$$B = \{\text{Son reaches one of the two highest classes}\}$$

From the table of probabilities,

$$P(A) = .48$$
$$P(B) = .05 + .01 = .06$$

What is the probability that the son of a Class 1 father does *not* remain in Class 1? This is

$$P(A^c) = 1 - P(A)$$
$$= 1 - .48 = .52$$

The events A and B are disjoint so the probability that the son of a lower class father either remains in the lower class or reaches one of the two top classes is

$$P(A \text{ or } B) = P(A) + P(B)$$
$$= .48 + .06 = .54 \qquad \blacksquare$$

Independence   Rule 3, the addition rule for disjoint events, describes the probability that *one or the other* of two events A and B will occur in the special situation when A and B cannot occur together because they are disjoint. Our final rule describes the probability that *both* events A and B occur, again only in a special situation. More general rules appear in Section 5.4, but in our study of statistics, we will need only those rules that apply to special situations.

Suppose that you toss a balanced coin twice. You are counting heads, so the events of interest are

$$A = \{\text{first toss is a head}\}$$
$$B = \{\text{second toss is a head}\}$$

The events A and B are not disjoint. They occur together whenever both tosses give heads. We want to compute the probability of the event $\{A$ and $B\}$ that *both* tosses are heads. Figure 5.4 is a Venn diagram that illustrates the event $\{A$ and $B\}$ as the overlapping area that is common to both A and B.

The coin-tossing experiments of Buffon, Pearson, and Kerrich described at the beginning of this chapter make us willing to assign probability 1/2 to a head when we toss a coin. So

$$P(A) = .5$$
$$P(B) = .5$$

What is $P(A$ and $B)$? Our common sense says that it is 1/4. The first coin will give a head half the time and then the second will give a head on half of those

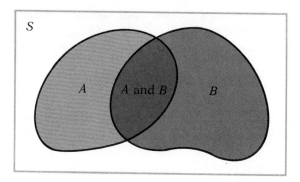

Figure 5.4 Venn diagram showing the event {A and B}.

trials, so both coins will give heads on 1/2 × 1/2 = 1/4 of all trials in the long run. This reasoning assumes that the second coin still has probability 1/2 of a head after the first has given a head. This is true—we can verify it by tossing two coins many times and observing the relative frequency of heads on the second toss after the first toss has produced a head. We say that the events "head on the first toss" and "head on the second toss" are independent.

Independence

> Events A and B are independent if knowing whether A occurs does not change the probability that B occurs. Two random variables X and Y are independent if any event involving only X is independent of any event involving only Y.

This definition is rather informal. A more exact definition appears in Section 5.4. In practice, though, we rarely need a precise definition of independence, because independence is usually *assumed* as part of a probability model when we want to describe random phenomena that seem to be physically independent of each other.

EXAMPLE 5.13

Because a coin has no memory and most coin-tossers cannot influence the fall of the coin, it is safe to assume that successive coin tosses are independent. For a balanced coin this means that after we see the outcome of the first toss, we still assign probability 1/2 to heads on the second toss.

On the other hand, the colors of successive cards dealt from a deck are not independent. A standard 52-card deck contains 26 red and 26 black cards. For the first card dealt from a shuffled deck, the probability of a red card is 26/52 = .50. Once we see that the first card is red, we know that there are only 25 reds left among the remaining 51 cards. The probability that the second card is red is therefore only 25/51 = .49. Knowing the outcome of the first deal changes the probabilities for the second.

If a doctor measures your blood pressure twice, it is reasonable to assume that the results X and Y are independent because the first result does not influence the instrument that takes the second reading. But if you take an IQ test or other mental test twice in succession, the test scores X and Y are not independent. The learning that occurs on the first attempt influences your second attempt. ■

As in the case of tossing two coins, independence leads to a multiplication rule for the probability that both of two events occur.

Multiplication rule for independent events

Rule 5. If events A and B are independent, then

$$P(A \text{ and } B) = P(A)P(B)$$

EXAMPLE 5.14

A person's blood type is inherited from the parents. Inheritance operates randomly. Each parent carries two genes for blood type, and each of these genes has probability 1/2 of being passed to the child. The two genes that the child receives, one from each parent, determine the child's blood type. The parents contribute their genes independently of each other.

Suppose that both parents carry the O and the A genes. The child will have blood type O if both parents contribute an O gene; otherwise, the child's blood type will be A. If M is the event that the mother contributes her O gene and F is the event that the father contributes his O gene, then the probability that the child has type O blood is

$$P(M \text{ and } F) = P(M)P(F)$$
$$= (.5)(.5) = .25$$

In the long run, 1/4 of all children born to such parents will have type O blood.
                                                                                                          ■

The multiplication rule $P(A \text{ and } B) = P(A)P(B)$ is valid if A and B are independent but not otherwise. The addition rule $P(A \text{ or } B) = P(A) + P(B)$ is valid if A and B are disjoint but not otherwise. Resist the temptation to use these simple formulas when the circumstances that justify them are not present. You must also be certain not to confuse disjointness and independence. If A and B are disjoint, then the fact that A occurs tells us that B cannot occur. So disjoint events are not independent. Unlike disjointness or complements, independence cannot be pictured by a Venn diagram because it involves the probabilities of the events rather than just the outcomes that make up the events.

Like the rule for complements, the multiplication rule for independence can actually be derived from the first three rules of probability. We are not concerned with such derivations, but it is useful to note that we need check only the first three rules to be sure that an assignment of probabilities to events is legitimate. All other probability laws follow from these three.*

The multiplication rule extends to collections of more than two events, provided that all are independent. Independence of events *A*, *B*, and *C* means that no information about any one or any two can change the probability of the remaining events. The formal definition is a bit messy. Fortunately, independence is usually assumed in setting up a probability model. We can then use the multiplication rule freely, as in this example.

EXAMPLE 5.15

A transatlantic telephone cable contains repeaters at regular intervals to amplify the signal. If a repeater fails it must be replaced by fishing the cable to the surface at great expense. Each repeater has probability 0.999 of functioning without failure for 10 years. Repeaters fail independently of each other. (This assumption means that there are no "common causes" such as current surges in the cable that would affect several repeaters at once.) Denote by A_i the event that the *i*th repeater operates successfully for 10 years.

The probability that two repeaters both last 10 years is

$$P(A_1 \text{ and } A_2) = P(A_1)P(A_2)$$
$$= .999 \times .999 = .998$$

For a cable with 10 repeaters the probability of no failures in 10 years is

$$P(A_1 \text{ and } A_2 \text{ and } \ldots \text{ and } A_{10}) = P(A_1)P(A_2) \cdots P(A_{10})$$
$$= .999 \times .999 \times \cdots \times .999$$
$$= .999^{10} = .990$$

Cables with 2 or 10 repeaters are quite reliable. Unfortunately, a transatlantic cable has 300 repeaters. The probability that all 300 work for 10 years is

$$P(A_1 \text{ and } \ldots \text{ and } A_{300}) = .999^{300} = .741$$

There is therefore about one chance in four that the cable will have to be fished up for replacement of a repeater sometime during the next 10 years. Repeaters are in fact designed to be much more reliable than 0.999 in 10 years. Some transatlantic cables have served for more than 20 years with no failures. ∎

By combining the rules we have learned, we can compute probabilities for rather complex events. Here is an example.

EXAMPLE 5.16

A diagnostic test for the presence of the AIDS virus has probability 0.005 of yielding a false positive. That is, when a person free of the AIDS virus is tested, the test has probability 0.005 of falsely indicating that the virus is present. If the 140

---

* Although probability is an old branch of mathematics, the idea of deriving all probability laws from a few basic rules is rather recent. The Russian mathematician A. N. Kolmogorov did this in 1933. (Actually, a more elaborate version of Rule 3 is needed to obtain all of the usual theory of probability.)

employees of a medical clinic are tested and all 140 are free of AIDS, what is the probability that at least one false positive will occur?

It is reasonable to assume as part of the probability model that the test results for different individuals are independent. The probability that the test is positive for a single person is 0.005, so the probability of a negative result is $1 - 0.005 = 0.995$ by the complement rule. The probability of at least one false positive among the 140 people tested is therefore

$$
\begin{aligned}
P(\text{at least one positive}) &= 1 - P(\text{no positives}) \\
&= 1 - P(140 \text{ negatives}) \\
&= 1 - .995^{140} \\
&= 1 - .496 = 0.504
\end{aligned}
$$

The probability is greater than 1/2 that at least one of the 140 people will test positive for AIDS, even though no one has the virus.   ■

SUMMARY

Probability can be viewed intuitively as relative frequency in many repeated trials of a random phenomenon. Alternatively, a probability can express a personal assessment of chance. Because we are familiar with sampling distributions, we emphasize the interpretation of probability as idealized long-run relative frequency.

A **probability model** for a random phenomenon consists of a **sample space** S and an assignment of probabilities P. S is the set of all possible outcomes of the random phenomenon. Sets of outcomes are called **events**. P assigns a number $P(A)$ to an event A as its probability.

Events A and B are called **disjoint** if they have no outcomes in common. The **complement** A^c of an event A consists of exactly the outcomes that are not in A. Events A and B are called **independent** if knowing whether one event occurs does not change the probability of the occurrence of the other event.

Any assignment of probability must obey the rules that state the basic properties of probability:

     1 $0 \leq P(A) \leq 1$ for any event A.

     2 $P(S) = 1$.

     3 **Addition rule**: If A and B are disjoint events, then
         $P(A \text{ or } B) = P(A) + P(B)$.

     4 **Complement rule**: For any event A, $P(A^c) = 1 - P(A)$.

     5 **Multiplication rule**: If events A and B are independent, then
         $P(A \text{ and } B) = P(A)P(B)$.

SECTION 5.1 EXERCISES

5.1 Suppose that instead of tossing a penny, you spin it on a smooth surface and observe which face is up when the coin finally falls. Assign approximate probabilities to the outcomes {heads, tails} by making at least 20 trials and using the relative frequencies to estimate the probabilities. (Hold the penny upright with a finger and snap it with a finger of the other hand to start it spinning. Do not count trials in which you fail to spin the coin well or in which the coin hits an object before falling.) If possible, combine your results with those of other students to obtain long-term relative frequencies that are closer to the probabilities.

5.2 In the game of *heads or tails*, Betty and Bob toss a coin four times. Betty wins a dollar from Bob for each head and pays Bob a dollar for each tail—that is, she wins or loses the difference between the number of heads and the number of tails. For example, if there are one head and three tails, Betty loses $2. You can check that Betty's possible outcomes are

$$\{-4, -2, 0, 2, 4\}$$

Assign probabilities to these outcomes by playing the game 20 times and using the relative frequencies of the outcomes as estimates of the probabilities. If possible, combine your trials with those of other students to obtain long-run relative frequencies that are closer to the probabilities.

5.3 In each of the following cases, describe a sample space S for the indicated random phenomenon. In some cases, you have some freedom in your choice of S.

(a) A seed is planted in the ground. It either germinates or fails to grow.

(b) A patient with a usually fatal form of cancer is given a new treatment. The response variable is the length of time that the patient lives after treatment.

(c) A student enrolls in a statistics course and at the end of the semester receives a letter grade.

(d) A basketball player shoots two free throws.

(e) A year after knee surgery, a patient is asked to rate the amount of pain in the knee. A seven-point scale is used, with 1 corresponding to no pain and 7 corresponding to extreme discomfort.

5.4 In each of the following cases, describe a sample space S for the indicated random phenomenon. In some cases you have some freedom in specifying S, especially in setting the largest and smallest value in S.

(a) Choose a student in your class at random. Ask how much time that student spent studying during the past 24 hours.

(b) The Physicians' Health Study asked 11,000 physicians to take an aspirin every other day and observed how many of them had a heart attack in a 5-year period.

(c) In a test of a new package design, a carton of a dozen eggs is dropped from a height of 1 foot. The number of broken eggs is counted.

(d) Choose a student in your class at random. Ask how much cash that student is carrying.

(e) A nutrition researcher feeds a new diet to a young male white rat. The response variable is the weight (in grams) that the rat gains in 8 weeks.

5.5 Let the random variable X be the maximum daily rainfall (in inches) recorded next year at South Bend, Indiana. Give a reasonable sample space S for the possible values of X. (Exercise 1.20 has past values of X to guide you.)

5.6 Let the random variable X be the number of calories in a hot dog. Give a reasonable sample space S for the possible values of X. (Table 1.4 contains some typical values to guide you.)

5.7 All human blood can be typed as one of O, A, B, or AB, but the distribution of the types varies a bit with race. Here is the distribution of the blood type of a randomly chosen black American. If this is to be a legitimate assignment of probabilities, what must be the probability of type AB blood?

Blood type	O	A	B	AB
Probability	.49	.27	.20	.04

5.8 The distribution of colors in M&M peanut candies differs from the distribution given in Example 5.10 for plain M&M's. The table below gives the probability that a randomly chosen peanut M&M has each color. What must the probability for tan candies be to make this a legitimate assignment of probabilities?

Color	Brown	Red	Yellow	Green	Orange	Tan
Probability	.3	.2	.2	.2	.1	

5.9 Here are several assignments of probabilities to the six faces of a die. We can learn which assignment is actually *accurate* for a particular die only by rolling the die many times. However, some of the assignments violate Rule 1 or 2, and so are not *legitimate* assignments of probability. Which are legitimate and which are not? In the case of the illegitimate models, explain what is wrong.

Outcome	Model 1	Model 2	Model 3	Model 4
⚀	$\frac{1}{3}$	$\frac{1}{6}$	$\frac{1}{7}$	$\frac{1}{3}$
⚁	0	$\frac{1}{6}$	$\frac{1}{7}$	$\frac{1}{3}$
⚂	$\frac{1}{6}$	$\frac{1}{6}$	$\frac{1}{7}$	$-\frac{1}{6}$
⚃	0	$\frac{1}{6}$	$\frac{1}{7}$	$-\frac{1}{6}$
⚄	$\frac{1}{6}$	$\frac{1}{6}$	$\frac{1}{7}$	$\frac{1}{3}$
⚅	$\frac{1}{3}$	$\frac{1}{6}$	$\frac{1}{7}$	$\frac{1}{3}$

5.10 In each of the following cases, state whether or not the given assignment of probabilities to individual outcomes is legitimate—that is, do they satisfy Rules 1 and 2? If not, give specific reasons for your answer.

(a) When a coin is spun, $P(H) = 0.55$ and $P(T) = 0.45$.

(b) When two coins are tossed, $P(HH) = 0.4$, $P(HT) = 0.4$, $P(TH) = 0.4$, and $P(TT) = 0.4$.

(c) The mixture of colors for M&M's given in Example 5.10 is quite new. Previously, there were no red candies and the other five colors had the same probabilities that are given in Example 5.10.

5.11 Las Vegas Zeke, when asked to predict the Atlantic Coast Conference basketball champion, follows the modern practice of giving probabilistic predictions. He says, "North Carolina's probability of winning is twice Duke's. North Carolina State and Virginia each have probability 0.1 of winning, but Duke's probability is three times that. Nobody else has a chance." Has Zeke given a legitimate assignment of probabilities to the eight teams in the conference?

5.12 You draw an M&M from a bag, with an outcome governed by the probabilities given in Example 5.10. Find the probability of each of the following events:

(a) You select brown or red.
(b) You select green, red, or tan.
(c) The M&M you draw is not yellow.
(d) The M&M you select is neither orange nor tan.
(e) You select brown, red, yellow, green, orange, or tan.

5.13 A sociologist studying social mobility in Denmark finds that the probability that the son of a lower class father remains in the lower class is 0.46. What is the probability that the son moves to one of the higher classes?

5.14 The probability that a randomly chosen American woman aged 20 to 24 is married is 0.38. What is the probability that she is not married?

5.15 Government data assign a single cause for each death that occurs in the United States. The data show that the probability is 0.48 that a randomly chosen death was due to cardiovascular (mainly heart) diseases, and 0.22 that it was due to cancer. What is the probability that a death was due either to cardiovascular disease or to cancer? What is the probability that the death was due to some other cause?

5.16 Choose an acre of land in the United States at random. The probability is 0.29 that it is forested and 0.26 that it is pasture. What is the probability that the land chosen is either forest or pasture? What is the probability that the land is anything other than forest or pasture?

5.17 Select a college freshman at random and ask what his or her academic rank was in high school. Here are the probabilities, based on relative frequencies from a large sample survey of freshmen.

Outcome	Top 20%	Second 20%	Third 20%	Fourth 20%	Lowest 20%
Probability	.41	.23	.29	.06	.01

Let A be the event that the student selected ranked in the top 40% in high school, and B be the event that the student fell in the lower 40%.

(a) Find $P(A)$ and $P(B)$.
(b) Describe the event A^c in words. Find $P(A^c)$ in two ways, first by adding the probabilities of outcomes and then by the complement rule.
(c) The events A and B are disjoint. Find $P(A \text{ or } B)$ in two ways, first by adding the probabilities of outcomes and then by the addition rule.

5.18 Choose an American farm at random and measure its size X in acres. Here are the probabilities that the farm chosen falls in several acreage categories.

Acres	<10	10–49	50–99	100–179	180–499	500–999	1000–1999	≥2000
Probability	.08	.20	.15	.16	.24	.09	.05	.03

$P(B) = .17$

$P(A) = .28$

$P(A \text{ or } B) = P(A) + P(B)$
$(- P(AB))$

Let A be the event that the farm is less than 50 acres in size, and B be the event that it is 500 acres or more.

(a) Find $P(A)$ and $P(B)$.

(b) Describe A^c in words and find $P(A^c)$ by the complement rule. $.72$

(c) Describe $\{A \text{ or } B\}$ in words and find its probability by the addition rule. <50 or >500 $.28 + .17$

5.19 Choose an American worker at random and classify his or her occupation into one of the following classes. These classes are used in government employment data.

A   Managerial and professional
B   Technical, sales, administrative support
C   Service occupations
D   Precision production, craft, and repair
E   Operators, fabricators, and laborers
F   Farming, forestry, and fishing

The following gives the probabilities that a randomly chosen worker falls into each of 12 sex-by-occupation classes:

	A	B	C	D	E	F
Male	.14	.11	.05	.11	.13	.025
Female	.10	.20	..08	.01	.04	.005

(a) Verify that this is a legitimate assignment of probabilities to these outcomes. $0 \le x \le 1$ $sum = 1$

(b) What is the probability that the worker is female?

(c) What is the probability that the worker is not engaged in farming, forestry, or fishing?

(d) Classes D and E include most mechanical and factory jobs. What is the probability that the worker holds a job in one of these classes?

(e) What is the probability that the worker does not hold a job in classes D or E?

5.20 A general can plan a campaign to fight one major battle or three small battles. He believes that he has probability 0.6 of winning the large battle and probability 0.8 of winning each of the small battles. Victories or defeats in the small battles are independent. The general must win either the large battle or all three small battles to win the campaign. Which strategy should he choose?

5.21 An automobile manufacturer buys computer chips from a supplier. The manufacturer has written quality control standards that will accept shipments containing an estimated 1% defective chips. Suppose that each chip has probability 0.01 of being defective and that each automobile uses 12 chips selected independently. What is the probability that all 12 chips in a car will work properly?

5.22 A college official claims that television sets are present in 70% of the dormitory rooms on campus and that stereo systems are found in 80% of the rooms. A student says because (0.7)(0.8) = 0.56, we can conclude that 56% of the rooms have *both* a TV and a stereo. Do you agree? Explain your answer.

5.23 A string of Christmas lights contains 20 lights. The lights are wired in series, so that if any light fails the whole string will go dark. Each light has probability 0.02 of failing during a 3-year period. The lights fail independently of each other. What is the probability that a string of lights will remain bright for 3 years?

5.24 A six-sided die has four green and two red faces, and is balanced so that each face is equally likely to come up. The die will be rolled several times. You must choose one of the following three sequences of colors; you will win $25 if the first rolls of the die give the sequence you have chosen.

RGRRR

RGRRRG

GRRRR

Which sequence should you choose? Explain your choice. (In a psychological experiment, 63% of 260 students who had not studied probability chose the second sequence. This is evidence that our intuitive understanding of probability is not very accurate. This and similar experiments are reported by A. Tversky and D. Kahneman, "Extensional versus intuitive reasoning: the conjunction fallacy in probability judgment," *Psychological Review*, 90 (1983), pp. 293–315.)

5.25 Looking back at Example 5.14 which dealt with inheritance of blood types, suppose that both parents carry genes for blood types A and B. The child will have blood type A if both parents pass their A genes, type B if both pass their B genes, and type AB if one A and one B gene are passed. What are the probabilities that a child of these parents has type A blood? Type B? Type AB?

5.26 The "random walk" theory of securities prices holds that price movements in disjoint time periods are independent of each other. Suppose that we record only whether the price is up or down each year, and that the probability that our portfolio rises in price in any one year is 0.65. (This probability is approximately correct for a portfolio containing equal dollar amounts of all common stocks listed on the New York Stock Exchange.)

(a) What is the probability that our portfolio goes up for 3 consecutive years?

(b) If you knew that the portfolio has risen in price 2 years in a row, what probability would you assign to the event that it will go down the next year?

(c) What is the probability that the portfolio's value moves in the same direction in both of the next 2 years?

5.27   Here is a two-way table of the composition of the 99th Congress (elected in 1986) by party and by seniority. The entries in the body of the table should be the probabilities that a randomly chosen member of Congress has both the stated seniority and party affiliation. Only the two marginal distributions of party alone and seniority alone are given. If party and seniority were independent, what would be the probabilities in the body of the table?

Seniority	Democrat	Republican	Total
< 2 years			.103
2–9 years			.556
10–19 years			.239
20–29 years			.074
≥ 30 years			.028
Total	.582	.418	

5.2   RANDOM VARIABLES

Our interest in assignments of probability is focused on the probability of numerical outcomes of random phenomena—that is, on random variables. In this section we will learn two ways of assigning probabilities to the values of a random variable. The two types of probability models that result will dominate our application of probability to statistical inference.

Discrete Random Variables

Up to this point we have learned several rules of probability but only one method of assigning probabilities: State the probabilities of the individual outcomes and assign probabilities to events by summing over the outcomes. The outcome probabilities must be between 0 and 1 and have sum 1. When the outcomes are numerical, they are values of a random variable. We will now attach a name to random variables having probability assigned in this way.[2]

Discrete random variable

A discrete random variable X takes a finite number of values, call them $x_1, x_2, \ldots, x_k$. A probability model for X is given by assigning probabilities p_i to these outcomes,

$$P(X = x_i) = p_i$$

The probabilities p_i must satisfy

1. $0 \leq p_i \leq 1$ for each i
2. $p_1 + p_2 + \cdots + p_k = 1$

(5.1)

The probability $P(X \text{ in } A)$ of any event is found by summing the p_i for the outcomes x_i making up A.

probability distribution

The assignment of probabilities to the values of a random variable X is called the *probability distribution* of X. When X is discrete, its probability distribution must satisfy requirements 1 and 2 of the definition 5.1. If a random variable takes only a few values, its probability distribution can be given as a table. This was done in Examples 5.9 and 5.12. Here is another example.

EXAMPLE 5.17

The instructor of a large class gives 15% each of A's and D's, 30% each of B's and C's, and 10% F's. If a student is selected at random from this course, his grade on a four-point scale (A = 4.0) is a discrete random variable X having the distribution

Outcome x_i	0	1	2	3	4
Probability p_i	.10	.15	.30	.30	.15

The probability of the event "the student selected got a B or better" is found from the table of probabilities as follows.

$$P(X \geq 3) = P(X = 3) + P(X = 4)$$
$$= .30 + .15 = .45 \qquad \blacksquare$$

probability histogram

The probability distribution of a discrete random variable can be presented graphically in a *probability histogram*. Figure 5.5 shows the distributions for the random digits in Example 5.9 and the grades in Example 5.17. The horizontal scale shows the possible values of X, and the height of each bar is the probability for the value at its base. A probability histogram is in effect a relative frequency histogram for a very large number of trials.

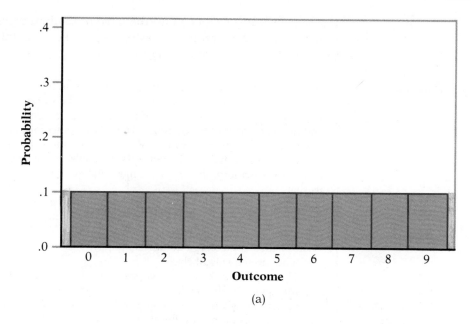

(a)

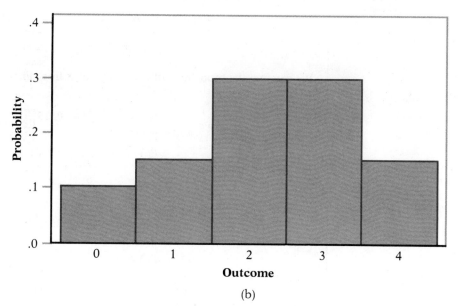

(b)

Figure 5.5   Probability histograms for the distributions of discrete random variables (a) generating a random digit (Example 5.9) and (b) grades in a large class (Example 5.17).

If a pencil point is dropped into the table of random digits many thousands of times, a relative frequency histogram of the outcomes will be close to the probability histogram in Figure 5.5(a).

EXAMPLE 5.18

What is the probability distribution of the discrete random variable X that counts the number of heads in four tosses of a coin? The sample space was discussed in Example 5.4. If we toss a coin four times, the outcome is a sequence of heads and tails such as HTTH. There are 16 possible outcomes in all. They are listed in Figure 5.6, along with the value of X for each outcome.

If the coin is balanced, each toss is equally likely to give H or T. Since the coin has no memory, the tosses are independent. The multiplication rule then tells us that

$$P(\text{HTTH}) = \frac{1}{2} \times \frac{1}{2} \times \frac{1}{2} \times \frac{1}{2} = \frac{1}{16}$$

Each of the 16 possible outcomes similarly has probability 1/16. That is, these outcomes are equally likely.

The number of heads X has possible values 0, 1, 2, 3, and 4. These values are *not* equally likely. As Figure 5.6 shows, there is only one way that $X = 0$ can occur, namely when the outcome is TTTT. So $P(X = 0) = 1/16$. But the event $\{X = 2\}$ can occur in six different ways. So

$$P(X = 2) = P(TTHH) + P(THHT) + \cdots + P(HTTH)$$

$$= \frac{1}{16} + \frac{1}{16} + \cdots + \frac{1}{16} = \frac{6}{16}$$

We can find the probability of each value of X from Figure 5.6 in the same way. Here is the result.

$$P(X = 0) = \frac{1}{16} = .0625$$

$$P(X = 1) = \frac{4}{16} = .25$$

$$P(X = 2) = \frac{6}{16} = .375$$

$$P(X = 3) = \frac{4}{16} = .25$$

$$P(X = 4) = \frac{1}{16} = .0625$$

These probabilities have sum 1, so this is a legitimate probability distribution. In table form the distribution is

Outcome	0	1	2	3	4
Probability	.0625	.25	.375	.25	.0625

Figure 5.7 is a probability histogram for this distribution. The probability distribution is exactly symmetric. It is an idealization of the relative frequency distri-

```
                        HTTH
                        HTHT
            HTTT   THTH   HHHT
            THTT   HHTT   HHTH
            TTHT   THHT   HTHH
    TTTT    TTTH   TTHH   THHH   HHHH

    X = 0   X = 1   X = 2   X = 3   X = 4
```

Figure 5.6 Possible outcomes in four tosses of a coin. The random variable X is the number of heads.

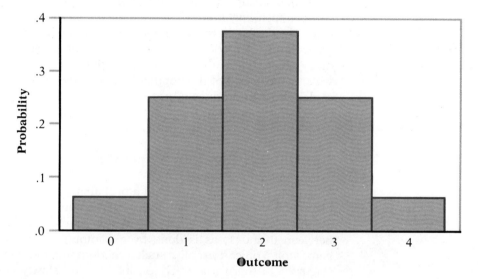

Figure 5.7 Probability histogram for the number of heads in four tosses of a coin.

bution of the number of heads after many tosses of four coins, which would be nearly symmetric but is unlikely to be exactly symmetric.

Any event involving the number of heads observed can be expressed in terms of X, and its probability can be found from the distribution of X. For example, the probability of tossing at least two heads is

$$P(X \geq 2) = .375 + .25 + .0625 = .6875$$

The probability of at least one head is found most simply by use of the complement rule:

$$P(X \geq 1) = 1 - P(X = 0)$$
$$= 1 - .0625 = .9375 \qquad \blacksquare$$

The assignment of probabilities to four tosses of a balanced coin gives the same probability to all 16 possible outcomes. Rules 1 and 2 then demand that each outcome have probability 1/16. In the same way, the 10 equally likely outcomes in Example 5.9 each had probability 1/10.

Equally likely outcomes

> If a random phenomenon has k possible outcomes, all equally likely, then each individual outcome has probability $1/k$.

If the coin in Example 5.18 is not balanced, say $P(H) = 0.4$, the outcomes are not equally likely and our work is a bit harder. Recall that tossing a coin n times is similar to choosing an SRS of size n from a large population and asking a yes-or-no question. If in fact 40% of the population favor reduced spending on student loans, the responses to a question about student loans are similar to repeated tosses of a coin with probability 0.4 of a head on each toss. A model with probability 0.5 of a head may be reasonable for many coins, but in the sampling situation it represents the special case in which exactly half of the population responds "Yes." The next chapter will explain how to produce a more general discrete distribution for a count X of "Yes" responses in a yes-or-no situation.

Continuous Random Variables

Examples 5.4 and 5.9 presented a probability model for selecting a random digit between 0 and 9. The model assigns equal probabilities to each of the 10 possible digits. In Example 5.5 we considered selecting a random number between 0 and 1, as is done by a computing system's random number generator. You can visualize such a random number by thinking of a spinner (Figure 5.8) that turns freely on its axis and slowly comes to a stop. The pointer can come to rest anywhere on a circle that is marked from 0 to 1. The most convenient choice of sample space allows the random number X to take any value between 0 and 1—that is, the possible values of X occupy

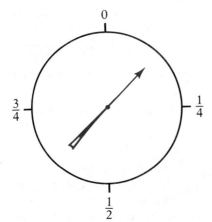

Figure 5.8 A spinner that generates a random number between 0 and 1.

an entire interval of numbers. How can we assign probabilities to such events as $\{0.3 \le X \le 0.7\}$? As in the case of selecting a random digit, we would like all possible outcomes to be equally likely. But we cannot assign probabilities to each individual value of X and then sum them, because the spinner can stop at any of an infinite number of points around the circle.

Instead, we use a new way of assigning probabilities directly to events—as *areas under a curve*. Probability histograms already picture probabilities as the area of their bars; we now use area to assign probability. Since the entire sample space must have probability 1, a curve with area 1 beneath it describes a probability distribution.

EXAMPLE 5.19

The random number generator will spread its output uniformly across the entire interval from 0 to 1 as we allow it to generate a long sequence of numbers. We therefore draw a line at height 1 above this interval (Figure 5.9). The area under this line is 1, and the probability of any event is the area under the line and above the event in question. Compare Figure 5.5(a), where 10 bars of equal height show the equal probabilities of the 10 digits, with Figure 5.9, where a line of fixed height shows probability spread evenly across an interval.

As Figure 5.9(a) illustrates, the probability that the idealized random number generator produces a number between 0.3 and 0.7 is

$$P(.3 \le X \le .7) = .4$$

because the area under the line and above the interval from 0.3 to 0.7 is 0.4. Since the height of the line is 1 and the area of a rectangle is the product of height and length, the probability of any interval of outcomes is just the length of the interval.

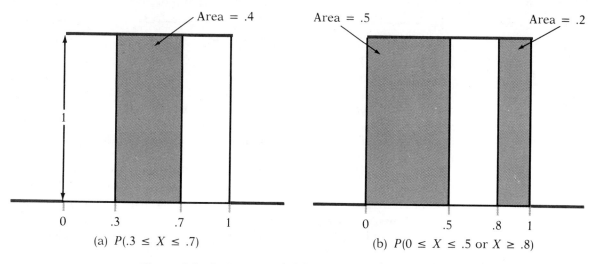

(a) $P(.3 \le X \le .7)$ (b) $P(0 \le X \le .5 \text{ or } X \ge .8)$

Figure 5.9 Assigning probability for generating a random number between 0 and 1. The probability of any interval of numbers is the area above the interval and under the curve.

Similarly,

$$P(X \le .5) = .5$$
$$P(X > .8) = .2$$
$$P(X \le .5 \text{ or } X > .8) = .7$$

You should draw a sketch for each of these areas. Notice in particular that the last event consists of two nonoverlapping intervals, so the total area above the event is found by adding the two areas, as illustrated by Figure 5.9(b). ■

The assignment of probabilities in Example 5.19 satisfies Rules 1 through 3, and therefore the other rules of probability. The entire sample space S has probability $P(0 \le X \le 1) = 1$ since the total area under the line is 1. For this same reason, the probability of any event is always a number between 0 and 1. Finally, the total area of two disjoint areas under the line is the sum of their individual areas. Probability as area under a curve is a second important way of assigning probabilities to events. Here is a general statement of the method.

Continuous random variable

A continuous random variable X takes all values in an interval of real numbers. A probability model for X is given by assigning to a set of outcomes A the probability $P(A)$ equal to the area above A and under a curve. The curve is the graph of a function $p(x)$ that satisfies

1. $p(x) \ge 0$ for all x
2. The total area under the graph of $p(x)$ is 1.

(5.2)

Figure 5.10 illustrates this definition. In Example 5.19, the function $p(x)$ takes the value 1 for $0 \le x \le 1$ and the value 0 everywhere else. In general, $p(x)$ is 0 in regions where the random variable X cannot take values. When the function $p(x)$ is at all complicated, we cannot find areas under it by simple geometry. In such cases, we find probabilities with the aid of tables or computer software.

The probability model for a continuous random variable assigns probabilities to intervals of outcomes, not to individual outcomes as we did in the case of discrete random variables. In fact, *all continuous probability distributions assign probability 0 to every individual outcome*. Only intervals of values have positive probability. To see why this is true, consider a specific outcome such as $P(X = 0.8)$ in Example 5.19. The probability of any interval is the same as its length. The point 0.8 has no length so its probability is 0. Although this fact may seem odd at first glance, it does make intuitive as well as mathematical sense. The random number generator produces a number between 0.79 and 0.81 with probability 0.02. An outcome between 0.799 and 0.801 has probability 0.002, and a result between 0.7999 and

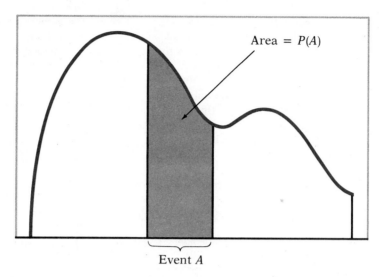

Area = $P(A)$

Event A

Figure 5.10   The probability distribution of a continuous random variable assigns probabilities as areas under a density curve.

0.8001 has probability 0.0002. Continuing to home in on 0.8, we can see clearly why an outcome *exactly* equal to 0.8 should have probability 0. Because there is no probability exactly at $X = 0.8$, the two events $\{X > 0.8\}$ and $\{X \geq 0.8\}$ have the same probability. We can ignore the distinction between $>$ and $\geq$ when finding probabilities for continuous (but not discrete) random variables.

Normal distributions   Continuous probability distributions have the same advantage over discrete distributions that density curves have over histograms—a single smooth curve is easier to work with than a large number of individual probabilities. In fact, the connection between continuous distributions and density curves is very close. Any density curve must lie on or above the horizontal axis and must have total area 1 underneath it. These are exactly the requirements on $p(x)$ in the definition 5.2 of a continuous random variable. So $p(x)$ is the equation of a density curve. *Any density curve describes the probability distribution of some continuous random variable.* The density curves that are most familiar to us are the normal curves. So *normal distributions are probability distributions.* That is, a normal curve assigns probabilities to intervals of outcomes, and these probabilities can be found by using Table A of areas under the standard normal curve.

EXAMPLE 5.20

An opinion poll asks an SRS of 1500 American adults what they consider to be the most serious problem facing our schools. Suppose that if all adults were asked this question, 30% would say "drugs." This proportion $p = 0.3$ is a population parameter. As the next chapter will show, the sample proportion who answer "drugs" is a random variable $\hat{p}$ that has approximately the $N(0.3, 0.0118)$ distribution. The mean 0.3 is the same as the population parameter because $\hat{p}$ is an unbiased estimate of p. The standard deviation is controlled mainly by the sample size, which is 1500 in this case.

What is the probability that the poll result differs from the truth about the population by more than two percentage points? Figure 5.11 shows this probability as an area under a normal density curve. By the addition rule for disjoint events, the desired probability is

$$P(\hat{p} < .28 \text{ or } \hat{p} > .32) = P(\hat{p} < .28) + P(\hat{p} > .32)$$

The individual probabilities are found by standardizing and using Table A.

$$P(\hat{p} < .28) = P\left(Z < \frac{.28 - .3}{.0118}\right)$$
$$= P(Z < -1.69) = .0455$$
$$P(\hat{p} > .32) = P\left(Z > \frac{.32 - .3}{.0118}\right)$$
$$= P(Z > 1.69) = .0455$$

Therefore,

$$P(\hat{p} < .28 \text{ or } \hat{p} > .32) = .0455 + .0455 = .0910$$

The probability that the sample result will miss the truth by more than two percentage points is 0.091. Compare the arrangement of this calculation with Examples 1.20 and 1.21. Only the language of probability is new.

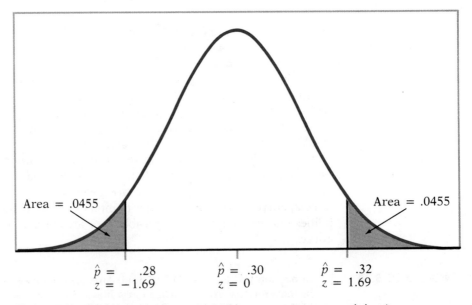

Figure 5.11   Probability in Example 5.20 as area under a normal density curve.

We could also do this calculation by first finding the probability of the complement,

$$P(.28 \leq \hat{p} \leq .32) = P\left(\frac{.28 - .3}{.0118} \leq Z \leq \frac{.32 - .3}{.0118}\right)$$
$$= P(-1.69 \leq Z \leq 1.69)$$
$$= .9545 - .0455 = .9090$$

Then by the complement rule,

$$P(\hat{p} < .28 \text{ or } \hat{p} > .32) = 1 - P(.28 \leq \hat{p} \leq .32)$$
$$= 1 - .9090 = .0910$$

There is often more than one correct way to use the rules of probability to answer a question.

■

We began this chapter with the idea that probability formalizes a few properties of relative frequencies. Two very useful ways of assigning probabilities are the distributions of discrete and continuous random variables. These two methods are adequate for our statistical uses.

SUMMARY

A **random variable** is a numerical outcome of a random phenomenon. Many probability models take S to be all possible values of a random variable X and define P from the **probability distribution** of X. A random variable X and its distribution can be discrete or continuous.

A **discrete random variable** takes a finite collection of values. The probability distribution assigns each of these values a probability between 0 and 1 such that the sum of all the probabilities assigned is 1. The probability of an event A is then the sum of the probabilities of all the values that make up A.

A **continuous random variable** takes all values in some interval of real numbers. The probability distribution assigns to an event A the area above A and under a **density curve**. The height of a density curve must be 0 or positive everywhere, and the curve must have total area 1 beneath it.

Normal distributions are one type of continuous probability distribution.

SECTION 5.2 EXERCISES

5.28 Return to Example 5.12 which dealt with social mobility in England. Let the random variable X be the class of a randomly chosen son of a lower class father. The distribution of X is as follows:

Son's class x_i	1	2	3	4	5
Probability p_i	.48	.38	.08	.05	.01

(a) Verify that the table gives a legitimate probability distribution for X. Draw a probability histogram to display the distribution.
(b) What is $P(X \leq 3)$?
(c) What is $P(X < 3)$?
(d) Write the event "a son of a lower class father reaches one of the two highest classes" in terms of values of X. What is the probability of this event?
(e) A father's social class is Class 1. Write the event "the son of a lower class father reaches a class higher than his father's" in terms of values of X. What is the probability of this event?

5.29 Choose an American household at random and let the random variable X be the number of persons living in the household. If we ignore the few households with more than seven inhabitants, the probability distribution of X is as follows:

Outcome	1	2	3	4	5	6	7
Probability	.237	.317	.178	.157	.070	.026	.015

(a) Verify that this is a legitimate discrete probability distribution and draw a probability histogram to display it.
(b) What is $P(X \geq 5)$? .111
(c) What is $P(X > 5)$? .41
(d) What is $P(2 < X \leq 4)$? .335
(e) What is $P(X \neq 1)$? .763
(f) Write the event that a randomly chosen household contains more than two persons in terms of the random variable X. What is the probability of this event?

$P(x > 2) = .446$

5.30 A study selected a sample of fifth grade pupils and recorded how many years of school they eventually completed. Let X be the highest year of school that a randomly chosen fifth grader completes. (Students who go on to college are included in the outcome $X = 12$.) The probability distribution of X is as follows:

x_i	4	5	6	7	8	9	10	11	12
p_i	.010	.007	.007	.013	.032	.068	.070	.041	.752

(a) Verify that this is a legitimate discrete probability distribution and draw a probability histogram to display it.

    (b) Find $P(X \geq 6)$. *.983*
    (c) Find $P(X > 6)$. *.976*

$P(x \geq 9)$ **(d)** What values of X make up the event "The student completed at least one year of high school?" (High school begins with the ninth grade.) What is the probability of this event?

$P(x \leq 12)$ **(e)** Express the event that a fifth grade student will not complete twelfth grade in terms of X and find its probability.

.248

5.31 A couple plans to have four children. If we consider only the sex of the children, having four children is much like tossing a coin four times. There are 16 possible arrangements of girls and boys, and all 16 are (approximately) equally likely.

    (a) Let the discrete random variable X be the number of girls among the four children. Following the pattern of Figure 5.6, list all 16 outcomes and give the value of X for each. The distribution of X is the same as was given in Example 5.18 for the number of heads in four tosses of a coin.

    (b) What is the probability that the couple has at least one girl?

    (c) What is the probability that the couple has two girls and two boys?

    (d) What is the probability that the first child is a girl? (Notice that this event cannot be described in terms of X. You must return to the original outcomes.)

5.32 A roulette wheel has 38 slots, numbered 0, 00, and 1 to 36. The slots 0 and 00 are colored green, 18 of the others are red, and 18 are black. The dealer spins the wheel and at the same time rolls a small ball along the wheel in the opposite direction. The wheel is carefully balanced so that the ball is equally likely to land in any slot when the wheel slows down. Gamblers can. bet on various combinations of numbers and colors.

$\frac{1}{38}$ **(a)** What is the probability that the ball lands in any one slot?

$\frac{18}{38}$ **(b)** A gambler betting on "red" wins if the ball lands in a red slot. What is the probability of winning?

$\frac{12}{38}$ **(c)** The slot numbers are laid out on a board on which gamblers place their bets. One column of numbers on the board contains all multiples of 3, that is, 3, 6, 9, . . . , 36. A gambler places a "column bet" that wins if any of these numbers comes up. What is the probability of winning?

5.33 Some games of chance rely on tossing two dice. Each die has six faces, marked with 1, 2, . . . , 6 spots called pips. The dice used in casinos are carefully balanced so that each face is equally likely to come up. When two dice are tossed, each of the 36 possible pairs of faces is equally likely to come up. The outcome of interest to a gambler is the sum of the pips on the two up faces. Call this random variable X.

(a) Write down all 36 possible pairs of faces.

$\frac{1}{36}$ (b) If all pairs have the same probability, what must be the probability of each pair?

(c) Write the value of X next to each pair of faces and use this information with the result of (b) to give the probability distribution of X. Draw a probability histogram to display the distribution.

(d) One bet available in craps wins if a 7 or an 11 comes up on the next roll of two dice. What is the probability of rolling a 7 or an 11 on the next roll?

(e) Several bets in craps lose if a 7 is rolled; if any outcome other than 7 occurs, these bets either win or continue to the next roll. What is the probability that anything other than a 7 is rolled?

5.34 Weary of the low turnout in student elections, a college administration decides to choose three students at random to form an advisory board that will represent student opinion. Suppose that 40% of all students oppose the use of student fees to fund student interest groups. The opinions of the three students on the board are independent and the probability is 0.4 that each opposes the funding of interest groups.

(a) Call the three students A, B, and C. What is the probability that A and B support funding and C opposes it?

(b) List all possible combinations of opinions that can be held by students A, B, and C. (Hint: There are eight possibilities.) Then give the probability of each of these outcomes. Note that they are not equally likely.

(c) Let the random variable X be the number of student representatives who oppose the funding of interest groups. Give the probability distribution of X.

(d) Express the event "a majority of the advisory board opposes funding" in terms of X and find its probability.

5.35 Let X be a random number between 0 and 1 produced by the idealized uniform random number generator described in Example 5.19 and Figure 5.9. Find the following probabilities:

(a) $P(0 \le X \le .4)$ .4

(b) $P(.4 \le X \le 1)$ .6

(c) $P(.3 \le X \le .5)$ .2

(d) $P(.3 < X < .5)$ .2

(e) $P(.226 \le X \le .713)$ .487

5.36 Let the random variable X be a random number with the density curve in Figure 5.9, as in the previous exercise. Find the following probabilities:

(a) $P(X \le .49)$

(b) $P(X \ge .27)$

(c) $P(.27 < X < 1.27)$

(d) $P(.1 \le X \le .2$ or $.8 \le X \le .9)$

(e) The probability that X is not in the interval .3 to .8

(f) $P(X = .5)$

5.37 Most random number generators allow users to specify the range of the random numbers to be produced. In MINITAB, for example, a random number X between 0 and 2 is placed in C1 by the command

```
MTB > RANDOM 1 C1;
   SUBC > UNIFORM 0 2.
```

The distribution of X spreads its probability uniformly between 0 and 2. The density curve of X should have constant height between 0 and 2, and height 0 elsewhere.

(a) What is the height of the density curve between 0 and 2? Draw a graph of the density curve of X.

(b) Use your graph from (a) and the fact that probability is area under the curve to find $P(X \le 1)$.

(c) Find $P(.5 < X < 1.3)$.

(d) Find $P(X \ge .8)$.

(e) Find $P(1.4 \le X \le 2.4)$.

(f) Find $P(X = 1)$.

5.38 Suppose that the random variable X is normally distributed with mean 0 and standard deviation 1. Use Table A to find the following probabilities:

$\dfrac{X - \mu}{\delta}$

(a) $P(X \ge 0)$ $P\left(z \ge \dfrac{0-0}{1}\right) = P(z \ge 0) = .5$

(b) $P(X \le 1)$ $\dfrac{1-0}{1} = 1$ $P(z \le 1) = .8413$

(c) $P(X \ge -1)$ $.8413$

(d) $P(X = 1)$ 0

(e) $P(-1 \le X \le 1) \approx 68\%$ $2(1 - .8413) = 2(.1587) = .3174$

(f) $P(1 \le X \le 2)$ $\dfrac{2-0}{1} = 2$ $P(z \le 2) - P(z \le 1) = .9772 - .8413$

(g) $P(1 < X < 2)$ Same $= .1359$

5.39 An SRS of 400 American adults is asked, "What do you think is the most serious problem facing our schools?" Suppose that in fact 30% of all adults would answer "drugs" if asked this question. That is, the population proportion is $p = 0.3$. The sample proportion $\hat{p}$ who answer "drugs" will vary in repeated sampling. The sampling distribution will be approximately normal with mean 0.3 and standard deviation 0.023. Using this approximation, find the probabilities of the following events:

(a) At least half of the sample believe that drugs are the schools' most serious problem.

(b) Less than 25% of the sample believe that drugs are the most serious problem.

(c) The sample proportion is between 0.25 and 0.35.

(d) $\{\hat{p} \leq .4 \text{ or } \hat{p} \geq .6\}$.

5.40 An opinion poll asks an SRS of 1500 adults, "Do you happen to jog?" Suppose (as is approximately correct) that the population proportion who jog is $p = 0.15$. Then the proportion $\hat{p}$ in the sample who answer "Yes" will be approximately normally distributed with $\mu = 0.15$ and standard deviation $\sigma = 0.0092$. Find the following probabilities:

(a) $P(\hat{p} = .2)$

(b) $P(\hat{p} \geq .2)$

(c) $P(.1 \leq \hat{p} \leq .2)$

5.3 DESCRIBING PROBABILITY DISTRIBUTIONS

We cannot foretell the number X of heads that we will get in tossing four balanced coins, nor the sample proportion $\hat{p}$ obtained in an SRS. We can, however, give probability distributions for these random variables that specify the relative frequencies of various outcomes in many repeated trials. The probability distribution of a random variable is similar to the relative frequency distribution of a set of data. Consider, for example, those mainstays of statistics, the normal distributions. We first met normal density curves as idealized descriptions of some distributions of data. Next we discovered by simulation that the sampling distributions of some important statistics are also well described by normal curves. Now we realize that a normal density curve specifies one type of continuous probability distribution.

Probability distributions can be described by many of the same numerical measures that apply to distributions of data. We can speak, for example, of the mean and the standard deviation of a probability distribution. In this section we will learn a bit more about how to compute such descriptive measures and about the laws they obey.

The Mean of a Random Variable

The probability distribution of a *continuous* random variable X is described by a density curve. Chapter 1 showed how to find the mean of the distribution: It is the point at which the density curve would balance if it were made out of solid material. The mean lies at the center of symmetric density curves such as the normal curves. Exact calculation of the mean of a distribution with a skewed density curve requires advanced mathematics. As in Chapter 1 we write the mean of a probability distribution as μ. It is convenient to also call μ the mean of the random variable X. To make clear

expected value

which random variable we have in mind we write μ_X. The mean of a random variable X is sometimes called the *expected value* of X.

The mean μ_X of a *discrete* random variable X is the point about which the probability histogram of the distribution would balance. In this case, μ_X can be computed quite simply from the probability distribution of X. Here is the method.

Mean of a random variable

> If X is a discrete random variable taking the values $x_1, x_2, \ldots, x_k$ with probabilities $p_1, p_2, \ldots, p_k$ then the mean of X is given by
>
> $$\mu_X = x_1 p_1 + x_2 p_2 + \cdots + x_k p_k \qquad (5.3)$$

To find the mean of a discrete random variable X, multiply each possible value of X by its probability and add the products. This method gives the balance point of a discrete distribution. Alternatively, we can think of Equation 5.3 as arising from the definition of the mean $\bar{x}$ of a data set. The mean $\bar{x}$ is the ordinary average of the observations, in which each observation has equal weight $1/n$. The values x_i of the random variable X are not equally likely to occur, so when we average them to produce the mean μ_X we weight each value by its probability.

EXAMPLE 5.21

The distribution of the count X of heads in four tosses of a balanced coin was found in Example 5.18 to be

Values x_i	0	1	2	3	4
Probability p_i	.0625	.25	.375	.25	.0625

The mean of X is therefore

$$\mu_X = (0)(.0625) + (1)(.25) + (2)(.375) + (3)(.25) + (4)(.0625)$$
$$= 2$$

This discrete distribution is symmetric, as the probability histogram of Figure 5.12(a) reminds us. The mean therefore falls at the center of symmetry. ■

EXAMPLE 5.22

What is the average size of an American family? Here is the distribution of the size of American families according to Census Bureau studies.

Persons in family	2	3	4	5	6	7	8
Fraction of families	.397	.231	.212	.097	.038	.017	.008

If we imagine selecting a single family at random, the size of the family drawn is a random variable X with probability distribution given by the table. The mean μ_X

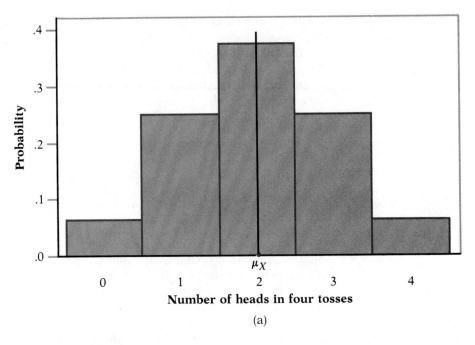

(a)

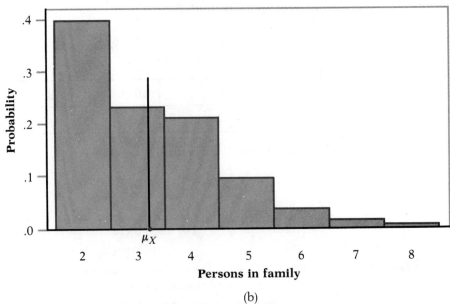

(b)

Figure 5.12 Locating the mean of a discrete random variable on a probability histogram for (a) the number of heads in four tosses of a coin (Example 5.21) and (b) the number of persons in a family (Example 5.22).

is the mean family size in the population. This mean is

$$\mu_X = (2)(.397) + (3)(.231) + (4)(.212) + (5)(.097) + (6)(.038) + (7)(.017) + (8)(.008)$$
$$= 3.23$$

Figure 5.12(b) locates the mean on a probability histogram. (In this example, we have ignored the few families with 9 or more members. The actual mean size of American families is therefore slightly larger than our result.) ∎

The Law of Large Numbers

The mean of a random variable is by definition the average outcome of the random variable as computed from the distribution. It is a striking fact that the mean of a random variable is also the average outcome in a different sense. If we actually observe a large number of outcomes of a random variable and calculate their mean (arithmetic average), this random mean will be close to the fixed mean of the distribution. This fact deserves closer study.

Suppose that we repeatedly toss four coins and record the number X of heads obtained on each trial. This was done in Example 5.1. The first toss gave $X = 3$, the second toss $X = 0$, then another 0, then 2, 4, 2, and so on. The relative frequency of each outcome in many repetitions will be close to its probability, as Figure 5.1 illustrates for the outcome $\{X = 2\}$. The probabilities of all possible outcomes make up the probability distribution of the discrete random variable X, which appears in Example 5.21. What happens to the average number of heads observed? The numbers of heads obtained on each of the first n trials form n observations. The average number of heads in the first n trials is the familiar mean $\bar{x}$ of these observations. The mean $\bar{x}$ is a random variable that takes different values if we repeat the tosses.[3] Here are the values of the observations and the mean $\bar{x}$ of all observations to date after each of the first few trials.

Trial	Outcome	Average number of heads per trial
1	3	$\bar{x} = \dfrac{3}{1} = 3.00$
2	0	$\bar{x} = \dfrac{3}{2} = 1.50$
3	0	$\bar{x} = \dfrac{3}{3} = 1.00$
4	2	$\bar{x} = \dfrac{5}{4} = 1.25$
5	4	$\bar{x} = \dfrac{9}{5} = 1.80$
6	2	$\bar{x} = \dfrac{11}{6} = 1.83$

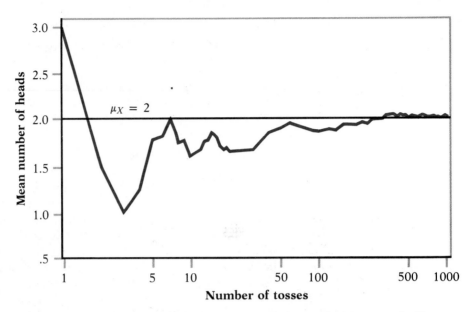

Figure 5.13 The long-term behavior of the mean outcome of many trials. The mean number of heads $\bar{x}$ observed when four coins are tossed many times approaches the mean $\mu_X = 2$ of the probability distribution.

Figure 5.13 shows the behavior of $\bar{x}$ as the number of trials n increases. At first the value of $\bar{x}$ is unstable, but *in the long run the observed mean $\bar{x}$ approaches and remains close to the distribution mean $\mu_X = 2$*. After 100 trials $\bar{x} = 1.86$; at the end of our series of 1000 tosses of four coins $\bar{x} = 2.012$. If we repeated our work, we would obtain a different sequence of outcomes and the graph of the progress of $\bar{x}$ would differ from Figure 5.13. But $\bar{x}$ will always approach $\mu_X = 2$ ever more closely as the number of trials grows.

Starting from the basic laws of probability, it can actually be *proved* that the sample mean outcome calculated from repeated independent observations of a random variable must approach the mean of the distribution. We will not give the proof. Be sure to notice, though, that the behavior of $\bar{x}$ is similar to the intuitive idea of probability—in the long run, relative frequencies of outcomes get close to the probability distribution, and the average outcome gets close to the distribution mean. These facts, especially when they are considered as mathematical results that can be derived from basic **law of large** laws of probability, are called the *law of large numbers*.* An essential re-
numbers quirement for the law of large numbers to apply is that the successive trials

* The earliest version of the law of large numbers was proved by the Swiss mathematician Jacob Bernoulli (1654–1705). The Bernoullis were a remarkably mathematical family; five of them contributed to the early study of probability.

of the random phenomenon be independent. If a coin is tossed by a magician who can influence whether it falls heads or tails, the law of large numbers may not apply. By looking at past outcomes and by manipulating the coin, the magician can change the long-term relative frequency of heads.

The law of large numbers is the foundation upon which such business enterprises as gambling casinos and insurance companies are built. The winnings (or losses) of a gambler on a single play are uncertain—that's why gambling is exciting. The average winnings of the house on tens of thousands of plays will be very close to the mean of the distribution of winnings—that's why gambling can be a business, for this distribution guarantees the house a profit.

EXAMPLE 5.23

Keno is a favorite game in the casinos of Nevada and Atlantic City, and similar games are popular with the states that operate lotteries. Balls numbered 1 to 80 are tumbled in a machine as the bets are placed, and then 20 of the balls are chosen at random. Players select numbers by marking a card. The simplest of the many wagers available is "Mark 1 Number." The payoff on a $1 bet is $3 if the number you select is one of those chosen. Since 20 of 80 numbers are chosen, your probability of winning is 20/80 or 0.25. The probability distribution of the payoff X on a single play is therefore

Payoff	$3	0
Probability	.25	.75

The mean payoff is

$$\mu_X = (\$3)(.25) + (0)(.75) = \$.75$$

That is, the average payoff on a $1 bet is 75 cents. The house keeps the other quarter. As hundreds of gamblers place thousands of bets, the casino is not gambling. In the long run, it will take in 25 cents for every dollar bet. ■

The law of averages The law of large numbers is the exact version of the popular "law of averages." Both say that random fluctuations eventually average out. But psychologists have discovered that the popular understanding of randomness is quite different from the true laws of chance.[4] Most people believe in an incorrect "law of small numbers." That is, we expect even short sequences of random events to show the kind of average behavior that in fact appears only in the long run.

Try this experiment: Write down a sequence of heads and tails that you think imitates 10 tosses of a balanced coin. How long was the longest string (called a *run*) of consecutive heads or consecutive tails in your tosses? Most people will write a sequence with no runs of more than two consecutive heads or tails. But in fact the probability of a run of three or more consecutive heads or tails in 10 tosses is greater than 0.8, and the probability of *both* a run of three or more heads and a run of three or more tails is almost 0.2.[5] This and other probability calculations suggest that a short sequence of coin tosses will often not appear random to us. The runs of consecutive heads or consecutive tails that appear in real coin tossing (and that are predicted by probability theory) seem surprising to us. Since we don't expect to see long runs, we may conclude that the coin tosses are not independent or that some influence is disturbing the random behavior of the coin.

Belief in the law of small numbers influences behavior. If a basketball player makes several consecutive shots, both the fans and his teammates believe that he has a "hot hand" and is more likely to make the next shot. This is wrong. Careful study has shown that runs of baskets made or missed are no more frequent in basketball than would be expected if each shot were independent of the player's previous shots. Players perform consistently, not in streaks. (Of course, some players make a higher percent of their shots in the long run than others.) Our perception of hot or cold streaks simply shows that we don't perceive random behavior very well.[6]

Gamblers often follow the hot hand theory, betting that a run will continue. At other times, however, they draw the opposite conclusion when confronted with a run of outcomes. If a coin gives 10 straight heads, some gamblers feel that it must now produce some extra tails to get back to the average of half heads and half tails. Not so. If the next 10,000 tosses give about 50% tails, those 10 straight heads will be swamped by the later thousands of heads and tails. No compensation is needed to get back to the average in the long run. Remember that it is *only* in the long run that the regularity described by probabilities and means takes over.

Our inability to accurately distinguish random behavior from systematic influences points out once more the need for statistical inference to supplement exploratory analysis of data. Probability calculations can help verify that what we see in the data is more than a random pattern.

Rules for Means

Since means of random variables are average outcomes (in several senses), it is not surprising that they obey laws that reflect the behavior of averages.

Rules for means

If X is a random variable and a and b are fixed numbers, then

$$\mu_{a+bX} = a + b\mu_X$$

If X and Y are random variables, then   (5.4)

$$\mu_{X+Y} = \mu_X + \mu_Y$$

The following example illustrates the use of these laws.

EXAMPLE 5.24

Gain Communications sells aircraft communications units to both the military and civilian markets. Next year's sales depend on market conditions that cannot be predicted exactly. Gain follows the modern practice of using probability estimates of sales. The military division estimates its sales as follows:

Units sold	1000	3000	5000	10,000
Probability	.1	.3	.4	.2

These are personal probabilities that express the informed opinion of Gain's executives. The corresponding sales estimates for the civilian division are as follows:

Units sold	300	500	750
Probability	.4	.5	.1

Take X to be the number of military units sold and Y the number of civilian units. From the probability distributions we compute that

$$\mu_X = (1000)(.1) + (3000)(.3) + (5000)(.4) + (10{,}000)(.2)$$
$$= 5000 \text{ units}$$

$$\mu_Y = (300)(.4) + (500)(.5) + (750)(.1)$$
$$= 445 \text{ units}$$

Gain makes a profit of $2000 on each military unit and $3500 on each civilian unit. Next year's profit will be

$$Z = 2000X + 3500Y$$

Using both rules in 5.4, the mean profit is

$$\mu_Z = 2000\mu_X + 3500\mu_Y$$
$$= (2000)(5000) + (3500)(445)$$
$$= \$11{,}557{,}500$$

This mean is the company's best estimate of next year's profits, combining the probability estimates of the two divisions. ∎

The Variance of a Random Variable

In addition to measures of the center of a distribution, we should be concerned with measures of spread that describe the variability of a random variable. The variance and the standard deviation are the measures of spread that accompany the choice of the mean to measure center. The definition of the variance of a random variable is similar to the definition of the sample variance s^2 that was given in Chapter 1. The variance is the average squared deviation of outcomes from their mean. It is computed as the mean of the squared deviations $(X - \mu_X)^2$. As usual, the computation for a continuous random variable requires advanced mathematics. The discrete case is easier.

Variance of a random variable

> The variance of any random variable X is the mean of the random variable $(X - \mu_X)^2$. If X is a discrete random variable taking values $x_1, x_2, \ldots, x_k$ with probabilities $p_1, p_2, \ldots, p_k$ the variance of X is given by
>
> $$\sigma_X^2 = (x_1 - \mu_X)^2 p_1 + (x_2 - \mu_X)^2 p_2 + \cdots + (x_k - \mu_X)^2 p_k \quad (5.5)$$
>
> The standard deviation σ_X of X is the square root of the variance.

EXAMPLE 5.25

In Example 5.24 we saw that the number Y of communications units sold by Gain Communications' civilian division has distribution

Units sold	300	500	750
Probability	.4	.5	.1

and mean $\mu_Y = 445$. The variance of Y can be found by arranging the sum in Equation 5.5 in the form of a table.

y_i	$(y_i - \mu_Y)^2$	p_i	$(y_i - \mu_Y)^2 p_i$
300	$(300 - 445)^2 = 21{,}025$	.4	$(21{,}025)(.4) = 8410$
500	$(500 - 445)^2 = 3025$	.5	$(3025)(.5) = 1512.5$
750	$(750 - 445)^2 = 93{,}025$	.1	$(93{,}025)(.1) = 9302.5$

Total   19,225

We see that $\sigma_Y^2 = 19{,}225$. The standard deviation of Y is $\sigma_Y = \sqrt{19{,}225} = 138.65$. The standard deviation is a measure of how variable the number of units sold is. As in the case of distributions for data, the standard deviation of a probability distribution is most meaningful when the distribution has a specific form, particularly when the distribution is normal. ∎

Rules for Variances

Rules for variances similar to the rules (5.4) for means are often needed to understand statistical questions. Although the mean of a sum of random variables is always the sum of their means, this addition rule is not true for variances. To understand why, suppose that X is the percent of a family's after-tax income that is spent and Y the percent that is saved. When X increases, Y decreases by the same amount. Though X and Y may vary widely from year to year, their sum $X + Y$ is always 100% and does not vary at all. It is the association between the variables X and Y that prevents their variances from adding. If random variables are *independent*, this kind of association between their values is ruled out and their variances do add.

Rules for variances

> If X is a random variable and a and b are fixed numbers, then
>
> $$\sigma^2_{a+bX} = b^2\sigma^2_X$$
>
> If X and Y are *independent* random variables, then   (5.6)
>
> $$\sigma^2_{X+Y} = \sigma^2_X + \sigma^2_Y$$
> $$\sigma^2_{X-Y} = \sigma^2_X + \sigma^2_Y$$

Notice that because a variance is the average of *squared* deviations from the mean, multiplying X by a constant b multiplies σ^2_X by the *square* of the constant. Adding a constant a to a random variable changes its mean but does not change its variability. The variance of $X + a$ is therefore the same as the variance of X. Because the square of -1 is 1, the addition rule says that the variance of a difference is the *sum* of the variances. Standard deviations are most easily combined by using the rules for variances rather than by giving separate rules for standard deviations. For example, the standard deviations of $2X$ and $-2X$ are both equal to $2\sigma_X$ because this is the square root of the variance $4\sigma^2_X$. The addition rule for variances implies that standard deviations do *not* generally add.

EXAMPLE 5.26

In Example 5.23 we found that the payoff X of a \$1 bet in Keno has mean $\mu_X = \$0.75$. The gambler's winnings on a \$1 bet are $W = X - 1$ since the dollar bet must be subtracted from the payoff. By the rules (5.4) for means, the mean amount won is

$$\mu_W = \mu_X - 1 = -\$.25$$

That is, the gambler loses an average of 25 cents on each dollar bet.

The variance of the outcome X is found as follows:

x_i	$(x_i - \mu_X)^2$	p_i	$(x_i - \mu_X)^2 p_i$
3	$(3 - .75)^2 = 5.0625$	.25	1.266
0	$(0 - .75)^2 = .5625$	.75	.422

Total   1.688

The standard deviation of the payoff X is $\sigma_X = \sqrt{1.688} = \$1.30$. The rules (5.6) for variances remind us that the variance and standard deviation of the winnings $W = X - 1$ are the same as those of X.

If the gambler makes a \$1 bet on two different Keno games, the payoffs X and Y are independent. The total payoff $X + Y$ has mean

$$\mu_{X+Y} = \mu_X + \mu_Y = \$.75 + \$.75 = \$1.50$$

and variance

$$\sigma_{X+Y}^2 = \sigma_X^2 + \sigma_Y^2 = (1.30)^2 + (1.30)^2 = 3.38$$

The standard deviation of the total payoff is

$$\sigma_{X+Y} = \sqrt{3.38} = \$1.84$$

This is not the same as the sum of the individual standard deviations, which is $\$1.30 + \$1.30 = \$2.60$. Variances add; standard deviations do not. ■

EXAMPLE 5.27

A college uses Scholastic Aptitude Test scores as one criterion for admission. Experience has shown that the distribution of SAT scores among the entire population of applicants is such that

$$\text{SAT math score } X \qquad \mu_X = 625 \qquad \sigma_X = 90$$
$$\text{SAT verbal score } Y \qquad \mu_Y = 590 \qquad \sigma_Y = 100$$

What are the mean and standard deviation of the total score $X + Y$ among students applying to this college?

The mean overall SAT score is

$$\mu_{X+Y} = \mu_X + \mu_Y = 625 + 590 = 1215$$

The variance and standard deviation of the total *cannot be computed* from the information given. The verbal and math scores are clearly not independent, since students who score high on one exam tend to score high on the other also. Therefore, the addition rule for variances does not apply. ■

EXAMPLE 5.28

Tom and George play golf at the same club. Tom's score X varies from round to round but has

$$\mu_X = 110 \quad \text{and} \quad \sigma_X = 10$$

George's score Y also varies, with

$$\mu_Y = 100 \quad \text{and} \quad \sigma_Y = 8$$

Tom and George are playing the first round of the club tournament. Because they are not playing together, we will assume that their scores vary independently of each other. The difference between their scores on this first round has mean

$$\mu_{X-Y} = 110 - 100 = 10$$

The variance of the difference between their scores is

$$\sigma_{X-Y}^2 = \sigma_x^2 + \sigma_Y^2 = 10^2 + 8^2 = 164$$

The standard deviation is found from the variance

$$\sigma_{X-Y} = \sqrt{164} = 12.8$$

The variation in the difference between the scores is greater than the variation in the score of either player. This might have been expected, since the difference contains two independent sources of variation. It also makes the tournament interesting, since George will not always finish ahead of Tom even though he has the lower mean score. Computations of the variance of sums and differences of statistics play an important role in statistical inference. ■

SUMMARY

The probability distribution of a random variable X, like a distribution of observations, has a **mean** μ_X and a **standard deviation** σ_X.

If X is a discrete random variable taking values $x_1, x_2, \ldots, x_k$ with probabilities $p_1, p_2, \ldots, p_k$ the mean and variance can be computed from its distribution as follows:

$$\mu_X = x_1 p_1 + x_2 p_2 + \cdots + x_k p_k$$
$$\sigma_X^2 = (x_1 - \mu_X)^2 p_1 + (x_2 - \mu_X)^2 p_2 + \cdots + (x_k - \mu_X)^2 p_k$$

The mean and variance of a continuous random variable can be computed from the density curve, but to do so requires more advanced mathematics.

The **law of large numbers** states that the actually observed mean outcome in many independent trials must approach the mean of the distribution of outcomes. This statement includes the fact that observed relative frequencies of events must approach the probabilities of the events. Short sequences of trials will often not display the regularity that appears in the long run.

The means and variances of random variables obey the following rules. If a and b are fixed numbers, then

$$\mu_{a+bX} = a + b\mu_X$$
$$\sigma_{a+bX}^2 = b^2 \sigma_X^2$$

If X and Y are any two random variables, then

$$\mu_{X+Y} = \mu_X + \mu_Y$$

and if X and Y are independent, then

$$\sigma_{X+Y}^2 = \sigma_X^2 + \sigma_Y^2$$
$$\sigma_{X-Y}^2 = \sigma_X^2 + \sigma_Y^2$$

SECTION 5.3 EXERCISES

5.41 Example 5.22 found that the mean number of people in an American family is 3.23. A family consists of two or more people related by blood, adoption, or marriage who live together, while a household consists of all people who live together in a dwelling unit. The mean size of a household should be less than the mean size of families.

The distribution of the size of American households is

[handwritten: 1(.237) + 2(.317) + 3(.178) + ···]

[handwritten: .2644]

Outcome	1	2	3	4	5	6	7
Probability	.237	.317	.178	.157	.070	.026	.015

Find the mean number of people living in a household.

5.42 The distribution of the highest grade X completed by schoolchildren who reach the fifth grade is

x_i	4	5	6	7	8	9	10	11	12
p_i	.010	.007	.007	.013	.032	.068	.070	.041	.752

Find the mean μ_X.

5.43 Example 5.17 gives the distribution of grades (A = 4, B = 3, and so on) in a large class as

Outcome	0	1	2	3	4
Probability	.10	.15	.30	.30	.15

Find the average (that is, the mean) grade in this course.

5.44 A life insurance company sells a term insurance policy to a 21-year-old male that pays $20,000 if the insured dies within the next 5 years. The probability that a randomly chosen male will die each year can be found in mortality tables. The company collects a premium of $100 each year as payment for the insurance. The amount X that the company earns on this policy is $100 per year, less the $20,000 that it must pay if the insured dies. Here is the distribution of X. Fill in the missing probability in the table and calculate the mean earnings μ_X.

[handwritten: $\mu_X = (-19,900)(.00183) + (-19,800)(.00186) + ···$]

[handwritten: $\mu_X =$ $309.74]

Age at death	21	22	23	24	25	≥26
Earnings	−$19,900	−$19,800	−$19,700	−$19,600	−$19,500	$500
Probability	.00183	.00186	.00189	.00191	.00193	*.99058*

5.45 It would be quite risky for you to insure the life of a 21-year-old friend under the terms of the previous exercise. There is a high probability that your friend would live and you would gain $500 in premiums. But if he were to die, you would lose almost $20,000. Explain carefully why selling insurance is not risky for an insurance company that insures many thousands of 21-year-old men.

[handwritten: law of large #'s says the company will earn $309.74 per person in the long run]

5.46 Green Mountain Numbers, a part of the Vermont state lottery, offers a choice of several bets. In each case, the player chooses a three-digit number. The lottery commission announces the winning number, chosen at random, at the end of each day. Find the expected payoff for a $1 bet for each of the following options in the lottery.

(a) The "triple" pays $500 if the number you choose matches the winning number exactly.

(b) The "box" pays $83.33 if the number you choose has the same digits as the winning number, in any order. (Assume that you chose a number having three distinct digits.)

5.47 For each of the following situations, would you expect the random variables X and Y to be independent? Explain your answers.

(a) X is the rainfall (in inches) on November 6 of this year and Y is the rainfall at the same location on November 6 of next year.

(b) X is the amount of rainfall today and Y is the rainfall at the same location tomorrow.

(c) X is today's rainfall at the Orlando, Florida airport, and Y is today's rainfall at Disney World just outside Orlando.

5.48 In which of the following games of chance would you be willing to assume independence of X and Y in making a probability model? Explain your answer in each case.

(a) In blackjack, you are dealt two cards and examine the total points X on the cards (face cards count 10 points). You can choose to be dealt another card and compete based on the total points Y on all three cards.

(b) In craps the betting is based on successive rolls of two dice. X is the sum of the faces on the first roll, and Y the sum of the faces on the next roll.

5.49 (a) A gambler knows that red and black are equally likely to occur on each spin of a roulette wheel. He observes five consecutive reds occur and bets heavily on red at the next spin. Asked why, he says that "red is hot" and that the run of reds is likely to continue. Explain to the gambler what is wrong with this reasoning.

(b) After hearing you explain why red and black remain equally probable after five consecutive reds on the roulette wheel, the gambler moves to a poker game. He is dealt five straight red cards. He remembers what you said and assumes that the next card dealt in the same hand is equally likely to be red or black. Is the gambler right or wrong? Why?

5.50 A time and motion study measures the time required for an assembly line worker to perform a repetitive task. The data show that the time required to bring a part from a bin to its position on an automobile chassis varies from car to car with mean 11 seconds and standard deviation 2 seconds. The time required to attach the part to the chassis varies with mean 20 seconds and standard deviation 4 seconds.

(a) What is the mean time required for the entire operation of positioning and attaching the part?

(b) If the variation in the worker's performance is reduced by better training, the standard deviations will decrease. Will this decrease change the mean you found in (a) if the mean times for the two steps remain as before?

(c) The study finds that the times required for the two steps are independent. A part that takes a long time to position, for example, does not take more or less time to attach than other parts. Would your answer in (a) change if the two variables were dependent?

5.51 Laboratory data show that the time required to complete two chemical reactions in a production process varies. The first reaction has a mean time of 40 minutes and a standard deviation of 2 minutes; the second has a mean time of 25 minutes and a standard deviation of 1 minute. The two reactions are run in sequence during production. There is a fixed period of 5 minutes between them as the product of the first reaction is pumped into the vessel where the second reaction will take place. What is the mean time required for the entire process?

5.52 Find the standard deviation σ_X of the distribution of grades in Exercise 5.42.

5.53 In an experiment on the behavior of young children, each subject is placed in an area with five toys. The response of interest is the number of toys that the child plays with. Past experiments with many subjects have shown that the probability distribution of the number X of toys played with is as follows:

Value x_i	0	1	2	3	4	5
Probability p_i	.03	.16	.30	.23	.17	.11

Calculate the mean μ_X and the standard deviation σ_X.

5.54 We will compare the distribution of the size of American families (Example 5.22) and households (Exercise 5.41).

(a) Draw probability histograms for both distributions, using the same scale for easy comparison. Mark the mean size on the axis in each case. Which distribution is more variable?

(b) Compute the standard deviation of the number of people in a family; then find the standard deviation of the number of people in a household. Does the distribution that appears more variable have a larger standard deviation?

5.55 Find the standard deviation of the time required to complete the chemical production process described in Exercise 5.51. The times for the two reactions are independent.

5.56 Find the standard deviation of the time required for the two-step assembly operation cited in Exercise 5.50.

5.57 Examples 5.24 and 5.25 concern a probabilistic projection of sales and profits by an electronics firm, Gain Communications.

(a) Find the variance and standard deviation of the estimated sales X of Gain's military unit, using the distribution and mean from Example 5.24.

(b) Because the military budget and the civilian economy are not closely linked, Gain is willing to assume that its military and civilian sales vary independently. Combine your result from (a) with the results for the civilian unit from Example 5.25 to obtain the standard deviation of the total sales $X + Y$.

(c) Find the standard deviation of the estimated profit, $Z = 2000X + 3500Y$.

5.58 The academic motivation and study habits of female students as a group are better than those of males. The Survey of Study Habits and Attitudes (SSHA) is a psychological test that measures these factors. Suppose that the distribution of SSHA scores among the women at a certain college has mean 120 and standard deviation 28, while the distribution of scores among men has mean 105 and standard deviation 35. A single male student and a single female student are selected at random and given the SSHA test.

(a) Explain why it is reasonable to assume that the scores of the two students are independent.

(b) What are the mean and standard deviation of the difference (female minus male) of their scores?

(c) From the information given, can you find the probability that the woman chosen scores higher than the man? If so, find this probability. If not, explain why you cannot.

5.59 The risk of an investment is often measured by the standard deviation of the return on the investment. The more variable the return is (the larger σ is), the riskier the investment. We can illustrate the great risk of insuring a single person's life in Exercise 5.44 by computing the standard deviation of the income X that the insurer will receive. Find σ_X, using the distribution and mean found in Exercise 5.44.

5.60 The risk of insuring one persons's life is reduced if we insure many people. Use the result of the previous exercise and rules 5.4 and 5.6 for means and variances to answer the following questions.

(a) Suppose that we insure two 21-year-old males, and that their ages at death are independent. If X_1 and X_2 are the insurer's income

$\mu_{X_1+X_2} = \mu_{X_1} + \mu_{X_2}$
$= 2(309.74) = 619.48$

$\delta^2_{X_1+X_2} = \delta^2_{X_1} + \delta^2_{X_2}$
$= 1951^2 + 1951^2$
$= 7,612,802$
$\delta_y \doteq 2759$

from the two insurance policies, the total income is $Y = X_1 + X_2$. The mean and variance of each of X_1 and X_2 are as you computed them in the previous exercise. What are the mean μ_Y and the standard deviation σ_Y of the total income Y?

(b) The insurer's *average income* on the two policies is

$\mu_2 = \dfrac{309.74 + 309.74}{2} = 309.74$

$$Z = \frac{X_1 + X_2}{2} = \frac{1}{2}X_1 + \frac{1}{2}X_2$$

$\delta_z = \sqrt{\delta^2_{\frac{1}{2}X_1} + \frac{1}{2}X_2}$

$= \sqrt{\delta^2_{\frac{1}{2}X_1} + \delta^2_{\frac{1}{2}X_2}}$

$= \sqrt{\frac{1}{2}^2(\delta^2_{X_1}) + \frac{1}{2}^2(\delta^2_{X_2})}$

$= \sqrt{(\frac{1}{4})1951^2 + (\frac{1}{4})1951^2}$

$= \sqrt{\frac{1}{2}1951^2}$

$\delta_z = \dfrac{1951}{\sqrt{2}} = 1379$

Find the mean and standard deviation of Z. You can see that the mean income is the same as for a single policy but the standard deviation is less.

(c) If four 21-year-old men are insured, the insurer's average income is

$$Z = \frac{1}{4}(X_1 + X_2 + X_3 + X_4)$$

μ still 309.74

δ_z $\dfrac{1951}{\sqrt{4}} = \dfrac{1951}{2} = 975.5$

where each income X_i has the same distribution as before. Find the mean and standard deviation of Z. Compare your results with the results of (b).

5.4 PROBABILITY LAWS*

Our study of probability has concentrated on random variables and their distributions. Now we return to the laws that govern any assignment of probilities. We have already met and used these five laws.

Laws of probability

Rule 1. $0 \leq P(A) \leq 1$ for any event A

Rule 2. $P(S) = 1$

Rule 3. Addition rule: If A and B are disjoint events, then

$$P(A \text{ or } B) = P(A) + P(B)$$

Rule 4. Complement rule: For any event A,

$$P(A^c) = 1 - P(A)$$

Rule 5. Multiplication rule: If A and B are independent events, then

$$P(A \text{ and } B) = P(A)P(B)$$

* The content of this section is important for an understanding of probability. However, it is not essential for understanding the statistical methods explained in later chapters, and can therefore be omitted without loss of continuity.

The first three laws state the fundamental properties of probability. The next two rules can be derived from these three laws and so can the other facts about probability that we will study. We will not emphasize the derivations, but it is important to note that the three common-sense facts in Rules 1 to 3 carry all of the other laws with them. Any assignment of probabilities to events that satisfies Rules 1 to 3 automatically satisfies the other laws. This is true in particular of the distributions of discrete and continuous random variables.

The goal of learning more laws of probability is to be able to give probability models for more complex random phenomena. Because of our interest in applying probability to statistical problems, we will use making decisions in the presence of uncertainty as an example of a more elaborate probability model.

Addition Rules

Probability has the property that if A and B are disjoint events, then $P(A$ or $B) = P(A) + P(B)$. But what if there are more than two events, or if the events are not disjoint? These circumstances are covered by more general addition rules for probability.

Union

> The union of any collection of events is the event that at least one of the collection occurs.

For two events A and B, the union is the event $\{A$ or $B\}$ that A or B or both occur. From the addition rule for disjoint events we can obtain rules for more general unions. Suppose first that we have several events—say, A, B, and C—that are disjoint in pairs. That is, no two can occur simultaneously. Figure 5.14 illustrates this case by a Venn diagram. The basic addition rule for two disjoint events extends to the following law:

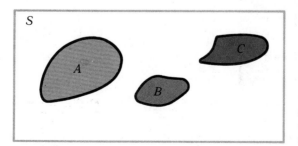

Figure 5.14 The addition rule for disjoint events: $P(A$ or B or $C) = P(A) + P(B) + P(C)$ when events A, B, and C are disjoint.

Addition rule for disjoint events

If events *A*, *B*, and *C* are disjoint in the sense that no two have any outcomes in common, then

$$P(\text{one or more of } A, B, C) = P(A) + P(B) + P(C)$$

This rule extends to any number of disjoint events.

EXAMPLE 5.29

What is the probability that the first digit of a random number will be odd? The probability model for generating a random number between 0 and 1 is given in Example 5.19. The random number *X* is a continuous random variable whose density curve has constant height 1 between 0 and 1, and is 0 elsewhere. The event that the first digit of *X* is odd is the union of five disjoint events. These events are

$$.10 \leq X < .20$$
$$.30 \leq X < .40$$
$$.50 \leq X < .60$$
$$.70 \leq X < .80$$
$$.90 \leq X < 1.00$$

Figure 5.15 illustrates the probabilities of these events as areas under the density curve. Each has probability 0.1 equal to its length. The union of the five therefore has probability equal to the sum, or 0.5. As we should expect, a random number is equally likely to begin with an odd or an even digit. ■

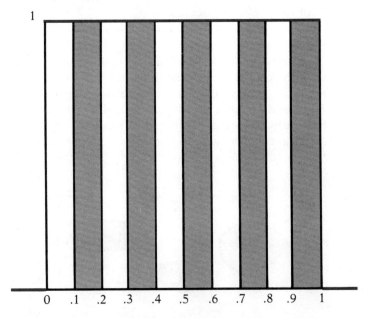

Figure 5.15 The probability that the first digit of a random number is odd is the sum of the probabilities of the five disjoint events shown.

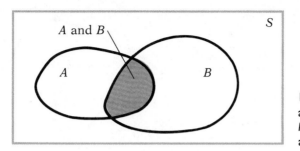

Figure 5.16 The general addition rule: $P(A \text{ or } B) = P(A) + P(B) - P(A \text{ and } B)$ for any events A and B.

If events A and B can occur simultaneously, the probability of their union is less than the sum of their probabilities. As Figure 5.16 suggests, the outcomes common to both are counted twice when we add probabilities, so we must subtract this probability once. Here is the addition rule for the union of any two events, disjoint or not.

General addition rule for unions of two events

> For any two events A and B
> $$P(A \text{ or } B) = P(A) + P(B) - P(A \text{ and } B)$$

empty event

If A and B are disjoint, the event $\{A \text{ and } B\}$ that both occur has no outcomes in it. As you might guess, this *empty event* must have probability 0. So the general addition rule includes the earlier addition rule for disjoint events.

EXAMPLE 5.30

Deborah and Matthew are anxiously awaiting word on whether they have been made partners in their law firm. Deborah guesses that her probability of making partner is 0.7 and that Matthew's is 0.5. (These are personal probabilities reflecting Deborah's assessment of chance.) This assignment of probabilities does not give enough information to compute the probability that at least one of the two is promoted. In particular, adding the individual probabilities of promotion gives the impossible result 1.2. If Deborah also guesses that the probability that *both* she and Matthew are made partners is 0.3, then by the addition rule for unions

$$P(\text{at least one is promoted}) = .7 + .5 - .3 = .9$$

The probability that *neither* is promoted is then 0.1 by the complement rule.

What is the probability that Deborah is promoted and Matthew is not? Notice that the event that Deborah is promoted (probability 0.7) is the union of the disjoint events that both are promoted (probability 0.3) and that Deborah is promoted and Matthew is not. So 0.7 is the sum of 0.3 and the unknown probability. The answer is therefore 0.4. ∎

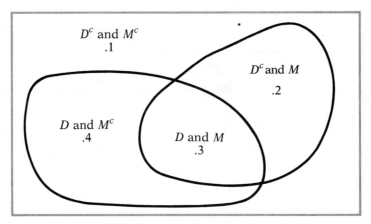

D = Deborah is made partner
M = Matthew is made partner

Figure 5.17 Venn diagram and probabilities for Example 5.30.

Venn diagrams are a great help in keeping events straight. Figure 5.17 shows the events and probabilities for Example 5.30. All but one of the probabilities in that figure have already been computed. You should be able to obtain the last probability shown.

Multiplication Rules

The sociologist Harry Edwards of the University of California at Berkeley is a leading authority on the sociology of sports. He has found that only 5% of male high school basketball, baseball, and football players go on to play at the college level. Of these, only 1.7% enter major league professional sports. The proportion of high school athletes who will have both a college career and a professional career is therefore 1.7% of 5%. That's $0.017 \times 0.05 = 0.00085$, or about eight out of 10,000.[7]

We will express this common-sense relative frequency calculation in terms of a new rule of probability theory. Our sample space consists of all male athletes in the three major high school sports. The events of interest are

$$A = \{\text{competes in college}\}$$
$$B = \{\text{competes professionally}\}$$

We wish to compute the probability of the event $\{A \text{ and } B\}$ that the athlete competes both in college and professionally. This is called the intersection of the events A and B.

Intersection

> The intersection of any collection of events is the event that *all* of the events occur.

conditional probability

We know that $P(A) = 0.05$. We also know that the probability that an athlete *who has already advanced to college sports* goes on to the pros is 0.017. This is not $P(B)$ but rather the probability that event B occurs given that the event A is known to occur. We call this the *conditional probability* of event B given the event A; we write it as $P(B|A) = 0.017$.

Whether one event has or has not occurred often influences the probability of another event—that is, the events are not independent. The conditional probability that a high school athlete plays professionally without having played in college, $P(B|A^c)$, is different from both $P(B|A)$ and the unconditional probability $P(B)$. Like unconditional probabilities, conditional probabilities can be thought of as relative frequencies. $P(B)$ is the proportion of a large group of high school athletes who go on to play professionally. $P(B|A)$ is found by looking only at the high school athletes who competed in college (event A) and finding the proportion of these who enter professional sports.

Be sure to keep in mind the distinct roles in $P(B|A)$ of the event B whose probability we are computing and the event A that represents the information we are given. The conditional probability $P(B|A)$ makes no sense if the event A can never occur, so we require that $P(A) > 0$ whenever we talk about $P(B|A)$. A rule for the probability of intersections can now be stated in terms of conditional probability.

Multiplication rule for intersections

> The probability of the intersection of two events is found by
> $$P(A \text{ and } B) = P(A)P(B|A)$$

Put in words, this rule says that for both of two events to occur, first one must occur [$P(A)$] and then, given that the first event has occurred, the second must occur [$P(B|A)$]. In the case of Edwards' study, the probability that an athlete competes both in college (event A) and professionally (event B) is

$$P(A \text{ and } B) = P(A)P(B|A)$$
$$= (.05)(.017) = .00085$$

as we computed earlier.

If $P(A)$ and $P(A$ and $B)$ are given, the multiplication rule becomes a definition of the conditional probability $P(B|A)$ in terms of unconditional probabilities.

Conditional probability

When $P(A) > 0$, the conditional probability of B given A is

$$P(B|A) = \frac{P(A \text{ and } B)}{P(A)}$$

EXAMPLE 5.31

Select at random a U.S. resident age 25 or older. The Census data in Table 3.4 gives us the following information about the age and education of this population. (Counts in the table are in thousands of persons.)

	<35	≥ 35	Total
Completed college	9771	17,091	26,862
Did not complete college	30,402	83,530	113,932
Total	40,173	100,621	140,794

Let

$$A = \text{the person selected is under 35}$$
$$B = \text{the person selected has completed college}$$

Then, because all 140,794,000 people are equally likely to be selected,

$$P(A) = \frac{40,173}{140,794} = .2853$$

$$P(B) = \frac{26,862}{140,794} = .1908$$

$$P(A \text{ and } B) = \frac{9771}{140,794} = .0694$$

The conditional probability $P(B|A)$ is the proportion of people who have completed college (event B) among those who are under 35 (event A). You can find this probability directly from the counts as

$$\frac{9771}{40,173} = .243$$

We dealt with conditional distributions in this manner in Section 3.4. Now we can also use the formal definition of conditional probability:

$$P(B|A) = \frac{P(A \text{ and } B)}{P(A)}$$

$$= \frac{.0694}{.2853} = .243$$

■

The definition of conditional probability reminds us that, in principle, all probabilities, including conditional probabilities, can be found from the assignment of probabilities to events that describes a random phenomenon. More often, however, conditional probabilities are part of the information given to us in a probability model, and the multiplication rule is used to compute $P(A$ and $B)$.

EXAMPLE 5.32

Slim is a professional poker player. At the moment, he wants very much to draw two diamonds in a row. As he sits at the table looking at his hand and at the up-turned cards on the table, Slim sees 11 cards. Of these, 4 are diamonds. The full deck contains 13 diamonds among its 52 cards so 9 of the 41 unseen cards are diamonds. Because the deck was carefully shuffled, each card that Slim draws is equally likely to be any of the cards that he has not seen.

To find Slim's probability of drawing two diamonds, first calculate

$$P(\text{first card diamond}) = \frac{9}{41}$$

$$P(\text{second card diamond} | \text{first card diamond}) = \frac{8}{40}$$

Both probabilities were found by counting cards. The probability that the first card drawn is a diamond is 9/41 because 9 of the 41 unseen cards are diamonds. If the first card is a diamond, that leaves 8 diamonds among the 40 remaining cards. So the *conditional* probability of another diamond is 8/40. The multiplication rule now says that

$$P(\text{both cards diamonds}) = \frac{9}{41} \times \frac{8}{40} = .044$$

Slim will need luck to draw his diamonds.   ■

The multiplication rule extends to finding the probability that all of several events occur. The key is to condition each event on the occurrence of *all* of the preceding events. For example, the intersection of three events A, B, and C has probability

$$P(A \text{ and } B \text{ and } C) = P(A)P(B|A)P(C|A \text{ and } B)$$

EXAMPLE 5.33

Call C the event that a major league professional athlete has a career that lasts longer than 3 years. Harry Edwards also finds that about 40% of the athletes who compete in college and then reach the pros have a career of more than 3 years. Taking A as the event that a high school athlete competes in college and B as the event that the athlete competes professionally, we know from Edwards' studies that

$$P(A) = .05$$
$$P(B|A) = .017$$
$$P(C|A \text{ and } B) = .4$$

The probability that a high school athlete competes in college and then goes on to have a pro career of more than 3 years is therefore

$$P(A \text{ and } B \text{ and } C) = P(A)P(B|A)P(C|A \text{ and } B)$$
$$= .05 \times .017 \times .40 = .00034$$

Only about three of every 10,000 high school athletes can expect to compete in college and have a professional career of more than 3 years. Edwards concludes that high school students should concentrate on studies rather than on unrealistic hopes of fortune from pro sports. ■

Probability problems often require us to combine several of the basic rules into a more elaborate calculation. Here is an example that combines the addition and multiplication rules.

EXAMPLE 5.34

What is the probability that a high school athlete will go on to professional sports? In the notation of Example 5.31, this is $P(B)$. Since every athlete either does or does not compete in college, B is the union of $\{B \text{ and } A\}$ (the athlete competes both in college and professionally) and $\{B \text{ and } A^c\}$ (the athlete competes professionally but not in college). These two events are disjoint because no one can both compete and not compete in college. Figure 5.18 shows the relations among events. By the addition rule for disjoint events

$$P(B) = P(B \text{ and } A) + P(B \text{ and } A^c)$$

Now apply the multiplication rule to obtain each term in the sum separately. We have already seen that

$$P(B \text{ and } A) = P(A)P(B|A)$$
$$= .05 \times .017 = .00085$$

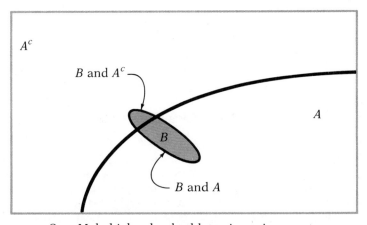

S = Male high school athletes in major sports
A = Competes in college
B = Competes professionally (colored)

Figure 5.18 Venn diagram for Example 5.34.

To apply the multiplication rule to the second term in the sum, we must know $P(B|A^c)$, the conditional probability that a high school athlete reaches a major professional league given that he does not compete in college. (Baseball players, for example, can reach the major leagues through the minor leagues without attending college.) If $P(B|A^c) = 0.0001$ then

$$P(B \text{ and } A^c) = P(A^c)P(B|A^c)$$
$$= .95 \times .0001 = .000095$$

The final result is

$$P(B) = .00085 + .000095 = .000945$$

About nine high school athletes out of 10,000 will play professional sports.   ■

tree diagram

There is another graphical device that is more helpful than a Venn diagram in organizing calculations that involve several successive stages. Figure 5.19 is a *tree diagram* for Example 5.34. Each segment in the tree is one stage of the problem. Each complete branch shows a path that an athlete can take. The probabilities written on the segments are conditional probabilities that an athlete follows that segment given that he has reached the point from which it branches. The multiplication rule says that the probability of reaching the end of any complete branch is the product of the probabilities written on its segments. The probability of any outcome, such as the event B that an athlete reaches professional sports, is then found by adding the probabilities of all branches that are part of that event.

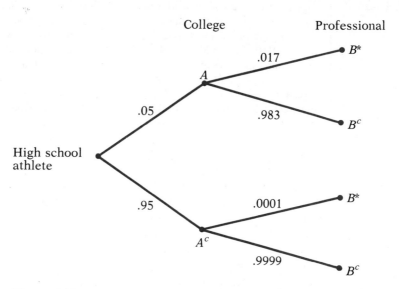

Figure 5.19   Tree diagram for Example 5.34. The probability $P(B)$ is the sum of the probabilities of the two marked branches.

Independence   The idea of the conditional probability $P(B|A)$ is that the occurrence of event A may cause a reassignment of probability to event B that differs from the original assignment $P(B)$. When the fact that A occurs gives no additional information about B, A and B are independent events. The precise definition of independence is expressed in terms of conditional probability.

Independent events

> Two events A and B that both have positive probability are independent if
>
> $$P(B|A) = P(B)$$

This is a more exact description of independence than was given in Section 5.3. The multiplication rule for independent events can now be seen to be a special case of the general multiplication rule, just as the addition rule for disjoint events is a special case of the general addition rule.

Decision analysis   Perhaps the most straightforward view of statistical inference regards the goal of inference as *making decisions in the presence of uncertainty*. One kind of decision-making in the presence of uncertainty seeks to make the probability of a favorable outcome as large as possible. Here is an example that illustrates how the multiplication and addition rules, organized with the help of a tree diagram, apply to a decision problem.

EXAMPLE 5.35   The kidneys of a patient with end-stage kidney disease will not support life. If the patient is to survive, the available choices are a kidney transplant or regular hemodialysis (use of a kidney machine several times a week). Lynn is faced with this choice. Both treatments are risky. Her doctor gives her the following information concerning patients in her condition: About 68% of dialysis patients survive for 5 years. Of transplant patients, about 48% survive with the transplanted kidney for 5 years, 43% must undergo regular dialysis because the transplanted kidney fails, and the remaining 9% do not survive the transplant. Of those transplant patients who return to dialysis, about 42% survive 5 years. Which treatment should Lynn choose to give herself the best chance of living for 5 years?[8]   ■

We can organize the information provided by the doctor in a tree diagram (Figure 5.20). Each path through the tree represents a possible outcome of Lynn's case. The probability written beside each branch after the first stage is the *conditional probability* of the next step given that Lynn has reached this point. For example, 0.68 is the probability that a dialysis patient survives 5 years. The conditional probability that a patient survives 5 years on dialysis given that he first had a transplant and then returned to dialysis because the transplant failed is different. It is 0.42, and appears on a different branch of the tree. Study the tree to convince yourself that it organizes all the information available.

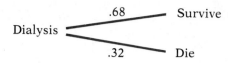

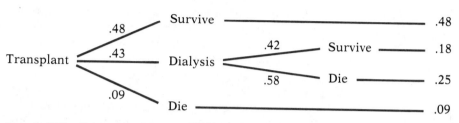

Figure 5.20 Tree diagram for the kidney failure decision problem in Example 5.35.

The multiplication rule for probabilities says that the probability of reaching the end of a branch in the tree is the product of the probabilities of all the steps along that branch. These probabilities are written at the end of each branch in Figure 5.20. For example, the event that a transplant patient returns to dialysis and then survives 5 years is

$$A = \{D \text{ and } S\}$$

where

$$D = \{\text{transplant patient returns to dialysis}\}$$
$$S = \{\text{patient survives 5 years}\}$$

Therefore,

$$P(A) = P(D \text{ and } S)$$
$$= P(S \mid D)P(D)$$
$$= (.42)(.43) = .18$$

Lynn can now compare her probability of surviving 5 years with regular dialysis with her probability of surviving 5 years with a transplant. For patients who choose to undergo dialysis, the survival probability is 0.68. A transplant patient can survive in either of two ways: with the transplant functioning or, if the transplant has failed, with a return to dialysis. These are *disjoint* events, represented by separate branches of the tree; they have probabilities 0.48 and 0.18, respectively. The addition rule for probability says that the overall probability of surviving 5 years is the sum, 0.48 + 0.18 = 0.66. This is slightly less than the survival probability with dialysis, so dialysis is the better choice.

The medical decision analysis for Example 5.35 illustrates a decision problem in which the goal is to maximize the probability of a desired outcome. The tree diagram organizes the different paths that can lead to the final outcome, and makes the calculation of outcome probabilities easy once all the necessary conditional probabilities are in place. We can then compare the probability of the desired outcome under each of the possible decisions. In Example 5.35, there are only two decisions to be compared, dialysis or transplant. Other decision problems may have more alternatives.

Where do the conditional probabilities in Example 5.35 come from? They are based in part on data—that is, on studies of many patients with kidney disease. But an individual's chances of survival depend on her age, general health, and other factors. Lynn's doctor considered her individual situation before giving her these particular probabilities. It is characteristic of most decision analysis problems that *personal probabilities* are used to describe the uncertainty of an informed decision maker.

SUMMARY

The **complement** A^c of an event A contains all outcomes that are not in A. The **union** $\{A \text{ or } B\}$ of events A and B contains all outcomes in A, in B, or in both A and B. The **intersection** $\{A \text{ and } B\}$ contains all outcomes that are in both A and B, but not outcomes in A alone or B alone.

The **conditional probability** $P(B|A)$ of an event B given an event A is defined by

$$P(B|A) = \frac{P(A \text{ and } B)}{P(A)}$$

when $P(A) > 0$, but in practice is most often found from directly available information.

The essential general laws of elementary probability are

Legitimate values: $0 \le P(A) \le 1$ for any event A, and $P(S) = 1$

Complement rule: $P(A^c) = 1 - P(A)$

Addition rule: $P(A \text{ or } B) = P(A) + P(B) - P(A \text{ and } B)$

Multiplication rule: $P(A \text{ and } B) = P(A)P(B|A)$

If A and B are **disjoint**, then $P(A \text{ and } B) = 0$. The general addition rule for unions then becomes the special addition rule, $P(A \text{ or } B) = P(A) + P(B)$.

A and B are **independent** when $P(B|A) = P(B)$. The multiplication rule for intersections then becomes $P(A \text{ and } B) = P(A)P(B)$.

In problems with several stages, use of the multiplication and addition rules is aided by drawing a **tree diagram**.

SECTION 5.4 EXERCISES

Handwritten notes in left margin:
$P(A \text{ or } B)$
$= P(A) + P(B) - P(A \text{ and } B)$
$= .148 + .209 - .075$
$= .282$

5.61 Call a household prosperous if its 1985 income exceeded $50,000. Call the household educated if the householder completed college. Select an American household at random and let A be the event that the selected household is prosperous and B be the event that it is educated. According to the Census Bureau, $P(A) = 0.148$, $P(B) = 0.209$, and the probability that a household is both prosperous and educated is $P(A \text{ and } B) = 0.075$. What is the probability $P(A \text{ or } B)$ that the household selected is either prosperous or educated?

5.62 Consolidated Builders has bid on two large construction projects. The company president believes that the probability of winning the first contract (event A) is 0.6, the probability of winning the second (event B) is 0.4, and the probability of winning both jobs (event $\{A \text{ and } B\}$) is 0.2. What is the probability of the event $\{A \text{ or } B\}$ that Consolidated will win at least one of the jobs?

5.63 Draw a Venn diagram that shows the relation between the events A and B in Exercise 5.61. Indicate each of the following events on your diagram and use the information in Exercise 5.61 to calculate the probability of each event. Finally, describe in words what each event is.

(a) $\{A \text{ and } B\}$
(b) $\{A \text{ and } B^c\}$
(c) $\{A^c \text{ and } B\}$
(d) $\{A^c \text{ and } B^c\}$

5.64 Draw a Venn diagram that illustrates the relation between events A and B in Exercise 5.62. Write each of the following events in terms of A, B, A^c, and B^c. Indicate the events on your diagram and use the information in Exercise 5.62 to calculate the probability of each.

(a) Consolidated wins both jobs.
(b) Consolidated wins the first job but not the second.
(c) Consolidated does not win the first job but does win the second.
(d) Consolidated does not win either job.

5.65 Here is the assignment of probabilities that describes the age (in years) and the sex of a randomly selected American college student as of 1985.

	14–17	18–24	25–34	≥ 35
Male	.01	.30	.12	.04
Female	.01	.30	.13	.09

(a) What is the probability that the student is female? *.53*
(b) What is the conditional probability that the student is female given that the student is at least 35 years old?
$\dfrac{.09}{.13}$
$= .69$

(c) What is the probability that the student is either female or at least 35 years old?

5.66 Exercise 3.67 presents a two-way table of suicides by sex of the victim and method used. Suppose that a suicide victim is selected at random. Answer the following questions from the table.

(a) What is the probability that the suicide victim is male?

(b) What is the probability that the victim used a firearm?

(c) What is the conditional probability that the suicide used a firearm given that the victim is male?

(d) Are the events "victim is male" and "a firearm was used in the suicide" independent? Why or why not?

5.67 Refer to the two-way table of households by education and income that appears in Exercise 3.63. Answer the following questions from this table.

(a) What is the probability that a randomly selected household has an income of at least $50,000?

(b) What is the conditional probability that a household earns over $50,000 given that the householder completed at least 4 years of college?

(c) What is the conditional probability that the householder completed at least 4 years of college given that the household income is at least $50,000?

(d) Are the random variables X (household income in dollars) and Y (years of education for the householder) independent? Why or why not?

(e) What is the probability that the household either has an income below $15,000 or is headed by a person who did not complete high school?

5.68 Common sources of caffeine in the American diet are coffee, tea, and cola drinks. Suppose that

55% of American adults drink coffee
25% of American adults drink tea
45% of American adults drink cola

and also that

15% drink both coffee and tea
5% drink all three beverages
25% drink both coffee and cola
5% drink only tea

Draw a Venn diagram marked with this information. Use it along with the addition rules to answer the following questions.

(a) What percent of American adults drink only cola?

(b) What percent drink none of these beverages?

5.69 Choose an employed person at random. Let A be the event that the person chosen is a woman, and B the event that the person holds a managerial or professional job. As of 1985, $P(A) = 0.441$ and the probability of managerial and professional jobs among women is $P(B|A) = 0.234$. Find the probability that a randomly chosen employed person is a woman holding a managerial or professional position.

5.70 Functional Robotics Corporation buys electrical controllers from a Japanese supplier. The company's treasurer feels that there is probability 0.4 that the dollar will fall in value against the Japanese yen in the next month. The treasurer also believes that *if* the dollar falls there is probability 0.8 that the supplier will demand renegotiation of the contract. What probability has the treasurer assigned to the event that the dollar falls and the supplier demands renegotiation?

5.71 In the language of government statistics, the "labor force" includes all civilians over 16 years of age who are working or looking for work. Select a member of the U.S. labor force at random. Let A be the event that the person selected is white and B be the event that he or she is employed. In 1985, 86.5% of the labor force was white. Of the whites in the labor force, 93.8% were employed. Among nonwhite members of the labor force, 86.1% were employed.

(a) Express each of the percents given as a probability involving the events A and B—for example, $P(A) = 0.865$.

(b) Draw a tree diagram for the outcomes of recording first the race (white or nonwhite) of a randomly chosen member of the labor force and then whether or not the person is employed.

(c) Find the probability that the person chosen is an employed white. Also find the probability that an employed nonwhite is chosen. What is the probability $P(B)$ that the person chosen is employed?

5.72 Suppose that in Exercise 5.70 the treasurer also feels that if the dollar does not fall there is probability 0.2 that the Japanese supplier will demand that the contract be renegotiated. What is the probability that the supplier will demand renegotiation?

5.73 Use your results from Exercise 5.71 and the definition of conditional probability to find the probability $P(A|B)$ that a randomly selected member of the labor force is white, given that he or she is employed.

5.74 An examination consists of multiple-choice questions, each having five possible answers. Linda estimates that she has probability 0.75 of knowing the answer to any question that may be asked. If she does not know the answer she will guess, with conditional probability 1/5 of being correct.

(a) What is the probability that Linda gives the correct answer to a question? (Draw a tree diagram to guide the calculation.)

(b) Use the result of (a) and the definition of conditional probability to find the conditional probability that Linda knows the answer, given that she supplies the correct answer.

5.75 The voters in a large city are 40% white, 40% black, and 20% Hispanic. (Hispanics may be of any race in official statistics, but in this case we are speaking of political blocs.) A black mayoral candidate anticipates attracting 30% of the white vote, 90% of the black vote, and 50% of the Hispanic vote. Draw a tree diagram with probabilities for the race (white, black, or Hispanic) and vote (for or against the candidate) of a randomly chosen voter. What percent of the overall vote does the candidate expect to get?

5.76 Choose a point at random in the square having sides $0 \le x \le 1$ and $0 \le y \le 1$. This means that the probability that the point falls in any region within the square is the area of that region. Let X be the x coordinate and Y the y coordinate of the point chosen. Find the conditional probability $P(Y < 1/2 \mid Y > X)$. (Hint: Draw a diagram of the square and the events $Y < 1/2$ and $Y > X$.)

5.77 Psychologists use probability models to describe learning in animals. In one experiment, a rat is placed in a dark compartment and a door leading to a light compartment is opened. The rat will not move to the light compartment without a reason. A bell is rung and if after 5 seconds the rat has not moved, it gets a shock through the floor of the dark compartment. To avoid the shock, rats soon learn to move when the bell is rung. A simple model for learning says that the rat can be in only one of two states:

In *state A* the rat will not move from the dark compartment until it receives a shock.
In *state B* the rat has learned to respond to the bell and moves immediately when the bell rings.

A rat starts in state A and eventually changes to state B. The change from state A to state B is the result of learning. A rat in state A gets a shock every time the bell rings; after it changes to state B it never gets a shock. Suppose that a rat has probability 0.2 of learning each time it is shocked. Let the random variable X be the number of shocks that this rat receives.[9]

(a) If a rat receives exactly 4 shocks, which state was the rat in at the end of each of trials 1, 2, 3, and 4?

(b) Use the result of (a) and the multiplication rule to find $P(X = 4)$, the probability that the rat receives exactly 4 shocks.

(c) Based on your work in (a) and (b), give the probability distribution of X. That is, for any positive whole number x, what is $P(X = x)$?

5.78 John has coronary artery disease. He and his doctor must decide between medical management of the disease and coronary bypass surgery. Because John has been quite active, he is concerned about his quality of life as well as the length of his life. He wants to make the decision that will maximize the probability of the event A that he survives for 5 years and is able to carry on moderate activity during that time. The doctor makes the following probability estimates for patients of John's age and condition.

- Under medical management, $P(A) = 0.7$.
- There is probability 0.05 that John will not survive bypass surgery, probability 0.10 that he will survive with serious complications, and probability 0.85 that he will survive the surgery without complications.
- If he survives with complications, the conditional probability of the desired outcome A is 0.73. If there are no serious complications, the conditional probability of A is 0.76.

Draw a tree diagram that summarizes this information. Then calculate $P(A)$ assuming that John chooses the surgery. Does surgery or medical management offer him the better chance of achieving his goal? (Based loosely on M. C. Weinstein, J. S. Pliskin, and W. B. Stason, "Coronary artery bypass surgery: decision and policy analysis," in J. P. Bunker, B. A. Barnes, and F. W. Mosteller (eds.), *Costs, Risks and Benefits of Surgery*, Oxford University Press, New York, 1977, pp. 342–371.)

5.79 Zipdrive, Inc. has developed a new disk drive for small computers. The demand for the new product is uncertain but it can be described as "high" or "low" in any one year. After 4 years the product is expected to be obsolete. Management must decide whether to build a plant or to contract with a factory in Hong Kong to manufacture the new drive. Building a plant will be profitable if demand remains high but could lead to a loss if demand drops in future years.

After careful study of the market and of all relevant costs, Zipdrive's planning office provides the following information. Let A be the event that the first year's demand is high, and B be the event that the following 3 years' demand is high. The marketing division's best estimate of the probabilities is

$$P(A) = .9$$
$$P(B|A) = .36$$
$$P(B|A^c) = 0$$

The probability that building a plant is more profitable than contracting the production to Hong Kong is 0.95 if demand is high all 4 years, 0.3 if demand is high only in the first year, and 0.1 if demand is low all 4 years.

Draw a tree diagram that organizes this information. The tree
will have three stages: first year demand, next 3 years' demand, and
whether building or contracting is more profitable. Which decision
has the higher probability of being more profitable? (When decision
analysis is used for investment decisions like this, firms in fact
compare the mean profits rather than the probability of a profit. We
ignore this complication.)

NOTES

1 An informative and entertaining account of the origins of probability theory is
Florence N. David, *Games, Gods and Gambling*, Charles Griffin, London, 1962.

2 We will consider only the case in which X takes a finite number of possible
values. The same ideas, implemented with more advanced mathematics, apply
to random variables with an infinite but still countable collection of values.

3 The notation $\bar{x}$ is used both for the random variable, which takes different
values in repeated sampling, and for the numerical value of the random variable
in a particular sample. Similarly, s and $\hat{p}$ stand both for random variables and
for specific values. This notation is mathematically imprecise but statistically
convenient.

4 See A. Tversky and D. Kahneman, "Belief in the law of small numbers,"
Psychological Bulletin, 76 (1971), pp. 105–110, and other writings of these
authors for a full account of our misperception of randomness.

5 Probabilities involving runs can be quite difficult to compute. That the
probability of a run of three or more heads in 10 independent tosses of a fair coin
is $(1/2) + (1/128) = 0.508$ can be found by clever counting, as can the other
results given in the text. A general treatment using advanced methods appears
in section XIII.7 of William Feller, *An Introduction to Probability Theory and
its Applications*, vol. 1, 3d ed., Wiley, New York, 1968.

6 R. Vallone and A. Tversky, "The hot hand in basketball: on the misperception
of random sequences," *Cognitive Psychology*, 17 (1985), pp. 295–314.

7 Professor Edwards' findings are reported in *The New York Times*, Feb. 25, 1986.

8 The probabilities in this example are based on Benjamin A. Barnes, "An
overview of the treatment of end-stage renal disease and a consideration of
some of the consequences," in J. P. Bunker, B. A. Barnes, and F. W. Mosteller
(eds.), *Costs, Risks and Benefits of Surgery*, Oxford University Press, New York,
1977, pp. 325–341. See this article for a more realistic analysis.

9 This model is too simple to fit experimental data well. Probability models for
learning that give more realistic descriptions of experiments with rats are
discussed in R. C. Atkinson, G. H. Bower, and E. J. Crothers, *An Introduction
to Mathematical Learning Theory*, Wiley, New York, 1965.

CHAPTER 5 EXERCISES

5.80 The original simple form of the Connecticut state lottery (ignoring a few gimmicks) awarded the following prizes for each 100,000 tickets sold. The winners were chosen by drawing tickets at random.

1 $5000 prize

18 $200 prizes

120 $25 prizes

270 $20 prizes

If you hold one ticket in this lottery, what is your probability of winning anything? What is the mean amount of your winnings?

5.81 The Rotter Brothers Company is planning a major investment. The amount of profit X is uncertain but a probabilistic estimate gives the following distribution (in millions of dollars):

Profit	1	1.5	2	4	10
Probability	.1	.2	.4	.2	.1

(a) Find the mean profit μ_X and the standard deviation of the profit.
(b) Rotter Brothers owes its source of capital a fee of $200,000 plus 10% of the profits X. So the firm actually retains

$$Y = .9X - .2$$

from the investment. Find the mean and standard deviation of Y.

5.82 A grocery store gives its customers cards that may win them a prize when matched with other cards. The back of the card announces the following probabilities of winning various amounts if a customer visits the store 10 times:

Amount	$1000	$200	$50	$10
Probability	1/10,000	1/1000	1/100	1/20

(a) What is the probability of winning nothing?
(b) What is the mean amount won?
(c) What is the standard deviation of the amount won?

5.83 The time required for a telephone operator to handle a directory assistance call is normally distributed with mean 25 seconds and standard deviation 5 seconds.

(a) What is the probability that it takes more than 30 seconds to handle such a call?

(b) The telephone company plans to monitor the longest 1% of directory assistance calls to study the reasons for the delay. You must instruct the computer how long to wait before beginning to monitor the call. What is the length of time that only 1% of directory assistance calls exceed?

5.84 A study of the weights of the brains of Swedish men found that the weight X was a random variable with mean 1400 grams and standard deviation 20 grams. Find positive numbers a and b such that $Y = a + bX$ has mean 0 and standard deviation 1.

5.85 In assigning a character's intelligence in the game "Dungeons and Dragons," a player rolls three dice and adds the spots on the up faces. Assume that the dice are all balanced so that each face is equally likely, and that the three dice fall independently.

(a) Give a sample space for the sum X of the spots.

(b) Find $P(X = 5)$.

(c) If X_1, X_2, and X_3 are the number of spots on the up faces of the three dice, then $X = X_1 + X_2 + X_3$. Use this fact to find the mean μ_X and the standard deviation σ_X without finding the distribution of X. (Start with the distribution of each of the X_i.)

5.86 The real return on an investment is its rate of increase corrected for the effects of inflation. You believe that the annual real return X on a portfolio of stocks will vary in the future with mean $\mu_X = 0.11$ and standard deviation $\sigma_X = 0.28$. (That is, you expect stocks to give an average return of 11% in the future, but with a large variation from year to year.) Further, you think that the annual real return Y on Treasury bills will vary with mean $\mu_Y = 0.02$ and standard deviation $\sigma_Y = 0.05$. Even though this is *not* realistic, assume that returns on stocks and Treasury bills vary independently.

(a) If you put half of your assets into stocks and half into Treasury bills, your overall return will be $Z = 0.5X + 0.5Y$. Calculate μ_Z and σ_Z.

(b) You decide that you are willing to take more risk (greater variation in the return) in exchange for a higher mean return. Choose an allocation of your assets between stocks and Treasury bills that will accomplish this and calculate the mean and standard deviation of the overall return. (If you put a proportion α of your assets into stocks, the total return is $Z = \alpha X + (1 - \alpha)Y$.)

5.87 The most popular game of chance in Roman times was tossing four *astragali*. An astragalus is a small bone from the heel of an animal that comes to rest on one of four sides when tossed. (The other two sides are rounded.) The following table gives the probabilities of the outcomes for a single astragalus based on modern experiments. The names "Broad convex" etc. describe the four sides of the heel bone.

The best throw was the "Venus," with all four uppermost sides different. What is the probability of rolling a Venus? (From Florence N. David, *Games, Gods and Gambling*, Charles Griffin, London, 1962, p. 7.)

Side	Broad convex	Broad concave	Narrow flat	Narrow hollow
Probability	.4	.4	.1	.1

5.88 Ann and Bob are playing the game "two-finger Morra." Each player shows either one or two fingers and at the same time calls out a guess for the number of fingers the other player will show. If a player guesses correctly and the other player does not, the player wins a number of dollars equal to the total number of fingers shown by both players. If both or neither guesses correctly, no money changes hands. On each play both Ann and Bob choose one of the following options:

Choice	Show	Guess
A	1	1
B	1	2
C	2	1
D	2	2

(a) Give the sample space S by writing all possible choices for both players on a single play of this game.

(b) Let X be Ann's winnings on a play. (If Ann loses $2, then $X = -2$; when no money changes hands, $X = 0$.) Write the value of the random variable X next to each of the outcomes you listed in (a). This is another choice of sample space.

(c) Now assume that Ann and Bob choose independently of each other. Moreover, they both play so that all four choices listed above are equally likely. Find the probability distribution of X.

(d) If the game is fair, X should have mean zero. Does it? What is the standard deviation of X?

The following exercises require familiarity with the material presented in the optional Section 5.4.

5.89 Here is the distribution of the adjusted gross income X (in thousands of dollars) reported on individual federal income tax returns in 1984.

Income	<11	11–24	25–49	50–99	≥100
Probability	.369	.316	.248	.057	.010

   (a) What is the probability that a randomly chosen return shows an adjusted gross income of $50,000 or more?
   (b) Given that a return shows an income of at least $50,000, what is the conditional probability that the income is at least $100,000?

5.90 You have torn a tendon and are facing surgery to repair it. The orthopedic surgeon explains the risks to you: Infection occurs in 3% of such operations, the repair fails in 14%, and both infection and failure occur together in 1%. What percent of these operations succeed and are free from infection?

5.91 The distribution of blood types among white Americans is approximately as follows: 37% type A, 13% type B, 44% type O, and 6% type AB. Suppose that the blood types of married couples are independent and that both the husband and wife follow this distribution.

   (a) An individual with type B blood can safely receive transfusions only from persons with type B or type O blood. What is the probability that the husband of a woman with type B blood is an acceptable blood donor for her?
   (b) What is the probability that in a randomly chosen couple the wife has type B blood and the husband has type A?
   (c) What is the probability that one of a randomly chosen couple has type A blood and the other has type B?
   (d) What is the probability that at least one of a randomly chosen couple has type O blood?

5.92 Exercise 5.19 gives the probability distribution of the sex and occupation of a randomly chosen American worker. Use this distribution to answer the following questions:

   (a) Given that the worker chosen holds a managerial (class A) job, what is the conditional probability that the worker is female?
   (b) Classes D and E include most mechanical and factory jobs. What is the conditional probability that a worker is female given that he or she holds a job in one of these classes?

5.93 It is difficult to conduct sample surveys on sensitive issues because many people will not answer questions if the answers embarrass them. "Randomized response" is an effective way to guarantee anonymity while collecting information on topics such as student cheating or sexual behavior. Here is the idea: To ask a sample of students whether they have plagiarized a term paper while in college, have each student toss a coin in private. If the coin lands "heads" *and* they have not plagiarized, they are to answer "No." Otherwise they are to give "Yes" as their answer. Only the student knows whether the answer reflects the truth or just the coin toss, but the researchers can use a proper random sample with followup for nonresponse and other good sampling practices.

Suppose that in fact the probability is 0.3 that a randomly chosen student has plagiarized a paper. Draw a tree diagram in which the first stage is tossing the coin and the second stage is the truth about plagiarism. The outcome at the end of each branch is the answer given to the randomized response question. What is the probability of a "No" answer in the randomized response poll? If the probability of plagiarism were 0.2, what would be the probability of a "No" response on the poll? Now suppose that you get 39% "No" answers in a randomized response poll of a large sample of students at your college. What do you estimate to be the percent of the population who have plagiarized a paper?

5.94 The ELISA test was introduced in the mid-1980s to screen donated blood for the presence of the AIDS virus. The test actually detects antibodies, substances that the body produces when the virus is present. When antibodies are present, ELISA is positive with probability about 0.98 and negative with probability 0.02. When the blood tested is not contaminated with antibodies, ELISA gives a positive result with probability about 0.07 and a negative result with probability 0.93. (Since ELISA was designed to keep the AIDS virus out of blood supplies, the higher probability 0.07 of a false positive is acceptable in exchange for the low probability 0.02 of failing to detect contaminated blood.) Suppose that 1% of a large population carry the AIDS antibody in their blood.

(a) Draw a tree diagram for selecting a person from this population (outcomes: the person does or does not carry the AIDS antibody) and for testing his or her blood (outcomes: positive or negative).

(b) What is the probability that the ELISA test for AIDS is positive for a randomly chosen person from this population?

(c) Use your results from (a) and (b) and the definition of conditional probability to find the probability that a person has the antibody given that the ELISA test is positive.
(This exercise illustrates a fact that is important when considering proposals for widespread testing for AIDS or illegal drugs: If the condition being tested is uncommon in the population, most positives will be false positives.)

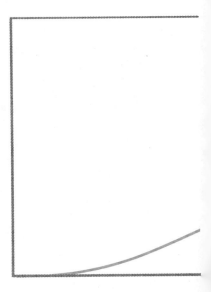

Prelude

Statistical inference is based on the same numerical measures we have used to describe data, along with probability calculations that tell us how reliable our conclusions are. This chapter discusses the probability distributions of some important statistics, especially counts, proportions, and means from sample data. We also discuss control charts, an important type of informal inference that combines graphical presentation of data with probability calculations.

- *How can the binomial probability distributions give us evidence that a local political random drawing is rigged?*

- *How can we express precisely the advantage of taking the average of several laboratory measurements rather than using a single measurement?*

- *How do manufacturers use control charts to improve the quality of their products by monitoring the manufacturing process?*

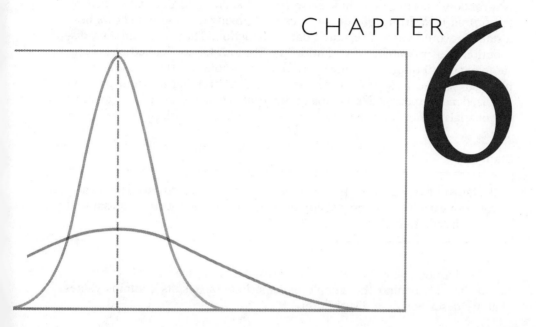

From Probability to Inference

Formal statistical inference is based on computing probabilities. Of course, it is also based on data that are summarized by *statistics* such as means, proportions, and the slopes of least squares regression lines. When the data are produced by random sampling or by randomized experimentation, a statistic is a numerical outcome of a random phenomenon—that is, a statistic is a random variable. The link between probability and data is formed by the *sampling distributions* of statistics. A sampling distribution shows how a statistic would vary in repeated data production. That is, a sampling distribution is a probability distribution. In Section 4.4 we looked at several sampling distributions empirically by actually simulating 1000 random samples. Considered as a probability distribution, a sampling distribution is an idealized mathematical description of the results of an indefinitely large number of trials rather than the results of a particular 1000 trials.

sampling distribution

Statistic

> A statistic from a probability sample or randomized experiment is a random variable. The probability distribution of the statistic is its sampling distribution.

Distributions play another role in statistical inference as well. The population that lies behind the sample or experimental results contains values that are described by a distribution.

EXAMPLE 6.1

The distribution of heights of women between the ages of 18 and 24 is approximately normal with mean 65.5 inches and standard deviation 2.5 inches. Suppose that we select one woman at random and measure her height. The resulting measurement X is a random variable. Its distribution in repeated sampling is the same $N(65.5, 2.5)$ distribution that describes the pattern of heights in the entire population. ∎

In statistical inference, probability distributions play two important roles.

- The distribution of a variable in the population is described by a probability distribution. This is the context in which distributions were first presented as models for data. The language of probability is used for convenience, asking the question "What if we selected one unit at random from the population?" We will simply speak of the *population distribution* without emphasizing that it is a probability distribution.
- The sampling distribution of a statistic is a probability distribution. This distribution describes the behavior of the statistic in repeated sampling or repeated experimentation.

To progress from discussing probability as a topic in itself to probability as a foundation for inference, we must study the sampling distributions of some common statistics. That is the topic of Sections 6.1 and 6.2. We will meet other sampling distributions in the study of inference. In each case, the nature of the sampling distribution depends on both the nature of the population distribution and the way in which the data were collected from the population.

Section 6.3 presents a simple and direct use of sampling distributions for inference: *control charts* for assessing whether a process remains in statistical control over time. Control charts are of considerable practical importance, and also illustrate the use of probability in inference in a simple setting.

6.1 COUNTS AND PROPORTIONS

An opinion poll selects a sample of 1785 persons and asks each if they attended church or synagogue in the past week; the number who say "No" is a random variable X. A quality control engineer selects a sample of 10 clutch assemblies from a supplier's shipment for detailed inspection; of these, a number X fail to meet the specifications. In both of these examples, the random variable X is a *count* of the occurrences of some outcome in a fixed number of trials. There are 1785 trials in the first example and 10 trials in the second. If the fixed number of trials is n, then the *sample proportion* is $\hat{p} = X/n$. For example, if 1000 of the 1785 people interviewed by the opinion poll say "No," the sample proportion is

sample proportion

$$\hat{p} = \frac{1000}{1785} = .56$$

This opinion poll example is the same one simulated in Example 4.17. Sample counts and sample proportions are common statistics. Finding their sampling distributions is the goal of this section.

The Binomial Distributions

The distribution of a count X depends on the data collection design. The simplest situation is a fixed number n of independent trials, on each of which the outcome being counted either occurs or does not occur. For the sake of convenience, we will call the outcome we are counting a "success." If the probability of a success has the same value p on each trial, the distribution of the count of successes X is completely determined by the number of trials n and the success probability p.

Binomial distribution

The distribution of the count X of successes in n independent trials with probability p of success on each trial is called the binomial distribution with parameters n and p. As an abbreviation, we say that X is $B(n, p)$.

The binomial distributions are an important class of discrete probability distributions. Later in this section we will obtain formulas for assigning probabilities to outcomes and for finding the mean and standard deviation of binomial distributions. But the first, and most important, aspect of using binomial distributions is the ability to recognize situations to which they apply. The key requirements are a fixed number n of trials, independent trials, and the same probability p of success on each trial.

- As Example 5.13 observed, independence is reasonable for describing successive tosses of a coin but not for describing the color of successive cards dealt from the same deck. So the number of heads in 10 tosses of a balanced coin has the binomial distribution $B(10, 0.5)$, but the number of red cards in 10 cards dealt from a deck does not.

- Genetics assumes that offspring receive genes chosen independently from those carried by each parent. If both parents carry genes for the O and A blood types, each offspring has probability 0.25 of getting two O genes and so of having blood type O. The number of O blood types among 5 children of these parents has the $B(5, 0.25)$ distribution.

- Engineers define reliability as the probability that an item will perform its function under specific conditions for a specific period of time. If an aircraft engine turbine has probability 0.999 of performing properly for an hour of flight, the number of turbines in a fleet of 350 engines that fly for an hour without failure has the $B(350, 0.999)$ distribution. This binomial distribution is obtained by assuming, as seems reasonable, that the turbines fail independently of each other. A common cause of failure, such as sabotage, would destroy the independence and make the binomial model inappropriate.

We will give a formula for the probability that a binomial random variable takes any of its values, but in practice you should avoid using this formula whenever possible. Some calculators and most statistical computing packages calculate binomial probabilities. If you do not have suitable computing facilities, you can still shorten the work of calculating binomial probabilities for some values of n and p by looking up probabilities in Table C in the back of this book. The entries in the table are the probabilities $P(X = k)$ of individual outcomes for a binomial random variable X.

EXAMPLE 6.2

A quality engineer chooses 10 clutch assemblies at random from a large shipment. Unknown to the engineer, 10% of the clutches in the shipment fail to meet the specifications. What is the probability that no more than 1 of the 10 clutches in the sample fail inspection?

Is this a binomial situation? Not quite. Removing one clutch changes the proportion of bad clutches in the remaining population, so the state of the second clutch chosen is not independent of the first. If the shipment is large, however, removing a few items has a very small effect on the composition of the remaining population; successive inspection results are very nearly independent. In this case we can act as if the count X of clutches in the sample that fail inspection has a binomial distribution. Since 10% of the population fail to meet the specifications, X is $B(10, 0.1)$. Figure 6.1 is a probability histogram for this distribution. The distribution is strongly skewed. Although X can take any whole number value from 0 to 10, the probabilities of values larger than 5 are so small that they do not appear in the histogram.

We want to calculate

$$P(X \leq 1) = P(X = 1) + P(X = 0)$$

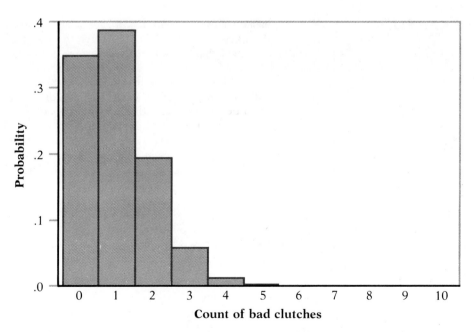

Figure 6.1 Probability histogram for the binomial distribution with $n = 10$ and $p = 0.1$.

when X is $B(10, 0.1)$. The probability $P(X = 1)$ can be obtained in MINITAB as follows:

```
MTB> PDF 1;
      BINOMIAL n = 10, p = .1.
```

Other computing systems have similar commands.

To use Table C for this calculation, look opposite $n = 10$ and under $p = 0.10$. The entry opposite each k is $P(X = k)$. So

		p
n	k	.10
10	0	0.3487
	1	0.3874
	2	0.1937
	3	0.0574

$$P(X \le 1) = P(X = 1) + P(X = 0)$$
$$= .3874 + .3487 = .7361$$

We see that about 74% of all samples will contain no more than 1 bad clutch. In fact, 35% of the samples will contain no bad clutches. A sample of size 10 cannot be trusted to alert the engineer to the presence of unacceptable items in the shipment. ∎

A look at Table C shows that the values of p that are present are all 0.5 or smaller. When the probability of a success is greater than 0.5, we must restate the problem in terms of the number of failures. The probability of a failure is less than 0.5 when the probability of a success exceeds 0.5. When using the table, we must stop to ask which outcomes should be counted.

EXAMPLE 6.3

Rick is a basketball player who makes 75% of his free throws over the course of a season. In a key game, Rick shoots 12 free throws and misses 5 of them. The fans think that he failed because he was nervous. Is it unusual for Rick to perform this poorly?

To answer this question, assume that free throws are independent trials with probability 0.75 of a success on each trial. (Studies of long sequences of free throws have found no evidence that they are dependent, so this is a reasonable assumption.) Because the probability of making a free throw is greater than 0.5, we count misses in order to use Table C. The probability of a miss is $1 - 0.75$, or 0.25. The number X of misses in 12 attempts has the $B(12, 0.25)$ distribution.

We want the probability of missing 5 or more. This is

$$P(X \ge 5) = P(X = 5) + P(X = 6) + \cdots + P(X = 12)$$
$$= .1032 + .0401 + \cdots + .0000 = .1576$$

Rick will miss 5 or more out of 12 free throws about 16% of the time, or roughly once in every six games. Although this performance is below his average level, it is well within the range of the usual chance variation in his shooting. ∎

Because sequences of independent success-or-failure trials like those in Example 6.3 are a common probability model, the binomial distributions occur frequently in applications of probability. Example 6.2, concerning quality control of clutch assemblies, illustrates why binomial distributions are also important in statistics. In that and similar examples, the objective is to estimate the proportion p of ''successes'' in a population. The population distribution in this setting is just the distribution of successes and failures in the population, described by the single parameter p.

Sampling distribution of a count

> When the population is much larger than the sample, a count of successes in an SRS of size n has approximately the $B(n, p)$ distribution if the population proportion of successes is p.

The accuracy of this approximation improves as the size of the population increases relative to the size of the sample. As a rule of thumb, you can use the binomial sampling distribution for counts when the population is at least 100 times as large as the sample.

EXAMPLE 6.4

An opinion poll asks an SRS of 1785 adults whether they attended church or synagogue during the past week. Suppose that 60% of the adult population did not attend. What is the probability that the sample proportion who did not attend is at least 58%?

The sample proportion does *not* have a binomial distribution because it is not a count. We therefore translate any question about the sample proportion $\hat{p}$ into a question about the sample count X, which *does* have the binomial distribution $B(1785, 0.6)$. Since 58% of 1785 is 1035.3,

$$P(\hat{p} \geq .58) = P(X \geq 1035.3)$$
$$= P(X = 1036) + P(X = 1037) + \cdots + P(X = 1785) \quad \blacksquare$$

To find the probability in Example 6.4 we must add over 700 binomial probabilities, each with $n = 1785$. Tables are not available for such large values of n, and most computer packages are unable to calculate binomial probabilities when n is this large. Even the formula for binomial probabilities given below is not practical in this case. We can state a probability question about the opinion poll in terms of the binomial distribution, but we cannot easily find the numerical value of the probability using this distribution. Recall from the simulations in Section 4.4 that a sample proportion $\hat{p}$ from a large sample follows a distribution that is close to normal. We will soon see that this empirical result can be stated exactly: Binomial distributions are nearly normal when n is large. So normal distributions can be used for binomial calculations for large n.

Binomial formulas*   We will illustrate by an example how to obtain a general formula for binomial probabilities. The argument uses the multiplication rule for the probability that all of a number of independent events occur.

---

* The formula for binomial probabilities is useful in many settings, but we will not use it in our study of statistical inference. This material can therefore be omitted if desired.

EXAMPLE 6.5

Nicholas Caputo is the clerk of Essex County, New Jersey. The top line on county ballots is supposed to be assigned by a random drawing to either the Republican or the Democratic candidate. Mr. Caputo, who is a Democrat, conducts the drawings. He has assigned the top line to the Democrats 40 out of 41 times. What is the probability that the Democrats would do so well in a truly random lottery?

In a properly conducted lottery, the successive drawings are independent and each has probability 1/2 of giving the top line to the Democrats. The count X of Democrats on the top line in 41 trials is therefore a binomial random variable with $n = 41$ and $p = 1/2$. The observed count of successes for Democrats was $X = 40$. We will calculate

$$P(X \geq 40) = P(X = 40) + P(X = 41)$$

which is the probability that the Democrats win at least 40 of the 41 drawings.

By the multiplication rule, the probability that the Democrats are awarded the top spot in all 41 drawings is

$$P(X = 41) = \left(\frac{1}{2}\right)^{41}$$

Similarly, the probability that the Democrats win the first 40 drawings is $(1/2)^{40}$. So 40 straight Democratic wins followed by a Republican win has probability

$$\left(\frac{1}{2}\right)^{40} \times \left(\frac{1}{2}\right) = \left(\frac{1}{2}\right)^{41} \tag{6.1}$$

This is *not* the probability $P(X = 40)$ because the event $\{X = 40\}$ allows the single Republican win to fall anywhere in the sequence of 41 trials. We argue as follows: Any specific sequence of 40 Democratic and 1 Republican wins has the same probability as that shown in Equation 6.1 by a similar use of the multiplication rule. There are 41 such outcomes because the single Republican success can occur on any one of the 41 trials. These 41 outcomes are disjoint. The event $\{X = 40\}$ therefore has probability given by the sum of 41 probabilities, each equal to Equation 6.1. We see at last that

$$P(X = 40) = 41 \times \left(\frac{1}{2}\right)^{40} \times \left(\frac{1}{2}\right)$$

These are very small probabilities. Their sum $P(X \geq 40)$ is about 0.00000000002, or 1 in 50 billion. There is almost no chance that the Democrats would have won a truly random drawing so consistently. The New Jersey Supreme Court made calculations like this one and told Mr. Caputo to change his ways.[1] ■

The idea of Example 6.5 extends to binomial situations in general. Suppose that we observe n independent trials, each of which leads to a "success" or a "failure." Suppose also that a success has the same probability p on each trial. Then any arrangement of k successes and $n - k$ failures in a specific order has probability $p^k(1 - p)^{n-k}$ by the multiplication rule. The probability $P(X = k)$ of exactly k successes is therefore $p^k(1 - p)^{n-k}$ times the number of different ways we can distribute k successes among n trials. We use the following fact to find this number.

Binomial coefficient

> The number of ways in which k successes can be chosen from among n trials is given by the binomial coefficient
>
> $$\binom{n}{k} = \frac{n!}{k!(n-k)!} \tag{6.2}$$
>
> for $k = 0, 1, 2, \ldots, n$.

factorial   The binomial coefficients are expressed using the *factorial* notation: For any positive whole number n,

$$n! = n \times (n-1) \times (n-2) \times \cdots \times 3 \times 2 \times 1$$

and $0! = 1$. Notice that the larger of the two factorials in the denominator of a binomial coefficient will cancel much of the $n!$ in the numerator. For example,

$$\binom{8}{3} = \frac{8!}{3!5!}$$

$$= \frac{(8)(7)(6)}{(3)(2)(1)} = 56$$

The notation $\binom{n}{k}$ is *not* related to the fraction $\frac{n}{k}$. A helpful way to remember its meaning is to read it as "binomial coefficient n choose k." Binomial coefficients have many uses in mathematics but we are interested in them only as an aid to finding binomial probabilities. The binomial coefficient $\binom{n}{k}$ counts the number of ways in which k successes can be distributed among n trials. The binomial probability $P(X = k)$ is this count multiplied by the probability of any specific arrangement of the k successes. Here is the formula we seek.

Binomial probability

> If X is $B(n, p)$, then
>
> $$P(X = k) = \binom{n}{k} p^k (1-p)^{n-k} \tag{6.3}$$
>
> for $k = 0, 1, \ldots, n$.

In Example 6.5, X is $B(41, 1/2)$ and $k = 1$. The probability

$$P(X = 40) = \binom{41}{40}\left(\frac{1}{2}\right)^{40}\left(\frac{1}{2}\right)^{1}$$

obtained from the binomial formula 6.3 agrees with the result obtained in Example 6.5 because

$$\binom{41}{40} = \frac{41!}{40!1!} = 41$$

Here is another example that illustrates the use of the binomial distributions.

EXAMPLE 6.6

The number X of clutches that fail inspection in Example 6.2 has the $B(10, 0.1)$ distribution. The probability that no more than 1 clutch fails is

$$P(X \le 1) = P(X = 1) + P(X = 0)$$

$$= \binom{10}{1}(.1)^{1}(.9)^{9} + \binom{10}{0}(.1)^{0}(.9)^{10}$$

$$= \frac{10!}{1!9!}(.1)(.3874) + \frac{10!}{0!10!}(1)(.3487)$$

$$= (10)(.1)(.3874) + (1)(1)(.3487)$$

$$= .3874 + .3487 = .7361$$

The calculation used the facts that $0! = 1$ and that $a^0 = 1$ for any number $a \ne 0$. The result agrees with that obtained from Table C in Example 6.2. ∎

Binomial Mean and Variance

If a count X is $B(n, p)$, what are the mean μ_X and the standard deviation σ_X? We can guess the mean. If a basketball player makes 80% of her free throws, the mean number made in 10 tries should be 80% of 10, or 8. That's μ_X when X is $B(10, 0.8)$. Intuition suggests more generally that the mean of the $B(n, p)$ distribution should be np. Can we show that this is correct and also obtain a short formula for the standard deviation? A nasty piece of algebra confronts us if we try to apply the definitions of μ and σ^2 for general discrete random variables to the binomial probabilities given by Equation 6.3. Here is an easier way.

A binomial random variable X is the count of successes in n independent trials with probability p of success on each trial. Let the random variable Z_i indicate whether trial i is a success or failure by taking the values $Z_i = 1$ if a success occurs and $Z_i = 0$ if the outcome is a failure. The Z_i are independent because the trials are, and each Z has the same simple distribution

Outcome	1	0
Probability	p	$1 - p$

You can use the expression 5.3 for the mean of a discrete random variable to calculate that

$$\mu_Z = (1)(p) + (0)(1 - p) = p$$

Similarly, the expression 5.5 for the variance shows that $\sigma_Z^2 = p(1 - p)$. The count X is just the number of 1s among the Z_i. That is,

$$X = Z_1 + Z_2 + \cdots + Z_n$$

We can now apply the addition rules for means and variances to this sum. The mean of X is the sum of the means of the variables Z_i,

$$\mu_X = \mu(Z_1) + \mu(Z_2) + \cdots + \mu(Z_n)$$
$$= n\mu_Z = np$$

Similarly, the variance is found as n times the variance of a single Z, so that $\sigma_X^2 = np(1 - p)$. The standard deviation σ_X is the square root of the variance. Here is the result.

Binomial mean and standard deviation

> If X has the $B(n, p)$ distribution, then
>
> $$\mu_X = np$$
> $$\sigma_X = \sqrt{np(1 - p)}$$
>
> (6.4)

EXAMPLE 6.7

The probability that a newborn child will be male is about $p = 0.52$. If a large hospital has 1290 live births in a given year, the number X of male children born has the $B(1290, 0.52)$ distribution. The mean number of males is

$$\mu_X = np = (1290)(.52) = 670.8$$

The standard deviation of the number of male children is

$$\sigma_X = \sqrt{np(1 - p)} = \sqrt{(1290)(.52)(.48)} = 17.9$$

Now consider instead the proportion $\hat{p} = X/1290$ of males among the 1290 children born in this hospital. By the rules for the mean and standard deviation of

bX we find that the mean of $\hat{p}$ is

$$\mu(\hat{p}) = \frac{1}{1290} \mu_X = .52$$

and that the standard deviation of $\hat{p}$ is

$$\sigma(\hat{p}) = \frac{1}{1290} \sigma_X = .014$$

■

The mean and standard deviation of the sample proportion can generally be calculated from the mean and standard deviation of a sample count just as in Example 6.7. Here is the result.

Mean and standard deviation of a sample proportion

> If $\hat{p}$ is the proportion of successes in an SRS of size n and p is the proportion of successes in the population from which the sample is drawn, then the mean and standard deviation of $\hat{p}$ are
>
> $$\mu(\hat{p}) = p$$
>
> $$\sigma(\hat{p}) = \sqrt{\frac{p(1-p)}{n}} \qquad\qquad (6.5)$$

The fact that $\mu(\hat{p}) = p$ states in statistical language that the sample proportion $\hat{p}$ in an SRS is an unbiased estimator of the population proportion p. We observed this fact empirically in Section 4.4 and have now verified it from the laws of probability. It is usual in statistics to describe lack of bias in terms of the mean of a sampling distribution.

Unbiased estimator

> A statistic is an unbiased estimator of a parameter if the mean of the statistic is equal to the true value of the parameter.

When a sample is drawn from a new population having a different value of the population proportion p, the sampling distribution of the unbiased estimator $\hat{p}$ changes so that its mean moves to the new value of p.

The variability of $\hat{p}$ about its mean, as described by the variance or standard deviation, decreases as the sample size increases. So a sample proportion from a large sample usually lies quite close to the population proportion p. We observed this in the simulation experiment of Example 4.17. Now we have discovered exactly how the variability decreases: The standard deviation is $\sqrt{p(1-p)/n}$. The $\sqrt{n}$ in the denominator of this expression means that the sample size must be multiplied by 4 if we wish to divide the standard deviation in half.

EXAMPLE 6.8

In the opinion poll of Example 6.4, the count of the 1785 persons in the sample who answer "No" has the $B(1785, 0.6)$ distribution. The proportion who answer "No" therefore has mean 0.6 (equal to the truth about the population) and standard deviation

$$\sigma(\hat{p}) = \sqrt{\frac{p(1-p)}{n}} = \sqrt{\frac{(.6)(.4)}{1785}}$$
$$= .0116$$

This is more exact than the result that we found by lengthy simulation in Example 4.17. To cut this standard deviation in half, the poll must take a sample of size 4×1785, or 7140. ∎

Normal Approximation to Binomial Distributions

We discovered empirically in Section 4.4 that the sampling distribution of a sample proportion $\hat{p}$ is close to normal. But now we have just deduced from basic probability that the distribution of $\hat{p}$ is the distribution of a binomial count divided by the sample size n. This seems to be a contradiction but it is not. The exact distribution is indeed based on the binomial but it approaches normality as n increases. Here are the facts.

Normal approximation for sample proportions

> Let $\hat{p}$ be the sample proportion of successes in an SRS of size n drawn from a population having population proportion p of successes. When n is large the sampling distribution of $\hat{p}$ is approximately normal with mean p and standard deviation $\sqrt{p(1-p)/n}$.

Since the binomial count X is just a fixed number n times the proportion, X is also approximately normal, with mean np and standard deviation $\sqrt{np(1-p)}$. Both of these normal approximations are easy to remember because they simply say that $\hat{p}$ and X are normal with their usual means and standard deviations given by Equations 6.4 and 6.5. The accuracy of the approximations improves as the sample size n increases. They are most accurate for any fixed n when p is close to $1/2$, and least accurate when p is near 0 or 1. Whether or not you use the normal approximations should depend on how accurate your calculations need to be. For most statistical purposes great accuracy is not required. We will therefore use the approximations for values of n and p that satisfy $np \geq 10$ and $n(1-p) \geq 10$.

EXAMPLE 6.9
The quality engineer of Example 6.2 realizes that a sample of 10 clutches is inadequate and decides to inspect 100 clutches from a large shipment. If in fact 10% of the clutches in the shipment fail to meet the specifications, then the count X of bad clutches in the sample has approximately the $B(100, 0.1)$ distribution. The distribution of the count of bad clutches in a sample of 10 is distinctly nonnormal, as Figure 6.1 showed. When the sample size is increased to 100, the shape of the binomial distribution becomes approximately normal. Figure 6.2 displays the $B(100, 0.1)$ distribution.

The probability $P(X \le 9)$ that no more than 9 of the clutches in the sample fail inspection can be found from the binomial probability formula, Equation 6.3, or from a computer program. The exact probability is $P(X \le 9) = 0.4513$.

According to the normal approximation to the binomial distributions, the count X is approximately normal with mean and standard deviation

$$\mu = np = (100)(.1) = 10$$
$$\sigma = \sqrt{np(1 - p)} = \sqrt{(100)(.1)(.9)} = 3$$

Figure 6.2 displays the probability histogram of the binomial distribution with the density curve of the approximating normal distribution superimposed. Both distributions have the same mean and standard deviation, and both the area under the histogram and the area under the curve are 1. (Although the binomial random variable X can take values as large as 100, the probabilities of outcomes larger than 20 are so small that the bars have no visible height in the figure.)

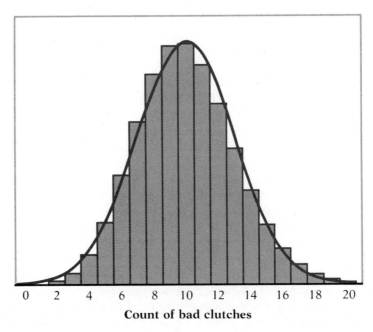

Figure 6.2 Probability histogram and normal approximation for the binomial distribution with $n = 100$ and $p = 0.1$.

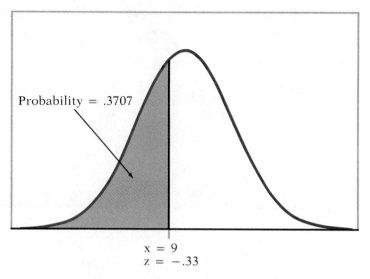

Probability = .3707

x = 9
z = −.33

Figure 6.3 Area under the normal approximation curve for the probability in Example 6.9.

The normal approximation to the probability of no more than 9 bad clutches is found from Table A, as illustrated by Figure 6.3.

$$P(X \le 9) = P\left(Z \le \frac{9 - 10}{3}\right)$$
$$= P(Z \le -.33) = .3707$$

The approximation 0.37 to the binomial probability 0.45 is not very accurate. Notice that $np = (100)(0.1) = 10$, so this combination of n and p is on the border of the values for which we are willing to use the approximation. ∎

EXAMPLE 6.10

The normal approximation makes the difficult calculation of Example 6.4 easy. We wished to calculate $P(\hat{p} \ge 0.58)$ when the sample size was $n = 1785$ and the population proportion was $p = 0.6$. In this case,

$$\mu(\hat{p}) = .6$$
$$\sigma(\hat{p}) = \sqrt{\frac{(.6)(.4)}{1785}} = .0116$$

So the approximate probability, as illustrated in Figure 6.4, is

$$P(\hat{p} \ge .58) = P\left(\frac{\hat{p} - .6}{.0116} \ge \frac{.58 - .6}{.0116}\right)$$
$$= P(Z \ge -1.72) = .9573$$

That is, 96% of all samples have a sample proportion that is at least 0.58. Since the sample was large, this normal approximation is quite accurate. ∎

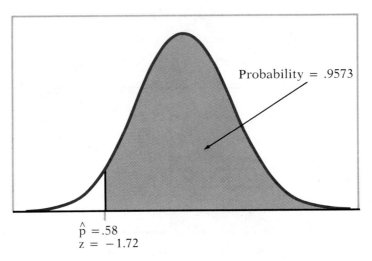

Probability = .9573

$\hat{p} = .58$
$z = -1.72$

Figure 6.4 The normal probability calculation for Example 6.10.

The continuity correction* Figure 6.2 suggests an idea that greatly improves the accuracy of the normal approximation to binomial probabilities. The discrete binomial distribution in Example 6.9 puts probability exactly on $X = 9$ and $X = 10$, and no probability between these whole numbers. The normal distribution spreads its probability continuously. The bar for $X = 9$ in the probability histogram of Figure 6.2 extends from 8.5 to 9.5, but the normal calculation of $P(X \leq 9)$ includes only the area below the middle of this bar. The normal approximation is more accurate if we consider $X = 9$ to extend from 8.5 to 9.5, $X = 10$ to extend from 9.5 to 10.5, and so on.

When we want to include the outcome $X = 9$, we include the entire interval from 8.5 to 9.5 that is the base of the $X = 9$ bar in the histogram. So $P(X \leq 9)$ is calculated as $P(X \leq 9.5)$. On the other hand, $P(X < 9)$ excludes the outcome $X = 9$, so we exclude the entire interval from 8.5 to 9.5 and calculate $P(X \leq 8.5)$ from the normal table. Here is the result of the normal calculation in Example 6.9 improved in this way.

$$P(X \leq 9) = P(X \leq 9.5)$$
$$= P\left(Z \leq \frac{9.5 - 10}{3}\right)$$
$$= P(Z \leq -.17) = .4325$$

The improved approximation 0.43 is much closer to the binomial probability 0.45. Acting as though a whole number occupies the interval from 0.5 below to 0.5 above the number is called the *continuity correction* to the normal

continuity
correction

* This material can be omitted if desired.

approximation. If you need accurate values for binomial probabilities, try to use computer software to do exact calculations but if no software is available, use the continuity correction unless n is very large. Because most statistical purposes do not require extremely accurate probability calculations, use of the continuity correction is not emphasized here.

SUMMARY

The number X of successes in n independent trials, each having probability p of success, has the **binomial distribution** $B(n, p)$. Binomial probabilities can be found in Table C or calculated from the formula

$$P(X = k) = \binom{n}{k} p^k (1 - p)^{n-k}$$

where the possible values of X are $k = 0, 1, \ldots, n$.

The binomial probability formula uses the **binomial coefficient**

$$\binom{n}{k} = \frac{n!}{k!(n-k)!}$$

Here the **factorial** $n!$ is

$$n! = n \times (n - 1) \times (n - 2) \times \cdots \times 3 \times 2 \times 1$$

for positive whole numbers n and $0! = 1$. The binomial coefficient counts the number of ways of distributing k successes among n trials.

The binomial distribution gives a good approximation to the sampling distribution of a count from an SRS of size n when the population is at least 100 times larger than the sample.

The mean and standard deviation of a binomial count X and a sample proportion of successes $\hat{p} = X/n$ are

$$\mu_X = np \qquad \qquad \mu(\hat{p}) = p$$
$$\sigma_X = \sqrt{np(1 - p)} \qquad \sigma(\hat{p}) = \sqrt{\frac{p(1 - p)}{n}}$$

The sample proportion $\hat{p}$ is therefore an unbiased estimator of the population proportion p.

If X has the binomial distribution $B(n, p)$ and n is large, then X has approximately the normal distribution $N(np, \sqrt{np(1 - p)})$ and the sample proportion $\hat{p} = X/n$ has approximately the normal distribution $N(p, \sqrt{p(1-p)/n})$. We will use these approximations when $np \geq 10$ and $n(1-p) \geq 10$. If the **continuity correction** is employed to improve accuracy, the normal approximations can be used more freely.

SECTION 6.1 EXERCISES

All of the binomial probability calculations required in these exercises can be done by using Table C. Your instructor may request that you use the binomial probability formula (Equation 6.3) or computer software. In exercises requiring the normal approximation, you should always use the continuity correction if you studied that topic.

6.1 For each of the following situations, indicate whether a binomial distribution is a reasonable probability model for the random variable X. Give your reasons in each case.

  (a) You observe the sex of the next 50 children born at a local hospital; X is the number of girls among them.

  (b) A couple decides to continue to have children until their first girl is born; X is the total number of children the couple has.

  (c) You want to know what percent of married people believe that mothers of young children should not be employed outside the home. You plan to interview 50 people, and for the sake of convenience you decide to interview both the husband and the wife in 25 married couples. The random variable X is the number among the 50 persons interviewed who think mothers should not be employed.

6.2 In each situation below, is it reasonable to use a binomial distribution for the random variable X? Give the reasons for your answer in each case.

  (a) An auto manufacturer chooses one car from each hour's production for a very detailed quality inspection. One variable that is recorded is the number X of finish defects (dimples, ripples, etc.) in the car's paint.

  (b) The pool of potential jurors for a murder case contains 100 persons chosen at random from the adult residents of a large city. Each person in the pool is asked whether he or she opposes the death penalty; X is the number who say "Yes."

  (c) Joe buys a state lottery ticket every week; X is the number of times in one year that he wins a prize.

6.3 In each of the following cases, decide whether or not a binomial distribution is an appropriate model and give your reasons.

  (a) Fifty students are taught about binomial distributions by a television program. After completing their study, all students take the same examination. The number of students who pass is counted.

(b) A student studies binomial distributions using computer assisted instruction. After the initial instruction is completed, the computer presents 10 problems. The student solves each problem and enters the answer; the computer gives additional instruction between problems if the student's answer is wrong. The number of problems that the student solves correctly is counted.

(c) A chemist repeats a solubility test 10 times on the same substance. Each test is conducted at a temperature 10° higher than the previous test.

6.4 A factory employs over 3000 workers, of whom 30% are black. If the 15 members of the union executive committee were chosen from the workers without regard to race, the number of blacks on the committee would have the $B(15, 0.3)$ distribution.

(a) What is the probability that 3 or fewer members of the committee are black?

(b) What is the probability that 10 or more members of the committee are white?

6.5 A university that is better known for its basketball program than for its academic strength claims that 80% of its basketball players get degrees. An investigation examines the fate of all 20 players who entered the program over a period of several years that ended 5 years ago. Of these players, 10 graduated and the remaining 10 are no longer in school. If the university's claim is true, the number of players who graduate among the 20 studied should have the $B(20, 0.8)$ distribution.

(a) Find the probability that exactly 10 players graduate under these assumptions.

(b) Find the probability that 10 or fewer players graduate. This probability is so small that it casts doubt on the university's claim.

6.6 Suppose that both parents in a family carry genes for blood types A and B. Then the blood types of their children are independent and each child has probability 1/4 of having blood type A. Let X be the number among the 4 children in the family who have blood type A. Compute the distribution of X (that is, the probability of each possible value) and draw a probability histogram for this distribution.

6.7 A firm believer in the "random walk" theory of the behavior of stock prices thinks that in the long run an index of stock prices has probability 0.65 of increasing in any given year; moreover, the direction of the index in any given year is not influenced by whether it rose or fell in earlier years. Let X be the number of years among the next 6 years in which the index rises. Give the possible values that X can take and the probability of each value. Draw a probability histogram for the distribution of X.

6.8 According to government data, 25% of employed women have never been married.

(a) If 10 employed women are selected at random, what is the probability that exactly 2 have never been married?

(b) What is the probability that 2 or fewer have never been married?

(c) What is the probability that at least 8 have been married?

6.9 Use the definition of binomial coefficients to show that each of the following facts is true. Then restate each fact in words in terms of the number of ways that k successes can be distributed among n trials.

(a) $\binom{n}{n} = 1$ for any whole number $n \geq 1$.

(b) $\binom{n}{n-1} = n$ for any whole number $n \geq 1$.

(c) $\binom{n}{k} = \binom{n}{n-k}$ for any n and k with $k \leq n$.

6.10 Find the mean number of children with type A blood in the family described in Exercise 6.6, and mark the location of the mean on your probability histogram for that problem. Then find the standard deviation of X.

6.11 Find the mean of the number X of years in which the stock price index rises according to the random walk stock price model of Exercise 6.7, and mark the mean on your probability histogram for this distribution. Then compute the standard deviation of X. What is the probability that X takes a value within one standard deviation of its mean?

6.12 In a test for detecting ESP (extrasensory perception), a subject is told that cards the experimenter can see but he cannot contain either a star, a circle, a wave, or a square. As the experimenter looks at each of 20 cards in turn, the subject names the shape on the card.

(a) If a subject simply guesses the shape on each card, what is the probability of a successful guess on a single card? Since the cards are independent, the count of successes in 20 cards has a binomial distribution.

(b) What is the probability that a subject correctly guesses at least 10 of the 20 shapes?

(c) In many repetitions of this experiment with a subject who is guessing, how many cards will the subject guess correctly on the average? What will be the standard deviation of the number of correct guesses?

(d) A standard ESP deck actually contains 25 cards, with five different shapes each appearing on five cards. The subject knows that the deck has this makeup. Is a binomial model still appropriate for the count of correct guesses in one pass through this deck? If so, what are *n* and *p*? If not, why not?

6.13 A college is considering a new core curriculum that would require all undergraduates to study a foreign language. The student newspaper plans to interview several faculty members and report their opinions. Suppose that in fact 60% of the faculty support the language requirement.

(a) If the newspaper interviews 5 faculty members, what is the probability that a majority (that is, 3 or more) oppose the language requirement? (Assume that the faculty members are chosen at random.)

(b) If 15 faculty members are interviewed instead, what is the probability that a majority (8 or more) oppose the requirement? Larger samples make it less likely that majority opinion in the population is incorrectly reported.

6.14 A study by a federal agency concludes that polygraph (lie detector) tests given to truthful persons have probability about 0.2 of suggesting that the person is deceptive. (Office of Technology Assessment, *Scientific Validity of Polygraph Testing: A Research Review and Evaluation*, Government Printing Office, Washington, DC, 1983.)

(a) A firm asks 12 job applicants about thefts from previous employers, using a polygraph test to assess their truthfulness. Suppose that all 12 answer truthfully. What is the probability that the polygraph says at least one is deceptive?

(b) What is the mean number among 12 truthful persons who will be classified as deceptive? What is the standard deviation of this number?

(c) What is the probability that the number classified as deceptive is less than the mean?

6.15 According to government data, 22% of American children under the age of six live in households with incomes lower than the official poverty level. A random sample of 300 children is selected for a study of learning in early childhood.

(a) What is the mean number of children in the sample who come from poverty-level households? What is the standard deviation of this number?

(b) Use the normal approximation to calculate the probability that at least 80 of the children in the sample live in poverty.

6.16 One way of checking for the effect of undercoverage, nonresponse, and other sources of error in a sample survey is to compare the sample with known demographic facts about the population. About 11% of American adults are black. The number X of blacks in a random sample of 1500 adults should therefore vary with the $B(1500, 0.11)$ distribution.

(a) What are the mean and standard deviation of X?

(b) Use the normal approximation to find the probability that the sample will contain 100 or fewer blacks.

6.17 A selective college would like to have an entering class of 1200 students. Since not all students who are offered admission accept, the college admits more than 1200 students. Past experience shows that about 70% of the students admitted will accept. The college decides to admit 1500 students. Assuming that students make their decisions independently, the number who accept has the $B(1500, 0.7)$ distribution. If this number is less than 1200, the college will admit students from its waiting list.

$\mu = np = 1050$

$\delta = 17.75 \quad \sqrt{np(1-p)}$

$p(x \geq 1000) = p(z \geq \frac{1000 - 1050}{17.75})$

$1 - .0024 = .9976$

(a) What are the mean and the standard deviation of the number X of students who accept?

(b) Use the normal approximation to find the probability that at least 1000 students accept.

(c) The college does not want more than 1200 students. What is the probability that more than 1200 will accept?

(d) If the college decides to increase the number of admission offers to 1700, what is the probability that more than 1200 will accept?

6.18 The Gallup Poll once asked a random sample of 1540 adults, "Do you happen to jog?" Suppose that in fact 15% of all American adults jog.

(a) Find the mean and standard deviation of the proportion of the sample who jog. (Assume that an SRS was used.)

(b) Use the normal approximation to find the probability that between 13% and 17% of the sample jog.

(c) What sample size would be required to reduce the standard deviation of the sample proportion to one-half the value you found in (a)?

6.19 Here is a simple probability model for multiple choice tests. Suppose that each student has probability p of correctly answering a question chosen at random from a universe of possible questions. (A strong student has a higher p than a weak student.) The correctness of answers to different questions are independent. Julie is a good student for whom $p = 0.75$.

(a) Use the normal approximation to find the probability that Julie scores 70% or lower on a 100-question test.

(b) If the test contains 250 questions, what is the probability that Julie will score 70% or lower?

(c) How many questions must the test contain in order to reduce the standard deviation of Julie's proportion of correct answers to half its value for a 100-item test?

(d) Laura is a weaker student for whom $p = 0.6$. Does the answer you gave in (c) for the standard deviation of Julie's score apply to Laura's standard deviation also?

6.20 When the ESP study described in Exercise 6.12 discovers a subject whose performance appears to be better than guessing, the study continues at greater length. Cards with five shapes (star, square, circle, wave, and cross) are used, and the order of the cards is determined by random numbers. Because the subject cannot see the experimenter as he looks at each card in turn, any possible nonverbal clues by the experimenter are eliminated. If sufficient care is taken, the answers of a subject who does not have ESP should be independent trials, each with probability 1/5 of success. One thousand trials are made.

(a) What are the mean and the standard deviation of the count of successes?

(b) What are the mean and standard deviation of the proportion of successes among the 1000 trials?

(c) What is the probability that a subject without ESP will be successful in at least 24% of 1000 trials?

(d) The researcher considers evidence of ESP to be a proportion of successes so large that there is only probability 0.01 that the subject could do this well or better by guessing. What proportion of successes must a subject have to meet this standard? (Hint: Example 1.22 shows how to do a normal calculation of the type required here.)

6.21 The ELISA test was introduced in the mid-1980s to screen donated blood for the presence of antibodies to the AIDS virus. When presented with AIDS-contaminated blood, ELISA gives a positive result (that is, it signals that an antibody is present) in about 98% of all cases. Suppose that among the many units of blood that pass through a blood bank in a year there are 20 units containing AIDS antibodies.

(a) What is the probability that ELISA will detect all of these cases?

(b) What is the probability that more than 2 of the 20 contaminated units will escape detection?

(c) What is the mean number of units among the 20 that will be detected by ELISA? What is the standard deviation of the number detected?

6.22 Since ELISA (see the previous exercise) is a screening test designed to keep the AIDS virus out of blood supplies, it is conservative in the sense of signaling the presence of AIDS antibodies too often. In fact, when presented with uncontaminated blood, ELISA is positive (claims that antibodies are present) about 7% of the time. Positive test results for uncontaminated blood are called *false positives.*

(a) A blood bank tests 12,000 units of uncontaminated blood. What is the mean number of false positives among these units? What is the standard deviation of the number of false positives?

(b) Suppose that the blood bank screens 12,000 units of uncontaminated blood and 20 units of blood that carry AIDS antibodies. What is the mean number of positive ELISA results among the entire 12,020 units? What is the standard deviation of the number of positive results? (Use the information in both this and the previous exercise along with the rules for the mean and variance of a sum of random variables.)

(c) What is the mean number of false positives as a percent of the overall mean number of positive results? This gives a rough answer to the following question: When ELISA is applied to 12,020 units of blood of which 20 carry AIDS antibodies, about what percent of positive test results are false positives? (A more exact way of answering such a question appears in Exercise 5.94. The conclusion is important: If a condition is uncommon, most positive test results will be false positives even if the test has a high probability of giving the correct result on any one trial. A detailed statistical analysis and more information appear in J. L. Gastwirth, "The statistical precision of medical screening procedures: application to polygraph and AIDS antibodies test data," *Statistical Science* 2 (1987), pp. 213–222.)

6.2   SAMPLE MEANS

Counts and proportions are discrete random variables that describe categorical data. The statistics most often used to describe measured data, on the other hand, are continuous random variables. The sample mean, percentiles and standard deviation are examples of statistics based on measured data. Statistical theory can describe the sampling distributions of these statistics. In this section we will concentrate on the sample mean.

The Distribution of a Sample Mean

The sample mean $\bar{x}$ from a sample or an experiment is an estimate of the mean μ of the underlying population, just as a sample proportion $\hat{p}$ is an estimate of a population proportion p. The sampling distribution of $\bar{x}$ is determined by the design used to produce the data, the sample size n, and the population distribution.

Suppose that we select an SRS of size n from a population, and measure a variable X on each unit in the sample. The data consist of observations on n random variables $X_1, X_2, \ldots, X_n$. A single X_i is a measurement on one unit selected at random from the population, so therefore it has the distribution of the population. If the population is large relative to the sample, we can consider $X_1, X_2, \ldots, X_n$ to be independent random variables each having the same distribution. This is our probability model for measurements on each unit in an SRS.

The sample mean of an SRS of size n is

$$\bar{x} = \frac{1}{n}(X_1 + X_2 + \cdots + X_n)$$

If the population has mean μ, then the mean of each X_i is $\mu(X_i) = \mu$. Therefore by the addition rule for means of random variables,

$$\mu(\bar{x}) = \frac{1}{n}\{\mu(X_1) + \mu(X_2) + \cdots + \mu(X_n)\}$$

$$= \frac{1}{n}n\mu = \mu$$

That is, *the mean of $\bar{x}$ is the same as the mean of the population.* The sample mean $\bar{x}$ is therefore an unbiased estimator of the unknown population mean μ.

Since the observations are independent, the addition rule for variances also applies.

$$\sigma^2(\bar{x}) = \left(\frac{1}{n}\right)^2\{\sigma^2(X_1) + \sigma^2(X_2) + \cdots + \sigma^2(X_n)\}$$

$$= \left(\frac{1}{n}\right)^2 n\sigma^2$$

$$= \frac{\sigma^2}{n}$$

Just as in the case of a sample proportion $\hat{p}$, the variability of the sampling distribution of a sample mean decreases as the sample size grows. *The mean of several observations is less variable than a single observation.* Since the standard deviation of $\bar{x}$ is $\sigma/\sqrt{n}$, it is again true that the standard deviation of the statistic decreases in proportion to the square root of the sample size. Here is a summary of these facts.

Mean and standard deviation of a sample mean

> If $\bar{x}$ is the mean of an SRS of size n from a population having mean μ and standard deviation σ, then
>
> $$\mu(\bar{x}) = \mu$$
>
> $$\sigma(\bar{x}) = \frac{\sigma}{\sqrt{n}}$$

EXAMPLE 6.11

The height X of a single randomly chosen young woman varies according to the $N(65.5, 2.5)$ distribution. If a medical study asked the height of an SRS of 100 young women, the sampling distribution of the sample mean height $\bar{x}$ would have mean and standard deviation

$$\mu(\bar{x}) = \mu = 65.5 \text{ inches}$$

$$\sigma(\bar{x}) = \frac{\sigma}{\sqrt{n}} = 0.25 \text{ inch}$$

The heights of individual women vary widely about the population mean, but the average height of a sample of 100 women has a standard deviation only one-tenth as large. ∎

The fact that the mean of several observations is less variable than a single observation is important in many endeavors. A basic principle of investment, for example, is that diversification reduces risk. That is, buying several securities rather than just one reduces the variability of the return on an investment. Figure 6.5 illustrates this principle in the case of common stocks listed on the New York Stock Exchange. Figure 6.5(a) shows the distribution of returns for all 1815 stocks listed on the Exchange for the entire year 1987. This was a year of extreme swings in stock prices, including a loss of over 20% in a single day. The mean return for all 1815 stocks was -3.5% and the distribution shows a very wide spread. Figure 6.5(b) shows the distribution of returns for all possible portfolios that invested equal amounts in each of 5 stocks. A portfolio is simply a sample of 5 stocks and its return is the average return for the 5 stocks chosen. The mean return for all portfolios is still -3.5% but the variation among portfolios is much less than the variation among individual stocks. For example, 11% of all individual stocks had a loss of more than 40% but only 1% of the portfolios had a loss that large.

EXAMPLE 6.12

The fact that a mean of several measurements is less variable than a single measurement is also important in science. When Simon Newcomb set out to measure the speed of light, he took repeated measurements of the time required for a beam of light to travel a known distance. Newcomb's 66 measurements appear in Table 1.1 We saw in Chapter 1 that the data contain two outliers that may not be observations from the same population. Disregarding these leaves 64 observa-

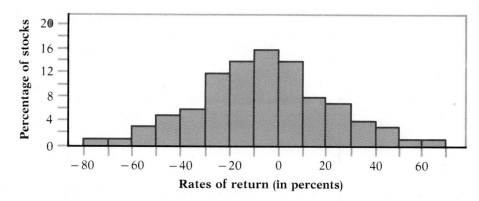

Figure 6.5(a) The distribution of returns for New York Stock Exchange common stocks in 1987.

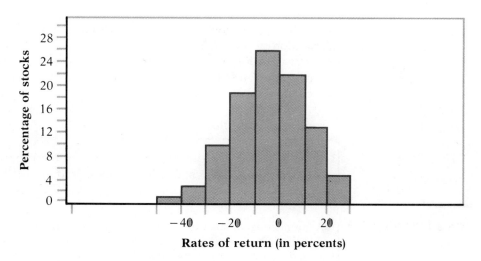

Figure 6.5(b) The distribution of returns for portfolios of five stocks in 1987. Figure 6.5 is taken with permission from John K. Ford, "A method for grading 1987 stock recommendations," *American Association of Individual Investors Journal*, March 1988, pp. 16–17.

tions. These are the values of 64 independent random variables, each with a probability distribution that describes the population of all measurements made with Newcomb's apparatus. If Newcomb's procedures were correct, the population mean μ is the true passage time of light. The population variability reflects the random variation in the measurements due to small changes in the environment, the equipment, and the procedure. Suppose that the standard deviation for this population is $\sigma = 5$. (The units of the coded data in Table 1.1 are billionths of a second, that is, seconds $\times 10^{-9}$.)

If Newcomb had taken a single measurement, the standard deviation of the result would be 5, so a second observation might have been quite different. Taking

10 measurements and reporting their mean $\bar{x}$ reduces the standard deviation to

$$\sigma(\bar{x}) = \frac{5}{\sqrt{10}} = 1.58$$

The sample mean of any number of measurements is an unbiased estimate of the population mean. Averaging over more measurements reduces the variability and makes it more likely that the result is close to the truth. The mean $\bar{x}$ of all 64 measurements has standard deviation

$$\sigma(\bar{x}) = \frac{5}{\sqrt{64}} = .625$$

Newcomb's observed value of this random variable, $\bar{x} = 27.75$, is a much more reliable estimate of the true passage time than is a single measurement. Increasing the number of measurements from 1 to 64 divides the standard deviation by 8, the square root of 64.   ■

We have described the mean and standard deviation of the probability distribution of a sample mean $\bar{x}$, but not the distribution itself. The shape of the distribution of $\bar{x}$ depends on the shape of the population distribution. In particular, if the population distribution is normal, then so is the distribution of the sample mean.

Sampling distribution of a sample mean

> If a population has the $N(\mu, \sigma)$ distribution, then the sample mean of n independent observations has the $N(\mu, \sigma/\sqrt{n})$ distribution.

EXAMPLE 6.13

A normal quantile plot of Newcomb's 64 observations in Figure 1.19(d) showed that they are close to normal. The relative frequency distribution of repeated observations reflects the population distribution, so we can be confident that the population of all measurements made with Newcomb's apparatus is close to normal. A single measurement has the $N(\mu, 5)$ distribution, where μ is the unknown population mean. By the 68–95–99.7 rule, 95% of these measurements will lie within 2×5, or 10, of the unknown μ.

Now we combine the calculations of $\sigma(\bar{x})$ from Example 6.12 with the fact that $\bar{x}$ follows a normal distribution. The mean $\bar{x}$ of 10 measurements follows the $N(\mu, 1.58)$ distribution, so 95% of the time the observed mean falls within $2 \times 1.58 = 3.16$ of μ. Figure 6.6 compares the distributions of a single measurement and of the sample mean of 10 measurements. The picture makes dramatically clear the advantages of averaging over many observations. Newcomb actually made 64 measurements. Their sample mean $\bar{x}$ has the $N(\mu, 0.625)$ distribution. In repetitions of the experiment, 95% of the values of $\bar{x}$ would lie within $2 \times 0.625 = 1.25$ of the population mean μ. Newcomb could be confident that his announced result was close to μ. (But if there was bias in his experimental method, the mean μ of the population of measurements may not be equal to the true passage time of light. This is a question of physics, not of statistics.)   ■

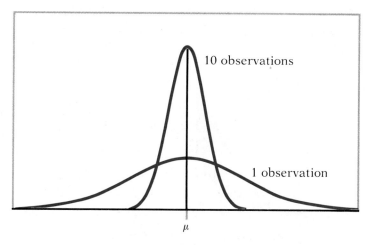

Figure 6.6   The sampling distribution of $\bar{x}$ for samples of size 10 compared with the distribution of a single observation.

The fact that the sample mean of an SRS from a normal population has a normal distribution is a special case of a more general fact: *Any linear combination of independent normal random variables is also normally distributed.* That is, if X and Y are independent normal random variables and a and b are constants, $aX + bY$ is also normally distributed, and so it is for any number of normal variables. In particular, the sum or difference of independent normal random variables has a normal distribution. The mean and standard deviation of $aX + bY$ are found as usual from the addition rules for means and variances. These facts are often used in statistical calculations.

EXAMPLE 6.14

If the golf scores of Tom and George (Example 5.28) vary normally and independently, what is the probability that Tom will score lower than George and thus do better in the tournament? Tom's score X has the $N(110, 10)$ distribution and George's score Y varies from round to round according to the $N(100, 8)$ distribution. We now know that the difference $X - Y$ between their scores is normally distributed, with mean and variance

$$\mu_{X-Y} = \mu_X - \mu_Y = 110 - 100 = 10$$
$$\sigma_{X-Y}^2 = \sigma_X^2 + \sigma_Y^2 = 10^2 + 8^2 = 164$$

Because $\sqrt{164} = 12.8$, $X - Y$ is $N(10, 12.8)$. Figure 6.7 illustrates the probability which can now be computed.

$$P(X < Y) = P(X - Y < 0)$$
$$= P\left(\frac{(X - Y) - 10}{12.8} < \frac{0 - 10}{12.8}\right)$$
$$= P(Z < -.78) = .2177$$

Although George's score is 10 strokes lower on the average, Tom will have the lower score in about one of every five matches.   ∎

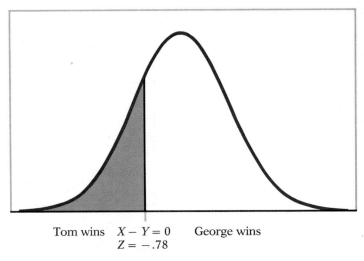

Tom wins $X - Y = 0$ George wins
                    $Z = -.78$

Figure 6.7   The normal probability calculation for Example 6.14.

The Central Limit Theorem

The sampling distribution of $\bar{x}$ is normal if the underlying population itself has a normal distribution. But what happens when the population distribution is not normal? It turns out that *as the sample size increases, the distribution of $\bar{x}$ becomes closer to a normal distribution*. This is true no matter what the population distribution may be, as long as the population has a finite standard deviation σ. This famous fact of probability theory is called the *central limit theorem.*[*]

central limit theorem

More generally, the central limit theorem states that the distribution of a sum or average of many random quantities is close to normal. This is true even if the quantities are not independent (as long as they are not too strongly associated) and even if they have different distributions (as long as no one random quantity is so large that it dominates the others). The central limit theorem suggests why the normal distributions are common models for observed data. Any variable that is a sum of many small influences will have approximately a normal distribution.

How large a sample size n is needed for $\bar{x}$ to be close to normal depends on the population distribution. More observations are required if the shape of the population distribution is far from normal.

EXAMPLE 6.15   Figure 6.8 shows the central limit theorem in action in the case of a strongly non-normal population. Figure 6.8(a) shows the density curve of a single observation, that is, of the population. The distribution is strongly right skewed and the most

* The first general version of the central limit theorem was established in 1810 by the French mathematician Pierre Simon Laplace (1749–1827).

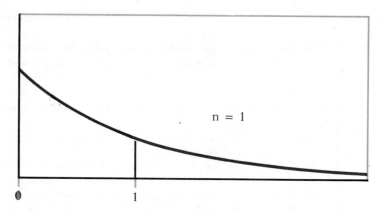

Figure 6.8(a) The central limit theorem in action: The density curve for a single observation from a strongly nonnormal population.

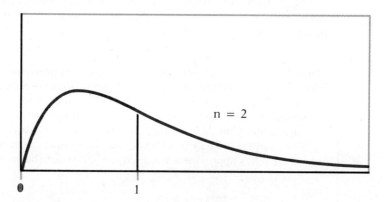

Figure 6.8(b) The central limit theorem in action: The density curve for the sample mean of two observations.

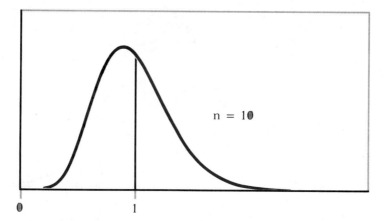

Figure 6.8(c) The central limit theorem in action: The density curve for the sample mean of 10 observations.

probable outcomes are near 0 at one end of the range of possible values. The mean μ of this distribution is 1 and its standard deviation σ is also 1. This particular continuous distribution is called an exponential distribution from the shape of its density curve. Exponential distributions are used as models for the lifetime in service of electronic components and for the time required to serve a customer or repair a machine.

Figure 6.8(b) and Figure 6.8(c) are the density curves of the sample mean of 2 and 10 observations from this population. As n increases, the shape becomes more normal. The mean remains at $\mu = 1$ and the standard deviation decreases, taking the value $1/\sqrt{n}$. The density curve for 10 observations is still somewhat skewed to the right but already resembles a normal curve with $\mu = 1$ and $\sigma = 1/\sqrt{10} = 0.32$. The contrast between the shape of the population distribution and the distribution of the mean of 10 observations is striking. ■

The central limit theorem allows us to use normal probability calculations to answer questions about sample means from many observations even when the population distribution is not normal.

EXAMPLE 6.16   The time X that a technician requires to perform preventive maintenance operations on an air conditioning unit is governed by the exponential distribution whose density curve appears in Figure 6.8(a). The mean time is $\mu = 1$ hour and the standard deviation is $\sigma = 1$ hour. Your company operates 70 of these units. What is the probability that their average maintenance time exceeds 50 minutes?

The central limit theorem says that the sample mean time $\bar{x}$ spent working on 70 units has approximately the $N(\mu, \sigma/\sqrt{70})$ distribution. In this case,

$$\frac{\sigma}{\sqrt{70}} = \frac{1}{\sqrt{70}} = .12$$

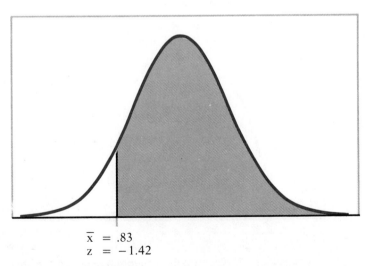

$$\bar{x} = .83$$
$$z = -1.42$$

Figure 6.9 The normal probability calculation for Example 6.16.

The distribution of $\bar{x}$ is therefore $N(1, 0.12)$. Since 50 minutes is 50/60 of an hour, or 0.83 hour, the probability we want is

$$P(\bar{x} > .83) = P\left(\frac{\bar{x} - 1}{.12} > \frac{.83 - 1}{.12}\right)$$
$$= P(Z > -1.42) = .9222$$

Figure 6.9 illustrates this calculation. ∎

The normal approximation to the binomial distributions is an important example of the central limit theorem. This is true because a sample proportion can be thought of as a sample mean. Recall the idea that we used to find the mean and variance of a $B(n, p)$ random variable X. We wrote the count X as a sum

$$X = Z_1 + Z_2 + \cdots + Z_n$$

of indicator variables. The variables Z_i take only the values 0 and 1 and are far from normal. The proportion $\hat{p} = X/n$ is the sample mean of the Z_i and like all sample means is approximately normal when n is large.

SUMMARY

The sample mean $\bar{x}$ of an SRS of size n drawn from a large population with mean μ and standard deviation σ has a sampling distribution with mean and standard deviation

$$\mu(\bar{x}) = \mu$$
$$\sigma(\bar{x}) = \frac{\sigma}{\sqrt{n}}$$

The sample mean $\bar{x}$ is therefore an unbiased estimator of the population mean μ.

Linear combinations of independent normal random variables have normal distributions. In particular, if the population has a normal distribution, so does $\bar{x}$.

The **central limit theorem** states that for large n the sampling distribution of $\bar{x}$ is approximately $N(\mu, \sigma/\sqrt{n})$ for any population with finite standard deviation σ.

SECTION 6.2 EXERCISES

6.23 Coal mine operators are required to test the amount of dust in the atmosphere of the mine. A laboratory carries out the test by weighing filters that have been exposed to the air in the mine. The test has a standard deviation of $\sigma = 0.08$ milligram in repeated weighings of the same filter. The laboratory weighs each filter 3 times and reports the mean result. What is the standard deviation of the reported result?

6.24 An automatic grinding machine in an auto parts plant prepares axles with a target diameter $\mu = 40.125$ millimeters (mm). The machine has some variability so the standard deviation of the diameters is $\sigma = 0.002$ mm. A sample of 4 axles is inspected each hour for process control purposes and records are kept of the sample mean diameter. What will be the mean and standard deviation of the numbers recorded?

6.25 The scores of individual students on the American College Testing (ACT) Program composite college entrance examination have a normal distribution with mean 18.6 and standard deviation 5.9. At Northside High, 76 seniors take the test. If the scores at this school have the same distribution as national scores, what are the mean and standard deviation of the average (sample mean) score for the 76 students?

6.26 In 1987 the mean return of all common stocks on the New York Stock Exchange was $\mu = -3.5\%$ (that is, these stocks lost 3.5% of their value in 1987). The standard deviation of the returns was about $\sigma = 26\%$. A student of finance forms all possible portfolios that invested equal amounts in 5 of these stocks and records the return for each portfolio. This return is the average of the returns of the 5 stocks chosen. What are the mean and the standard deviation of the portfolio returns?

6.27 The scores of students on the ACT college entrance examination in a recent year had the normal distribution with mean $\mu = 18.6$ and standard deviation $\sigma = 5.9$.

   (a) What is the probability that a single student randomly chosen from all those taking the test scores 21 or higher?
   (b) The average score of the 76 students at Northside High who took the test was $\bar{x} = 20.4$. What is the probability that the mean score for 76 students randomly selected from all who took the test nationally is 20.4 or higher?

6.28 A bottling company uses a filling machine to fill plastic bottles with a popular cola. The bottles are supposed to contain 300 milliliters (ml). In fact, the contents vary according to a normal distribution with mean $\mu = 298$ ml and standard deviation $\sigma = 3$ ml.

   (a) What is the probability that an individual bottle contains less than 295 ml?
   (b) What is the probability that the mean content of the bottles in a six-pack is less than 295 ml?

6.29 Judy's doctor is concerned that she may suffer from hypokalemia (low potassium in the blood). There is variation both in the actual potassium level and in the blood test that measures the level. Judy's measured potassium level varies according to the normal distribution with $\mu = 3.8$ and $\sigma = 0.2$. A patient is classified as hypokalemic if the level is below 3.5.

(a) If a single potassium measurement is made, what is the probability that Judy is diagnosed as hypokalemic?

(b) If measurements are made instead on 4 separate days and the mean result is compared with the criterion 3.5, what is the probability that Judy is diagnosed as hypokalemic?

6.30 A laboratory weighs filters from a coal mine to measure the amount of dust in the mine atmosphere. Repeated measurements of the weight of dust on the same filter vary normally with standard deviation $\sigma = 0.08$ mg because the weighing is not perfectly precise. The dust on a particular filter actually weighs 123 mg. Repeated weighings will have the $N(123, 0.08)$ distribution.

(a) The laboratory reports the mean of 3 weighings. What is the distribution of this mean?

(b) What is the probability that the laboratory reports a result of 124 mg or higher for this filter?

6.31 A company that owns and services a fleet of cars for its sales force has found that the service lifetime of disc brake pads varies from car to car according to a normal distribution with mean $\mu = 55,000$ miles and standard deviation $\sigma = 4500$ miles. A new brand of brake pads is installed on 8 cars.

(a) If the new brand has the same lifetime distribution as the previous type, what is the distribution of the sample mean lifetime for the 8 cars?

(b) The average life of the pads on these 8 cars turns out to be $\bar{x} = 51,800$ miles. What is the probability that the sample mean lifetime is 51,800 miles or less if the lifetime distribution is unchanged? The company takes this probability as evidence that the average lifetime of the new brand of pads is less than 55,000 miles.

6.32 The design of an electronic circuit calls for a 100-ohm resistor and a 250-ohm resistor connected in series so that their resistances add. The components used are not perfectly uniform, so the actual resistances vary independently according to normal distributions. The resistance of 100-ohm resistors has mean 100 ohms and standard deviation 2.5 ohms, while that of 250-ohm resistors has mean 250 ohms and standard deviation 2.8 ohms.

(a) What is the distribution of the total resistance of the two components in series?

(b) What is the probability that the total resistance lies between 345 and 355 ohms?

6.33 The clearance between a pin and the collar around it is important for the proper performance of a disc drive from small computers. The specifications call for the pin to have diameter 0.525 cm and for the collar to have diameter 0.526 cm. The clearance will then be 0.001 cm. In practice, both diameters vary from part to part independently of each other. The diameter X of the pin has the $N(0.525, 0.0003)$ distribution and the distribution of the diameter Y of the collar is $N(0.526, 0.0004)$.

(a) What is the distribution of the clearance $Y - X$?

(b) What is the probability $P(Y - X \leq 0)$ that the pin will not fit inside the collar?

6.34 An experiment to compare the nutritive value of normal corn and high-lysine corn (Exercise 1.28) divides 40 chicks at random into two groups of 20. One group is fed a diet based on normal corn while the other receives high-lysine corn. At the end of the experiment, inference about which diet is superior is based on the difference $\bar{y} - \bar{x}$ between the mean weight gain $\bar{y}$ of the 20 chicks in the high-lysine group and the mean weight gain $\bar{x}$ of the 20 chicks in the normal corn group. Because of the randomization, the two sample means are independent.

(a) Suppose that $\mu_X = 360$ grams (g) and $\sigma_X = 55$ g in the population of all chicks fed normal corn, and that $\mu_Y = 385$ g and $\sigma_Y = 56$ g in the high-lysine population. What are the mean and standard deviation of $\bar{y} - \bar{x}$?

(b) The weight gains are normally distributed in both populations. What is the distribution of $\bar{x}$? Of $\bar{y}$? What is the distribution of $\bar{y} - \bar{x}$?

(c) What is the probability that the mean weight gain in the high-lysine group exceeds the mean weight gain in the normal group by 25 g or more?

6.35 An experiment on the teaching of reading compares two methods, A and B. The response variable is the Degree of Reading Power (DRP) score. The experimenter uses method A in a class of 26 students and method B in a comparable class of 24 students. The classes are assigned to the teaching methods at random. Suppose that in the population of all children of this age the DRP score has the $N(34, 12)$ distribution if method A is used and the $N(37, 11)$ distribution if method B is used.

(a) What is the distribution of the mean DRP score $\bar{x}$ for the 26 students in the A group? (Assume that this group can be regarded as an SRS from the population of all children of this age.)

(b) What is the distribution of the mean score $\bar{y}$ for the 24 students in the B group?

(c) Use the results of (a) and (b), keeping in mind that $\bar{x}$ and $\bar{y}$ are independent, to find the distribution of the difference $\bar{y} - \bar{x}$ between the mean scores in the two groups.

(d) What is the probability that the mean score for the B group will be at least 4 points higher than the mean score for the A group?

6.36 The two previous exercises illustrate a common setting for statistical inference. This exercise gives the general form of the sampling distribution needed in this setting. We have a sample of n observations from a treatment group and an independent sample of m observations from a control group. Suppose that the response to the treatment has the $N(\mu_X, \sigma_X)$ distribution and that the response of control subjects has the $N(\mu_Y, \sigma_Y)$ distribution. Inference about the difference $\mu_Y - \mu_X$ between the population means is based on the difference $\bar{y} - \bar{x}$ between the sample means in the two groups.

(a) Under the assumptions given, what is the distribution of $\bar{y}$? Of $\bar{x}$?

(b) What is the distribution of $\bar{y} - \bar{x}$?

6.37 A mechanical assembly consists of a shaft with a bearing at each end. The total length of the assembly is the sum $X + Y + Z$ of the shaft length X and the lengths Y and Z of the bearings. These lengths vary from part to part in production, independently of each other and with normal distributions. The shaft length X has mean 11.2 inches and standard deviation 0.002 inch, while each bearing length Y and Z has mean 0.4 inch and standard deviation 0.001 inch.

(a) According to the 68–95–99.7 rule, about 95% of all shafts have lengths in the range $11.2 \pm d_1$ inches. What is the value of d_1? Similarly, about 95% of the bearing lengths fall in the range of $0.4 \pm d_2$ inches. What is the value of d_2?

(b) It is common practice in industry to state the "natural tolerance" of parts in the form used in (a). An engineer who knows no statistics thinks that tolerances add, so that the natural tolerance for the total length of the assembly (shaft and two bearings) is $12 \pm d$ inches, where $d = d_1 + 2d_2$. Find the standard deviation of the total length $X + Y + Z$. Then find the value d such that about 95% of all assemblies have lengths in the range $12 \pm d$. Was the engineer correct?

6.38 The amount of nitrogen oxides (NOX) present in the exhaust of a particular type of car varies from car to car according to the normal distribution with mean 1.4 grams per mile (g/mi) and standard deviation 0.3 g/mi. Two cars of this type are tested. One has 1.1 g/mi of NOX, the other 1.9. The test station attendant finds this much variation between two similar cars surprising. If X and Y are independent NOX levels for cars of this type, find the probability

$$P(X - Y \geq .8 \text{ or } X - Y \leq -.8)$$

that the difference is at least as large as the value the attendant observed.

6.39 Leona and Fred are friendly competitors in high school. Both are about to take the ACT college entrance examination. They agree that if one of them scores 5 or more points higher than the other, the loser will take the winner out to dinner. Suppose that in fact Fred and Leona have equal ability, so that each score varies normally with mean 24 and standard deviation 2. (The variation is due to luck in guessing and the accident of specific test questions being familiar to the student.) The two scores are independent. What is the probability that the scores differ by 5 or more points in either direction?

6.40 The study habits portion of the Survey of Study Habits and Attitudes (SSHA) psychological test consists of two sets of questions. One set measures "delay avoidance" and the other measures "work methods." A subject's study habits score is the sum $X + Y$ of the delay avoidance score X and the work methods score Y. The distribution of X in a broad population of college freshmen is close to $N(25, 10)$, and the distribution of Y in the same population is close to $N(25, 9)$.

(a) If a subject's X and Y scores were independent, what would be the distribution of the study habits score $X + Y$?

(b) Using the distribution you found in (a), what percent of the population have a study habits score of 60 or higher?

(c) In fact, the X and Y scores are probably strongly correlated. In this case, does the mean of $X + Y$ still have the value you found in (a)? Does the standard deviation still have the value you found in (a)?

yes

no

6.41 A study of working couples measures the income X of the husband and the income Y of the wife in a large number of couples in which both partners are employed. Suppose that you knew the means μ_X and μ_Y and the variances σ_X^2 and σ_Y^2 of both variables in the population.

yes, rules for means

No, rules for variances

(a) Is it reasonable to take the mean of the total income $X + Y$ to be $\mu_X + \mu_Y$? Explain your answer.

(b) Is it reasonable to take the variance of the total income to be $\sigma_X^2 + \sigma_Y^2$? Explain your answer. *need independence*

Variances add only if the two variables are uncorrelated (independent).

6.42 The number of flaws per square yard in a type of carpet material varies, with mean 1.6 flaws per square yard and standard deviation 1.2 flaws per square yard. The distribution is not normal—in fact, it is discrete. An inspector studies 200 square yards of the material, records the number of flaws found in each square yard, and calculates $\bar{x}$, the mean number of flaws per square yard inspected. Use the central limit theorem to find the approximate probability that the mean number of flaws exceeds 2 per square yard.

6.43 The number of accidents per week at a hazardous intersection varies with mean 2.2 and standard deviation 1.4. This distribution is discrete and so it is not normal.

(a) Let $\bar{x}$ be the mean number of accidents per week at the intersection during a year (52 weeks). What is the approximate distribution of $\bar{x}$ according to the central limit theorem?

(b) What is the approximate probability that $\bar{x}$ is less than 2?

(c) What is the approximate probability that there are fewer than 100 accidents at the intersection in a year? (Hint: Restate this event in terms of $\bar{x}$.)

6.44 The distribution of annual returns on common stocks is roughly symmetric but extreme observations are more frequent than in a normal distribution. Because the distribution is not strongly nonnormal, the mean return over even a moderate number of years is close to normal. In the long run, annual real returns on common stocks have varied with mean about 9% and standard deviation about 28%. Andrew plans to retire in 45 years, and is considering investing in stocks. What is the probability (assuming that the past pattern of variation continues) that the mean annual return on common stocks over the next 45 years will exceed 15%? What is the probability that the mean return will be less than 5%?

6.45 The level of nitrogen oxide (NOX) in the exhaust of a particular car model varies with mean 1.4 g/mi and standard deviation 0.3 g/mi. A company has 125 cars of this model in its fleet. If $\bar{x}$ is the mean NOX emission level for these cars, what is the level L such that the probability that $\bar{x}$ is greater than L is only 0.01?

6.46 Children in kindergarten are sometimes given the Ravin Progressive Matrices Test (RPMT) to assess their readiness for learning. Experience at Southwark Elementary School suggests that the RPMT scores for its kindergarten pupils have mean 13.6 and standard deviation 3.1. The distribution is close to normal. Mr. Lavin has 22 children in his kindergarten class this year. He suspects that their RPMT scores will be unusually low because the test was interrupted by a fire drill. To check this suspicion, he wants to find the level L such that the probability is only 0.05 that the mean score of his 22 children falls below L when the usual Southwark distribution remains true. What is the value of L?

6.3 CONTROL CHARTS*

statistical control

A variable that is described by the same distribution when observed at successive points in time is said to be *in statistical control*, or simply in control. The idea of control can be used to see whether your commuting time or your checking account balance is stable over time, as was done in Section 2.1. The most common application of statistical control is to monitor the performance of an industrial process. The same methods, however, can be used to check the stability of quantities as varied as the ratings of a television show and the level of ozone in the atmosphere. Control charts combine graphical and numerical descriptions of data with probability computations. They therefore provide a natural transition between exploratory analysis and the use of probability in formal statistical inference.

$\bar{x}$ Charts

Control charts were invented in the 1920s by Walter Shewhart at the Bell Telephone Laboratories.[2] The most common control chart is not a chart of individual observations such as the ones presented in Chapter 2 but of the sample means of groups of observations.

EXAMPLE 6.17

A manufacturer of high-resolution video terminals must control the tension on the mesh of fine wires that lies behind the surface of the viewing screen. Too much tension will tear the mesh while too little will allow wrinkles. The tension is measured by an electrical device with output readings in millivolts (mv). A careful study shows that the proper tension is 275 mv. Some variation is inherent in the production process and the study shows that when the production process is operating properly the standard deviation of the tension readings is $\sigma = 43$ mv. The quality engineer measures the tension on 4 terminals each hour. The mean $\bar{x}$ of each sample estimates the mean tension μ for the population of terminals produced that hour.

Table 6.1 shows the observed $\bar{x}$'s for 20 consecutive hours of production. How should these data be used to monitor the process? ∎

center line

A plot against time helps us see whether or not the process is stable. Figure 6.10 presents a plot of the successive sample means against the order in which the samples were taken. Since the desired value for the process mean is $\mu = 275$ mv, we draw a *center line* at that level across the plot. The means from the later samples fall above this line and are consistently higher than those from earlier samples. This suggests that the process mean μ may

---

* Control charts are important in industry and they also illustrate the use of sampling distributions in inference. Nonetheless, this section is not required for an understanding of later material.

Table 6.1 $\bar{x}$ from 20 samples of size 4

Sample	$\bar{x}$	Sample	$\bar{x}$
1	269.5	11	264.7
2	297.0	12	307.7
3	269.6	13	310.0
4	283.3	14	343.3
5	304.8	15	328.1
6	280.4	16	342.6
7	233.5	17	338.8
8	257.4	18	340.1
9	317.5	19	374.6
10	327.4	20	336.1

have shifted upward, away from its target value of 275 mv. We are aware of how easily our intuitive judgment can be misled by short random sequences. Can we back up our perception by calculation?

We expect $\bar{x}$ to have a distribution that is nearly normal. Not only are the tension measurements roughly normal, but also the central limit theorem effect suggests that the sample mean will be closer to normal than the individual measurements. Since a control chart is a warning device, it is not

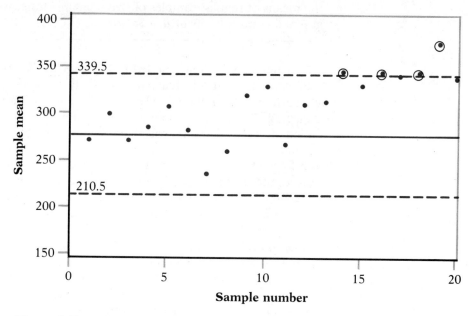

Figure 6.10 The $\bar{x}$ control chart for the data of Table 6.1.

necessary that our probability calculations be exactly correct. Approximate normality is good enough. In that same spirit, control charts use the approximate normal probabilities given by the 68–95–99.7 rule rather than more exact calculations from Table A.

If the standard deviation of the individual screens remains at $\sigma = 43$ mv, the standard deviation of $\bar{x}$ from 4 screens is

$$\sigma(\bar{x}) = \frac{\sigma}{\sqrt{n}}$$
$$= \frac{43}{\sqrt{4}} = 21.5$$

As long as the mean remains at its target value $\mu = 275$ mv, almost all values of $\bar{x}$ will lie between

$$\mu - 3\sigma(\bar{x}) = 275 - (3)(21.5) = 210.5$$
$$\mu + 3\sigma(\bar{x}) = 275 + (3)(21.5) = 339.5$$

control limits   We therefore draw dashed *control limits* at these two levels on the plot.

$\bar{x}$ control chart

> To evaluate the control of a process with given standards μ and σ, plot the means of regular samples of size n against time, with a center line at μ and control limits at $\mu \pm 3\sigma/\sqrt{n}$.

Four points, which are circled in Figure 6.10, lie above the upper control limit of the control chart. It is unlikely (probability less than 0.003) that a particular point would fall outside the control limits if μ and σ remain at their target values. These points are therefore good evidence that the distribution of the production process has changed. It appears that the process mean moved up at about sample number 12. In practice, the operators search for a change in the process as soon as the first out of control point is noticed, that is, after sample number 14. Lack of control might be caused by a new operator, a new batch of mesh, or a breakdown in the tensioning apparatus. The out of control signal alerts us to the change before a large number of defective screens are produced.

An $\bar{x}$ control chart is often called simply an $\bar{x}$ *chart*. Points $\bar{x}$ that vary between the control limits of an $\bar{x}$ chart represent the chance variation that is present in a normally operating process. Points that are out of control suggest that some source of additional variability has disturbed the stable operation of the process. Such a disturbance makes out of control points probable rather than unlikely. For example, if the process mean μ in Example 6.17 shifts from 275 mv to 339.5 mv, the probability that the next point falls above the upper control limit increases from about 0.0015 to 1/2.

Comments on control charts A control chart gets its name from the statistic that is plotted against time—an $\bar{x}$ chart plots $\bar{x}$, while a $\hat{p}$ chart plots the sample proportion $\hat{p}$. (By an accident of history that cannot now be undone, a $\hat{p}$ chart is called a p chart in most writing on quality control.) The center line and control limits are based on the sampling distribution of the statistic. Since $\hat{p}$ from a sample of size n has approximately the $N(p, \sqrt{p(1-p)/n})$ distribution, you should be able to set up a $\hat{p}$ chart to control a population proportion at a target value p. Such $\hat{p}$ charts are most often used to monitor the proportion p of items produced that fail to conform to a set of specifications.

There are many variations on even the $\bar{x}$ chart. We often have a target value for the mean μ but σ is not known. In this case we base the control limits on an estimate of σ obtained by combining the sample standard deviations from past samples. Each sample standard deviation s estimates σ and their average $\bar{s}$ is a better estimate. The control limits are then $\mu \pm 3\bar{s}/\sqrt{n}$. Note that the average $\bar{s}$ measures only the short-term variation that occurs during the period of a single sample (and then averages this variation over all of the samples). This $\bar{s}$ is usually much smaller than the sample standard deviation of all of the individual observations, which includes long-term variation over the entire period during which the samples were taken. The $\bar{x}$ chart with control limits based on $\bar{s}$ is more sensitive to disturbances in the process than the control charts presented in Chapter 2, which were based on a single observation at each time. There are some fine points in using control limits based on past data—for example, we must first check that the process was already in control when the data were gathered. We will content ourselves with the case in which values μ and σ are given.

The purpose of a control chart is not to ensure good quality by inspecting most of the items produced. Control charts focus on the manufacturing process itself rather than on the products. By checking the process at regular intervals, we can detect disturbances and correct them before they become serious. This is called *statistical process control*. Process control achieves high quality at a lower cost than inspecting all of the products. Small samples of 4 or 5 items are usually adequate for process control.

statistical process control

The probability calculations in a control chart assume that the individual observations are a random sample from the population of interest. If the usual 4 or 5 items in a sample are spread out over an hour's production, the population is all items produced that hour. It is more common, however, to sample 4 or 5 consecutive items once each hour. In that case the population exists only in our minds. It contains all items that would be produced by the process as it was operating at the time of sampling. The control chart monitors the state of the process once each hour to see if a change has taken place.

Out of Control Signals

The basic signal indicating lack of control in an $\bar{x}$ chart is a single point beyond the control limits. In practice, however, many other signals are used. For each case we can compute the probability that a process in control (that is, with μ and σ at their target values) gives a false signal. The probability of a false signal is about 0.003 for the "one point beyond the limits" signal. Figure 6.11 illustrates several popular signals.

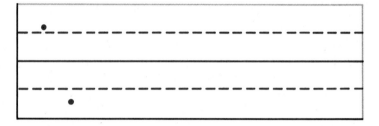

Figure 6.11(a)   Out of control signals for $\bar{x}$ charts, one point beyond the three sigma level.

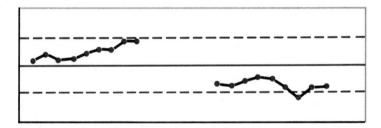

Figure 6.11(b)   Out of control signals for $\bar{x}$ charts, run of nine points on one side of the center line.

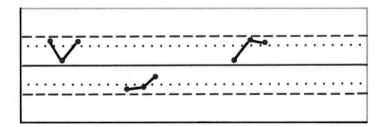

Figure 6.11(c)   Out of control signals for $\bar{x}$ charts, two out of three points beyond the two sigma level on the same side of the center line.

EXAMPLE 6.18 One common out of control signal is a *run* of 9 consecutive points above or below the center line (Figure 6.11b). What is the probability of a false signal of this type when the process is in fact in control?

When the process is in control, the center line is the true process mean μ. Because of the symmetry of the normal distributions, a sample mean $\bar{x}$ is equally likely to fall above or below the line. The means of successive samples are independent because the samples are separated in time. So by the multiplication rule, the probability of a run of 9 points above μ is

$$\left(\frac{1}{2}\right)^9 \approx .002$$

The probability of a run of 9 points below the center line is the same. The probability that the next 9 points show a run of either kind is therefore about 0.004. The probability that such a run occurs not in the next 9 points but somewhere in a longer sequence of observations would also be useful. Unfortunately, the calculation of such probabilities is quite difficult. ∎

In the $\bar{x}$ chart of Figure 6.10, the run criterion does not give an out of control signal until sample number 20. The "one point out" criterion alerts us at sample 14. In this case the run criterion was slow to detect lack of control. In cases where the process mean slowly drifts away from its target value, however, the run criterion will often give an out of control signal before any individual point falls outside the control limits. That is why it is common practice to use the "one point out" criterion and the run criterion simultaneously.

EXAMPLE 6.19 Another out of control signal is "2 out of 3 points beyond the two sigma level." That is, at least 2 of the next 3 $\bar{x}$'s fall either below $\mu - 2\sigma/\sqrt{n}$ or above $\mu + 2\sigma/\sqrt{n}$. Figure 6.11(c) illustrates several configurations that give this signal. The probability that a process in control produces an $\bar{x}$ larger than $\mu + 2\sigma/\sqrt{n}$ is about 0.025 (that's half of the 0.05 left over from the "95" part of the 68–95–99.7 rule). The number X of the next 3 samples beyond this level is a count of successes in 3 independent trials; X therefore has the $B(3, 0.025)$ distribution. The binomial formula gives the probability of a false signal as

$$P(X \geq 2) = P(X = 2) + P(X = 3)$$
$$= \binom{3}{2}(.025)^2(.975) + \binom{3}{3}(.025)^3(.975)^0$$
$$= .00183 + .00002 \approx .002$$

The probability that the next 3 samples give a false signal on the low side is the same. So the overall probability of a false signal by this criterion is about 0.004. ∎

These examples illustrate the application of probability calculations in inference. They confirm that each of these out of control signals is quite unlikely to occur so long as the process remains in control. So when a signal *does* occur, we suspect that the process has been disturbed. The probability calculations assumed that μ and σ were given. The results would be a bit different in the more common case in which σ is unknown and is estimated by $\bar{s}$. The same out of control signals are nonetheless used in both cases. Control charts are an informal tool designed for easy use. We do probability calculations in the simplest case to get a rough idea of how likely false signals are, and we do not attempt to refine these calculations when conditions change. We will see in Chapter 8 that more formal inference does adapt its probability calculations to each situation in turn.

SUMMARY

A process that can be measured over time is **in control** if the value at any time has the same probability distribution. A process that is in control is operating under stable conditions.

An $\bar{x}$ **control chart** is a graph of sample means plotted against the time order of the samples, with a solid center line at the target value μ of the process mean and dashed **control limits** at $\mu \pm 3\sigma/\sqrt{n}$. An $\bar{x}$ chart helps us decide if a process is in control with mean μ and standard deviation σ.

The probability that the next point lies outside the control limits on an $\bar{x}$ chart is about 0.003 if the process is in control. Such a point is evidence that the process is **out of control**—that is, that the distribution of the process has changed as a result of some disturbance. A cause for the change in the process should be sought.

There are other common signals for lack of control, such as a run of 9 consecutive points on the same side of the center line. In each case, we can compute the probability that a process in control will give the signal. If this probability is small, we can regard the signal as good evidence of lack of control.

SECTION 6.3 EXERCISES

6.47 A maker of auto air conditioners checks a sample of 4 thermostatic controls from each hour's production. The thermostats are set at 75° and then placed in a chamber where the temperature is raised gradually. The temperature at which the thermostat turns on the air conditioner is recorded. The standard for the process mean is $\mu = 75°$. Past experience indicates that the response temperature of properly adjusted thermostats varies with $\sigma = 0.5°$. The mean response temperature $\bar{x}$ for each hour's sample is plotted on an $\bar{x}$ control chart. Calculate the center line and control limits for this chart.

6.48 The width of a slot cut by a milling machine is important to the proper functioning of the hydraulic system of a farm tractor. The manufacturer checks the control of the milling process by measuring a sample of 5 consecutive items during each hour's production. The mean slot width for each sample is plotted on an $\bar{x}$ control chart. The target width for the slot is $\mu = 0.8750$ inch. When properly adjusted, the milling machine should produce slots with mean width equal to the target value and standard deviation $\sigma = 0.0012$ inch. What center line and control limits should be drawn on the $\bar{x}$ chart?

6.49 A pharmaceutical manufacturer forms tablets by compressing a granular material that contains the active ingredient and various fillers. The hardness of a sample from each lot of tablets is measured in order to control the compression process. The target values for the hardness are $\mu = 11.5$ and $\sigma = 0.2$. Table 6.2 gives three sets of data, each representing $\bar{x}$ for 20 successive samples of $n = 4$ tablets. One set remains in control at the target value. In a second set, the process mean μ shifts suddenly to a new value. In a third, the process mean drifts gradually.

(a) What are the center line and control limits for an $\bar{x}$ chart for this process?

Table 6.2 Three sets of $\bar{x}$ from 20 samples of size 4

Sample	Data set A	Data set B	Data set C
1	11.602	11.627	11.495
2	11.547	11.613	11.475
3	11.312	11.493	11.465
4	11.449	11.602	11.497
5	11.401	11.360	11.573
6	11.608	11.374	11.563
7	11.471	11.592	11.321
8	11.453	11.458	11.533
9	11.446	11.552	11.486
10	11.522	11.463	11.502
11	11.664	11.383	11.534
12	11.823	11.715	11.624
13	11.629	11.485	11.629
14	11.602	11.509	11.575
15	11.756	11.429	11.730
16	11.707	11.477	11.680
17	11.612	11.570	11.729
18	11.628	11.623	11.704
19	11.603	11.472	12.052
20	11.816	11.531	11.905

(b) Draw a separate $\bar{x}$ chart for each of the three data sets. Circle any points that are beyond the control limits. Also, check for runs of 9 points above or below the center line and mark the ninth point of any run as being out of control.

(c) Based on your work in (b) and the appearance of the control charts, which set of data comes from a process that is in control? In which case does the process mean shift suddenly and at about which observation do you think that the mean changed? Finally, in which case does the mean drift gradually?

6.50 The diameter of a bearing deflector in an electric motor is supposed to be 2.205 cm. Experience shows that when the manufacturing process is properly adjusted, it produces items with mean 2.2050 cm and standard deviation 0.0010 cm. A sample of 5 consecutive items is measured once each hour. The sample means $\bar{x}$ for the past 12 hours are given below.

Hour	1	2	3	4	5	6
$\bar{x}$	2.2047	2.2047	2.2050	2.2049	2.2053	2.2043

Hour	7	8	9	10	11	12
$\bar{x}$	2.2036	2.2042	2.2038	2.2045	2.2026	2.2040

Make an $\bar{x}$ control chart for the deflector diameter. Use both the "one point out" and "run of nine" signals to assess the control of the process. At what point should action have been taken to correct the process as the hourly point was added to the chart?

6.51 Ceramic insulators are baked in lots in a large oven. After the baking, 3 insulators are selected at random from each lot and tested for breaking strength. The mean breaking strength for these samples is plotted on a control chart. The specifications call for a mean breaking strength of at least 10 pounds per square inch (psi). Past experience suggests that if the ceramic is properly formed and baked the standard deviation in the breaking strength is about 1.2 psi. Here are the sample means from the last 15 lots.

Lot	1	2	3	4	5	6	7	8
$\bar{x}$	12.94	11.45	11.78	13.11	12.69	11.77	11.66	12.60

Lot	9	10	11	12	13	14	15
$\bar{x}$	11.23	12.02	10.93	12.38	7.59	13.17	12.14

(a) Find the center line and control limits for an $\bar{x}$ chart with standards $\mu = 10$ and $\sigma = 1.2$. Make a control chart with these lines drawn on it and plot the $\bar{x}$ points on it.

(b) In this case, a process mean breaking strength greater than 10 psi is acceptable, so points out of control in the high direction do not call for remedial action. With this in mind, use both the "one point out" and "run of nine" signals to assess the control of the process and to recommend action.

6.52 A manager who knows no statistics asks you, "What does it mean to say that a process is in control? Is being in control a guarantee that the quality of the product is good?" Answer these questions in plain language that the manager can understand.

6.53 There are other out of control signals that are sometimes used with $\bar{x}$ charts. One is "4 out of 5 points beyond the one sigma level" on the same side of the center line. That is, at least 4 of 5 consecutive points lie above $\mu + \sigma/\sqrt{n}$, or at least 4 of 5 lie below $\mu - \sigma/\sqrt{n}$. Find the probability that the next 5 points give this signal when the process remains in control at the given μ and σ. Use the binomial distribution and approximate probabilities from the 68–95–99.7 rule.

6.54 Another out of control signal that is sometimes used is "15 points in a row within the one sigma level." That is, 15 consecutive points fall between $\mu - \sigma/\sqrt{n}$ and $\mu + \sigma/\sqrt{n}$. This signal suggests either that the value of σ used for the chart is too large or that careless measurement is producing results that are suspiciously close to the target. Find the probability that the next 15 points will give this signal when the process remains in control with the given μ and σ.

6.55 The usual American and Japanese practice in making $\bar{x}$ charts is to place the control limits at $\mu \pm 3\sigma/\sqrt{n}$. The probability that a particular $\bar{x}$ falls outside these limits when the process is in control is about 0.003, using the 99.7 part of the 68–95–99.7 rule. British practice, on the other hand, places the control limits at $\mu \pm c\sigma/\sqrt{n}$, where the constant c is chosen to give exactly probability 0.001 of a point $\bar{x}$ falling above $\mu + c\sigma/\sqrt{n}$ when the target μ and σ remain true. (The probability that $\bar{x}$ falls below $\mu - c\sigma\sqrt{n}$ is also 0.001 because of the symmetry of the normal distributions.) Use Table A to find the value of c.

In Chapter 2 control charts were presented for a single observation at each time. These control charts for individual measurements are just $\bar{x}$ charts with the sample size $n = 1$. We cannot estimate the short-term process standard deviation σ from the individual samples because a sample of size 1 has no variation. Even with advanced methods[3] of combining the information in several samples, the estimated σ will include some long-term variation and so will be too large. To compensate for this, it is common to use "2σ" rather than "3σ" control limits, as was done in Chapter 2. The next two exercises deal with control charts for individual measurements.

6.56 Exercise 2.1 gives weekly measurements for Joe's weight for a period of 16 weeks. The short-term variation in Joe's weight, estimated from these measurements by advanced methods, is about $\sigma = 1.3$ lb. Joe has a target of $\mu = 162$ lb for his weight. Make a control chart for his measurements, using control limits $\mu \pm 2\sigma$. Comment on individual points out of control and on runs. Is Joe's weight stable or does it change systematically over this period?

6.57 Exercise 2.4 reports the amount spent on food by a family in each month over a 4-year period. The family budget calls for $\mu = \$325$ per month to be spent on food as a long-term average. The month-to-month standard deviation is about $\sigma = \$46$. Make a control chart for food spending with center line and control limits based on these target values. Comment on individual months that fall beyond the control limits and comment on runs.

6.58 A manufacturer of compact disc players uses statistical process control to monitor the quality of the circuit board that contains most of the player's electronic components. Every circuit board is tested for proper function by a computer-directed test bed after assembly. The plant produces 400 circuit boards per shift and the proportion $\hat{p}$ of the 400 boards that fail the test is recorded each day. Company standards call for a failure rate of no more than 10%, or $p = 0.1$.

(a) If the process is in control with proportion $p = 0.1$ of defective boards, what are the mean and standard deviation of the sample proportion $\hat{p}$?

(b) According to the normal approximation, what is the distribution of $\hat{p}$ when the process is in control?

(c) Give the center line (at the mean of $\hat{p}$) and control limits (at three standard deviations of $\hat{p}$ above and below the mean).

(d) Listed in the following chart are the proportions that are defective in 16 days' production. Make a $\hat{p}$ control chart with center line and control limits from (c). Plot the values of $\hat{p}$ on the chart. Are any points out of control?

Day	1	2	3	4	5	6	7	8
$\hat{p}$	.1150	.1600	.1300	.1225	.1000	.1225	.1900	.1150

Day	9	10	11	12	13	14	15	16
$\hat{p}$	.1000	.1600	.1675	.1225	.1375	.1975	.1525	.1675

6.59 In a process control setting, successive samples of n items are inspected. Each item is classified either as conforming to specifications or as failing to conform. The proportion $\hat{p}$ of nonconforming items in each sample is recorded and the values of $\hat{p}$ are plotted against the order of the samples. There is a target value p for the population proportion of nonconforming items. Generalize the results of the previous exercise to give formulas in terms of p and n for the center line and control limits for a $\hat{p}$ control chart. Use the normal approximation for $\hat{p}$ and follow the model of the $\bar{x}$ chart.

6.60 A producer of agricultural machinery purchases bolts from a supplier in lots of several thousand. The producer submits 80 bolts from each lot to a shear test and accepts the lot if no more than 3 of them fail the test. Past records show that an average of 1.8 bolts per lot failed the test. This is an opportunity to keep a control chart to monitor the quality level of the supplier over time. Since $1.8/80 = 0.0225$, a target proportion of nonconforming bolts is set at $p = 0.0225$ based on past data. The proportion of nonconforming bolts among the 80 tested from each incoming lot will be plotted on a control chart. Give the center line and control limits to be used on this chart. (Because of the informal nature of process control procedures, the chart is based on the normal approximation to the distribution of the sample proportion $\hat{p}$ even though the approximation is not accurate for this n and p.)

NOTES

1 Joseph F. Sullivan, "Ruling in Jersey upholds idea of equal odds for all," *The New York Times*, Aug. 13, 1985.

2 See his classic book, W. A. Shewhart, *Economic Control of Quality of Manufactured Product*, Van Nostrand, New York, 1931.

3 These methods are discussed in all texts on statistical quality control. See, for example, Irving W. Burr, *Statistical Quality Control Methods*, Marcel Dekker, New York, 1976, section 7.1.

CHAPTER 6 EXERCISES

6.61 Ray is a basketball player who has made about 70% of his free throws over several years. In a tournament game he makes only 2 of 6 free throws. Ray's coach says this was just bad luck. Suppose that Ray's free throws are independent trials with probability 0.7 of a success on each trial. What is the probability that he makes 2 or fewer in 6 attempts? Do you think that his tournament performance is just chance variation? *yes*

6.62 The distribution of scores for persons over 16 years of age on the Wechsler Adult Intelligence Scale (WAIS) is approximately normal with mean 100 and standard deviation 15. The WAIS is one of the most common "IQ tests" for adults.

   (a) What is the probability that a randomly chosen individual has a WAIS score of 105 or higher?

   (b) What are the mean and standard deviation of the average WAIS score $\bar{x}$ for an SRS of 60 people?

   (c) What is the probability that the average WAIS score of an SRS of 60 people is 105 or higher?

   (d) Would your answers to (a), (b), or (c) be affected if the distribution of WAIS scores in the adult population were distinctly nonnormal?

6.63 High school dropouts make up 14.1% of all Americans aged 18 to 24. A vocational school that wants to attract dropouts mails an advertising flyer to 25,000 persons between the ages of 18 and 24.

   (a) If the mailing list can be considered a random sample of the population, what is the mean number of high school dropouts who will receive the flyer? What is the standard deviation of this number?

   (b) What is the probability that at least 3500 dropouts will receive the flyer?

6.64 A political activist is gathering signatures on a petition by going door to door asking citizens to sign. She wants 100 signatures. Suppose that the probability of getting a signature at each household is 1/10 and let the random variable X be the number of households visited to collect exactly 100 signatures. Does X have a binomial distribution? If so, give n and p. If not, explain why not.

6.65 According to genetic theory, the blossom color in the second generation of a certain cross of sweet peas should be red or white in a 3:1 ratio. That is, each plant has probability 3/4 of having red blossoms and the blossom colors of separate plants are independent.

   (a) What is the probability that exactly 6 out of 8 of these plants have red blossoms?

(b) What is the mean number of red-blossomed plants when 80 plants of this type are grown from seeds?

(c) What is the probability of obtaining at least 50 red-blossomed plants when 80 plants are grown from seeds?

6.66 The weight of the eggs produced by a certain type of hen is normally distributed with mean 65 g and standard deviation 5 g. If cartons of such eggs can be considered to be SRSs of size 12 from the population of all eggs, what is the probability that the weight of a carton falls between 750 g and 825 g?

6.67 According to *USA Today* (Oct. 12, 1984), 46.4% of all residential telephone numbers in Las Vegas are unlisted. A telephone sales firm uses random digit dialing equipment that dials residential numbers at random, regardless of whether or not they are listed in the telephone directory. The firm calls 500 numbers in Las Vegas.

(a) What is the exact distribution of the number X of unlisted numbers that are called?

(b) Use a suitable approximation to calculate the probability that at least half of the numbers called are unlisted.

6.68 A study of rush hour traffic in Los Angeles records the number of people in each car entering a freeway at a suburban interchange. Suppose that this number X has mean 1.5 and standard deviation 0.75 in the population of all cars that enter at this interchange during rush hours.

(a) Does the count X have a binomial distribution? Why or why not?

(b) Could the exact distribution of X be normal? Why or why not?

(c) Traffic engineers estimate that the capacity of the interchange is 700 cars per hour. According to the central limit theorem, what is the approximate distribution of the mean number of persons $\bar{x}$ in 700 randomly selected cars at this interchange?

(d) The count of people in 700 cars is $700\bar{x}$. Use your result from (c) to give an approximate distribution for the count. What is the probability that 700 cars will carry more than 1075 people?

6.69 An opinion poll asks a sample of 500 adults whether they favor giving parents of school-age children vouchers that can be exchanged for education at any public or private school of their choice. Each school would be paid by the government on the basis of how many vouchers it collected. Suppose that in fact 45% of the population favor this idea. What is the probability that at least half of the sample are in favor of it? (Assume that an SRS was taken.)

6.70 A machine fastens plastic screw-on caps onto containers of motor oil. If the machine applies more torque than the cap can withstand, the cap will break. Both the torque applied and the strength of the caps vary. The capping machine torque has the normal distribution with mean 7 inch-pounds and standard deviation 0.9 inch-pound. The cap strength (the torque that would break the cap) has the normal distribution with mean 10 inch-pounds and standard deviation 1.2 inch-pounds.

(a) Explain why it is reasonable to assume that the cap strength and the torque applied by the machine are independent.

(b) What is the probability that a cap will break while being fastened by the capping machine?

6.71 The process that molds the plastic caps referred to in the previous exercise is monitored by testing a sample of 6 caps every 20 minutes. The breaking strength of the caps is measured and an $\bar{x}$ chart is kept. Use the information in the previous exercise to find the center line and control limits for this $\bar{x}$ chart.

6.72 An important step in the manufacture of integrated circuit chips is etching the lines that will conduct current between components on the chip. The chips contain a line width test pattern that is used for process control measurements. The target width is 3.0 micrometers (μm). Past data show that the line width varies in production according to a normal distribution with mean 2.829 μm and standard deviation 0.1516 μm.

(a) What is the probability that the line width of a randomly chosen chip falls outside the acceptable range 3.0 ± 0.2 μm?

(b) What are the control limits for an $\bar{x}$ chart for line width if samples of size 5 are selected at regular intervals during production? (Use the target value 3.0 as your center line.)

6.73 In an experiment to discover the best way to learn foreign languages, researchers studied the effect of delaying oral practice when beginning language study. The researchers randomly assigned 23 beginning students of Russian to an experimental group and another 23 to a control group. The control group began speaking practice immediately while the experimental group delayed speaking for 4 weeks. At the end of the semester both groups took a standard test of comprehension of spoken Russian. Suppose that in the population of all beginning students, the test scores under the control method vary according to the $N(32, 6)$ distribution. The population distribution when oral practice is delayed is $N(29, 5)$.

(a) What is the sampling distribution of the mean score $\bar{x}$ in the control group in many repetitions of the experiment? What is the sampling distribution of the mean score $\bar{y}$ in the experimental group?

(b) If the experiment was repeated many times, what would be the sampling distribution of the difference $\bar{y} - \bar{x}$ between the mean scores in the two groups?

(c) What is the probability that the experiment will find (misleadingly) that the experimental group has a mean at least as large as that of the control group?

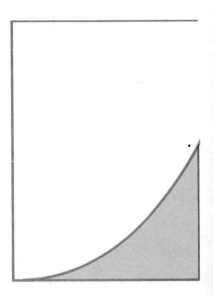

Prelude

Statistical inference not only draws conclusions from data but uses probability to state how reliable the conclusions are. In this chapter we discuss two basic types of inference, confidence intervals and tests of significance. Our emphasis is on the reasoning of inference and on the judgment that is required to use statistical inference wisely. In later chapters we will present specific methods of inference that apply the principles of this chapter in many practical settings.

- *How do confidence intervals allow us to estimate the mean SAT score of a large population of students from data on a few students and also to give a margin of error that says how precise our estimate is?*

- *How can a test of significance produce convincing evidence that the 1970 draft lottery was biased against men born late in the year?*

- *What does it mean to say that a result is "statistically significant"? What does statistical significance not tell us?*

7

Introduction to Inference

The purpose of statistical inference is to draw conclusions from data. Although we have examined data and arrived at conclusions many times previously, what is new here is an emphasis on substantiating our conclusions with probability calculations. An effect that appears systematic to our unaided judgment may in fact be the result of chance variation. The misperceptions that lie behind the popular belief in a law of small numbers, discussed on pages 351–352, are one example. Exploratory analysis of data can also err in the other direction by overlooking real effects when they are not large. For example, the scatterplot of the 1970 draft lottery (Figure 3.5) does not suggest any obvious bias in the lottery. We can look more closely by computing a numerical measure of the association. The correlation between birth date and draft number is $r = -0.226$. Since any two variables will have some chance association in practice, this rather small correlation may not seem convincing. Unaided judgment can easily fail to detect the unfairness inherent in the lottery. We were able to see the flaws in the 1970 draft lottery by clever graphics (Figures 3.6 and 3.7), but the pictures alone cannot convince us that more than bad luck is behind the trend that they show.

Control charts (Section 6.3) illustrate the use of probability calculations to aid our judgment in a rather informal way. In this chapter we will concentrate on the two most prominent types of formal statistical inference. Section 7.1 introduces *confidence intervals* for estimating the value of a population parameter. Section 7.2 presents *tests of significance*, which assess the evidence for a claim. Both types of inference are based on the sampling distributions of statistics. That is, both report probabilities that state *what would happen if we used the inference method many times*. This kind of probability statement is characteristic of standard statistical inference. Users of statistics should understand from the first the nature of the reasoning employed and the meaning of the probability statements that appear, for example, on computer output for statistical procedures.

Because the methods of formal inference are based on sampling distributions, they are most reliable when the data are produced by a properly randomized design. Use of these methods implies that we are acting as if the data are a random sample or result from a randomized experiment. If this is not true, the conclusions may be open to challenge. Do not be overly impressed by the complex details of formal inference. This elaborate machinery cannot remedy basic flaws in producing the data. Use the common sense developed in your study of the first four chapters of this book, and proceed to detailed formal inference only when you are satisfied that the data deserve such analysis.

The primary purpose of this chapter is to describe the reasoning used in statistical inference. We will discuss only a few specific inference techniques, and these require rather unrealistic assumptions. Later chapters will present inference methods for use in most of the settings we met in learning to explore data. There are libraries—both of books and of computer software—full of yet more elaborate statistical techniques. Informed use of any of these methods requires an understanding of the underlying reasoning. A computer will do the arithmetic, but you must still exercise judgment based on understanding.

7.1   ESTIMATING WITH CONFIDENCE

The Scholastic Aptitude Tests (SAT) are widely used measures of readiness for college study. There are two tests, one for verbal ability (SAT-V) and one for mathematical ability (SAT-M). The scores on each test are adjusted so that the mean is 500 and the standard deviation is 100 in a large "standardization group" on which the tests were developed. This scale is maintained from year to year so that scores have a constant interpretation. The scores actually achieved in a given year may have a mean that is higher or lower than 500. For example, the number of students taking the SATs peaked in 1976 at about 1 million; the mean scores that year were 431 on the verbal test and 472 for mathematics.

EXAMPLE 7.1   You want to estimate the mean SAT-M score for the more than 250,000 high school seniors in California. You know better than to trust data from the students who choose to take the SATs. Only about 41% of California students take the SATs. These self-selected students are all planning to attend college and are not representative of all California seniors. At considerable effort and expense, you give the test to a simple random sample (SRS) of 500 California high school seniors. The mean score for your sample is $\bar{x} = 461$. What can you say about the mean score μ in the population? ∎

The sample mean $\bar{x}$ is the natural estimator of the unknown population mean μ. We know that $\bar{x}$ is an unbiased estimator of μ. More important, the law of large numbers says that the sample mean must approach the population mean as the size of the sample grows. The value $\bar{x} = 461$ therefore appears to be a reasonable estimate of the mean score μ that all 250,000 students would achieve if they took the test. But how reliable is this estimate? A second sample would surely not give 461 again. Unbiasedness says only that there is no systematic tendency to underestimate or overestimate the truth. Could we plausibly get a sample mean of 530 or 410 on repeated samples? An estimate without an indication of its variability is of little value.

Statistical Confidence

Just as unbiasedness concerns the center of the sampling distribution, questions about variation are answered by looking at its spread. We know that if the entire population of SAT scores has mean μ and standard deviation σ, then in repeated samples of size 500 the sample mean $\bar{x}$ follows the $N(\mu, \sigma/\sqrt{500})$ distribution. Let us suppose that we know that the standard deviation σ of SAT-M scores in our California population is the same σ = 100 that is true of the standardization group. (This is not realistic. We will see in the next chapter how to proceed when σ is not known. For now, we are more interested in statistical reasoning than in details of realistic methods.) In repeated sampling the sample mean $\bar{x}$ follows the normal distribution centered at the unknown population mean μ and having standard deviation

$$\sigma(\bar{x}) = \frac{100}{\sqrt{500}} \approx 4.5$$

Now we are in business. Consider this line of thought, which is illustrated by Figure 7.1:

- We know by the 68–95–99.7 rule that the probability is about 0.95 that $\bar{x}$ will be within 9 points (two standard deviations of $\bar{x}$) of the population mean score μ.
- To say that $\bar{x}$ lies within 9 points of μ is the same as saying that μ is within 9 points of $\bar{x}$.
- So 95% of all samples will capture the true μ in the interval from $\bar{x} - 9$ to $\bar{x} + 9$.

We have simply restated a fact about the sampling distribution of $\bar{x}$. *The language of statistical inference uses this fact about what would happen in the long run to express our confidence in the results of any one sample.* Our

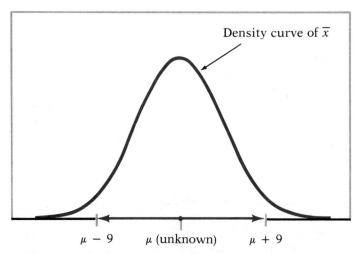

Figure 7.1 The sample mean $\bar{x}$ lies within ±9 of μ in 95% of all samples, so μ also lies within ±9 of $\bar{x}$ in those samples.

sample gave $\bar{x} = 461$. We say that we are *95% confident* that the unknown mean score for all California seniors lies between

$$\bar{x} - 9 = 461 - 9 = 452$$

and

$$\bar{x} + 9 = 461 + 9 = 470$$

Be sure you understand the grounds for our confidence. There are only two possibilities:

1 The interval between 452 and 470 contains the true μ.

2 Our SRS was one of the few samples for which $\bar{x}$ is not within 9 points of the true μ. Only 5% of all samples give such inaccurate results.

We cannot know whether our sample is one of the 95% for which the interval $\bar{x} \pm 9$ captures μ or one of the unlucky 5%. The statement that we are 95% confident that the unknown μ lies between 452 and 470 is shorthand for saying, ''We arrived at these numbers by a method that gives correct results 95% of the time.''

confidence interval

The interval of numbers between the values $\bar{x} \pm 9$ is called a *95% confidence interval* for μ. Figure 7.2 illustrates the behavior of 95% confidence intervals in repeated sampling. The center of each interval is at $\bar{x}$ and therefore varies from sample to sample. The sampling distribution of $\bar{x}$ appears at the top of the figure to show the long-term pattern of this variation. The 95% confidence intervals $\bar{x} \pm 9$ from 25 SRSs appear below. The center $\bar{x}$ of each interval is marked by a dot. The arrows on either side of the dot span the confidence interval. All except one of the 25 intervals cover the true value of μ. In a very large number of samples, 95% of the confidence intervals would contain μ.*

Confidence Intervals

confidence level

Statisticians have constructed confidence intervals for many different parameters based on a variety of designs for data collection. We will meet a number of these in later chapters. Any confidence interval has two aspects: an *interval* computed from the data and a *confidence level* giving the probability that the method produces an interval that covers the parameter. Users can choose the confidence level. It is most often 90% or higher and is usually written

* Confidence intervals as a systematic method were invented in 1937 by Jerzy Neyman (1894–1981). Neyman was a Pole who moved to England in 1934 and spent the last half of his long life at the University of California at Berkeley. He remained active until his death, almost doubling the list of his publications after his official retirement in 1961.

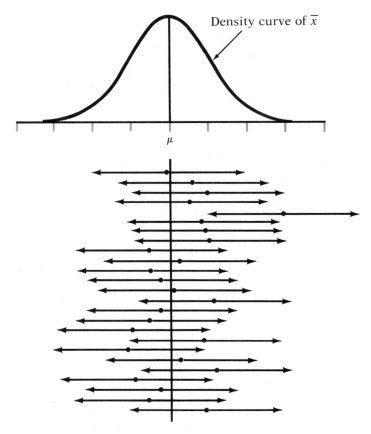

Figure 7.2 Twenty-five samples from the same population gave these 95% confidence intervals. In the long run, 95% of all samples give an interval that covers μ.

in formulas as $1 - \alpha$ in decimal form. For example, a 95% confidence level corresponds to $1 - \alpha = 0.95$, or to $\alpha = 0.05$. Here is the formal definition:

Confidence interval

A level $1 - \alpha$ confidence interval for a parameter is given by two statistics U and L such that when θ is the true value of the parameter,

$$P(L \le \theta \le U) = 1 - \alpha$$

In this definition, θ is any fixed (and unknown) true value of the parameter. The lower and upper endpoints L and U are random variables that vary in repeated sampling. These endpoints are chosen so that the probability that the interval from L to U contains θ is $1 - \alpha$ when θ describes the population.

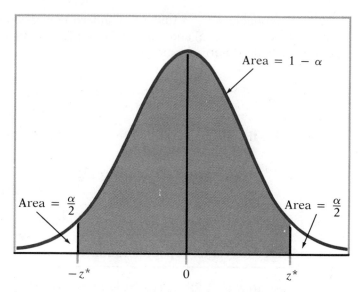

Figure 7.3 The upper $\alpha/2$ critical value z^* has area $\alpha/2$ above it under the standard normal curve. The area between $-z^*$ and z^* is $1 - \alpha$.

We will now construct a level $1 - \alpha$ confidence interval for the mean μ of a population when the data are an SRS of size n. The construction is based on our knowledge of the sampling distribution of the sample mean $\bar{x}$. This distribution is exactly $N(\mu, \sigma/\sqrt{n})$ when the population has the $N(\mu, \sigma)$ distribution. The central limit theorem says that this same sampling distribution is approximately correct for large samples whenever the population mean and standard deviation are μ and σ.

Our construction of a 95% confidence interval for the mean SAT score began by noting that any normal distribution has probability about 0.95 within ± 2 standard deviations of its mean. To construct a level $1 - \alpha$ confidence interval we first find the number z^* such that any normal distribution has probability $1 - \alpha$ within $\pm z^*$ standard deviations of its mean. We can find z^* from Table A of standard normal probabilities by following Figure 7.3. As the figure illustrates, the value z^* for confidence $1 - \alpha$ is the point with area $\alpha/2$ lying above it under the standard normal curve. For example, a 90% confidence interval has $1 - \alpha = 0.9$, and $\alpha/2 = 0.05$. In this case, z^* is the point with probability 0.05 above it and probability 0.95 below it. Table A shows that this z^* lies between 1.64 and 1.65. A more exact calculation of areas under the standard normal curve gives $z^* = 1.645$.

Critical value

The number z^* with probability p above it under the standard normal density curve is called the upper p critical value of the standard normal distribution.

There is probability $1 - \alpha$ under a standard normal curve between z^* (the upper $\alpha/2$ critical value) and $-z^*$, as Figure 7.3 shows. Standardizing $\bar{x}$ produces a standard normal random variable, so that

$$P\left(-z^* \leq \frac{\bar{x} - \mu}{\sigma/\sqrt{n}} \leq z^*\right) = 1 - \alpha$$

Rearranging the inequalities to write the same event in another form gives

$$P\left(\bar{x} - z^* \frac{\sigma}{\sqrt{n}} \leq \mu \leq \bar{x} + z^* \frac{\sigma}{\sqrt{n}}\right) = 1 - \alpha$$

This probability statement gives us our confidence interval.

Confidence interval for a population mean

> Suppose that an SRS of size n is drawn from a population having unknown mean μ and known standard deviation σ. A level $1 - \alpha$ confidence interval for μ is
>
> $$\bar{x} \pm z^* \frac{\sigma}{\sqrt{n}}$$
>
> where z^* is the upper $\alpha/2$ critical value for the standard normal distribution. This interval is exact when the population distribution is normal and is approximately correct for large n in other cases.

It is convenient to have a separate table of the critical values z^*. Table D in the back of the book gives the upper p critical values of the standard normal distribution for several values of p. Notice that for a level $1 - \alpha$ confidence interval you must use the $\alpha/2$ critical value. For a 90% confidence interval the $p = \alpha/2 = 0.05$ critical value is needed. Table D shows that this is $z^* = 1.645$. Here is a brief tabulation taken from Table D of the normal critical points needed for the most common levels of confidence.

Confidence level	p	z^*
90%	.05	1.645
95%	.025	1.960
99%	.005	2.576

EXAMPLE 7.2   A laboratory analyzes specimens of a pharmaceutical product to determine the concentration of the active ingredient. Such chemical analyses are not perfectly precise. Repeated measurements on the same specimen will give slightly different

results. The results of repeated measurements follow a normal distribution quite closely. The analysis procedure has no bias, so that the mean μ of the population of all measurements is the true concentration in the specimen. The standard deviation of this distribution is a property of the analytical procedure and is known to be $\sigma = 0.0068$ gram per liter. The laboratory analyzes each specimen three times and reports the mean result.

Three analyses of one specimen give concentrations

$$.8403 \quad .8363 \quad .8447$$

Give a 99% confidence interval for the true concentration μ.

The sample mean of these readings is

$$\bar{x} = \frac{.8403 + .8363 + .8447}{3} = .8404$$

For 99% confidence, $\alpha = 0.01$ and $\alpha/2 = 0.005$. We see from the table that $z^* = 2.576$. A 99% confidence interval for μ is therefore

$$\bar{x} \pm z^* \frac{\sigma}{\sqrt{n}} = .8404 \pm 2.576 \frac{.0068}{\sqrt{3}}$$

$$= .8404 \pm .0101$$

$$= (.8303, .8505)$$

margin of error

We are 99% confident that the true concentration lies between 0.8303 and 0.8505. The quantity ± 0.0101 is called the *margin of error* for the confidence interval. ∎

Suppose that a single measurement gave $x = 0.8404$, the same value that the sample mean took in Example 7.2. Repeating the calculation with $n = 1$ shows that the 99% confidence interval based on a single measurement is

$$\bar{x} \pm z^* \frac{\sigma}{\sqrt{1}} = .8404 \pm (2.576)\,(.0068)$$

$$= .8404 \pm .0175$$

$$= (.8229, .8579)$$

The mean of three measurements gives a smaller margin of error and hence a shorter interval than a single measurement. Figure 7.4 illustrates the gain from using three observations.

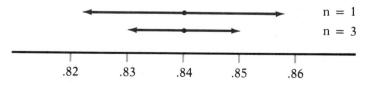

Figure 7.4 Confidence intervals for $n = 3$ and $n = 1$ for Example 7.2.

The argument leading to the form of confidence intervals for the population mean μ rested on the fact that the statistic $\bar{x}$ used to estimate μ has a normal distribution. Because many sample estimates have normal distributions (at least approximately), it is useful to notice that the confidence interval has the form

$$\text{estimate} \pm z^* \sigma_{\text{estimate}}$$

The desired confidence level determines z^* from Table D. The standard deviation of the estimate is found from a knowledge of the sampling distribution in a particular case. When the estimate is $\bar{x}$ from an SRS, the standard deviation of the estimate is $\sigma_{\bar{x}} = \sigma/\sqrt{n}$.

How Confidence Intervals Behave

The confidence interval $\bar{x} \pm z^* \sigma/\sqrt{n}$ for the mean of a normal population illustrates several important properties that are shared by all confidence intervals in common use. The user chooses the confidence level and the width of the interval follows from this choice. High confidence is desirable and so is a short interval. High confidence says that our method almost always gives correct answers. A short interval says that we have pinned down the parameter to within a small margin for error. There is a tradeoff between the confidence level and the width of the interval. To obtain higher confidence from the same data, you must be willing to accept a wider interval.

In the case of the confidence interval for a population mean, these facts reflect the behavior of the critical value z^*. A look at Figure 7.3 will convince you that z^* must be larger for higher confidence (smaller α). Table D shows that this is indeed the case. If n and σ are unchanged, a larger z^* leads to a wider interval. On the other hand, increasing the sample size n narrows the interval for any fixed confidence level. The square root in the formula means that we must multiply the number of observations by four in order to halve the length of the interval. The standard deviation σ measures the variation in the population. You can think of the variation among individuals in the population as noise that obscures the average value μ. It is harder to pin down the mean μ of a highly variable population; that is why the width of a confidence interval increases with σ.

EXAMPLE 7.3

Suppose that the laboratory in Example 7.2 is content with 90% confidence rather than 99%. Then $\alpha/2 = 0.05$. Table D gives the upper 0.05 critical value as $z^* = 1.645$. The 90% confidence interval for μ based on three repeated measurements is then

$$\bar{x} \pm z^* \frac{\sigma}{\sqrt{n}} = .8404 \pm 1.645 \frac{.0068}{\sqrt{3}}$$

$$= .8404 \pm .0065$$

$$= (.8339, .8469)$$

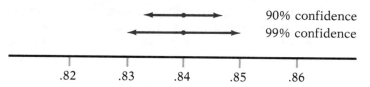

Figure 7.5 90% and 99% confidence intervals for Example 7.3.

Settling for 90% rather than 99% confidence has reduced the margin of error from ± 0.0101 to ± 0.0065. Figure 7.5 compares the two intervals.

Increasing the number of measurements from 3 to 5 will also reduce the width of the 99% confidence interval in Example 7.2. Check that replacing $\sqrt{3}$ by $\sqrt{5}$ reduces the ± 0.0101 margin of error to ± 0.0078. ∎

Choosing the sample size A wise user of statistics never plans data collection without at the same time planning the inference. In this way you can arrange to have both high confidence and a short interval. The width of the confidence interval $\bar{x} \pm z^* \sigma / \sqrt{n}$ for a normal mean is

$$2z^* \frac{\sigma}{\sqrt{n}}.$$

If your desired width is w, just set this expression equal to w, substitute the value of z^* for your desired confidence level, and solve for the sample size n. The result is as follows:

Sample size for desired width

> The confidence interval for a population mean will have a specified width w when the sample size is
>
> $$n = \left(\frac{2z^* \sigma}{w}\right)^2$$

This formula is not the proverbial free lunch. In practice, taking observations costs time and money. The sample size desired may be impossibly expensive.

EXAMPLE 7.4

A new customer of the laboratory of Example 7.2 wants results accurate to within ± 0.005 with 95% confidence. How many measurements must be averaged to comply with this request?

The desired width w is 2×0.005, or 0.01. For 95% confidence, $\alpha = 0.05$ and $\alpha/2 = 0.025$. Table D gives $z^* = 1.960$. (This is slightly more exact than the $z^* = 2$ from the 68–95–99.7 rule.) Therefore,

$$n = \left(\frac{2z^* \sigma}{w}\right)^2 = \left(\frac{2 \times 1.96 \times .0068}{.01}\right)^2 = 7.1$$

Since 7 measurements will give a slightly wider interval than desired and 8 measurements a slightly narrower interval, the lab must take 8 measurements on each specimen to meet the customer's demand. If this is too expensive, the lab can refuse the contract or raise the price. Signing a contract before doing the calculation of n would be foolish. ∎

Some cautions   We have already seen that short intervals and high confidence are not free. You should also be keenly aware that *any formula for inference is correct only in specific circumstances.* If the government required statistical procedures to carry warning labels like those on drugs, most inference methods would have long labels indeed. Our handy formula $\bar{x} \pm z^*\sigma/\sqrt{n}$ for estimating a normal mean comes with the following list of warnings for the user:

- The data must be an SRS from the population. We are completely safe if we actually did a randomization and drew an SRS. We are not in great danger if the data can plausibly be thought of as independent observations from a population. That is the case in Examples 7.2 to 7.4, where we have in mind the population resulting from a very large number of repeated analyses of the same specimen.
- The formula is not correct for probability sampling designs more complex than an SRS. Correct methods for other designs are available. We will not discuss confidence intervals based on multistage or stratified samples. If you plan such samples, be sure that you (or your statistical consultant) know how to carry out the inference you desire.
- There is no correct method for inference from data haphazardly collected with bias of unknown size. Fancy formulas cannot rescue badly produced data.
- Since $\bar{x}$ is not resistant, outliers can have a large effect on the confidence interval. You should search for outliers and try to correct them or justify their removal before computing the interval. If the outliers cannot be removed, ask your statistical consultant about procedures that are not sensitive to outliers.
- If the sample size is small and the population is not normal, the true confidence level will be different from the value $1 - \alpha$ used in computing the interval. Make a normal quantile plot to check normality. The interval relies only on the distribution of $\bar{x}$, which even for quite small sample sizes is much closer to normal than that of the individual observations. When $n \geq 15$, the confidence level is not greatly disturbed by nonnormal populations unless extreme outliers or quite strong skewness are present. We will discuss this issue in more detail in Chapter 8.

> ▪ You must know the standard deviation σ of the population. This unrealistic requirement renders the interval $\bar{x} \pm z^*\sigma/\sqrt{n}$ of little use in statistical practice. We will learn in the next chapter what to do when σ is unknown.

The most important caution concerning confidence intervals is a consequence of the first of these warnings. *The margin of error in a confidence interval covers only random sampling errors.* The margin of error is obtained from the sampling distribution and indicates how much error can be expected because of chance variation in randomized data production. Practical difficulties such as undercoverage and nonresponse in a sample survey can cause additional errors that may be larger than the random sampling error. Remember this unpleasant fact when reading the results of an opinion poll or other sample survey. The practical conduct of the survey influences the trustworthiness of its results in ways that are not described by the announced margin of error.

Every inference procedure that we will meet has its own list of warnings. Because many of the warnings are similar to those above, we will not print the full warning label each time. It is easy to state (from the mathematics of probability) conditions under which a method of inference is exactly correct. These conditions are *never* fully met in practice. For example, no population is exactly normal. Deciding when a statistical procedure should be used in practice often requires judgment assisted by exploratory analysis of the data. Mathematical facts are therefore only a part of statistics. The difference between statistics and mathematics can be stated thus: mathematical theorems are true; statistical methods are often effective when used with skill.

Finally, you should understand what statistical confidence does not say. We are 95% confident that the mean SAT-M score for the California students in Example 7.1 lies between 452 and 470. This says that these numbers were calculated by a method that gives correct results in 95% of all possible samples. It does *not* say that the probability is 95% that the true mean falls between 452 and 470. No randomness remains after we draw a particular sample and get from it a particular interval. The true mean either is or is not between 452 and 470. Probability in its interpretation as long-term relative frequency makes no sense in this situation. The probability calculations of standard statistical inference describe how often the *method* gives correct answers.

SUMMARY

A level $1 - \alpha$ **confidence interval** for a parameter θ is an interval computed from the data that contains the unknown true value of θ in a proportion $1 - \alpha$ of all samples.

A confidence level states the probability that the method will give a correct answer. That is, if you use 95% confidence intervals often, in the long run 95% of your intervals will contain the true parameter value. You cannot know whether the result of applying a confidence interval to a particular set of data is correct.

A level $1 - \alpha$ confidence interval for the mean μ of a normal population with known standard deviation σ, based on an SRS of size n, is given by

$$\bar{x} \pm z^* \frac{\sigma}{\sqrt{n}}$$

Here z^* is the **upper normal critical value** for $p = \alpha/2$, given in Table D.

Other things being equal, the width of a confidence interval decreases as

- the confidence level $1 - \alpha$ decreases,
- the sample size n increases, and
- the population standard deviation σ decreases.

The sample size required to obtain a confidence interval of specified width w for a normal mean is

$$n = \left(\frac{2z^*\sigma}{w}\right)^2$$

where z^* is the critical point for the desired level of confidence.

A particular form of confidence interval is correct only under specific conditions. The most important conditions concern the method used to produce the data. Other factors such as the form of the population distribution may also be important.

SECTION 7.1 EXERCISES

7.1 The weights of a random sample of 24 male runners are measured. The sample mean is $\bar{x} = 60$ kilograms (kg). Suppose that the standard deviation of the population is known to be $\sigma = 5$ kg.

    (a) What is $\sigma(\bar{x})$, the standard deviation of $\bar{x}$?

    (b) Give a 95% confidence interval for μ, the mean of the population from which the sample is drawn.

7.2 Fifty plots are planted with a new variety of corn. The average yield for these plots is $\bar{x} = 130$ bushels per acre. Assume that $\sigma = 10$.

    (a) Find the 90% confidence interval for the mean yield μ for this variety of corn.

(b) Find the 95% confidence interval.

(c) Find the 99% confidence interval.

(d) Calculate the widths of the intervals found in (a), (b), and (c). How do these widths change as the confidence level increases?

7.3 Find a 99% confidence interval for the mean weight μ of the population of male runners in Exercise 7.1. Is the 99% confidence interval wider or narrower than the 95% interval found in Exercise 7.1? Explain in plain language why this is true.

7.4 Refer to Exercise 7.2. Suppose that the same value of $\bar{x}$ were obtained, but from a sample of 100 plots rather than 50.

(a) Compute the 95% confidence interval for the mean yield μ.

(b) Find the width of this interval. Is the interval wider or narrower than the interval found for the sample of 50 plots in Exercise 7.2? Explain in plain language why this is true.

(c) Do you think that the 90% and 99% intervals for a sample of size 100 will be wider or narrower than those for $n = 50$? Verify your answer by calculating these intervals.

7.5 A test for the level of potassium in the blood is not perfectly precise. Moreover, the actual level of potassium in a person's blood varies slightly from day to day. Suppose that repeated measurements for the same person on different days vary normally with $\sigma = 0.2$.

(a) Julie's potassium level is measured once. The result is $x = 3.2$. Give a 90% confidence interval for her mean potassium level.

(b) If three measurements were taken on different days and the mean result is $\bar{x} = 3.2$, what is a 90% confidence interval for Julie's mean blood potassium level?

7.6 The Acculturation Rating Scale for Mexican Americans (ARSMA) is a psychological test developed to measure the degree of Mexican/ Spanish versus Anglo/English acculturation of Mexican-Americans. The distribution of ARSMA scores in a population used to develop the test was approximately normal, with mean 3.0 and standard deviation 0.8. A further study gave ARSMA to 42 first-generation Mexican-Americans. The mean of their scores was $\bar{x} = 2.13$. If the standard deviation for the first-generation population is also $\sigma = 0.8$, give a 95% confidence interval for the mean ARSMA score for first-generation Mexican Americans.

7.7 What value of z^* from Table D is required for a 98% confidence interval? For an 80% confidence interval?

7.8 What value of z^* from Table D is required for a 70% confidence interval? For a 99.9% confidence interval?

7.9 The crankshaft dimension data in Exercise 1.11 come from a production process that is known to have standard deviation

$\sigma = 0.060$ millimeter (mm). A normal quantile plot (Exercise 1.95) shows that the distribution is very close to normal. The process mean is supposed to be $\mu = 224$ mm but can drift away from this target during production. Give a 95% confidence interval for the process mean at the time these crankshafts were produced.

7.10 Exercise 1.23 gives the Degree of Reading Power (DRP) scores for a sample of 44 third grade students. The normal quantile plot in Figure 1.23 shows that the distribution is close to normal except for slightly short tails. Suppose that the standard deviation of the population of DRP scores is known to be $\sigma = 11$. Give a 99% confidence interval for the population mean score.

7.11 We wish to give a confidence interval for the mean weight in pounds of male runners, based on the interval given in Exercise 7.1 for the mean weight in kilograms. Note that 1 kg = 2.2 pounds. It seems reasonable to simply multiply the endpoints of the interval obtained in Exercise 7.1 by 2.2. The facts about changes in the unit of measurement discussed in Section 1.2 allow us to verify that this is in fact correct. Let X be the weight in kilograms, and $Y = 2.2X$, the same weight measured in pounds.

(a) Compute $\bar{y}$ from the fact that $\bar{x} = 60$.
(b) Compute σ_Y from the fact that $\sigma_X = 5$, using the rule $\sigma_{bX}^2 = b^2\sigma_X^2$.
(c) Compute $\sigma_{\bar{y}}$ from the value σ_Y that you found in (b). How could you have computed it more directly using $\sigma_{\bar{x}}$?
(d) Find the 95% confidence interval for μ_Y, the population mean weight expressed in pounds.
(e) Multiply by 2.2 the endpoints of the 95% confidence interval for μ_X obtained in Exercise 7.1. Is the result the same as the answer you found in (d)?

7.12 A pilot study was done to obtain information to be used in planning a large study of the reading ability of third grade children. The sample standard deviation s was found to be 12. The researchers would like to be able to construct a 95% confidence interval for the mean reading ability of the population of third graders that they will sample in the large study. They would like the width of the confidence interval to be 10 or less. Based on the pilot study, they take $\sigma = 12$ in preliminary calculations.

(a) The researchers' budget will allow as many as 100 students in the large study. Calculate the width of the 95% confidence interval for the population mean based on $n = 100$.
(b) There are many other demands on the research budget. If all of these demands were met, there would be funds to measure only 10 children. Calculate the width of the confidence interval based on $n = 10$ measurements.

(c) Find the smallest value of n that would satisfy the researchers' goal of a 95% confidence interval with width 10 or less. Is this value of n within the limits of the budget?

7.13 How large a sample of the crankshafts in Exercise 7.9 would be needed to estimate the mean μ within ± 0.020 mm with 95% confidence?

7.14 In Exercises 7.2 and 7.4, we compared confidence intervals based on corn yields from 50 and 100 small plots of ground. These are large sample sizes for an agricultural field experiment. How large a sample is required to estimate the mean yield within ± 5 bushels per acre with 90% confidence?

7.15 To assess the accuracy of a laboratory scale, a standard weight known to weigh 10 grams is weighed repeatedly. The scale readings are normally distributed, with unknown mean (this mean is 10 grams if the scale has no bias). The standard deviation of the scale readings is known to be 0.0002 gram.

(a) The weight is weighed five times. The mean result is 10.0023 grams. Give a 98% confidence interval for the mean of repeated measurements of the weight.

(b) How many measurements must be averaged to get a margin of error of ± 0.0001 with 98% confidence?

7.16 Family income was one of the items included only on the "long form" in the 1980 Census. This form was sent to a random sample of 20% of the nation's households. Suppose (alas, it is too simple to be true) that the households that returned the long form are an SRS of the population of all households in each district. In Middletown, a city of 45,000 persons, 2621 households reported their income. The mean of the responses was $\bar{x} = \$23,453$, and the standard deviation was $s = \$8721$. The sample standard deviation for so large a sample will be very close to the population standard deviation σ. Use these facts to give an approximate 99% confidence interval for the 1980 mean household income in Middletown.

7.17 A 1987 Gallup Poll asked 1571 adults what they considered to be the most serious problem facing the nation's public schools; 30% said drugs. This sample percent is an estimate of the percent of all adults who think that drugs are the schools' most serious problem. The news article reporting the poll result adds, "The poll has a margin of error—the measure of its statistical accuracy—of three percentage points in either direction; aside from this imprecision inherent in using a sample to represent the whole, such practical factors as the wording of questions can affect how closely a poll reflects the opinion of the public in general." (From *The New York Times*, August 31, 1987.)

The Gallup Poll uses a complex multistage sample design, but the sample percent has approximately a normal distribution. Moreover, it is standard practice to announce the margin of error for a 95% confidence interval when no other confidence level is stated.

(a) The announced poll result was 30% ± 3%. Can we be certain that the true population percent falls in this interval?

(b) Use the information given to state a 95% confidence interval for the percent of all adults who think that drugs are the schools' most serious problem.

(c) The confidence interval has the same form we have met earlier:

$$\text{estimate} \pm z^* \sigma_{\text{estimate}}$$

(Actually σ is estimated from the data, but we ignore this for now.) What is the standard deviation σ_{estimate} of the estimated percent?

(d) Does the announced margin of error include errors due to the practical problems mentioned in Section 4.3, such as undercoverage and nonresponse?

7.18 When the statistic that estimates an unknown parameter has a normal distribution, a confidence interval for the parameter has the form

$$\text{estimate} \pm z^* \sigma_{\text{estimate}}$$

In a complex sample survey design, the appropriate unbiased estimate of the population mean may not be the sample mean of all responses. Similarly, the standard deviation of the estimate may require an elaborate computation. But when the estimate is known to have a normal distribution and its standard deviation is given, we can calculate a confidence interval for μ from complex sample designs without knowing the formulas that led to the numbers given.

A report based on the Current Population Survey estimates the 1981 mean household income as $22,787 and also reports that the standard deviation of this estimate is $83. The Current Population Survey uses an elaborate multistage sampling design to select a sample of over 70,000 households. The sampling distribution of the estimated mean income is approximately normal. The value $83 computed from so large a sample will be very close to the true standard deviation of the estimate of the mean income. Give a 95% confidence interval for the 1981 mean income of all American households. (Remember that the mean income is certainly higher than the median income and for some purposes is a poorer descriptive measure.)

7.19 As we prepare to take a sample and compute a 95% confidence interval, we know that the probability that the interval we compute will cover the parameter is 0.95. That's the meaning of 95%

confidence. If we use several such intervals, however, our confidence that *all* give correct results is less than 95%.

In an agricultural field trial a corn variety is planted in seven separate locations, which may have different mean yields due to differences in soil and climate. At the end of the experiment, seven independent 95% confidence intervals will be calculated, one for the mean yield at each location.

(a) What is the probability that every one of the seven intervals covers the true mean yield at its location? This probability (expressed as a percent) is our overall confidence level for the seven simultaneous statements.

(b) What is the probability that at least six of the seven intervals cover the true mean yields?

7.2 TESTS OF SIGNIFICANCE

Confidence intervals are one of the two most common types of formal statistical inference. They are appropriate when our goal is to estimate a population parameter. The second common type of inference is directed at a quite different goal: to assess the evidence provided by the data in favor of some statement. An example will illustrate the reasoning that is used.

EXAMPLE 7.5

Does the result of the 1970 draft lottery provide strong evidence that the lottery was not truly random? We examined this question in Chapter 3. A scatterplot of draft numbers against birth dates (Figure 7.6) does not show a convincing non-random pattern. A statistic that measures the strength of the association between draft number (1 to 366) and birth date (1 to 366) is the correlation coefficient. The computer tells us that $r = -0.226$ for the 1970 lottery. Is this the evidence we seek?

> *Formal question*: Suppose for the sake of argument that the lottery were truly random. What is the probability that a random lottery would produce an r at least as far from 0 as the observed $r = -0.226$?

> *Answer*: The probability that a random lottery will produce an r this far from 0 is less than 0.001.

> *Conclusion*: Since an r as far from 0 as that observed in 1970 would almost never occur in a random lottery, we have strong evidence that the 1970 lottery was not random. ∎

In a random assignment of draft numbers to birth dates, we would expect the correlation to be close to 0. The observed correlation for the 1970 lottery was $r = -0.226$, showing that men born later in the year tended to get lower draft numbers. Common sense cannot decide if $r = -0.226$ means that the lottery was not random. After all, the correlation in a random lottery

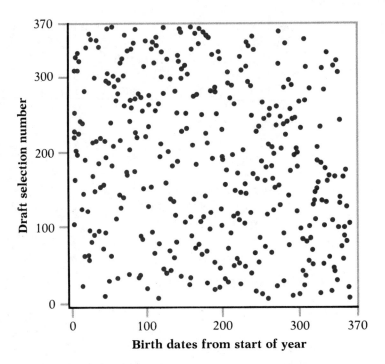

Figure 7.6 Scatterplot of birth date (1 to 366) and draft number (1 to 366) for the 1970 draft lottery.

will almost never be exactly 0. Perhaps $r = -0.226$ is within the range of values that could plausibly occur. As an aid to answering the informal question, "Is this good evidence of a nonrandom lottery?" we state a formal question about probability. We ask just how often a random lottery will produce an r as far from 0 as the r observed in 1970. We find that if a random draft lottery were run each year, a correlation as strong as that observed in 1970 would occur less than once in a thousand years.[1] This convinces us that the 1970 lottery was biased.

Be sure you understand why this probability calculation is convincing. There are two possible explanations for that notorious $r = -0.226$.

1 The lottery was random, and by bad luck a very unlikely outcome occurred.

2 The lottery was biased, so the outcome is about what would be expected from such a lottery.

We cannot be certain that the first explanation is untrue. The 1970 results *could* be due to chance alone. But the probability that such results will occur by chance in a random lottery is so small (0.001) that we are quite confident that the second explanation is correct. Here is another example of this reasoning.

EXAMPLE 7.6

Cobra Cheese Company buys milk from several suppliers as the essential raw material for its cheese. Cobra suspects that some producers are adding water to their milk to increase their profits. Excess water can be detected by determining the freezing point of the milk. The freezing temperature of natural milk varies normally, with a mean of $\mu = -0.545°$ Celsius (C) and a standard deviation of $\sigma = 0.008°$C. Added water raises the freezing temperature. Cobra's laboratory manager measures the freezing temperature of five consecutive lots of milk from one producer. The mean measurement is $\bar{x} = -0.538°$C. Is this good evidence that this producer is adding water to the milk?

> *Formal question*: Suppose for the sake of argument that no water has been added, so that the mean freezing point of the population of all milk from this producer is $\mu = -0.545°$C. What is the probability that five measurements would give a sample mean as high as $-0.538°$C or higher?
>
> *Answer*: An outcome this high or higher has probability 0.025 if natural milk is measured.
>
> *Conclusion*: Since a mean freezing temperature as high as that observed would occur only 2.5 times per 100 samples of natural milk, there is evidence that the producer is watering the milk. ■

The evidence in Example 7.6 is less strong than the evidence in Example 7.5, because the observed effect is more likely to occur simply by chance.

The Nature of Significance Testing

The reasoning used in Examples 7.5 and 7.6 is codified in *tests of significance*.* In both cases, we ask whether some effect is present—whether the draft lottery is biased in Example 7.5 and whether the freezing point of the milk is elevated in Example 7.6. To do this, we begin by supposing for the sake of argument that the effect is *not* present. In Example 7.5 we suppose that the lottery is random, not biased. In Example 7.6, we suppose that the freezing point of the suspect milk is the same as that of natural milk. We then see if the data provide evidence against the supposition we made. If so, we have evidence in favor of the effect we are seeking. The first step in a test of significance is to state a claim that we will try to find evidence *against*.

Null hypothesis

> The statement being tested in a test of significance is called the null hypothesis. The test of significance is designed to assess the strength of the evidence against the null hypothesis. Usually the null hypothesis is a statement of "no effect" or "no difference."

* The reasoning of significance tests has been used sporadically at least since Laplace in the 1820s. A clear exposition by the English statistician Francis Y. Edgeworth (1845–1926) appeared in 1885 in the context of a study of social statistics. Edgeworth introduced the term "significant" as meaning "corresponds to a real difference in fact."

Stating hypotheses   The term "null hypothesis" is abbreviated H_0. It is a statement about a population, expressed in terms of some parameter or parameters. For example, if μ is the mean freezing point of all milk shipped to the cheesemaker by the milk producer in Example 7.6, our null hypothesis is

$$H_0: \mu = -.545$$

This says that the mean freezing point of the milk shipped is the same as for unwatered milk.

It is convenient also to give a name to the statement we hope or suspect is true instead of H_0. This is called the *alternative hypothesis* and is abbreviated by H_a. In Example 7.6, the alternative hypothesis states that the mean freezing point of the suspect milk is higher than that of unwatered milk. We write this as

alternative
hypothesis

$$H_a: \mu > -.545$$

The hypotheses in Example 7.5 refer to the randomization procedure used in the 1970 draft lottery or, if you prefer, to the hypothetical population of all lottery outcomes that would arise from repeated use of this procedure. Let ρ be the correlation between birth date and draft number in this population, averaged over all of the outcomes. The null hypothesis states that the lottery is truly random, so that on the average there is no correlation. That is,

$$H_0: \rho = 0$$

The alternative is a biased lottery with

$$H_a: \rho \neq 0$$

Hypotheses always refer to some population, not to a particular outcome. For this reason, we must state H_0 and H_a in terms of population parameters.

Because H_a expresses the effect that we hope to find evidence *for*, we often begin with H_a and then set up H_0 as the statement that the hoped-for effect is not present. Nonetheless, stating H_a is often the more difficult task. It is not always clear, in particular, whether H_a should be *one-sided* or *two-sided*. In the draft lottery example, the alternative $H_a: \rho \neq 0$ is two-sided. That is, it allows the lottery to give men born later in the year either higher ($\rho > 0$) or lower ($\rho < 0$) draft numbers than men with earlier birth dates. This H_a simply says that the lottery procedure is biased without specifying the direction of the bias. The alternative $H_a: \mu > -0.545$ in the Cobra Cheese example is one-sided. Because watering milk always increases the freezing point, we are interested in detecting only an upward shift in μ. The alternative hypothesis should express the hopes or suspicions we bring to the data. It is cheating to first look at the data and then frame H_a to fit what the data show. Thus the fact that the 1970 draft lottery produced a negative correlation between birth date and draft number should not influence our choice of H_a. If you do not have a specific direction firmly in mind in advance, use a two-sided alternative.

The choice of the hypotheses in Example 7.6 as H_0: $\mu = -.545$ and H_a: $\mu > -.545$ deserves a final comment. The cheesemaker is not concerned with the possibility that the milk may have a *lower*-than-normal freezing point, perhaps due to an unusually high cream content. However, we can allow for the possibility that μ is less than $-0.545°C$ by including this case in the null hypothesis. Then we would write

$$H_0: \mu \leq -.545$$

Although this statement is logically satisfying because all possible values of μ are accounted for, only the parameter value in H_0 that is closest to H_a influences the form of the test in all common significance testing situations. We will therefore take H_0 to be the simpler statement that the parameter has a specific value, in this case H_0: $\mu = -0.545$.

test statistic　　**The form of the test**　A significance test is based on a *test statistic* that shows whether or not the data give evidence against the null hypothesis. The recipe for a significance test depends on the hypotheses and on the population distribution. We will learn specific recipes in a number of common situations. Here are some principles that apply to most tests and that help in understanding the form of tests.

- The test is based on a statistic that estimates the parameter that appears in the hypotheses. When H_0 is true, we expect the estimate to take a value near the parameter value specified by H_0.

- Values of the estimate far from the parameter value specified by H_0 give evidence against H_0. The alternative hypothesis determines which direction or directions count against H_0.

EXAMPLE 7.7　　In the draft lottery, Example 7.5, the hypotheses concern ρ, the population correlation

$$H_0: \rho = 0$$
$$H_a: \rho \neq 0$$

The test statistic is the sample correlation r, which estimates the population correlation. When H_0 is true, we expect r to be near zero. Because H_a is two-sided, values of r away from zero in either direction give evidence against the null hypothesis. In the Cobra Cheese case, Example 7.6, the hypotheses are stated in terms of the population mean freezing point:

$$H_0: \mu = -.545$$
$$H_a: \mu > -.545$$

The estimate of μ is the sample mean $\bar{x}$. Because H_a is one-sided on the high side, only large values $\bar{x}$ count as evidence against the null hypothesis. ∎

P-values A test of significance assesses the evidence against the null hypothesis in terms of probability. If the observed outcome is unlikely under the supposition that the null hypothesis is true and is more probable if the alternative hypothesis is true, that outcome is evidence against H_0 in favor of H_a. The less probable the outcome is, the stronger the evidence that H_0 is false. Usually, of course, any *specific* outcome has low probability. A random draft lottery is unlikely to give exactly $r = 0$, but if we observe $r = 0$, we certainly do not have evidence against the null hypothesis that the lottery is random. The alternative hypothesis determines what kinds of outcomes count as evidence against H_0 and in favor of H_a. In the draft lottery case, observed correlations r away from 0 in either direction count against the hypothesis of a random lottery. The farther from 0 the observed r is, the stronger the evidence. The probability that measures the strength of the evidence that the 1970 lottery was biased is therefore the probability that a random lottery would produce an r *at least as far from 0* as the 1970 lottery did. This is

$$P(r \leq -.226 \text{ or } r \geq .226)$$

The probability calculation is done assuming that H_0 is true.

In general, a test of significance finds the probability of getting an outcome *at least as extreme as the actually observed outcome*. "Extreme" means "far from what we would expect if H_0 were true." The direction or directions that count as "far from what we would expect" are determined by H_a as well as H_0. In Example 7.5, an observed r away from 0 in either direction is evidence of a nonrandom lottery, because H_a is two-sided. In Example 7.6, we want to know if the mean freezing point has been raised, a one-sided H_a. So the evidence against H_0 is measured by the probability

$$P(\bar{x} \geq -.538)$$

that the sample mean freezing point would be *at least as high* as the observed value. This probability is calculated under the assumption that the population mean remains at $\mu = -0.545$.

P-value

> The probability, computed assuming that H_0 is true, that the test statistic will take a value at least as extreme as that actually observed is called the *P*-value of the test. The smaller the *P*-value is, the stronger is the evidence against H_0 provided by the data.

Computer software that carries out tests of significance usually calculates the *P*-value for you. In some cases we can find *P*-values by using our knowledge of sampling distributions.

EXAMPLE 7.8

In Example 7.6 we observed $\bar{x} = -0.538$. We are given the information that the observations are an SRS of size $n = 5$ from a normal population with $\sigma = 0.008$. The P-value for testing

$$H_0: \mu = -.545$$
$$H_a: \mu > -.545$$

is

$$P(\bar{x} \geq -.538)$$

calculated assuming that H_0 is true. In this case, $\bar{x}$ has the normal distribution with

$$\mu_{\bar{x}} = \mu = -.545$$
$$\sigma_{\bar{x}} = \frac{\sigma}{\sqrt{n}} = \frac{.008}{\sqrt{5}}$$

The P-value is found by a normal probability calculation as follows:

$$P(\bar{x} \geq -.538) = P\left[\frac{\bar{x} - (-.545)}{.008/\sqrt{5}} \geq \frac{-.538 - (-.545)}{.008/\sqrt{5}}\right]$$
$$= P(Z \geq 1.96)$$
$$= 1 - .9750 = .025$$

This is the value that was reported in Example 7.6. Figure 7.7 illustrates the P-value in terms of the sampling distribution of $\bar{x}$ when H_0 is true. ∎

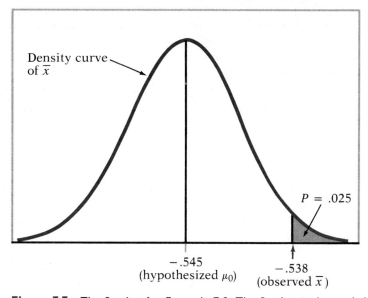

Figure 7.7 The P-value for Example 7.8. The P-value is the probability (when H_0 is true) that $\bar{x}$ takes a value as large as the actually observed value.

Statistical significance   One final step is sometimes taken to assess the evidence against H_0. We can compare the P-value we obtained to a fixed value that we regard as decisive. This amounts to announcing in advance how much evidence against H_0 we will insist on. The decisive value of P is **significance level** called the *significance level*. It is denoted by α. If we choose $\alpha = 0.05$, we are requiring that the data give evidence against H_0 so strong that it would happen no more than 5% of the time (one time in twenty) when H_0 is true. If we choose $\alpha = 0.01$, we are insisting on stronger evidence against H_0, evidence so strong that it would appear only 1% of the time (one time in a hundred) if H_0 is in fact true.

Statistical significance

> If the P-value is as small or smaller than α, we say that the data are statistically significant at level α.

"Significant" in the statistical sense does not mean "important." The original meaning of the word is "signifying something." In statistics the term is used to indicate only that the evidence against the null hypothesis reached the standard set by α. Significance at level 0.01 is often expressed by the statement "The results were significant ($P < 0.01$)." Here P stands for the P-value.

A recipe for testing the significance of the evidence against a null hy- **test of** pothesis is called a *test of significance*. The steps common to all tests of **significance** significance are as follows:

1 State the *null hypothesis* H_0 and the *alternative hypothesis* H_a. The test is designed to assess the strength of the evidence against H_0. H_a is the statement that we will accept if the evidence enables us to reject H_0.

2 (Optional) Specify the *significance level* α. This states how much evidence against H_0 we will regard as decisive.

3 Calculate the value of the *test statistic* on which the test will be based. This is a statistic that measures how well the data conform to H_0.

4 Find the *P-value* for the observed data. This is the probability, calculated assuming that H_0 is true, that the test statistic will weigh against H_0 at least as strongly as it does for these data. If the P-value is less than or equal to α, the test result is *statistically significant at level α*.

We will learn the details of many tests of significance in the following chapters. In most cases, Steps 3 and 4 of our outline are almost automatic once you have practiced a bit. The proper test statistic is determined by the hypotheses and the data collection design, and its numerical value is found by computer software or with a calculator. The computation of the *P*-value is done by computer software or from tables. The computer will not formulate your hypotheses for you, however. Nor will it decide if significance testing is appropriate or help you to interpret the *P*-value that it presents to you. The reasoning of significance tests is a bit subtle. We have now looked at this reasoning with little attention to the details of carrying out a test. Next we will examine the details of a simple significance test, the one that is appropriate in Example 7.6.

Tests for a Population Mean

The cheesemaker in Example 7.6 measured the freezing point of each of a sample of five shipments from a milk producer. If the milk is not watered, the mean freezing temperature should be $-0.545°C$, while watered milk has a higher freezing point. To test the suspicion that the producer is watering the milk, we take

$$H_0: \mu = -.545$$
$$H_a: \mu > -.545$$

We estimate the unknown μ by the sample mean $\bar{x}$. Because we are attempting to detect an increase in μ, values of $\bar{x}$ larger than -0.545 are evidence against H_0. We observed $\bar{x} = -0.538$. The *P*-value is the probability of an $\bar{x}$ at least this large. That is,

$$P = P(\bar{x} > -.538)$$

We calculated this probability in Example 7.8 by standardizing $\bar{x}$ in order to use the standard normal table. Because normal calculations require standardized variables, we will express the test statistic in standardized form to give a general recipe for the test.

We want to test the hypothesis that the mean of a normal population has a specified value. Call the specified value μ_0. (In the cheesemaking example, $\mu_0 = -0.545$.) The null hypothesis is

$$H_0: \mu = \mu_0$$

As the test statistic, we use the *standardized* sample mean

$$z = \frac{\bar{x} - \mu_0}{\sigma/\sqrt{n}}$$

The statistic z has the standard normal distribution when H_0 is true. If the alternative is one-sided on the high side

$$H_a: \mu > \mu_0$$

then the *P*-value is the probability that a standard normal random variable *Z* takes a value at least as large as the observed *z*. That is,

$$P = P(Z \geq z)$$

In Example 7.8, the standardized sample mean was $z = 1.96$ and the *P*-value was $P(Z \geq 1.96)$.

Similar reasoning applies when the alternative hypothesis states that the true μ lies below the hypothesized μ_0 (one-sided). When H_a states that μ is simply unequal to μ_0 (two-sided), values of *z* away from zero in either direction count against the null hypothesis. The *P*-value is the probability that a standard normal *Z* is at least as far from zero as the observed *z*. For example, if $z = 1.7$ is observed, the *P*-value is the probability that $Z \leq -1.7$ or $Z \geq 1.7$. We can write this more briefly in terms of the absolute value $|Z|$ of *Z* as $P(|Z| \geq 1.7)$. Because the standard normal distribution is symmetric, we calculate this probability by finding $P(Z \geq 1.7)$ and *doubling* it.

$$P(|Z| \geq 1.7) = 2P(Z \geq 1.7)$$
$$= 2(1 - .9554) = .0892$$

Here is a statement of the test in general terms.

z test for a population mean

> To test the hypothesis $H_0: \mu = \mu_0$ based on an SRS of size *n* from a population with unknown mean μ and known standard deviation σ, compute the standardized sample mean
>
> $$z = \frac{\bar{x} - \mu_0}{\sigma/\sqrt{n}}$$
>
> In terms of a standard normal random variable *Z*, the *P*-value for a test of H_0 against
>
> $$H_a: \mu > \mu_0 \quad \text{is} \quad P(Z \geq z)$$
> $$H_a: \mu < \mu_0 \quad \text{is} \quad P(Z \leq z)$$
> $$H_a: \mu \neq \mu_0 \quad \text{is} \quad P(|Z| \geq |z|)$$
>
> These *P*-values are exact if the population distribution is normal and are approximately correct for large *n* in other cases.

EXAMPLE 7.9

Do middle-aged male executives have higher or lower blood pressure than the general population? The mean systolic blood pressure for white males 35 to 44 years of age is 127 and the standard deviation in this population is 7. The medical director of a company looks at the medical records of 72 company executives in this age group and finds that the mean systolic blood pressure in this sample is 126.1. Is this evidence that executive blood pressures differ from the national average?

The null hypothesis is "no difference" from the national mean $\mu_0 = 127$. The alternative is two-sided, because the medical director did not have a particular direction in mind before examining the data. So the hypotheses about the unknown mean μ of the executive population are

$$H_0: \mu = 127$$
$$H_a: \mu \neq 127$$

As usual in this chapter, we make the unrealistic assumption that the population standard deviation is known, in this case that executives have the same $\sigma = 7$ as the general population of middle-aged white males. We also assume that the 72 executives in the sample are an SRS from the population of all middle-aged white male executives in the company. This assumption may be realistic but must be checked by asking how the data were produced. If medical records are available only for executives with recent medical problems, for example, the data are of little value for our purpose. It turns out that all executives are given a free annual medical exam and that the medical director selected 72 exam results at random.

We can now proceed to compute the standardized sample mean

$$z = \frac{\bar{x} - \mu_0}{\sigma/\sqrt{n}} = \frac{126.1 - 127}{7/\sqrt{72}}$$
$$= -1.09$$

Figure 7.8 illustrates the P-value, which is the probability that a standard normal variable Z takes a value at least 1.09 away from zero. From Table A we find that

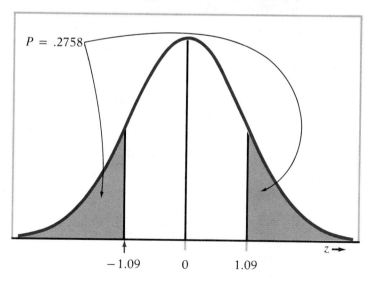

Figure 7.8   The P-value for Example 7.9. The P-value is the probability that the z statistic takes a value at least as far from 0 as the actually observed value.

this probability is

$$P(|Z| \geq 1.09) = 2P(Z \geq 1.09)$$
$$= 2(1 - 8621) = 2758$$

That is, an SRS of size 72 from the general male population would have a mean blood pressure at least as far from 127 as did the executive sample more than 27% of the time. The observed $\bar{x} = 126.1$ is therefore not good evidence that executives differ from other men. ∎

The data in Example 7.9 do *not* establish that the mean blood pressure μ for this company's executives is 127. We sought evidence that μ differed from 127 and failed to find convincing evidence. That is all we can say. No doubt the mean blood pressure of the entire executive population is not exactly equal to 127. A large enough sample would give evidence of the difference, even if it is very small. Tests of significance assess the evidence *against* H_0. If the evidence is strong, we can confidently reject H_0 in favor of the alternative. Failing to find evidence against H_0 means only that the data are consistent with H_0, not that we have clear evidence that H_0 is true.

EXAMPLE 7.10

In a discussion of SAT scores, someone comments: "Because only a minority of high school students take the test, the scores overestimate the ability of typical high school seniors. The mean SAT mathematics score is about 475, but I think that if all seniors took the test, the mean score would be no more than 450." You gave the test to an SRS of 500 seniors from California (Example 7.1). These students had a mean score of $\bar{x} = 461$. Is this good evidence against the claim that the mean for all California seniors is no more than 450?

The hypotheses are

$$H_0: \mu = 450$$
$$H_a: \mu > 450$$

As in Example 7.1, we assume that $\sigma = 100$. The z statistic is

$$z = \frac{\bar{x} - \mu_0}{\sigma/\sqrt{n}} = \frac{461 - 450}{100/\sqrt{500}}$$
$$= 2.46$$

Because H_a is one-sided on the high side, large values of z count against H_0. From Table A, we find that the P-value is

$$P(Z \geq 2.46) = 1 - .9931 = .0069$$

Figure 7.9 illustrates this P-value. It means that a mean score as large as that observed would occur fewer than seven times in 1000 samples if the population mean were no larger than 450. This is convincing evidence that the mean SAT-M score for all California high school seniors is higher than 450. ∎

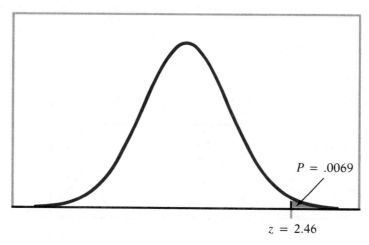

Figure 7.9   The P-value for Example 7.10.

Tests with Fixed Significance Level

We now know how to compute the P-value of a significance test for a mean if the standard deviation is known. Suppose that we require a specific degree of evidence, stated as a significance level α. In terms of the P-value, the outcome of a test is significant at level α if $P \le \alpha$. The following example illustrates an alternative way of assessing significance at a fixed level α that does not require calculating P.

EXAMPLE 7.11

In Example 7.10, we examined whether the mean SAT-M score of California high school seniors is higher than 450. The hypotheses are

$$H_0: \mu = 450$$
$$H_a: \mu > 450$$

Is the evidence against H_0 statistically significant at the 1% level? Compute the standardized value $z = 2.46$ of $\bar{x}$ as before. To be significant at the 1% level, z must lie in the upper 1% of the standard normal distribution. So to determine significance, we need only compare z with the 0.01 critical point $z^* = 2.326$ from Table D. Since z is larger than z^*, the observed $\bar{x}$ is significant at level $\alpha = 0.01$. Figure 7.10 illustrates the procedure. We say that we can *reject H_0* at the 1% level of significance. ∎

The number z^* with probability p falling to the right of it under the standard normal density curve is the *upper p critical value* for the standard normal distribution. Table D gives these critical values for several choices of p. We made use of upper critical values for confidence intervals. Example 7.11 shows how they are used in tests with a fixed level

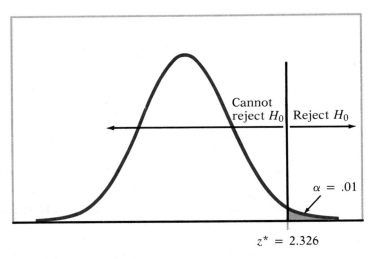

Figure 7.10   The rejection rule for an $\alpha = .01$ one-sided test, Example 7.11.

α with a one-sided alternative hypothesis. Similar reasoning applies to other choices of the alternative hypothesis. Here is a summary of the procedures.

Fixed level α z tests for a population mean

> To test the hypothesis $H_0: \mu = \mu_0$ based on an SRS of size n from a population with unknown mean μ and known standard deviation σ, compute the standardized sample mean
>
> $$ z = \frac{\bar{x} - \mu_0}{\sigma/\sqrt{n}} $$
>
> Reject H_0 at significance level α against a one-sided alternative
>
> $$ H_a: \mu > \mu_0 \quad \text{if} \quad z \geq z^* $$
> $$ H_a: \mu < \mu_0 \quad \text{if} \quad z \leq -z^* $$
>
> where z^* is the upper α critical value from Table D. Reject H_0 at significance level α against a two-sided alternative
>
> $$ H_a: \mu \neq \mu_0 \quad \text{if} \quad |z| \geq z^* $$
>
> where z^* is the upper $\alpha/2$ critical value from Table D.

EXAMPLE 7.12   The analytical laboratory of Example 7.2 is asked to evaluate the claim that the concentration of the active ingredient in a specimen is 0.86%. As in Example 7.2, the mean of three repeated analyses of the specimen is $\bar{x} = 0.8404$, and the

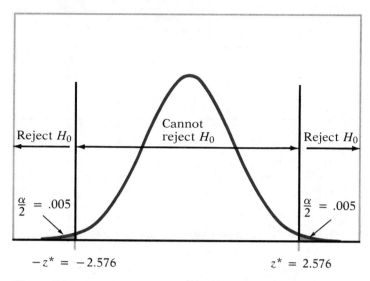

Figure 7.11 The rejection rule for an $\alpha = .01$ two-sided test, Example 7.12.

standard deviation of the analysis process is known to be $\sigma = 0.0068$. The true concentration is the mean μ of the population of repeated analyses. The hypotheses are

$$H_0: \mu = .86$$
$$H_a: \mu \neq .86$$

The lab chooses the 1% level of significance, $\alpha = 0.01$. The z statistic is

$$z = \frac{.8404 - .86}{.0068/\sqrt{3}} = -4.99$$

Because the alternative is two-sided, we compare $|z| = 4.99$ with the $\alpha/2 = 0.005$ critical value from Table D. This critical value is $z^* = 2.576$. The values of z that lead to rejection are illustrated in Figure 7.11. Since $|z| > z^*$, we reject H_0. ∎

The calculation in Example 7.12 for a 1% significance test is very similar to that in Example 7.2 for a 99% confidence interval. In fact, a two-sided test at significance level α can be carried out directly from a level $1 - \alpha$ confidence interval.

Confidence intervals and two-sided tests

A level α two-sided significance test rejects a hypothesis $H_0: \mu = \mu_0$ exactly when the value μ_0 falls outside a level $1 - \alpha$ confidence interval for μ.

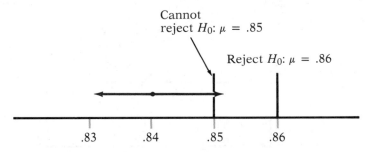

Figure 7.12 Values of μ falling outside a 99% confidence interval can be rejected at the 1% significance level; values falling inside the interval cannot be rejected.

EXAMPLE 7.13

The 99% confidence interval for μ in Example 7.2 was

$$\bar{x} \pm z^* \frac{\sigma}{\sqrt{n}} = .8404 \pm .0101$$

$$= (.8303, .8505)$$

The hypothesized value $\mu_0 = 0.86$ in Example 7.12 falls outside this confidence interval, and so

$$H_0: \mu = .86$$

is rejected at the 1% significance level. On the other hand,

$$H_0: \mu = .85$$

cannot be rejected at the 1% level in favor of the two-sided alternative $H_a: \mu \neq 0.85$, because 0.85 lies inside the 99% confidence interval for μ. Figure 7.12 illustrates both cases. ■

P-values versus fixed α The observed result in Example 7.12 was $z = -4.99$. The conclusion that this result is significant at the 1% level does not tell the whole story. The observed z is far beyond the 1% critical value, and the evidence against H_0 is far stronger than 1% significance suggests. The *P*-value

$$P(|Z| \geq 4.99) = .0000006$$

gives a better sense of how strong the evidence is. *The P-value is the smallest level α at which the data are significant.* Knowing the *P*-value therefore allows us to assess significance at any level.

EXAMPLE 7.14

In Example 7.10, we tested the hypotheses

$$H_0: \mu = 450$$

$$H_a: \mu > 450$$

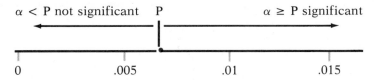

Figure 7.13 An outcome with P-value P is significant at all levels α at or above P and not significant at smaller levels α.

concerning the mean SAT mathematics score μ of California high school seniors. The test had the P-value $P = 0.0069$. This result is significant at the $\alpha = 0.01$ level because $0.0069 \leq 0.01$. It is not significant at the $\alpha = 0.005$ level, because the P-value is larger than 0.005. See (Figure 7.13) ∎

A *P*-value is more informative than a reject-or-not finding at a fixed significance level. But assessing significance at a fixed level α is easier, because no probability calculation is required. You need only look up a critical value in a table. Because the practice of statistics today almost always employs computer software that calculates P-values automatically, the use of tables of critical values is becoming outdated. We include the usual tables of critical values (such as Table D) at the end of the book for learning purposes and to rescue students without good computing facilities. The tables can be used directly to carry out fixed α tests. They also allow us to approximate P-values quickly without a probability calculation. The following example illustrates the use of Table D to find an approximate P-value.

EXAMPLE 7.15

Bottles of a popular cola drink are supposed to contain 300 milliliters (ml) of cola. There is some variation from bottle to bottle because the filling machinery is not perfectly precise. The distribution of the contents is normal with standard deviation $\sigma = 3$ ml. A student who suspects that the bottler is underfilling measures the contents of six bottles. The results are

299.4   297.7   301.0   298.9   300.2   297.0

Is this convincing evidence that the mean contents of cola bottles is less than the advertised 300 ml? The hypotheses are

$$H_0: \mu = 300$$
$$H_a: \mu < 300$$

The sample mean contents of the six bottles measured is $\bar{x} = 299.03$ ml. The z statistic is therefore

$$z = \frac{\bar{x} - \mu_0}{\sigma/\sqrt{n}} = \frac{299.03 - 300}{3/\sqrt{6}}$$
$$= -.792$$

Small values of z count against H_0, because H_a is one-sided on the low side. The *P*-value is

$$P(Z \leq -.792)$$

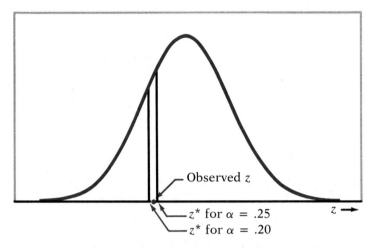

Figure 7.14 The *P*-value of a *z* statistic can be approximated by noting which α levels from Table D it falls between. Here, *P* lies between .20 and .25.

Rather than compute this probability, compare $z = -0.792$ with the critical values in Table D. According to this table, we would

- reject H_0 at level $\alpha = .25$ if $z \leq -.674$,
- reject H_0 at level $\alpha = .20$ if $z \leq -.841$.

As Figure 7.14 illustrates, the observed $z = -0.792$ lies between these two critical points. We can reject at $\alpha = 0.25$ but not at $\alpha = 0.20$. That is the same as saying that the *P*-value lies between 0.25 and 0.20. A sample mean at least as small as that observed will occur in between 20% and 25% of all samples if the population mean is $\mu = 300$ ml. There is no convincing evidence that the mean is below 300, and there is no need to calculate the *P*-value more exactly. ■

SUMMARY

A test of significance is intended to assess the evidence provided by data against a **null hypothesis** H_0 in favor of an **alternative hypothesis** H_a.

The hypotheses are stated in terms of population parameters. Usually H_0 is a statement that no effect is present and H_a says that a parameter differs from its null value, in a specific direction (**one-sided alternative**) or in either direction (**two-sided alternative**).

The test is based on a **test statistic**. The ***P*-value** is the probability, computed assuming that H_0 is true, that the test statistic will take a value at least as extreme as that actually observed. Small *P*-values indicate strong evidence against H_0. Calculating *P*-values requires knowledge of the sampling distribution of the test statistic when H_0 is true.

If the *P*-value is as small or smaller than a specified value α, the data are **statistically significant** at significance level α.

Significance tests for the hypothesis $H_0: \mu = \mu_0$ concerning the unknown mean μ of a population are based on the **z statistic**

$$z = \frac{\bar{x} - \mu_0}{\sigma/\sqrt{n}}$$

The z test assumes an SRS of size n, known population standard deviation σ, and either a normal population or a large sample. *P*-values are computed from the normal distribution (Table A). Fixed α tests use the table of **standard normal critical values** (Table D).

SECTION 7.2 EXERCISES

7.20 Each of the following situations requires a test of hypotheses about a population mean μ. State the appropriate null hypothesis H_0 and alternative hypothesis H_a in each case.

(a) The mean area of the several thousand apartments in a new development is advertised to be 1250 square feet. A tenants group thinks that the apartments are smaller than advertised. They hire an engineer to measure a sample of apartments to test their suspicion.

(b) Larry's car averages 32 miles per gallon on the highway. He now switches to a new motor oil that is advertised as increasing gas mileage. After driving 3000 highway miles with the new oil, he wants to determine if his gas mileage actually has increased.

(c) The diameter of a spindle in a small motor is supposed to be 5 mm. If the spindle is either too small or too large, the motor will not perform properly. The manufacturer measures the diameter in a sample of motors to determine whether the mean diameter has moved away from the target.

7.21 In each of the following situations, a significance test for a population mean μ is called for. State the null hypothesis H_0 and the alternative hypothesis H_a in each case.

(a) Experiments on learning in animals sometimes measure how long it takes a mouse to find its way through a maze. The mean time is 18 seconds for one particular maze. A researcher thinks that a loud noise will cause the mice to complete the maze faster. She measures how long each of 10 mice takes with a noise as stimulus.

(b) The examinations in a large psychology class are scaled after grading so that the mean score is 50. A self-confident teaching assistant thinks that his students have a higher mean score than the class as a whole. His students this semester can be considered a sample from the population of all students he might teach, so he compares their mean score with 50.

(c) A university gives credit in French language courses to students who pass a placement test. The language department wants to know if students who get credit in this way differ in their understanding of spoken French from students who actually take the French courses. Some faculty think the students who test out of the courses are better, but others argue that they are weaker in oral comprehension. Experience has shown that the mean score of students in the courses on a standard listening test is 24. The language department gives the same listening test to a sample of 40 students who passed the credit examination to see if their performance is different.

7.22 In each of the following situations, state an appropriate null hypothesis H_0 and alternative hypothesis H_a. Be sure to identify the parameters that you use to state the hypotheses. (We have not yet learned how to test these hypotheses.)

(a) A sociologist asks a large sample of high school students which academic subject they like best. She suspects that a higher percent of males than of females will name mathematics as their favorite subject.

(b) An educational researcher randomly divides sixth grade students into two groups for physical education class. He teaches both groups basketball skills, using the same methods of instruction in both classes. He encourages Group A with compliments and other positive behavior but acts cool and neutral toward Group B. He hopes to show that positive teacher attitudes result in a higher mean score on a test of basketball skills than do neutral attitudes.

(c) A political scientist hypothesizes that among registered voters there is a negative correlation between age and the percent who actually vote. To test this, she draws a random sample from voter registration records.

7.23 A randomized comparative experiment was conducted to examine whether a calcium supplement in the diet will reduce the blood pressure of healthy men. The subjects received either a calcium supplement or a placebo for 12 weeks. The statistical analysis was quite complex, but one conclusion was that "the calcium group had lower seated systolic blood pressure ($P = .008$) compared with the placebo group." Explain this conclusion, especially the P-value, as if you were speaking to a doctor who knows no statistics. (From R. M. Lyle et al., "Blood pressure and metabolic effects of calcium supplementation in normotensive white and black men," *Journal of the American Medical Association*, 257 (1987), pp. 1772–1776.)

7.24 A social psychologist reports that "in our sample, ethnocentrism was significantly higher ($P < 0.05$) among church attenders than among nonattenders." Explain what this means in language understandable to someone who knows no statistics. Do not use the word "significance" in your answer.

7.25 The financial aid office of a university asks a sample of students about their employment and earnings. The report says that "for academic year earnings, a significant difference ($P = 0.038$) was found between the sexes, with men earning more on the average. No difference ($P = 0.476$) was found between the earnings of black and white students." Explain both of these conclusions, for the effects of sex and of race on mean earnings, in language understandable to someone who knows no statistics. (From a study by M. R. Schlatter et al., Division of Financial Aids, Purdue University.)

7.26 An exercise researcher believes that the mean weight of competitive runners is about 140 pounds. (The mean weight of all men aged 25 to 34 is 173 pounds, but runners are lighter.) A sample of 24 elite distance runners has mean weight $\bar{x} = 136$ pounds. Suppose that these runners are an SRS from the population of elite distance runners and that weight varies normally in this population with standard deviation $\sigma = 11$ pounds. To see if this is evidence that the mean weight of elite distance runners is less than 140 pounds, test

$$H_0: \mu = 140$$
$$H_a: \mu < 140$$

Report the P-value of the test. Is there convincing evidence that the population mean is less than 140 pounds?

7.27 The Survey of Study Habits and Attitudes (SSHA) is a psychological test that measures the motivation, attitude toward school, and study habits of students. Scores range from 0 to 200. The mean score for U.S. college students is about 115, and the standard deviation is about 30. A teacher who suspects that older students have better attitudes toward school gives the SSHA to 25 students who are at least 30 years of age. The mean score is $\bar{x} = 125.2$.

(a) Assuming that $\sigma = 30$ for the population of older students, carry out a test of

$$H_0: \mu = 115$$
$$H_a: \mu > 115$$

Report the P-value of your test, and state your conclusion clearly.

(b) Your test in (a) required two important assumptions in addition to the assumption that the value of σ is known. What are they? Which of these assumptions is most important to the validity of your conclusion in (a)?

7.28 The mean yield of corn in the United States is about 120 bushels per acre. A survey of 50 farmers this year gives a sample mean yield of $\bar{x} = 123.6$ bushels per acre. We want to know whether this is good evidence that the national mean this year is not 120 bushels per acre. Assume that the farmers surveyed are an SRS from the population of all commercial corn growers and that the standard deviation of the yield in this population is $\sigma = 10$ bushels per acre. Give the *P*-value for the test of

$$H_0: \mu = 120$$
$$H_a: \mu \neq 120$$

Are you convinced that the population mean is not 120 bushels per acre? Is your conclusion correct if the distribution of corn yields is somewhat nonnormal? Why?

7.29 In the past, the mean score of the seniors at South High on the American College Testing (ACT) college entrance examination has been 20. This year a special preparation course is offered, and all 43 seniors planning to take the ACT test enroll in the course. The mean of their 43 ACT scores is 21.1. The principal believes that the new course has improved the students' ACT scores.

(a) Assume that ACT scores vary normally with standard deviation 6. Is the outcome $\bar{x} = 21.1$ good evidence that the population mean score is greater than 20? State H_0 and H_a, compute the *P*-value, and answer the question by interpreting your result.

(b) The results are in any case inconclusive because of the design of the study. The effects of the new course are confounded with any change from past years, such as other new courses or higher standards. Briefly outline the design of a better study of the effect of the new course on ACT scores.

7.30 The sample of crankshaft measurements in Exercise 1.11 comes from a production process that is known to vary normally with standard deviation $\sigma = 0.060$ mm. The process mean is supposed to be 224 mm. Do these data give evidence that the process mean is not equal to the target value 224 mm?

(a) State the H_0 and H_a that you will test.

(b) Give the *P*-value of the test. Are you convinced that the process mean is not 224 mm?

7.31 The level of calcium in the blood in healthy young adults varies with mean about 9.5 milligrams per deciliter and standard deviation about $\sigma = 0.4$. A clinic in rural Guatemala has measured the blood calcium level of 180 healthy pregnant women at their first visit for prenatal care. The mean is $\bar{x} = 9.57$. Is this an indication that the mean calcium level in the population from which these women come differs from 9.5?

(a) State H_0 and H_a.

(b) Carry out the test and give the P-value, assuming that $\sigma = 0.4$ in this population. Report your conclusion.

(c) Give a 95% confidence interval for the mean calcium level μ in this population. We are confident that μ lies quite close to 9.5. This illustrates the fact that a test based on a large sample ($n = 180$ here) will often declare even a small deviation from H_0 to be statistically significant.

7.32 Exercise 1.23 gives the Degree of Reading Power (DRP) scores for a sample of 44 third grade students. The students can be considered to be an SRS of the third graders in a suburban school district. DRP scores are approximately normal. Suppose that the standard deviation of scores in this school district is known to be $\sigma = 11$. The researcher suspects that the mean score μ of all third graders in this district is higher than the national mean, which is 32.

(a) State the appropriate H_0 and H_a to test this suspicion.

(b) Carry out the test. Give the P-value, and then interpret the result in plain language.

7.33 A computer has a random number generator designed to produce random numbers that are uniformly distributed within the interval from 0 to 1. If this is true, the numbers generated come from a population with $\mu = 0.5$ and $\sigma = 0.0833$. A command to generate 100 random numbers gives outcomes with mean $\bar{x} = 0.4817$. Assume that the population σ remains fixed. We want to test

$$H_0: \mu = .5$$
$$H_a: \mu \neq .5$$

(a) Calculate the value of the z statistic.

(b) Is the result significant at the 5% level ($\alpha = 0.05$)?

(c) Is the result significant at the 1% level ($\alpha = 0.01$)?

7.34 To determine whether the mean nicotine content of a brand of cigarettes is greater than the advertised value of 1.4 mg, a test of

$$H_0: \mu = 1.4$$
$$H_a: \mu > 1.4$$

is conducted. The calculated value of the test statistic is $z = 2.42$.

(a) Is the result significant at the 5% level?

(b) Is the result significant at the 1% level?

7.35 There are other z statistics that we have not yet studied. The significance of any z statistic is assessed from Table D. A study compares the habits of students who are on academic probation with students whose grades are satisfactory. One variable measured is the hours spent watching television last week. The null hypothesis is "no difference" between the means for the two populations. The alternative hypothesis is two-sided. The value of the test statistic $z = -1.37$.

(a) Is this result significant at the 5% level?

(b) Is the result significant at the 1% level?

7.36 Explain in plain language why a significance test that is significant at the 1% level must always be significant at the 5% level.

7.37 Use Table D to find the approximate *P*-value for the test in Exercise 7.33 without doing a probability calculation. That is, find from the table two numbers that contain the *P*-value between them.

7.38 Use Table D to find the approximate *P*-value for the test in Exercise 7.34. That is, between what two numbers obtained from the table must the *P*-value lie?

7.39 Between what values from Table D does the *P*-value for the outcome $z = -1.37$ in Exercise 7.35 lie? (Remember that H_a is two-sided.) Calculate the *P*-value using Table A, and verify that it lies between the values you found from Table D.

7.40 Exercise 7.1 asks you to find a 95% confidence interval for the mean weight of a population of male runners.

(a) Give the confidence interval from that exercise, or calculate the interval if you did not do the exercise.

(b) Based on this confidence interval, does a test of

$$H_0: \mu = 61.5 \text{ kg}$$
$$H_a: \mu \neq 61.5 \text{ kg}$$

reject H_0 at the 5% significance level?

(c) Would $H_0: \mu = 63$ be rejected at the 5% level if tested against a two-sided alternative?

7.41 Researchers studying the absorption of sugar by insects feed cockroaches a diet containing measured amounts of a particular sugar. After 10 hours, the cockroaches are killed and the concentration of the sugar in various body parts is determined by a chemical analysis. The paper that reports the research states that a 95% confidence interval for the mean amount (in milligrams) of the sugar in the hindgut of the cockroaches is 4.2 ± 2.3. (From D. L. Shankland et al., "The effect of 5-thio-D-glucose on insect development and its absorption by insects," *Journal of Insect Physiology*, 14 (1968), pp. 63–72.)

(a) Does this paper give evidence that the mean amount of sugar in the hindgut under these conditions is not equal to 7 mg? State H_0 and H_a and base a test on the confidence interval.

(b) Would the hypothesis that $\mu = 5$ mg be rejected at the 5% level in favor of a two-sided alternative?

7.42 An old farmer claims to be able to detect the presence of water with a forked stick. In a test of this claim, he is presented with five identical barrels, some containing water and some not. He is right in four of the five cases.

(a) Suppose the farmer has probability p of being correct. If he is just guessing, $p = 0.5$. State an appropriate H_0 and H_a in terms of p for a test of whether he does better than guessing.

(b) If the farmer is simply guessing, what is the distribution of the number X of correct answers in five tries?

(c) The observed outcome is $X = 4$. What is the P-value of the test that takes large values of X to be evidence against H_0?

7.3   USE AND ABUSE OF TESTS

Carrying out a test of significance is often quite simple, especially if a fixed significance level α is used or if the P-value is given effortlessly by a computer. Using tests wisely is not so simple. Each test is valid only in certain circumstances, with properly produced data being particularly important. The z test, for example, should bear the same warning label that was attached in Section 7.1 to the corresponding confidence interval. Similar warnings accompany the other tests that we will learn. There are additional caveats that concern tests more than confidence intervals, enough to warrant this separate section. Some hesitation about the unthinking use of significance tests is a sign of statistical maturity.

Using Significance Tests

The reasoning of significance tests has appealed to researchers in many fields, so that tests are widely used to report research results. In this setting H_a is a "research hypothesis" asserting that some effect or difference is present. The null hypothesis H_0 says that there is no effect or no difference. A low P-value represents good evidence that the research hypothesis is true. Here are some comments on the use of significance tests, with emphasis on their use in reporting scientific research.

Choosing a level of significance   The spirit of a test of significance is to give a clear statement of the degree of evidence provided by the sample against the null hypothesis. This is done by the P-value. But sometimes you will make some decision or take some action if your evidence reaches a certain standard. Such a standard is often set by giving a level of significance α. Perhaps you will announce a new scientific finding if your data are significant at the $\alpha = 0.05$ level. Or perhaps you will recommend using a new method of teaching reading if the evidence of its superiority is significant at the $\alpha = 0.01$ level.

Making a decision is different in spirit from testing significance, though the two are often mixed in practice. Choosing a level α in advance makes sense if you must make a decision, but not if you wish only to describe the strength of your evidence. Using tests with fixed level α for decision making is discussed at greater length at the end of this chapter.

If you do use a fixed level α significance test to make a decision, choose α by asking how much evidence is required to reject H_0. This depends first on how plausible H_0 is. If H_0 represents an assumption that everyone in your field has believed for years, strong evidence (small α) will be needed to reject it. Second, the level of evidence required to reject H_0 depends on the consequences of such a decision. If rejecting H_0 in favor of H_a means making an expensive changeover from one medical therapy or instructional method to another, strong evidence is needed. Both the plausibility of H_0 and H_a and the consequences of any action that rejection may lead to are somewhat subjective. Different persons may feel that different levels of significance are appropriate. When this is the case, it is better to report the P-value, which allows each of us to decide individually if the evidence is sufficiently strong.

Users of statistics have often emphasized certain standard levels of significance, such as 10%, 5%, and 1%. This emphasis reflects the time when tables of critical points rather than computer programs dominated statistical practice. The 5% level ($\alpha = 0.05$) is particularly common. Significance at that level is still a widely accepted criterion for meaningful evidence in research work. *There is no sharp border between "significant" and "insignificant," only increasingly strong evidence as the P-value decreases.* There is no practical distinction between the P-values 0.049 and 0.051. It makes no sense to treat $\alpha = 0.05$ as a universal rule for what is significant.

There is a reason for the common use of $\alpha = 0.05$—the great influence of Sir R. A. Fisher.* Fisher did not originate tests of significance. But because his writings organized statistics, especially as a tool of scientific research, his views on tests were enormously influential. Here is his opinion on choosing a level of significance:

> . . . it is convenient to draw the line at about the level at which we can say: "Either there is something in the treatment, or a coincidence has occurred such as does not occur more than once in twenty trials. . . . "
>
> If one in twenty does not seem high enough odds, we may, if we prefer it, draw the line at one in fifty (the 2 percent point), or one in a hundred (the 1 percent point). Personally, the writer prefers to set a low standard of significance at the 5 percent point, and ignore entirely all results which fail to reach that level. A scientific fact should be regarded as experimentally established only if a properly designed experiment *rarely fails* to give this level of significance.[2]

There you have it. Fisher thought 5% was about right, and who was to disagree with the master? Fisher was of course not an advocate of blind use of

---

* We have met Fisher as the inventor of randomized experimental designs. He also originated many other statistical techniques, derived the distributions of many common statistics, and introduced such basic terms as *parameter* and *statistic* into statistical writing.

significance at the 5% level as a yes-or-no criterion. The last sentence quoted above shows an experienced scientist's feeling for the repeated studies and variable results that mark the advance of knowledge.[3]

What statistical significance doesn't mean When a null hypothesis ("no effect" or "no difference") can be rejected at the usual levels, $\alpha = 0.05$ or $\alpha = 0.01$, there is good evidence that an effect is present. But that effect may be extremely small. When large samples are available, even tiny deviations from the null hypothesis will be significant. For example, suppose that we are testing the hypothesis of no correlation between two variables. With 1000 observations, an observed correlation of only $r = 0.08$ is significant evidence at the $\alpha = 0.01$ level that the correlation in the population is not zero but positive. The low significance level does not mean there is a strong association, only that there is strong evidence of some association. The true population correlation is probably quite close to the observed sample value, $r = 0.08$. We might well conclude that for practical purposes we can ignore the association between these variables, even though we are confident (at the 1% level) that the correlation is positive. Exercise 7.45 demonstrates in detail the effect on P of increasing the sample size. Remember the wise saying: *Statistical significance is not the same thing as practical significance.*

The remedy for attaching too much importance to statistical significance is to pay attention to the actual experimental results as well as to the P-value. Plot your data and examine them carefully. Are there outliers or other deviations from a consistent pattern? A few outlying observations can produce highly significant results if the data are run blindly through common tests of significance. Outliers can also destroy the significance of otherwise convincing data. The foolish user of statistics who feeds the data to a computer without exploratory analysis will often be embarrassed. Is the effect you are seeking visible in your plots? If not, ask yourself if the effect is large enough to be practically important. It is usually wise to give a confidence interval for the parameter in which you are interested. A confidence interval actually estimates the size of an effect, rather than simply asking if it is too large to reasonably occur by chance alone. Confidence intervals are not used as often as they should be, while tests of significance are perhaps overused.

Don't ignore lack of significance Researchers typically have in mind the research hypothesis that some effect exists. Following the peculiar logic of tests of significance, they set up as H_0 the null hypothesis that no such effect exists and try their best to get evidence against H_0. A perverse legacy of Fisher's opinion on $\alpha = 0.05$ is that research in some fields has rarely been published unless significance at that level is attained. For example, a survey of four journals of the American Psychological Association showed that of 294 articles using statistical tests, only 8 did not attain the 5% significance level.[4]

Such a publication policy impedes the spread of knowledge, and not only by declaring that a P-value of 0.051 is "not significant." If a researcher has good reason to suspect that an effect is present and then fails to find

significant evidence of it, that may be interesting news—perhaps more interesting than if evidence in favor of the effect at the 5% level had been found.
If you follow the history of science, you will recall examples such as the
Michelson-Morley experiment, which changed the course of physics by *not*
detecting an expected change in the speed of light. Keeping silent about negative results may condemn other researchers to repeat the attempt to find an
effect that isn't there.

Of course, an experiment that fails only causes a stir if it is clear that
the experiment would have detected the effect if it were really there. An
important aspect of planning a study is to verify that the test you plan to
use does have high probability of detecting an effect of the size you hope to
find. This probability is the *power* of the test. Power calculations are discussed later in this chapter.

Abuse of Significance Tests

Tests of statistical significance are routinely used to assess the results of research in agriculture, education, engineering, medicine, psychology, and
sociology and increasingly in other fields as well. Any tool used routinely is
often used unthinkingly. We therefore offer some comments for the thinking
researcher on possible abuses of this tool. Thinking consumers of research
findings (such as students) should also ponder these comments.

Statistical inference is not valid for all sets of data   We learned long
ago that badly designed surveys or experiments often produce invalid results.
Formal statistical inference cannot correct basic flaws in the design. There is
no doubt a significant difference in English vocabulary scores between high
school seniors who have studied a foreign language and those who have not.
But because the effect of actually studying a language is confounded with
the differences between students who choose language study and those who
do not, this statistical significance is hard to interpret. It does indicate that
the difference in English scores is greater than would often arise by chance
alone. That leaves unsettled the issue of *what* other than chance caused the
difference. The most plausible explanation is that students who were already
good at English chose to study another language. A randomized comparative
experiment would isolate the actual effect of language study and so make
significance meaningful.

Tests of significance and confidence intervals are based on the laws of
probability. Randomization in sampling or experimentation ensures that
these laws apply. When these statistical strategies for producing data cannot
be used, formal statistical inference should be done only with caution. It is
often necessary to analyze data produced without randomization. Tests of
significance are widely used to analyze such data. It is true that significance
at least points to the presence of an effect greater than would be likely by
chance. However, that indication alone is little evidence against H_0 and in

favor of the research hypothesis H_a. Do not allow the solid appearance of P-values on a computer printout to overrule your common sense understanding of data.

Beware of searching for significance Statistical significance is a commodity much sought after by researchers. It means (or ought to mean) that you have found an effect that you were looking for. *The reasoning behind statistical significance works well if you decide what effect you are seeking, design an experiment or sample to search for it, and use a test of significance to weigh the evidence you get.* But because a successful search for a new scientific phenomenon often ends with statistical significance, it is all too tempting to make significance itself the object of the search. There are several ways to do this, none of them acceptable in polite scientific society.

One tactic is to make many tests on the same data. Once upon a time there were three psychiatrists who studied a sample of schizophrenic persons and a sample of nonschizophrenic persons. They measured 77 variables for each subject—religion, family background, childhood experiences, and so on. Their goal was to discover what distinguishes persons who later become schizophrenic. Having measured 77 variables, they made 77 separate tests of the significance of the differences between the two groups of subjects. Pause for a moment of reflection. If you made 77 tests at the 5% level, you would expect a few of them to be significant by chance alone. After all, results significant at the 5% level do occur 5 times in 100 in the long run even when H_0 is true. The psychiatrists found 2 of their 77 tests significant at the 5% level and immediately published this exciting news.[5] Running one test and reaching the $\alpha = 0.05$ level is reasonably good evidence that you have found something; running 77 tests and reaching that level only twice is not.

The case of the 77 tests happened long ago. Such crimes are rarer now—or at least better concealed. The computer has freed us from the labor of doing arithmetic. This is surely a blessing in statistics, where the arithmetic can be long and complicated indeed. Comprehensive statistical software systems are everywhere available, so a few simple commands will set the machine to work performing all manner of complicated tests and operations on your data. The result can be much like the 77 tests of old. We will state it as a law that any large set of data—even several pages of a table of random digits—contains some unusual pattern. Sufficient computer time will discover that pattern, and when you test specifically for the pattern that turned up, the result will be significant. It also will mean exactly nothing.

One lesson here is not to be overawed by the computer. The computer has greatly extended the range of statistical inference, allowing us to handle larger sets of data and to carry out more complex analyses. But it has changed the logic of inference not one bit. Doing 77 tests and finding 2 of them significant at the $\alpha = 0.05$ level is not evidence of a real discovery. Neither is doing factor analysis followed by discriminant analysis followed by multiple regression and at last discovering a significant pattern in the data. Fancy computer programs are no remedy for bad scientific logic. It is convincing to hypothesize that an effect or pattern will be present, design a study to look for it, and find it at a low significance level. It is not convincing to search for any effect or pattern whatever and find one.

We do not mean that searching data for suggestive patterns is not proper scientific work. It certainly is. Many important discoveries have been made by accident rather than by design. Exploratory analysis of data is an essential part of statistics. We do mean that the usual reasoning of statistical inference does not apply when the search for a pattern is successful. You cannot legitimately test a hypothesis on the same data that first suggested that hypothesis. The remedy is clear. Once you have a hypothesis, design a study to search specifically for the effect you now think is there. If the result of this study is statistically significant, you have real evidence at last.

Power*

In examining the usefulness of a confidence interval, we are concerned with both the level of confidence and the width of the interval. The confidence level tells us how reliable the method is in repeated use. The width tells us how sensitive the method is, that is, how closely the interval pins down the parameter being estimated. Fixed level α significance tests are closely related to confidence intervals—in fact, we saw that a two-sided test can be carried out directly from a confidence interval. The significance level, like the confidence level, says how reliable the method is in repeated use. If we use 5% significance tests repeatedly when H_0 is in fact true, we will be wrong (the test will reject H_0) 5% of the time and right (the test will fail to reject H_0) 95% of the time.

High confidence is of little value if the interval is so wide that few values of the parameter are excluded. Similarly, it can happen that a test with a very small level α almost never rejects H_0 even when the true parameter value is far from the hypothesized value. We must be concerned with the ability of a test to detect that H_0 is false, just as we are concerned with the width of a confidence interval. This ability is measured by the probability that the test will reject H_0 when an alternative is true. The higher this probability is, the more sensitive the test is.

Power

> The probability that a fixed level α significance test will reject H_0 when a particular alternative value of the parameter is true is called the power of the test against that alternative.

* Although power is important in planning and interpreting significance tests, this section and later comments on power can be omitted without loss of continuity.

EXAMPLE 7.16

The cheesemaker of Examples 7.6 and 7.8 determines that milk so heavily watered that the freezing point is raised to $-0.53°C$ will damage the quality of the cheese. Will a 5% significance test of the hypotheses

$$H_0: \mu = -.545$$
$$H_a: \mu > -.545$$

based on a sample of five lots usually detect a mean freezing point this high?

The test measures the freezing point of five lots of milk from a producer and rejects H_0 when

$$z = \frac{\bar{x} - (-.545)}{.008/\sqrt{5}} \geq 1.645$$

This is the same as

$$\bar{x} \geq -.545 + 1.645\frac{.008}{\sqrt{5}}$$

or

$$\bar{x} \geq -.539$$

That the significance level is $\alpha = 0.05$ means that this event has probability 0.05 of occurring *when in fact the population mean μ is* -0.545. To help us keep in mind that the probability calculation assumes that $\mu = -0.545$, we will write

$$P(\bar{x} \geq -.539 | \mu = -.545) = .05$$

(This is not a conditional probability, because μ is not a random variable. It is just a notation to remind us what value of μ was assumed in calculating the probability.)

The power against the alternative -0.53 is the probability that H_0 will be rejected *when in fact $\mu = -0.53$*, or

$$P(\bar{x} \geq -.539 | \mu = -.53)$$

We can calculate this probability by standardizing $\bar{x}$, but we must use the value $\mu = -0.53$ for the population mean in doing so. The population standard deviation $\sigma = 0.008$ does not change.

$$P(\bar{x} \geq -.539 | \mu = -.53) = P\left[\frac{\bar{x} - (-.53)}{.008/\sqrt{5}} \geq \frac{-.539 - (-.53)}{.008/\sqrt{5}}\right]$$
$$= P(Z \geq -2.52) = .9941$$

Figure 7.15 illustrates the power in terms of the sampling distribution of $\bar{x}$ when $\mu = -0.53$ is true. This significance test will almost always (probability more than 99%) reject H_0 when in fact $\mu = -0.53$. It is sensitive enough for the cheesemaker's purpose. ∎

High power is desirable. The numerical value of the power depends on which particular parameter value in H_a we are interested in. Values of the mean μ that are in H_a but lie close to the hypothesized value μ_0 are harder to detect (lower power) than values of μ that are far from μ_0.

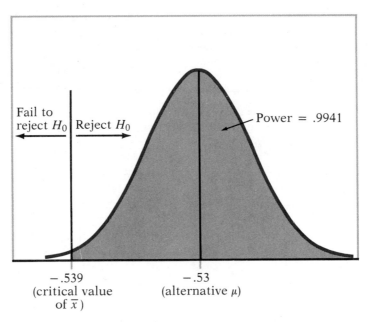

Figure 7.15 The power for Example 7.16. Power is the probability that the test rejects H_0 when the alternative is true.

In planning an investigation that will include a test of significance, a careful user of statistics should decide what alternatives the test should detect and check that the power is adequate. If the power is too low, a larger sample size will increase the power for the same significance level α. In order to calculate power, we must fix a level α so that there is a fixed criterion for rejecting H_0. We prefer to report P-values rather than to use a fixed significance level. The usual practice is to calculate the power at common significance levels such as $\alpha = 0.05$ even though you intend to report a P-value.

Power calculations are important in planning studies. The use of a significance test with low power makes it unlikely that you will find a significant effect even if the truth is far from the null hypothesis. A null hypothesis that is in fact false can become widely believed if repeated attempts to find evidence against it fail because of low power. Consider for example the "efficient market hypothesis" for the time series of stock prices. This hypothesis says that future stock prices (when adjusted for inflation) show only random variation. No information available now will help us predict stock prices in the future, because the efficient working of the stock market has already incorporated all available information in the present price. Many studies have tested the claim that one or another kind of information is helpful. In these studies, the efficient market hypothesis is H_0 and the claim that prediction is possible is H_a. The studies have almost all failed to find good evidence against H_0. As a result, the efficient market hypothesis is quite popular. But an examination of the significance tests employed finds that the power is generally low. Failure to reject H_0 when using tests of low power is not evidence that H_0 is true. As one expert says, "The widespread impression that there is strong evidence for market efficiency may be due just to a lack of appreciation of the low power of many statistical tests."[6]

The outline of a power calculation is as follows:

- Write down the event that the test rejects H_0 in a form that does not contain any unknown parameters.
- Find the probability of this event under an alternative value of the parameter. This probability is the power against that alternative.

In the case of tests on a population mean, we express the test in terms of z, the sample mean standardized assuming that H_0 is true. This form of the test contains the parameter value μ_0. To calculate power, first restate the test in terms of $\bar{x}$. Here is another example, this time for a two-sided z test.

EXAMPLE 7.17

Example 7.12 presented a test of

$$H_0: \mu = .86$$
$$H_a: \mu \neq .86$$

at the 1% level of significance. What is the power of this test against the specific alternative $\mu = 0.845$?

The test in Example 7.12 rejects H_0 when $|z| \geq 2.576$. Since

$$z = \frac{\bar{x} - .86}{.0068/\sqrt{3}}$$

some arithmetic shows that the test rejects when either of the following is true

$$z \geq \quad 2.576 \quad \text{(that is, } \bar{x} \geq .870)$$
$$z \leq -2.576 \quad \text{(that is, } \bar{x} \leq .850)$$

Since these are disjoint events, the power is the sum of their probabilities, *computed assuming that the alternative $\mu = 0.845$ is true.* We find that

$$P(\bar{x} \geq .87 \mid \mu = .845) = P\left(\frac{\bar{x} - .845}{.0068/\sqrt{3}} \geq \frac{.87 - .845}{.0068/\sqrt{3}}\right)$$
$$= P(Z \geq 6.37) \approx 0$$
$$P(\bar{x} \leq .85 \mid \mu = .845) = P\left(\frac{\bar{x} - .845}{.0068/\sqrt{3}} \leq \frac{.85 - .845}{.0068/\sqrt{3}}\right)$$
$$= P(Z \leq 1.27) = .8980$$

Figure 7.16 illustrates this calculation. Since the power is about 0.9, we are quite confident that the test will reject H_0 when this alternative is true.   ■

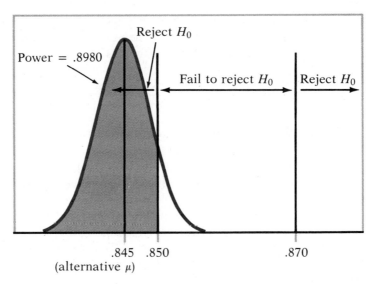

Figure 7.16 The power for Example 7.17.

Inference as Decision*

Tests of significance were presented in Section 7.2 as methods for assessing the strength of evidence against the null hypothesis. This assessment is made by the *P*-value, which is a probability computed under the assumption that the null hypothesis is true. The alternative hypothesis (the statement we seek evidence for) enters the test only to help us see what outcomes count against the null hypothesis. Such is the reasoning of tests of significance as advocated by Fisher and as practiced by many users of statistics.

But signs of another way of thinking were present in the discussion of significance tests with fixed level α. A level of significance α chosen in advance points to the outcome of the test as a *decision*. If the *P*-value is less than α, we reject H_0 in favor of H_a; otherwise we fail to reject H_0. The transition from measuring the strength of evidence to making a decision is not a small step. Many statisticians agree with Fisher's opinion that making decisions is too grand a goal, especially in scientific inference. A decision is reached only after the evidence of many studies is weighed. Indeed, the goal of research is not ''decision'' but a gradually evolving understanding. Statistical inference should content itself with confidence intervals and tests of significance. Many users of statistics are content with such methods. It is rare to set up a level α in advance as a rule for making a decision in a scientific problem. More commonly, users think of significance at level 0.05 as a description of good evidence. This is made clearer by giving the *P*-value.

---

* The purpose of this section is to clarify the reasoning of significance tests by contrasting them with a related type of reasoning. It can be omitted without loss of continuity.

acceptance sampling

Yet there are circumstances in which a decision or action is called for as the end result of inference. *Acceptance sampling* is one such circumstance. A producer of bearings and the consumer of the bearings agree that each carload lot shall meet certain quality standards. When a carload arrives, the consumer chooses a sample of bearings to be inspected. On the basis of the sample outcome, the consumer will either accept or reject the carload. Fisher agreed that this is a genuine decision problem. But he insisted that acceptance sampling is completely different from *scientific inference.* Other eminent statisticians have argued that if "decision" is given a broad meaning, almost all problems of statistical inference can be posed as problems of making decisions in the presence of uncertainty. We will not venture further into the arguments over how we ought to think about inference. We do want to show how a different concept—inference as decision—changes the reasoning used in tests of significance.

Two types of error   Tests of significance fasten attention on H_0, the null hypothesis. If a decision is called for, however, there is no reason to single out H_0. There are simply two hypotheses, and we must accept one and reject the other. It is convenient to call the two hypotheses H_0 and H_a, but H_0 no longer has the special status (the statement we try to find evidence against) that it had in tests of significance. In the acceptance sampling problem, we must decide between

$$H_0\text{: the lot of bearings meets standards}$$

$$H_a\text{: the lot does not meet standards}$$

on the basis of a sample of bearings.

We hope that our decision will be correct, but sometimes it will be wrong. There are two types of incorrect decisions, and it is important to distinguish between them. We can accept a bad lot of bearings, or we can reject a good lot. Accepting a bad lot injures the consumer, while rejecting a good lot hurts the producer. To help distinguish these two types of errors, we give them specific names.

Type I and Type II errors

> If we reject H_0 (accept H_a) when in fact H_0 is true, this is a Type I error. If we accept H_0 (reject H_a) when in fact H_a is true, this is a Type II error.

The possibilities are summed up in Figure 7.17. If H_0 is true, our decision is either correct (if we accept H_0) or is a Type I error. If H_a is true, our decision is either correct or is a Type II error. Only one error is possible at one time. Figure 7.18 applies these ideas to the acceptance sampling example.

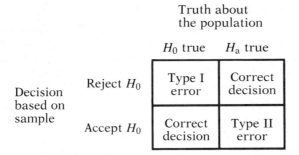

Figure 7.17 shows the following table:

Truth about
the population

		H_0 true	H_a true
Decision based on sample	Reject H_0	Type I error	Correct decision
	Accept H_0	Correct decision	Type II error

Figure 7.17 The two types of error in testing hypotheses.

Truth about the lot

		Does meet standards	Does not meet standards
Decision based on sample	Reject the lot	Type I error	Correct decision
	Accept the lot	Correct decision	Type II error

Figure 7.18 The two types of error in the acceptance sampling setting.

Error probabilities Any rule for making decisions is assessed in terms of the probabilities of the two types of error. This is in keeping with the idea that statistical inference is based on probability. We cannot (short of inspecting the whole lot) guarantee that good lots of bearings will never be rejected and bad lots never be accepted. But by random sampling and the laws of probability, we can say what the probabilities of both kinds of error are.

Significance tests with fixed level α give a rule for making decisions, because the test either rejects H_0 or fails to reject it. If we adopt the decision-making way of thought, failing to reject H_0 means deciding that H_0 is true. We can then describe the performance of a test by the probabilities of Type I and Type II error.

EXAMPLE 7.18

The mean diameter of a type of bearing is supposed to be 2.000 centimeters (cm). The bearing diameters vary normally with standard deviation $\sigma = 0.010$ cm. When a lot of the bearings arrives, the consumer takes an SRS of five bearings from the lot and measures their diameters. The consumer rejects the bearings if the sample mean diameter is significantly different from 2 at the 5% significance level.

This is a test of the hypotheses

$$H_0: \mu = 2$$
$$H_a: \mu \neq 2$$

To carry out the test, the consumer computes the standardized sample mean

$$z = \frac{\bar{x} - 2}{.01/\sqrt{5}}$$

and rejects H_0 if

$$z < -1.96 \quad \text{or} \quad z > 1.96$$

A Type I error is to reject H_0 when in fact $\mu = 2$.

What about Type II errors? Because there are many values of μ in H_a we will concentrate on one value. The producer and the consumer agree that a lot of bearings with mean 0.015 cm or more away from the desired mean 2.000 cm should be rejected. So a particular Type II error is to accept H_0 when in fact $\mu = 2.015$.

Figure 7.19 shows how the two probabilities of error are obtained from the two sampling distributions of $\bar{x}$, for $\mu = 2$ and for $\mu = 2.015$. When $\mu = 2$, H_0 is true and to reject H_0 is a Type I error. When $\mu = 2.015$, accepting H_0 is a Type II error. We will now calculate these error probabilities.   ■

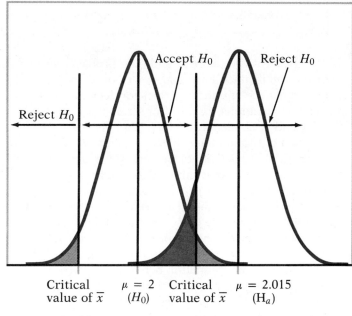

Figure 7.19   The two error probabilities for Example 7.18. The probability of a Type I error (*light shaded area*) is the probability of rejecting $H_0: \mu = 2$ when in fact $\mu = 2$. The probability of a Type II error (*dark shaded area*) is the probability of accepting H_0 when in fact $\mu = 2.015$.

The probability of a Type I error is the probability of rejecting H_0 when it is really true. In Example 7.18, this is the probability that $|z| \geq 1.96$ when $\mu = 2$. But this is exactly the significance level of the test. The critical value 1.96 was chosen to make this probability 0.05, so we do not have to compute it again. The definition of "significant at level 0.05" is that sample outcomes this extreme will occur with probability 0.05 when H_0 is true.

The significance level α of any fixed level test is the probability of a Type I error; that is, α is the probability that the test will reject the null hypothesis H_0 when H_0 is in fact true.

The probability of a Type II error for the particular alternative $\mu = 2.015$ in Example 7.18 is the probability that the test will fail to reject H_0 when μ has this alternative value. The *power* of the test against the alternative $\mu = 2.015$ is just the probability that the test *does* reject H_0. By following the method of Example 7.17, we can calculate that the power is about 0.92. The probability of a Type II error is therefore $1 - 0.92$, or 0.08.

The power of a fixed level test against a particular alternative is 1 minus the probability of a Type II error for that alternative.

The two types of error and their probabilities give another interpretation of the significance level and power of a test. The distinction between tests of significance and tests as rules for deciding between two hypotheses is not in the calculations. In a test of significance we focus on a single hypothesis (H_0) and a single probability (the *P*-value). The goal is to measure the strength of the sample evidence against H_0. Calculations of power are done to check the sensitivity of the test. If we cannot reject H_0, we conclude only that there is not sufficient evidence against H_0, not that H_0 is actually true. If the same inference problem is thought of as a decision problem, we focus on two hypotheses and give a rule for deciding between them based on the sample evidence. We therefore must focus equally on two probabilities, the probabilities of the two types of error. We must choose one or the other hypothesis and cannot abstain on grounds of insufficient evidence.

Such a clear distinction between the two ways of thinking is helpful for understanding. In practice, the two approaches often merge. We continued to call one of the hypotheses in a decision problem H_0. The common prac-

**testing
hypotheses**   tice of *testing hypotheses* mixes the reasoning of significance tests and decision rules as follows:

1 State H_0 and H_a just as in a test of significance.

2 Think of the problem as a decision problem, so that the probabilities of Type I and Type II error are relevant.

3 Because of Step 1, Type I errors are more serious. So choose an α (significance level) and consider only tests with probability of Type I error no greater than α.

4 Among these tests, select one that makes the probability of a Type II error as small as possible (that is, power as large as possible). If this probability is too large, you will have to take a larger sample to reduce the chance of an error.

Testing hypotheses may seem to be a hybrid approach. It was, historically, the effective beginning of decision-oriented ideas in statistics.* The decision-making approach came later (1940s) and grew out of the work of Neyman and Pearson. Because decision theory in its pure form leaves you with two error probabilities and no simple rule on how to balance them, it has been used less often than either tests of significance or tests of hypotheses. Decision ideas have been applied in testing problems mainly by way of the Neyman-Pearson hypothesis-testing theory. That theory asks you first to choose α, and the influence of Fisher often has led users of hypothesis testing comfortably back to $\alpha = 0.05$ or $\alpha = 0.01$. Fisher, who was exceedingly argumentative, violently attacked the Neyman-Pearson decision-oriented ideas, and the argument still continues.

SUMMARY

P-values are more informative than the reject-or-not result of a fixed level α test. Beware of placing too much weight on traditional values of α, such as $\alpha = 0.05$.

Very small effects can be highly significant (small *P*), especially when a test is based on a large sample. A statistically significant effect need not be practically significant. Plot the data to display the effect you are seeking, and use confidence intervals to estimate the actual value of parameters.

On the other hand, lack of significance does not imply that H_0 is true, especially when the test has low power.

Significance tests are not always valid. Faulty data collection, outliers in the data, and testing a hypothesis on the same data that suggested the hypothesis can invalidate a test. Many tests run at once will probably produce some significant results by chance alone, even if all the null hypotheses are true.

* An impressive mathematical theory of hypothesis testing was developed between 1928 and 1938 by Jerzy Neyman and the English statistician Egon Pearson.

The **power** of a significance test measures its ability to detect the truth of an alternative hypothesis. Power against a specific alternative is calculated as the probability that the test will reject H_0 when the alternative is true. This calculation requires knowledge of the sampling distribution of the test statistic under the alternative hypothesis. Increasing the size of the sample increases the power when the significance level remains fixed.

An alternative to significance testing regards H_0 and H_a as two statements of equal status that we must decide between. This **decision theory** point of view regards statistical inference in general as giving rules for making decisions in the presence of uncertainty.

In the case of testing H_0 versus H_a, decision theory chooses a decision rule on the basis of the probabilities of two types of error. A **Type I error** occurs if H_0 is rejected when it is in fact true. A **Type II error** occurs if H_0 is accepted when in fact H_a is true.

In a fixed level α significance test, the significance level α is the probability of a Type I error, and the power against a specific alternative is 1 minus the probability of a Type II error for that alternative.

SECTION 7.3 EXERCISES

7.43 Which of the following questions does a test of significance answer?

    (a) Is the sample or experiment properly designed?

    (b) Is the observed effect due to chance?

    (c) Is the observed effect important?

7.44 In a study of the suggestion that taking vitamin C will prevent colds, 400 subjects are assigned at random to one of two groups. The experimental group takes a vitamin C tablet daily, while the control group takes a placebo. At the end of the experiment, the difference in the percent of subjects in the two groups who were free of colds is calculated. This difference is statistically significant ($P = 0.03$) in favor of the vitamin C group. Can we conclude that vitamin C has a strong effect in preventing colds? Explain your answer.

7.45 Every user of statistics should understand the distinction between statistical significance and practical importance. A sufficiently large sample will declare very small effects statistically significant. Let us suppose that Scholastic Aptitude Test mathematics (SAT-M) scores in the absence of coaching vary normally with mean $\mu = 475$ and $\sigma = 100$. Suppose further that coaching may change μ but does not change σ. An increase in the SAT-M score from 475 to 478 is of no importance in seeking admission to college, but this unimportant change can be statistically very significant. To see this, calculate the

P-value for the test of

$$H_0: \mu = 475$$
$$H_a: \mu > 475$$

in each of the following situations:

(a) A coaching service coaches 100 students; their SAT-M scores average $\bar{x} = 478$.

(b) By the next year, the service has coached 1000 students; their SAT-M scores average $\bar{x} = 478$.

(c) An advertising campaign brings the number of students coached to 10,000; their average score is still $\bar{x} = 478$.

7.46 Give a 99% confidence interval for the mean SAT-M score μ after coaching in each part of the previous exercise. For large samples, the confidence interval says, "Yes, the mean score is higher after coaching, but only by a small amount."

7.47 As in the previous exercises, suppose that SAT-M scores vary normally with $\sigma = 100$. One hundred students go through a rigorous training program designed to raise their SAT-M scores by improving their mathematics skills. Carry out a test of

$$H_0: \mu = 475$$
$$H_a: \mu > 475$$

in each of the following situations:

(a) The students' average score is $\bar{x} = 491.4$. Is this result significant at the 5% level?

(b) The average score is $\bar{x} = 491.5$. Is this result significant at the 5% level?

(c) Do you think that the difference between the two outcomes in (a) and (b) is of any importance? Beware of attempts to treat $\alpha = 0.05$ as sacred.

7.48 A local television station announces a question for a call-in opinion poll on the six o'clock news and then gives the response on the eleven o'clock news. Today's question concerns a proposed gun control ordinance. Of the 2372 calls received, 1921 oppose the new law. The station, following standard statistical practice, makes a confidence statement: "81% of the Channel 13 Pulse Poll sample oppose gun control. We can be 95% confident that the proportion of all viewers who oppose the law is within 1.6% of the sample result." Is the station's conclusion justified? Explain your answer.

7.49 A researcher looking for evidence of ESP tests 500 subjects. (See Exercises 6.15 and 6.23 for the nature of ESP testing.) Four of these subjects do significantly better ($P < 0.01$) than random guessing.

(a) Is it proper to conclude that these four people have ESP? Explain your answer.

(b) What should the researcher now do to test whether any of these four subjects have ESP?

7.50 The text cites an example in which researchers carried out 77 separate significance tests, of which two were significant at the 5% level. Suppose that these tests are independent of each other. (In fact they were not independent, because all involved the same subjects.) If all of the null hypotheses are true, each test has probability 0.05 of being significant at the 5% level.

(a) What is the distribution of the number X of tests that are significant?

(b) Find the probability that two or more of the tests are significant.

The following exercises concern the optional sections beginning on p. 490:

7.51 Example 7.11 gives a test of a hypothesis about the SAT scores of California high school students based on an SRS of 500 students. The hypotheses are

$$H_0: \mu = 450$$
$$H_a: \mu > 450$$

Assume that the population standard deviation is $\sigma = 100$. The test rejects H_0 at the 1% level of significance when $z \geq 2.326$, where

$$z = \frac{\bar{x} - 450}{100/\sqrt{500}}$$

Is this test sufficiently sensitive to usually detect an increase of 10 points in the population mean SAT score? Answer this question by calculating the power of the test against the alternative $\mu = 460$.

7.52 Example 7.15 discusses a test about the mean contents of cola bottles. The hypotheses are

$$H_0: \mu = 300$$
$$H_a: \mu < 300$$

The sample size is $n = 6$, and the population is assumed to have a normal distribution with $\sigma = 3$. A 5% significance test rejects H_0 if $z \leq -1.645$, where the z statistic is

$$z = \frac{\bar{x} - 300}{3/\sqrt{6}}$$

Power calculations will help us determine how large a shortfall in the bottle contents the test can be expected to detect.

(a) Find the power of this test against the alternative $\mu = 299$.

(b) Find the power against the alternative $\mu = 295$.

(c) Is the power against $\mu = 290$ higher or lower than the value you found in (b)?

7.53 Increasing the sample size increases the power of a test when the level α is unchanged. Suppose that in the previous exercise a sample of n bottles had been measured. In that exercise, $n = 6$. The 5% significance test still rejects H_0 when $z \le -1.645$, but the z statistic is now

$$z = \frac{\bar{x} - 300}{3/\sqrt{n}}$$

(a) Find the power of this test against the alternative $\mu = 299$ when $n = 25$.

(b) Find the power against $\mu = 299$ when $n = 100$.

7.54 In Example 7.9, a company medical director failed to find significant evidence that the mean blood pressure of a population of executives differed from the national mean $\mu = 127$. The medical director now wonders if the test used would detect an important difference if one were present. For the SRS of size 72 from a population with standard deviation $\sigma = 7$, the z statistic is

$$z = \frac{\bar{x} - 127}{7/\sqrt{72}}$$

The two-sided test rejects

$$H_0\colon \mu = 127$$

at the 5% level of significance when $|z| \ge 1.96$.

(a) Find the power of the test against the alternative $\mu = 130$.

(b) Find the power of the test against $\mu = 124$. Can the test be relied on to detect a mean that differs from 127 by 3?

(c) If the alternative were farther from H_0, say $\mu = 135$, would the power be higher or lower than the values calculated in (a) and (b)?

7.55 You have an SRS of size $n = 9$ from a normal distribution with $\sigma = 1$. You wish to test

$$H_0\colon \mu = 0$$
$$H_a\colon \mu > 0$$

You decide to reject H_0 if $\bar{x} > 0$ and to accept H_0 otherwise.

(a) Find the probability of a Type I error, that is, the probability that your test rejects H_0 when in fact $\mu = 0$.

(b) Find the probability of a Type II error when $\mu = 0.3$. This is the probability that your test accepts H_0 when in fact $\mu = 0.3$.

(c) Find the probability of a Type II error when $\mu = 1$.

7.56 Use the result of Exercise 7.52 to give the probabilities of Type I and Type II errors for the test discussed there. Take the alternative hypothesis to be $\mu = 295$.

7.57 Use the result of Exercise 7.51 to give the probability of Type I error for the test in that exercise and its probability of Type II error when the alternative is $\mu = 460$.

7.58 You must decide which of two discrete distributions a random variable X has. We will call the distributions p_0 and p_1. Here are the probabilities they assign to the values x of X.

x	0	1	2	3	4	5	6
p_0	.1	.1	.1	.1	.2	.1	.3
p_1	.2	.1	.1	.2	.2	.1	.1

You have a single observation on X and wish to test

$$H_0: p_0 \text{ is correct}$$
$$H_a: p_1 \text{ is correct}$$

One possible decision procedure is to accept H_0 if $X = 4$ or $X = 6$ and reject H_0 otherwise.

(a) Find the probability of a Type I error, that is, the probability that you reject H_0 when p_0 is the correct distribution.
(b) Find the probability of a Type II error.

7.59 A computerized medical diagnostic program is being designed. The program will scan the results of routine medical tests (pulse rate, blood pressure, urinalysis, etc.) and either clear the patient or refer the case to a doctor. The program will be used as part of a preventive medicine system to screen many thousands of persons who do not have specific medical complaints. The program makes a decision about each patient.

(a) What are the two hypotheses and the two types of error that the program can make?
(b) The program can be adjusted to decrease one error probability, at the cost of an increase in the other error probability. Which error probability would you choose to make smaller, and why? (This is a matter of judgment. There is no single correct answer.)

7.60 The acceptance sampling test in Example 7.18 has probability 0.05 of rejecting a good lot of bearings and probability 0.08 of accepting a bad lot. The consumer of the bearings may imagine that acceptance sampling guarantees that most accepted lots are good. Alas, it is not so. Suppose that 90% of all lots shipped by the producer are bad.

(a) Draw a tree diagram for shipping a lot (the branches are "bad" and "good") and then inspecting it (the branches at this stage are "accept" and "reject").

(b) Write the appropriate probabilities on the branches, and find the probability that a lot shipped is accepted.

(c) Use the definition of conditional probability to find the probability that a lot is bad, given that the lot is accepted. This is the proportion of bad lots among the lots that the sampling plan accepts.

NOTES

1 The correlation here relates two orderings of men, by birth date and by draft number. We will not study the distribution of r in this case, but the probability calculation required is not hard. This and other analyses of the draft lottery data appear in S. E. Fienberg, "Randomization and social affairs: the 1970 draft lottery," *Science*, 171 (1971), pp. 255–261.

2 R. A. Fisher, "The arrangement of field experiments," *Journal of the Ministry of Agriculture of Great Britain*, 33 (1926), p. 504, quoted in Leonard J. Savage, "On rereading R. A. Fisher," *The Annals of Statistics*, 4 (1976), p. 471.

3 Fisher's work is described in a biography by his daughter: Joan Fisher Box, *R. A. Fisher: The Life of a Scientist*, Wiley, New York, 1978.

4 T. D. Sterling, "Publication decisions and their possible effects on inferences drawn from tests of significance—or vice versa," *Journal of the American Statistical Association*, 54 (1959), pp. 30–34. Related comments appear in J. K. Skipper, A. L. Guenther, and G. Nass, "The sacredness of 0.05: a note concerning the uses of statistical levels of significance in social science," *American Sociologist*, 1 (1967), pp. 16–18.

5 This example is cited by William Feller, "Are life scientists overawed by statistics?" *Scientific Research*, February 3, 1969, p. 26.

6 R. J. Schiller, "The volatility of stock market prices," *Science*, 235 (1987), pp. 33–36.

CHAPTER 7 EXERCISES

7.61 Patients with chronic kidney failure may be treated by dialysis, using a machine that removes toxic wastes from the blood, a function normally performed by the kidneys. Kidney failure and dialysis can cause other changes, such as retention of phosphorus, that must be corrected by changes in diet. A study of the nutrition of dialysis patients measured the level of phosphorus in the blood of several patients on six occasions. Here are the data for one patient (milligrams of phosphorus per deciliter of blood).

5.6   5.1   4.6   4.8   5.7   6.4

The measurements are separated in time and can be considered an SRS of the patient's blood phosphorus level. If this level varies normally with $\sigma = 0.9$ mg/dl, give a 90% confidence interval for the mean blood phosphorus level. (Data provided by Joan M. Susic.)

7.62 The normal range of phosphorus in the blood is considered to be 2.6 to 4.8 mg/dl. Is there strong evidence that the patient in the previous exercise has a mean phosphorus level that exceeds 4.8?

7.63 A government report gives a 99% confidence interval for the 1985 median family income as $27,735 \pm \$357$. This result was calculated by advanced methods from the Current Population Survey, a multi-stage random sample of over 70,000 households.

(a) Would a 95% confidence interval be wider or narrower? Explain your answer.

(b) Would the null hypothesis that the 1985 median family income was $28,000 be rejected at the 1% significance level in favor of the two-sided alternative?

(c) Can you determine whether this hypothesis would be rejected at the 5% level? Explain your answer.

7.64 An agronomist examines the cellulose content of a variety of alfalfa hay. Suppose that the cellulose content in the population has standard deviation $\sigma = 8$ mg/g. A sample of 15 cuttings has mean cellulose content $\bar{x} = 145$ mg/g.

(a) Give a 90% confidence interval for the mean cellulose content in the population.

(b) A previous study claimed that the mean cellulose content was $\mu = 140$ mg/g, but the agronomist believes that the mean is higher than that figure. State H_0 and H_a and carry out a significance test to see if the new data support this belief.

(c) The statistical procedures used in (a) and (b) are valid when several assumptions are met. What are these assumptions?

7.65 In a study of possible iron deficiency in infants, researchers compared several groups of infants who were following different feeding patterns. One group of 26 infants was being breast fed. At 6 months of age, these children had a mean hemoglobin level of $\bar{x} = 12.9$ grams per 100 milliliters of blood and a standard deviation of 1.6. Taking the standard deviation to be the population value σ, give a 95% confidence interval for the mean hemoglobin level of breast-fed infants. What assumptions are required for the validity of the method you used to get the confidence interval?

7.66 Statisticians prefer large samples. Describe briefly the effect of increasing the size of a sample (or the number of subjects in an experiment) on each of the following:

(a) The width of a level $1 - \alpha$ confidence interval

(b) The P-value of a test, when H_0 is false and all facts about the population remain unchanged as n increases

(c) The power of a fixed level α test, when α, the alternative hypothesis, and all facts about the population remain unchanged

7.67 A roulette wheel has 18 red slots among its 38 slots. You observe many spins and record the number of times that red occurs. Now you want to use these data to test whether the probability of a red has the value that is correct for a fair roulette wheel. State the hypotheses H_0 and H_a that you will test. (The test for this situation is described in Chapter 9.)

7.68 When asked to explain the meaning of "statistically significant at the $\alpha = 0.05$ level," a student says, "This means there is only probability 0.05 that the null hypothesis is true." Is this an essentially correct explanation of statistical significance? Explain your answer.

7.69 Another student, when asked why statistical significance appears so often in research reports, says, "Because saying that results are significant tells us that they cannot easily be explained by chance variation alone," Do you think that this statement is essentially correct? Explain your answer.

7.70 A study compares two groups of mothers with young children who were on welfare 2 years ago. One group had attended a voluntary training program offered free of charge at a local vocational school and advertised in the local news media. The other group had not chosen to attend the training program. The study finds a significant difference ($P < 0.01$) between the proportions of the mothers in the two groups who are still on welfare. The difference is not only significant but quite large. The report says that with 95% confidence the percent of the nonattending group still on welfare is 21% $\pm 4\%$ higher than that of the group who attended the program. You are on the staff of a member of Congress who is interested in the plight of welfare mothers, and who asks you about the report.

(a) Explain briefly and in nontechnical language what "a significant difference ($P < 0.01$)" means.

(b) Explain clearly and briefly what "95% confidence" means.

(c) Is this study good evidence that requiring job training of all welfare mothers would greatly reduce the percent who remain on welfare for several years?

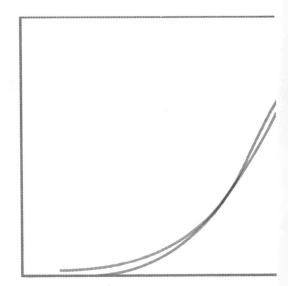

Prelude

Many statistical problems require that we answer a question about the mean of a population or compare the means of two populations—for example, the mean responses to two treatments in an experiment. In this chapter we learn how to give confidence intervals and carry out tests of significance in these situations. We will also consider some of the practical aspects of using these inference procedures in real problems.

- *Does attending a summer language institute improve the ability of high school French teachers to understand spoken French? How large is the improvement?*

- *Is it true that female college students as a group score higher than male students on a test of study habits and attitude toward learning?*

- *Can a significance test detect the effect of the pesticide DDT on the nervous system by comparing rats exposed to DDT with unexposed rats?*

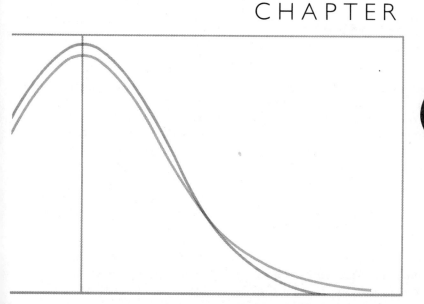

8

Inference for Distributions

With the principles in hand, we can proceed to practice. This chapter describes significance tests and confidence intervals for the mean of a single population and for comparing two populations. Later chapters will present procedures for categorical data, for studying relations among variables, and for comparing more than two populations.

8.1 INFERENCE FOR THE MEAN OF A POPULATION

Both confidence intervals and tests of significance for the mean μ of a normal population are based on the sample mean $\bar{x}$, which estimates the unknown μ. The sampling distribution of $\bar{x}$ depends on σ. This fact causes no difficulty when σ is known. When σ is unknown, however, it means that we must estimate σ even though we are primarily interested in μ. The sample standard deviation s is used to estimate the population standard deviation σ.

The One-Sample t Procedures

Suppose that we have a simple random sample (SRS) of size n from a normally distributed population with mean μ and standard deviation σ. The sample mean $\bar{x}$ then has the normal distribution with mean μ and standard deviation $\sigma/\sqrt{n}$. When σ is not known, we estimate the standard deviation of $\bar{x}$ by $s/\sqrt{n}$. This quantity is called the standard error of the sample mean $\bar{x}$.

Standard error

> When the standard deviation of a statistic is estimated from the data, the result is called the standard error of the statistic.

The term "standard error" is sometimes used for the actual standard deviation of a statistic, $\sigma/\sqrt{n}$ in the case of $\bar{x}$. The estimated value $s/\sqrt{n}$ is then called the "estimated standard error." In this book we will use the term "standard error" only when the standard deviation of a statistic is estimated from the data. The term has this meaning in the output of many statistical computer packages and in reports of research in many fields that apply statistical methods.

The standardized sample mean

$$z = \frac{\bar{x} - \mu}{\sigma/\sqrt{n}}$$

is the basis of the z procedures for use when σ is known. This statistic has the standard normal distribution $N(0, 1)$. When we substitute the standard error $s/\sqrt{n}$ for the standard deviation $\sigma/\sqrt{n}$ of $\bar{x}$, the statistic does *not* have a normal distribution. It has a distribution that is new to us, called a t distribution.

The t distributions

Suppose that an SRS of size n is drawn from an $N(\mu, \sigma)$ population. Then the one-sample t statistic

$$t = \frac{\bar{x} - \mu}{s/\sqrt{n}} \tag{8.1}$$

has the t distribution with $n - 1$ degrees of freedom.

degrees of freedom

There is a different t distribution for each sample size. A particular t distribution is specified by giving the *degrees of freedom*. The degrees of freedom for this t statistic come from the sample standard deviation s in the denominator of t. We saw in Chapter 1 that s has $n - 1$ degrees of freedom. This is true because the n deviations $x_i - \bar{x}$ that are used to calculate s always have sum 0. Therefore any $n - 1$ of the deviations determine the remaining deviation. We think of $n - 1$ of the deviations being free to change, and this number gives the degrees of freedom. There are other t statistics with different degrees of freedom, some of which we will meet later in this chapter. We will denote the t distribution with k degrees of freedom by $t(k)$ for short.*

The density curves of the $t(k)$ distributions are similar in shape to the standard normal curve. That is, they are symmetric about 0 and are bell-shaped. The spread of the t distributions is a bit greater than that of the standard normal distribution. This reflects the extra variability caused by substituting the random variable s for the fixed parameter σ. As the degrees of freedom k increase, the $t(k)$ density curve approaches the $N(0, 1)$ curve ever more closely. This reflects the fact that s approaches σ as the sample size increases. Figure 8.1 compares the density curves of the standard normal distribution and the t distribution with 5 degrees of freedom. The similarity in shape is apparent, as is the fact that the t distribution has more probability in the tails and less in the center than does the standard normal.

Table E in the back of the book gives upper p critical values for the t distributions. For comparison, the standard normal critical values from

---

* The t distributions were discovered in 1908 by William S. Gossett. Gossett was a statistician employed by the Guinness Brewing Company, which required that he not publish his discoveries under his own name. He therefore wrote under the pen name "Student." The t distribution is often called "Student's t" in his honor.

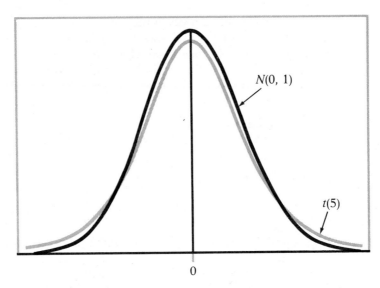

Figure 8.1 Density curves for the standard normal and $t(5)$ distributions. Both are symmetric with center 0. The t distributions have more probability in the tails than the standard normal distribution.

Table D are repeated in the bottom row of Table E. The t-values approach these normal values as the degrees of freedom increase. As in the case of the normal table, computer software often makes Table E unnecessary.

One-sample t procedures   With the t distributions to help us, we can analyze samples from normal populations with unknown σ by replacing the standard deviation $\sigma/\sqrt{n}$ of $\bar{x}$ with its standard error $s/\sqrt{n}$ in the z procedures of Chapter 7. The z statistic then becomes the one-sample t statistic of Equation 8.1. We must now employ P-values or critical values from t in place of the corresponding normal values.

The one-sample t procedures

Suppose that an SRS of size n is drawn from a population having unknown mean μ. A level $1 - \alpha$ confidence interval for μ is

$$\bar{x} \pm t^* \frac{s}{\sqrt{n}}$$

where t^* is the upper $\alpha/2$ critical value for the $t(n-1)$ distribution. This interval is exact when the population distribution is normal and is approximately correct for large n in other cases.

To test the hypothesis H_0: $\mu = \mu_0$ based on an SRS of size n, compute the one-sample t statistic

$$t = \frac{\bar{x} - \mu_0}{s/\sqrt{n}}$$

In terms of a random variable T having the $t(n-1)$ distribution, the P-value for a test of H_0 against

$$H_a: \mu > \mu_0 \quad \text{is} \quad P(T \geq t)$$
$$H_a: \mu < \mu_0 \quad \text{is} \quad P(T \leq t)$$
$$H_a: \mu \neq \mu_0 \quad \text{is} \quad P(|T| \geq |t|)$$

These P-values are exact if the population distribution is normal and approximately correct for large n in other cases.

The one-sample t procedures are similar in both reasoning and computational detail to the z procedures of Chapter 7. We can therefore put more emphasis on understanding their effective use in statistical practice.

EXAMPLE 8.1

In an experiment on the metabolism of insects, American cockroaches were fed measured amounts of a sugar solution after being deprived of food for a week and of water for 3 days. After 2, 5, and 10 hours, cockroaches were dissected and the amount of sugar in various tissues was measured.[1] Five cockroaches fed the sugar D-glucose and dissected after 10 hours had the following amounts (in micrograms) of D-glucose in their hindgut:

$$55.95 \qquad 68.24 \qquad 52.73 \qquad 21.50 \qquad 23.78$$

The researchers gave a 95% confidence interval for the mean amount of D-glucose in cockroach hindguts under these conditions.

First calculate that

$$\bar{x} = 44.44$$
$$s = 20.741$$

For 95% confidence, $\alpha = 0.05$ and we use the $\alpha/2 = 0.025$ critical value. The degrees of freedom are $n - 1 = 4$. From Table E we find that $t^* = 2.776$. The confidence interval is

$$\bar{x} \pm t^* \frac{s}{\sqrt{n}} = 44.44 \pm 2.776 \frac{20.741}{\sqrt{5}}$$
$$= 44.44 \pm 25.75$$
$$= (18.69, 70.19)$$

Comparing this estimate with those for other body tissues and different times before dissection gave some insight into cockroach metabolism. The large width of the confidence interval is due to the small sample and the rather large variation among the cockroaches, reflected in the large value of s. ■

The use of *t* procedures in Example 8.1 rests on assumptions that cannot easily be checked but are reasonable in this case. The carefully controlled treatment and random assignment of the cockroaches to different sugars and different times before dissection allow the researchers to consider this an SRS of all similarly treated American cockroaches. The assumption that the population distribution is normal cannot be effectively checked with only five observations. Experience with this and similar variables led the researchers to believe that approximate normality holds, despite the wide gap between the two smallest and the three largest observations. In observational data, this might suggest two different species of cockroach. In this case we know that all five cockroaches came from a homogeneous population grown in the laboratory for research purposes.

EXAMPLE 8.2

The first data that we looked at were Simon Newcomb's measurements of the passage time of light. Table 1.1 records his 66 measurements. A normal quantile plot (Figure 8.2) reminds us that when the 2 outliers to the left are omitted,

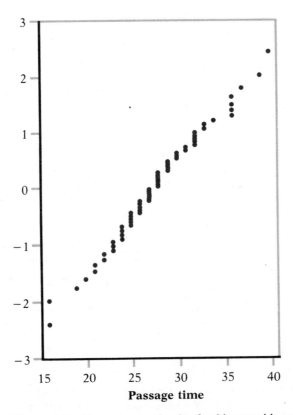

Figure 8.2 Normal quantile plot for Newcomb's passage time data with the outliers omitted, from Example 8.2.

the remaining 64 observations follow a normal distribution quite closely. What result should Newcomb report from these 64 observations?

We want to estimate the mean μ of the distribution of measurements from which Newcomb's 64 are a sample. The sample mean $\bar{x} = 27.750$ estimates μ, but to indicate the precision of this estimate we should give a confidence interval. The sample standard deviation is $s = 5.083$. A 99% confidence interval uses the 0.005 critical value of the $t(63)$ distribution. Table E contains no entry for 63 degrees of freedom. We therefore use the entry for the next smaller degrees of freedom, which is 60, and find that $t^* = 2.660$. (The next smaller entry gives a slightly wider interval than the exact degrees of freedom would, so this is the conservative choice.) The 99% confidence interval is

$$\bar{x} \pm t^* \frac{s}{\sqrt{n}} = 27.750 \pm 2.660 \frac{5.083}{\sqrt{64}}$$

$$= 27.75 \pm 1.69$$

$$= (26.06, 29.44)$$

The best modern measurements of the speed of light correspond to a passage time of 33.02 in Newcomb's experiment. Is his result significantly different from this modern value? To answer this question, we test

$$H_0: \mu = 33.02$$
$$H_a: \mu \neq 33.02$$

The t statistic is

$$t = \frac{\bar{x} - \mu_0}{s/\sqrt{n}} = \frac{27.75 - 33.02}{5.083/\sqrt{64}}$$

$$= -8.29$$

The P-value is $P(|T| \geq 8.29)$, where T has the $t(63)$ distribution. Because the t distributions are symmetric, P for a two-sided test is found from Table E by comparing t with the critical values t^* and *doubling* the corresponding levels p. Table E shows that 8.29 is beyond the 0.0005 critical value. The P-value of the two-sided test is therefore less than 2 times 0.0005, or 0.001. In fact, the P-value is 0 to many decimal places. It is in effect certain that Newcomb's result is not the same as the modern value. That the difference is significant at the 1% level could be seen from the confidence interval. Since 33.02 lies outside the 99% confidence interval, H_0 can be rejected at the 1% level. ∎

df = 60

p	.001	.0005
t^*	3.232	3.460

Newcomb's measurements can be regarded as independent observations drawn from the population of all measurements he might make. We verified the normality of the measurements with the normal quantile plot in Figure 8.2. Moreover, the sample mean of 64 observations will have a distribution that is nearly normal even if the population is not normal. Only normality of $\bar{x}$ is required by the t procedures. Use of these procedures in this example is very well justified.

Because the t procedures are so common, all statistical software systems will do the calculations for you. For example, if Newcomb's data are entered into the MINITAB system as column C1, the t procedures of Example 8.2 are carried out as follows:

```
MTB> TINTERVAL 99 C1

        N     MEAN    STDEV    SE MEAN      99.0 PERCENT C.I.
C1     64    27.750   5.083     0.635       (26.062, 29.438)


MTB> TTEST 33.02 C1

TTEST OF MU = 33.020 VS MU N.E. 33.020       p-value

        N     MEAN    STDEV    SE MEAN      99.0 PERCENT C.I.
C1     64    27.750   5.083     0.635       (26.062, 29.438)
```

Compare these results with those obtained in Example 8.2. The output includes $\bar{x}$, s, and the standard error $s/\sqrt{n}$ of the mean (SE MEAN), as well as the 99% confidence interval and the t statistic with its *P*-value.

Paired Comparisons

Newcomb wanted to estimate a constant of nature, and so only a single population was involved. We saw in discussing data production that comparative studies are usually preferred to single-sample investigations because of the protection they offer against confounding. For that reason, inference about a parameter of a single distribution is less common than comparative inference. One common comparative design, however, makes use of single-sample procedures. In a *paired comparisons* study, subjects are matched in pairs and the outcomes are compared within each matched pair. The experimenter can toss a coin to assign two treatments to the two subjects in each pair. Paired comparisons are also common when randomization is not possible. One situation calling for paired comparisons is before-and-after observations on the same subjects, as illustrated in the next example.

paired comparisons

EXAMPLE 8.3

The National Endowment for the Humanities sponsors summer institutes to improve the skills of high school teachers of foreign languages. One such institute hosted 20 French teachers for 4 weeks. At the beginning of the period, the teachers were given the Modern Language Association's (MLA) listening test of understanding of spoken French. After 4 weeks of immersion in French in and out of class, the listening test was given again. (The actual spoken French in the two tests was different, so that taking the first test should not improve the score on the second test.) Table 8.1 gives the pretest and posttest scores. The maximum possible score on the test is 36.[2]

To analyze these data, we first subtract the pretest score from the posttest score to obtain the improvement for each student. These 20 differences form a single sample. They are

2 0 6 6 3 3 2 3 −6 6 6 6 3 0 1 1 0 2 3 3

Table 8.1 MLA listening scores for French teachers

Teacher	Pretest	Posttest	Teacher	Pretest	Posttest
1	32	34	11	30	36
2	31	31	12	20	26
3	29	35	13	24	27
4	10	16	14	24	24
5	30	33	15	31	32
6	33	36	16	30	31
7	22	24	17	15	15
8	25	28	18	32	34
9	32	26	19	23	26
10	20	26	20	23	26

To assess whether the institute significantly improved the teachers' comprehension of spoken French, we test

$$H_0: \mu = 0$$
$$H_\alpha: \mu > 0$$

Here μ is the mean improvement that would be achieved if the entire population of French teachers attended a summer institute. The null hypothesis says that no improvement occurs, while H_α says that posttest scores are higher on the average. The 20 differences have

$$\bar{x} = 2.5 \quad \text{and} \quad s = 2.89$$

The one-sample t statistic of Equation 8.1 is therefore

$$t = \frac{\bar{x} - 0}{s/\sqrt{n}} = \frac{2.5}{2.89/\sqrt{20}}$$
$$= 3.87$$

df = 19

p	.001	.0005
t^*	3.579	3.883

The *P*-value is found from the $t(19)$ distribution (remember that the degrees of freedom are 1 less than the sample size). Table E shows that 3.87 lies between the upper 0.001 and 0.0005 critical values of the $t(19)$ distribution. The *P*-value therefore lies between these values. A computer statistical package gives the value $P = 0.00052$. The improvement in listening scores is very unlikely to be due to chance alone. We have strong evidence that the institute was effective in raising scores.

A 90% confidence interval for the mean improvement in the entire population requires the upper 0.05 critical value for the $t(19)$ distribution. Table E gives this value as 1.729. The confidence interval is

$$\bar{x} \pm t^* \frac{s}{\sqrt{n}} = 2.5 \pm 1.729 \frac{2.89}{\sqrt{20}}$$
$$= 2.5 \pm 1.12$$
$$= (1.38, 3.62)$$

The improvement, though statistically significant, is rather small. ∎

Example 8.3 illustrates how paired comparisons data are restated as single-sample data by taking differences within each pair. It is incorrect to ignore the pairs and analyze such data as if we had two samples, one from teachers who had attended an institute and a second from teachers who had not attended. Inference procedures for comparing two samples assume that the samples are selected independently of each other. This assumption does not hold when the same subjects are measured twice. The proper analysis depends on the design used to produce the data.

However, the use of the *t* procedures in Example 8.3 faces several difficulties. First, the teachers are not an SRS from the population of high school French teachers. There is some selection bias in favor of energetic, committed teachers who are willing to give up 4 weeks of their summer vacation. It is therefore not clear to what population the results apply. Second, a look at the data shows that several of the teachers had pretest scores close to the maximum of 36. They could not improve their scores very much even if their mastery of French increased substantially. This is a weakness in the listening test that is the measuring instrument in this study. The differences in scores may not adequately indicate the effectiveness of the institute. This is one reason why the average increase was small.

A final difficulty facing the *t* procedures in Example 8.3 is that the data show departures from normality. In a paired comparisons analysis, the population of differences must have a normal distribution because the *t* procedures are applied to the differences. In the language institute example, one teacher actually lost 6 points between the pretest and the posttest. This one subject lowered the sample mean from 2.95 for the other 19 subjects to 2.5 for all 20. A normal quantile plot (Figure 8.3) displays this outlier and

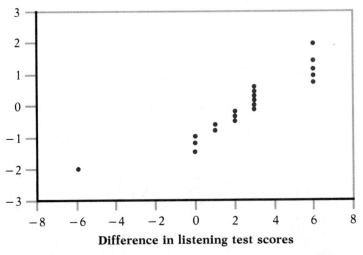

Figure 8.3 Normal quantile plot for the change in French listening scores from Example 8.3.

also granularity due to the fact that only whole-number scores are possible. The overall pattern of the plot is otherwise roughly straight. Does this non-normality forbid use of the t test? The behavior of the t procedures when the population does not have a normal distribution is one of their most important properties.

Robustness of t procedures   The one-sample t procedures are exactly correct only when the population is normal. Real populations are never exactly normal. The usefulness of the t procedures in practice therefore depends on how strongly they are affected by nonnormality.

Robust procedures

> A statistical inference procedure is called robust if the probability calculations required are insensitive to violations of the assumptions made.

The assumption that the population is normal rules out outliers, so the presence of outliers shows that this assumption is not valid. The t procedures are not robust against outliers, because $\bar{x}$ and s are not resistant to outliers. If we dropped the single outlier in Example 8.3, the test statistic would change from $t = 3.87$ to $t = 6.14$ and the P-value would be much smaller. In this case, the outlier makes the test result *less* significant and the confidence interval *wider* than they would otherwise be. The results of the t procedures in Example 8.3 are conservative in the sense that the conclusions show a smaller effect than would be the case if the outlier were not present.

Fortunately, the t procedures are quite robust against nonnormality of the population except in the case of outliers or strong skewness. Larger samples improve the accuracy of P-values and critical values from the t distributions when the population is not normal. This is true for two reasons. First, the sampling distribution of the sample mean $\bar{x}$ from a large sample is close to normal (that's the central limit theorem). We need be less concerned about the normality of the individual observations when the sample is large. Second, as the size n of a sample grows, the sample standard deviation s approaches the population standard deviation σ. This fact is closely related to the law of large numbers. In large samples, s will be an accurate estimate of σ whether or not the population has a normal distribution.

A normal quantile plot or other method to check for skewness and outliers is an important preliminary to the use of t procedures for small samples. For most purposes, the one-sample t procedures can be safely used when $n \geq 15$ unless an outlier or clearly marked skewness is present. Except in the case of small samples, the assumption that the data are an SRS from the population of interest is more crucial than the assumption that the population distribution is normal. Here are practical guidelines for inference on a single mean.[3]

- *Sample size less than 15.* Use t procedures if the data are close to normal. If the data are clearly nonnormal or if outliers are present, do not use t.
- *Sample size at least 15.* The t procedures can be used except in the presence of outliers or strong skewness.
- *Large samples.* The t procedures can be used even for clearly skewed distributions when the sample is large, roughly $n \geq 40$.

The Power of the t Test*

The power of a statistical test measures its ability to detect deviations from the null hypothesis. In practice we carry out the test in the hope of showing that the null hypothesis is false, so high power is practically important. The power of the one-sample t test against a specific alternative value of the population mean μ is the probability that the test will reject the null hypothesis when the alternative value of the mean is true. To calculate the power, we assume a fixed level of significance, usually $\alpha = 0.05$.

Calculation of the exact power of the t test takes into account the estimation of σ by s and is a bit complex. But an approximate calculation that acts as if σ were known is almost always adequate for planning a study. This calculation is very much like that for the z test, presented in Section 7.3. The method is: Write the event that the test rejects H_0 in terms of $\bar{x}$ and then find the probability of this event when the population mean has the alternative value.

EXAMPLE 8.4

It is the winter before the summer language institute of Example 8.3. The director, thinking ahead to the report he must write, hopes that the planned 20 students will enable him to be quite certain of detecting an average improvement of 2 points in the mean listening score. Is this realistic?

We wish to compute the power of the t test for

$$H_0: \mu = 0$$
$$H_a: \mu > 0$$

against the alternative $\mu = 2$ when $n = 20$. We must have a rough guess of the size of σ in order to compute the power. In planning a large study, a pilot study is often run for this and other purposes. In this case, listening-score improvements in past summer language institutes have had sample standard deviations of about 3. We therefore take both $\sigma = 3$ and $s = 3$ in our approximate calculation.

The t test with 20 observations rejects H_0 at the 5% significance level if the t statistic

$$t = \frac{\bar{x} - 0}{s/\sqrt{20}}$$

---

* This section can be omitted without loss of continuity.

exceeds the upper 5% point of $t(19)$, which is 1.729. Taking $s = 3$, the event that the test rejects H_0 is therefore

$$t = \frac{\bar{x}}{3/\sqrt{20}} \geq 1.729$$

$$\bar{x} \geq 1.729 \frac{3}{\sqrt{20}}$$

$$\geq 1.160$$

The power is the probability that $\bar{x} \geq 1.160$ when $\mu = 2$. Taking $\sigma = 3$, this probability is

$$P(\bar{x} \geq 1.160 \,|\, \mu = 2) = P\left(\frac{\bar{x} - 2}{3/\sqrt{20}} \geq \frac{1.160 - 2}{3/\sqrt{20}}\right)$$

$$= P(Z \geq -1.252)$$

$$= 1 - .1056 = .8944$$

A true difference of 2 points in the population mean scores will produce significance at the 5% level in 89% of all possible samples. The director can be reasonably confident of detecting a difference this large. ∎

Inference for Nonnormal Populations*

We have not discussed how to do inference about the mean of a clearly nonnormal distribution based on a small sample. If you face this problem, you should consult an expert. Three general strategies are available.

- In some cases a distribution other than a normal distribution will describe the data well. There are many nonnormal models for data, and inference procedures for these models are available.

- Since skewness is the chief barrier to the use of t procedures on data without outliers, you can attempt to transform skewed data so that the distribution is symmetric and as close to normal as possible. Confidence levels and P-values from the t procedures applied to the transformed data will be quite accurate for even moderate sample sizes.

distribution-free procedure

nonparametric procedures

- Another strategy is to use a *distribution-free* inference procedure. Such procedures do not assume that the population distribution has any specific form, such as normal. Distribution-free procedures are often called *nonparametric procedures*.

Each of these strategies can be effective, but each quickly carries us beyond the basic practice of statistics. We emphasize procedures based on normal distributions because they are the most common in practice, because their robustness makes them widely useful, and (most important)

* This section can be omitted without loss of continuity.

because we are first of all concerned with understanding the principles of inference. We will therefore not discuss procedures for nonnormal continuous distributions and will present only one of the many distribution-free procedures that do not require that the population have any specific type of distribution. We will be content with illustrating by example the use of a transformation and of a simple distribution-free procedure.

Transforming data   When the distribution of a variable is skewed, it often happens that a simple transformation results in a variable whose distribution is symmetric and even close to normal. The most common transfor-

logarithm transformation

mation is the *logarithm*. The logarithm tends to pull in the right tail of a distribution. For example, the data 2, 3, 4, 20 show an outlier in the right tail. Their logarithms 0.30, 0.48, 0.60, 1.30, are much less skewed. Taking logarithms is a possible remedy for right skewness. Instead of analyzing values of the original variable X, we first compute their logarithms and analyze the values of log X. Here is an example of this approach.

EXAMPLE 8.5

Table 3.6 presents data on the amounts (in grams per mile) of three pollutants in the exhaust of 46 vehicles of the same type, measured under standard conditions prescribed by the Environmental Protection Agency. We will concentrate on emissions of carbon monoxide (CO). We would like to give a confidence interval for the mean emissions μ for this vehicle type.[4]

A normal quantile plot of the CO data from Table 3.6 (Figure 8.4) shows that the distribution is skewed to the right. Because there are no extreme outliers, the sample mean of 46 observations will nonetheless have an approximately normal sampling distribution. The t procedures could be used for approximate inference. For more exact inference, we will transform the data so that the distribution

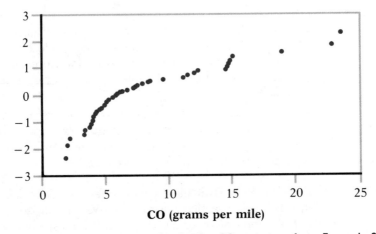

Figure 8.4 Normal quantile plot for CO emissions from Example 8.5. The distribution is skewed to the right.

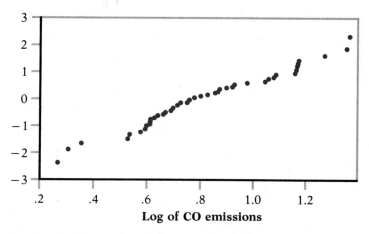

Figure 8.5 Normal quantile plot for the logarithms of the CO emissions from Example 8.5. This distribution is close to normal.

is more nearly normal. Figure 8.5 is a normal quantile plot of the logarithms of the CO measurements. The transformed data are very close to normal, so t procedures will give quite exact results. ∎

The application of the t procedures to the transformed data is straightforward. Call the original CO values from Table 3.6 values of the variable X. The transformed data are values of $X^* = \log X$. If the 46 values of X are entered into MINITAB as column C1, a 95% confidence interval for the mean μ^* of the transformed values X^* is found as follows:

```
MTB> LET C2 = LOGTEN(C1)

MTB> TINTERVAL 95 C2

        N      MEAN     STDEV    SE MEAN    95.0 PERCENT C.I.
C2      46    0.8198    0.2654    0.0391    ( 0.7409, 0.8986)
```

These commands put the logarithms of the CO measurements in column C2 and then find the 95% t confidence interval for μ^* from the transformed values. For comparison, the 95% t confidence interval for the original mean μ is found from the original data as follows:

```
MTB> TINTERVAL 95 C1

        N      MEAN     STDEV    SE MEAN    95.0 PERCENT C.I.
C1      46    7.960     5.261     0.776     ( 6.398, 9.523)
```

The advantage of analyzing transformed data is that use of procedures based on the normal distributions is better justified and the results are more exact. The disadvantage is that a confidence interval for the mean μ in the original scale of emissions measured in grams per mile cannot be recovered from the confidence interval for μ^*. The reason is that the mean μ^* of log X is *not* the logarithm of the mean μ of X. The MINITAB output above illustrates this annoying fact in the case of the sample mean. The mean $\bar{x}$ of the values of X is 7.960. The logarithm of 7.960 is 0.9009, which is not equal to the sample mean 0.8198 of the logarithms of the emissions. So we cannot transform the endpoints of the 95% confidence interval for μ^* back to the original scale and obtain a 95% confidence interval for μ.

In some cases we are content to abandon the original scale and do all our work in the scale that leads to a normal distribution. The use of logarithmic scales in particular is common in science. In Example 8.5, however, we would like to state our conclusions in terms of actual emissions (so many grams of CO per mile driven). If we are interested only in the mean, analysis of the original data is more attractive than working with the transformed data; this analysis is justified by the moderately large sample size. For other purposes that require normality of the individual observations rather than just normality of $\bar{x}$, the transformed data are superior.

The sign test   Perhaps the most straightforward way to cope with nonnormal data is to use a *distribution-free* procedure. As the name indicates, these procedures do not require the population distribution to have any specific form, such as normal. Distribution-free significance tests are quite simple and are available in most statistical software systems. Distribution-free tests have two drawbacks. First, they are generally less powerful than tests designed for use with a specific distribution, such as the t test. Second, we must often modify the statement of the hypotheses in order to use a distribution-free test. A distribution-free test concerning the center of a distribution, for example, is usually stated in terms of the median rather than the mean. This is sensible when the distribution may be skewed. But

sign test

the distribution-free test no longer asks the same question (Has the mean changed?) that the t test does. The simplest distribution-free test, and one of the most useful, is the *sign test*. The following example illustrates this test.

EXAMPLE 8.6

Return to the data of Example 8.3 showing the improvement in French listening scores after a summer institute. In that example we used the one-sample t test on these data, despite granularity and an outlier that make the P-value only roughly correct. The sign test is based on the following simple observation: of the 17 teachers whose scores changed, 16 improved and only 1 did more poorly. This is evidence that the institute improved French listening skills.

To give a significance test based on the count of teachers whose scores improved, let p be the probability that a randomly chosen teacher would improve if she attended the institute. The null hypothesis of "no effect" says that the posttest is just a repeat of the pretest with no change in ability, so a teacher is equally likely to do better on either test. We therefore want to test

$$H_0: p = 1/2$$
$$H_a: p > 1/2$$

The 17 teachers whose scores changed are 17 independent trials, so the number who improve has the binomial distribution $B(17, 1/2)$ if H_0 is true. The P-value for the observed count 16 is therefore $P(X \geq 16)$, where X has the $B(17, 1/2)$ distribution. You can compute this probability from computer software or from the binomial probability formula, Equation 6.3 on page 397:

$$P(X \geq 16) = P(X = 16) + P(X = 17)$$
$$= \binom{17}{16}\left(\frac{1}{2}\right)^{16}\left(\frac{1}{2}\right)^{1} + \binom{17}{17}\left(\frac{1}{2}\right)^{17}\left(\frac{1}{2}\right)^{0}$$
$$= (17)\left(\frac{1}{2}\right)^{17} + \left(\frac{1}{2}\right)^{17}$$
$$= 0.00014$$

(The normal approximation to the binomial gives the same result, even if the continuity correction is not used.) As in Example 8.3, there is very strong evidence that participation in the institute has improved performance on the listening test.

■

There are several varieties of sign test, all based on counts and the binomial distribution. The sign test for paired comparisons (Example 8.6) is the most useful. The null hypothesis of "no effect" is then always $H_0: p = 1/2$. The alternative can be one-sided in either direction or two-sided, depending on the type of change we are looking for. The test gets its name from the fact that we need look only at the signs of the differences, not their actual values.

Sign test for paired comparisons

> Ignore pairs with difference 0; the number of trials n is the count of the remaining pairs. The test statistic is the count X of pairs with a positive difference. P-values for X are based on the binomial $B(n, 1/2)$ distribution.

The paired comparisons t test in Example 8.3 tested the hypothesis that the mean of the distribution of differences (score after the institute minus score before) is 0. The sign test in Example 8.6 is in fact testing the hypothesis that the *median* of the differences is 0. If p is the probability that a difference is positive, then $p = 1/2$ when the median is 0. This is true because

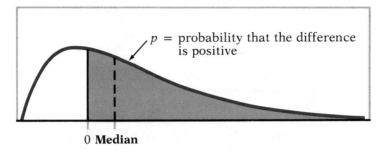

Figure 8.6 Why the sign test tests the median difference: when the median is greater than 0, the probability p of a positive difference is greater than 1/2, and vice versa.

the median of the distribution is the point with probability 1/2 above it. As Figure 8.6 illustrates, $p > 1/2$ when the median is greater than 0, again because the probability above the median is always just 1/2. Let η stand for the median of the difference in scores for an entire population of French teachers who might attend a summer institute. Then the sign test of $H_0 : p = 1/2$ against $H_a : p > 1/2$ is a test of

$$H_0 : \eta = 0$$
$$H_0 : \eta > 0$$

The sign test in Example 8.6 makes no use of the actual scores—it just counts how many teachers improved. The teachers whose scores did not change were ignored altogether. Because the sign test uses so little of the available information, it is much less powerful than the t test when the population is close to normal. There are other distribution-free tests that are more powerful than the sign test.[5]

SUMMARY

Tests and confidence intervals for the mean μ of a normal population are based on the sample mean $\bar{x}$ of an SRS. Because of the central limit theorem, the resulting procedures are approximately correct for other population distributions when the sample is large.

The standardized sample mean, or **one-sample z statistic**,

$$z = \frac{\bar{x} - \mu}{\sigma/\sqrt{n}}$$

has the $N(0, 1)$ distribution. If the standard deviation $\sigma/\sqrt{n}$ of $\bar{x}$ is replaced by the **standard error** $s/\sqrt{n}$, the **one-sample t statistic**

$$t = \frac{\bar{x} - \mu}{s/\sqrt{n}}$$

has the **t distribution** with $n - 1$ degrees of freedom.

An exact level $1 - \alpha$ confidence interval for the mean μ of a normal population is

$$\bar{x} \pm t^* \frac{s}{\sqrt{n}}$$

where t^* is the upper $\alpha/2$ critical value of the $t(n - 1)$ distribution.

Significance tests for $H_0: \mu = \mu_0$ are based on the t statistic. P-values or fixed significance levels are computed from the $t(n - 1)$ distribution.

These one-sample procedures are used to analyze **paired comparisons** data by first taking the difference within each matched pair to produce a single sample.

The t procedures are relatively **robust** against nonnormal populations, especially for large sample sizes. The t procedures are useful for nonnormal data when $n \geq 15$ unless the data show outliers or strong skewness. The $t(k)$ distribution approaches the $N(0, 1)$ distribution as k increases.

The power of the t test is calculated like that of the z test, using an approximate value for both σ and s.

Small samples from skewed populations can sometimes be analyzed by first applying a transformation (such as the logarithm) to obtain an approximately normally distributed variable. The t procedures then apply to the transformed data.

The **sign test** is a **distribution-free test** because it uses probability calculations that are correct for a wide range of population distributions.

The sign test for "no treatment effect" in paired comparisons counts the number of positive differences. The P-value is computed from the $B(n, 1/2)$ distribution, where n is the number of non-0 differences. The sign test is less powerful than the t test in cases where use of the t test is justified.

SECTION 8.1 EXERCISES

If you are not using computer software, find the P-values in these exercises by using Table E to give two values between which P lies.

8.1 What critical value t^* from Table E should be used for a confidence interval for the mean of the population in each of the following situations?

(a) A 95% confidence interval based on $n = 10$ observations

(b) A 99% confidence interval from an SRS of 20 observations

(c) An 80% confidence interval from a sample of size 7

8.2 The one-sample t statistic for testing

$$H_0: \mu = 0$$
$$H_a: \mu > 0$$

from a sample of $n = 15$ observations has the value $t = 1.82$.

(a) What are the degrees of freedom for this statistic?

(b) Give the two critical values t^* from Table E that bracket t. What are the upper-tail probabilities p for these two entries?

(c) Between what two values does the P-value of the test fall?

(d) Is the value $t = 1.82$ significant at the 5% level? Is it significant at the 1% level?

8.3 The one-sample t statistic from a sample of $n = 25$ observations for the two-sided test of

$$H_0: \mu = 64$$
$$H_a: \mu \neq 64$$

has the value $t = 1.12$.

(a) What are the degrees of freedom for t?

(b) Locate the two critical values t^* from Table E that bracket t. What are the upper-tail probabilities p for these two values?

(c) Between what two values does the P-value of the test fall? (Note that H_a is two-sided.)

(d) Is the value $t = 1.12$ statistically significant at the 10% level? At the 5% level?

8.4 The one-sample t statistic for a test of

$$H_0: \mu = 10$$
$$H_a: \mu < 10$$

based on $n = 10$ observations has the value $t = -2.25$.

(a) What are the degrees of freedom for this statistic?

(b) Between what two probabilities p from Table E does the P-value of the test fall?

8.5 The scores of four roommates on the Law School Aptitude Test have mean $\bar{x} = 589$ and standard deviation $s = 37$. What is the standard error of the mean?

8.6 A manufacturer of small appliances employs a market research firm to estimate retail sales of its products by gathering information from a sample of retail stores. This month an SRS of 75 stores in the Midwest sales region finds that these stores sold an average of 24 of the manufacturer's hand mixers, with standard deviation 11.

(a) Give a 95% confidence interval for the mean number of mixers sold by all stores in the region.

(b) The distribution of sales is strongly right-skewed, because there are many smaller stores and a few very large stores. The use of t in (a) is reasonably safe despite this violation of the normality assumption. Why?

8.7 A bank wonders whether omitting the annual credit card fee for customers who charge at least $2400 in a year would increase the amount charged on its credit card. The bank makes this offer to an SRS of 200 of its existing credit card customers. It then compares how much these customers charge this year with the amount that they charged last year. The mean increase is $332, and the standard deviation is $108.

(a) Is there significant evidence at the 1% level that the mean amount charged increases under the no-fee offer? State H_0 and H_a and carry out a t test.

(b) Give a 99% confidence interval for the mean amount of the increase.

(c) The distribution of the amount charged is skewed to the right, but outliers are prevented by the credit limit that the bank enforces on each card. Use of the t procedures is justified in this case even though the population distribution is not normal. Explain why.

(d) A critic points out that the customers would probably have charged more this year than last even without the new offer, because the economy is more prosperous and interest rates are lower. Briefly describe the design of an experiment to study the effect of the no-fee offer that would avoid this criticism.

8.8 The level of various substances in the blood of kidney dialysis patients is of concern because kidney failure and dialysis can lead to additional problems. A researcher did blood tests on several dialysis patients on six consecutive clinic visits. One variable measured was the level of phosphate in the blood. Phosphate levels for a single person tend to vary normally over time. The data on one patient, in milligrams of phosphate per deciliter (mg/dl) of blood, are given below. (The data are from Joan M. Susic, *Dietary Phosphorus Intakes, Urinary and Peritoneal Phosphate Excretion and Clearance in Continuous Ambulatory Peritoneal Dialysis Patients*, M.S. thesis, Purdue University, 1985.)

$$5.6 \quad 5.1 \quad 4.6 \quad 4.8 \quad 5.7 \quad 6.4$$

(a) Calculate the sample mean $\bar{x}$ and also its standard error.

(b) Use the t procedures to give a 90% confidence interval for this patient's mean phosphate level.

8.9 Poisoning by the pesticide DDT causes tremors and convulsions. In a study of DDT poisoning, researchers fed several rats a measured amount of DDT. They then measured electrical characteristics of the rats' nervous systems that might explain how DDT poisoning causes tremors. One important variable was the "absolutely refractory period," the time required for a nerve to recover after a stimulus. This period varies normally. Measurements on four rats gave the data below (in milliseconds). (Data from D. L. Shankland, "Involvement of spinal cord and peripheral nerves in DDT-poisoning syndrome in albino rats," *Toxicology and Applied Pharmacology*, 6 (1964), pp. 197–213.)

$$1.6 \quad 1.7 \quad 1.8 \quad 1.9$$

(a) Find the mean refractory period $\bar{x}$ and the standard error of the mean.

(b) Give a 90% confidence interval for the mean "absolutely refractory period" for all rats of this strain when subjected to the same treatment.

8.10 The normal range of values for blood phosphate levels is 2.6 to 4.8 mg/dl. The sample mean for the patient in Exercise 8.8 falls above this range. Is this good evidence that the patient's mean level in fact falls above 4.8? State H_0 and H_a and use the data in Exercise 8.8 to carry out a t test. Between which levels from Table E does the P-value lie? Are you convinced that the patient's phosphate level is higher than normal?

8.11 Suppose that the mean "absolutely refractory period" for unpoisoned rats is known to be 1.3 milliseconds. Do the data in Exercise 8.9 give good evidence that the mean is larger in rats poisoned with DDT? State H_0 and H_a and do a t test. Between what levels from Table E does the P-value lie? What do you conclude from the test?

8.12 In a randomized comparative experiment on the effect of calcium in the diet on blood pressure, 54 healthy white males were divided at random into two groups. One group received calcium; the other, a placebo. At the beginning of the study, the researchers measured many variables on the subjects. The paper reporting the study gives $\bar{x} = 114.9$ and $s = 9.3$ for the seated systolic blood pressure of the 27 members of the placebo group. (From R. M. Lyle et al., "Blood pressure and metabolic effects of calcium supplementation in normotensive white and black men," *Journal of the American Medical Association*, 257 (1987), pp. 1772–1776.)

(a) Give a 95% confidence interval for the mean blood pressure of the population from which the subjects were recruited.

(b) What assumptions about the population and the study design are required by the procedure you used in (a)? Which of these assumptions are important for the validity of the procedure in this case?

8.13 The Acculturation Rating Scale for Mexican Americans (ARSMA) is described in Exercise 7.6. During the development of ARSMA, the test was given to a group of 17 Mexicans. Their scores, from a possible range of 1.00 to 5.00, had $\bar{x} = 1.67$ and $s = 0.25$. Since low scores should indicate a Mexican cultural orientation, these results helped to establish the validity of the test. (Based on I. Cuellar, L. C. Harris, and R. Jasso, "An acculturation scale for Mexican American normal and clinical populations," *Hispanic Journal of Behavioral Sciences*, 2 (1980), pp. 199–217.)

(a) Give a 95% confidence interval for the mean ARSMA score of Mexicans.

(b) What assumptions does your confidence interval require? Which of these assumptions is most important in this case?

8.14 Exercise 1.11 gives a dimension measurement for 16 crankshafts. The mean dimension is supposed to be 224 millimeters (mm), and the variability of the manufacturing process is unknown. Is there evidence that the mean dimension is not 224 mm?

(a) Check the data graphically for outliers or strong skewness that might threaten the validity of the *t* procedures.

(b) State H_0 and H_a and carry out a *t* test. Use Table E to give two levels between which the *P*-value falls, or give the exact *P*-value if you use software. What do you conclude?

8.15 Table 3.6 gives the levels of three pollutants in the exhaust of 46 randomly selected vehicles of the same type. A normal quantile plot shows that the distribution of nitrogen oxide (NOX) emissions is approximately normal.

(a) Give a 99% confidence interval for the mean NOX level in vehicles of this type.

(b) Are the assumptions required by the procedure that you used in (a) valid in this case?

8.16 Gas chromatography is a sensitive technique used by chemists to measure small amounts of compounds. The response of a gas chromatograph is calibrated by repeatedly testing specimens containing a known amount of the compound to be measured. A calibration study for a specimen containing 1 nanogram (ng) (that's 10^{-9} gram) of a compound gave the following response readings:

$$21.6 \quad 20.0 \quad 25.0 \quad 21.9$$

The response is known from experience to vary according to a normal distribution unless an outlier indicates an error in the analysis. Estimate the mean response to 1 ng of this substance, and give a margin of error of your choosing. Then explain to a chemist who knows no statistics what your margin of error means. (Data from the appendix of D. A. Kurtz (ed.), *Trace Residue Analysis*, American Chemical Society Symposium Series, No. 284, 1985.)

8.17 The embryos of brine shrimp can enter a dormant phase in which metabolic activity drops to a low level. Researchers studying this dormant phase measured the level of several compounds important to normal metabolism. The results were reported in a table, with the note, "Values are means ± SEM for three independent samples." The table entry for the compound ATP was 0.84 ± 0.01. Biologists reading the article are presumed to be able to decipher this. (From S. C. Hand and E. Gnaiger, "Anaerobic dormancy quantified in *Artemia* embryos," *Science*, 239 (1988), pp. 1425–1427.)

(a) What does "SEM" stand for?

(b) The researchers made three measurements of ATP, which had $\bar{x} = 0.84$. What was the sample standard deviation s for these measurements?

(c) Give a 90% confidence interval for the mean ATP level in dormant brine shrimp embryos.

8.18 The design of controls and instruments has a large effect on how easily people can use them. A student project investigated this effect by asking 25 right-handed students to turn a knob (with their right hands) that moved an indicator by screw action. There were two identical instruments, one with a right-hand thread (the knob turns clockwise) and the other with a left-hand thread (the knob must be turned counter-clockwise). The table below gives the times required (in seconds) to move the indicator a fixed distance. (Data provided by Timothy Sturm.)

Subject	Right thread	Left thread	Subject	Right thread	Left thread
1	113	137	14	107	87
2	105	105	15	118	166
3	130	133	16	103	146
4	101	108	17	111	123
5	138	115	18	104	135
6	118	170	19	111	112
7	87	103	20	89	93
8	116	145	21	78	76
9	75	78	22	100	116
10	96	107	23	89	78
11	122	84	24	85	101
12	103	148	25	88	123
13	116	147			

(a) Each of the 25 students used both instruments. Discuss briefly how the experiment should be arranged and how randomization should be used.

(b) The project hoped to show that right-handed people find right-hand threads easier to use. State the appropriate H_0 and H_a about the mean time required to complete the task.

(c) Carry out a test of your hypotheses. Give the *P*-value and report your conclusions.

8.19 Give a 90% confidence interval for the mean time advantage of right-hand over left-hand threads in the setting of the previous exercise. Do you think that the time saved would be of practical importance if the task were performed many times, for example by an assembly line worker? To help answer this question, find the mean time for right-hand threads as a percent of the mean time for left-hand threads.

8.20 The table below gives the pretest and posttest scores on the MLA listening test in Spanish for 20 high school Spanish teachers who attended an intensive summer course in Spanish. The setting is identical to the French institute described in Example 8.3. (Data provided by Professor Joseph A. Wipf, Department of Foreign Languages and Literatures, Purdue University.)

Subject	Pretest	Posttest	Subject	Pretest	Posttest
1	30	29	11	30	32
2	28	30	12	29	28
3	31	32	13	31	34
4	26	30	14	29	32
5	20	16	15	34	32
6	30	25	16	20	27
7	34	31	17	26	28
8	15	18	18	25	29
9	28	33	19	31	32
10	20	25	20	29	32

(a) We hope to show that attending the institute improves listening skills. State the appropriate H_0 and H_a. Be sure to identify the parameters appearing in the hypotheses.

(b) Make a graphical check for outliers or strong skewness in the data that you will use in your statistical test, and report your conclusions on the validity of the test.

(c) Carry out a test. Can you reject H_0 at the 5% significance level? At the 1% significance level?

(d) Give a 90% confidence interval for the mean increase in listening score due to attending the summer institute.

8.21 The ARSMA test (Exercises 7.6 and 8.13) was compared with a similar test, the Bicultural Inventory (BI), by administering both tests to 22 Mexican-Americans. Both tests have the same range of scores (1.00 to 5.00) and are scaled to have similar means for the groups used to develop them. There was a high correlation between the two scores, giving evidence that both are measuring the same characteristics. The researchers wanted to know whether the population mean scores for the two tests are the same. The differences in scores (ARSMA − BI) for the 22 subjects had $\bar{x} = 0.2519$ and $s = 0.2767$.

(a) Describe briefly how the administration of the two tests to the subjects should be conducted, including randomization.

(b) Carry out a significance test for the hypothesis that the two tests have the same population mean. Give the *P*-value and state your conclusion.

(c) Give a 95% confidence interval for the difference between the two population mean scores.

8.22 It is claimed that a new filter for filter-tipped cigarettes leaves less nicotine in the smoke than does the current filter. Because cigarette brands differ in a number of ways, each filter is tested on one cigarette of each of nine brands. The difference between the nicotine content for the current filter and the new filter is recorded. The mean difference is $\bar{x} = 1.32$ milligrams (mg), and the standard deviation of the differences is $s = 2.35$ mg.

(a) How significant is the observed difference in means? State H_0 and H_a, and give a *P*-value.

(b) Give a 90% confidence interval for the mean amount of additional nicotine removed by the new filter.

8.23 The yield of two varieties of tomatoes for commercial use is compared in a field trial. Small plots of land in 10 locations are divided in half, and each tomato variety is planted on one-half of each plot. After harvest, the yields in pounds per plant are compared at each location. The 10 differences (variety A − variety B) give the following statistics: $\bar{x} = 0.34$ and $s = 0.83$. Is there convincing evidence that variety A has the higher mean yield? State H_0 and H_a and give a *P*-value to answer this question.

8.24 The following situations all require inference about a mean or means. Identify each as (1) a single sample; (2) paired comparisons; or (3) two independent samples. The procedures of this section apply to cases (1) and (2). We will learn procedures for (3) in the next section.

(a) An education researcher wants to learn whether inserting questions before or after introducing a new concept in an elementary school mathematics text is more effective. He prepares two text segments that teach the concept, one with motivating questions

ore and the other with review questions after. Each text
ment is used to teach a group of children, and their scores on
est over the material are compared.

other researcher approaches the same issue differently. She
epares text segments on two unrelated topics. Each segment
mes in two versions, one with questions before and the other
ith questions after. Each of a group of children is taught both
pics, one (chosen at random) with questions before and the
ther with questions after. Test scores for each child on the two
opics are compared to see which topic he or she learned better.

o check a new analytical method, a chemist obtains a reference
pecimen of known concentration from the National Bureau of
Standards. She then makes 20 measurements of the concentration
of this specimen with the new method and checks for bias by
comparing the mean result with the known concentration.
Another chemist is checking the same new method. He has no
reference specimen, but a familiar analytic method is available.
He wants to know if the new and old methods agree. He takes
a specimen of unknown concentration and measures the
concentration 10 times with the new method and 10 times with
the old method.

xercise 1.18 gives a table of the percent of residents 65 years of
ge and over in each of the 50 states. It does not make sense to use
he t procedures (or any other statistical procedures) to give a 95%
onfidence interval for the mean percent of residents of age 65 in
he population of American states. Explain why not.

ollowing exercises concern the optional material in the sections on the
r of the t test and on nonnormal populations.

The bank in Exercise 8.7 tested a new idea on a sample of 200
customers. Suppose that the bank wanted to be quite certain of
detecting a mean increase of $\mu = \$100$ in the amount charged, at the
$\alpha = 0.01$ significance level. Perhaps a sample of only $n = 50$
customers would accomplish this. Find the approximate power of the
test with $n = 50$ against the alternative $\mu = \$100$ as follows:

(a) What is the t critical value for the one-sided test with $\alpha = 0.01$
and $n = 50$?

(b) Write the criterion for rejecting H_0: $\mu = 0$ in terms of the t statistic.
Then take $s = 108$ (an estimate based on the data in Exercise 8.7)
and state the rejection criterion in terms of $\bar{x}$.

(c) Assume that $\mu = \$100$ (the given alternative) and that $\sigma = 108$ (an
estimate from the data in Exercise 8.7). The approximate power
is the probability of the event you found in (b), calculated under
these assumptions.

8.27 The tomato experts who carried out the field trial described in Exercise 8.23 suspect that the relative lack of significance there is due to low power. They would like to be able to detect a mean difference in yields of 0.5 pound per plant at the 0.05 significance level. Based on the previous study, use 0.83 as an estimate of both the population σ and the value of s in future samples.

    (a) What is the power of the test from Exercise 8.23 with $n = 10$ against the alternative $\mu = 0.5$?

    (b) If the sample size is increased to $n = 25$ plots of land, what will be the power against the same alternative?

8.28 Exercise 8.21 reports a small study comparing ARSMA and BI, two tests of the acculturation of Mexican-Americans. Would this study usually detect a difference in mean scores of 0.2? To answer this question, calculate the approximate power of the test (with $n = 22$ subjects and $\alpha = 0.05$) of

$$H_0: \mu = 0$$
$$H_a: \mu \neq 0$$

against the alternative $\mu = 0.2$. Note that this is a two-sided test.

    (a) From Table E, what is the critical value for $\alpha = 0.05$?

    (b) Write the criterion for rejecting H_0 at the $\alpha = 0.05$ level. Then take $s = 0.3$, the approximate value observed in Exercise 8.21, and restate the rejection criterion in terms of $\bar{x}$.

    (c) Find the probability of this event when $\mu = 0.2$ (the alternative given) and $\sigma = 0.3$ (estimated from the data in Exercise 8.21) by a normal probability calculation. This is the approximate power.

8.29 Apply the sign test to the data in Exercise 8.18 to assess whether the subjects can complete a task with right-hand thread significantly faster than with left-hand thread.

    (a) State the hypotheses two ways, in terms of a population median and in terms of the probability of completing the task faster with a right-hand thread.

    (b) Carry out the sign test. Find the P-value and draw a conclusion.

8.30 Use the sign test to assess whether the summer institute of Exercise 8.20 improves Spanish listening skills. State the hypotheses, give the P-value, and report your conclusion.

8.31 The paper reporting the results on ARSMA used in Exercise 8.21 does not give the raw data or any discussion of normality. You would like to replace the t procedure used in Exercise 8.21 by a sign test. Can you do this from the available information? Carry out the sign test and state your conclusion, or explain why you are unable to carry out the test.

8.32 In the tomato field trial of Exercise 8.23, variety A had the higher yield in 6 of the 10 locations. Variety B had the higher yield in the other 4 locations. Use the sign test to give a *P*-value for testing the hypothesis that the median difference in yields (A minus B) is positive.

8.33 The data below are the survival times of 72 guinea pigs after they were injected with tubercle bacilli in a medical experiment. Figure 1.24 is a normal quantile plot of these data. The distribution is strongly skewed to the right. (Data from T. Bjerkedal, "Acquisition of resistance in guinea pigs infected with different doses of virulent tubercle bacilli," *American Journal of Hygiene*, 72 (1960), pp. 130–148.)

43	45	53	56	56	57	58	66	67	73
74	79	80	80	81	81	81	82	83	83
84	88	89	91	91	92	92	97	99	99
100	100	101	102	102	102	103	104	107	108
109	113	114	118	121	123	126	128	137	138
139	144	145	147	156	162	174	178	179	184
191	198	211	214	243	249	329	380	403	511
522	598								

(a) Give a 95% confidence interval for the mean survival time by applying the *t* procedures to these data.
(b) Transform the data by taking the logarithm of each value. Display the transformed data by either a histogram or a normal quantile plot. The distribution of the logarithms remains somewhat right-skewed, but is much closer to symmetry than the original distribution. Probability values from the *t* distribution will be more accurate for the transformed data.
(c) Give a 95% confidence interval for the mean of the log survival time by applying the *t* procedures to the transformed data.

8.34 A manufacturer of electric motors tests insulation at a high temperature (250°C) and records the number of hours until the insulation fails. The data for five specimens are

300   324   372   372   444

(Data from Wayne Nelson, *Applied Life Data Analysis*, Wiley, New York, 1982, p. 471.)

(a) Make a stemplot of these data and also a normal quantile plot if your software permits. Then take the logarithm of the failure times and display the transformed data in the same way. Briefly compare the two distributions. The small sample size makes judgment from the data difficult, but engineering theory predicts that the logarithms of the failure times will have a normal distribution.

(b) Give a 90% confidence interval for the mean of the log failure times, using *t* methods.

8.2 COMPARING TWO MEANS

A medical researcher is interested in the effect on blood pressure of added calcium in our diet. She conducts a randomized comparative experiment in which one group of subjects is given a calcium supplement and a control group gets a placebo. A psychologist develops a test that measures social insight. He compares the social insight of male college students to that of female college students by giving the test to a large group of students of each sex. An educator hopes that involving elementary school children in directed reading activities in class will improve their reading ability. She incorporates such activities in one class, while a second class follows the same curriculum without the special activities. Two-sample problems such as these are among the most common situations encountered in statistical practice.

Two-sample problems

- Statistical inference must compare the responses in two groups.
- Each group is considered to be a sample from a distinct population.
- The responses in each group are independent of those in the other group.

A two-sample problem can arise from a randomized comparative experiment that randomly divides the subjects into two groups and exposes each group to a different treatment. Comparing random samples separately selected from two populations is also a two-sample problem. Unlike the paired comparisons designs studied earlier, there is no matching of the units in the two samples and the two samples may be of different sizes. Inference procedures for two-sample data differ from those for paired comparisons.

We can gain an overview of two-sample data from a back-to-back stemplot (for small samples) or from side-by-side boxplots (for larger samples). Now we will apply the ideas of formal inference in this setting. When both population distributions are symmetric, and especially when they are at least approximately normal, a comparison of the mean responses in the two populations is most often the goal of inference.

We have two independent samples, from two distinct populations (such as subjects given a treatment and those given a placebo). The same variable is measured for both samples. We will call the variable x_1 in the first population and x_2 in the second because the variable may have different distributions in the two populations. Here is the notation that we will use to describe the two populations:

Population	Variable	Mean	Standard deviation
1	x_1	μ_1	σ_1
2	x_2	μ_2	σ_2

We want to compare the two population means, either by giving a confidence interval for $\mu_1 - \mu_2$ or by testing the hypothesis of no difference, $H_0: \mu_1 = \mu_2$.

Inference is based on two independent SRSs, one from each population. Here is the notation used to describe the samples:

Population	Sample size	Sample mean	Sample standard deviation
1	n_1	$\bar{x}_1$	s_1
2	n_2	$\bar{x}_2$	s_2

Throughout this section, the subscripts 1 and 2 show the population to which a parameter or a sample statistic refers.

The Two-Sample z Statistic

The natural estimator of the difference $\mu_1 - \mu_2$ is the difference between the sample means, $\bar{x}_1 - \bar{x}_2$. If we are to base inference on this statistic, we must know its sampling distribution. Our knowledge of probability is equal to the task. We must recall the facts about the distribution of a sample mean and the addition rules for means and variances of a distribution. First, the mean of the sampling distribution is

$$\mu(\bar{x}_1 - \bar{x}_2) = \mu(\bar{x}_1) - \mu(\bar{x}_2)$$
$$= \mu_1 - \mu_2$$

Because the samples are independent, their sample means $\bar{x}_1$ and $\bar{x}_2$ are independent random variables. The variance of the difference is therefore the sum of their variances, so that

$$\sigma^2(\bar{x}_1 - \bar{x}_2) = \sigma^2(\bar{x}_1) + \sigma^2(\bar{x}_2)$$
$$= \frac{\sigma_1^2}{n_1} + \frac{\sigma_2^2}{n_2}$$

We now know the mean and variance of the distribution of $\bar{x}_1 - \bar{x}_2$ in terms of the parameters of the two populations. If the two population distributions are both normal, then the distribution of $\bar{x}_1 - \bar{x}_2$ is also normal. This is true because each sample mean alone is normally distributed and because a difference of independent normal random variables is again normal.

EXAMPLE 8.7

The Survey of Study Habits and Attitudes (SSHA) is a psychological test designed to measure the motivation, study habits, and attitudes toward learning of college students. These factors, along with ability, are important in explaining success in school. Scores on the SSHA range from 0 to 200. The mean score for women students on the SSHA is typically somewhat higher than the mean score for men at the same college. Suppose that the scores of all freshman women at Upper Wabash Tech have mean $\mu_1 = 120$ and standard deviation $\sigma_1 = 28$, while the scores for the population of freshman men have mean $\mu_2 = 105$ and standard deviation $\sigma_2 = 35$. A psychologist gives the SSHA test to an SRS of 10 women and an SRS of 12 men from the freshman class and compares the mean scores for these samples.

The difference $\bar{x}_1 - \bar{x}_2$ between the female and male mean scores varies in repeated sampling from the freshman class. The sampling distribution has mean

$$\mu(\bar{x}_1 - \bar{x}_2) = \mu_1 - \mu_2$$
$$= 120 - 105 = 15$$

and variance

$$\sigma^2(\bar{x}_1 - \bar{x}_2) = \frac{\sigma_1^2}{n_1} + \frac{\sigma_2^2}{n_2}$$
$$= \frac{28^2}{10} + \frac{35^2}{12} = 180.48$$

The standard deviation of the difference in sample means is therefore

$$\sigma(\bar{x}_1 - \bar{x}_2) = \sqrt{180.48} = 13.43$$

What is the probability that the particular 10 men chosen will have a higher mean SSHA score than the particular 12 women chosen? If scores vary normally, the difference in sample means is also normally distributed. We can standardize $\bar{x}_1 - \bar{x}_2$ by subtracting its mean 15 and dividing by its standard deviation 13.43. Therefore,

$$P[(\bar{x}_1 - \bar{x}_2) < 0] = P\left[\frac{(\bar{x}_1 - \bar{x}_2) - 15}{13.43} < \frac{0 - 15}{13.43}\right]$$
$$= P(Z < -1.12) = .1314$$

The men chosen will score higher than the women chosen in about 13% of all samples, despite the superior scores of women in the entire population. ∎

As Example 8.7 reminds us, any normal random variable has the $N(0, 1)$ distribution when standardized. We have arrived at a new z statistic.

Two-sample *z* statistic

Suppose that $\bar{x}_1$ is the mean of an SRS of size n_1 drawn from an $N(\mu_1, \sigma_1)$ population and that $\bar{x}_2$ is the mean of an independent SRS of size n_2 drawn from an $N(\mu_2, \sigma_2)$ population. Then the two-sample z statistic

$$z = \frac{(\bar{x}_1 - \bar{x}_2) - (\mu_1 - \mu_2)}{\sqrt{\dfrac{\sigma_1^2}{n_1} + \dfrac{\sigma_2^2}{n_2}}} \tag{8.2}$$

has the standard normal $N(0, 1)$ sampling distribution.

   In the unlikely event that both population standard deviations are known, the two-sample z statistic is the basis for inference about $\mu_1 - \mu_2$. Exact z procedures are little used, because σ_1 and σ_2 are rarely known. In Chapter 7, we discussed the one-sample z procedures in order to introduce the ideas of inference. Now we pass immediately to the more useful t procedures.

Two-Sample *t* Procedures

Suppose now that the population standard deviations σ_1 and σ_2 are not known. Following the pattern of the one-sample case, we substitute the standard errors $s_i/\sqrt{n_i}$ for the standard deviations $\sigma_i/\sqrt{n_i}$ in the two-sample z statistic Equation 8.2. The result is the *two-sample t statistic*

**two-sample
t statistic**

$$t = \frac{(\bar{x}_1 - \bar{x}_2) - (\mu_1 - \mu_2)}{\sqrt{\dfrac{s_1^2}{n_1} + \dfrac{s_2^2}{n_2}}} \tag{8.3}$$

Unfortunately, this statistic does *not* have a t distribution. A t distribution replaces a $N(0, 1)$ distribution only when a single standard deviation in a z statistic is replaced by a standard error. In this case, we replaced two standard deviations by the corresponding standard errors, which does not produce a statistic having a t distribution.

   Nonetheless, the statistic in Equation 8.3 is used with t critical values in inference for two-sample problems. There are two ways to do this.

- Option 1: The distribution of the statistic t is closely approximated by a t distribution with degrees of freedom computed from the data. The degrees of freedom are generally not a whole number.

- Option 2: Procedures based on the statistic t can be used with critical values from the t distribution with degrees of freedom equal to the smaller of $n_1 - 1$ and $n_2 - 1$. These procedures are always conservative for any two normal populations.

Most statistical software systems use the two-sample t statistic with option 1 for two-sample problems unless the user requests another method. Use of this option without software is a bit complicated. We will therefore present the second, simpler, option first. We recommend that you use option 2 when doing calculations without a computer. If you use a computer package, it will automatically do the calculations for option 1. Here is a statement of the option 2 procedures that includes an explanation of just how they are "conservative."

Two-sample t procedures

Suppose that an SRS of size n_1 is drawn from a normal population with unknown mean μ_1 and that an independent SRS of size n_2 is drawn from another normal population with unknown mean μ_2. The confidence interval for $\mu_1 - \mu_2$ given by

$$(\bar{x}_1 - \bar{x}_2) \pm t^* \sqrt{\frac{s_1^2}{n_1} + \frac{s_2^2}{n_2}}$$

has confidence level at least $1 - \alpha$ no matter what the population standard deviations may be. Here t^* is the upper $\alpha/2$ critical value for the $t(k)$ distribution with k the smaller of $n_1 - 1$ and $n_2 - 1$.

To test the hypothesis H_0: $\mu_1 = \mu_2$, compute the two-sample t statistic

$$t = \frac{\bar{x}_1 - \bar{x}_2}{\sqrt{\dfrac{s_1^2}{n_1} + \dfrac{s_2^2}{n_2}}}$$

and use P-values or critical values for the $t(k)$ distribution. The true P-value or fixed significance level will always be equal to or less than the value calculated from $t(k)$ no matter what values the unknown population standard deviations have.

These two-sample t procedures always err on the safe side, reporting higher P-values and lower confidence than may be true. The gap between what is reported and the truth is quite small unless the sample sizes are both small and unequal. As the sample sizes increase, probability values based on t with the smaller of $n_1 - 1$ and $n_2 - 1$ as the degrees of freedom become more accurate.[6] The following examples illustrate the two-sample t procedures.

EXAMPLE 8.8

An educator believes that new directed reading activities in the classroom will help elementary school pupils improve some aspects of their reading ability. She arranges for a third grade class of 21 students to follow these activities for an 8-week period. A control classroom of 23 third graders follows the same curricu-

Table 8.2   DRP scores for third
graders

Treatment group		Control group	
24	56	42	46
43	59	43	10
58	52	55	17
71	62	26	60
43	54	62	53
49	57	37	42
61	33	33	37
44	46	41	42
67	43	19	55
49	57	54	28
53		20	48
		85	

lum without the activities. At the end of the 8 weeks, all students are given a
Degree of Reading Power (DRP) test, which measures the aspects of reading ability
that the treatment is designed to improve. The data appear in Table 8.2.[7]
     From the table we calculate the following summary statistics:

Group	n	$\bar{x}$	s
Treatment	21	51.48	11.01
Control	23	41.52	17.15

Separate normal quantile plots for both groups (Figure 8.7) show that both are
approximately normal. Since we hope to show that the treatment (group 1) is
better than the control (group 2), the hypotheses are

$$H_0: \mu_1 = \mu_2$$
$$H_a: \mu_1 > \mu_2$$

The two-sample t statistic is

$$t = \frac{\bar{x}_1 - \bar{x}_2}{\sqrt{\dfrac{s_1^2}{n_1} + \dfrac{s_2^2}{n_2}}}$$

$$= \frac{51.48 - 41.52}{\sqrt{\dfrac{11.01^2}{21} + \dfrac{17.15^2}{23}}} = 2.31$$

df = 20

p	.02	.01
t^*	2.197	2.528

The P-value comes from the $t(k)$ distribution with degrees of freedom k equal to
the smaller of

$$n_1 - 1 = 21 - 1 = 20 \quad \text{and} \quad n_2 - 1 = 23 - 1 = 22$$

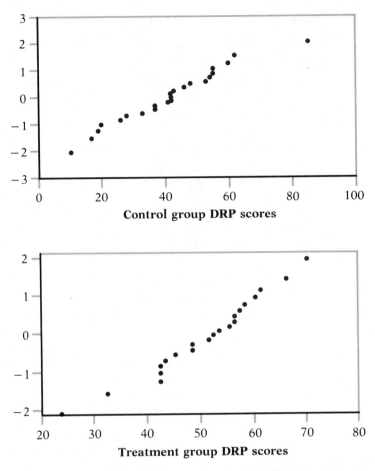

Figure 8.7   Normal quantile plots for the DRP scores in Table 8.2.

The *P*-value for the one-sided test is $P(T \geq 2.31)$. Comparing 2.31 with the entries in Table E for 20 degrees of freedom, we see that *P* lies between 0.02 and 0.01. The data give strong support to the thesis that directed reading activity improves the DRP score.

A 95% confidence interval for the mean amount of the improvement in the entire population of third graders uses the 0.025 critical value $t^* = 2.086$ of the $t(20)$ distribution. The interval is

$$(\bar{x}_1 - \bar{x}_2) \pm t^* \sqrt{\frac{s_1^2}{n_1} + \frac{s_2^2}{n_2}} = (51.48 - 41.52) \pm 2.086 \sqrt{\frac{11.01^2}{21} + \frac{17.15^2}{23}}$$

$$= 9.96 \pm 8.99$$

$$= (.97, 18.95)$$

Although we have good evidence of some improvement, the data do not allow a very precise estimate of the size of the average improvement. ∎

y in Example 8.8 is not ideal. Random assignment
e in a school environment, so existing third grade
t of the reading programs is therefore confounded
between the two classes. The classes were chosen
—for example, in the social and economic status
pretesting showed that the two classes were on
in reading ability at the beginning of the experi-
of two different teachers, the researcher herself
sses during the 8-week period of the experiment.
what confident that the two-sample test is detecting
and not some other difference between the classes.
many situations in which an experiment is carried
s not possible.

ht Test is a psychological test designed to measure how
ppraises other people. The possible scores on the test range
e development of the Chapin Test, it was given to several
pl. Here are the results for male and female college students
al arts:[8]

Sex	n	$\bar{x}$	s
Male	133	25.34	5.05
Female	162	24.94	5.44

rt the contention that female and male students differ in av-
Because no specific direction for the male/female difference
efore looking at the data, a two-sided alternative is chosen.

$$H_0: \mu_1 = \mu_2$$
$$H_a: \mu_1 \neq \mu_2$$

tistic is

$$t = \frac{\bar{x}_1 - \bar{x}_2}{\sqrt{\dfrac{s_1^2}{n_1} + \dfrac{s_2^2}{n_2}}}$$

$$= \frac{25.34 - 24.94}{\sqrt{\dfrac{5.05^2}{133} + \dfrac{5.44^2}{162}}}$$

$$= .654$$

d by comparing 0.654 to critical values for the $t(132)$ distri-
ubling p because the alternative is two-sided. Table E (for 100
) shows that 0.654 does not reach the 0.25 critical value, which

is the largest upper-tail probability in Table E. The *P*-value is therefore greater than 0.50. The data give no evidence of a male/female difference in mean social insight score. ∎

The researcher in Example 8.9 did not do an experiment but compared samples from two populations. The large samples imply that the assumption that the populations have normal distributions is of little importance. The sample means will be nearly normal in any case. The major question concerns the population to which the conclusions apply. The student subjects are certainly not an SRS of all liberal arts majors. If they are volunteers from a single college, the sample results may not extend to a wider population.

Comments The two sample *t* procedures are more robust than the one-sample *t* methods. When the sizes of the two samples are equal and the two populations being compared have similar shapes, probability values from the *t* table are quite accurate for a broad range of distributions when the sample sizes are as small as $n_1 = n_2 = 5$.[9] When the two population distributions have different shapes, larger samples are needed. As a guide to practice, the guidelines given on page 520 for the use of one-sample *t* procedures can be adapted to two-sample procedures by replacing "sample size" with the "sum of the sample sizes" $n_1 + n_2$. These guidelines are rather conservative, especially when the two samples are of equal size. In planning a two-sample study, you should usually plan equal sample sizes. The two-sample *t* procedures are most robust against nonnormality in this case, and the conservative probability values are most accurate.

As in the one-sample case, the power of the *t* test of $H_0: \mu_1 = \mu_2$ can be calculated approximately by a normal probability calculation. The calculation uses approximate values (often guesses based on past data) for both the σ_i and the s_i. Exercise 8.56 leads you through an example.

More accurate levels in the t procedures* The two-sample *t* statistic 8.3 does not have a *t* distribution. Moreover, the exact distribution changes as the unknown population standard deviations σ_1 and σ_2 change. However, the distribution can be approximated by a *t* distribution with degrees of freedom given by

$$df = \frac{\left(\dfrac{s_1^2}{n_1} + \dfrac{s_2^2}{n_2}\right)^2}{\dfrac{1}{n_1 - 1}\left(\dfrac{s_1^2}{n_1}\right)^2 + \dfrac{1}{n_2 - 1}\left(\dfrac{s_2^2}{n_2}\right)^2} \tag{8.4}$$

This approximation is quite accurate when both sample sizes n_1 and n_2 are 5 or larger. The *t* procedures remain exactly as before except that the *t*

---

* This material can be omitted unless you are using statistical software and wish to understand what the software does.

distribution with *df* degrees of freedom is used to give critical values and *P*-values.

EXAMPLE 8.10

In the DRP study of Example 8.8 we had for the data in Table 8.2

Group	n	$\bar{x}$	s
1	21	51.48	11.01
2	23	41.52	17.15

For greatest accuracy, we will use critical points from the *t* distribution with degrees of freedom given by Equation 8.4:

$$df = \frac{\left(\dfrac{11.01^2}{21} + \dfrac{17.15^2}{23}\right)^2}{\dfrac{1}{20}\left(\dfrac{11.01^2}{21}\right)^2 + \dfrac{1}{22}\left(\dfrac{17.15^2}{23}\right)^2}$$

$$= \frac{344.486}{9.099} = 37.86$$

Notice that the degrees of freedom *df* is not a whole number. The conservative 95% confidence interval in Example 8.8 used the 0.025 critical value $t^* = 2.086$ based on 20 degrees of freedom. A more exact confidence interval replaces this critical value with the 0.025 critical value for $df = 37.86$ degrees of freedom. We cannot find this critical value exactly without using a computer package. A close approximation can be found by interpolating between the two closest entries (for 30 and 40 degrees of freedom) in Table E. Instead, we just use the 30 degrees of freedom entry in Table E, $t^* = 2.042$. The 95% confidence interval is now

$$(\bar{x}_1 - \bar{x}_2) \pm t^* \sqrt{\frac{s_1^2}{n_1} + \frac{s_2^2}{n_2}} = (51.48 - 41.52) \pm 2.042 \sqrt{\frac{11.01^2}{21} + \frac{17.15^2}{23}}$$

$$= 9.96 \pm 8.80$$

$$= (1.16, 18.76)$$

This confidence interval is a bit shorter (margin of error 8.80 rather than 8.99) than the conservative interval in Example 8.8. ∎

As Example 8.10 illustrates, the two-sample *t* procedures are exactly as before, except that a *t* distribution with more degrees of freedom is used. The number *df* given by Equation 8.4 is always at least as large as the smaller of $n_1 - 1$ and $n_2 - 1$. On the other hand, *df* is never larger than the sum $n_1 + n_2 - 2$ of the two individual degrees of freedom. The number of degrees of freedom *df* is generally not a whole number. There is a *t* distribution with any positive degrees of freedom, even though Table E contains entries only for whole-number degrees of freedom. When *df* is small and is not a whole number, interpolation between entries in Table E may be needed to obtain an accurate critical value or *P*-value. Because of this and the need to calculate *df*, we do not recommend regular use of Equation 8.4 if a computer is not doing the arithmetic. With a computer, the more accurate procedures are painless, as the following example illustrates.

EXAMPLE 8.11

The pesticide DDT causes tremors and convulsions if it is ingested by humans or other mammals. Researchers seek to understand how the convulsions are caused. In a randomized comparative experiment, 6 white rats poisoned with DDT were compared with a control group of 6 unpoisoned rats. Electrical measurements of nerve activity are the main clue to how DDT poisoning works. When a nerve is stimulated, its electrical response shows a sharp spike followed by a much smaller, second spike. In rats fed DDT, the second spike was found to be larger than in normal rats. This observation helps biologists understand how DDT causes tremors.[10]

The researchers measured the amplitude of the second spike as a percentage of the first spike when a nerve in the rat's leg was stimulated. For the poisoned rats the results were

$$12.207 \quad 16.869 \quad 25.050 \quad 22.429 \quad 8.456 \quad 20.589$$

The control group data were

$$11.074 \quad 9.686 \quad 12.064 \quad 9.351 \quad 8.182 \quad 6.642$$

The difference in means is quite large, but in such small samples the sample mean is highly variable. A significance test can help confirm that we are seeing a real effect. Since the increase in the size of the second spike was not suspected in advance but was discovered in the course of the experiment, a two-sided alternative is appropriate. We want to test

$$H_0: \mu_1 = \mu_2$$
$$H_a: \mu_1 \neq \mu_2$$

Normal quantile plots (Figure 8.8) show no evidence of outliers or strong skewness. Both populations are reasonably normal, as far as can be judged from six observations.

If the treatment group data are entered into MINITAB as column C1 and the control group data as column C2, the two-sample *t* test is carried out as follows:

```
MTB> TWOSAMPLE T C1 C2

TWOSAMPLE T FOR C1 VS C2

        N     MEAN    STDEV   SE MEAN
C1      6     17.60   6.34    2.6
C2      6     9.50    1.95    0.80


95 PCT CI FOR MU C1 - MU C2: (1.1, 15.06)
TTEST MU C1 = MU C2 (VS NE): T=2.99 P=0.030 DF=5.9
```

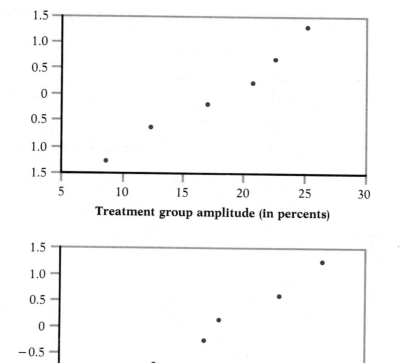

Figure 8.8 Normal quantile plots for the amplitude of the second spike as a percent of the first spike in Example 8.11.

Notice that in the absence of other instructions, MINITAB carries out a two-sided test and gives a 95% confidence interval. The two-sample t statistic has the value $t = 2.99$. The test is based on the t distribution with $df = 5.9$ degrees of freedom, obtained from Equation 8.4. The *P*-value is 0.030. There is good evidence that the mean size of the secondary spike differs in the two groups of rats. ■

Would the conservative test based on 5 degrees of freedom (both $n_1 - 1$ and $n_2 - 1$ are 5) have given a different result in Example 8.11? The statistic is exactly the same, so $t = 2.99$ as in the example. The conservative *P*-value is $P(|T| \geq 2.99)$ where T has the $t(5)$ distribution. Table E shows that 2.99 lies between the 0.02 and 0.01 upper critical values of the $t(5)$ distribution.

We double these values of p for a two-sided test. So P lies between 0.04 and 0.02. For practical purposes this is the same result as that given by MINITAB. As this example and Example 8.10 suggest, the difference between the t procedures using the conservative and approximately correct distributions is rarely of practical importance. That is why we recommend the simpler approximate procedure for inference without a computer.

The Pooled Two-Sample t Procedures*

There is one situation in which a t statistic for comparing two means has exactly a t distribution. Suppose that the two normal population distributions have the *same* standard deviation. In this case we need only substitute a single standard error in a z statistic, and the resulting t statistic has a t distribution. We will develop the z statistic first, as usual, and from it the t statistic.

Call the common—and still unknown—standard deviation of both populations σ. Both sample variances s_1^2 and s_2^2 estimate σ^2. The best way to combine these two estimates is to average them with weights equal to their degrees of freedom. This gives more weight to the information from the larger sample, which is reasonable. The resulting estimator of σ^2 is

$$s_p^2 = \frac{(n_1 - 1)s_1^2 + (n_2 - 1)s_2^2}{n_1 + n_2 - 2} \tag{8.5}$$

pooled estimator

This is called the *pooled estimator* of σ^2 because it combines the information in both samples.

When both populations have variance σ^2, the addition rule for variances says that $\bar{x}_1 - \bar{x}_2$ has variance

$$\sigma^2(\bar{x}_1 - \bar{x}_2) = \sigma^2(\bar{x}_1) + \sigma^2(\bar{x}_2)$$
$$= \frac{\sigma^2}{n_1} + \frac{\sigma^2}{n_2} = \sigma^2\left(\frac{1}{n_1} + \frac{1}{n_2}\right)$$

The standardized difference of means in this equal-variance case is therefore

$$z = \frac{(\bar{x}_1 - \bar{x}_2) - (\mu_1 - \mu_2)}{\sigma\sqrt{\dfrac{1}{n_1} + \dfrac{1}{n_2}}}$$

This is a special two-sample z statistic for the case in which the populations have the same σ. Replacing the unknown σ by the estimate s_p gives a t sta-

---

* This material can be omitted if desired, but should be read if you plan to read Chapter 11.

tistic. The degrees of freedom are $n_1 + n_2 - 2$, the sum of the degrees of freedom of the two sample variances. This statistic is the basis of the pooled two-sample t inference procedures.

Pooled two-sample t procedures

Suppose that an SRS of size n_1 is drawn from a normal population with unknown mean μ_1 and that an independent SRS of size n_2 is drawn from another normal population with unknown mean μ_2. Suppose also that the two populations have the same standard deviation. A level $1 - \alpha$ confidence interval for $\mu_1 - \mu_2$ is

$$(\bar{x}_1 - \bar{x}_2) \pm t^* s_p \sqrt{\frac{1}{n_1} + \frac{1}{n_2}}$$

Here t^* is the upper $\alpha/2$ critical value for the $t(n_1 + n_2 - 2)$ distribution.

To test the hypothesis $H_0: \mu_1 = \mu_2$, compute the pooled two-sample t statistic

$$t = \frac{\bar{x}_1 - \bar{x}_2}{s_p \sqrt{\frac{1}{n_1} + \frac{1}{n_2}}} \tag{8.6}$$

In terms of a random variable T having the $t(n_1 + n_2 - 2)$ distribution, the P-value for a test of H_0 against

$$H_a: \mu_1 > \mu_2 \quad \text{is} \quad P(T \geq t)$$
$$H_a: \mu_1 < \mu_2 \quad \text{is} \quad P(T \leq t)$$
$$H_a: \mu_1 \neq \mu_2 \quad \text{is} \quad P(|T| \geq |t|)$$

EXAMPLE 8.12

Does increasing the amount of calcium in our diet reduce blood pressure? Examination of a large sample of people revealed a relationship between calcium intake and blood pressure, but such observational studies do not establish causation. Animal experiments then showed that calcium supplements do reduce blood pressure in rats, justifying an experiment with human subjects. A randomized comparative experiment gave one group of 10 black men a calcium supplement for 12 weeks. The control group of 11 black men received a placebo that appeared identical. (In fact, a block design with black and white men as the blocks was used. We will look only at the results for blacks, since the earlier survey suggested that calcium is more effective for blacks.) The experiment was double-blind. Table 8.3 gives the seated systolic (heart contracted) blood pressure for all subjects at the beginning and end of the 12-week period. Because the researchers were interested in decreasing blood pressure, Table 8.3 also shows the decrease for each subject. An increase appears as a negative entry.[11]

To compare the effects of the two treatments, take the response variable to be the amount of the decrease in blood pressure. Group 1 is the calcium group and group 2 the placebo group. The evidence that calcium lowers blood pressure

Table 8.3 Seated systolic blood pressure

Calcium group			Placebo group		
Begin	End	Decrease	Begin	End	Decrease
107	100	7	123	124	−1
110	114	−4	109	97	12
123	105	18	112	113	−1
129	112	17	102	105	−3
112	115	−3	98	95	3
111	116	−5	114	119	−5
107	106	1	119	114	5
112	102	10	112	114	2
136	125	11	110	121	−11
102	104	−2	117	118	−1
			130	133	−3

more than a placebo is assessed by testing

$$H_0: \mu_1 = \mu_2$$
$$H_0: \mu_1 > \mu_2$$

Normal quantile plots (Figure 8.9) show no outliers or strong skewness. The calcium group has a somewhat short left tail, but there are no departures from normality that will prevent use of t procedures.

The following tabulation gives the summary statistics for the decrease in blood pressure:

Group	Treatment	n	$\bar{x}$	s
1	Calcium	10	5.000	8.743
2	Placebo	11	−.273	5.901

The calcium group shows a drop in blood pressure, while the placebo group has a small increase. The sample standard deviations do not rule out equal population standard deviations. A difference this large will often arise by chance in samples this small. We are willing to assume equal population standard deviations. The pooled sample variance is

$$s_p^2 = \frac{(n_1 - 1)s_1^2 + (n_2 - 1)s_2^2}{n_1 + n_2 - 2}$$
$$= \frac{(9)(8.743)^2 + (10)(5.901)^2}{10 + 11 - 2} = 54.536$$

so that

$$s_p = \sqrt{54.536} = 7.385$$

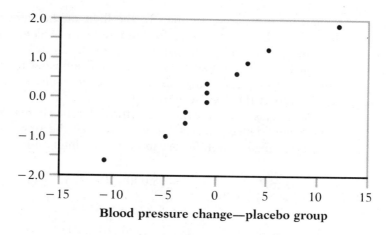

Blood pressure change—placebo group

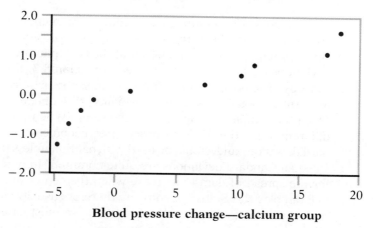

Blood pressure change—calcium group

Figure 8.9 Normal quantile plots for the decrease in blood pressure from Table 8.3.

The pooled two-sample t statistic is

$$t = \frac{\bar{x}_1 - \bar{x}_2}{s_p\sqrt{\dfrac{1}{n_1} + \dfrac{1}{n_2}}} = \frac{5.000 - (-.273)}{7.385\sqrt{\dfrac{1}{10} + \dfrac{1}{11}}}$$

$$= \frac{5.273}{3.227} = 1.634$$

df = 19

p	.10	.05
t^*	1.328	1.729

The P-value is $P(T \geq 1.634)$, where T has the $t(19)$ distribution. From Table E we can see that P falls between the $\alpha = 0.10$ and $\alpha = 0.05$ levels. Statistical software gives the exact value $P = 0.059$. The experiment found evidence that calcium reduces blood pressure, but the evidence falls a bit short of the traditional 5% and 1% levels. ∎

The P-value of a test is strongly influenced by the sample size. An effect that fails to be significant at a specified level α in a small sample will be significant in a larger sample. In the light of the rather small samples in Example 8.12, the evidence for some effect of calcium on blood pressure is rather good. The published account of the study combined these results for blacks with the results for whites and adjusted for pretest differences among the subjects. Using this more detailed analysis, the researchers were able to report the P-value $P = 0.008$.

We can also estimate the size of the difference in means in Example 8.12. A 90% confidence interval for $\mu_1 - \mu_2$ uses the upper 0.05 critical value $t^* = 1.729$ from the $t(19)$ distribution. The interval is

$$(\bar{x}_1 - \bar{x}_2) \pm t^* s_p \sqrt{\frac{1}{n_1} + \frac{1}{n_2}} = [5.000 - (-.273)] \pm (1.729)(7.385)\sqrt{\frac{1}{10} + \frac{1}{11}}$$

$$= 5.273 \pm 5.579$$

$$= (-.306,\ 10.852)$$

The pooled two-sample t procedures are anchored in statistical theory, and so have long been the standard version of two-sample t in textbooks. But they require the assumption that the two unknown population standard deviations be equal. As we shall see in Section 8.3, this assumption is hard to verify. The pooled t procedures are therefore a bit risky. They are reasonably robust against both nonnormality and unequal standard deviations when the sample sizes are nearly the same. When the samples are of quite different sizes, the pooled t procedures become sensitive to unequal standard deviations and should be used with caution unless the samples are large. Unequal standard deviations are quite common. In particular, it is common for the spread of data to increase when the center moves up, as happened in Example 8.11. Statistical software therefore usually allows you to calculate both the pooled and unpooled t statistics. In most cases, the practical conclusions from the two statistics are similar.

SUMMARY

Tests and confidence intervals for the difference of the means μ_1 and μ_2 of two normal populations are based on the difference $\bar{x}_1 - \bar{x}_2$ of the sample means from two independent SRSs. Because of the central limit theorem, the resulting procedures are approximately correct for other population distributions when the sample sizes are large.

When independent SRSs of sizes n_1 and n_2 are drawn from two normal populations with parameters μ_1, σ_1 and μ_2, σ_2 the **two-sample z statistic**

$$z = \frac{(\bar{x}_1 - \bar{x}_2) - (\mu_1 - \mu_2)}{\sqrt{\dfrac{\sigma_1^2}{n_1} + \dfrac{\sigma_2^2}{n_2}}}$$

has the $N(0, 1)$ distribution.

The **two-sample t statistic**

$$t = \frac{(\bar{x}_1 - \bar{x}_2) - (\mu_1 - \mu_2)}{\sqrt{\dfrac{s_1^2}{n_1} + \dfrac{s_2^2}{n_2}}}$$

does *not* have exactly a t distribution.

Conservative inference procedures for comparing μ_1 and μ_2 are obtained from the two-sample t statistic by using the $t(k)$ distribution with degrees of freedom k equal to the smaller of $n_1 - 1$ and $n_2 - 1$. More accurate probability values can be obtained by estimating the degrees of freedom from the data. This is the usual procedure in statistical software.

The confidence interval for $\mu_1 - \mu_2$ given by

$$(\bar{x}_1 - \bar{x}_2) \pm t^* \sqrt{\frac{s_1^2}{n_1} + \frac{s_2^2}{n_2}}$$

has confidence level at least $1 - \alpha$ if t^* is the upper $\alpha/2$ critical value for $t(k)$ with k the smaller of $n_1 - 1$ and $n_2 - 1$.

Significance tests for $H_0: \mu_1 = \mu_2$ based on

$$t = \frac{\bar{x}_1 - \bar{x}_2}{\sqrt{\dfrac{s_1^2}{n_1} + \dfrac{s_2^2}{n_2}}}$$

have a true P-value no higher than that calculated from $t(k)$.

The guidelines for practical use of two-sample t procedures are similar to those for one-sample t procedures. Equal sample sizes are recommended.

If it is known that the two populations have equal variances, **pooled two-sample t procedures** can be used. These are based on the **pooled estimator**

$$s_p^2 = \frac{(n_1 - 1)s_1^2 + (n_2 - 1)s_2^2}{n_1 + n_2 - 2}$$

of the unknown common variance and the $t(n_1 + n_2 - 2)$ distribution.

SECTION 8.2 EXERCISES

In exercises that call for two-sample t procedures, you may use as the degrees of freedom either the smaller of $n_1 - 1$ and $n_2 - 1$ or the more exact value *df* given by Equation 8.4. We recommend the first choice unless you are using a computer.

8.35 In a study of cereal leaf beetle damage on oats, researchers measured the number of beetle larvae per stem in small plots of oats after randomly applying one of two treatments: no pesticide or malathion at the rate of 0.25 pound per acre. The data gave the summary statistics below. (Based on M. C. Wilson et al., "Impact of cereal leaf beetle larvae on yields of oats," *Journal of Economic Entomology*, 62 (1969), pp. 699–702.)

Group	Treatment	n	$\bar{x}$	s
1	Control	13	3.47	1.21
2	Malathion	14	1.36	.52

Is there significant evidence at the 1% level that the mean number of larvae per stem is reduced by malathion? Be sure to state H_0 and H_a.

8.36 Physical fitness is related to personality characteristics. In one study of this relationship, middle-aged college faculty who had volunteered for a fitness program were divided into low-fitness and high-fitness groups based on a physical examination. The subjects then took the Cattell Sixteen Personality Factor Questionnaire. Here are the data for the "ego strength" personality factor. (From A. H. Ismail and R. J. Young, "The effect of chronic exercise on the personality of middle-aged men," *Journal of Human Ergology*, 2 (1973), pp. 47–57.)

Group	Fitness	n	$\bar{x}$	s
1	Low	14	4.64	.69
2	High	14	6.43	.43

(a) Is the difference in mean ego strength significant at the 5% level? At the 1% level? Be sure to state H_0 and H_α.

(b) You should be hesitant to generalize these results to the population of all middle-aged men. Explain why.

8.37 A market research firm supplies manufacturers with estimates of the retail sales of their products from samples of retail stores. Marketing managers are prone to look at the estimate and ignore sampling error. Suppose that an SRS of 75 stores this month shows mean sales of 52 units of a small appliance, with standard deviation 13 units. During the same month last year, an SRS of 53 stores gave mean sales of 49 units, with standard deviation 11 units. An increase from 49 to 52 is a rise of 6%. The marketing manager is happy, because sales are up 6%.

(a) Use the two-sample t procedure to give a 95% confidence interval for the difference in mean number of units sold at all retail stores.

(b) Explain in language that the manager can understand why he cannot be certain that sales rose by 6%, and that in fact sales may even have dropped.

8.38 A bank compares two proposals to increase the amount that its credit card customers charge on their cards. (The bank earns a percentage of the amount charged, paid by the stores that accept the card.) Proposal A offers to eliminate the annual fee for customers who charge $2400 or more during the year. Proposal B offers a small percent of the total amount charged as a cash rebate at the end of the year. Each proposal is offered to an SRS of 150 of the bank's existing credit card customers. At the end of the year, the total amount charged by each customer is recorded. Here are the summary statistics.

Group	n	$\bar{x}$	s
A	150	$1987	$392
B	150	$2056	$413

(a) Do the data show a significant difference between the mean amounts charged by customers offered the two plans? Give the null and alternative hypotheses, and calculate the two-sample t statistic. Obtain the P-value (either approximately from Table E or more accurately from software). State your practical conclusions.

(b) The distributions of amounts charged are skewed to the right, but outliers are prevented by the limits that the bank imposes on credit balances. Do you think that skewness threatens the validity of the test that you used in (a)? Explain your answer.

8.39 In a study of iron deficiency among infants, samples of infants following different feeding regimens were compared. One group contained breast-fed infants, while the children in another group were fed a standard baby formula without any iron supplements. Here are summary results on blood hemoglobin levels at 12 months of age. (From M. F. Picciano and R. H. Deering, "The influence of feeding regimens on iron status during infancy," *The American Journal of Clinical Nutrition*, 33 (1980), pp. 746–753.)

Group	n	$\bar{x}$	s
Breast-fed	23	13.3	1.7
Formula	19	12.4	1.8

(a) Is there significant evidence that the mean hemoglobin level is higher among breast-fed babies? State H_0 and H_a and carry out a t test. Give the P-value. What is your conclusion?

    (b) Give a 95% confidence interval for the mean difference in hemoglobin level between the two populations of infants.

    (c) State the assumptions that your procedures in (a) and (b) require in order to be valid.

8.40 In a study of heart surgery, one issue was the effect of drugs called β-blockers on the pulse rate of patients during surgery. The available subjects were divided at random into two groups of 30 patients each. One group received a β-blocker; the other, a placebo. The pulse rate of each patient at a critical point during the operation was recorded. The treatment group had mean 65.2 and standard deviation 7.8. For the control group, the mean was 70.3 and the standard deviation 8.3.

    (a) Do β-blockers reduce the pulse rate? State the hypotheses and do a t test. Is the result significant at the 5% level? At the 1% level?

    (b) Give a 99% confidence interval for the difference in mean pulse rates.

8.41 What aspects of rowing technique distinguish between novice and skilled competitive rowers? Researchers compared two groups of female competitive rowers: a group of skilled rowers and a group of novices. The researchers measured many mechanical aspects of rowing style as the subjects rowed on a Stanford Rowing Ergometer. One important variable is the angular velocity of the knee (roughly the rate at which the knee joint opens as the legs push the body back on the sliding seat). This variable was measured when the oar was at right angles to the machine. The measurements gave the following statistics:

Group	n	$\bar{x}$	s
Novice	8	3.01	.96
Skilled	10	4.18	.48

The data showed no outliers or strong skewness. (From W. N. Nelson and C. J. Widule, "Kinematic analysis and efficiency estimate of intercollegiate female rowers," unpublished manuscript, 1983.)

    (a) The researchers believed that the knee velocity would be higher for skilled rowers. State H_0 and H_a.

    (b) Carry out the two-sample t test and give a P-value (either from Table E or from software). What do you conclude?

    (c) Give a 90% confidence interval for the mean difference between the knee velocities of skilled and novice female rowers.

8.42 The novice and skilled rowers in the previous exercise were also compared with respect to several physical variables. For weight in kilograms, the data gave the following statistics:

Group	n	$\bar{x}$	s
Novice	8	68.45	9.04
Skilled	10	70.37	6.10

Is there significant evidence of a difference in the mean weights of skilled and novice rowers? State H_0 and H_a, calculate the two-sample t statistic, give a P-value, and state your conclusion.

8.43 The Johns Hopkins Regional Talent Searches give the Scholastic Aptitude Tests (intended for high school juniors and seniors) to 13-year-olds. In all, 19,883 males and 19,937 females took the tests between 1980 and 1982. The mean scores of males and females on the verbal test are nearly equal, but there is a clear difference between the sexes on the mathematics test. The reason for this difference is not understood. Here are the data. (From a news article in *Science*, 224 (1983), pp. 1029–1031.)

Group	$\bar{x}$	s
Males	416	87
Females	386	74

Give a 99% confidence interval for the difference between the mean score for males and the mean score for females in the population that Johns Hopkins searches.

8.44 The SSHA is described in Example 8.7. A selective private college gives the SSHA to an SRS of both male and female freshmen. The data for the women are as follows:

154	109	137	115	152	140	154	178	101
103	126	126	137	165	165	129	200	148

Here are the scores of the men:

108	140	114	91	180	115	126	92	169	146
109	132	75	88	113	151	70	115	187	104

(a) Examine each sample graphically, with special attention to outliers and skewness. Is use of a t procedure acceptable for these data?

(b) Most studies have found that the mean SSHA score for men is lower than the mean score in a comparable group of women. Test this supposition here. That is, state hypotheses, carry out the test and obtain a P-value, and give your conclusions.

(c) Give a 90% confidence interval for the mean difference between the SSHA scores of male and female freshmen at this college.

8.45 Exercise 1.28 gives the data from an experiment that compared the growth of chicks fed normal corn with that of chicks fed high-lysine corn. The purpose of the experiment was to see if high-lysine corn led to faster growth.

(a) Present the data graphically. Are there outliers or strong skewness that might prevent the use of *t* procedures?

(b) State the hypotheses for a statistical test of the claim that chicks fed high-lysine corn gain weight faster. Carry out the test. Is the result significant at the 10% level? At the 5% level? At the 1% level?

(c) Give a 95% confidence interval for the mean extra weight gain in chicks fed high-lysine corn.

8.46 Table 8.3 gives data on the blood pressure before and after treatment for two groups of black males. One group took a calcium supplement, and the other group received a placebo. Example 8.12 compares the decrease in blood pressure in the two groups using pooled two-sample *t* procedures.

(a) Repeat the significance test using two-sample *t* procedures that do not require equal population standard deviations. Compare your *P*-value with the result $P = 0.059$ for the pooled *t* test.

(b) Give a 90% confidence interval for the difference in means, again using a procedure that does not require equal standard deviations. How does the width of your interval compare with that in the discussion following Example 8.12?

8.47 Table 1.4 gives data on the calories in a sample of brands of each of three kinds of hot dogs. The boxplot in Figure 1.9 shows that poultry hot dogs have fewer calories than either beef or meat hot dogs.

(a) Give a 95% confidence interval for the difference in mean calorie content between beef and poultry hot dogs.

(b) Based on your confidence interval, can the hypothesis that the population means are equal be rejected at the 5% significance level? Explain your answer.

(c) What assumptions does your statistical procedure in (a) require? Which of these assumptions are justified or not important in this case? Are any of the assumptions doubtful in this case?

8.48 Researchers studying the learning of speech often compare measurements made on the recorded speech of adults and children. One variable of interest is called the voice onset time (VOT). Here are the results for 6-year-old children and adults asked to pronounce the word ''bees.'' The VOT is measured in milliseconds and can be either positive or negative. (From M. A. Zlatin and

R. A. Koenigsknecht, "Development of the voicing contrast: a comparison of voice onset time in stop perception and production," *Journal of Speech and Hearing Research*, 19 (1976), pp. 93–111).

Group	n	$\bar{x}$	s
Children	10	−3.67	33.89
Adults	20	−23.17	50.74

(a) What is the standard error of the sample mean VOT for the 20 adult subjects? What is the standard error of the difference $\bar{x}_1 - \bar{x}_2$ between the mean VOT for children and adults?

(b) The researchers were investigating whether VOT distinguishes adults from children. State H_0 and H_a and carry out a two-sample t test. Give a P-value and report your conclusions.

(c) Give a 95% confidence interval for the difference in mean VOTs when pronouncing the word "bees." Explain why you knew from your result in (b) that this interval would contain 0 (no difference).

8.49 The researchers in the study discussed in the previous problem looked at VOTs for adults and children pronouncing many different words. Explain why they should not do a separate two-sample t test for each word and conclude that the words with a significant difference (say $P < 0.05$) distinguish children from adults. (The researchers did not make this mistake.)

8.50 College financial aid offices expect students to use summer earnings to help pay for college. But how large are these earnings? One college studied this question by asking a sample of students how much they earned. Omitting students who were not employed, 1296 responses were received. Here are the data in summary form. (Data for 1982, provided by Marvin Schlatter, Division of Financial Aid, Purdue University.)

Group	n	$\bar{x}$	s
Males	675	$1884.52	$1368.37
Females	621	$1360.39	$1037.46

(a) Use the two-sample t procedures to give a 90% confidence interval for the difference between the mean summer earnings of male and female students.

(b) The distribution of earnings is strongly skewed to the right. Nevertheless, use of t procedures is justified. Why?

(c) Once sample size was decided, the samples were chosen by taking every kth name from an alphabetical list of undergraduates. Is it reasonable to consider the samples as SRSs chosen from the male and female undergraduate populations?

(d) What other information about the study would you request before accepting the results as describing all undergraduates?

8.51 Example 8.11 reports the MINITAB analysis of some biological data. MINITAB uses the two-sample t test with degrees of freedom given by Equation 8.4. Starting from MINITAB's results for $\bar{x}_i$ and s_i, show the calculations that verify MINITAB's values for the test statistic $t = 2.99$ and the degrees of freedom $df = 5.9$.

8.52 Example 8.9 analyzed the Chapin Social Insight Test scores of male and female college students. The sample standard deviations s_i for the two samples are almost identical. This suggests that the two populations have very similar standard deviations σ_i. The use of the pooled two-sample t procedures is justified in this case. Repeat the analysis in Example 8.9 using the pooled procedures. Compare your results with those obtained in the example.

8.53 The pooled two-sample t procedures can be used to analyze the hemoglobin data in Exercise 8.39, because the very similar s-values suggest that the assumption of equal population standard deviations is justified. Repeat (a) and (b) of Exercise 8.39 using the pooled t procedures. Compare your results with those you obtained with the two-sample t procedures in Exercise 8.39.

8.54 The data on weights of skilled and novice rowers in Exercise 8.42 can be analyzed by the pooled t procedures. Do the pooled t test to answer the question asked in Exercise 8.42. (The pooled procedures should not be used for the more important comparison of knee velocities in Exercise 8.41, because the s-values in the two groups are different enough to cast doubt on the assumption of a common standard deviation.)

8.55 Repeat the comparison of mean VOTs for children and adults in Exercise 8.48 using a pooled t procedure. (We prefer not to pool in this case, because the data suggest some difference in the population standard deviations.)

(a) Carry out the significance test, and give a P-value.

(b) Give a 95% confidence interval for the difference in population means.

(c) How similar are your results to those you obtained in Exercise 8.48 from the two-sample t procedures?

8.56 In Example 8.12, a small study of black men suggested that a calcium supplement can reduce blood pressure. Now we are planning a larger clinical trial of this effect. We plan to use 100 subjects in each of the two groups. Are these sample sizes large enough to make it very likely that the study will give strong evidence ($\alpha = 0.01$) of the effect of calcium if in fact calcium lowers blood pressure by 5 more than a placebo? To answer this question, we will compute the power of the two-sample t test of

$$H_0: \mu_1 = \mu_2$$
$$H_a: \mu_1 > \mu_2$$

against the specific alternative $\mu_1 - \mu_2 = 5$. (We plan to use the two-sample t because we are not convinced that the calcium and placebo groups have equal standard deviations.) Based on the pilot study reported in Example 8.12, we take 8, the larger of the two observed s-values, as a rough estimate of both the population σ's and future sample s's.

(a) What is the approximate value of the $\alpha = 0.01$ critical value t^* for the two-sample t statistic when $n_1 = n_2 = 100$?

(b) The test rejects H_0 when

$$\frac{\bar{x}_1 - \bar{x}_2}{\sqrt{\dfrac{s_1^2}{n_1} + \dfrac{s_2^2}{n_2}}} \geq t^*$$

Take both s_1 and s_2 to be 8, and n_1 and n_2 to be 100. What is the number c such that the test rejects H_0 when $\bar{x}_1 - \bar{x}_2 \geq c$?

(c) Suppose that $\mu_1 - \mu_2 = 5$ and that both σ_1 and σ_2 are 8. The power we seek is the probability that $\bar{x}_1 - \bar{x}_2 \geq c$ under these assumptions. Calculate the power.

8.57 You are planning a larger study of VOTs, based on the pilot study reported in Exercise 8.48. Not all words distinguish children (group 1) from adults (group 2) as well as "bees," so you want high power against the alternative that $\mu_1 - \mu_2 = 10$ in a *two-sided* t test. From the pilot study, take 30 as an estimate of σ_1 and s_1, and 50 as an estimate of σ_2 and s_2.

(a) State H_0 and H_a, and write the formula for the test statistic.

(b) Give the $\alpha = 0.05$ critical value of the test when $n_1 = 100$ and $n_2 = 200$.

(c) Find the approximate power against the given alternative.

8.58 A major bank asks you to design a study of ways to increase the use of their credit cards. One plan (plan A) would offer customers a cash-back rebate of a small percent of the total amount charged each 6 months. Plan B would reduce the interest rate charged by an amount that would cost the bank as much as the rebate. The response variable is the total amount each customer charges during the test period. You decide to offer plan A and plan B each to a separate SRS of the bank's existing charge card customers. In the past, the mean amount charged in a 6-month period has been about \$1100, with a standard deviation of \$400. Will a two-sample t test based on SRSs of 350 customers each detect a difference of \$100 in the mean amounts spent under the two plans?

(a) State H_0 and H_a, and write the formula for the test statistic.
(b) Give the $\alpha = 0.05$ critical value for the test when $n_1 = n_2 = 350$.
(c) Calculate the power of the test with $\alpha = 0.05$, using \$400 as a rough estimate of all standard deviations.

8.3 INFERENCE FOR POPULATION SPREAD*

The two most basic descriptive features of a distribution are its center and spread. In a normal population, these aspects are measured by the mean and the standard deviation. We have described procedures for inference about population means for normal populations and found that these procedures are often useful for nonnormal populations as well. It is natural to turn next to inference about the standard deviations of normal populations. Our advice here is short and clear: don't do it.

There are indeed inference procedures appropriate for the standard deviations of normal populations. We will describe the most common such procedure, the F test for comparing the spread of two normal populations. Unlike the t procedures for means, the F test and other procedures for standard deviations are extremely sensitive to nonnormal distributions. This lack of robustness does not improve in large samples. It is difficult in practice to tell whether a significant F-value is evidence of unequal population spreads or simply evidence that the populations are not normal.

The deeper difficulty that underlies the very poor robustness of normal population procedures for inference about spread already appeared in our work on describing data. The standard deviation is a natural measure of spread for normal distributions, but not for distributions in general. In fact, since skewed distributions have unequally spread tails, no single numerical measure is adequate to describe the spread of a skewed distribution. Thus, the standard deviation is not always a useful parameter, and even when it is (in the normal case), the results of inference are not trustworthy. Consequently, we do not recommend use of inference about population standard deviations in basic statistical practice.[12]

Sometimes equality of standard deviations is tested as a preliminary to performing the pooled two-sample t test for equality of two population means. It is better practice to check the distributions graphically, with

---

* This section can be omitted without loss of continuity.

special attention to skewness and outliers. The pooled t test is reasonably robust against unequal population standard deviations, particularly when the population distributions are roughly symmetric and the two sample sizes are similar. On the other hand, the test for equal standard deviations is often misleading because of its extreme sensitivity to departures from normality (see below).

The situation is similar when the means of more than two populations are being compared. Chapter 11 discusses the analysis of variance procedures for comparing several means. These procedures are extensions of the pooled t test and require that the populations have a common standard deviation. Like the t test, analysis of variance comparisons of means are quite robust. (Analysis of variance uses F statistics, but these are not the same as the F statistic for comparing two population standard deviations.) Formal tests for the assumption of equal standard deviations are very nonrobust and will often give misleading results. In the words of one distinguished statistician, "To make a preliminary test on variances is rather like putting to sea in a rowing boat to find out whether conditions are sufficiently calm for an ocean liner to leave port!"[13]

The F Test

Because of the limited usefulness of procedures for inference on the standard deviations of normal distributions, we will present only one such procedure. Suppose that we have independent SRSs from two normal populations, a sample of size n_1 from $N(\mu_1, \sigma_1)$ and a sample of size n_2 from $N(\mu_2, \sigma_2)$. The population means and standard deviations are all unknown. The hypothesis of equal spread,

$$H_0: \sigma_1 = \sigma_2$$
$$H_a: \sigma_1 \neq \sigma_2$$

is tested by a natural statistic, the ratio of sample variances.

The F statistic and F distributions

When s_1^2 and s_2^2 are sample variances from independent SRSs of sizes n_1 and n_2 drawn from normal populations, the F statistic

$$F = \frac{s_1^2}{s_2^2}$$

has the F distribution with $n_1 - 1$ and $n_2 - 1$ degrees of freedom when $H_0: \sigma_1 = \sigma_2$ is true.

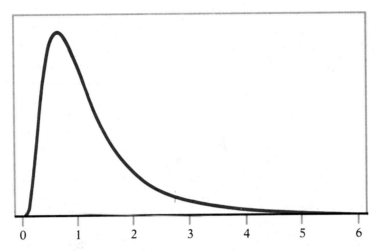

Figure 8.10 The density curve for the $F(9, 10)$ distribution. The F distributions are skewed to the right.

F distributions   The F distributions are a family of distributions with two parameters, the degrees of freedom of the sample variances in the numerator and denominator of the F statistic.* The numerator degrees of freedom are always mentioned first. Interchanging the degrees of freedom changes the distribution, so the order is important. Our brief notation will be $F(j, k)$ for the F distribution with j degrees of freedom in the numerator and k in the denominator. The F distributions are not symmetric, but right-skewed. The density curve in Figure 8.10 illustrates the shape. Because sample variances cannot be negative, the F statistic takes only positive values and the F distribution has no probability below 0. The peak of the F density curve is near 1; values far from 1 in either direction provide evidence against the hypothesis of equal standard deviations.

Tables of F critical points are awkward, since a separate table is needed for every pair of degrees of freedom j and k. Table F in the back of the book gives upper p critical points of the F distributions for $p = 0.10$, 0.05, 0.025, 0.01, and 0.001. For example, these critical points for the $F(9, 10)$ distribution shown in Figure 8.10 are

p	.10	.05	.025	.01	.001
F^*	2.35	3.02	3.78	4.94	8.96

* The F distributions are another of R. A. Fisher's contributions to statistics and are called F in his honor. Fisher introduced F statistics for comparing several means. We will meet these useful statistics in later chapters.

The skewness of the F distributions causes additional complications. In the symmetric normal and t distributions, the point with probability 0.05 below it is just the negative of the point with probability 0.05 above it. This is not true for F distributions. We therefore need either tables of both the upper and lower tails or means of eliminating the need for lower-tail critical values. Statistical software that eliminates the need for tables is plainly very convenient.

If you do not use statistical software, arrange the F test as follows:

1 Take the test statistic to be

$$F = \frac{\text{larger } s^2}{\text{smaller } s^2}$$

This amounts to naming the populations so that s_1^2 is the larger of the observed sample variances. The resulting F is always 1 or greater.

2 Compare the value of F to critical values from Table F. Then *double* the significance levels from the table to obtain the significance level for the two-sided F test.

The idea is that we calculate the probability in the upper tail and double to obtain the probability of all ratios on either side of 1 that are at least as improbable as that observed. Remember that the order of the degrees of freedom is important in using Table F.

EXAMPLE 8.13

Example 8.12 recounts a medical experiment comparing the effects of calcium and a placebo on the blood pressure of black men. The pooled two-sample t procedures were employed in the analysis. Since these procedures require equal population standard deviations, it is tempting to first test

$$H_0: \sigma_1 = \sigma_2$$
$$H_a: \sigma_1 \neq \sigma_2$$

The larger of the two sample standard deviations is $s = 8.743$ from 10 observations. The other is $s = 5.901$ from 11 observations. The two-sided test statistic is therefore

$$F = \frac{\text{larger } s^2}{\text{smaller } s^2} = \frac{8.743^2}{5.901^2} = 2.20$$

We compare the calculated value $F = 2.20$ with critical points for the $F(9, 10)$ distribution. Table F shows that 2.20 is *less* than the 0.10 critical value of the $F(9, 10)$ distribution, which is $F^* = 2.35$. Doubling 0.10, we know that the observed F falls short of the 0.20 critical value. The results are not significant at the 20% level (or any lower level). Statistical software shows that the exact upper-tail probability is 0.13, and hence $P = 0.26$. *If* the populations were normal, the observed standard deviations would give little reason to suspect unequal population standard deviations. Since one of the populations shows some nonnormality, we cannot be fully confident of this conclusion. ∎

Robustness of Normal Inference Procedures

We have claimed that

- The *t* procedures for inference about means are quite robust against nonnormal population distributions. These procedures are particularly robust when the population distributions are symmetric and (for the two-sample case) when the two sample sizes are equal.
- The *F* test and other procedures for inference about variances are so lacking in robustness as to be of little use in practice.

Figure 8.11 presents evidence for these claims. The paper from which the figure is reproduced contains more information.[14] This figure contains a great deal of information in compact form, so careful attention is needed. It displays the results of a number of large simulations somewhat like those we studied in Section 4.4. All were carried out with samples of size 25, and all concern significance tests with fixed level $\alpha = 0.05$. The four types of tests studied are the one-sample and pooled two-sample *t* tests, the *F* test (called "two-sample *F*"), and the test for the variance of a single normal population. We have not discussed this last test.

Figure 8.11 reports results for one-sided tests ("above" and "below" the null hypothesis of no difference) and two-sided tests ("beyond"). The 12 small boxes in each half of the figure show the behavior of the 12 combinations of test statistic (four choices) and alternative hypothesis (three choices). Figure 8.11(a) depicts the test results when the population is in fact nonnormal but strongly skewed to the right. Figure 8.11(b) describes performance for symmetric populations, both normal and nonnormal. The horizontal scale for each box is a "heavy tails" parameter that describes whether the particular population is strongly peaked or has heavy tails. Normal populations lie at 3 on this scale in Figure 8.11(b), nonnormal populations, on either side of 3. All of the skewed populations of Figure 8.11(a) are nonnormal.

All of the tests are carried out at the fixed significance level 5% ($\alpha = 0.05$) using critical values from tables such as *t* or *F* that are correct for normal populations. If the population distribution is not normal, the actual significance level (probability of rejecting the hypothesis when it is true) will generally be higher or lower than 5%. A test is robust if the actual level stays close to 5% even when the population is not normal. The colored dots in Figure 8.11 show the actual significance level in each case. The significance levels appear on the vertical scale, with 5% as the target value. The results for normal populations appear in Figure 8.11(b) above 3 in each box within the figure. These circles are all close to 5% on the vertical scale, as they should be. The other circles show what happens when the tests are used on nonnormal data. The faster and farther the circles move from the 5% level, the less robust the test is.

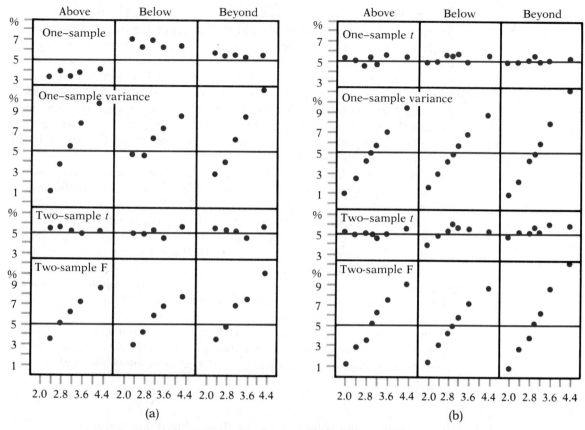

Figure 8.11 The robustness of common tests of significance. The points plotted are the actual significance levels for tests made at the 5% level. The tests assume normality, while the actual distributions are mostly nonnormal. (a) Right-skewed distributions; (b) symmetric distributions. From E. S. Pearson and N. W. Please, "Relation between the shape of population distribution and the robustness of four simple test statistics," *Biometrika*, 62 (1975), pp. 223–241. Reproduced by permission of the Biometrika trustees.

EXAMPLE 8.14

A symmetric population has a heavy tails parameter equal to 4.4, a quite extreme value. We carry out a two-sample t test for equality of means against a two-sided alternative using 5% critical values. Although the population is not at all normal, the actual significance level of this test is about 5.9%. Look at the rightmost dot in the "two-sample t" box of Figure 8.11(b), above 4.4 on the horizontal scale.

Now consider the F test in the same situation. The actual significance level of what is supposed to be a 5% test is about 11.2%. This is the rightmost dot in the "two-sample F" box in Figure 8.11(b). ∎

The robustness of the two-sample t is remarkable. The true significance level remains between about 4% and 6% for the entire range of populations studied. This test and the corresponding confidence interval are among the most reliable tools in the statistician's shop. Do remember, however, that outliers can greatly disturb the t procedures and that they are less robust when the sample sizes are not similar. Note also that the happy results of Figure 8.11 are obtained when both populations have the same shape, even though that shape is not normal. Dissimilar shapes, especially skewness in only one of the populations, can cause the type of results that occur in the one-sample t discussed below.

The lack of robustness of the tests for variances is equally remarkable. The true levels depart rapidly from the target 5% as the population distribution departs from normality. The two-sided F test carried out with 5% critical values can have a true level of less than 1% or greater than 11% even in symmetric populations with no outliers. Look at the lower right box in Figure 8.11(b). Results such as these are the basis for our recommendation that these procedures not be used.

The one-sample t is pleasingly robust in symmetric populations, but the top row of Figure 8.11(a) shows that one-sided tests are systematically disturbed by skewed distributions. For example, the t test of $H_0: \mu = \mu_0$ against $H_a: \mu > \mu_0$ rejects H_0 too seldom (the true significance level is less than 5%) when the population distribution is right-skewed. Recall that the guidelines on page 520 cautioned against skewness as well as outliers.

SUMMARY

Inference procedures for comparing the standard deviations of two normal populations are based on the **F statistic**, which is the ratio of sample variances

$$F = \frac{s_1^2}{s_2^2}$$

If an SRS of size n_1 is drawn from the x_1 population and an independent SRS of size n_2 is drawn from the x_2 population, the F statistic has the **F distribution** $F(n_1 - 1, n_2 - 1)$ if the two population standard deviations σ_1 and σ_2 are in fact equal.

The two-sided test of $H_0: \sigma_1 = \sigma_2$ uses the statistic

$$F = \frac{\text{larger } s^2}{\text{smaller } s^2}$$

and doubles the upper-tail probability to obtain the P-value.

The F tests and other procedures for inference on the spread of one or more normal distributions are so strongly affected by nonnormality that we do not recommend them for regular use.

SECTION 8.3 EXERCISES

In all exercises calling for use of the F test, assume that both population distributions are very close to normal. The actual data are not always sufficiently normal to justify use of the F test.

8.59 The F statistic $F = s_1^2/s_2^2$ is calculated from samples of size $n_1 = 10$ and $n_2 = 8$. (Remember that n_1 is the numerator sample size.)

    (a) What is the upper 5% critical value for this F?

    (b) In a test of equality of standard deviations against the two-sided alternative, this statistic has the value $F = 3.45$. Is this value significant at the 10% level? Is it significant at the 5% level?

8.60 The F statistic for equality of standard deviations based on samples of sizes $n_1 = 21$ and $n_2 = 16$ takes the value $F = 2.78$.

    (a) Is this significant evidence of unequal population standard deviations at the 5% level? At the 1% level?

    (b) Between which two values obtained from Table F does the P-value of the test fall?

8.61 The sample variance for the treatment group in the DDT experiment of Example 8.11 is more than 10 times as large as the sample variance for the control group. Calculate the F statistic. Can you reject the hypothesis of equal population standard deviations at the 5% significance level? At the 1% level?

8.62 Exercise 8.41 records the results of comparing an important measure of rowing style for skilled and novice female competitive rowers. Is there significant evidence of inequality between the standard deviations of the two populations?

    (a) State H_0 and H_a.

    (b) Calculate the F statistic. Between which two levels does the P-value lie?

8.63 Answer the same questions for the weights of the two groups, recorded in Exercise 8.42.

8.64 The data for VOTs of children and adults in Exercise 8.48 show quite different sample standard deviations. How statistically significant is the observed inequality?

8.65 Return to the SSHA data in Exercise 8.44. SSHA scores are generally less variable among women than among men. We want to know whether this is true for this college.

    (a) State H_0 and H_a. Note that H_a is one-sided in this case.

    (b) Because Table F contains only upper critical values for F, a one-sided test requires that in calculating F the numerator s^2 belong to the group that H_a claims to have the larger σ. Calculate this F.

(c) Compare F to the entries in Table F (no doubling of p) to obtain the P-value. Be sure the degrees of freedom are in the proper order. What do you conclude about the variation in SSHA scores?

8.66 The observed inequality between the sample standard deviations of male and female SAT scores in Exercise 8.43 is clearly significant. You can say this without doing any calculations. Find F and look in Table F. Then explain why the significance of F could be seen without arithmetic.

8.67 The Pearson and Please paper from which Figure 8.11 was taken also investigated several sets of industrial data. One variable, the failure time of a metal part when repeatedly bent, had the same degree of right skewness as the simulated distributions in Figure 8.11(a). The heavy tails measure that appears on the horizontal scales in Figure 8.11 was about 3.6 for this distribution. Answer the following questions from Figure 8.11.

(a) You are doing a two-sample t test for the equality of mean failure times for two similar types of part. You measure 25 specimens of each type of part, the same sample sizes as in Figure 8.11. The alternative is two-sided. You do the test at fixed significance level 5% by using the 0.025 critical value from Table E of the t distribution. About what will be the actual significance level of your test?

(b) You do an F test for equality of standard deviations against a two-sided alternative for the same data. Again you choose the 5% level and carry out the test using Table F. About what will be the actual significance level?

8.68 In the setting of the previous exercise, suppose that you were doing inference on a single sample. You carry out a one-sample t test at the $\alpha = 0.05$ significance level.

(a) If the alternative is one-sided on the high side, about what will be the actual significance level of your test?

(b) About what will be the actual significance level for the test against a two-sided alternative?

(c) Instead of a t test for the mean, you carry out the test for a specific value of the variance, or standard deviation, of a single distribution, again at the 5% level. About what will be the actual level of this test if the alternative is two-sided?

NOTES

1 This example is based on information in D. L. Shankland et al., "The effect of 5-thio-D-glucose on insect development and its absorption by insects," *Journal of Insect Physiology*, 14 (1968), pp. 63–72.

2 Data provided by Professor Joseph Wipf, Department of Foreign Languages and Literatures, Purdue University.

3 These recommendations are based on extensive computer work. See, for example, Harry O. Posten, "The robustness of the one-sample t-test over the Pearson system," *Journal of Statistical Computation and Simulation,* 9 (1979), pp. 133–149, and E. S. Pearson and N. W. Please, "Relation between the shape of population distribution and the robustness of four simple test statistics," *Biometrika,* 62 (1975), pp. 223–241.

4 The data, and the use of the log transformation, come from Thomas J. Lorenzen, "Determining statistical characteristics of a vehicle emissions audit procedure," *Technometrics,* 22 (1980), pp. 483–493.

5 You can find a practical discussion of distribution-free inference in Myles Hollander and Douglas A. Wolfe, *Nonparametric Statistical Methods,* Wiley, New York, 1973.

6 Detailed information about the conservative t procedures can be found in Paul Leaverton and John J. Birch, "Small sample power curves for the two sample location problem," *Technometrics,* 11 (1969), pp. 299–307, and in Henry Scheffé, "Practical solutions of the Behrens-Fisher problem," *Journal of the American Statistical Association,* 65 (1970), pp. 1501–1508.

7 This example is adapted from Maribeth C. Schmitt, *The Effects of an Elaborated Directed Reading Activity on the Metacomprehension Skills of Third Graders,* Ph.D. dissertation, Purdue University, 1987.

8 From H. G. Gough, *The Chapin Social Insight Test,* Consulting Psychologists Press, Palo Alto, Calif., 1968.

9 See the extensive simulation studies in Harry O. Posten, "The robustness of the two-sample t test over the Pearson system," *Journal of Statistical Computation and Simulation,* 6 (1978), pp. 295–311.

10 This example is loosely based on D.L. Shankland, "Involvement of spinal cord and peripheral nerves in DDT-poisoning syndrome in albino rats," *Toxicology and Applied Pharmacology,* 6 (1964), pp. 197–213.

11 This study is reported in Roseann M. Lyle et al., "Blood pressure and metabolic effects of calcium supplementation in normotensive white and black men," *Journal of American Medical Association,* 257 (1987), pp. 1772–1776. The individual measurements in Table 8.3 were provided by Dr. Lyle.

12 The problem of comparing spreads is difficult even with advanced methods. The common distribution-free procedures do not offer a satisfactory alternative to the F test, as they are sensitive to unequal shapes when comparing two distributions. If the samples are reasonably large (say 25 or more observations), modern jackknife and bootstrap methods are quite satisfactory, at least for symmetric distributions. The jackknife is described in section 5.5 of Hollander and Wolfe (see Note 5).

13 G. E. P. Box, "Non-normality and tests on variances," *Biometrika,* 40 (1953), pp. 318–335. The quote appears on p. 333.

14 Figure 8.11 is from the paper of Pearson and Please cited in Note 3.

CHAPTER 8 EXERCISES

8.69 As health care costs continue to rise, many employers are offering preventive screening and education to their employees. One such program screened all employees for health risks on a voluntary basis. Employees with identified risks were invited to voluntary risk-reduction programs. Of the 518 employees identified as having high blood cholesterol, 48 regularly attended a nutrition education program that showed how to control cholesterol levels through weight reduction and changes in the diet. The change in cholesterol level for these 48 employees after 6 weeks had $\bar{x} = -12.23$ and $s = 34.53$. We want to know if the program was effective. (Based on a study by John Sciacca, Purdue University.)

    (a) Use a t test to assess the significance of the observed drop in cholesterol levels. State H_0 and H_a, give the P-value (from Table E or software), and state your conclusions.

    (b) To what population does your conclusion apply? Be as specific as possible.

    (c) The raw data are not available. The author of the study did use a t test. Is the normality of the population distribution essential to the validity of the test in this case? Explain your answer.

8.70 Nitrites are often added to meat products as preservatives. In a study of the effect of these chemicals on bacteria, the rate of uptake of a radio-labeled amino acid was measured for a number of cultures of bacteria, some growing in a medium to which nitrites had been added. Here are the summary statistics from this study.

Group	n	$\bar{x}$	s
Nitrite	30	7880	1115
Control	30	8112	1250

Carry out a test of the research hypothesis that nitrites decrease amino acid uptake, and report your results.

8.71 The one-hole test is used to test the manipulative skill of job applicants. This test requires subjects to grasp a pin, move it to a hole, insert it, and return for another pin. The score on the test is the number of pins inserted in a fixed time interval. In one study, male college students were compared with experienced female industrial workers. Here are the data for the first minute of the test. (Based on G. Salvendy, "Selection of industrial operators: the one-hole test," *International Journal of Production Research*, 13 (1973), pp. 303–321.)

Group	n	$\bar{x}$	s
Students	750	35.12	4.31
Workers	412	37.32	3.83

(a) It was expected that the experienced workers would outperform the students, at least during the first minute, before learning occurs. State the hypotheses for a statistical test of this expectation and perform the test. Give a *P*-value and state your conclusions.

(b) The distribution of scores is slightly skewed to the left. Explain why the procedure you used in (a) is nonetheless acceptable.

(c) One purpose of the study was to develop performance norms for job applicants. Based on the data above, what is the range that covers the middle 95% of experienced workers? (Be careful! This is not the same as a 95% confidence interval for the mean score of experienced workers.)

(d) The five-number summary of the distribution of scores among the workers is

$$23 \quad 33.5 \quad 37 \quad 40.5 \quad 46$$

for the first minute, and

$$32 \quad 39 \quad 44 \quad 49 \quad 59$$

for the fifteenth minute of the test. Display these facts graphically, and describe briefly the differences between the distributions of scores in the first and the fifteenth minute.

8.72 The composition of the earth's atmosphere may have changed over time. One attempt to discover the nature of the atmosphere long ago studies the gas trapped in bubbles inside ancient amber. Amber is tree resin that has hardened and been trapped in rocks. The gas in bubbles within amber should be a sample of the atmosphere at the time the amber was formed. Measurements on specimens of amber from the late Cretaceous era (75 to 95 million years ago) give these percents of nitrogen:

$$63.4 \quad 65.0 \quad 64.4 \quad 63.3 \quad 54.8 \quad 64.5 \quad 60.8 \quad 49.1 \quad 51.0$$

These values are quite different from the present 78.1% of nitrogen in the atmosphere. Assume (this is not yet agreed on by experts) that these observations are an SRS from the late Cretaceous atmosphere. (Data from R. A. Berner and G. P. Landis, "Gas bubbles in fossil amber as possible indicators of the major gas composition of ancient air," *Science*, 239 (1988), pp. 1406–1409.)

(a) Graph the data, and comment on skewness and outliers.

(b) The *t* procedures will be only approximate in this case. Give a 95% *t* confidence interval for the mean percent of nitrogen in ancient air.

8.73 Do various occupational groups differ in their diets? In a British study of this question, two of the groups compared were 98 drivers and 83 conductors of London double-decker buses. The conductors' jobs require more physical activity. The article reporting the study gives the data as "Mean daily consumption ($\pm$ s. e.)." Some of the study results appear below. (From J. W. Marr and J. A. Heady, "Within- and between-person variation in dietary surveys: number of days needed to classify individuals," *Human Nutrition: Applied Nutrition*, 40A (1986), pp. 347–364.)

	Drivers	Conductors
Total calories	2821 $\pm$ 44	2844 $\pm$ 48
Alcohol (g)	.24 $\pm$ .06	.39 $\pm$ .11

(a) What does "s. e." stand for? Give $\bar{x}$ and s for each of the four sets of measurements.
(b) Is there significant evidence at the 5% level that conductors consume more calories per day than do drivers? Use the two-sample *t* method to give a *P*-value, and then assess significance.
(c) How significant is the observed difference in mean alcohol consumption? Use two-sample *t* methods to obtain the *P*-value.
(d) Give a 90% confidence interval for the mean daily alcohol consumption of London double-decker bus conductors.
(e) Give an 80% confidence interval for the difference in mean daily alcohol consumption between drivers and conductors.

8.74 The pooled two-sample *t* test is justified in part (b) of the previous exercise. Explain why. Find the *P*-value for the pooled *t* statistic, and compare with your result in the previous exercise.

8.75 The report cited in Exercise 8.73 says that the distribution of alcohol consumption among the individuals studied is "grossly skew."
(a) Do you think that this skewness prevents the use of the two-sample *t* test for equality of means? Explain your answer.
(b) Do you think that the skewness of the distributions prevents the use of the *F* test for equality of standard deviations? Explain your answer.

8.76 Do the data in Example 8.9 provide evidence of different standard deviations in Chapin Test scores in the populations of female and male college liberal arts majors?

(a) State the hypotheses and carry out the test. You will need software or a statistical calculator to assess significance because of the large sample sizes.

(b) Do the large sample sizes allow us to ignore the assumption that the population distributions are normal?

8.77 Exercise 1.99 gives the populations of all 92 counties in the state of Indiana. Is it proper to apply the one-sample t method to these data to give a 95% confidence interval for the mean population of an Indiana county? Explain your answer.

8.78 A pharmaceutical manufacturer checks the potency of products during manufacture by chemical analysis. The standard release potency for cephalothin crystals is set at 910. An assay of the previous 16 lots gives the following potency data:

897	914	913	906	916	918	905	921
918	906	895	893	908	906	907	901

(a) Check the data for outliers or strong skewness that might threaten the validity of the t procedures.

(b) Give a 95% confidence interval for the mean potency.

(c) Is there significant evidence at the 5% level that the mean potency is not equal to the standard release potency?

8.79 The amount of lead in a certain type of soil, when released by a standard extraction method, averages 86 parts per million (ppm). A new extraction method is tried on 40 specimens of the soil, yielding a mean of 83 ppm lead and a standard deviation of 10 ppm.

(a) Is there significant evidence at the 1% level that the new method frees less lead from the soil?

(b) A critic argues that because of variations in the soil, the effectiveness of the new method is confounded with characteristics of the particular soil specimens used. Briefly describe a better data production design that avoids this criticism.

8.80 High levels of cholesterol in the blood are not healthy in either humans or dogs. Because a diet rich in saturated fats raises the cholesterol level, it is plausible that dogs owned as pets have higher cholesterol levels than dogs owned by a veterinary research clinic. "Normal" levels of cholesterol based on the clinic's dogs would then be misleading. A clinic compared healthy dogs it owned with healthy pets brought to the clinic to be neutered. The summary statistics for blood cholesterol levels (milligrams per deciliter of blood) appear below. (From V. D. Bass, W. E. Hoffmann, and J. L. Dorner, "Normal canine lipid profiles and effects of experimentally induced pancreatitis and hepatic necrosis on lipids," *American Journal of Veterinary Research*, 37 (1976), pp. 1355–1357.)

Group	n	$\bar{x}$	s
Pets	26	193	68
Clinic	23	174	44

(a) Is there strong evidence that pets have higher mean cholesterol level than clinic dogs? State H_0 and H_a and carry out an appropriate test. Give the P-value and state your conclusion.

(b) Give a 95% confidence interval for the difference in mean cholesterol levels between pets and clinic dogs.

(c) Give a 95% confidence interval for the mean cholesterol level in pets.

(d) What assumptions must be satisfied to justify the procedures you used in (a), (b), and (c)? Assuming that the cholesterol measurements have no outliers and are not strongly skewed, what is the chief threat to the validity of the results of this study?

8.81 Exercise 3.25 gives data on the concentration of airborne particulate matter in a rural area upwind from a small city and in the center of the city. In that exercise, the focus was on using the rural readings to predict the city readings for the same day. Now we want to compare the mean level of particulates in the city and in the rural area. We suspect that pollution is higher in the city and hope to find evidence for this suspicion.

(a) State H_0 and H_a.

(b) Which type of t procedure is appropriate: one-sample, paired comparisons, or two-sample?

(c) Make a graph to check for outliers or strong skewness that might prevent the use of t procedures. Your graph should reflect the type of procedure that you will use.

(d) Carry out the appropriate t test. Give the P-value and report your conclusion.

(e) Give a 90% confidence interval for the mean amount by which the city particulate level exceeds the rural level.

8.82 The sign test allows us to assess whether city particulate levels are higher than nearby rural levels on the same day without the use of normal distributions. Carry out a sign test for the data used in the previous exercise. State H_0 and H_a and give the P-value and your conclusion.

8.83 Exercise 1.19 gives data on the daily egg production of a population of 25 female and 10 male *Ctenocephalides felis* fleas. Provide the flea experts with an estimate of the mean daily production together with a margin of error.

8.84 Exercise 1.25 gives 29 measurements of the density of the earth, made in 1798 by Henry Cavendish. Display the data graphically to check for skewness and outliers. Then give an estimate for the density of the earth from Cavendish's data and a margin of error for your estimate.

8.85 Elite distance runners are thinner than the rest of us. Here are data on skinfold thickness, which indirectly measures body fat, for 20 elite runners and 95 ordinary men in the same age group. The data are in millimeters and are given in the form "mean (standard deviation)." (From M. L. Pollock, et al., "Body composition of elite class distance runners," in P. Milvey (ed.), *The Marathon: Physiological, Medical, Epidemiological, and Psychological Studies*, New York Academy of Sciences, 1977, p. 366.)

	Runners	Others
Abdomen	7.1 (1.0)	20.6 (9.0)
Thigh	6.1 (1.8)	17.4 (6.6)

Use confidence intervals to describe the difference between runners and typical young men.

Prelude

For categorical variables we summarize data using percents or proportions. Confidence intervals and tests of significance are the basic tools that we use for inference in this setting. We start with a single proportion and then treat the problem of comparing two proportions. Finally general methods for analyzing count data classified in a two-way table are presented.

- *From a sample of 500 Indiana households, what can be inferred about the proportion of all Indiana households having a Christmas tree?*

- *Is there a difference between urban and rural Christmas tree users in the preference for natural versus artificial trees?*

- *Is there a relationship between smoking behavior and socioeconomic status?*

- *Do male and female college students have different reasons for participating in sports?*

9

Inference for Count Data

An opinion poll asks a sample of adults whether or not they approve of the president's conduct of his office. An experiment is performed to compare the effectiveness of two treatments to prevent the common cold. For each treatment the number of subjects who catch a cold during the next month is recorded. A similar experiment is performed to compare several treatments. A sample of students is classified in a two-way table according to their major field of study and their political preference. All of these studies produce counts rather than measurements. In this chapter we will present statistical inference procedures for such problems.

We start with the problem of one proportion. The statistical model is the binomial distribution, which we studied in Section 6.1. Comparing two proportions is discussed in Section 9.2. Again, binomial distributions play an important role. Finally, in Section 9.3, we study a general method for analyzing count data when the observations are classified in a two-way table. We have already studied such tables descriptively in Section 3.4.

9.1 INFERENCE FOR A SINGLE PROPORTION

Confidence Intervals and Significance Tests

When a simple random sample (SRS) is drawn from a population that is much larger than the sample, a sample count has a binomial distribution. If the sample is small, tests and confidence intervals must be based on the binomial distributions. These are difficult to compute because of the discreteness of the distributions.[1] We will consider only procedures based on the normal approximation to the binomial. Recall that this approximation is adequate for most purposes when the sample size n is so large that $np \geq 10$ and $n(1 - p) \geq 10$. The resulting procedures are similar to those for the mean of a normal distribution.

Suppose that we count "successes" (such as "yes" answers to a question) in an SRS of size n from a population containing an unknown proportion p of successes. The count X has the $B(n, p)$ distribution. The unknown population proportion p is estimated by the sample proportion $\hat{p} = X/n$. We know (Chapter 6) that $\hat{p}$ has approximately the normal distribution with mean $\mu(\hat{p}) = p$ and standard deviation $\sigma(\hat{p}) = \sqrt{p(1 - p)/n}$. If p were known, standardizing $\hat{p}$ would produce a z statistic

$$z = \frac{\hat{p} - p}{\sqrt{\dfrac{p(1 - p)}{n}}}$$

that has approximately the $N(0, 1)$ distribution. In order to test $H_0: p = p_0$, we calculate P-values by assuming that H_0 is true. Substituting p_0 for p in the z statistic allows such calculations.

To find a confidence interval for p, the standard deviation of $\hat{p}$ must be estimated from the data. To do this, we replace p with $\hat{p}$ in the expression for $\sigma(\hat{p})$. This gives the standard error of $\hat{p}$, which is

$$s(\hat{p}) = \sqrt{\frac{\hat{p}(1 - \hat{p})}{n}}$$

Two approximations are used: the normal approximation to the binomial distribution and the approximation of p by $\hat{p}$ in the standard deviation. Here is a summary of the inference procedures that result.

Large-sample inference for a population proportion

Suppose that an SRS of size n is drawn from a large population with unknown proportion p of successes. An approximate level $1 - \alpha$ confidence interval for p is

$$\hat{p} \pm z^* \sqrt{\frac{\hat{p}(1 - \hat{p})}{n}}$$

where z^* is the upper $\alpha/2$ standard normal critical value. To test the hypothesis H_0: $p = p_0$, compute the z statistic

$$z = \frac{\hat{p} - p_0}{\sqrt{\dfrac{p_0(1 - p_0)}{n}}}$$

In terms of a standard normal random variable Z, the approximate P-value for a test of H_0 against

H_a: $p > p_0$	is	$P(Z \geq z)$
H_a: $p < p_0$	is	$P(Z \leq z)$
H_a: $p \neq p_0$	is	$P(\lvert Z \rvert \geq \lvert z \rvert)$

Note that for confidence intervals we use the standard error $s(\hat{p})$, whereas for hypothesis testing we use the hypothesized value p_0 in the expression for $\sigma(\hat{p})$. These confidence intervals are quite accurate when n is so large that $n\hat{p} \geq 10$ and $n(1 - \hat{p}) \geq 10$. For the tests we will use this normal approximation whenever $np_0 \geq 10$ and $n(1 - p_0) \geq 10$.

EXAMPLE 9.1

An association of Christmas tree growers in Indiana sponsored a sample survey of Indiana households to help improve the marketing of Christmas trees.[2] An SRS of 500 households was contacted by telephone and asked several questions in a 2-minute interview. One question was, "Did you have a Christmas tree this year?" Of the 500 respondents, 421 answered "Yes." The sample proportion who had a tree is therefore

$$\hat{p} = \frac{421}{500} = .842$$

We compute a 95% confidence interval for the true proportion of all Indiana households who displayed a Christmas tree. From Table D we find the value of z^* to be 1.96. The interval is

$$\hat{p} \pm z^* \sqrt{\frac{\hat{p}(1 - \hat{p})}{n}} = .842 \pm 1.960 \sqrt{\frac{(.842)(.158)}{500}}$$

$$= .842 \pm .032$$

$$= (.810, .874)$$

We are 95% confident that between 81% and 87% of Indiana homes displayed Christmas trees.   ■

Remember that the margin of error in this confidence interval includes only random sampling error. There are other sources of error that are not accounted for. For example, the sample was chosen from telephone directories and so omits households without telephones or with unlisted numbers. Moreover, households were called at random until 500 responded. About 700 households were called in all, with no second call if the number dialed was busy or no one answered. These facts of real statistical life introduce some bias into the survey. The bias was checked by comparing demographic data collected from the respondents with census data for the state as a whole.

EXAMPLE 9.2

Of the 500 respondents in the Christmas tree market survey, 38% were from rural areas (including small towns) and the other 62% were from urban areas (including suburbs). According to the 1980 Census, 36% of Indiana residents live in rural areas and the remaining 64% live in urban areas. To examine how well the sample represents the state population in regard to rural versus urban residence, we perform a hypothesis test of

$$H_0: p = .36$$

versus the alternative

$$H_a: p \neq .36$$

where p represents the proportion of rural households that would be obtained by the telephone sampling procedure if it were repeated over and over again. The test statistic is

$$z = \frac{\hat{p} - .36}{\sqrt{\dfrac{(.36)(.64)}{500}}} = \frac{.38 - .36}{\sqrt{\dfrac{(.36)(.64)}{500}}} = .93$$

From Table A, we find that the probability that a Z is less than or equal to 0.93 is 0.8238. Figure 9.1 illustrates the calculation of the P-value. The probability in each tail is $1 - 0.8238 = 0.1762$. Therefore, the P-value is $2(0.1762) = 0.35$. There is a 35% chance of getting a value of Z larger than 0.93 or smaller than -0.93 if H_0 is true. We therefore have no reason to reject the hypothesis that the sampling procedure is unbiased with respect to rural versus urban residence.   ■

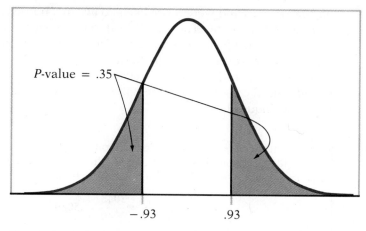

Figure 9.1   The P-value in Example 9.2.

In this example we arbitrarily chose to formulate the question of interest in terms of the proportion of rural respondents. We could just as easily have chosen to use the proportion of urban respondents. Since the sum of the two proportions is 1, our procedure should give us the same results.

EXAMPLE 9.3

For the Christmas tree market survey, we let p represent the proportion of urban households that would be obtained by the telephone sampling procedure if it were repeated over and over again. To check agreement with the 1980 Census results, we test

$$H_0: p = .64$$

versus the alternative

$$H_a: p \neq .64$$

The test statistic is

$$z = \frac{\hat{p} - .64}{\sqrt{\dfrac{(.64)(.36)}{500}}} = \frac{.62 - .64}{\sqrt{\dfrac{(.64)(.36)}{500}}} = -.93$$

Since only the sign of z has changed, the P-value remains the same.   ■

The fact demonstrated in Example 9.3 is true in general. When performing hypothesis tests on single proportions, we obtain the same value of the z statistic but with a reversed sign when we interchange the labels of success and failure in the binomial model. This corresponds to interchanging the probabilities p and $1 - p$.

For some of the other demographic variables recorded for the Christmas tree market survey, statistically significant differences between the sample and the 1980 Census values were found. The respondents had somewhat higher incomes and were more likely to live in a house rather than an apartment or mobile home. These facts lead to some overestimation of Christmas tree usage, since well-off people living in a house are more likely to have a tree. In other words, the estimate that we have calculated is slightly biased upward. Using more advanced statistical techniques, the bias can be assessed, and the estimate can be corrected for the bias. We will not discuss these techniques here.

EXAMPLE 9.4

As was mentioned in the introduction to Chapter 5, the French naturalist Buffon once tossed a coin 4040 times and obtained 2048 heads. This is a binomial experiment with $n = 4040$. The sample proportion is

$$\hat{p} = \frac{2048}{4040} = .5069$$

If Buffon's coin was balanced, then the probability of obtaining heads on any toss is 0.5. To assess whether the data provide evidence that the coin was not balanced, we test

$$H_0: p = .5$$
$$H_a: p \neq .5$$

The test statistic is

$$z = \frac{\hat{p} - .5}{\sqrt{\frac{(.5)(.5)}{4040}}} = \frac{.5069 - .5}{\sqrt{\frac{(.5)(.5)}{4040}}} = .88$$

Figure 9.2 illustrates the calculation of the *P*-value. From Table A we find

$$P(Z \leq .88) = .8106$$

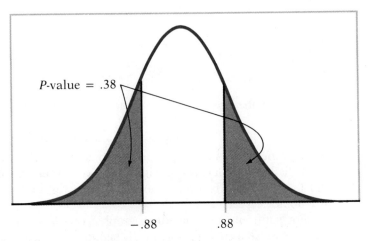

P-value = .38

$-.88$ $.88$

Figure 9.2 The *P*-value in Example 9.4.

Therefore, the probability in each tail is $1 - 0.8106 = 0.1894$ and the *P*-value is $P = 2(0.1894) = 0.38$. We do not have evidence that the coin was unbalanced.

■

Unless a coin is perfectly balanced, its true probability of heads will not be *exactly* 0.5. What has been demonstrated in the above example is that Buffon's coin is statistically indistinguishable from a fair coin on the basis of 4040 tosses. To see what other values of p are compatible with the sample results, a confidence interval is used.

EXAMPLE 9.5

For Buffon's coin-tossing experiment, the 99% confidence interval for the probability that his coin comes up heads is

$$\hat{p} \pm z^* \sqrt{\frac{\hat{p}(1 - \hat{p})}{n}} = .5069 \pm 2.576 \sqrt{\frac{(.5069)(.4931)}{4040}}$$

$$= .5069 \pm .0203$$

$$= (.4866, .5272)$$

We are 99% confident that the probability of a head is between 0.4866 and 0.5272.

■

The confidence interval of Example 9.5 is much more informative than the results of the hypothesis test of Example 9.4. It indicates the range of values of p that are consistent with observed results. We would not be surprised if the true probability of heads for Buffon's coin was something like 0.51. The width of the 99% confidence interval for the 4040 trials is $2(0.0203) = 0.0406$. To obtain a shorter interval, we would need to increase the sample size.

Choosing a Sample Size

In Chapter 7, the problem of choosing the sample size to obtain a confidence interval of prespecified width for a normal mean was discussed. Since we are using the normal approximation to the binomial for the analysis of proportions, the considerations for sample size selection are very similar.

Recall that an approximate level $1 - \alpha$ confidence interval for p is

$$\hat{p} \pm z^* \sqrt{\frac{\hat{p}(1 - \hat{p})}{n}}$$

where z^* is the upper $\alpha/2$ normal critical value. The width of the confidence interval is therefore

$$2z^* \sqrt{\frac{\hat{p}(1 - \hat{p})}{n}}$$

Since the value of $\hat{p}$ is not known before the experiment is conducted, we must guess a value to use in the calculations. We let p^* denote the guessed value. The width of the confidence interval is largest when the value of $\hat{p} = 0.5$. Therefore, a conservative approach to the problem is to use $p^* = 0.5$ as the guessed value. This ensures that the calculated confidence interval will have width less than or equal to the prespecified value. If w is the desired width, set

$$w = 2z^* \sqrt{\frac{p^*(1 - p^*)}{n}}$$

and solve this equation for n to find the sample size required. Here is the result.

Sample size for desired width

> The level $1 - \alpha$ confidence interval for a proportion p will have width approximately equal to a specified value w when the sample size is
>
> $$n = \left(\frac{2z^*}{w}\right)^2 p^*(1 - p^*)$$
>
> where p^* is a guessed value for the true proportion.
>   The width will be less than or equal to w if p^* is chosen to be 0.5. This gives
>
> $$n = \left(\frac{z^*}{w}\right)^2$$

The value of n obtained by this method is not particularly sensitive to the choice of p^* when p^* is fairly close to 0.5. However, if the value of p is, in fact, close to 0 or 1, then the use of $p^* = 0.5$ will select a sample size that is much larger than needed.

EXAMPLE 9.6

You are doing a sample survey to determine which of two candidates a group of voters prefers. Let p be the true proportion of voters who prefer the first candidate. Since it is judged that the value of p is not very different from 0.5, the choice $p^* = 0.5$ is used for planning the sample size.
    Suppose that you want to estimate p with a 95% confidence interval and a margin of error less than or equal to 3%, or 0.03. This corresponds to a 95% confidence interval for p with a width less than or equal to 0.06. The sample size required is

$$n = \left(\frac{1.96}{.06}\right)^2 = 1067$$

Similarly for a 2.5% margin of error we have

$$n = \left(\frac{1.96}{.05}\right)^2 = 1537$$

and for a 2% margin of error the required sample size is

$$n = \left(\frac{1.96}{.04}\right)^2 = 2401$$

∎

You will frequently encounter news reports that describe the results of surveys with sample sizes between 1500 and 1600 and a 2.5% or 3% margin of error. These surveys often use sampling procedures more complicated than simple random sampling, so that the calculation of confidence intervals in such cases is more involved than we have studied in this section. The calculations in Example 9.6 nonetheless show in principle how such surveys are planned.

In practice the choice of a sample size depends upon many factors. In the Christmas tree market survey, the researchers consulted a statistician when planning the study. It was estimated that each telephone interview would take about 2 minutes. Nine trained students in agribusiness marketing were to make the phone calls between 1:00 P.M. and 8:00 P.M. on a Sunday. After discussing problems related to people not being at home or being unwilling to answer the questions, a sample size of 500 was proposed. The widths of 95% confidence intervals for various values of $\hat{p}$ were then calculated for evaluation by the researchers.

EXAMPLE 9.7

In planning the Christmas tree market survey, the widths of 95% confidence intervals for various values of $\hat{p}$ and $n = 500$ were calculated using the following formula:

$$w = 2z^* \sqrt{\frac{\hat{p}(1 - \hat{p})}{n}}$$

$$= 2(1.96) \sqrt{\frac{\hat{p}(1 - \hat{p})}{500}}$$

$$= .175\sqrt{\hat{p}(1 - \hat{p})}$$

The results are shown in the following tabulation:

$\hat{p}$	w	$\hat{p}$	w
.05	.038	.60	.086
.10	.053	.70	.080
.20	.070	.80	.070
.30	.080	.90	.053
.40	.086	.95	.038
.50	.088		

These widths were judged to be acceptable by the researchers, and the sample size of 500 was then used in their study.   ∎

Several interesting things can be seen in the table presented in Example 9.7. First, note that the widths for $\hat{p} = 0.05$ and $\hat{p} = 0.95$ are the same. In fact, the widths will always be the same for $\hat{p}$ and $1 - \hat{p}$. This is a direct consequence of the form of the confidence interval.

Second, the width of the interval varies only between 0.080 and 0.088 as $\hat{p}$ varies from 0.3 to 0.7, and the width is greatest when $\hat{p} = 0.5$, as we claimed earlier. It is true in general that the width of the interval will vary relatively little for values of $\hat{p}$ between 0.3 and 0.7. Therefore, when planning a study, it is not necessary to have a very precise guess for p. If $p^* = 0.5$ is used and the observed $\hat{p}$ is between 0.3 and 0.7, the actual interval will be a little shorter than needed but the difference will be small.

Finally, note that the widths w for very small and very large values of $\hat{p}$ have not been included in the table. Suppose, for example, that we expect p to be about 0.01. Then for $n = 500$, we have $np = 5$. Recall that for the normal approximation to be reasonably accurate, we need to have $np \geq 10$ and $n(1 - p) \geq 10$. Therefore, we would not be justified in using the normal approximation to the binomial to calculate the confidence interval. More advanced methods are needed in this case.

SUMMARY

Inference about a population proportion p is based on the sample proportion $\hat{p} = X/n$ from an SRS of size n. When n is large, $\hat{p}$ has approximately the normal distribution, with mean p and standard deviation $\sqrt{p(1 - p)/n}$.

The level $1 - \alpha$ confidence interval for p is

$$\hat{p} \pm z^* \sqrt{\frac{\hat{p}(1 - \hat{p})}{n}}$$

where z^* is the upper $\alpha/2$ standard normal critical value.

Tests of H_0: $p = p_0$ are based on the **z statistic**

$$z = \frac{\hat{p} - p_0}{\sqrt{\frac{p_0(1 - p_0)}{n}}}$$

with P-values calculated from the $N(0, 1)$ distribution.

The sample size required to obtain a confidence interval of approximate width w for a proportion is

$$n = \left(\frac{2z^*}{w}\right)^2 p^*(1 - p^*)$$

where p^* is a guessed value for the proportion and z^* is the standard normal critical point for the desired level of confidence. To ensure that the width of the interval is less than or equal to w no matter what p may be, use

$$n = \left(\frac{z^*}{w}\right)^2$$

SECTION 9.1 EXERCISES

9.1 For each of the following cases state whether or not the normal approximation to the binomial should be used for a significance test on the population proportion p.

 (a) $n = 10$ and H_0: $p = 0.4$.
 (b) $n = 100$ and H_0: $p = 0.6$.
 (c) $n = 1000$ and H_0: $p = 0.996$.
 (d) $n = 500$ and H_0: $p = 0.3$.

9.2 For each of the following cases state whether or not the normal approximation to the binomial should be used for a confidence interval for the population proportion p.

 (a) $n = 30$ and we observe $\hat{p} = 0.9$.
 (b) $n = 25$ and we observe $\hat{p} = 0.5$.
 (c) $n = 100$ and we observe $\hat{p} = 0.04$.
 (d) $n = 600$ and we observe $\hat{p} = 0.6$.

9.3 Use the results of the Christmas tree market survey described in Example 9.1 to find a 99% confidence interval for the proportion of Indiana homes that displayed Christmas trees.

9.4 Use the results of the Christmas tree market survey described in Example 9.1 to find a 90% confidence interval for the proportion of Indiana homes that *did not* display a Christmas tree.

9.5 In recent years over 70% of freshman college students responding to a national survey have identified "being well-off financially" as an important personal goal. A liberal arts college finds that in an SRS of 200 of its freshman, 132 consider this goal important. Give a 95% confidence interval for the proportion of all freshman at the college who would identify being well-off as an important personal goal.

9.6 The Gallup Poll asked a sample of 1785 U.S. adults, "Did you, yourself, happen to attend church or synagogue in the last 7 days?" Of the respondents, 750 said "Yes." Suppose (it is not, in fact, true) that Gallup's sample was an SRS.

(a) Give a 99% confidence interval for the proportion of all U.S. adults who attended church or synagogue during the week preceding the poll.

(b) Do the results provide good evidence that less than half of the population attended church or synagogue?

(c) How large a sample would be required to obtain a margin of error of ± 0.01 in a 99% confidence interval for the proportion who attend church or synagogue? (Use Gallup's result as the guessed value of p.)

9.7 A national opinion poll found that 44% of all American adults agree that parents should be given vouchers good for education at any public or private school of their choice. The result was based on a small sample. How large an SRS is required to obtain a margin of error of ± 0.03 (that is, $\pm 3\%$) in a 95% confidence interval? (Use the previous poll's result to obtain the guessed value p^*.)

9.8 An entomologist samples a field for egg masses of a harmful insect by placing a yard-square frame at random locations and carefully examining the ground within the frame. An SRS of 75 locations selected from a county's pasture land found egg masses in 13 locations. Give a 95% confidence interval for the proportion of all possible locations that are infested.

9.9 In the Christmas tree market survey, 43% of the 500 respondents had incomes of $25,000 or less. The 1980 Census found that 64% of Indiana households had incomes of $25,000 or less. Test the null hypothesis that the telephone survey technique has a probability of selecting a household with an income of $25,000 or less that is equal to the value obtained by the 1980 Census. State H_0 and H_a and calculate the z statistic. Give the P-value and report your conclusions.

9.10 Of the 500 respondents in the Christmas tree market survey, 44% had no children at home and 56% had at least one child at home. The corresponding figures for the 1980 Census are 48% with no children and 52% with at least one child. Test the null hypothesis that the telephone survey technique has a probability of selecting a household with no children that is equal to the value obtained by the 1980 Census. Give the z statistic and the P-value. What do you conclude?

9.11 The introduction to Chapter 5 mentions a coin-tossing experiment carried out by the English statistician Karl Pearson. Pearson tossed a coin 24,000 times and obtained 12,012 heads.

(a) Find the z statistic for testing the null hypothesis that Pearson's coin had probability 0.5 of coming up heads versus the two-sided alternative. For $\alpha = 0.01$ do you reject H_0? Find the P-value.

(b) Find a 99% confidence interval for the probability of heads for Pearson's coin.

9.12 Another coin-tossing experiment mentioned in the introduction to Chapter 5 was performed by the English mathematician John Kerrich. Kerrich tossed a coin 10,000 times and obtained 5067 heads.

   (a) Find the z statistic for testing the null hypothesis that Kerrich's coin had probability 0.5 of coming up heads versus the two-sided alternative. For $\alpha = 0.05$ do you reject H_0? Find the P-value.

   (b) Find a 95% confidence interval for the probability of heads for Kerrich's coin.

9.13 An experiment was performed to compare the taste of instant versus freshly brewed coffee. Each subject was presented with two unmarked cups of coffee, one of each type. They were asked to state which they preferred. Of the 50 subjects who participated in the study, 19 stated that they preferred the cup containing the instant coffee. Let p denote the probability that a randomly chosen subject selects instant coffee in preference to freshly brewed coffee.

   (a) The claim that instant coffee tastes just as good as freshly brewed coffee is formulated as a test of the null hypothesis, H_0: $p = 0.5$, against the alternative H_a: $p < 0.5$. Find the z statistic for testing this H_0. Do you reject H_0 for $\alpha = 0.05$? What is the P-value?

   (b) Find a 90% confidence interval for p.

9.14 One of the starting players for a major college basketball team made only 40% of his free throws during the last season. During the summer he worked on developing a softer shot in the hopes of improving his free throw accuracy. In the first eight games of this season he made 25 free throws in 40 attempts. Let p denote the probability that he makes a free throw this season.

   (a) In terms of p, write the null hypothesis H_0 that his free throw probability has remained the same as last year and the alternative H_a that his work in the summer resulted in an improved free throw probability of success.

   (b) Calculate the z statistic for testing H_0 versus H_a.

   (c) Do you accept or reject H_0 for $\alpha = 0.05$? Find the P-value.

   (d) Give a 90% confidence interval for his free throw success probability for the new season.

   (e) What assumptions are needed for the validity of the test and confidence interval calculations that you performed.

9.15 You want to estimate the proportion of students at your college or university who are employed for 10 or more hours per week while classes are in session. A 95% confidence interval will be used to present the results. Using a guessed value of $p^* = 0.3$, find the sample size required if the interval is to have approximate width $w = 0.1$.

9.16 A magazine publisher would like to know the proportion of subscribers who have annual household incomes in excess of $25,000. To do this they will survey a sample of their subscribers. They would like the width of the 99% confidence interval for the proportion to be about 0.05 or less. Use a guessed value of $p^* = 0.5$ to find the required sample size.

9.17 A student organization wants to start a nightclub for students under the age of 21. To assess support for this proposal, a random sample of students will be drawn. Each student will be asked if he or she would patronize this type of establishment. They expect that about 70% of the student body will respond favorably. What sample size is required to obtain a 90% confidence interval of approximate width 0.075 for this problem? Suppose that 50% of the sample respond favorably. Calculate the width of the 90% confidence interval in this case.

9.18 An automobile manufacturer would like to know the proportion of its customers who are dissatisfied with the service received from their local automobile dealer. A random sample of customers will be surveyed, and a 99% confidence interval for the proportion dissatisfied will be computed. From past studies, they believe that the proportion will be about 0.2. Find the sample size needed if the width of the confidence interval is to be about 0.03. Suppose 10% of the sample say that they are dissatisfied. What is the width of the 99% confidence interval in this case?

9.19 You have been asked to survey students at a large college to determine the proportion of students who favor an increase in the student fees to support an expansion of the student newspaper. Each student will be asked to state whether or not he or she is in favor of the proposed increase. Using records provided by the registrar you can select a random sample of students from the college. After careful consideration of your resources, you decide that a study with a sample of 100 students would be reasonable to undertake.

(a) For this sample size, construct a table of the widths of the 95% confidence intervals that you would calculate if you obtained $\hat{p}$-values of 0.1, 0.2, 0.3, 0.4, 0.5, 0.6, 0.7, 0.8, and 0.9.

(b) For your sample size, would you use the normal approximation to the binomial if $\hat{p} = 0.04$? Why or why not?

9.20 Refer to the previous exercise. A former editor of the student newspaper has just agreed to underwrite your study because she believes that the results will demonstrate that a large proportion of the students will support the increase in the fees. She is willing to provide funds for a sample of size 500. Answer all of the questions posed in the previous exercise and write a short summary for your benefactor of why the increased sample size will provide better results.

9.2 COMPARING TWO PROPORTIONS

Inference procedures for comparing proportions in two populations are similar to those for the comparison of means. Suppose that we have two independent SRSs, of size n_1 from population 1 and size n_2 from population 2. We wish to compare the two population proportions of successes, p_1 and p_2. Each of these population proportions is estimated by the corresponding sample proportion. If the sample counts of successes are X_1 and X_2, p_1 is estimated by $\hat{p}_1 = X_1/n_1$ and p_2 by $\hat{p}_2 = X_2/n_2$.

Using the addition rule for means, we find that

$$\mu(\hat{p}_1 - \hat{p}_2) = \mu(\hat{p}_1) - \mu(\hat{p}_2) = p_1 - p_2$$

That is, the difference $\hat{p}_1 - \hat{p}_2$ between the sample proportions is an unbiased estimate of the population difference $p_1 - p_2$. Similarly, the addition rule for variances tells us that

$$\sigma^2(\hat{p}_1 - \hat{p}_2) = \sigma^2(\hat{p}_1) + \sigma^2(\hat{p}_2)$$
$$= \frac{p_1(1 - p_1)}{n_1} + \frac{p_2(1 - p_2)}{n_2}$$

Therefore, when n_1 and n_2 are large, the difference $D = \hat{p}_1 - \hat{p}_2$ is approximately normal with mean $p_1 - p_2$ and standard deviation

$$\sigma(D) = \sqrt{\frac{p_1(1 - p_1)}{n_1} + \frac{p_2(1 - p_2)}{n_2}}$$

Confidence Intervals

For both confidence intervals and hypothesis tests, we need to estimate the unknown variance, $\sigma^2(D)$. A different estimate will be used in each of these two cases. First, consider the problem of constructing a confidence interval for $p_1 - p_2$. Here we simply substitute the sample values $\hat{p}_1$ and $\hat{p}_2$ for p_1 and p_2 in the above formula for $\sigma(D)$ to obtain

$$s(D) = \sqrt{\frac{\hat{p}_1(1 - \hat{p}_1)}{n_1} + \frac{\hat{p}_2(1 - \hat{p}_2)}{n_2}}$$

This is the standard error of D that we will use in our confidence interval calculations.

Confidence intervals for comparing two proportions

Suppose that an SRS of size n_1 is drawn from a large population having proportion p_1 of successes, and that an independent SRS of size n_2 is drawn from another population having proportion p_2 of successes. When n_1 and n_2 are large, an approximate level $1 - \alpha$ confidence interval for $p_1 - p_2$ is

$$(\hat{p}_1 - \hat{p}_2) \pm z^* s(D)$$

where

$$s(D) = \sqrt{\frac{\hat{p}_1(1 - \hat{p}_1)}{n_1} + \frac{\hat{p}_2(1 - \hat{p}_2)}{n_2}}$$

and z^* is the upper $\alpha/2$ standard normal critical value.

This interval is approximately correct when the sample sizes n_1 and n_2 are large. As a general rule, we will use this method when $n_1\hat{p}_1$, $n_1(1 - \hat{p}_1)$, $n_2\hat{p}_2$, and $n_2(1 - \hat{p}_2)$ are all 5 or more.

EXAMPLE 9.8

If a respondent to the Christmas tree survey introduced in Example 9.1 did display a tree during the last holiday season, the next question asked was whether the tree was natural or artificial. Respondents were also asked if they lived in an urban area or in a rural area. Of the 421 households displaying a Christmas tree, 261 were urban and 160 lived in rural areas.

Take population 1 to be urban tree users and population 2 to be rural tree users. We can consider that we have independent SRSs of size $n_1 = 261$ from population 1 and size $n_2 = 160$ from population 2. The tree growers want to know if there is a difference in preference for natural trees versus artificial trees between urban and rural households. The survey results are tabulated below.

Population	n	Count X
1 (urban)	261	89
2 (rural)	160	64

The proportion of urban trees users displaying a natural tree is

$$\hat{p}_1 = \frac{89}{261} = .341$$

and the proportion for the rural group is

$$\hat{p}_2 = \frac{64}{160} = .400$$

To compute a 90% confidence interval for the difference between urban and rural tree users in the proportions of households that prefer a natural tree, we first calculate the standard error of the difference:

$$s(D) = \sqrt{\frac{\hat{p}_1(1 - \hat{p}_1)}{n_1} + \frac{\hat{p}_2(1 - \hat{p}_2)}{n_2}}$$

$$= \sqrt{\frac{(.341)(.659)}{261} + \frac{(.400)(.600)}{160}}$$

$$= .04859$$

The 90% confidence interval is

$$(\hat{p}_1 - \hat{p}_2) \pm z^* s(D) = (.341 - .400) \pm (1.645)(.04859)$$

$$= -.059 \pm .080$$

$$= (-.139, .021)$$

With 90% confidence we can say that the difference in the proportions is between -0.14 and $+0.02$. Because the interval contains 0, we are not confident that either group has a stronger preference for natural trees than the other group.

■

The two samples in this example were not drawn separately and did not have sizes chosen in advance. The sample sizes are actually random, arising from asking the 500 randomly chosen respondents where they lived and whether they had a tree last holiday season. If the SRS were repeated, we would no doubt not have 261 urban and 160 rural tree users among the 500 new respondents. It is in fact true—we do not say that it is obvious— that we can analyze the data just as if the sample sizes had been fixed in the design of the study. This fact extends the usefulness of procedures for the comparison of two proportions. The issue of fixed versus random sample sizes will be discussed in Section 9.3, where we consider more elaborate analyses of count data.

Significance Tests

Although we prefer to compare two proportions by giving a confidence interval for the difference between the two population proportions, it is sometimes useful to perform a test of the null hypothesis that the two population proportions are the same.

For the testing problem, we will not use the same estimate $s(D)$ of $\sigma(D)$ that we used for the confidence intervals. Although $s(D)$ would lead to a valid test, we instead adopt the more common practice of using an estimate of $\sigma(D)$ based on the fact that under the null hypothesis being tested, $p_1 = p_2$.

Specifically, if the two proportions are equal, then we can view all of the data as coming from a single population. Let p denote the common value of p_1 and p_2; then the standard deviation of $D = \hat{p}_1 - \hat{p}_2$ is

$$\sigma(D) = \sqrt{\frac{p_1(1 - p_1)}{n_1} + \frac{p_2(1 - p_2)}{n_2}}$$

$$= \sqrt{p(1 - p)\left(\frac{1}{n_1} + \frac{1}{n_2}\right)}$$

We estimate the common value of p by the overall proportion of successes in both samples,

$$\hat{p} = \frac{X_1 + X_2}{n_1 + n_2}$$

This estimate of p is called the *pooled* estimate because we are combining, or pooling, the information from both samples for our estimate.

To estimate $\sigma(D)$ under the null hypothesis, we substitute $\hat{p}$ for p in the above expression for $\sigma(D)$. The result is a standard error for D that assumes the truth of H_0: $p_1 = p_2$:

$$s_p(D) = \sqrt{\hat{p}(1 - \hat{p})\left(\frac{1}{n_1} + \frac{1}{n_2}\right)}$$

The subscript on s_p reminds us that we pooled data from the two samples to construct the estimate.

Significance tests for comparing two proportions

To test the hypothesis

$$H_0: p_1 = p_2$$

compute the z statistic

$$z = \frac{\hat{p}_1 - \hat{p}_2}{s_p(D)}$$

where

$$s_p(D) = \sqrt{\hat{p}(1 - \hat{p})\left(\frac{1}{n_1} + \frac{1}{n_2}\right)}$$

and

$$\hat{p} = \frac{X_1 + X_2}{n_1 + n_2}$$

In terms of a standard normal random variable Z, the P-value for a test of H_0 against

$$H_a: p_1 > p_2 \quad \text{is} \quad P(Z \geq z)$$
$$H_a: p_1 < p_2 \quad \text{is} \quad P(Z \leq z)$$
$$H_a: p_1 \neq p_2 \quad \text{is} \quad P(|Z| \geq |z|)$$

This test is based on the normal approximation to the binomial distribution. As a general rule, we will use it when $n_1\hat{p}$, $n_1(1 - \hat{p})$, $n_2\hat{p}$, and $n_2(1 - \hat{p})$ are all 5 or more.

EXAMPLE 9.9

For the Christmas tree market survey we will compare the urban tree users and the rural tree users with respect to their preference for natural versus artificial trees. Other studies have shown that there is a tendency toward greater use of natural Christmas trees in rural areas. A one-sided test will be used to examine whether or not this sample supports this previous finding. Again, take population 1 to consist of the urban households that use a tree and population 2 to be the rural tree users. We want to test the hypotheses

$$H_0: p_1 = p_2$$
$$H_a: p_1 < p_2$$

The survey responses show that 89 of the urban households and 64 of the rural households who displayed a tree chose a natural tree. So

$$\hat{p}_1 = \frac{89}{261} = .341$$

$$\hat{p}_2 = \frac{64}{160} = .400$$

For the combined sample, the proportion of respondents who chose a natural tree was

$$\hat{p} = \frac{89 + 64}{261 + 160} = .363$$

The test statistic is calculated as follows:

$$s_p(D) = \sqrt{(.363)(.637)\left(\frac{1}{261} + \frac{1}{160}\right)} = .04828$$

$$z = \frac{\hat{p}_1 - \hat{p}_2}{s_p(D)} = \frac{.341 - .400}{.04828}$$

$$= -1.22$$

The calculation of the *P*-value is illustrated in Figure 9.3. From Table A we find

$$P(Z \le -1.22) = .1112$$

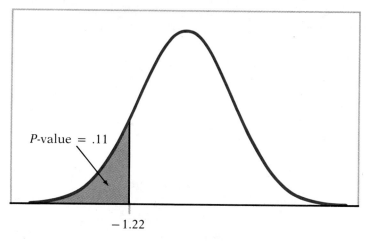

P-value = .11

−1.22

Figure 9.3 The *P*-value in Example 9.9.

Since we are doing a one-sided test, the *P*-value is 0.11. Even though rural house-holds in the survey chose natural Christmas trees more often than the urban households, our calculations indicate that there is not sufficient evidence in the data to conclude that this difference in preferences is true in the population of Indiana tree users. If the preferences of rural and urban households were identical, rural usage would exceed urban usage by an amount leading to a *z* statistic at least as large as the one observed in 11% of all samples of this size. ∎

SUMMARY

Comparison of two population proportions from independent SRSs of sizes n_1 and n_2 is based on the difference of sample proportions $\hat{p}_1 - \hat{p}_2$. The level $1 - \alpha$ confidence interval for $p_1 - p_2$ is

$$(\hat{p}_1 - \hat{p}_2) \pm z^* s(D)$$

where

$$s(D) = \sqrt{\frac{\hat{p}_1(1 - \hat{p}_1)}{n_1} + \frac{\hat{p}_2(1 - \hat{p}_2)}{n_2}}$$

Significance tests of H_0: $p_1 = p_2$ use the **z statistic**

$$z = \frac{\hat{p}_1 - \hat{p}_2}{s_p(D)}$$

with $N(0, 1)$ probabilities. In this statistic,

$$s_p(D) = \sqrt{\hat{p}(1 - \hat{p})\left(\frac{1}{n_1} + \frac{1}{n_2}\right)}$$

and $\hat{p}$ is the **pooled estimate** of the common value of p_1 and p_2,

$$\hat{p} = \frac{X_1 + X_2}{n_1 + n_2}$$

SECTION 9.2 EXERCISES

9.21 As of July 1, 1988, the New York Mets baseball team had played 37 games at home and 40 games away. They won 26 of their home games and 23 of the games played away.

   (a) Find the proportion of wins for the home games. Do the same for the away games.

(b) Find the standard error needed to compute a confidence interval for the difference in the proportions.

(c) Compute a 90% confidence interval for the difference between the probability that the Mets win at home and the probability that they win when on the road.

9.22 In 1988 random samples of Indiana farmers were asked whether or not they favored a mandatory corn checkoff program to pay for corn product marketing and research. In Tippecanoe County, 263 farmers were in favor of the program and 252 were not. In neighboring Benton County, 260 were in favor while 377 were not.

(a) Find the proportions of farmers in favor of the program in each of the two counties.

(b) Find the standard error needed to compute a confidence interval for the difference in the proportions.

(c) Compute a 99% confidence interval for the difference in the proportions of farmers favoring the program in Tippecanoe County versus Benton County.

9.23 Refer to the New York Mets baseball data given in Exercise 9.21.

(a) Combining all of the games played, what is the proportion of wins?

(b) Find the standard error needed for testing that the probability of winning is the same at home and away.

(c) It is generally believed that it is easier to win at home than away. Formulate appropriate null and alternative hypotheses to examine this idea.

(d) Compute the z statistic and its P-value. What conclusion do you draw?

9.24 Refer to the survey of farmers described in Exercise 9.22.

(a) Combine the two samples and find the proportion of farmers who favor the corn checkoff program.

(b) Find the standard error needed for testing that the population proportions of farmers favoring the program are the same in the two counties.

(c) Formulate appropriate null and alternative hypotheses for comparing the two counties.

(d) Compute the z statistic and its P-value. What conclusion do you draw?

9.25 Refer to the New York Mets baseball data and the confidence interval computed in Exercise 9.21. Redo the exercise using the proportion of losses in place of the proportion of wins. Compare these results with those obtained in Exercise 9.21.

9.26 Refer to the survey of farmers described in Exercise 9.22. Redo the exercise using the proportion of farmers not in favor of the program in place of the proportion in favor. Compare these results with those obtained in Exercise 9.22.

9.27 Refer to the New York Mets baseball data described in Exercise 9.21 and the significance test computed in Exercise 9.23. Redo Exercise 9.23 using the proportion of losses in place of the proportion of wins. Compare these results with those obtained in Exercise 9.23.

9.28 Refer to the survey of farmers described in Exericse 9.22 and the significance test you performed in Exercise 9.24. Redo Exercise 9.24 using the proportion of farmers not in favor of the program in place of the proportion in favor. Compare these results with those obtained in Exercise 9.24.

9.29 The 1958 Detroit Area Study was an important sociological investigation of the influence of religion on everyday life. It is described in Gerhard Lenski, *The Religious Factor*, Doubleday, New York, 1961. The sample "was basically a simple random sample of the population of the metropolitan area." Of the 656 respondents, 267 were white Protestants and 230 were white Catholics. One question asked whether the government was doing enough in areas such as housing, unemployment, and education; 161 of the Protestants and 136 of the Catholics said "No." Is there evidence that white Protestants and white Catholics differed on this issue?

9.30 The respondents in the Detroit Area Study (see the previous exercise) were also asked whether they believed that the right of free speech included the right to make speeches in favor of communism. Of the white Protestants, 104 said "Yes," while 75 of the white Catholics said "Yes." Give a 95% confidence interval for the amount by which the proportion of Protestants who agree that communist speeches are protected exceeds the proportion of Catholics who hold this opinion.

9.31 A university financial aid office surveyed an SRS of undergraduate students to study their summer employment. Not all students were employed the previous summer. Here are the results for men and women:

	Men	Women
Employed	718	593
Not employed	79	139
Total	797	732

(a) Is there evidence that the proportion of male students employed during the summer differs from the proportion of female students who were employed? State H_0 and H_a, compute the test statistic, and give its *P*-value.

(b) Give a 99% confidence interval for the difference between the proportions of male and female students who were employed during the summer.

9.32 The power takeoff driveline on tractors used in agriculture is a potentially serious hazard to operators of farm equipment. The driveline is covered by a shield in new tractors, but for a variety of reasons, the shield is often missing on older tractors. Two types of shield are the bolt-on and the flip-up. In a study initiated by the National Safety Council, a sample of older tractors was taken to examine the proportions of types of shields removed. Of 83 tractors designed to have bolt-on shields, 35 had been removed. For the 136 tractors with flip-up shields, 15 were removed. (Data taken from W. E. Sell and W. E. Field, "Evaluation of PTO master shield usage on John Deere tractors," presented at the American Society of Agricultural Engineers 1984 Summer Meeting.)

(a) Test the null hypothesis that there is no difference between the proportions of the two types of shields removed. Give the *z* statistic and the *P*-value. State your conclusion in words.

(b) Give a 90% confidence interval for the difference in the proportions of removed shields for the bolt-on and the flip-up types.

9.33 A study was performed to test the effectiveness of aspirin in the treatment of cerebral ischemia (stroke). Patients were randomized into treatment and control groups. The study was double-blind in the sense that neither the patients nor the physicians who evaluated the patients knew whether the patient received aspirin or the placebo—a preparation that appeared to be identical to aspirin but that contained no active ingredients. After 6 months of treatment, patients were evaluated and their progress was recorded as either favorable or unfavorable. Of the 78 patients in the aspirin group, 63 had favorable outcomes; 43 of the 77 control patients had favorable outcomes. (From William S. Fields et al., "Controlled trial of aspirin in cerebral ischemia," *Stroke*, 8 (1977), pp. 301–315.)

(a) Compute the sample proportions of patients having favorable outcomes in the two groups.

(b) Give a 95% confidence interval for the difference between the favorable proportions in the treatment and control groups.

(c) The physicians conducting the study had concluded from previous research that aspirin was likely to increase the chance of a favorable outcome. Formulate appropriate null and alternative hypotheses for this study.

(d) Test the null hypothesis and report the *P*-value. What do you conclude about the effectiveness of aspirin as a treatment for cerebral ischemia?

9.34 Infestations of the German cockroach, *Blatella germanica*, are commonly treated by the pesticide diazinon. In one study the persistence of this pesticide on various types of surfaces was investigated. A 0.5% emulsion of diazinon was applied to glass and plasterboard. After 14 days, 18 cockroaches were placed on each surface, and the number that died within 48 hours was recorded. On the glass, 9 cockroaches died, while on the plasterboard, 13 died. (Based on Elray M. Roper and Charles G. Wright, "German cockroach (Orthoptera: Blatellidae) mortality on various surfaces following application of diazinon," *Journal of Economic Entomology*, 78 (1985), pp. 733–737.)

(a) Calculate the mortality rates (sample proportion that died) for the two surfaces.

(b) Find a 90% confidence interval for the difference in the two proportions.

(c) Chemical analysis of the residues of diazinon suggests that it may persist longer on plasterboard than on glass because it binds to the paper covering on the plasterboard. The researchers therefore expected the mortality rate to be greater on plasterboard than on glass. State appropriate null and alternative hypotheses for comparing the two surfaces.

(d) Test the null hypothesis and report the *P*-value. What do you conclude?

9.35 Refer to the study of aspirin in the treatment of cerebral ischemia described in Exercise 9.33. Suppose that a smaller number of patients had been used in each group and that similar proportions of favorable and unfavorable responses were obtained. Specifically, suppose that 10 patients were used in each group and the number of favorable responses was 8 in the aspirin group and 5 in the control group.

(a) Recompute the *z* statistic for these data and report the *P*-value. What do you conclude?

(b) Compare the results of this significance test with those you gave in Exercise 9.33. What do you conclude about the effect of the sample size on the results of these significance tests?

9.36 Refer to the cockroach study described in Exercise 9.34. Suppose that more cockroaches were used on each surface and that similar mortality rates were observed. Specifically, suppose that 36 cockroaches were used on each surface and that 26 died on the plasterboard while 18 died on the glass.

(a) Recompute the *z* statistic for these data and report the *P*-value. What do you conclude?

(b) Compare the results of this significance test with those you gave in Exercise 9.34. What do you conclude about the effect of the sample size on the results of these significance tests?

9.3 INFERENCE FOR TWO-WAY TABLES

In the previous section, we discussed inference procedures for comparing two proportions. We now turn our attention to the more general problem of analyzing count data that can be displayed in an $r \times c$ table. In some applications the rows (r) correspond to possible outcomes and the columns (c) correspond to different sampled populations. When comparing two proportions, we have $c = 2$ populations and $r = 2$ possible outcomes (success and failure). Thus, the problem of comparing two proportions is a special case of the class of problems that we study in this section. In other applications, the rows and columns correspond to two different ways of classifying observations sampled from a single population. The statistical methods used to analyze these two types of application are the same. Here is an example in which a single sample is classified in two ways.

EXAMPLE 9.10

In a study of heart disease in male federal employees, volunteer subjects were classified according to their socioeconomic status, abbreviated as SES, and their smoking habits.[3] There were three categories of SES—high, middle, and low. For the smoking classification, individuals were asked whether they were current smokers, former smokers, or had never smoked. The data table for this two-way classification is shown below.

	SES			
Smoking	High	Middle	Low	Total
Current	51	22	43	116
Former	92	21	28	141
Never	68	9	22	99
Total	211	52	93	356

Descriptive Tables

A two-way table with r rows and c columns contains $r \times c$ sample counts. Note that the table in Example 9.10 is a 3×3 table with 9 counts. We let N_{ij} denote the number of observations in the ith row and the jth column. Here i runs from 1 to r and j runs from 1 to c. The general form of the data table giving the sample counts is as follows:

Row	1	2	$\cdots$	j	$\cdots$	c	Total
			Column				
1	N_{11}	N_{12}	$\cdots$	N_{1j}	$\cdots$	N_{1c}	R_1
2	N_{21}	N_{22}	$\cdots$	N_{2j}	$\cdots$	N_{2c}	R_2
$\vdots$	$\vdots$	$\vdots$	$\vdots$	$\vdots$	$\vdots$	$\vdots$	$\vdots$
i	N_{i1}	N_{i2}	$\cdots$	N_{ij}	$\cdots$	N_{ic}	R_i
$\vdots$	$\vdots$	$\vdots$	$\vdots$	$\vdots$	$\vdots$	$\vdots$	$\vdots$
r	N_{r1}	N_{r2}	$\cdots$	N_{rj}	$\cdots$	N_{rc}	R_r
Total	C_1	C_2	$\cdots$	C_j	$\cdots$	C_c	n

In this table C_j are the column totals and R_i are the row totals. Thus

$$C_j = \sum_{i=1}^{r} N_{ij} \quad \text{and} \quad R_i = \sum_{j=1}^{c} N_{ij}$$

As in the previous section, the total sample size is denoted by n. This value can be found by adding either the column totals or the row totals:

$$n = \sum_{j=1}^{c} C_j = \sum_{i=1}^{r} R_i$$

column percent

For many $r \times c$ data tables, we can get a better understanding of the information contained in the table by computing *column percents*. To find the column percents, first determine the value of each entry as a proportion of the column total using

$$j\text{th column proportion} = \frac{N_{ij}}{C_j}$$

conditional distribution

Then multiply by 100 to convert the proportion to a percent. The resulting entries in each column form the *conditional distribution* of the row variable given that value of the column variable. These conditional distributions can be displayed using the method presented in Section 3.4.

EXAMPLE 9.11

To calculate the column percents for the SES-smoking example, we proceed as follows. For the high-SES group, there are 51 current smokers out of a total of 211 people. The column proportion for this cell is

$$\frac{N_{11}}{C_1} = \frac{51}{211} = .242$$

We see that 24.2% of the high-SES group are current smokers. Similarly, 92 of the 211 people in this group are former smokers. The column proportion is

$$\frac{N_{21}}{C_1} = \frac{92}{211} = .436$$

Therefore, 43.6% of the high-SES group are former smokers. The results are summarized in the following table:

Smoking	SES			
	High	Middle	Low	All
Current	24.2	42.3	46.2	32.6
Former	43.6	40.4	30.1	39.6
Never	32.2	17.3	23.7	27.8
Total	100.0	100.0	100.0	100.0

Of course, the sum of the percents in each column should be 100. In general, we expect a slight round-off error in these calculations, and so the sum may differ from this value slightly. It is good practice to calculate the percents for each entry in the table and then sum each column as a check. In this way we can find arithmetic errors that would not be uncovered if, for example, we calculated the column percent for the Never row by subtracting the sum of the percents for Current and Former from 100.

Inspection of the table reveals that there is an apparent increase in the percent of current smokers as the SES group varies from high to middle to low. There is a similar, but decreasing, trend for the former smokers. The percents of people who never smoked show a different pattern, however, with the highest percent in the high-SES group and the lowest in the middle-SES group. The column percents suggest that there is a negative association between smoking and SES: higher-SES people tend to smoke less. The test statistic that we present later will allow us to assess whether this observed association is sufficiently large for us to conclude that it is due to underlying differences within the population we have sampled.

Tables giving entries as percents of row totals or as percents of the total sample are also sometimes useful for understanding count data. Which description is best depends upon the particular set of data being analyzed and is a matter of judgment. In particular, if there is a natural explanatory variable, such as SES in Example 9.10, percents using the marginal totals for this variable in the denominator are generally very useful.

Models and Hypotheses

The test procedure for $r \times c$ count data is sufficiently general that it is valid for different assumptions regarding the data. However, the analysis of the two models presented here requires *each experimental unit or subject to be counted only once in the data table*. In the SES-smoking example, each federal employee fits into one SES-smoking cell. Procedures for data that do not satisfy this assumption have been developed, and many of them can be performed using standard statistical computer programs. However, we will restrict our attention to problems for which this assumption holds.

EXAMPLE 9.12

In the SES-smoking example, 356 federal employees were classified according to two categorical variables—SES and smoking. Because there are nine possible categories in the 3×3 table, we view the process of selecting a federal employee at random and then classifying the individual according to the two categorical variables as a random phenomenon with nine possible outcomes. Each possible outcome has a probability. We can also think of the population as having corresponding proportions of individuals in each of the nine cells in the table. Rather than denoting the probabilities or population proportions for the possible outcomes as $p_1, \ldots, p_9$, it is more convenient to use the notation p_{ij} for the population proportion in row i and column j of the table. In this case, $i = 1, 2, 3$ describes the three smoking possibilities and $j = 1, 2, 3$ describes the three SES groups. Since each individual falls into one of the nine cells, it follows that the sum of the nine probabilities is 1. ∎

Example 9.10 represents a special case of a general statistical model where a single SRS is taken from a single population and each observation is classified into one cell of an $r \times c$ table. A summary of the population proportions is given in the following table:

Row	1	2	$\cdots$	j	$\cdots$	c	Total
			Column				
1	p_{11}	p_{12}	$\cdots$	p_{1j}	$\cdots$	p_{1c}	r_1
2	p_{21}	p_{22}	$\cdots$	p_{2j}	$\cdots$	p_{2c}	r_2
$\vdots$	$\vdots$	$\vdots$	$\vdots$	$\vdots$	$\vdots$	$\vdots$	$\vdots$
i	p_{i1}	p_{i2}	$\cdots$	p_{ij}	$\cdots$	p_{ic}	r_i
$\vdots$	$\vdots$	$\vdots$	$\vdots$	$\vdots$	$\vdots$	$\vdots$	$\vdots$
r	p_{r1}	p_{r2}	$\cdots$	p_{rj}	$\cdots$	p_{rc}	r_r
Total	c_1	c_2	$\cdots$	c_j	$\cdots$	c_c	1

The marginal proportions in this table are the sums of the proportions in the rows and columns. We use the notation r_i for the row sums and c_j for the column sums. In other words, r_i is the sum of the proportions across row i, and c_j is the sum down column j.

The marginal proportions are easily interpreted as probabilities. Each r_i is the probability that a randomly selected member of the population falls in the ith row category. Similarly, each c_j is the probability of the jth column category.

EXAMPLE 9.13

The unknown population proportions for the SES-smoking example are tabulated below.

	SES			
Smoking	High	Middle	Low	Total
Current	p_{11}	p_{12}	p_{13}	r_1
Former	p_{21}	p_{22}	p_{23}	r_2
Never	p_{31}	p_{32}	p_{33}	r_3
Total	c_1	c_2	c_3	1

A natural H_0 in this example is that smoking habits are unrelated to SES. Although this hypothesis is relatively easy to state (and understand) in words, it is more difficult to express in terms of the population proportions in our statistical model. The multiplication rule for independent events is the tool we need.

The probability that a member of the population is a current smoker is r_1, and the probability that the person has high SES is c_1. If smoking and SES are independent, then the probability that a federal employee is *both* a current smoker and high SES is $r_1 \times c_1$. In other words, $p_{11} = r_1 c_1$. In general, the hypothesis that smoking and SES are independent says that $p_{ij} = r_i c_j$. ■

Example 9.13 illustrates the first model for which our analysis of count data in an $r \times c$ table is valid.

First model for $r \times c$ tables

An SRS of size n is drawn from a population. Each individual in the sample is classified according to two categorical variables. The probabilities for the row classification are r_i, and the probabilities for the column classification are c_j.

The null hypothesis is that the row and column classifications are independent; that is, there is no relationship between the row and column classifications. Letting p_{ij} denote the probability of an observation being classified in row i and column j, the null hypothesis is

$$H_0: p_{ij} = r_i c_j \quad \text{for all } i \text{ and } j$$

The alternative hypothesis is that the row and column classifications are dependent; that is, the row and column classifications are related in some way. We write this alternative as

$$H_a: p_{ij} \neq r_i c_j \quad \text{for some } i \text{ and } j$$

The second model for a two-way table is illustrated by the comparison of two proportions discussed in Section 9.2. We assumed that there were independent samples from each of two populations. The possible outcomes were classified into two categories—success and failure. The natural generalization of this situation is the case where c populations are independently sampled and the number of possible outcomes in each population is r, where r can be 2 or greater.

EXAMPLE 9.14

People have different goals when participating in recreational sports. One goal is social comparison—the desire to win or to do better than other people. Another is mastery—the desire to improve one's skills or to try one's best. In a study on why students participate in sports,[4] data on 67 male and 67 female undergraduates from a large university were collected. Each student was classified into one of four categories based on the responses to a questionnaire about their sports goals. The four categories were high social comparison–high mastery (HSC-HM), high social comparison–low mastery (HSC-LM), low social comparison–high mastery (LSC-HM), and low social comparison–low mastery (LSC-LM). One purpose of the study was to compare the goals of male and female students. There are therefore $c = 2$ populations. Females are population 1 and males are population 2. The table of counts has the usual form.

	Sex	
Goal	Female	Male
HSC-HM	N_{11}	N_{12}
HSC-LM	N_{21}	N_{22}
LSC-HM	N_{31}	N_{32}
LSC-LM	N_{41}	N_{42}
Total	67	67

Notice that the column totals $C_1 = 67$ and $C_2 = 67$ are known before the data are collected. They are just the sample sizes. To describe the population proportions for females, we use

$$p_{1(1)}, \ p_{2(1)}, \ p_{3(1)}, \ \text{and } p_{4(1)}$$

where $p_{i(1)}$ denotes the probability that a randomly selected female student will fall into goals category i. Since each female student must fall into exactly one of these categories, the sum of the four population proportions is 1. Similarly, the population proportions for the males (column 2) are denoted by

$$p_{1(2)}, \ p_{2(2)}, \ p_{3(2)}, \ \text{and } p_{4(2)}$$

These four proportions also sum to 1. The table of population proportions is given at the top of the next page.

	Sex	
Goal	Female	Male
HSC-HM	$p_{1(1)}$	$p_{1(2)}$
HSC-LM	$p_{2(1)}$	$p_{2(2)}$
LSC-HM	$p_{3(1)}$	$p_{3(2)}$
LSC-LM	$p_{4(1)}$	$p_{4(2)}$
Total	1	1

This example represents a special case of a general statistical model where an independent SRS is taken from each of c populations and the outcomes are classified into one of r categories. A summary of the c population proportions is given in the following table:

	Column (population)					
Row	1	2	$\cdots$	j	$\cdots$	c
1	$p_{1(1)}$	$p_{1(2)}$	$\cdots$	$p_{1(j)}$	$\cdots$	$p_{1(c)}$
2	$p_{2(1)}$	$p_{2(2)}$	$\cdots$	$p_{2(j)}$	$\cdots$	$p_{2(c)}$
$\vdots$	$\vdots$	$\vdots$	$\vdots$	$\vdots$	$\vdots$	$\vdots$
i	$p_{i(1)}$	$p_{i(2)}$	$\cdots$	$p_{i(j)}$	$\cdots$	$p_{i(c)}$
$\vdots$	$\vdots$	$\vdots$	$\vdots$	$\vdots$	$\vdots$	$\vdots$
r	$p_{r(1)}$	$p_{r(2)}$	$\cdots$	$p_{r(j)}$	$\cdots$	$p_{r(c)}$
Total	1	1	1	1	1	1

Since the sum of the probabilities for each of the c populations is 1, the column sums in this table are all 1. The row sums are not equal to 1 and are usually not meaningful.

EXAMPLE 9.15

A natural H_0 for the sports goals example is that the sports goals for males and females are the same. For females, $p_{1(1)}$ is the proportion in the HSC-HM group. Similarly, $p_{1(2)}$ is the proportion in this group for the males. H_0 states that these two proportions are equal. In other words, the values in the first row of the population proportions table are equal. The same is true for each of the other three goals. H_0 is that the two proportions in each row of the population proportions table are equal. ∎

This example illustrates the second model for which our analysis of count data in an $r \times c$ table is valid.

Second model for $r \times c$ tables

For each of c populations, independent SRSs of sizes $C_1, C_2, \ldots, C_c$ are drawn. Each individual in a sample is classified according to a categorical outcome variable with r possible values. For the jth population the probability that an individual will fall into category i is $p_{i(j)}$.

The null hypothesis is that the distributions of the outcome variable are the same in all c populations. Letting $p_{i(j)}$ denote the proportion of population j in category i, the null hypothesis is

$$H_0: p_{1(1)} = p_{1(2)} = \cdots = p_{1(c)}$$
$$p_{2(1)} = p_{2(2)} = \cdots = p_{2(c)}$$
$$\vdots$$
$$p_{r(1)} = p_{r(2)} = \cdots = p_{r(c)}$$

The alternative hypothesis is

$$H_a: \text{at least one of the equalities in } H_0 \text{ does not hold}$$

The sample sizes from each of the populations are the column totals in the sample count table. Since we will be using the same statistical analysis for both the first and the second models, we call these sample sizes C_j. In the first model, the C_j are random variables. The total sample size n is set by the researcher, and the column sums are known only when the data are analyzed. For the second model, on the other hand, the column sums are the sample sizes selected at the design phase of the research. The null hypothesis in both models says that there is "no relationship" between the column variable and the row variable. The way in which the hypothesis is expressed differs, depending upon the sampling design. Fortunately, the test of the hypothesis of "no relationship" is the same for both models.

Expected Counts

The statistic that tests H_0 in $r \times c$ tables compares the sample counts with *expected* counts that are calculated under the assumption that H_0 is true. We use the notation E_{ij} for the expected count in the ijth cell of the table. Thus, the expected count table has the following general form:

Row	Column					
	1	2	...	j	...	c
1	E_{11}	E_{12}	...	E_{1j}	...	E_{1c}
2	E_{21}	E_{22}	...	E_{2j}	...	E_{2c}
$\vdots$	$\vdots$	$\vdots$	$\vdots$	$\vdots$	$\vdots$	$\vdots$
i	E_{i1}	E_{i2}	...	E_{ij}	...	E_{ic}
$\vdots$	$\vdots$	$\vdots$	$\vdots$	$\vdots$	$\vdots$	$\vdots$
r	E_{r1}	E_{r2}	...	E_{rj}	...	E_{rc}

Calculation of expected counts

> For an $r \times c$ table the expected counts are calculated from the marginal totals in the sample count table using the formula
>
> $$E_{ij} = \frac{R_i C_j}{n} \qquad (9.1)$$

This formula works for both of the statistical models we have presented. As with most statistical calculations, the formula for expected counts is intuitively reasonable when viewed in the proper way. The intuitive idea behind these calculations depends upon which of the two models is the basis for our calculations. The ideas are best illustrated using our examples.

EXAMPLE 9.16

The sample count table for the SES-smoking example is reproduced below as an aid to our calculations. These counts arise from classifying a single sample of 356 federal employees.

Smoking	SES			
	High	Middle	Low	Total
Current	51	22	43	116
Former	92	21	28	141
Never	68	9	22	99
Total	211	52	93	356

What is the expected count E_{11} in the (1, 1) cell under the H_0 that smoking and SES are independent? ∎

Under H_0, N_{11} is a binomial random variable based on n trials, each with probability

$$p_{11} = r_1 c_1$$

of producing an outcome in this cell. The mean of N_{11} is therefore np_{11}. The expected count E_{11} is an estimate of this mean. To find E_{11} we must estimate r_1 and c_1 from the data.

The number of current smokers in the data is $R_1 = 116$. The estimate of the proportion of current smokers in the population is therefore

$$\hat{r}_1 = \frac{R_1}{n} = \frac{116}{356}$$

Similarly, the number of high-SES individuals is $C_1 = 211$, and the estimate of the proportion of high-SES individuals is

$$\hat{c}_1 = \frac{C_1}{n} = \frac{211}{356}$$

Combining these two estimates gives the estimate of p_{11} under H_0 as

$$\hat{p}_{11} = \hat{r}_1 \hat{c}_1 = \frac{R_1}{n} \frac{C_1}{n}$$

Finally, to calculate the expected count for the sample of $n = 356$ individuals, we multiply our estimate of p_{11} by n:

$$E_{11} = n \frac{R_1}{n} \frac{C_1}{n} = \frac{R_1 C_1}{n}$$

This is the same as the expression given in Equation 9.1 for expected counts. For our example,

$$E_{11} = \frac{R_1 C_1}{n} = \frac{(116)(211)}{356} = 68.75$$

The expected counts for the other cells are calculated similarly. Here is the table of expected counts, calculated from the table of sample counts using Equation 9.1.

		SES		
Smoking	High	Middle	Low	Total
Current	68.75	16.94	30.30	115.99
Former	83.57	20.60	36.83	141.00
Never	58.68	14.46	25.86	99.00
Total	211.00	52.00	92.99	355.99

The row and column totals in this table should agree with those in the sample count table. The differences in the last column and the first row are due to round-off error. As we have done with previous calculations for $r \times c$ tables, we prefer to use the marginal sums as a check on our calculations rather than to force the totals to be equal to those in the sample count table.

EXAMPLE 9.17

> The second model for $r \times c$ tables is illustrated by the sports goals example. The computer output for this example generated by the SAS procedure FREQ is given in Figure 9.4. Similar output is produced by most statistical software packages. In each cell of the 4×2 table, there are five entries. An index is given in the upper left corner. The first entry, labeled FREQUENCY, is the sample count, and the second, labeled EXPECTED, is the expected count. The other three entries, labeled PERCENT, ROW PCT, and COL PCT, are the total percent, row percent, and column percent. The row and column totals and marginal percents are also given. ■

To understand why the formula

$$E_{ij} = \frac{R_i C_j}{n}$$

is true in this model, we need to examine the margins of the table given in Figure 9.4. Let us first consider computing the expected count for the HSC-HM females.

We see that for the combined sample, 45 of 134 students (33.58%) are in the HSC-HM group. (Find these numbers on the output.) In terms of our formulas, this combined proportion is

$$\frac{R_1}{n} = \frac{45}{134} = .3358$$

Under the null hypothesis that the probability of observing a student in this group does not depend upon sex, we expect 33.58% of both the females and the males in our sample to be HSC-HM. Since there are $C_1 = 67$ females in the study, the expected number of females in the HSC-HM group is

$$E_{11} = C_1 \frac{R_1}{n} = 67 \frac{45}{134} = 22.5$$

This agrees with the EXPECTED entry in the female HSC-HM cell. Under the null hypothesis we expect 22.5 females to be in the HSC-HM group. Since the number of males C_2 is the same as the number of females in this study, the expected number of males in the HSC-HM group is also 22.5. The other expected counts are calculated similarly.

```
                   TABLE OF GOAL BY SEX

        GOAL          SEX

        FREQUENCY |
        EXPECTED  |
         PERCENT  |
         ROW PCT  |
         COL PCT  | FEMALE  |  MALE   |  TOTAL
        ----------+---------+---------+
        HSC-HM    |      14 |      31 |     45
                  |    22.5 |    22.5 |
                  |   10.45 |   23.13 |   33.58
                  |   31.11 |   68.89 |
                  |   20.90 |   46.27 |
        ----------+---------+---------+
        HSC-LM    |       7 |      18 |     25
                  |    12.5 |    12.5 |
                  |    5.22 |   13.43 |   18.66
                  |   28.00 |   72.00 |
                  |   10.45 |   26.87 |
        ----------+---------+---------+
        LSC-HM    |      21 |       5 |     26
                  |    13.0 |    13.0 |
                  |   15.67 |    3.73 |   19.40
                  |   80.77 |   19.23 |
                  |   31.34 |    7.46 |
        ----------+---------+---------+
        LSC-LM    |      25 |      13 |     38
                  |    19.0 |    19.0 |
                  |   18.66 |    9.70 |   28.36
                  |   65.79 |   34.21 |
                  |   37.31 |   19.40 |
        ----------+---------+---------+
        TOTAL            67        67      134
                      50.00     50.00   100.00


        STATISTICS FOR TABLE OF GOAL BY SEX

        STATISTIC          DF      VALUE     PROB
        -------------------------------------------
        CHI-SQUARE          3     24.898    0.000
        SAMPLE SIZE=134
```

Figure 9.4 Computer output for the sports goals study, Example 9.1.

Significance Tests

X² statistic

To test H_0 that there is no relationship between the row and column classifications, we use a statistic, called the *chi-square statistic*, that compares the sample counts with the expected counts. Specifically, we take the difference between each sample count and its corresponding expected count, square these values, divide by the expected count, and sum over all entries.

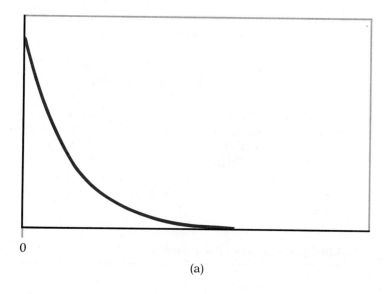

0

(a)

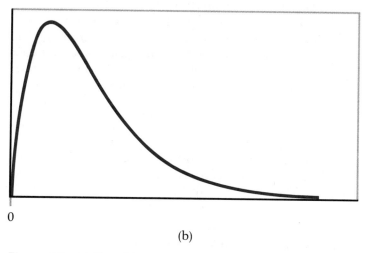

0

(b)

Figure 9.5 (a) The $\chi^2(2)$ density function. (b) The $\chi^2(4)$ density function.

Chi-square statistic

To compare the sample and expected counts we use a statistic X^2, called the chi-square statistic. It is calculated from the following formula:

$$X^2 = \sum \frac{(\text{sample} - \text{expected})^2}{\text{expected}} \tag{9.2}$$

where *sample* represents the sample counts, *expected* represents the expected counts, and the sum is over all $r \times c$ entries in the sample and expected count tables.[5]

χ^2 **distribution**

To test H_0, we need the sampling distribution of X^2 under the assumption that H_0 is true. This leads us to a new distribution, the *chi-square distribution*, which we denote by χ^2.

Like the t distributions, the χ^2 distributions form a family described by a single parameter, the degrees of freedom. We use $\chi^2(\text{df})$ to indicate a particular member of this family. Figure 9.5 depicts the density curves for the $\chi^2(2)$ and $\chi^2(4)$ distributions. Notice that these distributions take only positive values and are skewed to the right. These facts are true for all χ^2 distributions. Table G in the back of the book gives critical values for the χ^2 distributions.

Chi-square test for $r \times c$ tables

The data for an $r \times c$ table can be obtained by random sampling as described by either of the two models discussed in the Models and Hypotheses section of this chapter.

The null hypothesis to be tested is that the row and column classifications are independent (first model) or that the row classification proportions for the c populations are all equal (second model). The alternative hypothesis is that the null hypothesis is not true.

The test statistic is the X^2 statistic

$$X^2 = \sum_{i,j} \frac{(N_{ij} - E_{ij})^2}{E_{ij}}$$

If H_0 is true, the statistic X^2 has approximately a χ^2 distribution with $(r-1)(c-1)$ degrees of freedom.

The P-value for the test is

$$P(\chi^2 \geq X^2)$$

where χ^2 is a random variable having the $\chi^2[(r-1)(c-1)]$ distribution.

The approximation of the distribution of X^2 by a χ^2 distribution is based on having large samples. We will use this approximation whenever the average of the expected counts is 5 or more and the smallest expected count is 1 or more.

EXAMPLE 9.18

In Example 9.11, we examined the column percentages for the SES-smoking example. There appeared to be an association between the two categorial variables in the sample data. We now perform a significance test to determine whether the patterns seen are evidence for concluding that there is an association in the population. The sample count and expected count tables for the SES-smoking example are reproduced below to aid us in the calculation of the X^2 statistic for this problem.

Sample Counts for SES and Smoking

Smoking	SES			
	High	Middle	Low	Total
Current	51	22	43	116
Former	92	21	28	141
Never	68	9	22	99
Total	211	52	93	356

Expected Counts for SES and Smoking

Smoking	SES			
	High	Middle	Low	Total
Current	68.75	16.94	30.30	115.99
Former	83.57	20.60	36.83	141.00
Never	58.68	14.46	25.86	99.00
Total	211.00	52.00	92.99	355.99

For the high-SES current smokers, the sample count is $N_{11} = 51$ and the expected count is $E_{11} = 68.75$. The contribution to the X^2 statistic for this cell is therefore

$$\frac{(N_{11} - E_{11})^2}{E_{11}} = \frac{(51 - 68.75)^2}{68.75} = 4.583$$

Similarly, the calculation for the middle-SES current smokers is

$$\frac{(N_{12} - E_{12})^2}{E_{12}} = \frac{(22 - 16.94)^2}{16.94} = 1.511$$

To calculate X^2, we perform this computation for each of the nine cells in the table and sum the results:

$$
\begin{aligned}
X^2 &= \sum_{i,j} \frac{(N_{ij} - E_{ij})^2}{E_{ij}} \\
&= \frac{(51 - 68.75)^2}{68.75} + \frac{(22 - 16.94)^2}{16.94} + \frac{(43 - 30.30)^2}{30.30} \\
&\quad + \frac{(92 - 83.57)^2}{83.57} + \frac{(21 - 21.60)^2}{21.60} + \frac{(28 - 36.83)^2}{36.83} \\
&\quad + \frac{(68 - 58.68)^2}{58.68} + \frac{(9 - 14.46)^2}{14.46} + \frac{(22 - 25.86)^2}{25.86} \\
&= 4.583 + 1.511 + 5.323 + .850 + .017 + 2.117 \\
&\quad + 1.480 + 2.062 + .576 \\
&= 18.52
\end{aligned}
$$

Since there are $r = 3$ smoking categories and $c = 3$ SES groups, the degrees of freedom for this table are

$$(r - 1)(c - 1) = (3 - 1)(3 - 1) = 4$$

df = 4

p	.001	.0005
χ^2	18.47	20.00

Therefore, under the null hypothesis assumption that smoking and SES are independent, the test statistic X^2 has a $\chi^2(4)$ distribution.

To calculate a P-value, refer to the row in Table G corresponding to 4 df. The calculated value $X^2 = 18.52$ lies between upper critical points corresponding to probabilities of 0.001 and 0.0005. The P-value is therefore between 0.001 and 0.0005. This is very strong evidence against H_0. The analysis indicates that there is an association between smoking and SES in the population of federal workers studied. The size and nature of this association are described by the table of percents examined in Example 9.11. The chi-square test assesses only the significance of the association, so the descriptive percents are essential to an understanding of the data.

EXAMPLE 9.19

To illustrate the use of the computer for the analysis of $r \times c$ tables, we return to the sports goals example and the output given in Figure 9.4. Below the table, the value of the chi-square statistic X^2 is given with its df and P-value. As a check we calculate the df:

$$(r - 1)(c - 1) = (4 - 1)(2 - 1) = 3$$

From the output we see that $X^2 = 24.898$, df = 3, and the P-value is given as 0.000 under the heading **PROB**. Since the P-value is rounded, we know that it is less than 0.0005. ■

The chi-square test tells us that the data contain clear evidence against the null hypothesis that female and male students have the same sports goals. Under H_0, the chance of obtaining a value of X^2 greater than or equal to the calculated value of 24.898 is very small—less than 0.0005.

The nature of the difference between females and males in their sports goals is best understood by examining the column percents. The percent of males in both of the HSC goal classes is more than twice the percent of females. For the HSC-HM group we have 46.27% of the males versus 20.90% of the females, while for the HSC-LM group we have 26.87% of the males and 10.45% of the females. The pattern is reversed for the LSC goal classes. Thus, males are more likely to be motivated by social comparison goals than females.

SUMMARY

Two different models for generating $r \times c$ tables lead to the same analysis of two-way count data. In the first model, an SRS of size n is drawn from a population, and samples are classified according to two categorical variables having r and c possible values. In the second model, independent SRSs of size C_j are drawn from each of c populations, and each sample is classified according to a categorical variable with r possible values.

The **null hypothesis** is that there is no relationship between the column variable and row variable. In the first model this means that the two variables are independent. In the second it means that the distributions of the row categorical variable are the same for all c populations.

Expected counts are computed using the formula

$$E_{ij} = \frac{R_i C_j}{n}$$

where R_i is the ith row total and C_j is the jth column total.

The null hypothesis is tested by the **chi-square statistic**

$$X^2 = \sum_{i,j} \frac{(N_{ij} - E_{ij})^2}{E_{ij}}$$

Under the null hypothesis, X^2 has approximately the $\chi^2[(r-1)(c-1)]$ distribution. The *P*-value for the test is

$$P(\chi^2 \geq X^2)$$

where χ^2 is a random variable having the $\chi^2[(r-1)(c-1)]$ distribution.

SECTION 9.3 EXERCISES

9.37 There are many "indicators" that investors use to predict the behavior of the stock market. One of these is the "January indicator." Some investors believe that if the market is up in January, then it will be up for the rest of the year. On the other hand, if it is down in January, then it will be down for the rest of the year. The following table gives the data for the 72 years from 1916 to 1987:

	Jan.	
Feb.–Dec.	Up	Down
Up	33	13
Down	13	13

(a) Calculate the column percents for this table. Give a short explanation of what they express.

(b) Repeat (a) for the row percents.

(c) Give a description of appropriate null and alternative hypotheses for this problem. Use words rather than symbols.

(d) Find the table of expected counts under H_0. In what cells do the expected counts exceed the sample counts? In what cells are they less than the sample counts? Explain.

(e) Test H_0. Give the values of the X^2 statistic, the df, and the P-value. What do you conclude?

(f) Write a short explanation of what the January indicator means based on your analysis of these data.

9.38 In Exercise 3.69, Reggie Jackson's times at bat and hits are given for regular season and World Series play. During regular seasons, Reggie was at bat 9864 times and had 2584 hits. During World Series games, he was at bat 98 times and had 35 hits.

(a) Display these data in a 2 × 2 table of sample counts with regular season and World Series as the column headings and fill in the marginal sums.

(b) Find the row and column percents. Which of these gives the proportions of hits to times at bat?

(c) Calculate the table of expected counts under the null hypothesis that Reggie hits equally well in regular season games and World Series play. How does the expected number of hits in World Series play compare with the actual number of hits?

(d) Test H_0 versus the H_a that Reggie's hitting probability is not the same in these two circumstances. Give the X^2 statistic, df, and P-value.

(e) Summarize your conclusions.

9.39 In Exercise 3.61 a survey on the severity of rodent problems in egg and turkey poultry houses is described. In this study a random sample of poultry houses was surveyed, and the houses were classified according to the type of operation and the extent of the rodent problems. The results are summarized in the following table:

Rodent problem	Type	
	Egg	Turkey
Mild	34	22
Moderate	33	22
Severe	7	4

(a) Since the type of poultry operation is a natural explanatory variable for this problem, compute the table of column percents. Include the marginal percents for the three levels of rodent problems. Give a short summary of the results.

(b) The population proportions for this problem are denoted by p_{ij}. Explain in words the meaning of p_{11}. Do the same for p_{32}.

(c) In terms of the p_{ij}, express the H_0 that the two ways of classifying the poultry operations are independent. What is H_a?

(d) Compute the table of expected counts under the H_0.

(e) Give the X^2 statistic, df, and the P-value for the test. What do you conclude?

9.40 In Exercise 3.60 a survey of businesses of different sizes is described. Questionnaires were sent to 200 randomly selected businesses of each of three sizes. The question of interest was whether or not the response rates were the same for the three sizes of businesses. The data are summarized in the following table. Note that the column sums are fixed by the design of the survey.

	Business size		
	Small	Medium	Large
Response	125	81	40
No Response	75	119	160
Total	200	200	200

(a) For each size of business find the column percents for the response and no response categories. Do the same for all of the data combined and present the results in a table. In what way do the response rates appear to vary with the size of the business?

(b) The population proportions are given by $p_{i(j)}$ for this problem. Explain in words what $p_{1(2)}$ represents. Do the same for $p_{2(3)}$.

(c) In terms of the $p_{i(j)}$, write an H_0 that specifies that there are no differences among the three sizes of businesses. What is an appropriate H_a?

(d) Calculate the table of expected counts.

(e) Give the X^2 statistic, df, and P-value for testing H_0. What do you conclude?

9.41 In January 1975, the Committee on Drugs of the American Academy of Pediatrics recommended that tetracycline drugs not be used for children under the age 8. A 2-year study was conducted in Tennessee to investigate the extent to which physicians prescribed this drug between 1973 and 1975. In one part of the study, family practice physicians were categorized according to whether the county of their practice was urban, intermediate, or rural. The proportions of doctors in each of these categories who prescribed tetracycline to at least one patient under the age of 8 were compared. The table of sample counts is given below. (Data taken from Wayne A. Ray et al., "Prescribing of tetracycline to children less than 8 years old," *Journal of the American Medical Association*, 237 (1977), pp. 2069–2074.)

		County type	
	Urban	Intermediate	Rural
Tetracycline	65	90	172
No tetracycline	149	136	158

(a) Find the row and column sums and put them in the margins of the sample count table.

(b) For each type of county find the percent of physicians who prescribed tetracycline and the percent of those who did not. Do the same for the combined sample. Display the results in a table.

(c) In this study the 770 family practice physicians were classified according to the type of county where they practiced and whether or not they prescribed tetracycline. Write null and alternative hypotheses that can be used to assess whether or not these two ways of classifying physicians are independent.

(d) Calculate a table of expected counts.

(e) Find the X^2 statistic for testing H_0 and report the df and the P-value.

(f) Summarize the results.

9.42 Refer to the previous exercise. The part of the study described in that exercise concerned only physicians in family practice. In another part of the study, tetracycline prescribing was compared for physicians engaged in different types of practices. The practices were classified as family practice, pediatrics, and other. The data are given in the following table.

	Family practice	Pediatrics	Other
Tetracycline	327	32	159
No tetracycline	443	122	808

(a) Find the row and column sums and put them in the margins of the sample count table.
(b) For each type of practice find the percent of physicians who prescribed tetracycline and the percent of those who did not. Do the same for the combined sample. Display the results in a table.
(c) In this study 1891 physicians were classified according to type of practice and whether or not tetracycline was prescribed. Write null and alternative hypotheses that can be used to assess whether or not these two ways of classifying physicians are independent.
(d) Calculate a table of expected counts.
(e) Find the X^2 statistic for testing H_0 and report the df and the P-value.
(f) Summarize the results.

9.43 Most colleges and universities have annual campaigns in which they ask former graduates to contribute money. For the 1984 to 1985 Providence College fund-raising campaign, statistics were recorded for the number of people contacted and the number of donors categorized by class year. Some of the data are summarized in the following table. (Data taken from the *Providence College Fund Year Report 1984–85*, Providence College, Providence, Rhode Island, 1985.)

	Class				
	1961	1966	1971	1976	1981
Contributed	192	240	182	263	325
Did not contribute	174	258	260	339	586

(a) Find the row and column totals. Put these in the margins of the sample count table.
(b) For each class, find the percent of alumni who contributed and the percent who did not contribute. Do the same for all of the classes combined. Present the results in a table.
(c) For each class the number of alumni contributing to the annual fund can be viewed as a binomial random variable with probability

$p_{1(j)}$ as the probability that a randomly selected individual from class j will contribute and probability $p_{2(j)} = 1 - p_{1(j)}$ as the probability that the individual will not contribute. Write a null hypothesis that states that the probabilities of contributing and not contributing are the same for all of the classes. Specify an appropriate H_a.

(d) Calculate a table of expected counts under the assumption that H_0 is true.

(e) Calculate the value of the X^2 statistic and give the df and the P-value. What do you conclude? In what way does the proportion of alumni contributing depend upon class year?

9.44 Refer to the previous exercise. The corresponding data for the 1986 to 1987 campaign are given in the table below. (Data taken from the *Providence College Fund Year Report 1986–87*, Providence College, Providence, Rhode Island, 1987.)

	Class				
	1961	1966	1971	1976	1981
Contributed	196	266	194	276	333
Did not contribute	123	226	241	322	568

(a) Complete the sample count table by putting the row and column totals in the margins.

(b) Calculate column percents. Include the values in the margin. Explain in words the meaning of the entry in the first row and the second column.

(c) For each class the number of alumni contributing to the annual fund can be viewed as a binomial random variable with probability $p_{1(j)}$ as the probability that a randomly selected individual from class j will contribute and probability $p_{2(j)} = 1 - p_{1(j)}$ as the probability that the individual will not contribute. Write a null hypothesis that states that the probabilities of contributing and not contributing are the same for all of the classes. Specify an appropriate H_a.

(d) Calculate a table of expected counts under the assumption that H_0 is true.

(e) Calculate the value of the X^2 statistic and give the df and the P-value. State your conclusion. How does the proportion of alumni contributing depend upon class year?

9.45 Alcohol and nicotine consumption during pregnancy are believed to be associated with certain characteristics of children. Since drinking and smoking behaviors may be related, it is important to understand the nature of this relationship when assessing the possible effects of these variables on children. In one study, 452 mothers were classified according to their alcohol intake prior to pregnancy recognition and their nicotine intake during pregnancy. The data are summarized in the following table. (Data taken from Ann P. Streissguth et al., "Intrauterine alcohol and nicotine exposure: attention and reaction time in 4-year-old children," *Developmental Psychology*, 20 (1984), pp. 533–541.)

	Nicotine (mg/day)		
Alcohol (ounces/day)	None	1–15	16 or more
None	105	7	11
.01–.10	58	5	13
.11–.99	84	37	42
1.00 or more	57	16	17

(a) Calculate the column percents. In what way does the pattern of alcohol consumption vary with nicotine consumption?

(b) Calculate the row percents. In what way does the pattern of nicotine consumption vary with alcohol consumption?

(c) Write H_0 and H_a for assessing whether or not alcohol consumption and nicotine consumption are independent.

(d) Compute the table of expected counts.

(e) Find the X^2 statistic. Report the df and the *P*-value.

(f) What do you conclude from your analysis of these data?

9.46 Nutrition and illness are related in a complex way. If the diet is inadequate, the ability to resist infections can be impaired and illness results. On the other hand, certain types of illness cause lack of appetite, so that poor nutrition can be the result of illness. In a study of morbidity and nutritional status in 1165 preschool children living in poor conditions in Delhi, India, data were obtained on nutrition and illness. Nutrition was categorized by a standard method as normal or one of four levels of inadequate: I, II, III, and IV. For the purpose of analysis, the two most severely undernourished groups, III and IV, were combined. In one part of the study, illness during the past year was categorized into one of four categories: upper respiratory infection (URI), diarrhea, URI and diarrhea, and none. The data are summarized in the following table. (Data taken from Vimlesh Seth et al., "Profile of morbidity and nutritional status and their effect on the growth potentials in preschool children in Delhi, India," *Tropical Pediatrics and Environmental Health*, 25 (1979), pp. 23–29.)

Illness	Nutritional status			
	Normal	I	II	III and IV
URI	95	143	144	70
Diarrhea	53	94	101	48
URI and diarrhea	27	60	76	27
None				
Total	288	345	365	167

(a) Complete the table by calculating the entries for the illness category None and the row sums.

(b) Calculate the column percents. Do the percents of children in the different illness categories appear to vary with nutritional status? Explain your answer.

(c) Calculate the table of row percents. Do the percents in the different nutritional status categories appear to vary with the illness group? Explain your answer.

(d) Of the 1165 children in this study, what percent have normal nutritional status and no illness? In what kind of table would you find this percent?

(e) Write null and alternative hypotheses for testing that there is no relationship between nutritional status and illness.

(f) Calculate the table of expected counts.

(g) Report the value of the X^2 statistic, the df, and the *P*-value. What do you conclude?

NOTES

1 Details of exact binomial procedures can be found in chapter 2 of Myles Hollander and Douglas Wolfe, *Nonparametric Statistical Methods*, Wiley, New York, 1973.

2 This example is adapted from a survey directed by Professor Joseph N. Uhl of the Department of Agricultural Economics, Purdue University. The survey was sponsored by the Indiana Christmas Tree Growers Association and was conducted in April 1987.

3 These data were taken from Ray H. Rosenman et al., "A 4-year prospective study of the relationship of different habitual vocational physical activity to risk and incidence of ischemic heart disease in volunteer male federal employees," in P. Milvey (ed.), *The Marathon: Physiological, Medical, Epidemiological and Psychological Studies*, New York Academy of Sciences, 301 (1977), pp. 627–641.

4 This study is reported in Joan L. Duda, "The relationship between goal perspectives, persistence and behavioral intensity among male and female recreational sport participants," *Leisure Sciences*, 10 (1988), pp. 95–106.

5 An alternative formula that can be used for hand or calculator computations is

$$X^2 = \sum \frac{(\text{sample})^2}{\text{expected}} - n$$

CHAPTER 9 EXERCISES

9.47 Several years ago many colleges that enrolled only male or only female students became coeducational. One concern of administrators and alumni was that the academic standards of the institution might decrease with the change. At one such institution, a study of the first freshman class to have women was undertaken. The class consisted of 851 students, 214 of whom were women. An examination of the first semester grades revealed that 15 of the top 30 students were female.

   (a) What is the proportion of women in the freshman class? Call this value p_0.

   (b) Assume that the number of females in the top 30 is approximately a binomial random variable with $n = 30$ and probability p of success. In this case success corresponds to the student being female. What is the value of $\hat{p}$?

   (c) Formulate a null hypothesis and an alternative hypothesis in terms of p and p_0 that expresses the question of interest.

   (d) Test H_0 and summarize your conclusion.

9.48 In 1982, 75% of the 20 fatal accidents in Tippecanoe County, Indiana, were alcohol related. The national average is 50%. Is there evidence in these data to conclude that alcohol is involved in a different proportion of fatal accidents in Tippecanoe County than in the nation as a whole? Give a summary of your analysis and reasons for your conclusion. It would be tempting to use a one-sided alternative for this problem. Why is this choice not appropriate?

9.49 For a study on blood pressure and diet, a random sample of Seventh Day Adventists was taken at a national meeting. Since many people who belong to this denomination are vegetarians, they are a very useful group for studying the effects of this type of diet. It is known that blacks in the population as a whole have a higher average blood pressure than whites. Therefore, in a study of this type it is important to take race into account in the analysis. In one part of the study, the 312 people in the sample were categorized by race and whether or not they were vegetarians. The data are given in the following table. (Data provided by Chris Melby and David Goldflies, Department of Physical Education, Health, and Recreation Studies, Purdue University, West Lafayette, Ind.)

Diet	Black	White
Vegetarian	42	135
Not vegetarian	47	88

The question of interest is whether or not the proportion of vegetarians is the same among black and white Seventh Day Adventists who attended this meeting. Analyze the data, paying particular attention to this question. Summarize your analysis and conclusions. What can you infer about the proportions of vegetarians among black and white Seventh Day Adventists in general? What about blacks and whites in general?

9.50 In Exercise 3.66 the results of a study on high blood pressure and cardiovascular disease are discussed. Of the 2676 men with low blood pressure, 21 died from cardiovascular disease. In the high blood pressure group, there were 3338 men and 55 deaths.

(a) Calculate the sample proportions of deaths for the low and high blood pressure groups.

(b) It is known that high blood pressure is associated with increased risk of death from cardiovascular disease. Formulate appropriate null and alternative hypotheses for this study.

(c) Test H_0 using the z statistic. Report the P-value and state your conclusion.

(d) Give a 95% confidence interval for the difference between the death proportions for the low and high blood pressure groups.

9.51 Traditional practices in Egypt result in food being withheld from children with diarrhea. Since it is known that feeding children with this illness reduces mortality, a nationwide program designed to promote this practice was undertaken in 1985. To evaluate the impact of the program, surveys were taken before and after the program was implemented. In the first survey, 457 of 1003 surveyed mothers followed the practice of feeding children with diarrhea. For the second survey, 437 of 620 surveyed followed this practice. (Data taken from O. M. Galal et al., "Feeding the child with diarrhea: a strategy for testing a health education message within the primary health care system in Egypt," *Socio-Economic Planning Sciences*, 21 (1987), pp. 139–147.)

(a) Assume that the data come from two independent samples. Test the hypothesis that the program was effective, that is, that the practice of feeding children with diarrhea increased between the time of the first study and the time of the second. State H_0 and H_a, give the test statistic and its P-value, and summarize your conclusion.

(b) Present the results using a 95% confidence interval for the difference in proportions.

9.52 In a study of chromosome abnormalities and criminality, data on 4124 Danish males born in Copenhagen were collected. They were classified as having criminal records or not, using the penal registers maintained in the offices of the local police chiefs. They were also classified as having the normal male XY chromosome pair or one of the abnormalities XYY or XXY. Of the 4096 men with normal chromosomes, 381 had criminal records, while 8 of the 28 men with chromosome abnormalities had criminal records. Some experts believe that chromosome abnormalities are associated with increased criminality. Do these data lend support to this belief? Summarize your analysis. (Data taken from H. A. Witkin et al., "Criminality in XYY and XXY men," *Science*, 193 (1976), pp. 547–555.)

9.53 The progress of male and female Ph.D. graduate students at a major university was the subject of a study designed to see if sex differences existed. All students entering Ph.D. programs in a given year were classified as to their status 6 years later. The following categories were used: completed the degree, still enrolled, and dropped out. The data are given in the following table:

Status	Men	Women
Completed	423	98
Still enrolled	134	33
Dropped out	238	98

Describe the data using whatever percents are appropriate. State and test a null hypothesis and alternative that address the question of sex differences. Summarize the results of your test. Describe any other factors not given that might be relevant to this study.

9.54 PTC is a compound that has a strong bitter taste for some people and is tasteless for others. The ability to taste this compound is an inherited trait. Many studies have assessed the proportions of people in different populations who can taste PTC. Some of the results for different countries are given in the following table. (Data taken from A. E. Mourant et al., *The Distribution of Human Blood Groups and Other Polymorphisms*, Oxford University Press, London, 1976.)

	Ireland	Portugal	Norway	Italy
Tasters	558	345	185	402
Nontasters	225	109	81	134

Complete the table and describe the data. Do they provide evidence that the proportion of PTC tasters varies among the four countries? Give a complete summary of your analysis.

9.55 There are four major blood types in humans: O, A, B, and AB. In a study conducted using blood samples from the Blood Bank of Hawaii in 1950, individuals were classified according to blood type and ethnic group. The ethnic groups were Hawaiian, Hawaiian-white, Hawaiian-Chinese, and white. (Data taken from the book cited in exercise 9.54.)

Blood type	Hawaiian	Hawaiian-white	Hawaiian-Chinese	White
O	1903	4469	2206	53,759
A	2490	4671	2368	50,008
B	178	606	568	16,252
AB	99	236	243	5001

Summarize the data. Is there evidence to conclude that blood type and ethnic group are related? Explain how you arrived at your conclusion.

9.56 In *The New York Times* of January 30, 1988, an article described the results of a study of 5139 male British medical doctors who participated in a study of the effects of aspirin on cardiovascular disease. The doctors who participated were randomly assigned to two groups. One group of 3429 doctors took one aspirin daily, and the other did not take aspirin. After 6 years, there were 148 deaths from heart attack or stroke in the first group and 79 in the second group. A similar study for male American medical doctors was reported in the January 27, 1988, *New York Times*. In this study the doctors were also randomly assigned to one of two groups. The 11,037 doctors in the first group took one aspirin every other day, and the 11,034 doctors in the second group took no aspirin. After nearly 5 years there were 104 deaths from heart attacks in the first group and 189 in the second. Analyze the data from these two studies and summarize the results. In what ways do the two studies differ?

9.57 Refer to the gastric freezing experiment described in Example 4.5. It was reported that 28 of the 82 patients who were subjected to gastric freezing improved, while 30 of the 78 patients in the control group improved. The hypothesis of "no difference" between the two groups can be tested in two ways: using a z statistic or using a X^2 statistic.

(a) State the appropriate hypothesis with a two-sided alternative and carry out a z-test. What is the P-value?

(b) Present the data in a 2×2 table. State the appropriate hypothesis and carry out the X^2 test. What is the P-value? Verify that the X^2 statistic is the square of the z statistic.

(c) What do you conclude about the effectiveness of gastric freezing as a treatment for ulcers?

9.58 In Example 3.20 data concerning the survival rates of surgery patients for two hospitals are presented.

(a) Analyze the data for all patients combined and summarize the results.

(b) Run separate analyses for the patients in good condition and for those in poor condition. Summarize these results.

(c) What conclusions do you draw from the analyses performed in (a) and (b)?

9.59 In this exercise we examine the effect of the sample size on the significance test for comparing two proportions. In each case suppose that $\hat{p}_1 = .6$, $\hat{p}_2 = .4$, and n represents the common value of n_1 and n_2. Use the z statistic to test $H_0: p_1 = p_2$ versus the alternative H_a: $p_1 \neq p_2$. Compute the statistic and the associated P-value for the following values of n: 15, 25, 50, 75, 100, and 500. Summarize the results in a table. Explain what you have learned about the effect of the sample size on the statistical significance of the same sample proportions $\hat{p}_1$ and $\hat{p}_2$.

9.60 In the first section of this chapter, we studied the effect of the sample size on the width of the confidence interval for a single proportion. In this exercise we perform some calculations to observe this effect for the two sample problem. As in the exercise above, suppose that $\hat{p}_1 = .6$, $\hat{p}_2 = .4$, and n represents the common value of n_1 and n_2. Compute the 95% confidence intervals for the difference in the two proportions for $n = $ 15, 25, 50, 75, 100, and 500. For each interval calculate the width. Summarize and explain your results.

9.61 For a single proportion we noted that the width of a confidence interval is largest for any given sample size n and confidence level $1 - \alpha$ when $\hat{p} = .5$. This led us to use $p^* = .5$ for planning purposes. The same kind of result is true for the two sample problem. Specifically, the width of the confidence interval for the difference in two proportions is largest when $\hat{p}_1 = \hat{p}_2 = .5$. Use these conservative values in the following calculations. Assume that the common value of n_1 and n_2 is n. Calculate the widths of the 99% confidence intervals for the difference in two proportions for the following choices of n: 10, 30, 50, 100, 200, and 500. Present the results in a table or with a graph. Summarize your conclusions.

9.62 Refer to the previous problem for background information. You are planning a survey and will calculate a 95% confidence interval for the difference in two proportions when the data are collected. You would like the width of the interval to be less than or equal to .1. Assume that n represents the common value of n_1 and n_2. To guarantee that the width of your interval will be less than or equal to .1, use $\hat{p}_1 = \hat{p}_2 = .5$ in your calculations.

(a) How large a value of n is needed?
(b) Give a general formula for n in terms of the width w and z^*.

9.63 You are planning a survey and you will use a 90% confidence interval for the difference in two proportions to present the results. Assume that it is reasonable to use .5 as the value for $\hat{p}_1$ and $\hat{p}_2$ in your planning. You would like the width of the confidence interval to be less than or equal to .2. Suppose that it is very difficult to sample from the first population and that it will be impossible for you to obtain more than 20 observations from this population. Assuming that $n_1 = 20$, can you find a value of n_2 that will guarantee the desired result? If so, report the value; if not, explain why not.

9.64 Refer to Exercise 9.43. Display the data for the 1984–85 fund raising campaign using appropriate charts or figures suitable for inclusion in a report on the campaign. Do the same for the data on the 1986–87 campaign presented in Exercise 9.44.

9.65 A statistically significant association between alcohol consumption and nicotine exposure was found using the data presented in Exercise 9.45. Present one or more charts or figures which clearly display this association.

9.66 The proportions of individuals with each of the four major blood types were compared across four ethnic groups in Exercise 9.55 and significant differences were found. Display the data using charts or figures in such a way that the differences among the four ethnic groups can be easily seen.

9.67 You are asked to evaluate a proposal for a study of the effects of aspirin on cardiovascular disease similar to those described in Exercise 9.56. The researchers are planning to use two groups, one treatment and one control. The proposed sample sizes are 200 in each group. Using any information in Exercise 9.56 that you consider relevant, write a short evaluation of this proposal.

9.68 The sports goals study described in Example 9.14 actually involved three categorical variables: sex, social comparison and mastery. Methods for analyzing count data classified according to more than two variables have been developed but are beyond the scope of this text. The analyses and discussion presented in this chapter suggest that the major sex difference is in the social comparison variable. To investigate this idea, rearrange the data into a 2 × 2 table that classifies the students by sex and social comparison. Analyze this table and summarize the results. Do the same using sex and mastery as the classification variables. Considering these analyses and the analysis presented in the text, what conclusions do you draw?

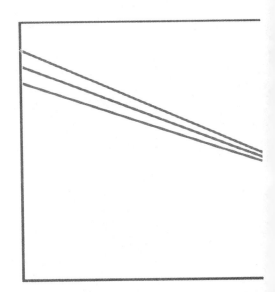

Prelude

When the relationship between an explanatory variable and a
response variable is linear, we summarize the data with a line. In
this chapter we discuss confidence intervals and significance tests
for the slope and intercept. Confidence intervals for the mean of the
response variable for a given value of the explanatory variable are
presented as well as a method for finding a prediction interval for a
future value of the response variable. Finally a case study illustrates
the type of analysis used when there are several explanatory
variables and one response variable.

- *How can we construct an interval to predict the mean gas consumption
  for a house from the number of degree days?*
- *Body density is difficult to measure, but skinfold thickness is easy to
  determine. How can we use skinfold measures to predict body density?*
- *Do high school grades predict academic performance in the first year
  of college? Are they better predictors than SAT scores?*

10

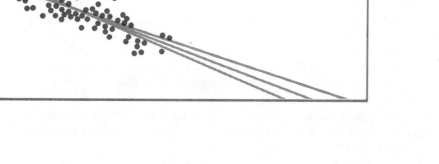

Inference for Regression

In Chapter 1 the idea of using the sample mean $\bar{x}$ as a measure of the center for a collection of observations was introduced. Later we learned that when the data are viewed as a random sample from a population, the sample mean is an estimate of the population mean μ. In Chapters 7 and 8 $\bar{x}$ was used to make inferences about μ using confidence intervals and tests of significance.

In Section 10.1, we follow the same program for the problem of fitting straight lines to data. In Chapter 3 the method of least squares was used to fit a line when a scatterplot suggests that the dependence of a response variable y on an explanatory variable x can be summarized by a straight line. The fitted line can be used to predict y for a given value of x. To do statistical inference for regression, we think of a population with a *true* regression line. The least-squares line is an estimate of this true line. By extending the ideas in Chapters 7 and 8, we can give confidence intervals and significance tests for the slope and intercept. Confidence intervals for the mean response and prediction intervals for an individual future observation are also presented. Correlation was introduced in Chapter 3 as a measure of the strength of the linear association between two variables when we do not have a clear explanatory-response structure. Statistical inference for correlation is presented in the last part of Section 10.1.

multiple linear regression

Some general ideas about *multiple linear regression* are introduced in Section 10.2. These problems involve one response variable and *several* explanatory variables. A case study is presented that illustrates this technique. The important case of a single explanatory variable, which is the topic of Section 10.1, is often called *simple linear regression* to distinguish it from multiple regression.

simple linear regression

In discussing the least-squares line $a + bx$ in Chapter 3, we did not need to distinguish between sample and population. That distinction is essential to formal statistical inference. We emphasize it by introducing new notation. Following the usual practice of using Greek letters for the population parameters to be estimated from the data, the population line is written as $\beta_0 + \beta_1 x$. The fitted line estimates the intercept β_0 by b_0 and the slope β_1 by b_1.

10.1 SIMPLE LINEAR REGRESSION

Statistical Model for Linear Regression

In Section 8.2, methods for comparing two means were introduced. The statistical model for the blood pressure study described in Example 8.12 is displayed in Figure 10.1. In each population the observed blood pressure change is assumed to be normally distributed around a mean μ. Since there are two populations, we use μ_1 and μ_2 to represent the population means.

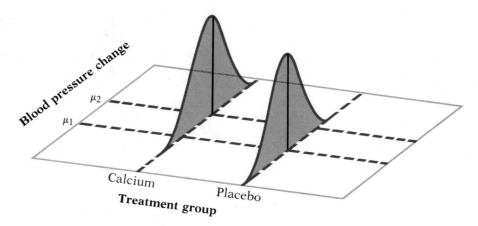

Figure 10.1 Illustration of statistical model for Example 8.12.

In the figure the two normal distributions have the same spread, which indicates that the population standard deviations are assumed to be equal. We can think of the treatment (Placebo or Calcium) as being an explanatory variable in this example.

For linear regression the explanatory variable x can have many different values. For each value of x, the observed values of the response variable y are assumed to be normally distributed about a mean that depends upon x. We use $\mu_{y(x)}$ to represent these means. The variation about the mean as expressed by σ is assumed to be the same for all values of x.

Sometimes it is convenient to think of the values of x as defining different subpopulations, one for each possible value of x. Each subpopulation consists of all individuals or items having the same value of x. In this view, the response variable y is assumed to be normally distributed with mean $\mu_{y(x)}$ and standard deviation σ in each of the subpopulations.

In general the means $\mu_{y(x)}$ can vary according to any sort of pattern as x changes. In linear regression we assume that they all lie on a line when plotted against x. The equation of the line is given by

$$\mu_{y(x)} = \beta_0 + \beta_1 x$$

population regression line

where β_0 is the intercept and β_1 is the slope. This *population regression line* describes how the mean response varies with x. A display of this statistical model is given in Figure 10.2.

The data for a linear regression are the observed values of y and x. In the model, x is viewed as a fixed known quantity defining the subpopulations. On the other hand, y is a random variable with a mean and standard deviation that can be estimated from the data.

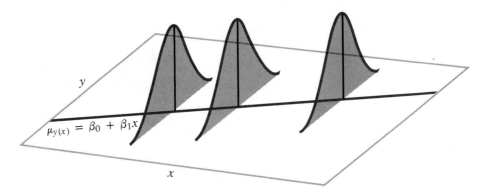

Figure 10.2   Illustration of statistical model for linear regression.

EXAMPLE 10.1

In Example 3.6 the relationship between an explanatory variable, heating degree days, and a response variable, gas consumption, is described by a regression line. There are nine different values for heating degree days. The data are given in the following table:

	Oct.	Nov.	Dec.	Jan.	Feb.	Mar.	Apr.	May	June
x	15.6	26.8	37.8	36.4	35.5	18.6	15.3	7.9	0.0
y	5.2	6.1	8.7	8.5	8.8	4.9	4.5	2.5	1.1

Heating degree days x is an average for all of the days in the given month, and gas consumption is a similar daily average measured in hundreds of cubic feet. For the sake of brevity, we will refer to these two variables as degree days and gas consumption. The data and the least-squares line are plotted in Figure 10.3.

To do statistical inference for this problem, we assume that for each fixed value of degree days, the gas consumption has a normal distribution. For example, consider the observation for October. The value of degree days is $x = 15.6$. We can think of measuring gas consumption for many months, each with 15.6 degree days. The regression model assumes that these gas consumption measurements would follow a normal distribution with mean $\mu_{y(15.6)}$ and standard deviation σ.

Similarly, for any other value of degree days, gas consumption is assumed to follow a normal distribution with mean $\mu_{y(x)}$ and standard deviation σ. The means for different values of x are related by the population regression line

$$\mu_{y(x)} = \beta_0 + \beta_1 x \qquad \blacksquare$$

The value $x = 20$ is not present in the data collected. However, since this value is within the range of the data available, we may want to predict mean gas consumption for the subpopulation of months with 20 degree days. For this reason, our model assumes a linear relationship for the means over a *range* of values of degree days where inference is desired. In our

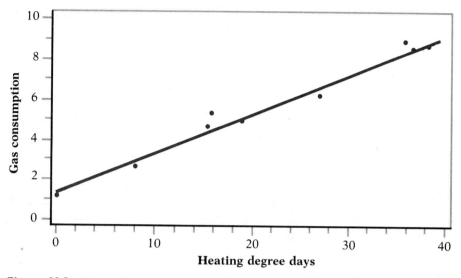

Figure 10.3 Data and regression line for gas consumption example.

example, it would be reasonable to make inferences for values of degree days from 0 to about 40. Negative values of degree days do not make sense, and for very large values we would face the dangers of extrapolation.

EXAMPLE 10.2

"The fat content of the human body has physiological and medical importance. It may influence morbidity and mortality, it may alter the effectiveness of drugs and anesthetics, and it may affect the ability to withstand exposure to cold and starvation."[1] Fat content is generally calculated from body density, with higher fat values corresponding to lower body density.

Body density is very difficult to measure directly. The standard method requires that subjects be weighed under water. For this reason, scientists have sought variables that are easier to measure and that can be used to predict body density. Research has shown that skinfold thickness measures can be used to accurately predict body density. At four body locations, a fold of skin is pinched between calipers and its thickness is determined. A linear relationship between body density and the logarithm of the sum of the four skinfold measures has been established. We use the names DEN for body density and LSKIN for the logarithm of the sum of the four skinfold measures in discussing this problem.

For the statistical model that is the foundation of this prediction method, the subpopulations are defined by the variable LSKIN. All individuals with the same value of LSKIN are in the same subpopulation. Their body densities are assumed to be normally distributed about a mean density. The mean densities are related by the equation

$$\mu_{\text{DEN(LSKIN)}} = \beta_0 + \beta_1\text{LSKIN}$$ ■

The population regression line gives the means for all of the subpopulations. The statistical model for linear regression consists of this line and a description of the variation of individual values about the means. In Chapter 2 the principle of seeking an overall pattern and deviations from it when examining data was expressed in the equation

$$DATA = FIT + RESIDUAL$$

This principle also can be used to express the statistical model for linear regression. The FIT part of the model consists of the subpopulation means, given by the expression

$$\beta_0 + \beta_1 x$$

The RESIDUAL part represents deviations of the data from the line of population means. The assumptions state that these deviations are normally distributed with standard deviation σ. We use ϵ to represent the RESIDUAL part of the statistical model.

Simple linear regression model

> Given n observations on the explanatory variable x and the response variable y
>
> $$(x_1, y_1), (x_2, y_2), \ldots, (x_n, y_n)$$
>
> the statistical model for simple linear regression states that
>
> $$y_i = \beta_0 + \beta_1 x_i + \epsilon_i$$
>
> where $i = 1, 2, \ldots, n$.
>   The ϵ_i are assumed to be independent and normally distributed with mean 0 and standard deviation σ. In other words, they are a simple random sample (SRS) from an $N(0, \sigma)$ distribution.
>   The parameters of the model are β_0, β_1, and σ.

Because the means $\mu_{y(x)}$ lie on the line

$$\mu_{y(x)} = \beta_0 + \beta_1 x$$

they are parameters determined by β_0 and β_1. With data we can find estimates of the model parameters. From the estimates of β_0 and β_1, the linear relationship determines the estimates of the $\mu_{y(x)}$. Thus, the model for linear regression allows us to do inference not only on subpopulations for which we have data but also on those corresponding to x's not present in the data.

Estimation of Regression Parameters

The method of least squares presented in Chapter 3 can be used as a data summary procedure in a variety of circumstances. However, the statistical model presented in the preceding section is not applicable to all problems that are amenable to least-squares methods. When using the least-squares estimates as statistics that estimate the parameters of the simple linear regression model, we use b_0 for the intercept and b_1 for the slope. This reminds us that b_0 is an estimate of β_0 and b_1 is an estimate of β_1.

Using the formulas from Chapter 3 and our new notation, the slope of the least-squares line is

$$b_1 = \frac{\sum(x_i - \bar{x})(y_i - \bar{y})}{\sum(x_i - \bar{x})^2}$$

and the intercept is

$$b_0 = \bar{y} - b_1\bar{x}$$

Some algebra based on the rules for means of random variables (Section 5.3) shows that b_0 and b_1 are unbiased estimates of β_0 and β_1.

The predicted value of y for a given x is

$$\hat{y} = b_0 + b_1 x$$

and is an unbiased estimate of the population mean $\mu_{y(x)}$. The sample residuals are

$$\begin{aligned} e_i &= y_i - \hat{y}_i \\ &= y_i - b_0 - b_1 x_i \end{aligned}$$

The sample residuals e_i correspond to the model residuals ϵ_i. The e_i sum to 0, and the ϵ_i come from a population with a mean of 0.

The remaining parameter to be estimated is σ. Since this parameter is the standard deviation of the model residuals, it should come as no surprise that we use the sample residuals to estimate it. As usual, we work first with the variance and take the square root to obtain the standard deviation.

For linear regression the estimate of σ^2 is

$$\begin{aligned} s^2 &= \frac{\sum e_i^2}{n - 2} \\ &= \frac{\sum(y_i - \hat{y}_i)^2}{n - 2} \end{aligned}$$

The denominator $n - 2$ in this formula plays the same role as the quantity $n - 1$ in the formula for a sample variance given in Equation 1.2 of Chapter 1. That is, s^2 is an unbiased estimate of σ^2. The quantity $n - 2$ is called the *degrees of freedom* for s^2. The estimate of σ is given by

$$s = \sqrt{s^2}$$

Regression calculations The calculations required for a regression problem are done in two steps:

Step 1 Calculate the means and standard deviations of x and y and obtain the least-squares line.

Step 2 Calculate the residuals e_i and the estimated standard deviation s about the line.

Step 1 consists of the descriptive calculations that are familiar from Chapter 3. If step 2 is to be carried out successfully, the step 1 calculations must be done very accurately. A statistical calculator or computing system is very helpful. If you do regression calculations with a basic calculator, you must carry many significant digits and check your work carefully. We illustrate the calculations using a basic calculator for the data from Example 10.1 and using a computer regression program for the data from Example 10.2.

EXAMPLE 10.3   The data in Example 10.1 were used in Chapter 3 to find the least-squares regression line

$$\hat{y} = 1.23 + .202x$$

for the gas consumption problem. The line is plotted with the data in Figure 10.3. The number of digits given for the intercept and slope are sufficient to describe the data and plot the line. However, the calculations that follow require greater accuracy. The work is organized in Table 10.1 for step 1 and Table 10.2 for step 2. ∎

We start with step 1; let us first examine Table 10.1. The first three columns give the observation number, x, and y. From the sums given in the

Table 10.1 Calculations for gas consumption data (step 1)

Case	x_i	y_i	$x_i - \bar{x}$	$y_i - \bar{y}$	$(x_i - \bar{x})^2$	$(y_i - \bar{y})^2$	$(x_i - \bar{x})(y_i - \bar{y})$
1	15.6	5.2	−5.94	−.389	35.284	.151	2.311
2	26.8	6.1	5.26	.511	27.668	.261	2.688
3	37.8	8.7	16.26	3.111	264.388	9.678	50.585
4	36.4	8.5	14.86	2.911	220.820	8.474	43.257
5	35.5	8.8	13.96	3.211	194.882	10.311	44.826
6	18.6	4.9	−2.94	−.689	8.644	.475	2.026
7	15.3	4.5	−6.24	−1.089	38.938	1.186	6.795
8	7.9	2.5	−13.64	−3.089	186.050	9.542	42.134
9	0.0	1.1	−21.54	−4.489	463.972	20.151	96.693
Sum	193.9	50.3	.04	−.001	1440.646	60.229	291.315

bottom row, we calculate

$$\bar{x} = \frac{193.9}{9} = 21.54$$

$$\bar{y} = \frac{50.3}{9} = 5.589$$

The next two columns give the deviations of x and y from their means. For the first case, for example, we have

$$x_1 - \bar{x} = 15.6 - 21.54 = -5.94$$
$$y_1 - \bar{y} = 5.2 - 5.589 = -.389$$

The columns of deviations should be inspected for any unusual values. We note that case 9 has a very low value of x and that this value appears to be reasonable since it corresponds to June. The sums for these columns should be 0 or, because of round-off error, near 0. The values 0.04 and -0.001 appear to be sufficiently small. Had the sums been far from 0, we would recheck our calculations. If no errors were found, we would need to include more digits in the means used to calculate the deviations.

The next two columns give the squares of the x deviations and the squares of the y deviations. For case 1 these values are

$$(x_1 - \bar{x})^2 = (-5.94)^2 = 35.284$$
$$(y_1 - \bar{y})^2 = (-.389)^2 = .151$$

We use the sums for these columns to calculate the variance and standard deviation for x and y. These quantities are not used in the regression calculations but should be determined to aid in inspecting the data.

$$s_x^2 = \frac{\sum (x_i - \bar{x})^2}{n - 1} = \frac{1440.646}{9 - 1} = 180.08$$

$$s_y^2 = \frac{\sum (y_i - \bar{y})^2}{n - 1} = \frac{60.229}{9 - 1} = 7.529$$

The standard deviations are

$$s_x = \sqrt{180.08} = 13.42$$
$$s_y = \sqrt{7.529} = 2.744$$

The last column of Table 10.1 gives the products of the x and y deviations. The entry for case 1 is

$$(x_1 - \bar{x})(y_1 - \bar{y}) = (-5.94)(-.389) = 2.311$$

The slope of the least-squares line is calculated from the column sums for the squares of the x deviations and the products of the x and the y deviations:

$$b_1 = \frac{\sum(x_i - \bar{x})(y_i - \bar{y})}{\sum(x_i - \bar{x})^2} = \frac{291.315}{1440.646} = .20221$$

The intercept is

$$b_0 = \bar{y} - b_1\bar{x} = 5.589 - (.20221)(21.54) = 1.233$$

Our more exact calculation of the least-squares regression line therefore gives

$$\hat{y} = 1.233 + .20221x$$

The quantities needed for step 2 are presented in Table 10.2. The first three columns are the same as those in Table 10.1. The next three columns give the predicted values, the residuals, and the squares of the residuals.

For case 1,

$$\hat{y}_1 = b_0 + b_1 x_1$$
$$= 1.233 + (.20221)(15.6)$$
$$= 4.387$$

Note that we have used the more accurate values of b_1 and b_0 in this calculation. The residual for the first case is

$$e_1 = y_1 - \hat{y}_1 = 5.2 - 4.387 = .813$$

Table 10.2 Calculations for gas consumption data (step 2)

Case	x_i	y_i	$\hat{y}_i$	e_i	e_i^2
1	15.6	5.2	4.387	.813	.661
2	26.8	6.1	6.652	−.552	.305
3	37.8	8.7	8.876	−.176	.031
4	36.4	8.5	8.593	−.093	.009
5	35.5	8.8	8.411	.389	.151
6	18.6	4.9	4.994	−.094	.009
7	15.3	4.5	4.327	.173	.030
8	7.9	2.5	2.830	−.330	.109
9	0.0	1.1	1.233	−.133	.018
Sum	193.9	50.3	50.303	−.003	1.323

and the squared residual is

$$e_1^2 = .813^2 = .661$$

In Chapter 3 we learned of the importance of examining residuals. Before proceeding with statistical inference for regression, we apply the same principles and inspect the residuals for any unusual values. The first case has the largest deviation, but it does not appear to be particularly large relative to the other values given. The residuals are plotted versus case in Figure 10.4 and versus degree days in Figure 10.5. No particular patterns or unusual observations are evident.

The sum of the residuals is always 0. In Table 10.2 the sum of the residuals is given as -0.003. This is small enough for our purposes. Since the residuals are recorded with three places after the decimal point, we expect some round-off error in the last digit of the sum. If the sum were far from 0, we would need to compute $\hat{y}$ more accurately.

Another useful check on the accuracy of our calculations is based on the fact that the fitted regression line goes through the point corresponding to $\bar{y}$ and $\bar{x}$. To do this, calculate

$$b_0 + b_1\bar{x}$$

The result should agree with $\bar{y}$. The reader is invited to perform this calculation for our example.

To find s^2, the sample estimate of the population σ^2, we find in the bottom row of Table 10.2 that the sum of squares of the residuals is

$$\sum e_i^2 = 1.323$$

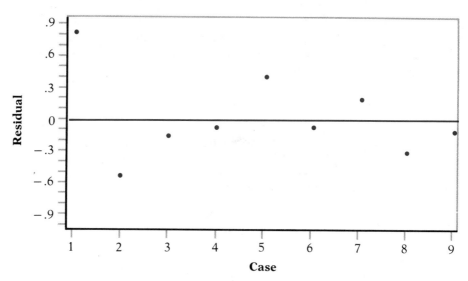

Figure 10.4 Plot of residuals versus case for gas consumption example.

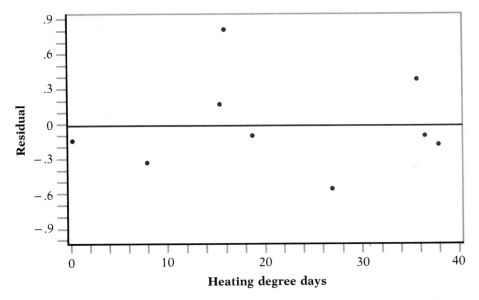

Figure 10.5 Plot of residuals versus degree days for gas consumption example.

Therefore,

$$s^2 = \frac{\sum e_i^2}{n-2} = \frac{1.323}{9-2} = .189$$

The estimate of σ is

$$s = \sqrt{.189} = .435$$

These calculations are tedious even though there are only nine observations. We must often manipulate a large number of digits to obtain accurate results. Checking that the sums for the deviations and residuals are close to 0 helps to ensure accuracy. Nevertheless, we prefer to use a computer to perform the calculations, and when there are many observations, a computer is essential.

EXAMPLE 10.4   We use the SAS statistical software package to calculate the regression of DEN on LSKIN for Example 10.2. (Other software packages will yield a similar output.) Because the relationship between body density, DEN, and the logarithm of the sum of the four skinfold measures, LSKIN, is known to vary slightly with age and sex, separate regressions are estimated for males and females in different age groups. In the report on this study, the estimated regression line for a sample of size $n = 92$ males aged 20 to 29 is given as

$$\widehat{\text{DEN}} = 1.1631 - .0632\text{LSKIN}$$

The value of s was reported to be 0.0084.   ∎

VARIABLE	N	MEAN	STANDARD DEVIATION	VARIANCE
LSKIN	92	1.56800000	0.21590000	0.04661281
DEN	92	1.06400240	0.01602319	0.00025674

Figure 10.6 Descriptive statistics for LSKIN and DEN.

In much of the discussion that follows, the numbers from the output are rounded off. Computers often give us many more digits than are meaningful or useful. Means and standard deviations for the two variables are given in the output from the SAS procedure MEANS in Figure 10.6. We see that

$$\overline{\text{LSKIN}} = 1.568 \qquad s_{\text{LSKIN}} = .2159$$
$$\overline{\text{DEN}} = 1.064 \qquad s_{\text{DEN}} = .0160$$

The data are plotted with the least-squares line in Figure 10.7. The relationship appears to be linear, and no unusual observations are evident.

The output of the SAS regression procedure REG for this example is given in Figure 10.8. The values of b_0 and b_1 are in the column labeled PARAMETER ESTIMATE. The column labeled VARIABLE tells us that the

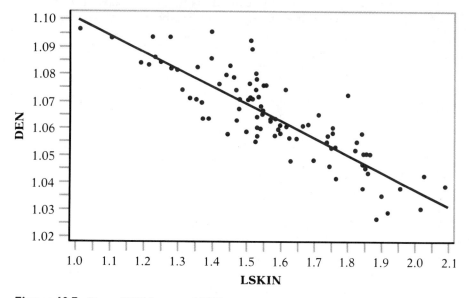

Figure 10.7 Plot of DEN versus LSKIN with regression line.

```
DEP VARIABLE: DEN
                              ANALYSIS OF VARIANCE


                         SUM OF             MEAN
         SOURCE     DF   SQUARES           SQUARE      F VALUE    PROB>F

         MODEL      1    0.01694263     0.01694263     237.478    0.0001
         ERROR     90    0.006420960    0.000071344
         C TOTAL   91    0.02336359


             ROOT MSE    0.008446538       R-SQUARE      0.7252
             DEP MEAN    1.064002          ADJ R-SQ      0.7221
             C.V.        0.7938457


                             PARAMETER ESTIMATES


                     PARAMETER        STANDARD      T FOR H0:
         VARIABLE    ESTIMATE         ERROR         PARAMETER=0    PROB > |T|

         INTERCEP    1.16310000     0.006490615       179.197       0.0001
         LSKIN      -0.06320000     0.004101147       -15.410       0.0001
```

Figure 10.8 Regression output for body density example.

first entry, INTERCEP, is b_0 and the second, LSKIN, is b_1. The values agree with those presented in Example 10.4.

The sum of the squared residuals is found in the **SUM OF SQUARES** column in the row marked **ERROR**. The value given is 0.00642. The degrees of freedom for this sum of squares are found in the same row in the column marked **DF**. Since $n = 92$, we have $n - 2 = 90$ for this entry. The value of s^2 is found in the same row in the column labeled **MEAN SQUARE**. It is given as 0.0000713. The estimated standard deviation s is labeled **ROOT MSE** and is given on the output as 0.0084.

Computer outputs often give more information than we want or need. On the other hand, it is very frustrating to find that a computer output does not contain the particular statistics that we want for an analysis. The experienced user of statistical software packages ignores the parts of the output that are not needed for the problem at hand.

A plot of the residuals versus case number is displayed in Figure 10.9. No unusual values or patterns are evident in this plot. Similarly, the plot of the residuals versus LSKIN given in Figure 10.10 reveals nothing that would cause us to question the validity of our analysis. The normal quantile plot for the residuals is displayed in Figure 10.11. Since it is fairly straight, the assumption of normally distributed errors appears to be reasonable.

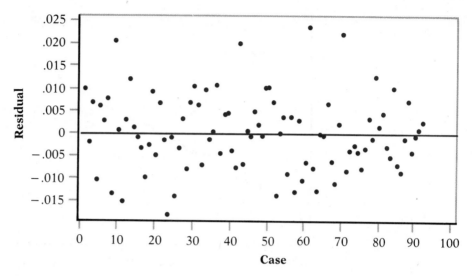

Figure 10.9 Plot of residuals versus case for body density example.

For the 92 males aged 20 to 29 in the study, the average body density was 1.064, with a standard deviation of 0.016. By using a regression model and defining subpopulations according to skinfold measures, the common sub-population standard deviation is estimated to be only 0.0084. In other words, the standard deviation of the body density observations about their mean is

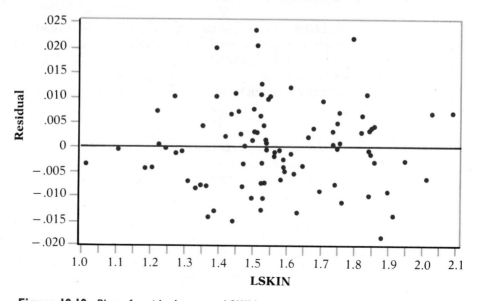

Figure 10.10 Plot of residuals versus LSKIN.

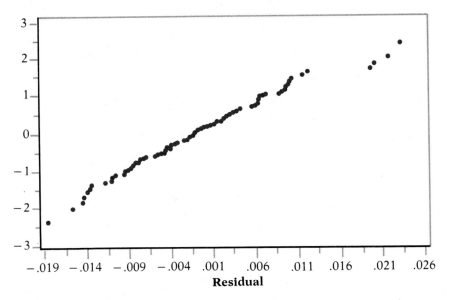

Figure 10.11   Normal quantile plot of residuals for body density example.

0.0160, while the standard deviation of these same observations about the regression line is only half as large, 0.0084. This decrease in the standard deviation is due to the use of skinfolds in the regression model.

Confidence Intervals and Significance Tests

In Chapter 8, confidence intervals and significance tests for means and differences in means were developed using the standard errors for these statistics and t distributions. Similar procedures are applicable to the intercept and slope in a linear regression although some of the formulas are a little more complicated. All of the confidence intervals, for example, have the form

$$\text{estimate} \pm t^* s(\text{estimate})$$

where t^* is a critical point of the t distribution.

Computation based on the rules for variances (Section 5.3) establishes that the standard deviation of b_1 is

$$\sigma(b_1) = \frac{\sigma}{\sqrt{\sum(x_i - \bar{x})^2}}$$

To estimate $\sigma(b_1)$, we need only replace σ with its estimate s. Therefore, the standard error for the slope b_1 in a simple linear regression is

$$s(b_1) = \frac{s}{\sqrt{\sum(x_i - \bar{x})^2}}$$

Similarly, the standard deviation of b_0 is

$$\sigma(b_0) = \sigma \sqrt{\frac{1}{n} + \frac{\bar{x}^2}{\sum(x_i - \bar{x})^2}}$$

which is estimated by the standard error of b_0

$$s(b_0) = s \sqrt{\frac{1}{n} + \frac{\bar{x}^2}{\sum(x_i - \bar{x})^2}}$$

The estimated regression coefficients b_0 and b_1 are each normally distributed, with means β_0 and β_1 and standard deviations $\sigma(b_0)$ and $\sigma(b_1)$. Furthermore, a general version of the central limit theorem implies that even if the model residuals are not normally distributed, the statistics b_0 and b_1 will tend to be normal as the sample size increases.

Confidence intervals and tests for the slope and intercept are based on t distributions. If the standard deviation were known, we would use the normal z statistic in place of the t statistic. The degrees of freedom used in estimating the standard deviation are $n - 2$. We therefore use the $t(n - 2)$ distribution in our calculations.

Confidence intervals and hypothesis tests for β_0 and β_1

A level $1 - \alpha$ confidence interval for the intercept β_0 is

$$b_0 \pm t^* s(b_0)$$

A level $1 - \alpha$ confidence interval for the slope β_1 is

$$b_1 \pm t^* s(b_1)$$

In these expressions t^* is the upper $\alpha/2$ critical value for the $t(n - 2)$ distribution.

To test the hypothesis $H_0: \beta_1 = 0$, compute

$$t = \frac{b_1}{s(b_1)}$$

In terms of a random variable T having the $t(n - 2)$ distribution, the P-value for a test of H_0 against

$$H_a: \beta_1 > 0 \quad \text{is} \quad P(T \geq t)$$
$$H_a: \beta_1 < 0 \quad \text{is} \quad P(T \leq t)$$
$$H_a: \beta_1 \neq 0 \quad \text{is} \quad P(|T| \geq |t|)$$

Similar formulas can be given for a significance test of β_0 using $s(b_0)$ and the $t(n - 2)$ distribution. Although computer outputs sometimes include tests for $H_0: \beta_0 = 0$, this information usually has little practical value. From our subpopulation means equation

$$\mu_{y(x)} = \beta_0 + \beta_1 x$$

we see that β_0 is the subpopulation mean corresponding to $x = 0$. In many practical situations, this subpopulation is not particularly interesting or it does not exist.

On the other hand, the test $H_0: \beta_1 = 0$ is quite useful. Consider the model when $\beta_1 = 0$. The term involving x drops out of the formula for the means, and we are left with

$$y_i = \beta_0 + \epsilon_i$$

With this model, the mean of y does not vary with x. All of the y's come from a single population with mean β_0, which we would estimate by $\bar{y}$. Thus, the significance test of β_1 indicates, on the basis of the data available, whether or not x is useful for predicting y in a linear equation.

EXAMPLE 10.5

In Example 10.3, we found the least-squares slope

$$b_1 = .20221$$

and intercept

$$b_0 = 1.233$$

for the regression of gas consumption on heating degree days. We now construct confidence intervals for β_1 and β_0 and test $H_0: \beta_1 = 0$ versus $H_a: \beta_1 > 0$. The one-sided alternative is used because we expect gas consumption to be positively related to degree days. ■

We begin by finding the standard error of b_1. From Example 10.3,

$$s = .435$$

and from Table 10.1,

$$\sum(x_i - \bar{x})^2 = 1440.646$$

Thus, the standard error of b_1 is

$$s(b_1) = \frac{s}{\sqrt{\sum(x_i - \bar{x})^2}}$$

$$= \frac{.435}{\sqrt{1440.646}}$$

$$= .01146$$

To test

$$H_0: \beta_1 = 0$$
$$H_a: \beta_1 > 0$$

we calculate

$$t = \frac{b_1}{s(b_1)}$$
$$= \frac{.20221}{.01146}$$
$$= 17.64$$

The degrees of freedom for s are $n - 2 = 7$, so the value of t is compared with the critical values for the $t(7)$ distribution in Table E. The largest entry for 7 degrees of freedom is 5.408, corresponding to $\alpha = 0.0005$. Therefore, we conclude that b_1 is significant with $P < 0.0005$ because $t = 17.64$ is greater than 5.408.

For the 95% confidence interval for β_1 we need the upper 0.025 critical value for the $t(7)$ distribution. In Table E this value is given as $t^* = 2.365$. The 95% confidence interval for β_1 is

$$b_1 \pm t^*s(b_1) = .20221 \pm (2.365)(.01146)$$
$$= .20221 \pm .02710$$
$$= (.17511, .22931)$$

From our data we can be 95% confident that each additional degree day results in an increased consumption of between 0.175 and 0.229 hundreds of cubic feet of gas.

For this example, the confidence interval for β_0 also has a meaningful interpretation. Since β_0 is the mean of y when $x = 0$, β_0 represents the gas for cooking, hot water, and other uses that are present when there is no demand for heating. To construct a confidence interval for β_0, we need to first calculate the standard error

$$s(b_0) = s\sqrt{\frac{1}{n} + \frac{\bar{x}^2}{\sum(x_i - \bar{x})^2}}$$
$$= .435\sqrt{\frac{1}{9} + \frac{21.54^2}{1440.646}}$$
$$= .2863$$

The 95% confidence interval for β_0 is

$$b_0 \pm t^*s(b_0) = 1.233 \pm (2.36)(.2863)$$
$$= 1.233 \pm .6757$$

The consumption for purposes other than heating is estimated to between 0.56 and 1.91 hundreds of cubic feet of gas.

EXAMPLE 10.6 | Refer to the computer output for the body density problem given in Figure 10.8. In the column labeled STANDARD ERROR, the value of $s(b_1)$ is given as 0.0041. The estimated slope $b_1 = -0.0632$ is given to the left of this value. (As usual, we have rounded the values from the output.)

The t statistic and P-value for the test of H_0: $\beta_1 = 0$ against the two-sided alternative H_a: $\beta_1 \neq 0$ are given in the columns labeled T FOR H0: PARAMETER = 0 and PROB > |T|. We can verify the t calculation with the formula

$$t = \frac{b_1}{s(b_1)} = \frac{-.0632}{.0041} = -15.41$$

The P-value is 0.0001. The null hypothesis is clearly rejected, and we conclude that the logarithm of the skinfold measures can be used in a linear equation to predict body density.

Since body density and skinfold measures are expected to be negatively related, we could use H_a: $\beta_1 < 0$. This choice makes no difference in our conclusion for this example, although the P-value for the one-sided alternative is half of the value for the two-sided alternative.

To construct a confidence interval for β_1 we need the value of t^* from the $t(n-2) = t(90)$ distribution. In Table E there are entries for 80 and 100 degrees of freedom. Note that the values for these rows are very similar. To be conservative, we will use the larger value given for 80 degrees of freedom. For a 95% confidence interval we use the value in the 0.025 column, which is 1.99.

The 95% confidence interval for β_1 is

$$-.0632 \pm (1.99)(.0041) = -.0632 \pm .0082$$
$$= (-.0714, -.0550)$$

From the analysis, we estimate that an increase of 1 in LSKIN is associated with a decrease of between 0.0714 and 0.0550 in DEN. ∎

In contrast to Example 10.5, the intercept in this example has no natural interpretation. It corresponds to a subpopulation with LSKIN = 0, which is impossible. The values of LSKIN in the data range from 1.20 to 2.26. For this reason, we do not compute a confidence interval for β_0.

Confidence Intervals for Mean Response

For any given value of x, say x^*, the mean of the response y in the subpopulation is given by

$$\mu_{y(x^*)} = \beta_0 + \beta_1 x^*$$

To estimate this mean from the sample, we substitute the estimates b_0 and b_1 for β_0 and β_1:

$$\hat{\mu}_{y(x*)} = b_0 + b_1 x*$$

To construct a confidence interval for $\mu_{y(x*)}$, we need the standard error of $\hat{\mu}$. It is given by the following formula:

$$s(\hat{\mu}) = s\sqrt{\frac{1}{n} + \frac{(x* - \bar{x})^2}{\sum(x_i - \bar{x})^2}}$$

Confidence interval for $\mu_{y(x*)}$

A level $1 - \alpha$ confidence interval for the mean response $\mu_{y(x*)}$ when x takes the value $x*$ is

$$\hat{\mu}_{y(x*)} \pm t* s(\hat{\mu})$$

where $t*$ is the upper $\alpha/2$ critical value for the $t(n - 2)$ distribution.

The computations for these confidence intervals are very similar to those for the parameters of the regression line that we performed in the previous section. The basic form is the estimate plus or minus a critical value times a standard error.

We will illustrate the computations using a calculator for the data in Example 10.1 and using a computer program for the problem in Example 10.2.

EXAMPLE 10.7

Consider the gas consumption problem. We compute a 95% confidence interval for the mean gas consumption in months with 20 degree days.

The estimate of the mean is

$$\begin{aligned}
\hat{\mu}_{y(20)} &= b_0 + b_1 x \\
&= 1.233 + (.20221)(20) \\
&= 5.277
\end{aligned}$$

The standard error is

$$\begin{aligned}
s(\hat{\mu}) &= s\sqrt{\frac{1}{n} + \frac{(x* - \bar{x})^2}{\sum(x_i - \bar{x})^2}} \\
&= .435\sqrt{\frac{1}{9} + \frac{(20 - 21.54)^2}{1440.646}} \\
&= .435\sqrt{\frac{1}{9} + \frac{2.372}{1440.646}} \\
&= .146
\end{aligned}$$

The numbers in this computation are taken from the previous calculations performed for the gas consumption example.

Finally, we need a value for t^*. For a 95% confidence interval we use the $\alpha/2 = 0.025$ critical value $t^* = 2.36$, as we did in Example 10.5. The 95% confidence interval is

$$\hat{\mu}_{y(20)} \pm t^*s(\hat{\mu}) = 5.277 \pm (2.36)(.146)$$
$$= 5.277 \pm .345$$
$$= (4.932, 5.622)$$

For months averaging 20 degree days, we estimate the mean consumption to be between 4.93 and 5.62 hundreds of cubic feet of gas, with 95% confidence. ∎

Many computer programs calculate confidence intervals for the mean response corresponding to each of the x values in the data. Some[2] can cal-

OBS	LSKIN	ACTUAL	PREDICT VALUE	STD ERR PREDICT	LOWER95% MEAN	UPPER95% MEAN
1	1.26594	1.0930	1.0831	.00152	1.0801	1.0861
2	1.55720	1.0627	1.0647	.00088	1.0629	1.0664
3	1.45025	1.0783	1.0714	.00100	1.0694	1.0734
4	1.52207	1.0564	1.0669	.00090	1.0651	1.0687
5	1.51477	1.0734	1.0674	.00091	1.0656	1.0692
6	1.50624	1.0706	1.0679	.00092	1.0661	1.0697
7	1.49688	1.0761	1.0685	.00093	1.0667	1.0703
8	1.61624	1.0474	1.0610	.00090	1.0592	1.0627
9	1.50313	1.0886	1.0681	.00092	1.0663	1.0699
10	1.75201	1.0531	1.0524	.00116	1.0501	1.0547
11	1.43343	1.0573	1.0725	.00104	1.0704	1.0746
12	1.81344	1.0513	1.0485	.00134	1.0458	1.0511
13	1.60096	1.0738	1.0619	.00089	1.0601	1.0637
14	1.49381	1.0699	1.0687	.00093	1.0668	1.0705
15	1.28574	1.0808	1.0818	.00145	1.0790	1.0847
16	1.51957	1.0636	1.0671	.00090	1.0653	1.0689
17	1.83410	1.0371	1.0472	.00140	1.0444	1.0500
18	1.58400	1.0603	1.0630	.00088	1.0612	1.0647
19	1.69952	1.0649	1.0557	.00103	1.0536	1.0577
20	1.58701	1.0577	1.0628	.00088	1.0610	1.0646
21	2.02173	1.0421	1.0353	.00206	1.0312	1.0394
22	1.84178	1.0451	1.0467	.00143	1.0439	1.0495
23	1.87409	1.0265	1.0447	.00153	1.0416	1.0477
24	1.83495	1.0460	1.0471	.00140	1.0443	1.0499

Figure 10.12 Confidence intervals for subpopulation means for body density example.

culate an interval for any value of x. As the previous example illustrates, the computations required are not particularly difficult and can easily be performed with a calculator.

EXAMPLE 10.8

For the problem of predicting body density from skinfold measures, Figure 10.12 gives part of the output generated by the SAS procedure REG. Only output for the first 24 cases is shown.

The observed, or actual, values y_i are given in the column labeled ACTUAL, the predicted values $\hat{\mu}_{y(x_i)}$ are given in the column labeled PREDICT VALUE, and the standard errors of the predicted values $s(\hat{\mu})$ are given in the column labeled STD ERR PREDICT. The lower and upper limits for the 95% confidence intervals are in the columns labeled LOWER95% MEAN and UPPER95% MEAN.

Consider the first line of the output. The value of LSKIN is 1.266, and

$$\hat{\mu}_{\text{DEN}(1.266)} = 1.0831$$

with a standard error of

$$s(\hat{\mu}) = .0015$$

The 95% confidence interval is (1.0801, 1.0861). We conclude with 95% confidence that the mean body density is between 1.0801 and 1.0861 for men aged 20 to 29 whose logarithm of skinfold measure is 1.266. ∎

The upper and lower confidence limits for all values of x within a given range can be plotted on a graph with the data and the least-squares line. Such a plot is shown in Figure 10.13. Observe that the width of the confidence interval for any value of x is the vertical distance between the upper and lower confidence limits. The widths are smallest for values of x^*

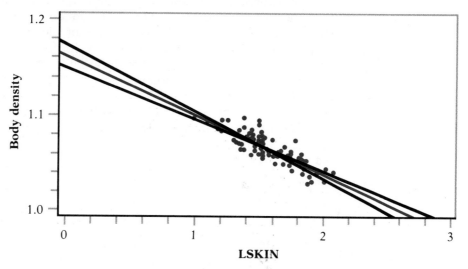

Figure 10.13 Regression line with 95% confidence limits for body density example.

near the mean and increase as x^* moves from $\bar{x}$. This phenomenon is quite general and is a consequence of the term $(x^* - \bar{x})^2$ in the expression for $s(\hat{\mu})$.

Prediction Intervals

A regression equation also can be used to predict a future value of the response variable y corresponding to a particular value x^* of the explanatory variable x. The predicted value is

$$\hat{y} = b_0 + b_1 x^*$$

This is the same as the expression for $\hat{\mu}_{y(x^*)}$. That is, the fitted line is used both to estimate the mean response when $x = x^*$ and to predict a single future response. The two notations $\hat{\mu}_{y(x^*)}$ and $\hat{y}$ are used to remind us of these two distinct uses.

The distinction between estimating a subpopulation mean and predicting a future observation becomes clearer when we use intervals that provide a margin of error. The margin of error is larger when we are predicting a single response than when we are estimating a mean response. The confidence interval for the subpopulation mean was discussed in the last section. The interval used for predicting a future observation is called a *prediction interval*.

prediction interval

Although the response that is being predicted is a random variable, the interpretation of a prediction interval is similar to that for a confidence interval. Consider an experiment in which a sample of size n and one additional observation are drawn. Each time the experiment is run, the 95% prediction interval for the additional observation is calculated. If the experiment is repeated many times, 95% of the prediction intervals will contain the y-value of the additional observation. In other words, the probability that the prediction interval contains the value of a future observation is 0.95.

The form of the prediction interval is very similar to that of the confidence interval. The difference is that the standard error used must include both (1) the variability due to the fact that the least-squares line is not exactly equal to the true regression line and (2) the variability of the future response variable y around the subpopulation mean. The standard error for the prediction interval is

$$s(\hat{y}) = s \sqrt{1 + \frac{1}{n} + \frac{(x^* - \bar{x})^2}{\sum (x_i - \bar{x})^2}}$$

Note that the only difference between this standard error and that given for the confidence interval for the mean response is the extra 1 under the square root sign.

Prediction interval for a future observation

A $1 - \alpha$ prediction interval for a future observation on the response variable y from the subpopulation corresponding to x^* is

$$\hat{y} \pm t^* s(\hat{y})$$

where t^* is the upper $\alpha/2$ critical value for the $t(n - 2)$ distribution.

We again illustrate the computations using a calculator for the data given in Example 10.1 and using a computer program for the problem in Example 10.2.

EXAMPLE 10.9

Consider the gas consumption problem. We will construct a 95% prediction interval for the gas consumption during a future month with 20 degree days. For $x^* = 20$, the predicted value of y is

$$\hat{y} = b_0 + b_1 x^*$$
$$= 1.233 + (.20221)(20)$$
$$= 5.277$$

This is the same value that we found for $\hat{\mu}_y$ in Example 10.7.

To compute the standard deviation we use many of the same quantities that were used for the computation of $s(\hat{\mu})$ in Example 10.7:

$$s(\hat{y}) = s\sqrt{1 + \frac{1}{n} + \frac{(x^* - \bar{x})^2}{\sum(x_i - \bar{x})^2}}$$

$$= .435\sqrt{1 + \frac{1}{n} + \frac{(20 - 21.54)^2}{1440.646}}$$

$$= .435\sqrt{1 + \frac{1}{9} + \frac{2.372}{1440.646}}$$

$$= .459$$

The 95% prediction interval is

$$\hat{y} \pm t^* s(\hat{y}) = 5.277 \pm (2.36)(.459)$$
$$= 5.277 \pm 1.083$$
$$= (4.194, 6.360)$$

For a future month with 20 degree days, we predict the consumption to be between 4.19 and 6.36 hundreds of cubic feet of gas with 95% confidence. ∎

In Example 10.7 we found the 95% confidence interval for the mean gas consumption for months with 20 degree days to be 5.277 ± 0.345. Comparing this interval with the 95% prediction interval, we note that the prediction interval is centered at the same value but is substantially wider.

OBS	LSKIN	ACTUAL	PREDICT VALUE	STD ERR PREDICT	LOWER95% PREDICT	UPPER95% PREDICT
1	1.26594	1.0930	1.0831	.00152	1.0660	1.1001
2	1.55720	1.0627	1.0647	.00088	1.0478	1.0816
3	1.45025	1.0783	1.0714	.00100	1.0545	1.0883
4	1.52207	1.0564	1.0669	.00090	1.0500	1.0838
5	1.51477	1.0734	1.0674	.00091	1.0505	1.0842
6	1.50624	1.0706	1.0679	.00092	1.0510	1.0848
7	1.49688	1.0761	1.0685	.00093	1.0516	1.0854
8	1.61624	1.0474	1.0610	.00090	1.0441	1.0778
9	1.50313	1.0886	1.0681	.00092	1.0512	1.0850
10	1.75201	1.0531	1.0524	.00116	1.0354	1.0693
11	1.43343	1.0573	1.0725	.00104	1.0556	1.0894
12	1.81344	1.0513	1.0485	.00134	1.0315	1.0655
13	1.60096	1.0738	1.0619	.00089	1.0450	1.0788
14	1.49381	1.0699	1.0687	.00093	1.0518	1.0856
15	1.28574	1.0808	1.0818	.00145	1.0648	1.0989
16	1.51957	1.0636	1.0671	.00090	1.0502	1.0839
17	1.83410	1.0371	1.0472	.00140	1.0302	1.0642
18	1.58400	1.0603	1.0630	.00088	1.0461	1.0799
19	1.69952	1.0649	1.0557	.00103	1.0388	1.0726
20	1.58701	1.0577	1.0628	.00088	1.0459	1.0797
21	2.02173	1.0421	1.0353	.00206	1.0181	1.0526
22	1.84178	1.0451	1.0467	.00143	1.0297	1.0637
23	1.87409	1.0265	1.0447	.00153	1.0276	1.0617
24	1.83495	1.0460	1.0471	.00140	1.0301	1.0641

Figure 10.14 Prediction intervals for body density example.

EXAMPLE 10.10

For the problem of predicting body density from skinfold measures, 95% prediction intervals can be calculated by the SAS procedure REG. Output for the first 24 cases is given in Figure 10.14. The lower and upper 95% limits are in the last two columns.

Consider observation 1, which has LSKIN = 1.266. The 95% prediction interval is 1.066 to 1.100. For a male aged 20 to 29 with LSKIN = 1.266, we predict the body density to be between 1.066 and 1.100. ∎

Figure 10.15 shows the upper and lower prediction limits for all values of x within a given range plotted on a graph with the data and the least-squares line. The 95% prediction limits are indicated by the black curves. Compare this figure with Figure 10.13, which shows the 95% confidence limits. The black curves for the prediction intervals are farther from the

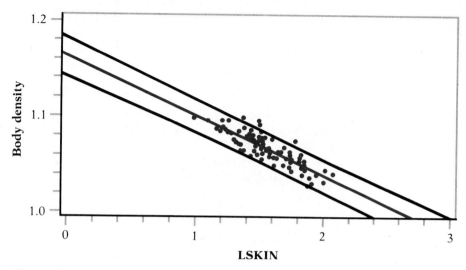

Figure 10.15 Regression line with 95% prediction limits for body density example.

least-squares line than are the confidence limits. This is a graphic illustration of the idea that the interval for a single future observation must be larger than an interval for the mean of its subpopulation.

Analysis of Variance for Regression*

In the preceding sections we have covered the basic material that is important for using regression analysis to examine the effect of a single explanatory variable on a response variable. The standard computer output for this type of problem includes additional calculations called *analysis of variance*. They are essential for multiple regression, treated briefly in Section 10.2, and for comparing several means, treated in Chapter 11.

Analysis of variance summarizes information related to the sources of variation in the data. To see this, we return to the

$$DATA = FIT + RESIDUAL$$

framework and apply this idea to examine variation. The total variation in the response y is expressed by the deviations $y_i - \bar{y}$. If these deviations were all 0, all observations would be equal and there would be no variation in the response.

There are two reasons why the individual observations y_i are not all equal to $\bar{y}$. First, $\bar{y}$ does not, in general, estimate the subpopulation mean

---

* This optional section presents material that should be studied if multiple regression is to be covered.

for y_i. Rather, it is estimated by the fitted value $\hat{y}_i$. The difference is $(\hat{y}_i - \bar{y})$. Second, because of variation within the subpopulation, individual observations will vary about their mean. This difference is given by the residual $(y_i - \hat{y}_i)$. We express these ideas with the following equation:

$$(y_i - \bar{y}) = (\hat{y}_i - \bar{y}) + (y_i - \hat{y}_i)$$

For deviations, this equation expresses the DATA = FIT + RESIDUAL conceptualization.

Variation is quantified by squaring and summing deviations. It is an algebraic fact that the sums of squares add:

$$\sum(y_i - \bar{y})^2 = \sum(\hat{y}_i - \bar{y})^2 + \sum(y_i - \hat{y}_i)^2$$

We rewrite this equation as

$$\text{SST} = \text{SSM} + \text{SSE}$$

where

$$\text{SST} = \sum(y_i - \bar{y})^2$$
$$\text{SSM} = \sum(\hat{y}_i - \bar{y})^2$$
$$\text{SSE} = \sum(y_i - \hat{y}_i)^2$$

The SS in each abbreviation stands for sum of squares and the T, M, and E stand for total, model, and error, respectively. Thus the total variation, as expressed by SST, is composed of the variation due to the model (SSM) and the variation due to the error (SSE).

If $H_0: \beta_1 = 0$ were true, then there would be no subpopulations and all of the data should be viewed as coming from a single population with mean μ. To calculate a variance in this circumstance, we would use the formula from Chapter 1:

$$\frac{\sum(y_i - \bar{y})^2}{n - 1}$$

The numerator in this expression is SST. The denominator is the total *degrees of freedom*, or simply DFT.

Just as the total sum of squares SST is the sum of SSM and SSE, the total degrees of freedom DFT is the sum of DFM and DFE, the degrees of freedom for the model and for the error.

$$\text{DFT} = \text{DFM} + \text{DFE}$$

Because the model has one explanatory variable x, DFM = 1. Since DFT = $n - 1$, this leaves DFE = $n - 2$ as the degrees of freedom for the error.

mean square

For each source, the ratio of the sum of squares to the degrees of freedom is called the *mean square*, or simply MS. Thus, the general formula for a mean square is

$$MS = \frac{\text{sum of squares}}{\text{degrees of freedom}}$$

We have seen above that MST is the sample variance that we would calculate if all of the data came from a single population. Earlier in this chapter, the estimate of the parameter σ^2 for the regression model was given as

$$s^2 = \frac{\sum (y_i - \hat{y}_i)^2}{n - 2}$$

We now see that this estimate is also MSE.

Sums of squares, degrees of freedom, and mean squares

> Sums of squares represent variation present in the data. They are calculated by summing squared deviations. In linear regression there are three sources of variation: model, error, and total.
>   The sums of squares are related by the formula
>
> $$SST = SSM + SSE$$
>
> Thus, the total variation is composed of two parts, one due to the model and one due to error.
>   Degrees of freedom are associated with each sum of squares. They are related in the same way:
>
> $$DFT = DFM + DFE$$
>
>   To calculate mean squares, use the formula
>
> $$MS = \frac{\text{sum of squares}}{\text{degrees of freedom}}$$

The null hypothesis $H_0: \beta_1 = 0$ that y is not linearly related to x can be tested by comparing MSM with MSE. The test statistic is

$$F = \frac{MSM}{MSE}$$

When H_0 is true, this statistic has an F distribution with 1 degree of freedom in the numerator and $n - 2$ degrees of freedom in the denominator. These degrees of freedom are those of MSM and MSE. The F distributions were introduced in Section 8.3; Figure 8.10 shows a typical F density curve. Just as there are many t statistics, there are many F statistics. The analysis of variance F statistic is not the same as the F statistic of Section 8.3.

When $\beta_1 \neq 0$, MSM tends to be large relative to MSE. So large values of F give evidence against H_0 in favor of the two-sided alternative.

Analysis of variance F test

In the simple linear regression model, the hypotheses

$$H_0: \beta_1 = 0$$
$$H_a: \beta_1 \neq 0$$

are tested by the F statistic

$$F = \frac{\text{MSM}}{\text{MSE}}$$

The P-value is the probability that a random variable having the $F(1, n-2)$ distribution is greater than or equal to the calculated value of the F statistic.

Since the F statistic tests the same null hypothesis as the t statistic that we encountered earlier in this chapter, it is not surprising that the two are related. It is an algebraic fact that $t^2 = F$ in this case. For linear regression with one explanatory variable, we prefer the t form of the test because it more easily allows the possibility of testing one-sided alternatives and is closely related to the confidence interval for β_1.

analysis of variance table

The analysis of variance calculations are displayed in an *analysis of variance table*, often abbreviated ANOVA table. Here is the table for simple linear regression.

Source	Sum of squares	Degrees of freedom	Mean square	F
Model	$\sum(\hat{y}_i - \bar{y})^2$	1	SSM/DFM	MSM/MSE
Error	$\sum(y_i - \hat{y}_i)^2$	$n-2$	SSE/DFE	
Total	$\sum(y_i - \bar{y})^2$	$n-1$	SST/DFT	

EXAMPLE 10.11

For the gas consumption problem of Example 10.1 we have already computed most of the entries for the ANOVA table. In the bottom row of Table 10.1 we computed

$$\text{SST} = \sum(y_i - \bar{y})^2 = 60.299$$

and in the bottom row of Table 10.2, we find

$$\text{SSE} = \sum(y_i - \hat{y}_i)^2 = \sum e_i^2 = 1.323$$

To obtain SSM we rearrange the equation

$$SST = SSM + SSE$$

to obtain

$$
\begin{aligned}
SSM &= SST - SSE \\
&= 60.229 - 1.323 \\
&= 58.906
\end{aligned}
$$

Since $n = 9$, we have

$$
\begin{aligned}
DFM &= 1 \\
DFE &= n - 2 = 7 \\
DFT &= n - 1 = 8
\end{aligned}
$$

The mean squares are computed as follows:

$$MSM = \frac{SSM}{DFM} = \frac{58.906}{1} = 58.906$$

$$MSE = \frac{SSE}{DFE} = \frac{1.323}{7} = .189$$

$$MST = \frac{SST}{DFT} = \frac{60.229}{8} = 7.529$$

Finally, the F statistic is

$$F = \frac{MSM}{MSE} = \frac{58.906}{.189} = 312$$

■

The results of these computations are summarized in the following ANOVA table:

Source	Sum of squares	Degrees of freedom	Mean square	F
Model	58.906	1	58.906	312
Error	1.323	7	.189	
Total	60.229	8	7.529	

To test

$$
\begin{aligned}
H_0 &: \beta_1 = 0 \\
H_a &: \beta_1 \neq 0
\end{aligned}
$$

using the F statistic, we refer to the critical values given in Table F. Under H_0, the F statistic has an $F(1, n - 2) = F(1, 7)$ distribution. Thus, we find the table entries corresponding to 1 degree of freedom in the numerator and 7 degrees of freedom in the denominator. The largest critical value is 29.25, corresponding to $\alpha = 0.001$. Since the calculated $F = 312$ exceeds this number, we report the P-value as $P < 0.001$.

In Example 10.5, we calculated the t statistic for $H_0: \beta_1 = 0$ as $t = 17.64$. If we square this number we get $17.64^2 = 311$, which is quite close to the value of F that we calculated. The difference is due to round-off error in our calculations. Had we kept more digits in our calculations, we would find $t = 17.663$ and $F = 311.972$.

The critical values for the F distributions with 1 degree of freedom in the numerator are related to the critical values for the t distributions. For 7 degrees of freedom, the t upper 0.025 critical value is given as $t^* = 2.365$ in Table E. If we square this value, we get $2.365^2 = 5.59$, which agrees with the $\alpha = 0.05$ F critical value for 1 degree of freedom in the numerator and 7 degrees of freedom in the denominator given in Table F.

EXAMPLE 10.12 | For the problem of predicting body density from skinfold measures, Figure 10.16 gives the ANOVA table provided by the SAS procedure REG. Recall that we discussed part of this output in Example 10.4.

```
DEP VARIABLE: DEN
                          ANALYSIS OF VARIANCE

                        SUM OF             MEAN
     SOURCE     DF      SQUARES           SQUARE      F VALUE    PROB>F

     MODEL       1    0.01694263       0.01694263    237.478    0.0001
     ERROR      90    0.006420960      0.000071344
     C TOTAL    91    0.02336359


           ROOT MSE    0.008446538       R-SQUARE      0.7252
           DEP MEAN    1.064002          ADJ R-SQ      0.7221
           C.V.        0.7938457


                          PARAMETER ESTIMATES

                    PARAMETER        STANDARD       T FOR HO:
     VARIABLE       ESTIMATE          ERROR       PARAMETER=0    PROB > |T|

     INTERCEP     1.16310000      0.006490615      179.197        0.0001
     LSKIN       -0.06320000      0.004101147      -15.410        0.0001
```

Figure 10.16 Regression output for body density example.

The output corresponds very well to the ANOVA table given earlier. The only difference in the labels is that C TOTAL is used in the output rather than TOTAL.

The calculated value of the F statistic is 237.478 and the P-value is $P = 0.0001$. The null hypothesis is therefore rejected. Note that the t statistic for LSKIN is -15.410. If we square this, we get 237.468, which agrees reasonably well with the F statistic given on the output. Whenever numbers are rounded off in calculations such as this, there will be round-off error. ∎

Computers generally keep many digits in the calculations and round off only when printing the results. When doing computations by hand, we often need to write down intermediate results and then do more computations with them later. This is a major disadvantage in using a basic calculator to perform complex computations such as those needed for linear regression.

Correlation*

In Section 3.3 the correlation coefficient was introduced and formulas were given for its calculation. Correlation is a measure of linear association between two variables that do not necessarily fit into an explanatory-response relationship.

If we assume that two random variables x and y are each normally distributed and that all of the dependence between them can be summarized by a population correlation that we call ρ (technically, the assumption is that x and y are jointly normal), statistical theory provides a procedure for testing the null hypothesis that the population correlation is 0.

Test for a zero population correlation

To test the hypothesis $H_0: \rho = 0$, compute

$$t = \frac{r\sqrt{n-2}}{\sqrt{1-r^2}}$$

where n is the sample size and r is the sample correlation.

In terms of a random variable T having the $t(n-2)$ distribution, the P-value for a test of H_0 against

$$H_a: \rho > 0 \quad \text{is} \quad P(T \geq t)$$
$$H_a: \rho < 0 \quad \text{is} \quad P(T \leq t)$$
$$H_a: \rho \neq 0 \quad \text{is} \quad P(|T| \geq |t|)$$

We illustrate the calculation with our two examples.

* This material is optional and can be omitted without loss of continuity.

EXAMPLE 10.13

For the gas consumption problem we test the $H_0: \rho = 0$. Since we expect gas consumption to be positively related to degree days, it is appropriate to test H_0 versus the one-sided alternative $H_a: \rho > 0$.

In Example 3.12 the sample correlation between gas consumption and degree days is calculated as

$$r = .989$$

Since the sample size is $n = 9$, the value of t is

$$t = \frac{.989\sqrt{9 - 2}}{\sqrt{1 - .989^2}}$$

$$= \frac{2.6166}{.1479}$$

$$= 17.69$$

We compare the result to the values for the $t(7)$ distribution given in Table E. The critical value for 7 degrees of freedom and $\alpha = 0.0005$ is 5.408. We therefore have strong evidence that gas consumption is positively related to degree days. ∎

Most computer packages have routines for calculating and testing correlations.

EXAMPLE 10.14

For the body density and skinfold measures problem, the output of the SAS procedure CORR is given in Figure 10.17. The sample correlation between DEN and LSKIN is $r = -0.85157$ and is called a Pearson correlation coefficient in this output. The P-value for a two-sided test is given below the correlation and is 0.0001. If we wanted to test the one-sided alternative that the population correlation is negative, we would divide the P-value in the output by 2, after checking that the sample coefficient is, in fact, negative. We conclude that there is a negative correlation between DEN and LSKIN. ∎

In Section 3.3 the connection between correlation and regression is discussed. It is noted there that the sample correlation and the slope of the

```
PEARSON CORRELATION COEFFICIENTS / PROB > |R| UNDER HO: RHO=0 / N = 92


                            LSKIN              DEN


           LSKIN         1.00000          -0.85157
                         0.0000            0.0001


           DEN          -0.85157           1.00000
                         0.0001            0.0000
```

Figure 10.17 Correlation output for body density example.

least-squares regression line are related by the equation

$$b_1 = r \frac{s_y}{s_x}$$

From this fact we can see that if the slope is 0, so is the correlation and vice versa. Therefore, it should come as no surprise to learn that the procedures for testing $H_0: \beta_1 = 0$ and $H_0: \rho = 0$ are closely related. In fact, the t statistics for testing these hypotheses are numerically equal. This means that

$$\frac{b_1}{s(b_1)} = \frac{r\sqrt{n-2}}{\sqrt{1-r^2}}$$

For the two examples discussed above, the result that there is a significant correlation between the two variables would not come as a surprise to anyone familiar with the meaning of these variables. The statistical test simply tells us whether or not there is evidence in the data to conclude that the population correlation is different from 0. The actual size of the correlation is of considerably more interest. To approach the problem from this point of view we would like to report the sample correlation with a confidence interval. Unfortunately, most software packages do not perform this calculation. Since the steps needed to do the calculation by hand involve some rather tedious manipulations, we do not include them here.[3]

In Section 3.3 it is also noted that r^2 is the fraction of variation in the values of y that is explained by the least-squares regression of y on x. By writing an alternative formula for r^2, we can clarify this statement:

$$r^2 = \frac{\text{SSM}}{\text{SST}} = \frac{\sum(\hat{y}_i - \bar{y})^2}{\sum(y_i - \bar{y})^2}$$

Here, SSM represents the variation explained by the model and SST represents the total variation in y.

SUMMARY

The **statistical model for simple linear regression** is

$$y_i = \beta_0 + \beta_1 x_i + \epsilon_i$$

where $i = 1, 2, \ldots, n$. The ϵ_i are assumed to be independent and normally distributed with mean 0 and standard deviation σ. The **parameters** of the model are β_0, β_1, and σ.

The β's are estimated by the intercept and slope of the least-squares regression line, b_0 and b_1. The parameter σ is estimated by

$$s = \sqrt{\frac{\sum e_i^2}{n-2}}$$

where the e_i are the sample residuals:

$$e_i = y_i - \hat{y}_i$$

The **standard errors** for b_0 and b_1 are

$$s(b_0) = s\sqrt{\frac{1}{n} + \frac{\bar{x}^2}{\sum(x_i - \bar{x})^2}}$$

$$s(b_1) = \frac{s}{\sqrt{\sum(x_i - \bar{x})^2}}$$

A level $1 - \alpha$ confidence interval for β_1 is

$$b_1 \pm t^* s(b_1)$$

where t^* is the upper $\alpha/2$ critical value for the $t(n - 2)$ distribution. The test of the hypothesis H_0: $\beta_1 = 0$ is based on the statistic

$$t = \frac{b_1}{s(b_1)}$$

and the $t(n - 2)$ distribution. Similar formulas are used for confidence intervals and tests for β_0, but these are meaningful only in special cases.

The **estimated mean response** for the subpopulation corresponding to the value x^* of the explanatory variable is

$$\hat{\mu}_{y(x^*)} = b_0 + b_1 x^*$$

A level $1 - \alpha$ confidence interval for the mean response is

$$\hat{\mu}_{y(x^*)} \pm t^* s(\hat{\mu})$$

where

$$s(\hat{\mu}) = s\sqrt{\frac{1}{n} + \frac{(x^* - \bar{x})^2}{\sum(x_i - \bar{x})^2}}$$

is the standard error and t^* is the upper $\alpha/2$ critical value for the $t(n - 2)$ distribution.

The estimated value of the response variable y for a future observation from the subpopulation corresponding to the value x^* of the explanatory variable is

$$\hat{y} = b_0 + b_1 x^*$$

A $1 - \alpha$ **prediction interval** for the estimated response is

$$\hat{y} \pm t^* s(\hat{y})$$

where

$$s(\hat{y}) = s \sqrt{1 + \frac{1}{n} + \frac{(x^* - \bar{x})^2}{\sum (x_i - \bar{x})^2}}$$

and t^* is the upper $\alpha/2$ critical value for the $t(n - 2)$ distribution.

The **ANOVA table** for a linear regression gives the sum of squares, degrees of freedom, and mean squares for the model, error, and total sources of variation. The F statistic is the ratio MSM/MSE. Under H_0: $\beta_1 = 0$, this statistic has an $F(1, n - 2)$ distribution and is used to test H_0 versus the two-sided alternative.

When the variables y and x are jointly normal, the sample correlation is an estimate of the population correlation ρ. The test of H_0: $\rho = 0$ is based on the statistic

$$t = \frac{r\sqrt{n - 2}}{\sqrt{1 - r^2}}$$

which has a $t(n - 2)$ distribution under H_0. This test statistic is numerically identical to the t statistic used to test H_0: $\beta_1 = 0$.

The **square of the sample correlation** can be expressed as

$$r^2 = \frac{\text{SSM}}{\text{SST}}$$

and is interpreted as the proportion of the variability in the response variable y that is explained by the explanatory variable x in the linear regression.

SECTION 10.1 EXERCISES

10.1 A study similar to that described in Examples 3.6 and 10.1 was run for a different house. The study was run for 2 years, giving $n = 18$ observations. The fitted equation for these data is $\hat{y} = 2.405 + 0.26896x$, where y represents gas consumption and x represents degree days. The standard error of the slope is $s(b_1) = 0.00815$.

(a) Find a 95% confidence interval for β_1.

(b) Since this interval is based on twice as many observations as the one calculated in Example 10.5, how would you expect its width to compare to the one calculated in the example? Check your answer by comparing the two widths.

10.2 Suppose the standard error of the intercept is $s(b_0) = 0.20351$ for the problem described in the previous exercise.

(a) Recall that the intercept represents gas for cooking, hot water, and other uses that are present when there is no demand for heating. Find a 95% confidence interval for the intercept.

(b) Compare the width of your interval to the one calculated in Example 10.5. Explain why it is shorter.

10.3 Refer to the gas consumption problem given in Exercise 10.1.

(a) Calculate the t statistic for testing $H_0: \beta_1 = 0$.

(b) For the alternative $H_1: \beta_1 > 0$, what critical value would you use for the test corresponding to $\alpha = 0.05$?

(c) Do you reject H_0 using this value?

(d) How would you report the P-value for the test?

10.4 Students in an introductory statistics class were given a pretest on mathematics skills at the beginning of the semester. This pretest x is used in a regression equation to predict the score on the final exam y. The estimated regression equation is $\hat{y} = 10.5 + 0.82x$. The standard error of b_1 is 0.38 and the sample size is 55. Test the null hypothesis that there is no linear relationship between the pretest and the score on the final exam versus the two-sided alternative.

10.5 Use the procedure illustrated in Example 10.7 for the gas consumption problem to answer the following questions:

(a) Find a 99% confidence interval for the mean gas consumption for months with degree days = 30.

(b) Repeat the calculation for degree days = 80.

(c) Which interval has greater width? Explain why.

(d) Do any of the calculations that you performed in this exercise involve extrapolation? Explain your answer.

10.6 Answer all parts of the previous exercise using 90% confidence intervals for mean gas consumption for months with degree days = 31.54 and degree days = 11.54.

10.7 In Example 10.9 a prediction interval for the gas consumption problem is illustrated. Use this method to answer the following questions:

(a) Find a 99% prediction interval for gas consumption in a future month having degree days = 30.

(b) Repeat the calculation for degree days = 80.

(c) Compare the width of the interval found in (a) of this exercise with the width of the confidence interval found in (a) of Exercise 10.5. Do the same for (b) of the two exercises. Give an explanation for the results of these comparisons.

10.8 Find 90% prediction intervals for the gas consumption in a future month having degree days = 31.54 and repeat the calculation for degree days = 11.54. Compare the results with those for the confidence intervals that you found in Exercise 10.6.

10.9 In Exercise 3.23 data are given for a study on blood flow in the stomachs of dogs. Assume that the assumptions needed for simple linear regression hold. The estimated regression line is $\hat{y} = 1.031 + 0.902x$, where y is the response variable vein and x is the explanatory variable spheres. The estimate of σ is $s = 1.757$.

(a) Find the value of $\bar{x}$ and $\sum(x_i - \bar{x})^2$ using the raw data.
(b) Test the null hypothesis that the slope is 0 versus the one-sided alternative that it is positive. What conclusion do you draw?
(c) Find a 99% confidence interval for the slope.
(d) What is the estimate of the mean value of vein corresponding to spheres = 12.0? Give a 95% confidence interval for this mean.
(e) Suppose that a value of spheres equal to 15.0 is observed. Calculate a 90% prediction interval for the response variable vein.

10.10 Exercise 3.25 describes data on air pollution measurements in two locations—rural and city. The regression equation for predicting the city reading from the rural one is given as $\hat{y} = -2.580 + 1.0935x$.

(a) Only cases for which both readings are present are used in the regression calculations. What is the sample size for these computations? Calculate $\bar{x}$ and $\sum(x_i - \bar{x})^2$ for these cases.
(b) State appropriate null and alternative hypotheses for assessing whether or not there is a linear relationship between the city and rural readings. Give the test statistic and report the P-value for testing your null hypothesis. Summarize your conclusion.
(c) A radio report tells you that the rural reading is 43 today. Give a 95% prediction interval for today's city reading.

10.11 Ohm's law states that the current I in a metal wire is proportional to the potential difference V applied to its ends and is inversely proportional to the resistance R in the wire. Expressed in an equation, Ohm's law is $I = V/R$. In a high school physics lab students performed several experiments to study this law. The voltage was varied, and for each setting the current was read from a meter. The problem was to determine the value of R for the wire. We can rewrite Ohm's law in the form of a linear regression as $I = \beta_0 + \beta_1 V$, where $\beta_0 = 0$ and $\beta_1 = 1/R$. Since V was set by the experimenter, we think of this variable as the explanatory variable. The current I is viewed as the response. The data for an experiment on one wire are as follows (data provided by Sara McCabe):

V	.5	1.0	1.5	1.8	2.0
I	.52	1.19	1.62	2.00	2.40

(a) Plot the data. Are there any outliers or unusual points?

(b) Find the least-squares fit to the data. What is the estimate of $1/R$ for this wire?

(c) Find a 95% confidence interval for $1/R$.

(d) Ohm's law states that β_0 in the model is 0. Calculate the test statistic for this hypothesis and give an approximate P-value.

(e) If b_1 is an estimate of $1/R$, then $1/b_1$ is an estimate of R. Calculate this estimate of the resistance R. Similarly, if L and U represent the lower and upper confidence limits for $1/R$, then the corresponding limits for R are given by $1/U$ and $1/L$, as long as L and U are positive. Use this fact and your answer to (c) to find a 95% confidence interval for R.

10.12 Refer to the previous exercise on Ohm's law. The data for an experiment on another wire are as follows.

V	.5	1.0	2.0	3.0	4.0
I	.12	.31	.69	1.10	1.30

Answer the questions given in the previous exercise for this set of data.

10.13 Refer to the Ohm's law data given in Exercise 10.11. Some computer packages have an option for doing regressions in which the intercept is set in advance to 0. In SAS this is the NOINT option. If you have access to such software, run the analysis with this option and report the estimate of R. The output should also include an estimated standard error for $1/R$. Use this to calculate the 95% confidence interval for R. Note that with this option the degrees of freedom for t^* will be 1 greater than for the model with the intercept.

10.14 Answer the questions posed in the previous exercise for the data given in Exercise 10.12.

10.15 The human body takes in more oxygen when exercising than when it is at rest, and to deliver the oxygen to the muscles, the heart must beat faster. Heart rate is easy to measure but the measurement of oxygen uptake requires elaborate equipment. If oxygen uptake (VO2) can be accurately predicted from heart rate (HR) under a particular set of exercise conditions, then predicted, rather than measured, values can be used for various research purposes. Unfortunately, not all human bodies are the same, and a single equation that works for all people cannot be found. But individuals can be measured for both variables under varying sets of exercise conditions for a short time, and regression equations for predicting oxygen uptake from heart rate can be calculated for each person. The predicted oxygen uptakes can then be used in place of measured uptakes for the same individuals in later experiments. The data for one individual are as follows. (Data provided by Paul Waldsmith from experiments conducted in Don Corrigan's lab at Purdue University, West Lafayette, Ind.)

HR	94	96	95	95	94	95	94	104	104	106
VO2	.473	.753	.929	.939	.832	.983	1.049	1.178	1.176	1.292

HR	108	110	113	113	118	115	121	127	131
VO2	1.403	1.499	1.529	1.599	1.749	1.746	1.897	2.040	2.231

(a) Plot the data. Are there any outliers or unusual points?

(b) Compute the least-squares regression line for predicting oxygen uptake from heart rate for this individual.

(c) Test the null hypothesis that the slope of the regression line is 0.

(d) Since the regression is to be used for prediction, we calculate prediction intervals for future observations. Calculate 95% intervals for heart rates of 95 and 110.

(e) From what you have learned in (a), (b), (c), and (d) of this exercise, do you think that the researchers should use predicted VO2 in place of measured VO2 for this individual under similar experimental conditions? Explain your answer.

10.16 Refer to the previous exercise. These data were collected as part of a study designed to assess the effects of wearing different protective masks on worker productivity. In the previous example, the individual was wearing a half-mask; when the data below were collected, he was wearing no mask.

HR	103	112	113	110	111	112	115	120	118	126
VO2	.768	.953	1.090	1.167	1.025	1.130	1.173	1.296	1.358	1.507

HR	123	128	132	136	137	138	140	145	147
VO2	1.565	1.740	1.807	1.891	1.958	2.023	2.148	2.395	2.507

Using the questions given in the previous exercise as a guide, analyze these data.

10.17 The data given in Exercise 10.15 were taken while the individual performed progressively more vigorous exercise. Therefore, the observations are ordered by time (each observation corresponds to a consecutive 1-minute period). The statistical model for inference in linear regression assumes that the errors are independent and all have the same variance. Plot the residuals versus the order in which the data were taken. Do you see any patterns that would lead you to question the validity of the regression methodology that we have used in Exercise 10.15?

10.18 Answer the questions given in the previous exercise for the data on the individual with no mask given in Exercise 10.16.

10.19 Premature infants are often kept in intensive care nurseries after they are born, and it is common practice to measure their blood pressure frequently. The noninvasive oscillometric method is one technique that is relatively easy to use and has been automated. The traditional procedure, called the direct intra-arterial method, is believed to be more accurate but is much more difficult to perform. Several studies have reported that the correlation between the two methods is high, ranging from 0.49 to 0.98, and that the results are statistically significant. For one study designed to investigate the relation between the two methods, the regression equation $\hat{y} = 15 + 0.83x$ was reported. Here x represents the easy method and y represents the difficult one. (From John A. Wareham et al., "Prediction of arterial blood pressure on the premature neonate using the oscillometric method," *American Journal of Diseases of Children*, 141 (1987), pp. 1108–1110.)

(a) The standard error of the slope is 0.065 and the sample size is 81. Calculate the t statistic for testing $H_0: \beta_1 = 0$. Specify an appropriate alternative hypothesis for this problem and give an approximate P-value for the test.

(b) Give a 99% confidence interval for β_1.

(c) Do you conclude that the easy method should be used in place of the more difficult one for measuring the blood pressures of premature infants? The authors of the report on this study calculated and plotted prediction intervals. They found the widths to be unacceptably large and concluded that statistical significance does not imply that results are clinically useful.

10.20 Mechanisms for soil aeration and soil water evaporation involve the exchange of gases between the soil and the atmosphere. To investigate the effect of the air flow above the soil on this process, a series of experiments was carried out. In one such experiment, the speed of the air x was varied and the rate of evaporation y was measured. The fitted regression equation based on 18 observations was $\hat{y} = 5.0 + 0.00665x$. The standard error of the slope was given as 0.00182.

(a) Calculate the t statistic for testing $H_0: \beta_1 = 0$. It is reasonable to assume that greater air flow would cause more evaporation. Formulate an appropriate alternative hypothesis for this problem. Find an approximate P-value for the significance test.

(b) Construct a 95% confidence interval for β_1.

10.21 Refer to the experiment on Ohm's law given in Exercise 10.11.

(a) Compute all values for the ANOVA table.

(b) State the null hypothesis associated with the F statistic.

(c) What is the distribution of the F statistic when H_0 is true? Find an approximate P-value for the test of H_0.

10.22 A second set of data for an Ohm's law experiment is given in Exercise 10.12. Answer the questions given in the previous exercise for this set of data.

10.23 Refer to the oxygen uptake and heart rate data given in Exercise 10.15.

(a) Construct the ANOVA table.

(b) What null hypothesis is tested by the F statistic?

(c) Give the degrees of freedom for the F statistic and an approximate P-value for the test of H_0.

(d) Verify that the square of the t statistic that you calculated in Exercise 10.15 is equal to the F statistic in your ANOVA table. Any difference found is due to round-off error.

(e) What proportion of the variation in oxygen uptake is explained by heart rate for this set of data?

10.24 A second set of oxygen uptake and heart rate data is given in Exercise 10.16. Answer the questions given in the previous exercise for this set of data.

10.25 In a study conducted in the Egyptian village of Kalama, the relationship between the birth weights of infants and various socioeconomic variables was examined. The sample size was 40. (This study is reported in M. El-Kholy, F. Shaheen, and W. Mahmoud, "Relationship between socioeconomic status and birth weight, a field study in a rural community in Egypt," *Journal of the Egyptian Public Health Association*, 61 (1986), pp. 349–358.)

(a) The correlation between monthly income and birth weight was reported to be 0.39. Calculate the t statistic for testing the null hypothesis that the population correlation is 0.

   (b) The researchers expected that higher birth weights would be
         associated with higher incomes. Express this idea in terms of an
         alternative hypothesis for this correlation.

   (c) Determine a *P*-value for H_0 versus the alternative that you
         specified in (b).

10.26 Chinese third grade students from public schools in Hong Kong
         were the subjects of a study designed to investigate the relationship
         between various measures of parental behavior and other variables.
         The sample size was 713. Information was obtained from
         questionnaires administered to the students. One of the variables
         examined was parental control—an indication of the amount of
         control that the parents exercised in relation to the behavior of
         the students. Another was the self-esteem of the students. (This study
         is reported in S. Lau and P. C. Cheung, "Relations between Chinese
         adolescents' perception of parental control and organization and
         their perception of parental warmth," *Developmental Psychology,*
         23 (1987), pp. 726–729.)

   (a) The correlation between these two variables was reported to be
         -0.19. Calculate the *t* statistic for testing the null hypothesis
         that the population correlation is 0.

   (b) Find an approximate *P*-value for testing H_0 versus the two-sided
         alternative.

10.2   MULTIPLE LINEAR REGRESSION*

The methods presented in the first section of this chapter are useful for
examining a linear relationship between a *single* explanatory variable *x* and
a response variable *y*. With multiple linear regression we use *several* explanatory variables to explain or predict a single response variable. Many of
the ideas that we encountered in our study of simple linear regression carry
over to this more general case. However, the introduction of several explanatory variables leads to many additional considerations. In this short
section we cannot explore all of these issues. Rather, we will discuss some
general ideas and illustrate them with a case study.

Statistical Model for Multiple Linear Regression

For simple linear regression we assume that the subpopulation means of
the response variable *y* depend upon the explanatory variable *x* according

---

* This section is optional.

to the equation

$$\mu_{y(x)} = \beta_0 + \beta_1 x$$

For multiple regression we will once again use y for the response variable. The several explanatory x variables will be denoted by $x_1, x_2, \ldots, x_p$.

With this notation we describe the model for the subpopulation means of y as

$$\mu_{y(x_1, x_2, \ldots, x_p)} = \beta_0 + \beta_1 x_1 + \beta_2 x_2 + \cdots + \beta_p x_p$$

For simple linear regression the subpopulations correspond to different values of the explanatory variable x. We now think of subpopulations corresponding to a particular set of values for *all* of the explanatory variables $x_1, x_2, \ldots, x_p$.

EXAMPLE 10.15

As part of a study at a large university,[4] data were collected on all freshmen computer science majors in a particular year. Of particular interest was the cumulative grade point average (GPA) after three semesters. Other variables were recorded at the time the students enrolled in the university. These included average high school grades in mathematics (HSM), science (HSS), and English (HSE).

We use high school grades to predict the response variable GPA. Thus there are $p = 3$ explanatory variables: $x_1 = $ HSM, $x_2 = $ HSS, and $x_3 = $ HSE. The high school grades were coded on a scale from 1 to 10, with 10 corresponding to A, 9 to A$-$, 8 to B$+$, etc. These grades define the subpopulations. For example, the straight-C students are the subpopulation defined by HSM $= 4$, HSS $= 4$, and HSE $= 4$.

The model for the subpopulation means is

$$\mu_{\text{GPA(HSM,HSS,HSE)}} = \beta_0 + \beta_1 \text{HSM} + \beta_2 \text{HSS} + \beta_3 \text{HSE}$$

For the straight-C subpopulation of students, the model gives the subpopulation mean as

$$\mu_{\text{GPA(4,4,4)}} = \beta_0 + 4\beta_1 + 4\beta_2 + 4\beta_3$$ ■

The subpopulation means describe the FIT part of our statistical model. The RESIDUAL part represents the variation of observations about the means. We will use the same assumptions for the RESIDUAL that we used for simple linear regression. The symbol ϵ represents the deviation of an individual observation from its subpopulation mean. We assume that these deviations are normally distributed with mean 0 and some unknown standard deviation σ.

Since there are several explanatory variables, the notation needed to describe the data is a little more complicated. Each observation or case consists of a value for the response variable and for each of the explanatory variables.

We use x_{ij} to represent the value of the jth explanatory variable for the ith case. Thus, the data are

$$(x_{11}, x_{12}, \ldots, x_{1p}, y_1)$$
$$(x_{21}, x_{22}, \ldots, x_{2p}, y_2)$$
$$\vdots$$
$$(x_{n1}, x_{n2}, \ldots, x_{np}, y_n)$$

Here, n is the number of cases and p is the number of explanatory variables. Data are often entered into computer regression programs in this format. Each row is a case and each column corresponds to a different variable. Appendix A at the end of this section gives the data for Example 10.15 in this format with Scholastic Aptitude Test (SAT) scores and sex as additional explanatory variables.

Multiple linear regression model

> The statistical model for a multiple linear regression is given by
>
> $$y_i = \beta_0 + \beta_1 x_{i1} + \beta_2 x_{i2} + \cdots + \beta_p x_{ip} + \epsilon_i$$
>
> where $i = 1, 2, \ldots, n$.
>   The ϵ_i are assumed to be independent and normally distributed with mean 0 and standard deviation σ. In other words, they are an SRS from the $N(0, \sigma)$ distribution.
>   The parameters of the model are $\beta_0, \beta_1, \beta_2, \ldots, \beta_p$ and σ.

The assumption that the subpopulation means are related to the β parameters of the model by the equation

$$\mu_{y(x_1, x_2, \ldots, x_p)} = \beta_0 + \beta_1 x_1 + \beta_2 x_2 + \cdots + \beta_p x_p$$

implies that we can estimate all subpopulation means from estimates of the β's. To the extent that this equation is accurate, we have a very useful tool for answering many interesting questions.

Estimation, Confidence Intervals, and Hypothesis Testing

For simple linear regression we used the principle of least squares described in Section 3.2 to obtain estimates of the intercept and slope of the regression line. For multiple regression the principle is the same but the details are much more complicated.

We let

$$b_0, b_1, b_2, \ldots, b_p$$

denote the estimates of the parameters

$$\beta_0, \beta_1, \beta_2, \ldots, \beta_p$$

For the ith observation the predicted value is

$$\hat{y}_i = b_0 + b_1 x_{i1} + b_2 x_{i2} + \cdots + b_p x_{ip}$$

The residual is therefore

$$\begin{aligned} e_i &= y_i - \hat{y}_i \\ &= y_i - b_0 - b_1 x_{i1} - b_2 x_{i2} - \cdots - b_p x_{ip} \end{aligned}$$

The method of least squares chooses the values of the b's that make the sum of the squares of the residuals as small as possible. In other words, the parameter estimates $b_0, b_1, b_2, \ldots, b_p$ minimize the quantity

$$\sum (y_i - b_0 - b_1 x_{i1} - b_2 x_{i2} - \cdots - b_p x_{ip})^2$$

We will not give the formulas for the least-squares estimates here. Rather, we will be content to understand the principle on which they are based (i.e., least squares) and to let a computer package do the computations.

The parameter σ^2 is estimated from the sample residuals using the formula

$$\begin{aligned} s^2 &= \frac{\sum e_i^2}{n - p - 1} \\ &= \frac{\sum (y_i - \hat{y}_i)^2}{n - p - 1} \end{aligned}$$

Note that when $p = 1$, the denominator is $n - 2$, the value given for simple linear regression. The quantity $n - p - 1$ is the degrees of freedom associated with s^2. To estimate σ we use

$$s = \sqrt{s^2}$$

Confidence intervals and significance tests for each of the regression coefficients are constructed in a manner similar to that described for simple linear regression. The formulas for the standard errors of the b's are complicated, and we rely on computer software for the calculations. Let $s(b_j)$ denote the standard error of b_j.

Confidence intervals and hypothesis tests for β_j

A level $1 - \alpha$ confidence interval for β_j is

$$b_j \pm t^* s(b_j)$$

Here t^* is the upper $\alpha/2$ critical value for the $t(n - p - 1)$ distribution. To test the hypothesis H_0: $\beta_j = 0$, compute

$$t = \frac{b_j}{s(b_j)}$$

In terms of a random variable T having the $t(n - p - 1)$ distribution, the *P*-value for a test of H_0 against

$$H_a: \beta_j > 0 \quad \text{is} \quad P(T \geq t)$$
$$H_a: \beta_j < 0 \quad \text{is} \quad P(T \leq t)$$
$$H_a: \beta_j \neq 0 \quad \text{is} \quad P(|T| \geq |t|)$$

Confidence intervals for a mean response and prediction intervals for a future observation can be computed for multiple regression models. Once again, the basic ideas are the same as in the simple linear regression case. Computer software that will perform these calculations for simple linear regression will generally do the same for multiple regression. The commands used will usually be the same. The only difference is that we specify a list of explanatory variables in place of a single variable. It is an interesting commentary on modern computer technology that we can easily perform these rather complex calculations without an intimate knowledge of all of the computational details. This frees us to concentrate on the meaning and appropriate use of the results.

In simple linear regression the *F* test from the ANOVA table is equivalent to the two-sided test of the hypothesis that the slope of the regression line is 0. For multiple regression there is a corresponding *F* test that is used to test the hypothesis that *all* of the regression coefficients (with the exception of the intercept) are 0. Here is the general form of the ANOVA table for multiple regression.

Source	Sum of squares	Degrees of freedom	Mean square	F
Model	$\sum(\hat{y}_i - \bar{y})^2$	p	SSM/DFM	MSM/MSE
Error	$\sum(y_i - \hat{y}_i)^2$	$n - p - 1$	SSE/DFE	
Total	$\sum(y_i - \bar{y})^2$	$n - 1$	SST/DFT	

This table is very similar to the one that we encountered for simple linear regression. The degrees of freedom for the model have changed from 1 to p to reflect the fact that we now have p explanatory variables in place of just one. As a consequence, the degrees of freedom for error have been decreased by a corresponding amount. The estimate of the variance σ^2 for our model is given by the MSE in the ANOVA table.

Sums of squares, degrees of freedom, and mean squares

Sums of squares represent variation present in the data. They are calculated by summing squared deviations. In multiple regression there are three sources of variation: model, error, and total.

The sums of squares are related by the formula

$$SST = SSM + SSE$$

Thus, the total variation is composed of two parts, one due to the model and one due to error.

Degrees of freedom are associated with each sum of squares. They are related in the same way:

$$DFT = DFM + DFE$$

To calculate mean squares, use the formula,

$$MS = \frac{\text{sum of squares}}{\text{degrees of freedom}}$$

The ratio MSM/MSE is an F statistic that is used to test

$$H_0: \beta_1 = \beta_2 = \cdots = \beta_p = 0$$

versus

$$H_a: \beta_j \neq 0 \text{ for at least one } j = 1, 2, \ldots, p$$

The null hypothesis means that none of the explanatory variables are predictors of the response variable when used in the form expressed by the multiple regression equation. The alternative states that at least one of them is linearly related to the response. As in simple linear regression, large values of F give evidence against H_0. When H_0 is true, F has the $F(p, n - p - 1)$ distribution. Note that the degrees of freedom for the F distribution are those associated with the model and error.

Analysis of variance *F* test

> In the multiple regression model, the hypothesis
>
> $$H_0: \beta_1 = \beta_2 = \cdots = \beta_p = 0$$
>
> is tested by the *F* statistic
>
> $$F = \frac{\text{MSM}}{\text{MSE}}$$
>
> The *P*-value is the probability that a random variable having the $F(p, n - p - 1)$ distribution is greater than or equal to the calculated value of the *F* statistic.

For simple linear regression we noted that the square of the sample correlation could be written as the ratio of SSM to SST and could be interpreted as the proportion of variation in *y* explained by *x*. A similar statistic is routinely calculated for multiple regression.

The squared multiple correlation

> The statistic
>
> $$R^2 = \frac{\text{SSM}}{\text{SST}} = \frac{\sum(\hat{y}_i - \bar{y})^2}{\sum(y_i - \bar{y})^2}$$
>
> is the proportion of the variation of the response variable *y* that is explained by the explanatory variables $x_1, x_2, \ldots, x_p$ in a multiple linear regression.

Often, R^2 is multiplied by 100 and expressed as a percent. The square root of R^2 is the correlation between the observations y_i and the predicted values $\hat{y}_i$.

A Case Study

In this section we illustrate multiple regression by analyzing the data from the study described in Example 10.15. The response variable is the cumulative GPA after three semesters for a group of computer science majors at a large university. The explanatory variables previously mentioned are average high school grades, represented by HSM, HSS, and HSE. We also examine the SAT mathematics and SAT verbal scores as explanatory variables. The data for the $n = 224$ students in the study are listed in Appendix A of this chapter.

VARIABLE	N	MEAN	STANDARD DEVIATION	VARIANCE	MINIMUM VALUE	MAXIMUM VALUE
GPA	224	4.6352232	0.77939493	0.607456	2.1200000	6.000000
SATM	224	595.2857143	86.40144374	7465.209481	300.0000000	800.000000
SATV	224	504.5491071	92.61045907	8576.697129	285.0000000	760.000000
HSM	224	8.3214286	1.63873672	2.685458	2.0000000	10.000000
HSS	224	8.0892857	1.69966270	2.888853	3.0000000	10.000000
HSE	224	8.0937500	1.50787358	2.273683	3.0000000	10.000000

Figure 10.18 Descriptive statistics for case study.

Preliminary analysis   The first step in the analysis is to carefully examine each of the variables. Means, standard deviations, and minimum and maximum values are given by the SAS procedure MEANS and are displayed in Figure 10.18. The minimum value for the SAT mathematics (SAT-M) variable appears to be rather extreme; it is $(595 - 300)/86 = 3.43$ standard deviations below the mean. We do not discard this case at this time but will take care in our subsequent analyses to see if it has an excessive influence on our result.

The mean for the SAT mathematics score is higher than the mean for the verbal score (SAT-V) as we might expect for a group of computer science majors. The two standard deviations are about the same. The high school grade variables have similar means, with the mathematics grades being a bit higher. The standard deviations for the three high school grade variables are very close to each other. We also note that the mean GPA is 4.635, with a standard deviation of 0.779.

Since the variables GPA, SAT-M, and SAT-V have many possible values, we could use stemplots, histograms, and boxplots to examine the shape of their distributions. Normal quantile plots indicate whether or not the distributions look normal. It is very important to note that the multiple regression model *does not* require any of these distributions to be normal. Only the residuals are assumed to be normal. The purpose of examining these plots is to understand something about each variable alone before attempting to use it in a complicated model. Extreme values for any variable should be noted and checked for accuracy. If found to be correct, the cases with these values should be carefully examined to see if they are truly exceptional and perhaps do not belong in the same analysis with the other cases. When these data were examined in this way, no obvious problems were evident.

The high school grade variables HSM, HSS, and HSE have relatively few values and are best summarized by giving the relative frequencies for each possible value. The output of the SAS procedure FREQ, given in Figure 10.19, provides these summaries. Note that the distributions are all skewed,

HSM	FREQUENCY	PERCENT	CUMULATIVE FREQUENCY	CUMULATIVE PERCENT
2	1	0.4	1	0.4
3	1	0.4	2	0.9
4	4	1.8	6	2.7
5	6	2.7	12	5.4
6	23	10.3	35	15.6
7	28	12.5	63	28.1
8	36	16.1	99	44.2
9	59	26.3	158	70.5
10	66	29.5	224	100.0

HSS	FREQUENCY	PERCENT	CUMULATIVE FREQUENCY	CUMULATIVE PERCENT
3	1	0.4	1	0.4
4	7	3.1	8	3.6
5	9	4.0	17	7.6
6	24	10.7	41	18.3
7	42	18.8	83	37.1
8	31	13.8	114	50.9
9	50	22.3	164	73.2
10	60	26.8	224	100.0

HSE	FREQUENCY	PERCENT	CUMULATIVE FREQUENCY	CUMULATIVE PERCENT
3	1	0.4	1	0.4
4	4	1.8	5	2.2
5	5	2.2	10	4.5
6	23	10.3	33	14.7
7	43	19.2	76	33.9
8	49	21.9	125	55.8
9	52	23.2	177	79.0
10	47	21.0	224	100.0

Figure 10.19 Frequencies for high school grade variables.

with a large proportion of high grades ($10 = $ A and $9 = $ A$-$). Again we emphasize the fact that these distributions need not be normal.

Relationships between pairs of variables The second step in our analysis is to examine the relationships between all pairs of variables. One way to do this is to calculate the correlations. These are provided by the SAS procedure CORR and are displayed in Figure 10.20. Note that the output gives the P-value for the test of the null hypothesis that the correlation is 0 versus the two-sided alternative for each pair. Thus, we see that the correlation between GPA and HSM is 0.44, with a P-value of 0.0001, whereas the correlation between GPA and SAT-V is 0.11, with a P-value of 0.087. The first is statistically significant by any reasonable standard, while the second is rather marginal.

The high school grades all have higher correlations with GPA than the SAT scores. As we might expect, the mathematics grades (0.44) have the highest correlation, followed by the science grades (0.33) and then the English grades (0.29). The two SAT scores have a rather high correlation with each other (0.46), while the high school grades also correlate well with each other (0.45 to 0.58). The SAT mathematics score correlates well with HSM

```
PEARSON CORRELATION COFFICIENTS / PROB > |R| UNDER H0: RHO=0 / N = 224

              GPA        SATM       SATV        HSM        HSS        HSE

GPA        1.00000    0.25171    0.11449    0.43650    0.32943    0.28900
           0.0000     0.0001     0.0873     0.0001     0.0001     0.0001

SATM       0.25171    1.00000    0.46394    0.45351    0.24048    0.10828
           0.0001     0.0000     0.0001     0.0001     0.0003     0.1060

SATV       0.11449    0.46394    1.00000    0.22112    0.26170    0.24371
           0.0873     0.0001     0.0000     0.0009     0.0001     0.0002

HSM        0.43650    0.45351    0.22112    1.00000    0.57569    0.44689
           0.0001     0.0001     0.0009     0.0000     0.0001     0.0001

HSS        0.32943    0.24048    0.26170    0.57569    1.00000    0.57937
           0.0001     0.0003     0.0001     0.0001     0.0000     0.0001

HSE        0.28900    0.10828    0.24371    0.44689    0.57937    1.00000
           0.0001     0.1060     0.0002     0.0001     0.0001     0.0000
```

Figure 10.20 Correlations for case study variables.

(0.45), less well with HSS (0.24), and rather poorly with HSE (0.11). The correlations of the SAT verbal score with the three high school grades are about equal, ranging from 0.22 to 0.26.

It is important to keep in mind that by examining pairs of variables we are seeking a better understanding of the data. The fact that the correlation of a particular explanatory variable with the response variable does not achieve statistical significance does not *necessarily* imply that it will not be a useful (and significant) predictor in a multiple regression.

Numerical summaries such as correlations are useful, but plots are generally more informative when seeking to understand real data. Plots tell us whether or not the numerical summary gives a fair representation of the data. For a multiple regression, each pair of variables should be plotted. For the six variables in our case study, this means that we should examine 15 plots. In general there are $p + 1$ variables in a multiple regression analysis with p explanatory variables, so that $p(p + 1)/2$ plots are required. Multiple regression is a complicated procedure. If we do not do the necessary preliminary work, we are in serious danger of producing useless or misleading results. We leave the task of making these plots to the reader as an exercise.

Regression on high school grades   To explore the relationship between the explanatory variables and our response variable GPA, we will run several multiple regressions. The explanatory variables fall into two classes. High school grades are represented by HSM, HSS, and HSE, and standardized tests represented by the SAT scores. We start our analysis by using the high school grades to predict GPA. The output generated by the SAS procedure REG for this problem is given in Figure 10.21.

The output contains an ANOVA table, some additional descriptive statistics, and information about the parameter estimates. When examining any ANOVA table, it is always a good idea to first check the degrees of freedom. This assures that we have not made some serious error in specifying the model for the program or in entering the data. Since there are $n = 224$ cases, we have DFT $= n - 1 = 223$. The three explanatory variables give DFM $= p = 3$, and DFE $= n - p - 1 = 224 - 3 - 1 = 220$.

The F statistic is 18.86, with a P-value of 0.0001. Under the null hypothesis

$$H_0: \beta_1 = \beta_2 = \beta_3 = 0$$

the F statistic has an $F(3, 220)$ distribution. According to this distribution, the chance of obtaining an F statistic of 18.86 or larger is 0.0001. We therefore conclude that at least one of the three regression coefficients for the high school grades is different from 0.

In the descriptive statistics that follow the ANOVA table we find that ROOT MSE is 0.6998. This value is the square root of the MSE given in the ANOVA table and is the estimate of the parameter σ of our model. The value of R^2 is given as 0.20. Thus, 20% of the variation in the GPA scores is explained by the high school grades.

```
DEP VARIABLE: GPA
                        ANALYSIS OF VARIANCE

                        SUM OF           MEAN
    SOURCE      DF      SQUARES         SQUARE      F VALUE    PROB>F

    MODEL       3     27.71233132     9.23744377    18.861     0.0001
    ERROR     220      107.75046      0.48977481
    C TOTAL   223      135.46279

        ROOT MSE         0.6998391     R-SQUARE     0.2046
        DEP MEAN         4.635223      ADJ R-SQ     0.1937
        C.V.            15.09828

                        PARAMETER ESTIMATES

                  PARAMETER        STANDARD        T FOR H0:
    VARIABLE      ESTIMATE          ERROR        PARAMETER=0    PROB > |T|

    INTERCEP     2.58987662       0.29424324        8.802        0.0001
    HSM          0.16856664       0.03549214        4.749        0.0001
    HSS          0.03431557       0.03755888        0.914        0.3619
    HSE          0.04510182       0.03869585        1.166        0.2451
```

Figure 10.21 Multiple regression output for regression using high school grades to predict GPA.

From the **PARAMETER ESTIMATES** section we obtain the fitted regression equation

$$\widehat{\text{GPA}} = 2.590 + .169\text{HSM} + .034\text{HSS} + .045\text{HSE}$$

Recall that the t statistics for testing each regression coefficient are obtained by dividing the estimates by their standard errors. Thus, for the coefficient of HSM we obtain the t-value given in the output by calculating

$$t = \frac{.16856664}{.03549214} = 4.749$$

The P-values are given in the last column. Thus for HSM we have a P-value of 0.0001, and we conclude that the regression coefficient for this explanatory variable is significantly different from 0.

The P-values for the other explanatory variables (0.36 for HSS and 0.25 for HSE) do not achieve statistical significance. This result seems to contradict the impression that we obtained by examining the correlations in

Figure 10.20. In that display we see that the correlation between GPA and HSS is 0.33 and the correlation between GPA and HSE is 0.29. The *P*-values for both of these correlations are 0.0001. In other words, if we used HSS alone in a regression to predict GPA or if we used HSE alone, we would obtain statistically significant regression coefficients. This sort of phenomenon is not unusual in multiple regression analysis. Part of the explanation is contained in the correlations between HSM and the other two explanatory variables. These are rather high (at least compared with the other correlations in Figure 10.20). The correlation between HSM and HSS is 0.58, and that between HSM and HSE is 0.45. Thus, when we have a regression model that contains all three high school grades as explanatory variables, there is considerable overlap of the predictive information contained in these variables. *The significance tests calculated for individual regression coefficients assess the significance of each predictor variable assuming that all other predictors are included in the regression equation.* Given that we used a model with HSM and HSS as predictors, the coefficient of HSE is not statistically significant. Similarly, given that we have HSM and HSE in the model, HSS does not have a significant regression coefficient.

Unfortunately, we cannot conclude from this analysis that the *pair* of explanatory variables HSS and HSE contribute nothing significant to our model for predicting GPA. Correlations among explanatory variables in a multiple regression can cause many complicated problems.

As in simple linear regression we should always examine the residuals. Because there are several explanatory variables in multiple regression, there are several plots to examine. It is usual to plot the residuals versus predicted values and versus each of the explanatory variables. Look for outliers, influential observations, evidence for a curved, rather than linear, relation, and anything else unusual. We leave the task of making these plots to the reader as an exercise. The plots in this case all appear to show more or less random noise around the center value of 0.

A normal quantile plot of the residuals is given in Figure 10.22. The distribution of the residuals appears to be approximately normal. It should be noted that there are many other plots providing a great deal of information that can aid us in performing a multiple regression. Discussion of these, however, is more than we can undertake in this chapter.

Since the variable HSS has the largest *P*-value of the three explanatory variables and therefore appears to contribute the least to our explanation of GPA, we rerun the regression using only HSM and HSE as explanatory variables. The SAS output is given in Figure 10.23. The *F* statistic indicates that we can reject the null hypothesis that the regression coefficients for the two explanatory variables are both 0. The *P*-value is still 0.0001. The value of R^2 has dropped a very small amount compared with our previous run—from 0.2046 to 0.2016. The *t* statistics for the individual regression coefficients indicate that HSM is still clearly significant ($P = 0.0001$) while the coefficient of HSE is larger than before (1.747 versus 1.166) and approaches the traditional 0.05 level of significance ($P = 0.082$).

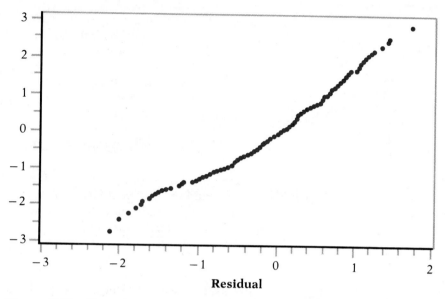

Figure 10.22 Normal quantile plot for residuals from high school grade model.

```
DEP VARIABLE: GPA
                        ANALYSIS OF VARIANCE

                      SUM OF         MEAN
   SOURCE      DF     SQUARES       SQUARE     F VALUE    PROB>F

   MODEL        2   27.30349117   13.65174558   27.894    0.0001
   ERROR      221   108.15930      0.48940859
   C TOTAL    223   135.46279


        ROOT MSE      0.6995774     R-SQUARE    0.2016
        DEP MEAN      4.635223      ADJ R-SQ    0.1943
        C.V.         15.09264


                    PARAMETER ESTIMATES

              PARAMETER      STANDARD      T FOR HO:
   VARIABLE    ESTIMATE       ERROR      PARAMETER=0    PROB > |T|

   INTERCEP   2.62422848    0.29172204      8.996        0.0001
   HSM        0.18265442    0.03195581      5.716        0.0001
   HSE        0.06067015    0.03472914      1.747        0.0820
```

Figure 10.23 Multiple regression output for regression using HSM and HSE to predict GPA.

Comparison of the fitted equations for the two multiple regression analyses tells us something more about the intricacies of this procedure. For the first run we have

$$\widehat{GPA} = 2.590 + .169HSM + .034HSS + .045HSE$$

while the second gives us

$$\widehat{GPA} = 2.624 + .183HSM + .061HSE$$

By eliminating HSS from the model, the regression coefficients for *all* of the remaining variables and the intercept have changed. This is a phenomenon that occurs quite generally with multiple regression. *Individual regression coefficients and their standard errors and significance tests are meaningful only when interpreted in the context of the other explanatory variables in the model.*

Regression on SAT scores   We now turn to the problem of predicting GPA using the two SAT scores. The output from the SAS procedure REG is given in Figure 10.24. The degrees of freedom are as expected: 2, 221, and 223. The F statistic is 7.476, with a P-value of 0.0007. We conclude that the regression coefficients for SAT-M and SAT-V are not both 0. Recall that we obtained a P-value of 0.0001 when we used the high school grades to predict GPA. We could state that both multiple regression equations are *highly* significant. This view obscures the fact that the results for the two models are quite different. For the SAT regression, $R^2 = 0.0634$, while for the high school grades model (see Figure 10.21) we have $R^2 = 0.2046$, a value about three times as large. Stating that we have a statistically significant result is quite different from saying that an effect is large or important.

Further examination of the output in Figure 10.24 reveals that the coefficient of SAT-M is significant ($t = 3.44$, $P = 0.0007$), and that for SAT-V is not ($t = -0.04$, $P = 0.9684$). For a complete analysis we should carefully examine the residuals. Also, we might want to run the analysis with SAT-M as the only explanatory variable.

Regression using all variables   So far we have learned that the high school grades and the SAT scores give highly significant regression equations. The mathematics component of each of these groups of explanatory variables appears to be the key predictor. Comparing the values of R^2 for the two models indicates that the high school grade variables are better predictors than the SAT scores. We still do not know whether or not we can get a better prediction equation using all of the explanatory variables together in one multiple regression.

To address this question we run the regression with all five explanatory variables. The output is given in Figure 10.25. The F statistic is 11.69, with a P-value of 0.0001. We conclude that at least one of our explanatory vari-

DEP VARIABLE: GPA

ANALYSIS OF VARIANCE

SOURCE	DF	SUM OF SQUARES	MEAN SQUARE	F VALUE	PROB>F
MODEL	2	8.58383910	4.29191955	7.476	0.0007
ERROR	221	126.87895	0.57411289		
C TOTAL	223	135.46279			

ROOT MSE	0.7577024	R-SQUARE	0.0634	
DEP MEAN	4.635223	ADJ R-SQ	0.0549	
C.V.	16.34662			

PARAMETER ESTIMATES

VARIABLE	PARAMETER ESTIMATE	STANDARD ERROR	T FOR H0: PARAMETER=0	PROB > |T|
INTERCEP	3.28867737	0.37603684	8.746	0.0001
SATM	0.002282834	0.000662914	3.444	0.0007
SATV	-0.000024562	0.000618470	-0.040	0.9684

Figure 10.24   Multiple regression output for regression using SAT scores to predict GPA.

ables has a nonzero regression coefficient. This result is not particularly surprising given that we have already seen that HSM and SAT-M are strong predictors of GPA. The value of R^2 is 0.2115, not much higher than the value of 0.2046 that we found for the high school grades equation.

Examination of the t statistics and the associated P-values for the individual regression coefficients reveals that HSM is the only one that is significant ($P = 0.0003$). Once again it is important to be aware of the fact that this result alone is not sufficient grounds to conclude that the regression coefficients for the four other explanatory variables are *all* 0.

Many statistical software packages provide the capability for testing whether a collection of regression coefficients in a multiple regression model are *all* 0. We use this approach to address two very interesting questions for this set of data. In the previous section of this chapter we have not discussed this procedure but the basic idea is quite simple.

In the context of the multiple regression model with all five predictors, we ask first whether or not the coefficients for the two SAT scores are both 0. In other words, do the SAT scores add any significant predictive information to that already contained in the high school grades? To be fair, we also

```
DEP VARIABLE: GPA
                         ANALYSIS OF VARIANCE

                         SUM OF          MEAN
          SOURCE    DF    SQUARES        SQUARE      F VALUE    PROB>F

          MODEL     5    28.64364489   5.72872898   11.691    0.0001
          ERROR    218   106.81914     0.48999607
          C TOTAL  223   135.46279


              ROOT MSE     0.6999972    R-SQUARE     0.2115
              DEP MEAN     4.635223     ADJ R-SQ     0.1934
              C.V.        15.10169


                         PARAMETER ESTIMATES

                     PARAMETER       STANDARD      T FOR HO:
          VARIABLE    ESTIMATE        ERROR      PARAMETER=0    PROB > |T|

          INTERCEP    2.32671874    0.39999643       5.817      0.0001
          SATM        0.000943593   0.000685657      1.376      0.1702
          SATV       -0.000407850   0.000591893     -0.689      0.4915
          HSM         0.14596108    0.03926097       3.718      0.0003
          HSS         0.03590532    0.03779841       0.950      0.3432
          HSE         0.05529258    0.03956869       1.397      0.1637


    TEST: SAT    NUMERATOR:     0.465657   DF:    2   F VALUE:    0.9503
                 DENOMINATOR:   0.489996   DF:  218   PROB >F:    0.3882


    TEST: HS     NUMERATOR:     6.6866     DF:    3   F VALUE:   13.6462
                 DENOMINATOR:   0.489996   DF:  218   PROB >F:    0.0001
```

Figure 10.25   Multiple regression output for regression using all variables to predict GPA.

ask the complementary question—do the high school grades add any significant predictive information to that already contained in the SAT scores?

The answers are given in the last two sections of the output in Figure 10.25. For the first test we see an *F* statistic of 0.9503. Under the null hypothesis that the two SAT coefficients are 0, this statistic has an $F(2, 218)$ distribution and the *P*-value is 0.95. We conclude that the SAT scores are not significant predictors of GPA in a regression that already contains the high school scores as predictor variables. Recall that the model with just SAT scores has a highly significant *F* statistic. This result means that whatever predictive information is in the SAT scores can also be found in the high school grades. In this sense, the SAT scores are superfluous.

The test for the three high school grade variables has $F = 13.6462$. Under the null hypothesis that these three regression coefficients are 0, the statistic has an $F(3, 218)$ distribution and the P-value is 0.0001. We conclude that the high school grades contain useful information for predicting GPA that is not contained in the SAT scores.

Of course, our statistical analysis of these data does not lead us to conclude that SAT scores are less useful than high school grades for predicting college grades for all groups of students. We have studied a select group of students—computer science majors—from a specific university. Generalizations to other situations based on these data alone are beyond the scope of inferential statistics.

SUMMARY

The statistical model for **multiple linear regression** with one response variable y and p explanatory variables $x_1, x_2, \ldots, x_p$ is

$$y_i = \beta_0 + \beta_1 x_{i1} + \beta_2 x_{i2} + \cdots + \beta_p x_{ip} + \epsilon_i$$

where $i = 1, 2, \ldots, n$. The ϵ_i are assumed to be independent and normally distributed with mean 0 and standard deviation σ. The parameters of the model are $\beta_0, \beta_1, \beta_2, \ldots, \beta_p$, and σ.

The β's are estimated by $b_0, b_1, b_2, \ldots, b_p$, which are obtained by the principle of least squares. The parameter σ is estimated by

$$s = \sqrt{\mathrm{MSE}} = \sqrt{\frac{\sum e_i^2}{n - p - 1}}$$

where the e_i are the sample residuals:

$$e_i = y_i - \hat{y}_i$$

A level $1 - \alpha$ confidence interval for β_j is

$$b_j \pm t^* s(b_j)$$

where t^* is the upper $\alpha/2$ critical value for the $t(n - p - 1)$ distribution.

The test of H_0: $\beta_j = 0$ is based on the statistic

$$t = \frac{b_j}{s(b_j)}$$

and the $t(n - p - 1)$ distribution.

The estimate b_j of β_j and the test and confidence interval for β_j are all based on a specific multiple linear regression model. The results of all of these procedures change if other explanatory variables are added to or deleted from the model.

The **ANOVA table** for a multiple linear regression gives the sum of squares, degrees of freedom, and mean squares for the model, error, and total sources of variation. The **F statistic** is the ratio MSM/MSE and is used to test the null hypothesis

$$H_0: \beta_1 = \beta_2 = \cdots = \beta_p = 0$$

If H_0 is true, this statistic has an $F(p, n - p - 1)$ distribution.

The **squared multiple correlation** is given by the expression

$$R^2 = \frac{\text{SSM}}{\text{SST}}$$

and is interpreted as the proportion of the variability in the response variable y that is explained by the explanatory variables $x_1, x_2, \ldots, x_p$ in a multiple linear regression.

SECTION 10.2 EXERCISES

10.27 The model for subpopulation means for the computer science study is described in Example 10.15.

   (a) Give the model for the subpopulation mean GPA for students having high school grade scores of HSM = 9 (A−), HSS = 8 (B+), and HSE = 7 (B).

   (b) Using the parameter estimates given in Figure 10.21, calculate the estimate of this subpopulation mean.

10.28 Use the model for subpopulation means described in Example 10.15 to answer the following questions:

   (a) For students having high school grade scores of HSM = 6 (B−), HSS = 7 (B), and HSE = 8 (B+), express the subpopulation mean in terms of the parameters β_j.

   (b) Calculate the estimate of this subpopulation mean using the b_j given in Figure 10.21.

10.29 Consider the regression problem of predicting GPA from the three high school grade variables. The computer output is given in Figure 10.21.

   (a) Use information given in the output to calculate a 95% confidence interval for the regression coefficient of HSM.

   (b) Do the same for the explanatory variable HSE.

10.30 The output for the regression run for predicting GPA from the high school grade variables HSM and HSE is given in Figure 10.23. Answer (a) and (b) of the previous exercise for this regression. Why are the results for this exercise different from those that you calculated for the previous exercise?

10.31 Consider the regression problem of predicting GPA from the three high school grade variables. The computer output is given in Figure 10.21.

(a) Write the fitted regression equation.
(b) What is the value of s, the estimate of σ?
(c) State the H_0 and H_a tested by the F statistic for this problem.
(d) What is the distribution of the F statistic under H_0?
(e) What percent of the variation in GPA is explained by these three high school grade variables?

10.32 Figure 10.24 gives the output for the regression analysis in which the two SAT scores are used to predict GPA. Answer the questions given in the previous exercise for this analysis.

10.33 Use the computer to make a plot of GPA versus SAT-M. Do the same for GPA versus SAT-V. Describe the general patterns. Are there any unusual values?

10.34 Make a plot of GPA versus HSM. Do the same for the other two high school grade variables. Describe the three plots. Are there any outliers or influential points?

10.35 Using a computer, run a regression using the three high school grade variables to predict GPA. Plot the residuals versus each of the three predictors and versus the predicted value of GPA. Are there any unusual points or patterns in these four plots?

10.36 Use the two SAT scores in a multiple regression to predict GPA. Plot the residuals versus each of the explanatory variables and versus the predicted GPA. Describe the plots.

10.37 Use a computer plotting routine to plot HSM versus HSS. Since there are relatively few possible values for these variables, many observations will be plotted on the same point. To handle this problem, find the two-way table of counts for this pair of high school grades. (For example, use the SAS procedure FREQ.) Then label each point on the plot with the number of observations at that point.

10.38 Refer to the previous exercise and plot HSM versus HSE.

10.39 Run a multiple regression using HSM and SAT-M to predict GPA using the data given in Appendix A of this chapter.

(a) Give the fitted regression equation.

   (b) State the H_0 and H_a tested by the F statistic. Give the value of the F statistic and its P-value. Do you reject H_0?

   (c) Give 95% confidence intervals for the regression coefficients of HSM and SAT-M. Do either of these include the point 0?

   (d) Report the t statistics and P-values for the tests of the regression coefficients of HSM and SAT-M. What conclusions do you draw from these tests?

   (e) What is the value of s, the estimate of σ?

   (f) What percent of the variation in GPA is explained by HSM and SAT-M in your model?

10.40 Using the data given in Appendix A of this chapter, perform a multiple regression analysis to predict GPA from HSE and SAT-V. Summarize the results and compare them with those obtained in the previous exercise.

NOTES

1 The quote and the example are taken from J. V. G. A. Durnin and J. Womersley, "Body fat assessed from total body density and its estimation from skinfold thicknesses: measurements on 481 men and women aged 16 to 72 years," *British Journal of Nutrition*, 32 (1974), pp. 77–97. The data analyzed in the examples that follow have been generated by computer simulation in such a way that they give the same statistical results as are reported in the article.

2 This calculation can be done in some software packages, such as SAS, by defining an additional observation with LSKIN equal to the value of x^* and DEN missing. The additional case will be ignored for the computation of the least-squares line. However, the predicted value and associated statistics will be computed and printed. Confidence intervals for any number of additional cases can be produced by this technique.

3 The method is described in G. W. Snedecor and W. G. Cochran, *Statistical Methods*, 7th ed., Iowa State University Press, Ames, Iowa, 1980.

4 Results of the study are reported in P. F. Campbell and G. P. McCabe, "Predicting the success of freshmen in a computer science major," *Communications of the ACM*, 27 (1984), pp. 1108–1113.

CHAPTER 10 EXERCISES

10.41 In Exercise 3.16 of Chapter 3, an experiment is described in which the strengths of 50 strips of yellow poplar wood were measured two times. The data are given in that exercise. In this problem we look at this experiment as one in which the process of measuring strength is to be evaluated. Specifically, we ask how well a repeat measurement T2 can be predicted from the first measurement T1 for strips of wood of this type.

(a) Plot T2 versus T1 and describe the pattern you see.

(b) Run the linear regression using T1 as the explanatory variable and T2 as the response variable. Write the fitted equation and give s, the estimate of the standard deviation σ of the model.

(c) What is the correlation between T1 and T2? Sometimes this quantity is called the *reliability* of the strength-measuring procedure. What proportion of the variability in T2 is explained by T1?

(d) Give the t statistic for testing the null hypothesis that the slope is 0. State an appropriate alternative hypothesis for this problem and state the P-value for the significance test.

(e) Verify that the square of the t statistic for the slope hypothesis is equal to the F statistic given in the ANOVA table.

10.42 Refer to the previous exercise. Rerun the regression for the 25 observations with odd-numbered strips; that is $S = 1, 3, \ldots, 49$. We want to compare the results of this run with those for the full data set given in the previous exercise. In particular, we want to see which quantities remain approximately the same and which change.

(a) Construct a table giving b_0, b_1, s, and $s(b_1)$ for the two runs. Which values are approximately the same and which have changed?

(b) Summarize the differences between the two ANOVA tables.

(c) Compare the two correlations.

(d) Summarize your results.

(e) Suppose that we had available another 50 samples of wood strips from the same population measured under the same conditions. If all 100 observations were run in one analysis, approximately what values would you expect to see for b_0, b_1, s, and r?

10.43 For this exercise you will use the Gesell score data discussed in Section 3.3 and given in Table 3.3 of Chapter 3. Run a linear regression to predict the Gesell score from the age at first word. Use all 21 cases for this analysis.

(a) Write the fitted regression equation.

(b) What is the value of the t statistic for testing the null hypothesis that the slope is 0? What is the distribution of this statistic under the assumption that H_0 is true? Give an approximate P-value for the test of H_0 versus the alternative that the slope is negative.

(c) Give a 95% confidence interval for the slope.

(d) What percentage of the variation in Gesell scores is explained by the age at first word?

(e) What is s, the estimate of the model standard deviation σ?

10.44 In Chapter 3 it was noted that cases 18 and 19 in the Gesell data set were somewhat unusual. Rerun the analysis described in the previous exercise with these two cases excluded. Compare your results with those obtained in the previous exercise.

10.45 Refer to the data given in Appendix A of this chapter. In the listing there is a variable SEX, which has the value 1 for males and 2 for females. Create a data set containing the values for males only. Run a multiple regression analysis for predicting GPA from the three high school grade variables for this group. Using the case study as a guide, interpret the results and state what conclusions can be drawn from this analysis.

10.46 Refer to the previous exercise. Perform the analysis using the data for females only.

10.47 Refer to the data given in Example 10.1. Suppose that the gas consumption for December was incorrectly recorded as 87 instead of 8.7.

(a) Calculate the least-squares regression line for the incorrect set of data.

(b) Find the standard error of b_1.

(c) Compute the test statistic for H_0: $\beta_1 = 0$ and find the P-value. How do the results compare with those for the correct set of data?

Exercises 10.48 to 10.56 refer to the data in Table 10.3. As cheddar cheese matures a variety of chemical processes take place. The taste of matured cheese is related to the concentration of several chemicals in the final product. In a study of cheddar cheese from the LaTrobe Valley of Victoria, Australia, samples of cheese were analyzed for their chemical composition and were subjected to taste tests. Table 10.3 presents data for one type of cheese manufacturing process. "Taste" is the response variable of interest. The taste scores were obtained by combining the scores from several tasters. Three of the chemicals whose concentrations were measured were acetic acid, hydrogen sulfide and lactic acid. For acetic acid and hydrogen sulfide (natural) log transformations were taken. Thus the explanatory variables in Table 10.3 are the transformed concentrations of acetic acid ("acetic") and hydrogen sulfide ("H2S") and the untransformed concentration of lactic acid ("lactic"). (Data based on experiments performed by G. T. Lloyd and E. H. Ramshaw of the CSIRO Division of Food Research, Victoria, Australia. Some results of the statistical analyses of these data are given in G. McCabe, L.

Table 10.3 Cheese data

Case	Taste	Acetic	H2S	Lactic
1	12.3	4.543	3.135	0.86
2	20.9	5.159	5.043	1.53
3	39.0	5.366	5.438	1.57
4	47.9	5.759	7.496	1.81
5	5.6	4.663	3.807	0.99
6	25.9	5.697	7.601	1.09
7	37.3	5.892	8.726	1.29
8	21.9	6.078	7.966	1.78
9	18.1	4.898	3.850	1.29
10	21.0	5.242	4.174	1.58
11	34.9	5.740	6.142	1.68
12	57.2	6.446	7.908	1.90
13	0.7	4.477	2.996	1.06
14	25.9	5.236	4.942	1.30
15	54.9	6.151	6.752	1.52
16	40.9	6.365	9.588	1.74
17	15.9	4.787	3.912	1.16
18	6.4	5.412	4.700	1.49
19	18.0	5.247	6.174	1.63
20	38.9	5.438	9.064	1.99
21	14.0	4.564	4.949	1.15
22	15.2	5.298	5.220	1.33
23	32.0	5.455	9.242	1.44
24	56.7	5.855	10.199	2.01
25	16.8	5.366	3.664	1.31
26	11.6	6.043	3.219	1.46
27	26.5	6.458	6.962	1.72
28	0.7	5.328	3.912	1.25
29	13.4	5.802	6.685	1.08
30	5.5	6.176	4.787	1.25

McCabe and A. Miller, "Analysis of taste and chemical composition of cheddar cheese 1982–83 experiments," CSIRO Division of Mathematics and Statistics Consulting Report VT85/6.)

10.48 For each of the four variables in Table 10.3, find the mean, median, standard deviation and interquartile range. Display the distribution of each with a stemplot and use a normal quantile plot to assess normality of the data. Summarize your findings. Note that when doing regressions with these data, we do not assume that these distributions are normal. Only the residuals from our model need be (approximately) normal. The careful study of each variable to be analyzed is nonetheless an important first step in any statistical analysis.

10.49 Make a scatterplot for each pair of variables in Table 10.3 (you will have 6 plots). Describe the relationships. Calculate the correlation for each pair of variables and report the *P*-value for the test of zero correlation in each case.

10.50 Perform a regression analysis using taste as the response variable and acetic as the explanatory variable. Be sure to examine the residuals carefully. Summarize the results. Include a plot of the data with the least squares regression line. Plot the residuals versus each of the other two chemicals. Are any patterns evident?

10.51 Repeat the analysis of Exercise 10.50 using taste as the response variable and H2S as the explanatory variable.

10.52 Repeat the analysis of Exercise 10.50 using taste as the response variable and lactic as the explanatory variable.

10.53 Compare the results of the regressions performed in the three previous exercises. Give a table with values of the *F* statistic, *P*-value, R^2, and the estimate of the standard deviation for each model. Report the three regression equations. Why are the intercepts in these three equations not all the same?

10.54 Carry out a multiple regression using acetic and H2S to predict taste. Summarize the results of your analysis. Compare the statistical significance of acetic in this model with its significance in the model with acetic alone as a predictor (Exercise 10.50). Which model do you prefer? Give a simple explanation for the fact that acetic alone appears to be a good predictor of taste but with H2S in the model, it is not.

10.55 Carry out a multiple regression using H2S and lactic to predict taste. Comparing the results of this analysis to the simple linear regressions using each of these explanatory variables alone, it is evident that a better result is obtained by using both predictors in a model. Support this statement with explicit information obtained from your analysis.

10.56 Use the three explanatory variables acetic, H2S and lactic in a multiple regression to predict taste. Write a short summary of the results including an examination of the residuals. Based on all of the regression analyses you have carried out on the data in Table 10.3, which model do you prefer and why?

10.57 Economists know that the consumption of a product increases when the price drops. In a study of this relationship, researchers looked at the price and consumption of textiles in the Netherlands in the 17 years from 1923 to 1939. During this period, the price of textiles was falling and consumption was rising. The explanatory variable x is the price of textiles, adjusted for changes in the overall cost of living and the response variable y is per person consumption of textile goods. Both x and y are reported as index numbers with 1925 = 100. That is, they are given as percents of their values in the year 1925. A computer least squares regression routine gives the following output.

	Coeff	Std Err	t Value
Intercept	235.4897	9.079104	25.93755
Price	-1.323306	0.1163298	-11.37547

Residual Standard Error = 7.84818
R-Square = 0.896123   N = 17

"Residual Standard Error" is the standard error s about the fitted line. A scatterplot of the data shows a negative linear association with no outliers or influential observations. (From a larger set of data in Henry Theil, *Principles of Econometrics*, Wiley, New York, 1971, p. 102.)

(a) Explain in plain language what the slope β_1 of the true regression line means. This is an important economic quantity. At your request, the computer tells you that the mean and variance of the explanatory variable are

$$\bar{x} = 76.3118 \text{ and } s_x^2 = 284.47$$

Give a 90% confidence interval for β_1. (Hint: remember that $(n-1)s_x^2 = \sum(x - \bar{x})^2$.)

(b) Explain why the intercept β_0 of the true regression line is of no interest in this case.

(c) In the year 1925, $x = 100$ and $y = 100$ because of the way the two variables are defined. Predict the mean consumption in years when the price of textiles is the same as the 1925 price. Then give a 90% confidence interval for this mean.

(d) Suppose you are told that the 1940 price of textiles was 75. Give a 90% prediction interval for textile consumption based on the regression line fitted to the 1923 to 1939 data. (In fact this prediction is very inaccurate. In 1940 the Germans invaded and conquered the Netherlands. Extrapolation by even one year is risky in economics because outside events can change economic conditions quickly.)

Appendix A Listing of Data for Case Study

OBS	GPA	HSM	HSS	HSE	SAT-M	SAT-V	SEX
1	5.32	10	10	10	670	600	1
2	5.14	9	9	10	630	700	2
3	3.84	9	6	6	610	390	1
4	5.34	10	9	9	570	530	2
5	4.26	6	8	5	700	640	1
6	4.35	8	6	8	640	530	1
7	5.33	9	7	9	630	560	2
8	4.85	10	8	8	610	460	2
9	4.76	10	10	10	570	570	2
10	5.72	7	8	7	550	500	1
11	4.08	9	10	7	670	600	1
12	5.38	8	9	8	540	580	1
13	2.40	6	6	7	560	690	1
14	5.50	8	7	8	630	500	1
15	5.69	10	10	8	710	470	2
16	5.35	9	9	9	580	540	2
17	5.29	10	8	8	760	630	1
18	3.80	7	7	7	620	470	2
19	4.83	6	7	7	690	440	1
20	4.57	9	10	10	417	518	2
21	5.80	10	9	8	560	530	2
22	4.88	9	7	6	690	460	1
23	4.28	8	10	10	600	600	2
24	5.06	8	6	5	540	400	1
25	5.21	8	8	7	600	400	1
26	3.60	4	7	7	460	460	2
27	5.47	10	10	9	720	680	2
28	4.00	3	7	6	460	530	1
29	5.18	9	10	8	670	450	1
30	4.77	6	5	9	590	440	1
31	4.38	9	9	10	650	570	2
32	4.58	10	10	9	440	430	2
33	4.34	7	7	6	570	480	1
34	4.26	5	7	7	530	440	1
35	5.44	10	10	9	640	590	2
36	4.03	6	7	9	540	610	1
37	5.08	9	10	6	491	488	1
38	5.34	5	9	7	600	600	1
39	3.40	6	8	8	510	530	1
40	4.00	2	4	6	300	290	2
41	3.43	10	9	9	750	610	1

Appendix A (Continued)

OBS	GPA	HSM	HSS	HSE	SAT-M	SAT-V	SEX
42	4.48	8	9	6	650	460	1
43	5.73	10	10	9	720	630	1
44	4.43	7	10	10	530	560	1
45	3.69	7	6	7	560	480	2
46	5.80	10	10	9	760	500	1
47	5.18	10	10	10	570	750	2
48	6.00	9	10	10	640	480	1
49	6.00	9	9	8	800	610	1
50	4.00	9	6	5	640	670	1
51	5.06	9	10	9	590	420	2
52	5.74	9	10	9	750	700	1
53	4.32	9	7	8	520	440	1
54	4.63	10	10	6	640	500	1
55	4.79	8	8	7	610	530	1
56	5.70	8	10	8	520	410	1
57	3.66	8	4	3	590	470	1
58	5.41	9	9	9	520	490	1
59	5.21	7	9	8	505	435	1
60	5.08	9	10	8	559	607	1
61	4.75	8	9	8	559	435	2
62	4.81	9	7	4	559	488	1
63	4.12	7	7	8	559	545	1
64	5.33	7	6	7	500	460	1
65	5.75	10	9	9	760	620	1
66	3.69	8	7	7	490	390	1
67	4.87	8	8	7	476	576	2
68	5.16	8	9	8	680	700	2
69	3.93	8	6	8	590	510	1
70	4.62	9	10	8	550	440	2
71	5.16	10	9	8	640	490	1
72	4.73	9	8	7	520	360	1
73	4.46	6	7	7	490	370	1
74	5.06	8	10	10	580	460	1
75	2.39	7	10	9	550	660	2
76	4.44	10	9	9	650	350	2
77	3.07	7	8	6	700	520	1
78	5.46	9	9	8	610	520	2
79	5.35	10	10	10	620	570	1
80	4.37	8	7	9	530	480	2
81	3.25	7	8	6	480	360	2
82	3.82	6	8	6	490	550	1
83	3.59	8	9	7	670	480	1

(*continued*)

Appendix A (Continued)

OBS	GPA	HSM	HSS	HSE	SAT-M	SAT-V	SEX
84	4.80	10	9	9	550	450	2
85	3.14	10	10	7	720	610	1
86	5.07	9	8	9	490	480	2
87	2.65	9	7	7	640	520	1
88	4.12	7	6	7	520	380	1
89	5.70	10	10	10	580	580	1
90	5.68	10	8	9	590	490	2
91	4.82	4	5	7	400	470	1
92	5.12	10	10	7	640	520	1
93	5.90	10	10	10	650	500	2
94	4.14	5	4	8	560	420	2
95	5.34	10	9	10	590	580	2
96	4.45	7	7	8	430	330	2
97	4.71	9	7	10	490	400	2
98	4.59	10	10	10	590	470	2
99	4.25	9	7	4	550	290	1
100	4.93	10	10	10	600	520	1
101	4.34	8	9	7	480	410	1
102	5.16	9	7	7	400	390	1
103	4.11	6	9	9	480	390	1
104	3.34	6	7	8	530	470	1
105	4.96	9	7	6	670	440	1
106	4.83	10	9	9	710	530	1
107	5.10	9	10	9	750	670	1
108	4.93	9	9	10	690	510	2
109	4.53	7	6	9	570	480	2
110	3.95	6	6	9	550	600	2
111	5.07	7	4	7	660	480	1
112	5.39	9	9	10	510	570	2
113	3.93	10	8	8	650	490	2
114	4.53	8	9	8	550	500	1
115	4.87	9	9	9	620	480	1
116	4.75	10	10	10	720	500	1
117	5.61	10	10	9	630	440	1
118	5.14	9	8	9	640	630	1
119	4.69	8	6	9	470	420	2
120	4.25	10	10	10	690	580	1
121	3.00	8	9	10	640	600	1
122	4.43	9	5	9	480	520	2
123	4.79	9	6	7	690	400	1
124	5.65	9	9	9	640	430	2

Appendix A (Continued)

OBS	GPA	HSM	HSS	HSE	SAT-M	SAT-V	SEX
125	4.17	10	7	7	650	450	1
126	3.95	7	8	9	550	570	1
127	3.94	8	8	8	470	330	2
128	5.00	10	8	9	510	360	2
129	4.39	6	5	6	470	330	1
130	4.15	6	6	6	480	460	1
131	4.09	9	7	8	450	460	2
132	3.85	10	8	10	530	550	2
133	5.28	10	10	10	670	490	2
134	2.75	7	6	6	540	590	1
135	5.35	10	10	10	730	650	1
136	5.34	10	9	10	490	410	2
137	5.58	10	7	8	710	400	1
138	5.06	5	9	9	510	380	1
139	4.75	10	7	5	770	720	1
140	4.25	6	9	10	530	510	2
141	6.00	10	10	10	580	490	2
142	3.32	9	8	9	740	460	2
143	4.50	9	9	9	560	500	1
144	5.66	10	10	10	710	600	2
145	4.78	9	9	10	600	510	1
146	4.72	6	5	7	580	490	2
147	4.64	9	9	8	620	590	1
148	5.26	10	10	9	610	560	1
149	5.09	10	10	8	540	470	1
150	5.41	9	4	7	600	360	1
151	4.29	7	6	8	570	570	2
152	4.44	8	8	8	690	490	1
153	3.11	7	7	7	590	480	1
154	5.00	4	3	4	620	560	1
155	5.67	10	10	10	640	570	1
156	5.12	10	10	10	580	340	1
157	5.81	10	10	7	750	540	1
158	4.17	8	7	8	650	500	1
159	2.12	4	6	6	630	490	1
160	5.30	10	10	9	650	480	1
161	4.30	9	10	10	590	420	1
162	4.00	6	5	6	530	320	1
163	5.22	9	7	9	650	490	1
164	5.62	10	10	8	660	630	1
165	4.61	9	7	8	620	420	2
166	4.32	6	6	7	430	460	2

(continued)

Appendix A (Continued)

OBS	GPA	HSM	HSS	HSE	SAT-M	SAT-V	SEX
167	5.39	10	10	10	500	390	2
168	5.64	8	6	8	590	580	2
169	3.88	10	6	6	620	430	1
170	5.20	9	5	7	570	570	1
171	4.55	7	8	8	570	480	1
172	4.04	8	7	7	690	440	1
173	4.82	10	9	9	660	550	1
174	5.25	9	7	8	690	550	1
175	4.97	10	10	10	770	540	1
176	4.21	7	7	8	670	500	1
177	4.81	10	10	10	620	570	1
178	4.19	6	5	6	540	460	2
179	5.06	10	10	10	620	510	2
180	3.80	8	7	9	620	600	2
181	4.50	10	9	9	660	460	1
182	5.03	8	8	7	600	630	1
183	3.92	9	10	8	447	320	1
184	4.58	10	9	9	720	740	1
185	3.52	9	9	10	520	570	2
186	5.40	6	9	9	480	480	2
187	4.16	6	6	6	590	440	1
188	6.00	7	6	6	600	410	1
189	4.86	9	4	8	640	470	2
190	4.50	7	10	10	630	500	1
191	3.81	9	9	9	620	580	1
192	3.85	10	8	7	700	480	1
193	4.70	6	8	6	580	470	1
194	4.96	9	7	8	630	630	1
195	5.23	10	10	10	600	660	2
196	3.46	7	7	8	630	540	1
197	4.86	10	9	10	650	430	2
198	5.32	10	9	10	660	560	1
199	4.76	10	10	10	600	560	1
200	4.71	8	7	9	700	440	1
201	5.32	10	9	10	640	560	2
202	5.40	9	10	9	550	560	1
203	4.95	9	9	8	620	400	1
204	4.65	8	10	8	680	450	1
205	4.51	8	9	10	510	440	2
206	4.07	9	7	6	600	440	2
207	2.85	7	7	9	510	480	2

Appendix A (Continued)

OBS	GPA	HSM	HSS	HSE	SAT-M	SAT-V	SEX
208	4.48	8	8	7	630	500	1
209	2.80	8	10	9	470	410	1
210	5.86	10	10	10	750	760	1
211	5.40	7	8	4	710	500	1
212	4.86	8	9	8	550	510	2
213	2.91	6	5·	7	586	697	1
214	4.67	9	9	10	586	670	1
215	4.51	9	8	7	700	500	1
216	3.79	7	7	5	550	570	1
217	3.86	7	9	7	356	350	2
218	4.59	5	4	7	630	470	2
219	4.42	6	6	8	505	518	1
220	5.41	8	6	8	630	470	2
221	2.58	5	7	7	515	285	1
222	5.00	10	10	9	774	688	1
223	4.28	9	8	9	559	488	2
224	4.62	9	8	7	491	391	1

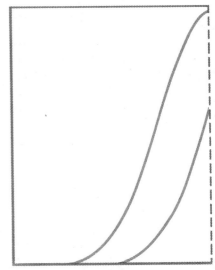

Prelude

Methods based on *t* distributions are used to compare two means. In this chapter we develop analysis of variance procedures to make comparisons among several means. Specific research questions can be incorporated into the analysis. Power calculations allow us to assess the chance of obtaining statistically significant results. When the groups to be compared are classified by two categorical variables, we use two-way analysis of variance to perform the inference.

- *Students who started as computer science majors in their freshman year are classified into three groups on the basis of their major in the sophomore year. Are the mean SAT mathematics scores different for the three groups?*

- *Do new methods of instruction result in higher mean reading comprehension than the traditional method?*

- *When comparing the salaries of men and women scientists, is the difference the same across fields?*

11

Analysis of Variance

---

analysis of variance

one-way ANOVA

two-way ANOVA

Procedures for comparing the means of two populations were described in Chapter 8. We are now ready to extend those methods to problems involving more than two populations. When comparing several means, we will use a methodology called *analysis of variance*, or simply *ANOVA*.

In this chapter we will consider two ANOVA techniques. When there is only one way to classify the populations of interest, we use a *one-way ANOVA* to analyze the data. This technique is discussed in detail in the first section. However, in many practical situations there is more than one way to classify the populations. This necessitates two-way, three-way, and higher ANOVA techniques. Most of the interesting ideas related to these techniques can be illustrated by studying *two-way ANOVA*, which we discuss in Section 11.2.

11.1   ONE-WAY ANALYSIS OF VARIANCE

When comparing two means we compute a two-sample t statistic and its P-value to assess the statistical significance of the difference in the sample means. For several means the situation is quite similar. An F statistic and its P-value are used to evaluate the null hypothesis that several population means are equal.

In the sections that follow, we will examine the basic ideas and assumptions that are needed for the analysis of variance. Many of these concepts are very similar to those discussed in the two-sample case.

Comparing Means

One-way ANOVA is a statistical method for comparing several population means. We draw a simple random sample (SRS) from each population and use the data to test the null hypothesis that the population means are all equal.

EXAMPLE 11.1

A medical researcher is interested in comparing the effectiveness of three different treatments to lower the cholesterol of patients with high values. Sixty individuals are randomly assigned to the three treatments (20 to each) and the reduction in cholesterol is recorded for each patient.   ■

EXAMPLE 11.2

An ecologist is interested in comparing the amount of a certain pollutant in five streams. Fifty water specimens are collected from each stream and the amount of the pollutant in each sample is measured.   ■

These two examples have some similarities and some differences. In both there is a single response variable measured on several units: people in the first and water samples in the second. We might expect the data to be approximately normal and would consider a transformation if needed. A comparison among several populations is desired (three treatments in the first and five streams in the second). In the first example patients are randomly assigned to treatments. In the second example, this kind of randomization does not make sense. We assume, instead, that the water specimens obtained from a stream are a random sample of all water samples that could be obtained from the stream. In both cases we can use ANOVA to make the desired comparisons.

The same ANOVA methods are used for sampling designs and randomized experiments. The data collection method, however, must be kept in mind when interpreting the results. A strong case for causation is best made with a randomized experiment.

factor

We use the term *factor* for the categorical variable that is used to distinguish the populations to be compared. Thus, in Example 11.1 the factor is treatments, while in Example 11.2 it is streams. In general we will use the term *groups* for the factor in a one-way ANOVA.

group

level

Different values of a factor are called *levels*. In Example 11.1 there are three levels for treatments and in Example 11.2 there are five levels for streams.

To assess whether several populations all have the same mean, we compare the means of samples drawn from each population. Figure 11.1 displays the sample means for Example 11.1. Of course, we do not expect sample means to be equal even if the population means are all the same. The purpose of ANOVA is to assess whether the observed differences among sample means are *significant*. In other words, could a variation this large be plausibly due to chance, or is it good evidence for a difference among

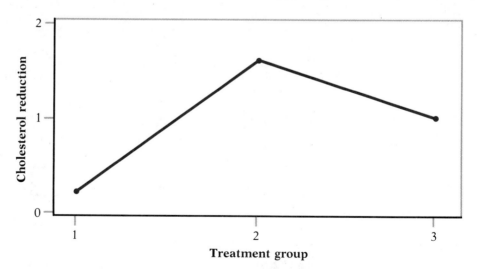

Figure 11.1 Serum cholesterol reduction in three groups.

the population means? The answer depends upon both the variation within the groups and the sizes of the samples used to compute the means.

Side-by-side boxplots help us to see the within-group variation. Compare Figures 11.2(a) and 11.2(b). The sample medians are the same in both cases, but the large variation within the groups in Figure 11.2(a) suggests that the differences among the sample medians could be due simply to chance variation. To assess this possibility we must know the size of the samples, information that is not contained in the boxplots. In ANOVA we compare means rather than medians. If the distributions are symmetric, these two

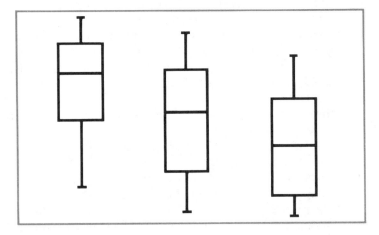

Figure II.2(a) Side-by-side boxplots for three groups with large within-group variation.

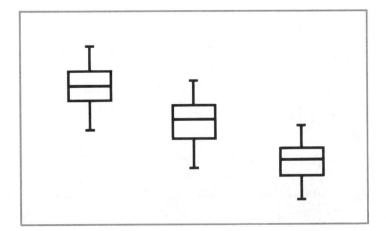

Figure II.2(b) Side-by-side boxplots for three groups with small within-group variation.

measures of the center of a distribution will be the same. We use boxplots as a preliminary step in ANOVA to get a general summary of the data in each group.

To compare the means of two populations, we developed two-sample t statistics. If the two populations are assumed to have equal but unknown standard deviations and the sample sizes are both equal to n, the t statistic is

$$t = \frac{\bar{x}_1 - \bar{x}_2}{s_p\sqrt{\dfrac{1}{n} + \dfrac{1}{n}}} = \frac{\sqrt{\dfrac{n}{2}}(\bar{x}_1 - \bar{x}_2)}{s_p}$$

We can rewrite the square of this t statistic as

$$t^2 = \frac{\dfrac{n}{2}(\bar{x}_1 - \bar{x}_2)^2}{s_p^2}$$

If we use ANOVA to compare two populations, the F statistic that we obtain is exactly equal to the statistic t^2 given above. Therefore, we can learn something about how ANOVA works by looking carefully at the statistic in this form.

The numerator in the t^2 statistic measures the variation *between* the groups. It includes a factor for the common sample size n. It can be large because of a large difference in the sample means or because the sample sizes are large. The denominator measures the variation *within* groups using s_p^2, the pooled estimate of the common variance. If the within-group variation is small, this will tend to make the statistic large.

Although the general form of the F statistic is more complicated, the idea is basically the same. To assess whether several populations all have the same mean, we compare the variation *among* the means of several groups with the variation *within* groups. Because we are comparing variation, the method is called *analysis of variance*.

The null hypothesis that is tested in an ANOVA is that the population means are *all* equal. The alternative is that they are not all equal. Note that the alternative could be true because all of the means are different or simply because one of them differs from the rest. Because of this framework, the situation is a bit more complex than we encountered in the case of two populations. If we reject the null hypothesis, we need to perform some further analysis to make statements about which populations appear to differ from which others.

The computations needed for an ANOVA are more lengthy than those for the t test. For this reason we generally use computer programs to perform the calculations. This frees us from the burden of arithmetic and allows us to concentrate on the basic concepts.

The next example illustrates the practical use of ANOVA in analyzing real data. Later we will explore the technical details.

EXAMPLE 11.3

In the computer science department of a large university, many students change their major after the first year. A detailed study[1] of the 256 students enrolled as freshmen computer science majors in one year was undertaken to help understand this phenomenon. Students were classified on the basis of their status at the beginning of their sophomore year, and a variety of variables measured at the time of their entrance to the university were obtained. Summary data for the Scholastic Aptitude Test (SAT) mathematics scores are given in the following table:

Second-year major	n	$\bar{x}$	s
Computer science	103	619	86
Engineering or other science	31	629	67
Other	122	575	83

An ANOVA is performed on the data. The null hypothesis is that the population mean SAT scores for the three groups are equal, and the alternative is that they are not all equal. The F statistic is 10.35 and the P-value is reported as less than 0.001. ∎

Note that when P-values are very small, it is common practice to report them as simply being less than some small number. In this case 0.001 was chosen. This is sufficiently small for most purposes. If the null hypothesis is true, we would obtain an F statistic of 10.35 or larger less than 0.1% of the time. Since the variation in the sample means is much larger than we would expect to occur by chance, we conclude that the mean SAT scores in the three populations are not all identical.

Although we rejected the null hypothesis, we have not demonstrated that all three populations are different. The means are plotted in Figure 11.3. Inspection of this plot suggests that the first two groups are approximately equal and that the third group differs from these two. We would need to do some additional analysis to clarify the situation.

In this particular study the researchers expected the mathematics scores of the engineering or other science group to be similar to those of the computer science group. They thought that perhaps the students in the third (other) group would have poorer mathematics scores and that this might be related to the fact that they chose to transfer to a program of study that would require less mathematics. Since these comparisons were planned before the data were examined, *contrasts* were used to explore these questions. Had no such ideas been entertained before looking at the data, a *multiple comparison* procedure would have been used to determine which pairs of populations differed significantly. In later sections we will explore the use of contrasts and multiple comparisons in detail.

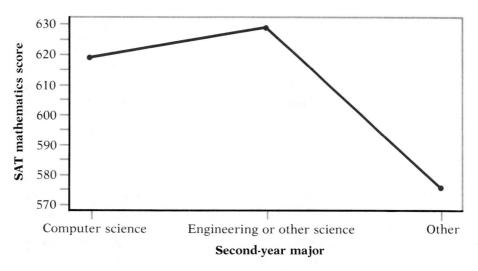

Figure 11.3 SAT mathematics score means for a computer science study.

A purist could argue that using ANOVA for the situation described in Example 11.3 is inappropriate. *All* first-year computer science majors in this particular university for the year studied were included in the analysis. There was no random sampling. On the other hand, one could think of these three groups of students as samples from three populations of students in similar circumstances. They may be reasonably similar to students in the next few years at the same university or other universities with similar programs. Judgments such as this are very important and must be made by experts who are knowledgeable about such matters. They are inferences that are beyond the realm of statistical inference.

Statistical Models

In Chapter 2 we learned that in examining data we seek an overall pattern and deviations from it. We expressed this idea through an equation:

$$DATA = FIT + RESIDUAL$$

When studying the method of least squares in Chapter 3, we again encountered this idea and learned to look carefully at residuals. We now apply this framework to discuss the statistical models used in ANOVA. These models provide a convenient way to summarize the assumptions that are the foundation for our analysis. They also give us the necessary notation to describe the calculations that we use.

First, we recall the statistical model for a random sample of observations from a single normal population with mean μ and standard deviation σ. If we let

$$x_1, x_2, \ldots, x_n$$

denote the observations, then we can describe this model by stating that the x_j are an SRS from the $N(\mu, \sigma)$ distribution.

Another way to describe the same model is to think of the x's varying about their population mean. To do this, write each observation x_j as

$$x_j = \mu + \epsilon_j$$

where the ϵ_j are an SRS from the $N(0, \sigma)$ distribution. Since μ is unknown, the ϵ's cannot actually be observed. This form more closely corresponds to our DATA = FIT + RESIDUAL way of thinking. The FIT part of the model is represented by μ. It is the systematic part of the model, like the line in a regression. The RESIDUAL part is represented by ϵ_j. It represents the deviations of the data from the fit and is due to random, or chance, variation.

In our statistical model there are two unknown parameters: μ and σ. We have already learned that we estimate μ by $\bar{x}$, the sample mean, and σ by s, the sample standard deviation. The differences $e_j = x_j - \bar{x}$ are the sample residuals and correspond to the ϵ_j of our statistical model.

The model for one-way ANOVA is very similar. Suppose we take observations from I different populations. For the ith population, the sample size is n_i. We let x_{ij} represent the jth observation from the ith population. The population means are the FIT part of the model and are represented by μ_i. The random variation, or RESIDUAL, part of the model is represented by ϵ_{ij}. Thus, the model is

$$x_{ij} = \mu_i + \epsilon_{ij}$$

for $i = 1, \ldots, I$ and $j = 1, \ldots, n_i$. The ϵ_{ij} are assumed to be an SRS from an $N(0, \sigma)$ distribution. The sample sizes n_i may differ, but the standard

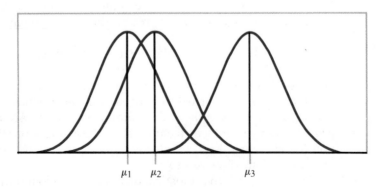

Figure 11.4 Model for one-way ANOVA with three groups.

deviation σ is the same in all of the populations. A picture of the model is given in Figure 11.4 for $I = 3$. Note that the population means μ_i are different. The shapes of the three normal distributions are the same, reflecting the assumption that all three populations have the same standard deviation.

EXAMPLE 11.4

A survey of college students was conducted to determine how much money they spend per year on textbooks and to compare students by year of study. From lists of students provided by the registrar, SRSs of size 50 were chosen from each of the four classes (freshmen, sophomores, juniors, and seniors). The students selected were asked how much they spent on textbooks during the previous year.

Here we have $I = 4$ populations. The population means μ_1, μ_2, μ_3, and μ_4 represent the yearly average amounts spent on textbooks by *all* freshmen, sophomores, juniors, and seniors at this college for the previous year. The sample sizes n_i are 50, 50, 50, and 50.

Suppose the first freshman sampled is Eve Brogden. Then $x_{1,1}$ represents the amount spent by Eve. The data for the other freshmen sampled are denoted by $x_{1,2}, x_{1,3}, \ldots, x_{1,50}$. Similarly, the data for the other groups have a first subscript indicating the group and a second subscript indicating the student in the group.

The difference between Eve's spending and the average for *all* of the students in the freshman class is $x_{1,1} - \mu_1 = \epsilon_{1,1}$. We are assuming that the ϵ_{ij} are independent and normally distributed (at least approximately) with mean 0 and standard deviation σ. ∎

The unknown parameters in the statistical model for ANOVA are the population means μ_i and the common population standard deviation σ. To estimate μ_i we use the sample mean for the ith group

$$\bar{x}_i = \frac{1}{n_i} \sum_{j=1}^{n_i} x_{ij}$$

The sample residuals are $e_{ij} = x_{ij} - \bar{x}_i$. Thus, for the *sample* data we represent the FIT by $\bar{x}_i$ and the RESIDUAL by e_{ij}.

We have assumed that the population standard deviations are all equal. If we have unequal standard deviations, we generally try to transform the data so that they are approximately equal. Fortunately, the same transformation that makes a single population look more normal often will make the standard deviations among populations more alike. If the standard deviations are markedly different and cannot be made similar by a transformation, we need methods that are beyond the scope of this text.

Unfortunately, formal tests for the equality of standard deviations in several groups share the lack of robustness against nonnormality that we noted in Chapter 8 for the case of two groups. Because ANOVA procedures are not extremely sensitive to unequal standard deviations, we do not recommend a formal test of equality of standard deviations as a preliminary to ANOVA. Instead, we will simply use the following general rule:

> If the ratio of the largest sample standard deviation to the smallest sample standard deviation is less than 2, we can use methods based on the assumption of equal standard deviations and our results will still be approximately correct.[2]

When we assume that the population variances are equal, each sample standard deviation is an estimate of σ. To combine these into one estimate, we use a generalization of the pooling method introduced in Chapter 8.

Pooled estimate of variance

> Suppose we have variance estimates $s_1^2, s_2^2, \ldots, s_I^2$ from I independent samples of size $n_1, n_2, \ldots, n_I$ from populations with common variance σ^2. The pooled sample variance
>
> $$s_p^2 = \frac{(n_1 - 1)s_1^2 + (n_2 - 1)s_2^2 + \cdots + (n_I - 1)s_I^2}{(n_1 - 1) + (n_2 - 1) + \cdots + (n_I - 1)}$$
>
> is an unbiased estimator of σ^2.

This formula gives more weight to groups with larger sample sizes. If the sample sizes are equal, this computation is equivalent to averaging the sample variances. To compute the pooled standard deviation s_p, we take the square root of the pooled variance s_p^2. Note that this is *not* the same as averaging the standard deviations.

EXAMPLE 11.5

Consider the computer science study described in Example 11.3. There are $I = 3$ groups and the sample sizes are $n_1 = 103$, $n_2 = 31$, and $n_3 = 122$. The sample means for the SAT mathematics scores are $\bar{x}_1 = 619$, $\bar{x}_2 = 629$, and $\bar{x}_3 = 575$. The sample standard deviations are $s_1 = 86$, $s_2 = 67$, and $s_3 = 83$. Since the ratio of the largest to the smallest standard deviation is

$$\frac{86}{67} = 1.28$$

which is less than 2, our general rule indicates that we can use the assumption of equal population standard deviations. The pooled variance estimate is

$$s_p^2 = \frac{(102)(86)^2 + (30)(67)^2 + (121)(83)^2}{102 + 30 + 121}$$

$$= \frac{1,722,631}{253}$$

$$= 6809$$

The pooled standard deviation is

$$s_p = \sqrt{6809} = 82.5$$

This is our estimate of the common standard deviation σ of the SAT mathematics scores in the three populations of students. ∎

The ANOVA Table

ANOVA table

Comparison of several means is accomplished by using an F statistic to compare the variation among groups with the variation within groups. We now show how the F statistic is calculated. The calculations are organized in an *ANOVA table*, which contains numerical measures of the variation among groups and within groups.

First we will summarize our assumptions and hypotheses for one-way ANOVA. As usual, I represents the number of populations to be compared, and x_{ij} represents the jth observation from the ith population.

Assumptions and hypotheses for one-way ANOVA

Suppose we draw independent SRSs of size n_i from each of I normal populations. The population means are μ_i and the standard deviations are all assumed to be equal to σ. The μ_i and σ are unknown.

The statistical model is

$$x_{ij} = \mu_i + \epsilon_{ij}$$

for $i = 1, \ldots, I$ and $j = 1, \ldots, n_i$, where the ϵ_{ij} are an SRS from an $N(0, \sigma)$ distribution.

The null and alternative hypotheses are

$$H_0: \mu_1 = \mu_2 = \cdots = \mu_I$$
$$H_a: \text{not all of the } \mu_i \text{ are equal}$$

The example that follows illustrates how to do a one-way ANOVA. Since the calculations are generally performed using statistical software on a computer, we focus our attention on interpretation of the output.

EXAMPLE 11.6

Three methods of instruction were compared in a study on reading comprehension in children.[3] As is common in such studies, several pretest variables were measured before any instruction was given. One purpose of the pretest is to see if the three groups of children are similar in their comprehension skills. One of the pretest variables was an "intruded sentences" measure, which is related to one type of reading comprehension skill. The data for the 22 subjects in each group are given in Table 11.1. The three methods are called basal, DRTA, and strategies. We use Basal, DRTA, and Strat as values for the categorical variable indicating which method of instruction each student received. ∎

Table 11.1   Pretest reading scores

Group	Subject	Score	Group	Subject	Score	Group	Subject	Score
Basal	1	4	DRTA	23	7	Strat	45	11
Basal	2	6	DRTA	24	7	Strat	46	7
Basal	3	9	DRTA	25	12	Strat	47	4
Basal	4	12	DRTA	26	10	Strat	48	7
Basal	5	16	DRTA	27	16	Strat	49	7
Basal	6	15	DRTA	28	15	Strat	50	6
Basal	7	14	DRTA	29	9	Strat	51	11
Basal	8	12	DRTA	30	8	Strat	52	14
Basal	9	12	DRTA	31	13	Strat	53	13
Basal	10	8	DRTA	32	12	Strat	54	9
Basal	11	13	DRTA	33	7	Strat	55	12
Basal	12	9	DRTA	34	6	Strat	56	13
Basal	13	12	DRTA	35	8	Strat	57	4
Basal	14	12	DRTA	36	9	Strat	58	13
Basal	15	12	DRTA	37	9	Strat	59	6
Basal	16	10	DRTA	38	8	Strat	60	12
Basal	17	8	DRTA	39	9	Strat	61	6
Basal	18	12	DRTA	40	13	Strat	62	11
Basal	19	11	DRTA	41	10	Strat	63	14
Basal	20	8	DRTA	42	8	Strat	64	8
Basal	21	7	DRTA	43	8	Strat	65	5
Basal	22	9	DRTA	44	10	Strat	66	8

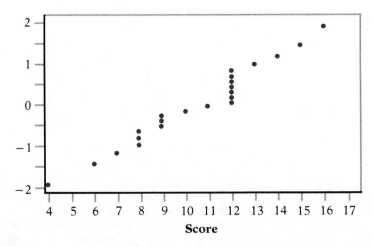

Figure 11.5(a)   Normal quantile plot for the pretest in the Basal group for the reading comprehension study.

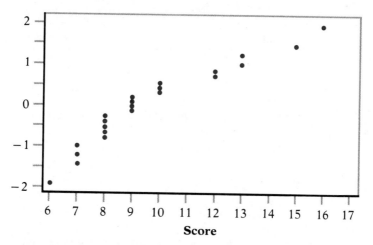

Figure 11.5(b) Normal quantile plot for the pretest in the DRTA group for the reading comprehension study.

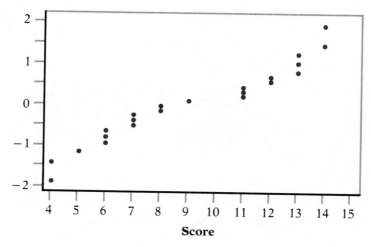

Figure 11.5(c) Normal quantile plot for the pretest in the Strat group for the reading comprehension study.

Before proceeding with ANOVA, several important preliminaries need attention. Normal quantile plots for the three groups are given in Figures 11.5(a), 11.5(b), and 11.5(c). The data look reasonably normal. Although there are a relatively small number of possible values for the score, this should not cause difficulties in using ANOVA.

Summary statistics generated by the SAS procedure MEANS are shown in Figure 11.6. The ratio of the largest standard deviation to the smallest standard deviation is $3.34/2.69 = 1.24$, which is less than 2. Our general rule

VARIABLE	N	MEAN	STANDARD DEVIATION
------------GROUP=Basal------------			
SCORE	22	10.50000000	2.97209242
------------GROUP=DRTA------------			
SCORE	22	9.72727273	2.69358706
------------GROUP=Strat------------			
SCORE	22	9.13636364	3.34230395

Figure 11.6   Descriptive statistics for pretest in reading comprehension study.

tells us that we can assume that the three populations have the same standard deviation. The pooled variance is

$$s_p^2 = \frac{(21)(2.972)^2 + (21)(2.694)^2 + (21)(3.342)^2}{21 + 21 + 21}$$

$$= \frac{572.45}{63}$$

$$= 9.09$$

Since the sample sizes are equal, this s_p^2 is simply the average of the three variances. The pooled standard deviation is

$$s_p = \sqrt{9.09} = 3.01$$

Note that we would *not* get the same answer if we averaged the three standard deviations.

Since the data look reasonably normal and the assumption of equal variances can be used, we proceed with ANOVA. The results produced by the **SAS** General Linear Models (GLM) procedure are shown in Figure 11.7. On this computer output, the calculated value of the F statistic is given under the heading F VALUE and the P-value is given under the heading PR $> F$. The value of F is 1.13 with a P-value of 0.3288. That is, an F of 1.13 or larger would occur about 33% of the time by chance when the population means are equal. Since the P-value is large, the observed varia-

```
                    GENERAL LINEAR MODELS PROCEDURE

DEPENDENT VARIABLE: SCORE

SOURCE                   DF     SUM OF SQUARES        MEAN SQUARE     F VALUE

MODEL                     2       20.57575758        10.28787879        1.13

ERROR                    63      572.45454545         9.08658009      PR > F

CORRECTED TOTAL          65      593.03030303                         0.3288


R-SQUARE               C.V.          ROOT MSE           SCORE MEAN

0.034696              30.7972       3.01439548          9.78787879


SOURCE                   DF          TYPE I SS      F VALUE    PR > F

GROUP                     2         20.57575758       1.13     0.3288
```

Figure 11.7 Analysis of variance output for the pretest in the reading comprehension study.

tion in the sample means can be attributed to chance, and we have no evidence to reject the null hypothesis that the three populations have equal means.

The information in an analysis of variance is organized in an *ANOVA table*. In the computer output depicted in Figure 11.7, the columns of this table are labeled SOURCE, DF, SUM OF SQUARES, MEAN SQUARE, and *F* VALUE. The rows are labeled MODEL, ERROR, and CORRECTED TO-TAL. These are the three sources of variation in the one-way ANOVA.

The MODEL row in the table corresponds to the FIT term in our DATA = FIT + RESIDUAL way of thinking. It gives information related to the variation *among* group means. In writing ANOVA tables we will use the generic label *groups* or some other term that describes the factor being studied for this row.

The ERROR row in the table corresponds to the RESIDUAL term in DATA = FIT + RESIDUAL. It gives information related to the variation *within* groups. The term "error" is most appropriately used for experiments in the physical sciences where all observations within a group are not the same because of measurement error. However, in the biological and social sciences, the within-group variation is often due to the fact that not all plants or people are the same. This sort of variation is not really error and is more appropriately described as "residual."

Finally, the CORRECTED TOTAL row in the table corresponds to the DATA term in our DATA = FIT + RESIDUAL framework. Often this row is simply labeled TOTAL.

sum of squares   As you might expect, each *sum of squares* is a sum of squared deviations. We use SSG, SSE, and SST for the entries in this column corresponding to groups, error, and total. To see how the sums of squares are calculated, it is helpful to write the data in the following form:

$$(x_{ij} - \bar{x}) = (\bar{x}_i - \bar{x}) + (x_{ij} - \bar{x}_i)$$

where

$$\bar{x} = \frac{\sum_{\text{obs}} x_{ij}}{N}$$

is the overall mean of all of the observations and

$$N = \sum n_i$$

is the total number of observations in all of our samples.

Each of these three terms is a different type of deviation. The first,

$$x_{ij} - \bar{x}$$

corresponds to the total variation. If we calculate this deviation for each observation, square, and sum, we obtain SST. Note that this is the type of calculation that we would use to compute a variance for the data if all observations came from a single population. Thus, SST represents the total variation in all of the data. Some of this variation is due to the fact that the group means are not all the same, while the rest is due to within-group variation.

The deviation

$$\bar{x}_i - \bar{x}$$

corresponds to the variation among the groups. If we calculate this deviation for each observation, square, and sum, we obtain SSG. Since this deviation is the same for all observations in a group, we calculate the square once for each group and multiply by n_i. SSG represents that part of SST that is due to the variation among group means. If all groups had the same $\bar{x}_i$, then SSG would be 0.

In a similar way,

$$x_{ij} - \bar{x}_i$$

corresponds to the error or residual variation within the groups. Again we calculate this deviation for each observation, square, and sum to obtain SSE. Recall that the sample residuals are $e_{ij} = x_{ij} - \bar{x}_i$. Thus, SSE is the sum of the squared residuals. It represents the part of SST that is due to the within-group variation.

In Figure 11.7 we can find the values for the three sums of squares. Note that the output gives many more digits than we need. Rounded-off values are SSG = 21, SSE = 572, and SST = 593. In this example it appears that much of the variation is coming from within groups or error. You can verify that SST = SSG + SSE for this example. In other words, the total variation is equal to the among-group variation plus the within-group variation. This is true in general.

degrees of freedom

The entries in the DF column are *degrees of freedom.* They are a little more difficult to understand than the sums of squares but are much easier to calculate. Each entry in the DF column is associated with a sum of squares and can be found by examining the deviations that make up the sum of squares. As we saw in earlier chapters, degrees of freedom represent the number of terms that are free to vary independently of the other terms in the sum of squares. The sums of squares in general all grow larger as the number of terms in the sums increases. The degrees of freedom supply a number by which to divide a sum of squares to yield a measure of the "average" variation.

First consider SSG. It is based on the deviations $\bar{x}_i - \bar{x}$. There are I sample means being compared with one overall mean. There are $I - 1$ degrees of freedom for groups; that is, DFG $= I - 1$.

Similarly, consider SSE. It is based on the deviations $x_{ij} - \bar{x}_i$. Here we have N observations being compared with I sample means. There are $N - I$ degrees of freedom for error; that is, DFE $= N - I$.

Finally, SST is based on the deviations $x_{ij} - \bar{x}$. The N observations are being compared with one mean and DFT $= N - 1$.

In our example, we have $I = 3$ and $N = 66$. Therefore,

$$\text{DFG} = 3 - 1 = 2$$
$$\text{DFE} = 66 - 3 = 63$$
$$\text{DFT} = 66 - 1 = 65$$

You can find these values in the DF column of Figure 11.7. Note that the degrees of freedom add in the same way that the sums of squares add; that is DFT = DFG + DFE.

mean square

For each source of variation, the *mean square* is the ratio of the sum of squares and the degrees of freedom. You can verify this by doing the divisions for the values given on the output in Figure 11.7.

Sums of squares, degrees of freedom, and mean squares

Sums of squares represent variation present in the data. They are calculated by summing squared deviations. In the one-way ANOVA there are three sources of variation: groups, error, and total. The sums of squares are related by the formula

$$SST = SSG + SSE$$

Thus the total variation is composed of two parts: one due to groups and one due to error.

Degrees of freedom are related to the deviations that are used in the sums of squares. The degrees of freedom are related in the same way as the sums of squares:

$$DFT = DFG + DFE$$

To calculate each mean square, we divide the corresponding sum of squares by its degrees of freedom.

At the beginning of this section we found the pooled variance to be $572.45/63 = 9.09$. Now look at the ERROR row on the computer output in Figure 11.7. Observe that

$$s_p^2 = MSE = \frac{SSE}{DFE}$$

The values in the ratio are SSE and DFE. Thus, when we divide the sum of squares by the degrees of freedom for the error row, we obtain an estimate of the within-group variance σ^2. Note that s_p is given in the output under the heading ROOT MSE.

The MEAN SQUARE entry in the CORRECTED TOTAL row of the SAS output is empty because this mean square (MST) is not needed in later calculations. If we wanted to fill it in, we would calculate

$$MST = \frac{593}{65} = 9.12$$

If H_0 is true, all three mean squares are estimates of σ^2. In this case we can view all of the observations as coming from a single population, so that there is only one source of variability present in the data. MST is the usual variance we would calculate in this situation. It is the sum of squared deviations about the mean divided by the sample size minus 1. In our example $MSG = 10.29$, $MSE = 9.09$, and $MST = 9.12$. These values are reasonably close, and this is why we did not reject H_0.

On the other hand, if H_a is true, both MSG and MST tend to overestimate σ^2 because they include group-to-group variation as well as within-group variation. Because MSE is based only on the within-group variation, it is a valid estimate of σ^2 both when H_0 is true *and* when H_a is true.

The ratio MSG/MSE is a statistic that is approximately 1 if H_0 is true and tends to be larger if H_a is true. This is the F statistic. In our example,

$$F = \frac{10.29}{9.09} = 1.13$$

If H_0 is true, the F statistic has an F distribution that depends upon two numbers: the *degrees of freedom for the numerator* and the *degrees of freedom for the denominator*. These degrees of freedom are those associated with the mean squares in the numerator and denominator of the F statistic. Thus, for the one-way ANOVA, the degrees of freedom for the numerator are DFG $= I - 1$, and the degrees for freedom for the denominator are DFE $= N - I$. We use the notation $F(\text{DFG, DFE})$ for this distribution.

The F statistic

$$F = \frac{\text{MSG}}{\text{MSE}}$$

If H_0 is true, the F statistic has an $F(\text{DFG, DFE})$ distribution. If H_a is true, the F statistic tends to be large. Therefore, we reject H_0 in favor of H_a if the F statistic is sufficiently large.

The P-value is the probability that a random variable having the $F(\text{DFG, DFE})$ distribution is greater than or equal to the calculated value of the F statistic.

We have already seen several examples where the F statistic and its P-value were used to choose between H_0 and H_a. Critical values for the F distribution are given in Table F for $\alpha = 0.100$, 0.050, 0.025, 0.010, and 0.001. For the one-way ANOVA we use critical values in the table corresponding to DFG degrees of freedom in the numerator and DFE degrees of freedom in the denominator.

EXAMPLE 11.7

P	Critical value
.100	2.33
.050	3.04
.025	3.76
.010	4.71
.001	7.15

In the study of computer science students in Example 11.3, $F = 10.35$. Since there were three populations sampled, the degrees of freedom in the numerator are DFG $= I - 1 = 2$. For this example the degrees of freedom in the denominator are $N - I = 256 - 3 = 253$. In Table F we first find the column corresponding to 2 degrees of freedom in the numerator. For the degrees of freedom in the denominator, we note that there are entries for 200 and 1000. The values for each α are very close. To be conservative we use the values corresponding to 200 degrees of freedom in the denominator since these are slightly larger. Since 10.35 is larger than all of the tabulated values, we reject H_0 and conclude that the differences in means are statistically significant with $P < 0.001$.   ■

EXAMPLE 11.8

P	Critical value
.100	2.39
.050	3.15
.025	3.93
.010	4.98
.001	7.77

In the comparison of teaching methods of Example 11.6, $F = 1.13$. Since there were three populations sampled, the degrees of freedom in the numerator are DFG $= I - 1 = 2$ and the degrees of freedom in the denominator are DFE $= N - I = 66 - 3 = 63$. We use Table F to find the critical values. There are entries for 2 degrees of freedom in the numerator. Since we do not find entries for 63 degrees of freedom in the denominator, we use the slightly larger values corresponding to 60. Since 1.13 is less than any of the tabulated values, we do not have evidence in the data to reject H_0 at any reasonable level of significance. We conclude that the differences in sample means are not significant at the 10% level. ■

The ANOVA F test shares the robustness of the two-sample t test. It is relatively insensitive to moderate nonnormality and unequal variances, especially when the sample sizes are similar.

ANOVA table

The following display shows the general form of a one-way *ANOVA table*. The formulas given in the sum of squares column can be used for calculations in small problems. There are other formulas that are more convenient for hand or calculator use, but ANOVA calculations are usually performed by computer software.

Source	Sum of squares	Degrees of freedom	Mean square	F
Groups	$\sum_{\text{groups}} n_i(\bar{x}_i - \bar{x})^2$	$I - 1$	SSG/DFG	MSG/MSE
Error	$\sum_{\text{groups}}(n_i - 1)s_i^2$	$N - I$	SSE/DFE	
Total	$\sum_{\text{obs}}(x_{ij} - \bar{x})^2$	$N - 1$	SST/DFT	

The value of the F statistic tends to be large when H_a is true.

EXAMPLE 11.9

In Example 11.8 we could not reject H_0 that the three groups in Example 11.6 had equal mean scores on a pretest of reading comprehension. We now change the data in a way that favors H_a. For each of the observations in the Basal group, add 1 point to the score. Similarly, for each of the observations in the Strat group, subtract 1. We leave the observations in the DRTA group as they were. We call the new variable SCOREX. The new summary statistics are given in Figure 11.8. Compared with the original means, given in Figure 11.6, the SCOREX means are farther apart. Note that we have not changed any of the standard deviations.

The ANOVA results appear in Figure 11.9. Compare this table with the one for the variable SCORE given in Figure 11.7. The values of the degrees of freedom are the same. The values of SSG and SST have increased, while SSE has not changed. In constructing SCOREX, we have added more total variation to the data, and all of this added variation is reflected in the added variation between groups. The increase in SSG is reflected in MSG. The new value is about six times the old. Since MSE has not changed, our F statistic is also about six times as large. The P-value has decreased dramatically and we now have evidence to reject H_0 in favor of H_a. ■

```
VARIABLE        N         MEAN       STANDARD
                                     DEVIATION

- - - - - - - - - GROUP=Basal - - - - - - - - -

SCOREX         22    11.50000000    2.97209242

- - - - - - - - - GROUP=DRTA - - - - - - - - -

SCOREX         22     9.72727273    2.69358706

- - - - - - - - - GROUP=Strat - - - - - - - - -

SCOREX         22     8.13636364    3.34230395
```

Figure 11.8 Descriptive statistics for SCOREX in the reading comprehension study.

```
              GENERAL LINEAR MODELS PROCEDURE

   DEPENDENT VARIABLE: SCORE

SOURCE              DF     SUM OF SQUARES      MEAN SQUARE     F VALUE

MODEL                2      124.57575758      62.28787879       6.85

ERROR               63      572.45454545       9.08658009      PR > F

CORRECTED TOTAL     65      697.03030303                       0.0020

R-SQUARE          C.V.        ROOT MSE         SCOREX MEAN

0.178724        30.7972      3.01439548         9.78787879


SOURCE              DF       TYPE I SS     F VALUE    PR > F

GROUP                2      124.57575758     6.85     0.0020
```

Figure 11.9 Analysis of variance output for SCOREX in the reading comprehension study.

coefficient of
determination

For an analysis of variance, we define the *coefficient of determination* as

$$R^2 = \frac{\text{SSG}}{\text{SST}}$$

This coefficient is very similar to the squared multiple correlation R^2 in a multiple regression. In Figure 11.7, we find R^2 listed under the heading R-SQUARE. The value is 0.035. One way to describe this result is to say that the FIT part of our model (that is, groups) explains 3.5% of the total variation in our data. Compare this value with that for the changed data described in Example 11.9 and given in Figure 11.9. The change has increased the coefficient of determination to 17.9%.

Contrasts

The ANOVA *F* test gives a general answer to a general question: are the differences among observed group means significant? Unfortunately, although a small *P*-value tells us that the group means are different, it does not tell us specifically which means differ from each other. Of course, inspection and plotting of the means give us some indication of where the differences lie.

In the ideal situation, particular questions regarding comparisons among means are specified before the data are collected. Our analysis is capable of providing specific answer to specific questions of this kind. We now explore these ideas with an example.

In Example 11.6, we discussed an experiment in which three methods for teaching reading comprehension were compared. We analyzed the pretest scores and found no reason to reject H_0 that the three groups had similar population means on this measure. This was the desired outcome. We now consider an analysis of the response variable, another measure of reading comprehension. This variable, which we call COMP, was measured by a test taken after the instruction was given.

EXAMPLE 11.10

For the experiment described in Example 11.6, Figure 11.10 gives the summary statistics computed by the SAS procedure MEANS for COMP. The ANOVA results generated by the SAS procedure GLM are given in Figure 11.11

The ANOVA null hypothesis is

$$H_0: \mu_B = \mu_D = \mu_S$$

where the subscripts correspond to the group labels. For these data we have $F = 4.48$, with degrees of freedom 2 and 63. The *P*-value is 0.0152 and the value of R^2 is 0.12. The MSE is 39.87 and $s_p = 6.31$. We have good evidence against H_0.

What have we learned from this analysis that the researchers can interpret and discuss? The alternative hypothesis is true if $\mu_B \neq \mu_D$ or if $\mu_B \neq \mu_S$ or if $\mu_D \neq \mu_S$ or if any combination of these statements is true. ■

```
VARIABLE      N        MEAN        STANDARD
                                   DEVIATION

- - - - - - - - -GROUP=Basal- - - - - - - - -

COMP         22    41.04545455    5.63557808

- - - - - - - - - GROUP=DRTA - - - - - - - - -

COMP         22    46.72727273    7.38841963

- - - - - - - - -GROUP=Strat

COMP         22    44.27272727    5.76675049
```

Figure 11.10 Descriptive statistics for comprehension score in the reading comprehension study.

```
                GENERAL LINEAR MODELS PROCEDURE

DEPENDENT VARIABLE: COMP

SOURCE              DF     SUM OF SQUARES      MEAN SQUARE    F VALUE

MODEL                2      357.30303030      178.65151515      4.48

ERROR               63     2511.68181818       39.86796537    PR > F

CORRECTED TOTAL     65     2868.98484848                       0.0152


R-SQUARE          C.V.         ROOT MSE          COMP MEAN

0.124540        14.3453       6.31410844        44.01515152


SOURCE              DF        TYPE I SS    F VALUE    PR > F

GROUP                2      357.30303030      4.48    0.0152


CONTRAST            DF              SS    F VALUE    PR > F

B VS D AND S         1      291.03030303      7.30    0.0088
D VS S               1       66.27272727      1.66    0.2020
```

Figure 11.11 Analysis of variance and contrasts output for comprehension score in the reading comprehension study.

In this experiment, the researchers had in mind a particular theory about reading comprehension that they wanted to investigate. The instruction for the Basal group was the standard method commonly used in schools. The DRTA and Strat groups received innovative methods of teaching that were specifically designed to increase the reading comprehension of the children. The DRTA and Strat methods were not identical, but they both involved teaching the students to use certain comprehension strategies in their reading.

It is clear that the F test from ANOVA did not directly answer the researchers' questions. We have demonstrated that there is some sort of difference in the means but more is needed to complete the analysis. The primary question that the researchers would like answered can be formulated as a null hypothesis:

$$H_{01}: \frac{1}{2}(\mu_D + \mu_S) = \mu_B$$

with the alternative

$$H_{a1}: \frac{1}{2}(\mu_D + \mu_S) > \mu_B$$

The hypothesis H_{01} compares the standard method (Basal) with the average of the two innovative methods (DRTA and Strat). The alternative is one-sided because the researchers are interested in demonstrating that the new methods are better than the old.

A secondary question involves a comparison of the two new methods We formulate this as the hypothesis that the methods DRTA and Strat are equally effective:

$$H_{02}: \mu_D = \mu_S$$

versus the alternative

$$H_{a2}: \mu_D \neq \mu_S$$

Note that we have used numerical subscripts on the hypotheses to distinguish them.

contrasts

To answer the specific questions posed by H_{01} and H_{02}, we express the combinations of means that we want to study as *contrasts*. We use the Greek letter ψ for these quantities. In this case, we have two contrasts:

$$\psi_1 = -\mu_B + \frac{1}{2}(\mu_D + \mu_S)$$
$$= (-1)\mu_1 + (.5)\mu_D + (.5)\mu_S$$

and

$$\psi_2 = \mu_D - \mu_S$$

In writing the contrasts we express each combination of means in such a way that the value of the contrast is 0 when H_0 is true.

A contrast expresses an effect in the population as a combination of population means. To estimate the contrast, form the corresponding *sample contrast* by using sample means in place of population means. Under the ANOVA assumptions, a sample contrast is a linear combination of independent normal variables and therefore has a normal distribution. We can obtain the standard error of a contrast by using the rules for variances given in Section 5.3. Inference is based on the t distributions.

Contrasts

A contrast is a combination of means of the form

$$\psi = \sum_{i=1}^{I} c_i \mu_i$$

where the c_i have sum 0. The corresponding sample contrast is

$$C = \sum_{i=1}^{I} c_i \bar{x}_i$$

The standard error of C is

$$s(C) = s_p \sqrt{\sum_{i=1}^{I} \frac{c_i^2}{n_i}}.$$

To test the null hypothesis,

$$H_0: \psi = 0$$

use the t statistic,

$$t = \frac{C}{s(C)}$$

The degrees of freedom are those associated with s_p, i.e. DFE. The alternative hypothesis may be one-sided or two-sided. A level $1 - \alpha$ confidence interval for ψ is

$$C \pm t^* s(C)$$

where t^* is the upper $\alpha/2$ critical value for the $t(\text{DFE})$ distribution.

Since each $\bar{x}_i$ estimates the corresponding μ_i, the addition rule for means tells us that $\mu_C = \psi$. In other words, C estimates ψ.

Testing the hypothesis that a contrast is 0 assesses the significance of the effect measured by the contrast. It is often more informative to estimate the size of the effect using a confidence interval for the population contrast.

For our examples we have $(c_1, c_2, c_3) = (-1, 0.5, 0.5)$ for ψ_1 and $(0, 1, -1)$ for ψ_2, where the subscripts 1, 2 and 3 correspond to B, D, and S. In each case the sum of the c_i is 0.

The sample contrast that estimates ψ_1 is

$$C_1 = -\bar{x}_B + \frac{1}{2}(\bar{x}_D + \bar{x}_S)$$

$$= -41.05 + \frac{1}{2}(46.73 + 44.27)$$

$$= 4.45$$

with standard error

$$s(C_1) = 6.314 \sqrt{\frac{(-1)^2}{22} + \frac{(.5)^2}{22} + \frac{(.5)^2}{22}}$$

$$= 1.65$$

To test $H_{01}: \psi_1 = 0$ versus $H_{a1}: \psi_1 > 0$, we use

$$t_1 = \frac{4.45}{1.65} = 2.70$$

Since s_p has 63 degrees of freedom, we use the $t(63)$ distribution and calculate the one-sided P-value as 0.0044. We used a computer to calculate this P-value, but if we used Table E, we would conclude that $P < 0.005$. Since the P-value is small, we reject H_{01}. The researchers have demonstrated that the new methods produce higher mean scores than the old. The size of the improvement can be described with a confidence interval. The 95% confidence interval is

$$C_1 \pm t^* s(C_1) = 4.45 \pm 2.00(1.65)$$

$$= 4.45 \pm 3.30$$

$$= (1.15, 7.75)$$

With 95% confidence we ascertain that the mean improvement obtained by using the innovative methods as compared to the old method is between 1.15 and 7.75 points.

Similarly, for the second contrast comparing the two new methods, we have

$$C_2 = \bar{x}_D - \bar{x}_S$$
$$= 46.73 - 44.27$$
$$= 2.46$$

with standard error

$$s(C_2) = 6.314 \sqrt{\frac{(1)^2}{22} + \frac{(-1)^2}{22}}$$
$$= 1.90$$

The t statistic is

$$t_2 = \frac{2.46}{1.90} = 1.29$$

Here we have a two-sided alternative. The P-value is 0.2020, and we conclude that either the two new methods have the same population mean or the sample sizes are not sufficiently large to distinguish them. The confidence interval helps us to be more specific. The 95% confidence interval for ψ_2 is

$$C_2 \pm t^* s(C_2) = 2.46 \pm 2.00(1.90)$$
$$= 2.46 \pm 3.80$$
$$= (-1.34, 6.26)$$

With 95% confidence we state that the difference between the population means for the two new methods is between -1.34 and 6.26.

Many computer software packages report the statistics associated with contrasts as F statistics rather than t statistics. These F statistics are the squares of the t statistics described above. The SAS output from the GLM procedure is given at the bottom of Figure 11.11. Note the F statistics 7.30 and 1.66 given in the F VALUE column. The degrees of freedom for each of these are 1 and 63. The P-values are correct for two-sided alternative hypotheses. They are given as 0.0088 and 0.2020. To convert the computer-generated results to apply to our one-sided hypothesis concerning ψ_1, we can simply divide the reported P-value by 2 after checking that the value of C_1 is in the direction of H_a (that is, that C_1 is positive).

In this section we showed how to translate questions about population means into hypotheses involving contrasts. Each contrast in our example addressed a specific question that the researchers had in mind when designing the experiment. For this reason, the analysis of contrasts as described is valid whether or not the ANOVA H_0 of equality of means is rejected. Since the F test answers a very general question, it is not particularly powerful when compared with the use of contrasts designed to answer specific questions. Specifying the important questions before the analysis is undertaken enables us to use this powerful statistical technique.

Multiple Comparisons*

For many experiments, specific questions either are not or cannot be specified in advance of the analysis. If H_0 is not rejected, we conclude that the population means are indistinguishable on the basis of the data given. On the other hand, if H_0 is rejected, we would like to know which pairs of means differ. *Multiple comparisons* are methods that address this question. It is important to keep in mind that they are used only after rejecting the ANOVA H_0.

multiple comparisons

We start by returning to the reading comprehension data given in Example 11.10. Since there are three samples, we can construct three t statistics to compare the population means. Specifically, we have

$$t_{12} = \frac{\bar{x}_1 - \bar{x}_2}{s_p \sqrt{\dfrac{1}{n_1} + \dfrac{1}{n_2}}}$$

$$= \frac{41.05 - 46.73}{6.31 \sqrt{\dfrac{1}{22} + \dfrac{1}{22}}}$$

$$= -2.99$$

to compare populations 1 and 2. We have used subscripts on the t to specify which groups are being compared. For the other two pairs, we have

$$t_{13} = \frac{\bar{x}_1 - \bar{x}_3}{s_p \sqrt{\dfrac{1}{n_1} + \dfrac{1}{n_3}}}$$

$$= \frac{41.05 - 44.27}{6.31 \sqrt{\dfrac{1}{22} + \dfrac{1}{22}}}$$

$$= -1.69$$

and

$$t_{23} = \frac{\bar{x}_2 - \bar{x}_3}{s_p \sqrt{\dfrac{1}{n_2} + \dfrac{1}{n_3}}}$$

$$= \frac{46.73 - 44.27}{6.31 \sqrt{\dfrac{1}{22} + \dfrac{1}{22}}}$$

$$= 1.29$$

* This section is optional.

Note that we performed the last calculation when we analyzed the contrast ψ_2 in the previous section. These statistics are very similar to the two-sample t statistics that we encountered in Chapter 8. The only real difference is that each statistic uses the variance information from all groups as expressed in s_p rather than only the information from the two groups being compared. As a consequence, we increase the power of the procedures. The degrees of freedom for all of these statistics are DFE = 63, those associated with s_p.

Since we do not have any particular ordering of the means in mind as an alternative to equality, we must use a two-sided approach to the problem of deciding which pairs of means are significantly different.

Multiple comparisons

> To perform a multiple comparisons procedure, we compute t statistics for all pairs of means using the formula
>
> $$t_{ij} = \frac{\bar{x}_i - \bar{x}_j}{s_p \sqrt{\dfrac{1}{n_i} + \dfrac{1}{n_j}}}$$
>
> If
>
> $$|t_{ij}| \geq t^{**}$$
>
> we declare that the population means μ_i and μ_j are different. Otherwise, we conclude that they are not distinguishable. The value of t^{**} depends upon which multiple comparisons procedure we choose.

One obvious choice for t^{**} is the upper $\alpha/2$ critical value for the $t(\text{DFE})$ distribution. This choice simply carries out as many separate significance tests of fixed level α as there are pairs of means to be compared. The procedure based on this choice is called the least-significant differences method, or simply the LSD. This method has some undesirable properties, particularly if the number of means being compared is large. Suppose, for example, we had $I = 20$ groups and used the LSD with $\alpha = 0.05$. The number of pairs of means to consider is 190. If we perform 190 tests, each with a Type I error rate of 5%, our overall error rate will be unacceptably large. We would expect about 5% of the 190 to be significant even if the corresponding population means were the same. Since 5% of 190 is 9.5, we would expect 9 or 10 of these false conclusions.

The LSD procedure fixes the probability of a false rejection for each single pair of means being compared. It puts no control over the overall probability of some false rejection among all pairs. Other choices of t^{**} control possible errors in other ways. The choice of t^{**} is therefore a complex problem. A detailed discussion of it is beyond the scope of this text. Many choices for t^{**} have been suggested and are used in practice. One major statistical package allows selection from a list of at least a dozen choices.

Bonferroni procedure

We will discuss one of these, called the *Bonferroni procedure*. Use of this procedure with $\alpha = 0.05$, for example, guarantees that the probability of *any* false rejection among all comparisons made is no greater than 0.05. This is much stronger protection than controlling the probability of a false rejection at 0.05 for *each* separate comparison.

With many computer programs the output for multiple comparison procedures takes the form of a display of the means with a value for the least- or minimum-significant difference. Any pair of sample means that differ by more than this amount are significantly different. That is, we conclude that the corresponding population means are different. In our notation, this minimum-significant difference (MSD) is simply

$$\text{MSD} = t^{**} s_p \sqrt{\frac{1}{n_i} + \frac{1}{n_j}}$$

EXAMPLE 11.11

We apply the Bonferroni multiple comparison procedure with $\alpha = 0.05$ to the data from the reading comprehension study used in Example 11.6. The value of t^{**} for this procedure is 2.46. Referring to the statistics t_{12}, t_{13}, and t_{23} that we calculated in the beginning of this section, we see that only t_{12} is significant. The output generated by the MEANS statement in the SAS procedure GLM is given in Figure 11.12(a). Note that the value of t^{**} is given on the output as CRITICAL

```
            GENERAL LINEAR MODELS PROCEDURE

BONFERRONI (DUNN) T TESTS FOR VARIABLE: COMP
NOTE: THIS TEST CONTROLS THE TYPE I EXPERIMENTWISE ERROR RATE
      BUT GENERALLY HAS A HIGHER TYPE II ERROR RATE THAN REGWQ

        ALPHA=0.05 DF=63 MSE=39.868
        CRITICAL VALUE OF T=2.45958
        MINIMUM SIGNIFICANT DIFFERENCE=4.6825

  MEANS WITH THE SAME LETTER ARE NOT SIGNIFICANTLY DIFFERENT.

      BON        GROUPING          MEAN      N    GROUP

                         A        46.727    22    DRTA
                         A
                 B       A        44.273    22    Strat
                 B
                 B                41.045    22    Basal
```

Figure 11.12(a) Bonferroni multiple comparisons output for comprehension scores in the reading comprehension study.

```
                    GENERAL LINEAR MODELS PROCEDURE


BONFERRONI (DUNN) T TESTS FOR VARIABLE: COMP
NOTE:  THIS TEST CONTROLS THE TYPE I EXPERIMENTWISE ERROR RATE
       BUT GENERALLY HAS A HIGHER TYPE II ERROR RATE THAN TUKEY'S


       ALPHA=0.05  CONFIDENCE=0.95  DF=63  MSE=39.868
       CRITICAL VALUE OF T=2.45958
       MINIMUM SIGNIFICANT DIFFERENCE=4.6825


COMPARISONS SIGNIFICANT AT THE 0.05 LEVEL ARE INDICATED BY '***'
```

GROUP COMPARISON	SIMULTANEOUS LOWER CONFIDENCE LIMIT	DIFFERENCE BETWEEN MEANS	SIMULTANEOUS UPPER CONFIDENCE LIMIT	
DRTA — Strat	-2.228	2.455	7.137	
DRTA — Basal	0.999	5.682	10.364	***
Strat — DRTA	-7.137	-2.455	2.228	
Strat — Basal	-1.455	3.227	7.910	
Basal — DRTA	-10.364	-5.682	-0.999	***
Basal — Strat	-7.910	-3.227	1.455	

Figure 11.12(b)   Bonferroni multiple comparisons output for confidence limits for differences in comprehension scores in the reading comprehension study.

VALUE OF T. The MSD is 4.6825. The letters A under the heading **GROUPING** extend from the DRTA group to the Strat group. This means that these two groups are not distinguishable. Similarly, the letters B extend from the Strat group to the Basal group, indicating that these two groups also cannot be distinguished. There is no collection of letters that connects the DRTA and Basal groups. This means that we can declare these two groups to be significantly different.   ■

Unfortunately, this is not a very satisfactory conclusion. It appears to be illogical. If μ_1 is the same as μ_3, and μ_2 is the same as μ_3, then doesn't it follow that μ_1 is the same as μ_2? Logically, the answer must be yes.

This situation points out dramatically the limitations of any statistical procedure for inference. Some of the difficulty can be resolved by noting the choice of words used above. In describing the inferences, we talk about being either able or unable to distinguish between pairs of means. In making the logical statements, we say, "is the same as." There is a big difference between the two modes of thought. If we think about the problem in another way, we realize that it is really very unlikely that any two methods of teaching reading comprehension would give *exactly* the same population means. Statistical procedures infer differences and similarities from the data collected rather than reveal the ultimate truth about the population means.

One way to deal with the difficulties described above is to use confidence intervals for the differences. Again, we must face the problem that we have more than one procedure—in this case, more than one method of obtaining intervals—to consider.

Confidence intervals for all differences $\mu_i - \mu_j$ between means have the form

$$(\bar{x}_i - \bar{x}_j) \pm t^{**} s_p \sqrt{\frac{1}{n_i} + \frac{1}{n_j}}$$

where the values of t^{**} are the same as those used for the multiple comparison procedure chosen.

The confidence intervals generated in this way are closely related to the multiple comparison result. If the confidence interval includes the value 0, then the means will not be declared different, and vice versa.

EXAMPLE 11.12 | Adding the CLDIFF option to the MEANS statement in the SAS GLM procedure gives the output in Figure 11.12b for the data in Example 11.5. ∎

Power*

Recall that the power of a test is the probability of rejecting H_0 when H_a is in fact true. When planning a study in which ANOVA is to be used for the analysis of data, it is important to perform power calculations to check that the sample sizes are adequate to detect differences among means that are judged to be important. Power calculations can also be used to evaluate and interpret the results of studies in which H_0 is not rejected. In such cases we sometimes find that we have an unacceptably small chance of obtaining a significant F when the means differ as might be expected.

noncentrality parameter
To find the power for the ANOVA F test, we proceed in three steps. First we calculate a quantity called the *noncentrality parameter*, λ. It measures the variation among population means relative to the within-group

---

* This section is optional.

variance σ^2. Specifically,[4]

$$\lambda = \frac{\sum_{i=1}^{I} n_i(\mu_i - \bar{\mu})^2}{\sigma^2}$$

where

$$\bar{\mu} = \sum_{i=1}^{I} w_i\mu_i$$

and

$$w_i = \frac{n_i}{\sum_{i=1}^{I} n_i}$$

Note that if the n_i are all equal with common value n, then

$$\lambda = \frac{n \sum_{i=1}^{I} (\mu_i - \bar{\mu})^2}{\sigma^2}$$

and $\bar{\mu}$ is the usual average of the μ_i.

    The second step in the calculation of power is to find the critical value that would be used to determine whether or not H_0 is rejected. This value, which we denote by F^*, is the upper α critical value for the $F(\text{DFG}, \text{DFE})$ distribution. As before, power calculations are done using a standard value of α, such as 0.05, even though P-values will be reported in the actual analysis.

    The last step is to find the desired probability. We prefer to use a computer for this calculation, although tables and charts are available. Under **noncentral F** H_a, the F statistic has a distribution known as the *noncentral F distribution*. **distribution** Many software packages have routines that can be used to find probabilities for this distribution. In SAS the function PROBF $(x, \text{dfn}, \text{dfd}, \lambda)$ gives the probability of falling below x for the noncentral F distribution with noncentrality parameter λ and degrees of freedom dfn and dfd. Using this routine, we find the power as

$$\text{Power} = 1 - \text{PROBF}(F^*, \text{DFG}, \text{DFE}, \lambda)$$

EXAMPLE 11.13 | Suppose that a study on reading comprehension similar to the one described in Example 11.10 had been performed with sample sizes of 10 in each group. Let us suppose that the population means and the standard deviation are approximately the same as we observed in that example. For our calculation we will assume $\mu_1 = 41$, $\mu_2 = 47$, $\mu_3 = 44$, and $\sigma = 7$. Since the n_i are equal, $\bar{\mu}$ is simply the average of the μ_i:

$$\bar{\mu} = \frac{41 + 47 + 44}{3} = 44$$

Therefore,

$$\lambda = \frac{10[(41 - 44)^2 + (47 - 44)^2 + (44 - 44)^2]}{49}$$

$$= \frac{(10)(18)}{49}$$

$$= 3.67$$

Since there are three groups with 10 observations per group, we have DFG = 2 and DFE = 27. The critical value for $\alpha = 0.05$ is $F^* = 3.35$. Our power is therefore

$$1 - \text{PROBF}(3.35, 2, 27, 3.67) = .35$$

The chance that we reject the ANOVA H_0 is only about 35%.   ■

If the assumed values of the μ_i in this example describe differences among the groups that the experimenter wants to detect, then there is little point in conducting the experiment with only 10 subjects per group. Although H_0 is assumed to be false, the chance of rejecting it is only about 35%. This chance can be increased to acceptable levels by increasing the sample sizes.

EXAMPLE 11.14

To decide on an appropriate sample size for the experiment described in the previous example, we repeat the power calculation for different values of n, the number of subjects in each group. The results are summarized in the following table:

n	DFG	DFE	F^*	λ	Power
20	2	57	3.16	7.35	.65
30	2	87	3.10	10.02	.80
40	2	117	3.07	14.69	.93
50	2	147	3.06	18.37	.97
100	2	297	3.03	36.73	≈ 1

■

With $n = 40$ the experimenters have a 93% chance of rejecting H_0 with $\alpha = 0.05$ and thereby demonstrating that the groups have different means. In the long run, 93 out of every 100 such experiments would reject H_0 at the $\alpha = 0.05$ level of significance.

Note that if they use 50 subjects per group, they increase their chance of finding significance to 97%. For 100 subjects per group, they are virtually certain to reject H_0. The actual power for $n = 100$ is 0.99989. This is a bit excessive. In most cases the cost added to the experiment by increasing the sample size from 50 to 100 subjects per group would not be justified by the relatively small increase in the chance of obtaining statistically significant results.

SUMMARY

One-way analysis of variance is used to compare several population means. The categorical variable that distinguishes the populations is a **factor** and the different values for a factor are **levels**.

The populations are assumed to be normal with possibly different means and the same variance. Independent SRSs of size n_i are drawn from each population.

To do an analysis of variance, first compute sample means and standard deviations. Examine the normal quantile plots to detect outliers or extreme deviations from normality. Compute the ratio of the largest to the smallest sample standard deviation. If this ratio is less than 2 and the normal quantile plots are satisfactory, proceed with the analysis.

The **null hypothesis** is that the population means are *all* equal. The **alternative hypothesis** is true if there are *any* differences among the population means.

The analysis is based on separating the total variation observed in the data into two parts: variation *among* group means and variation *within* groups. If the variation among groups is large relative to the variation within groups, we have evidence against the null hypothesis.

An **analysis of variance table** is used to organize the calculations. **Sums of squares**, **degrees of freedom**, and **mean squares** are part of the table. The F statistic and its P-value are used to test the null hypothesis.

Specific questions formulated before examination of the data can be expressed as **contrasts**. Tests and confidence intervals for contrasts provide answers to these questions.

If no specific questions are formulated before examination of the data and the null hypothesis of equality of population means is rejected, methods called **multiple comparisons** are used to assess the statistical significance of the differences between pairs of means.

The **power** of the F test depends upon the sample sizes, the variation among population means, and the within-group standard deviation.

SECTION 11.1 EXERCISES

11.1 For each of the following situations, identify the response variable and the factor, and state the number of levels of the factor.

   (a) Four varieties of tomato plants are compared. Ten plants of each variety are grown and the yield in pounds of tomatoes is recorded.

(b) In a marketing experiment six different types of packaging are used for a laundry detergent. Each package is shown to 120 different potential consumers who rate the attractiveness of the product on a 1 to 10 scale.

(c) To compare the effectiveness of three different weight loss programs, 10 people are randomly assigned to each. At the end of the program, the weight loss for each of the participants is recorded.

11.2 For each of the following situations, identify the response variable and the factor, and state the number of levels of the factor.

(a) In a study on smoking, subjects are classified as nonsmokers, moderate smokers, or heavy smokers. A sample of size 200 is drawn from each group. Every person is asked to report the number of hours of sleep they get on a typical night.

(b) The strength of cement depends upon the formula used to prepare it. In one study, five different mixtures are to be compared. Six batches of each mixture are prepared, and the strength of each sample is measured.

(c) Four methods for teaching sign language are to be compared. Ten students are randomly assigned to each of the methods, and the score on a final exam is recorded.

11.3 Three programs designed to help people lose weight are to be compared. There are 20 subjects in each program. The sample standard deviations for the amount of weight lost (in pounds) are 5.2, 8.9, and 10.1. Can you use the assumption of equal standard deviations to analyze these data? Compute the pooled variance. Find s_p.

11.4 Data were collected on the weight (in kilograms) of men in four different age groups. The sample standard deviations for the groups were 12.2, 10.4, 9.2, and 11.7. The sample sizes for the groups were 92, 34, 35, and 24. Can you use the asssumption of equal standard deviations to analyze these data? Compute the pooled variance. Find s_p.

11.5 Refer to Exercise 11.1. For each part, outline the ANOVA table, giving the sources of variation and the degrees of freedom. (Do not compute the numerical values for the sums of squares and mean squares.)

11.6 Refer to Exercise 11.2. For each part, outline the ANOVA table, giving the sources of variation and the degrees of freedom. (Do not compute the numerical values for the sums of squares and mean squares.)

11.7 Refer to the computer science study described in Example 11.3.

(a) State H_0 and H_a.

(b) Outline the ANOVA table, giving the sources of variation and the degrees of freedom.

(c) What is the distribution of the F statistic under the assumption that H_0 is true?

(d) Find the critical value for an $\alpha = 0.05$ test using Table F.

11.8 Refer to the survey of college students described in Example 11.4.

(a) State H_0 and H_a.

(b) Outline the ANOVA table, giving the sources of variation and the degrees of freedom.

(c) What is the distribution of the F statistic under the assumption that H_0 is true?

(d) Find the critical value for an $\alpha = 0.05$ test using Table F.

11.9 In a study of the effects of exercise on physiological and psychological variables, four groups of male subjects were studied. The treatment group (T) consisted of 10 participants in an exercise program. A control group (C) of 5 subjects volunteered for the program but were unable to attend for various reasons. Subjects in the other two groups were selected to be similar to those in the first two groups in age and other characteristics. These were 11 joggers (J) and 10 sedentary people (S) who did not regularly exercise. (Data were provided by Dennis Lobstein, from his Ph.D. dissertation "A multivariate study of exercise training effects on beta-endorphin and emotionality in psychologically normal, medically healthy men," Purdue University, 1983.) One of the variables measured at the end of the program was a physical fitness score. Part of the ANOVA table used to analyze these data is given below.

Source	Sum of squares	Degrees of freedom	Mean square	F
Groups	104,855.87	3		
Error	70,500.59	32		
Total				

(a) Fill in the missing entries in the ANOVA table.

(b) State H_0 and H_a for this experiment.

(c) What is the distribution of the F statistic under the assumption that H_0 is true? Using Table F, give an approximate P-value for the ANOVA test. What do you conclude?

(d) What is s_p^2, the estimate of the within-group variance? What is s_p?

11.10 Refer to the previous exercise for details of the experiment. Another variable measured was a depression score. Higher values of this score indicate more depression. Part of the ANOVA table for these data is given below.

Source	Sum of squares	Degrees of freedom	Mean square	F
Groups	476.87	3		
Error	2009.88	32		
Total				

(a) Fill in the missing entries in the ANOVA table.
(b) State H_0 and H_a for this experiment.
(c) What is the distribution of the F statistic under the assumption that H_0 is true? Using Table F, give an approximate P-value for the ANOVA test. What do you conclude?
(d) What is s_p^2, the estimate of the within-group variance? What is s_p?

11.11 The weight gain of women during pregnancy has an important effect on the birth weight of their children. If the weight gain is not adequate, the infant is more likely to be small and will tend to be less healthy. In a study conducted in three countries, weight gains (in kilograms) of women during the third trimester of pregnancy were measured. The results are summarized in the following table. (Data are taken from Collaborative Research Support Program in Food Intake and Human Function, *Management Entity Final Report*, University of California, Berkeley, 1988.)

Country	n	$\bar{x}$	s
Egypt	46	3.7	2.5
Kenya	111	3.1	1.8
Mexico	52	2.9	1.8

(a) Find the pooled estimate of the within-country variance s_p^2. What entry in the ANOVA table gives this quantity?
(b) The sum of squares for countries (groups) is 17.22. Use this information and that given above to complete all entries in an ANOVA table.
(c) State H_0 and H_a for this study.
(d) What is the distribution of the F statistic under the assumption that H_0 is true? Use Table F to find an approximate P-value for the significance test. Report your conclusion.
(e) Calculate R^2, the coefficient of determination.

11.12 In another part of the study described in the previous exercise, measurements of food intake in kilocalories were taken on many individuals several times during the period of a year. From these data, average yearly food intake values were computed for each individual. The results for toddlers aged 18 to 30 months are summarized in the following table:

Country	n	$\bar{x}$	s
Egypt	88	1217	327
Kenya	91	844	184
Mexico	54	1119	285

(a) Find the pooled estimate of the within-country variance s_p^2. What entry in the ANOVA table gives this quantity?

(b) The sum of squares for countries (groups) is 6,572,551. Use this information and that given above to complete all entries in an ANOVA table.

(c) State H_0 and H_a for this study.

(d) What is the distribution of the F statistic under the assumption that H_0 is true? Use Table F to find an approximate P-value for the significance test. Report your conclusion.

(e) Calculate R^2, the coefficient of determination.

11.13 Refer to the computer science study described in Example 11.3. Let μ_1, μ_2, and μ_3 represent the population mean SAT scores for the three groups.

(a) Since the first two groups (computer science and engineering or other science) are majoring in areas that require mathematical skills, it is natural to compare the average of these two with the third group. Write an expression for a contrast ψ_1 in terms of the μ_i that represents this comparison.

(b) Let ψ_2 be a contrast that compares the first two groups. Write an expression for ψ_2 in terms of the μ_i that represents this comparison.

11.14 Refer to the survey of college students described in Example 11.4. Let μ_1, μ_2, μ_3, and μ_4 represent the population mean expenditures on textbooks for the freshmen, sophomores, juniors, and seniors.

(a) Since juniors and seniors take higher-level courses, which might use more expensive textbooks, we want to compare the average of the freshmen and sophomores with the average of the juniors and seniors. Write a contrast for this comparison.

(b) Write a contrast for comparing the freshmen with the sophomores.

(c) Write a contrast for comparing the juniors with the seniors.

11.15 Refer to the SAT mathematics scores for the computer science study given in Example 11.3. Answer the following questions for the two contrasts that you defined in Exercise 11.13.

(a) For each contrast give H_0 and an appropriate H_a. Note that in deciding upon the alternatives you can use information given in the description of the problem, but you cannot consider any impressions that you have obtained by inspection of the sample means.

(b) Find the value of the corresponding sample contrasts C_1 and C_2.

(c) Using the value $s_p = 82.5$, calculate the standard errors $s(C_1)$ and $s(C_2)$ for the sample contrasts.

(d) Give the test statistics and approximate P-values for the two significance tests. What do you conclude?

(e) Compute 95% confidence intervals for the two contrasts.

11.16 Refer to the computer science study described in Example 11.3. Data for the high school mathematics grades of these students is presented in the following table. The sample sizes in this exercise are different from those in Example 11.3 because complete information was not available on all students. The values have been coded so that $10 = $ A, $9 = $ A$-$, etc.

Second-year major	n	$\bar{x}$	s
Computer science	90	8.77	1.41
Engineering or other science	28	8.75	1.46
Other	106	7.83	1.74

Answer the following questions for the two contrasts that you defined in (a) and (b) of Exercise 11.13.

(a) For each contrast give H_0 and an appropriate H_a. Note that in deciding upon the alternatives you can use information given in the description of the problem, but you cannot consider any impressions that you have obtained by inspection of the sample means.

(b) Find the value of the corresponding sample contrasts C_1 and C_2.

(c) Using the value $s_p = 1.581$, calculate the standard errors $s(C_1)$ and $s(C_2)$ for the sample contrasts.

(d) Give the test statistics and approximate P-values for the two significance tests. What do you conclude?

(e) Compute 95% confidence intervals for the two contrasts.

11.17 Refer to the exercise program study described in Exercise 11.9. The summary statistics for physical fitness scores are as follows:

Group	n	$\bar{x}$	s
Treatment (T)	10	291.91	38.17
Control (C)	5	308.97	32.07
Joggers (J)	11	366.87	41.19
Sedentary (S)	10	226.07	63.53

In planning the experiment, the researchers wanted to address the following questions about the physical fitness scores. In these questions "better" means a higher fitness score. (1) Is T better than C? (2) Is T better than the average of C and S? (3) Is J better than the average of the other three groups?

(a) For each of the three questions, define an appropriate contrast. Translate the questions into null and alternative hypotheses about the contrasts.
(b) Calculate the sample contrasts and the associated standard errors.
(c) Test the contrasts and give approximate P-values.
(d) Summarize your conclusions. Do you think that the use of contrasts in this way gave a fair summary of the results?
(e) You found that the groups differ significantly in the physical fitness scores. Does this study allow conclusions about causation—for example, that a sedentary lifestyle causes people to be less physically fit? Explain your answer.

11.18 Refer to the exercise program study described in Exercise 11.9 and the ANOVA for the depression scores given in Exercise 11.10. The summary statistics for the depression scores are as follows:

Group	n	$\bar{x}$	s
Treatment (T)	10	51.90	6.42
Control (C)	5	57.40	10.46
Joggers (J)	11	49.73	6.27
Sedentary (S)	10	58.20	9.49

In planning the experiment, the researchers wanted to address the following questions about the depression scores. In these questions "better" means a lower depression score. (1) Is T better than C? (2) Is T better than the average of C and S? (3) Is J better than the average of the other three groups?

(a) For each of the three questions, define an appropriate contrast. Translate the questions into null and alternative hypotheses about the contrasts.

(b) Calculate the sample contrasts and the associated standard errors.

(c) Test the contrasts and give approximate P-values.

(d) Summarize your conclusions. Do you think that the use of contrasts in this way gave a fair summary of the results?

(e) You found that the groups differ significantly in the depression scores. Does this study allow conclusions about causation—for example, that a sedentary lifestyle causes people to be more depressed? Explain your answer.

11.19 Refer to the weight gains of pregnant women in Egypt, Kenya, and Mexico given in Exercise 11.11. Computer software gives the critical value for the Bonferroni multiple comparison procedure with $\alpha = 0.05$ as $t^{**} = 2.41$. Use this procedure to compare the mean weight gains for the three countries. Summarize your conclusions.

11.20 Refer to the food intake values for toddlers in Egypt, Kenya, and Mexico given in Exercise 11.12. Computer software gives the critical value for the Bonferroni multiple comparison procedure with $\alpha = 0.05$ as $t^{**} = 2.41$. Use this procedure to compare the toddler food intake means for the three countries.

11.21 Refer to the physical fitness scores for the four groups in the exercise program study discussed in Exercises 11.9 and 11.17. Computer software gives the critical value for the Bonferroni multiple comparison procedure with $\alpha = 0.05$ as $t^{**} = 2.53$. Use this procedure to compare the mean fitness scores for the four groups. Summarize your conclusions.

11.22 Refer to the depression scores for the four groups in the exercise program study discussed in Exercises 11.10 and 11.18. Computer software gives the critical value for the Bonferroni multiple comparison procedure with $\alpha = 0.05$ as $t^{**} = 2.53$. Use this procedure to compare the mean depression scores for the four groups. Summarize your conclusions.

11.23 Suppose that you want to plan a study of the weight gains of pregnant women during the third trimester of pregnancy similar to that described in Exercise 11.11. The standard deviations given in that exercise range from 1.8 to 2.5. Perform the following calculations assuming a standard deviation $\sigma = 2.3$. Assume that you have three groups and you would like to reject the ANOVA H_0 when the alternative $\mu_1 = 2.5$, $\mu_2 = 3.0$, and $\mu_3 = 3.5$ is true. Prepare a table similar to the one in Example 11.14 in the text for the following numbers of women in each group: $n = 50, 100, 150, 175,$ and 200. What sample size would you choose for your study?

11.24 Refer to the previous exercise. Repeat the exercise for the alternative $\mu_1 = 2.7$, $\mu_2 = 3.0$, and $\mu_3 = 3.3$

11.25 In Exercise 3.13 an experiment conducted to assess the effects of nematodes on the growth of tomato seedlings is described. There were four treatments corresponding to 0, 1000, 5000, and 10,000 nematodes per plant. Each treatment was applied to four plants, and the growth of the plants was recorded. The raw data are given in the following table:

Nematodes	Seedling growth			
0	10.8	9.1	13.5	9.2
1000	11.1	11.1	8.2	11.3
5000	5.4	4.6	7.4	5.0
10,000	5.8	5.3	3.2	7.5

(a) Make a table of means and standard deviations for the four treatments.
(b) State H_0 and H_a for an ANOVA on these data.
(c) Using computer software, run the ANOVA. What are the F statistic and its P-value? Report your conclusion. Give the values of s_p and R^2.

11.26 In Exercise 3.14 an experiment on the attractiveness of different colors to cereal leaf beetles is described. Four colors were tested, and six traps were used for each color. The raw data are given in the following table:

Color	Insects trapped					
Lemon yellow	45	59	48	46	38	47
White	21	12	14	17	13	17
Green	37	32	15	25	39	41
Blue	16	11	20	21	14	7

(a) Make a table of means and standard deviations for the four colors.
(b) State H_0 and H_a for an ANOVA on these data.
(c) Using computer software, run the ANOVA. What are the F statistic and its P-value? Do you reject H_0? Give the values of s_p and R^2.

11.27 Refer to the nematode problem described in Exercise 11.25.

(a) Define the contrast that compares the 0 treatment with the average of the other three.

(b) State H_0 and H_a for using this contrast to test whether or not the presence of nematodes causes decreased growth in tomato seedlings.

(c) Perform the significance test and give the *P*-value. Do you reject H_0?

(d) Define the contrast that compares the 0 treatment with the treatment with 10,000 nematodes. This contrast is a measure of the decrease in growth due to a very large nematode infestation. Give a 95% confidence interval for this decrease in growth.

11.28 Refer to the color attractiveness problem described in Exercise 11.26. For the Bonferroni procedure with $\alpha = 0.05$, the value of t^{**} is 2.61. Use this multiple comparison procedure to decide which pairs of colors are significantly different. Summarize your results.

11.2 TWO-WAY ANALYSIS OF VARIANCE

When we want to compare means for populations that are classified in two ways or mean responses in two-factor experiments, we use two-way analysis of variance. Many of the ideas and key concepts are similar to those that we studied for one-way analysis of variance. We assume that the data are approximately normal, we pool to estimate the variance, and we use *F* statistics for our significance tests. A major difference between the one-way and two-way ANOVAs is in the FIT part of our model. We will carefully study this term, and we will see that it is both interesting and useful.

Advantages of Two-Way ANOVA

In the one-way ANOVA our populations were classified according to one categorical variable, or factor. For the two-way ANOVA model, we have two factors, each with its own number of levels.

EXAMPLE 11.15

In an experiment on the influence of dietary minerals on blood pressure, rats were fed diets prepared with varying amounts of calcium and magnesium but with all other ingredients of the diets the same. Calcium and magnesium are the two factors in this experiment.

As is common in such experiments, high, normal, and low values for each of the two minerals were selected for study. Thus, there were three levels for each of the factors and a total of nine diets, or treatment combinations. The following diagram summarizes the factors and levels for this experiment:

Magnesium	Calcium		
	L	M	
L	1	2	3
M	4	5	6
H	7	8	9

In this diagram the diets are numbered from 1 to 9. Each diet was fed to 9 rats, giving a total of 81 rats used in this experiment. (This experiment was conducted by Connie Weaver, Department of Foods and Nutrition, Purdue University.) ■

In this example, the researchers were primarily interested in the effect of calcium on the blood pressure. They could have used a one-way ANOVA with magnesium set at the normal value and three levels of calcium. Had they then wanted to study the effect of magnesium, they could have performed another experiment with calcium set at the normal value and three levels of magnesium. Suppose that their budget allowed for only 80 to 90 rats for experimentation. For the first experiment, they could use 15 rats for each of the three calcium diets. Similarly, they could use 15 rats for each of the three magnesium diets in the second experiment. However, this would require a total of 90 rats, 9 more than the design in Example 11.15.

Let us compare the two-way experiment with the two one-way experiments. In the two-way experiment there are 27 rats on each of the three calcium diets and 27 rats on each of the three magnesium diets. In the one-way experiments there are only 15 rats per diet. Therefore, in the two-way experiment we gain more information to estimate the effects of both calcium and magnesium.

The effects can also be expressed in terms of standard deviations. If σ represents the standard deviation of blood pressure for rats fed the same diet, then the standard deviation of a sample mean is $\sigma/\sqrt{n}$. For the two-way experiment this is $\sigma/\sqrt{27} = 0.19\sigma$, whereas for the two one-way experiments we have $\sigma/\sqrt{15} = 0.26\sigma$.

When the actual experiment was analyzed using a two-way ANOVA, no effect of magnesium on blood pressure was found. On the other hand, it was clearly demonstrated that the high calcium diet had the beneficial effect of producing lowered blood pressure measurements. Means for the three calcium levels were reported with standard errors based on 27 rats per group.

In comparing this two-way experiment with our hypothetical one-way experiments, there is an important additional consideration. In the one-way experiments calcium was varied only for the normal level of magnesium, and magnesium was varied only for the normal level of calcium. In the two-way experiment all combinations were included in the experiment. If magnesium affected blood pressure only at high levels of calcium, for example, only the two-way analysis would have been capable of detecting this reaction.

EXAMPLE 11.16

A group of students and university administrators concerned about alcohol abuse is planning an education and awareness program for the student body at a large university. To aid in the design of their program they decide to gather information on the attitudes of undergraduate students toward alcohol.

Initially, they considered taking an SRS of all the students on the campus. However, it was pointed out that the attitudes of freshmen, sophomores, juniors, and seniors might be quite different. A stratified sample is therefore more appropriate. The study planners decided to use year in school as a factor with four levels and to take separate random samples from each class of students.

Upon further discussion, they realized that the alcohol attitudes might be different for men and women. They therefore decided to add sex as a second factor. In the final study design there were two factors: year with four levels and sex with two levels. An SRS of size 50 was drawn from each of the eight populations. The eight groups for this study are shown in the following diagram:

		Year		
Sex	1	2	3	4
Male	1	2	3	4
Female	5	6	7	8

■

This example illustrates another advantage of the two-way ANOVA. In the one-way design where year is the only factor, the random samples would obviously include both men and women. If a sex difference exists, the one-way ANOVA would assign this variation to the RESIDUAL (within groups) part of the model. In the two-way ANOVA, sex is included as a factor, and therefore this variation would be included in the FIT part of the model. Whenever we can take variation and move it from RESIDUAL to FIT, we reduce the σ of our model and increase the power of our tests.

EXAMPLE 11.17

A textile researcher is interested in how four different colors of dye affect the durability of fabrics. Since the effects of the dyes may be different for different types of cloth, each dye is applied to five different kinds of cloth. The two factors in this experiment are dyes and cloths, with four levels and five levels, respectively. For each of the 20 dye-cloth combinations, six samples were dyed.   ■

In this experiment the researchers could have used only one type of cloth and compared the four dyes using a one-way analysis. However, based on knowledge about the chemistry of textiles and dyes, they thought that some of the dyes might have a chemical reaction with some of the cloths.

Such a reaction could have a strong negative effect on the durability of the cloth. They wanted their results to be useful to manufacturers who would use the dyes on a variety of cloths. Therefore, they chose cloth types for the study that would represent the different kinds that would be used with these dyes.

If one or more dye-cloth combinations produced exceptionally bad or exceptionally good durability measures, they wanted to be able to find this out with their experiment. The idea that the effects of the dyes might be different for different types of cloths is represented in the FIT part of a two-way model as an *interaction*. In contrast, the average values for dyes and cloths are represented as *main effects*. Thus, for the two-way model, we will represent FIT as a main effect for each of the two factors and an interaction. These concepts are discussed more fully in a later section.

interaction

main effect

These examples illustrate several reasons why experimenters choose to use two-way ANOVAs in preference to one-way ANOVAs. We summarize these ideas as follows.

Advantages of two-way ANOVAs

> **1** Valuable resources can be spent more efficiently by studying two factors simultaneously rather than separately.
>
> **2** The residual variation in a model can be reduced by including a factor thought to influence the response.
>
> **3** We can investigate interactions.

Although these considerations also apply to study designs with more than two factors, we will be content to explore only the two-way case. The choice of a study design for an experiment is an important consideration in any research problem using these methods. Careful selection of factors and levels must be done by an individual or team who understands both the statistical models as well as the research questions to be addressed.

The Two-Way Model

When discussing two-way models in general, we will use the labels A and B for the two factors. For particular examples and when using statistical software, it is better to use names for these categorical variables that suggest their meaning. Thus, in Example 11.15 we would say that the factors are calcium and magnesium.

The number of levels of the factors are often used to describe the model. Again using Example 11.15 as an illustration, we would call this a 3×3 ANOVA. Similarly, Example 11.16 illustrates a 4×2 ANOVA. In general, factor A will have I levels and factor B will have J levels. Therefore, we call the general two-way problem an $I \times J$ ANOVA.

The sample size[5] for level i of factor A and level j of factor B is n_{ij}, and the total number of observations is

$$N = \sum n_{ij}$$

Assumptions for the two-way ANOVA

Independent SRSs of size n_{ij} are drawn from each of the $I \times J$ normal populations. The variances are assumed to be the same in all populations, and the means are μ_{ij}.

We let x_{ijk} represent the kth observation from the population having factor A at level i and factor B at level j. The statistical model is

$$x_{ijk} = \mu_{ij} + \epsilon_{ijk}$$

for $i = 1, \ldots, I$ and $j = 1, \ldots, J$ and $k = 1, \ldots, n_{ij}$, where the ϵ_{ijk} are an SRS from an $N(0, \sigma)$ distribution.

The FIT part of our model is represented by the μ_{ij}, and the RESIDUAL part by ϵ_{ijk}. To estimate the μ_{ij} we use the sample averages

$$\bar{x}_{ij} = \frac{1}{n_{ij}} \sum_k x_{ijk}$$

In this formula, the k below the $\sum$ means that we sum the n_{ij} observations that belong to the (i, j)th group.

For the RESIDUAL part we calculate the sample variances for each SRS and pool these to estimate σ^2:

$$s_p^2 = \frac{\sum (n_{ij} - 1)s_{ij}^2}{\sum (n_{ij} - 1)}$$

Just as in the one-way ANOVA, the numerator in this fraction is SSE and the denominator is DFE. Because the n_{ij} have sum N, DFE $= N - IJ$, the overall number of observations minus the number of groups.

Main Effects and Interactions

In this section we will concentrate on the FIT part of the two-way ANOVA, which is represented in the model as μ_{ij} and in the data as $\bar{x}_{ij}$. Since we have $I \times J$ groups, we can think of the problem initially as a one-way ANOVA with $I \times J$ groups.

From this point of view, we can calculate sums of squares and degrees of freedom. In accordance with the conventions used by many computer software packages, we use the word *model* when discussing the sums of squares and degrees of freedom calculated in this way. Thus, SSM is a model sum of squares constructed from deviations of the form $\bar{x}_{ij} - \bar{x}$, where $\bar{x}$ is the average of all of the observations and $\bar{x}_{ij}$ is the mean of the (i, j)th group. Similarly, DFM is simply $IJ - 1$.

In the two-way ANOVA, SSM and DFM are broken down into terms corresponding to a main effect for A, a main effect for B, and an AB interaction. We use the following notation to express this idea:

$$SSM = SSA + SSB + SSAB$$

and

$$DFM = DFA + DFB + DFAB$$

The term SSA represents variation among the means for the different levels of the factor A. Since there are I such means, we have DFA $= I - 1$. Similarly, SSB represents variation of the means for the different levels of the factor B, and DFB $= J - 1$.

Interactions are a little more difficult to describe. From the above we can see that SSAB represents the variation in the model that is not accounted for by the main effects. By subtraction we see that DFAB $= (IJ - 1) - (I - 1) - (J - 1) = (I - 1)(J - 1)$. Interactions come in many different varieties. The easiest way to study them is through examples.

EXAMPLE 11.18

In a National Science Foundation survey of 1985 salaries of men and women scientists in the United States,[6] summary statistics for scientists in different fields were reported. Some of these results are given in the following table. Salaries are expressed in thousands of dollars on a yearly basis.

Field	Women	Men	Mean
Physics	41.2	48.6	44.90
Math	34.7	42.3	38.50
Biology	34.5	42.0	38.25
Mean	36.8	44.3	40.55

The table supplies averages of the means in the rows and columns. Thus, the first entry in the far right column is the average of the salaries for women and men in physics:

$$\frac{41.2 + 48.6}{2} = 44.90$$

Similarly, the average of the three salaries for women is

$$\frac{41.2 + 34.7 + 34.5}{3} = 36.8$$

marginal mean   These averages are called *marginal means* (because of their location at the *margins* of such tabulations).

A plot of these data is given in Figure 11.13. From the plot it is clear that the men earn more than the women. We can also see that the physicists earn more than the mathematicians and the biologists, whose salaries are approximately the same. In statistical language we say that there is a main effect for sex and a main effect for field. The main effect for field is due to the higher salaries of the physicists.

What about the interaction? We have an interaction if the main effects provide an incomplete description of our data. Note that according to the marginal means for sex (36.8 and 44.3), we would say that the salary difference for men versus women is 7.5. Taking the salary difference for each of the three fields, we get 7.4, 7.6, and 7.5. These values are remarkably similar. Therefore, if we want to describe the salary differential between men and women for these three fields, the difference between the marginal means is quite adequate.

Similarly, if we want to discuss the differences among the fields, the marginal means provide an adequate summary. For each of the two sexes, the differences in salary among fields are very much alike.

Because the important information in the above example can be explained by the main effects or marginal means, we do not need an interaction to summarize the salaries. The absence of interaction can be seen in our plot. Note that the lines connecting the means for women and the lines connecting the means for men are roughly parallel. This happens because

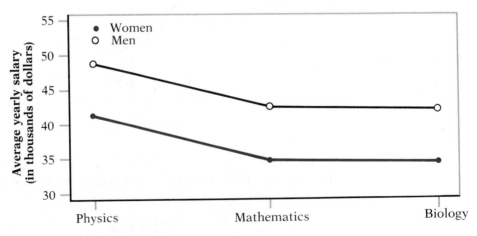

Figure 11.13 Plot of salaries of women and men scientists in three fields.

the difference between the lines is equal to the difference between the men's and women's salaries for each field, and as we have seen, this difference is approximately the same for all three fields.

In the next example, we add another field to our data. We will see that in this case an interaction is present.

EXAMPLE 11.19

The survey described in the previous example also gave salaries for medical scientists. Adding this field gives the following table:

Field	Women	Men	Mean
Physics	41.2	48.6	44.90
Math	34.7	42.3	38.50
Biology	34.5	42.0	38.25
Medicine	36.2	50.4	43.30
Mean	36.65	45.83	41.24

The means are plotted in Figure 11.14. ∎

Note that the marginal means for sex and the mean of all groups have changed. In our plot of the means, we no longer see a simple parallel pattern in the lines. The sex difference in the marginal means is now 9.18, a rather substantial increase from the previous value of 7.5.

In this example we have an interaction. The sex difference in salary for the medical scientists is 14.2, nearly double the amount for the other fields. To fully explain the table we need more than the marginal means.

Note that the presence of an interaction does not necessarily mean that our main effects are useless. It is still clear that the men earn more than the

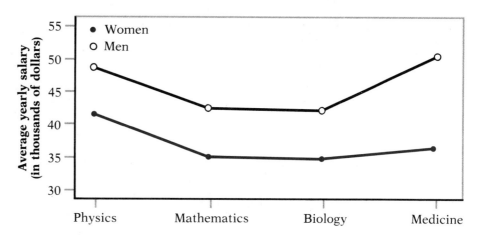

Figure 11.14 Plot of salaries of women and men scientists in four fields.

women, and this is evident in the marginal means for the sex factor. On the other hand, the marginal means for fields are a little less informative, as we see when we compare the medical scientists to the other fields. The marginal means for the physicists and the medical scientists are approximately the same. However, if we compare women, the physicists make more, whereas if we compare men, the medical scientists make more.

Interactions come in many shapes and forms. When we find them, a careful examination of the means is needed to properly interpret the data. Simply stating that interactions are significant tells us very little. Plots of the group means are very helpful.

The next example has an interaction that is a little different from the one we saw in the previous example.

EXAMPLE 11.20 | In a study of the energy expenditure of farmers in the Upper Volta, data were collected during the wet and dry seasons.[7] In this area millet is grown during the wet season. In the dry season, there is relatively little activity because the ground is too hard to grow crops. The mean energy expended (in calories) by men and women in the Upper Volta during the wet and dry seasons is given in the following table:

Season	Men	Women	Mean
Dry	2310	2320	2315
Wet	3460	2890	3175
Mean	2885	2605	2745

The means are plotted in Figure 11.15.   ∎

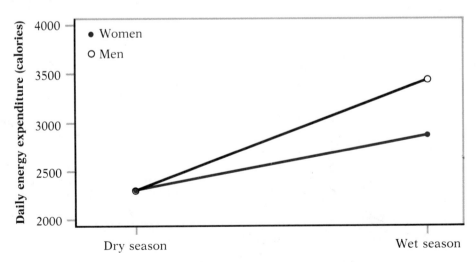

Figure 11.15 Plot of energy expenditures of farmers in the Upper Volta.

During the dry season relatively little activity takes place. As a consequence, the men and women use up about the same number of kilocalories in their daily activities. When the wet season arrives, there is a great deal of activity. Both sexes use more energy. Since this season is rather short, it is necessary to work very hard to survive the next dry season. The men work long hours in the field, and as a result they expend much more energy than the women. The amount of energy used by the men in the wet season is very high by any reasonable standard. Such values are sometimes found in developed countries for people engaged in coal mining and lumberjacking.

In a statistical analysis, the pattern of means shown in Figure 11.15 produced significant main effects for season and sex in addition to a season-by-sex interaction. For the main effects we would say that men use more energy than women and that the wet season is associated with greater activity than the dry season. This clearly does not tell the whole story. We need to discuss the men and women in each of the two seasons to fully understand how energy is used in the Upper Volta. In particular, if there is not enough food to support the activities of both the men and the women during the wet season, serious problems may arise.

A different kind of interaction is present in the next example. Here, we must be very cautious in our interpretation of the main effects since one of them can lead to a distorted conclusion.

EXAMPLE 11.21

An experiment was designed to assess the effects of noise on the performance of children on certain tasks.[8] Second grade hyperactive students were tested as well as a control group of students who were not hyperactive. One of the tasks involved solving mathematical problems. The tasks were performed under high- and low-noise conditions. Mean scores are given in the following table:

Group	High noise	Low noise	Mean
Control	214	170	192
Hyperactive	120	140	130
Mean	167	155	161

As in the previous examples, the marginal means are given in the table. The means are plotted in Figure 11.16. In the analysis of this experiment, both main effects and the interaction were statistically significant. How are we to interpret these results? ∎

From the plot of the means it is clear that the control children performed better than the hyperactive children under both conditions. However, the main effect for noise could be easily misinterpreted if we do not look at the means carefully. Under high-noise conditions the marginal mean was 167, compared with a low-noise mean of 155. This suggests that high noise was, on the average, a more favorable condition than low noise.

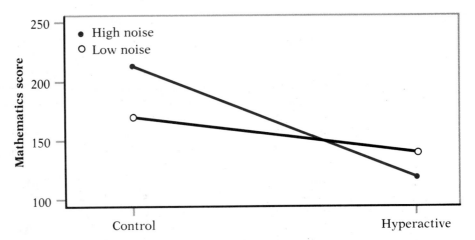

Figure 11.16 Plot of mathematics scores of control and hyperactive children under low and high noise conditions.

This interpretation is clearly not consistent with the data presented. The high noise was a favorable condition for the control children, whereas the hyperactive children performed better under the low-noise conditions. To understand the effects of high and low noise on the performance of the children in this experiment, we must know whether or not they are hyperactive. The interaction is essential to our understanding of these data.

ANOVA Table for the Two-Way ANOVA

The results of a two-way ANOVA are summarized in an ANOVA table. Our general rules regarding sums of squares and degrees of freedom apply. For the two-way ANOVA, we have

$$SST = SSA + SSB + SSAB + SSE$$

and

$$DFT = DFA + DFB + DFAB + DFE$$

For the mean squares, we use the general formula

$$\text{mean square} = \frac{\text{sum of squares}}{\text{degrees of freedom}}$$

To compute *F* statistics we divide the mean square for the source of interest by MSE. The following is the general form of the two-way ANOVA table:

Source	Sum of squares	Degrees of freedom	Mean square	F
A	SSA	$I-1$	SSA/DFA	MSA/MSE
B	SSB	$J-1$	SSB/DFB	MSB/MSE
AB	SSAB	$(I-1)(J-1)$	SSAB/DFAB	MSAB/MSE
Error	SSE	$N-IJ$	SSE/DFE	
Total	SST	$N-1$	SST/DFT	

With the two-way ANOVA there are three null hypotheses. We can test for significance of the main effect of A, the main effect of B, and the AB interaction. It is generally a good practice to examine the test for interaction first, since the presence of a strong interaction may influence the interpretation of the main effects. Of course, it is assumed that the means have been carefully examined and plotted before the results of any significance tests are interpreted.

Significance tests for the two-way ANOVA

To test the main effect of A, we use the F statistic

$$F_A = \frac{\text{MSA}}{\text{MSE}}$$

If there is no main effect of A, F_A has the $F(\text{DFA, DFE})$ distribution. The P-value for the null hypothesis of no main effect of A is the probability that a random variable having the $F(\text{DFA, DFE})$ distribution is greater than or equal to the calculated value.

To test the main effect of B, we use the F statistic

$$F_B = \frac{\text{MSB}}{\text{MSE}}$$

If there is no main effect of B, F_B has the $F(\text{DFB, DFE})$ distribution. The P-value for the null hypothesis of no main effect of B is the probability that a random variable having the $F(\text{DFB, DFE})$ distribution is greater than or equal to the calculated value.

To test the interaction of A and B, we use the F statistic

$$F_{AB} = \frac{\text{MSAB}}{\text{MSE}}$$

If there is no interaction, F_{AB} has the $F(\text{DFAB, DFE})$ distribution. The P-value for the null hypothesis of no interaction is the probability that a random variable having the $F(\text{DFAB, DFE})$ distribution is greater than or equal to the calculated value.

The following example illustrates how to do a two-way ANOVA. As with the one-way ANOVA, we focus our attention on interpretation of the computer output.

EXAMPLE 11.22 In a study of cardiovascular risk factors, runners who averaged at least 15 miles per week were compared with a control group described as "generally sedentary." Both men and women were included in the study.[9] The design is a 2×2 ANOVA with the factors group and sex. In each of the four combinations, 200 people were studied. One of the variables measured was the heart rate after 6 minutes of exercise on a treadmill. A computer analysis produces the outputs in Figure 11.17 (from the SAS procedure MEANS) and Figure 11.18 (from the SAS procedure GLM). ■

We start with the usual preliminaries. From Figure 11.17 we see that the ratio of the largest to the smallest standard deviation is less than 2. Therefore, we are not concerned about violating the assumption of equal population standard deviations. Normal quantile plots (not shown) do not reveal any outliers, and the data appear to be reasonably normal.

The ANOVA table at the top of the output in Figure 11.18 gives an analysis that is the same as we would obtain by running a one-way ANOVA with the four groups female control, female runner, male control, and male runner. In this analysis MODEL has 3 degrees of freedom, and there are 796 degrees of freedom for ERROR. The F test and its associated P-value

```
 VARIABLE       N         MEAN        STANDARD
                                      DEVIATION


 ----------GROUP=CONTROL  SEX=FEMALE----------

 HR           200    150.16768826    15.90882113


 ----------GROUP=CONTROL  SEX=MALE----------

 HR           200    129.40308425    16.91896949


 ----------GROUP=RUNNERS  SEX=FEMALE----------

 HR           200    117.86423502    15.21759686


 ----------GROUP=RUNNERS  SEX=MALE----------

 HR           200    103.09755344    12.66789092
```

Figure 11.17 Descriptive statistics for heart rates in 2×2 ANOVA.

```
                    GENERAL LINEAR MODELS PROCEDURE

DEPENDENT VARIABLE: HR

SOURCE                  DF      SUM OF SQUARES        MEAN SQUARE      F VALUE

MODEL                   3       236673.01696169     78891.00565390    338.81

ERROR                   796     185347.17372171       232.84820819    PR > F

CORRECTED TOTAL         799     422020.19068340                       0.0001


R-SQUARE            C.V.            ROOT MSE              HR MEAN

0.560810          12.1945         15.25936461          125.13314024


SOURCE                  DF          TYPE I SS     F VALUE    PR > F

GROUP                   1       171750.65050977    737.61    0.0001
SEX                     1        63123.61278194    271.09    0.0001
GROUP*SEX               1         1798.75366998      7.73    0.0056
```

Figure 11.18 Analysis of variance output for heart rates.

for this analysis refer to the hypothesis that all four groups have the same population mean. Since we are interested in the main effects and interaction, we ignore this test.

In the bottom part of the output, the sums of squares for the group and sex main effects and the group-by-sex interaction are given under the heading TYPE I SS. Note that these sum to the sum of squares for MODEL. Similarly, the degrees of freedom for these sums of squares sum to the degrees of freedom for MODEL.

SAS does not provide mean square calculations for the main effects and interaction. If they were needed, we would divide the sum of squares by the degrees of freedom. Since these degrees of freedom are all 1 in this case, the mean squares are the same as the sums of squares. Thus, for example, the mean square for the group main effect is SSG/DFG = 171,751/1 = 171,751. The F statistics for the three effects are given in the column labeled F VALUE and the P-values are under the heading PR > F. For the group main effect, we verify the calculation of F as follows:

$$F = \frac{\text{MSG}}{\text{MSE}} = \frac{171,751}{232.848} = 737.61$$

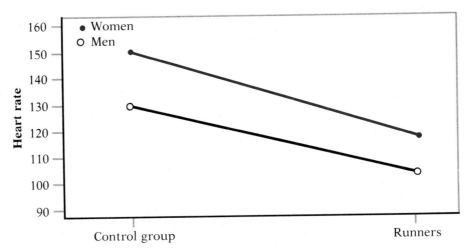

Figure 11.19 Plot of means for heart rates in 2 × 2 ANOVA.

All three effects are statistically significant. The group effect has the largest *F*, followed by the sex effect and then the group-by-sex interaction. To interpret these results, we examine the plot of means given in Figure 11.19. The significance of the main effect for group is due to the fact that the controls have higher average heart rates than the runners for both sexes. This is clearly the largest effect evident in the plot.

Similarly, the significance of the main effect for sex is due to the fact that the females have higher heart rates than the men in both groups. The differences are not as large as those for the group effect, and this observation is reflected in the smaller value of the *F* statistic.

The analysis indicates that a complete description of the average heart rates requires consideration of the interaction in addition to the main effects. Note that the two lines in the plot are not parallel. This can be explained in two ways. First, the female-male difference in average heart rates is greater for the controls than for the runners. Another view is that the difference in average heart rates between controls and runners is greater for the females than for the males.

SUMMARY

When populations are classified according to two factors, **two-way analysis of variance** is used to compare the means.

The populations are assumed to be normal with possibly different means and the same variance. Independent SRSs of size n_{ij} are drawn from each population.

As with one-way ANOVA, preliminary analysis includes examination of means, standard deviations, and normal quantile plots. **Marginal means** are calculated by taking averages of the cell means across rows and columns. Pooling is used to estimate the within-group variance.

The total variation is separated into parts for the model and error. The model variation is separated into parts for each of the main effects and the interaction.

The calculations are organized into an **ANOVA table**. F statistics and P-values are used to test hypotheses about the main effects and the interaction.

Careful inspection of the means is necessary to interpret significant main effects and interactions. Plots are a useful aid.

SECTION 11.2 EXERCISES

11.29 Each of the following situations is a two-way study design. For each case, identify the response variable and both factors, and state the number of levels for each factor.

(a) Four varieties of tomato plants are compared. Two types of fertilizers are used with each variety. Five plants are grown at each combination of variety and fertilizer and the yield in pounds of tomatoes is recorded for each plant.

(b) In a marketing experiment six different types of packaging are used for a laundry detergent. A survey is conducted to determine the attractiveness of the packaging in four different parts of the country. Each type of packaging is shown to 30 different potential customers in each part of the country who are asked to rate the attractiveness of the product on a 1 to 10 scale.

(c) To compare the effectiveness of three different weight loss programs, five men and five women are randomly assigned to each. At the end of the program, the weight loss for each of the participants is recorded.

11.30 Each of the following situations is a two-way study design. For each case, identify the response variable and both factors, and state the number of levels for each factor.

(a) In a study on smoking, subjects are classified as nonsmokers, moderate smokers, or heavy smokers. Samples of 100 men and 100 women are drawn from each group. Every person is asked to report the number of hours of sleep they get on a typical night.

(b) The strength of cement depends upon the formula used to prepare it. In one study, five different mixtures are to be compared. Six batches of each mixture are prepared. For each mixture two batches are subjected to 0 cycles of freezing and thawing, two batches are subjected to 100 cycles, and two batches are subjected to 500 cycles. The strength of each sample is then measured.

(c) Four methods for teaching sign language are to be compared. Five students in special education and five students in linguistics are randomly assigned to each of the methods and the score on a final exam is recorded.

11.31 Refer to Exercise 11.29. For each part, outline the ANOVA table, giving the sources of variation and the degrees of freedom. (Do not compute the numerical values for the sums of squares and mean squares.)

11.32 Refer to Exercise 11.30. For each part, outline the ANOVA table, giving the sources of variation and the degrees of freedom. (Do not compute the numerical values for the sums of squares and mean squares.)

11.33 As part of a clinical trial designed to reduce coronary heart disease, blood pressure measurements were taken on 12,866 men. Individuals were classified by age group and race. The means for systolic blood pressure are given in the following table. (Data from W. M. Smith et al., "The multiple risk factor intervention trial," in H. M. Perry, Jr., and W. M. Smith (eds.) *Mild Hypertension: To Treat or Not to Treat*, New York Academy of Sciences, 1978, pp. 293–308.)

	35–39	40–44	45–49	50–54	55–59
White	131.0	132.3	135.2	139.4	142.0
Nonwhite	132.3	134.2	137.2	141.3	144.1

(a) Make a plot of the mean systolic blood pressures for these groups, with age on the x axis and blood pressure on the y axis. For each racial group connect the points for the different ages.

(b) Describe the patterns you see. Does there appear to be a difference in the two racial groups? Does systolic blood pressure appear to vary with age? If so, how does it vary? Is there an interaction?

(c) Compute the marginal means. Compute the differences between the white and nonwhite blood pressures for each age group. Use this information to summarize numerically the patterns in the plot.

11.34 Refer to the previous exercise. The means for diastolic blood pressure are given in the following table:

	35–39	40–44	45–49	50–54	55–59
White	89.4	90.2	90.9	91.6	91.4
Nonwhite	91.2	93.1	93.3	94.5	93.5

(a) Make a plot of the mean diastolic blood pressures for these groups, with age on the x axis and blood pressure on the y axis. For each racial group connect the points for the different ages.

(b) Describe the patterns you see. Does there appear to be a difference in the two racial groups? Does diastolic blood pressure appear to vary with age? If so, how does it vary? Is there an interaction?

(c) Compute the marginal means. Compute the differences between the white and nonwhite blood pressures for each age group. Use this information to summarize numerically the patterns in the plot.

11.35 It is known that the amount of chromium in the diet has an effect on the way the body processes insulin. In an experiment designed to study this phenomenon, four diets were fed to male rats. There were two factors. Chromium had two levels: low (L) and normal (N). Either the rats were allowed to eat as much as they wanted (M) or the total amount that they could eat was restricted (R). We call the second factor Eat. One of the variables measured was the amount of an enzyme called GITH. The means for this variable are given in the following table. (Data provided by Julie Hendricks and V. J. K. Liu of the Department of Foods and Nutrition, Purdue University.)

	Eat	
Chromium	M	R
L	4.545	5.175
N	4.425	5.317

(a) Make a plot of the mean GITH for these diets, with the factor Chromium on the x axis and GITH on the y axis. For each Eat group connect the points for the two Chromium means.

(b) Describe the patterns you see. Does the amount of chromium in the diet appear to affect the GITH mean? Does restricting the diet as compared to letting the rats eat as much as they want appear to have an effect? Is there an interaction?

(c) Compute the marginal means. Compute the differences between the M and R diets for each level of the factor Chromium. Use this information to summarize numerically the patterns in the plot.

11.36 The Chapin Social Insight Test is designed to measure how well people can appraise others and predict what they will say or do. In one study this test was administered to different groups of people and the means were compared. Some of the results are given in the table below. Means for males and females who were either psychology graduate students (PG) or liberal arts undergraduates (LU) are presented. The two factors are labeled Sex and Group. (This example is based on results reported in H. G. Gough, *The Chapin Social Insight Test*, Consulting Psychologists Press, Palo Alto, Calif., 1968.)

Sex	Group	
	PG	LU
Males	27.56	25.34
Females	29.25	24.94

(a) Make a plot of the mean Chapin scores, with Sex on the x axis and Chapin scores on the y axis. For each Group connect the points for the male and female means.
(b) Describe the patterns you see. Is there a sex difference? Are the two groups of students different? Is there an interaction?
(c) Compute the marginal means. Compute the differences between the two groups of students for each sex. Use this information to summarize numerically the patterns in the plot.

11.37 Refer to the computer science study described in Example 11.3. For the one-way ANOVA described in that example, students were classified into one of three groups depending upon their major in the sophomore year. In this exercise we also consider the sex of the students. Therefore, there are two factors: Major, with three levels; and Sex, with two levels. The mean SAT mathematics scores for the six groups that result appear in the following table. For convenience we use the labels CS for computer science majors, EO for engineering or other science majors, and O for other majors.

Sex	Major		
	CS	EO	O
Male	628	618	589
Female	582	631	543

(a) Make a plot of the mean SAT mathematics scores, with Major on the x axis and SAT mathematics scores on the y axis. For each Sex connect the points for the three majors.
(b) Describe the patterns you see. Is there a sex difference? Are the mean scores different for the different majors? Is there an interaction?
(c) Compute the marginal means. Compute the differences between the two sexes for each major. Use this information to summarize numerically the patterns in the plot.

11.38 Refer to the previous exercise. The mean high school mathematics grades for these students are summarized in the following table. The grades have been coded so that $10 = A$, $9 = A-$, etc.

		Major	
Sex	CS	EO	O
Male	8.68	8.35	7.65
Female	9.11	9.36	8.04

(a) Make a plot of the mean high school mathematics grades, with Major on the x axis and high school mathematics grades on the y axis. For each sex connect the points for the three majors.
(b) Describe the patterns you see. Is there a sex difference? Are the mean scores different for the different majors? Is there an interaction?
(c) Compute the marginal means. Compute the differences between the two sexes for each major. Use this information to summarize numerically the patterns in the plot.

11.39 Refer to the chromium-insulin problem described in Exercise 11.35. Part of the ANOVA table for these data is given below.

Source	Sum of squares	Degrees of freedom	Mean square	F
A (Chromium)	.00121			
B (Eat)	5.79121			
AB	.17161			
Error	1.08084			
Total				

(a) The total number of rats used in this study was 40. Fill in the missing values in the ANOVA table.
(b) What is the value of the F statistic used to test the null hypothesis that there is no interaction? What is its distribution

when the null hypothesis is true? Using Table F, find an
approximate P-value for this test.

(c) Answer the questions in part (b) for the main effect of
Chromium and the main effect of Eat.

(d) What is s_p^2, the within-group variance? What is s_p?

(e) Using what you have learned in this exercise and your answer to
Exercise 11.35, summarize the results of this experiment.

11.40 Refer to the Chapin Social Insight Test study described in Exercise
11.36. Part of the ANOVA table for these data is given below.

Source	Sum of squares	Degrees of freedom	Mean square	F
A (Sex)	62.40			
B (Group)	1599.03			
AB				
Error	13633.29			
Total	15458.52			

(a) There were 150 individuals tested in each of the groups. Fill in
the missing values in the ANOVA table.

(b) What is the value of the F statistic used to test the null
hypothesis that there is no interaction? What is its distribution
when the null hypothesis is true? Using Table F, find an
approximate P-value for this test.

(c) Answer the questions in part (b) for the main effect of Sex and
the main effect of Group.

(d) What is s_p^2, the within-group variance? What is s_p?

(e) Using what you have learned in this exercise and your answer
to Exercise 11.36, summarize the results of this study.

NOTES

1 Results of the study are reported in P. F. Campbell and G. P. McCabe,
"Predicting the success of freshmen in a computer science major,"
Communications of the ACM, 27 (1984), pp. 1108–1113.

2 This rule is intended to provide a general guideline for deciding when serious
errors may result from applying ANOVA procedures. When the sample sizes in
each group are very small, the sample variances will tend to vary much more
than when the sample sizes are large. In this case, the rule may be a little too
conservative. For unequal sample sizes, particular difficulties can arise when a
relatively small sample size is associated with a population having a relatively
large standard deviation. Careful judgment is needed in all cases. By considering

P-values rather than fixed level α testing, judgments in ambiguous cases can more easily be made; for example, if the *P*-value is very small, say 0.001, then it is probably safe to reject H_0 even if there is a fair amount of variation in the sample standard deviations.

3 This example is based on data provided from a study conducted by Jim Baumann and Leah Jones of the Purdue University Education Department.

4 There are different definitions for the noncentrality parameter of the noncentral *F* distribution. Many authors prefer $\phi = \sqrt{\lambda/I}$. We have chosen to use λ because it is the form needed for the SAS function PROBF.

5 We have presented the two-way ANOVA model and analysis for the general case in which the sample sizes may be unequal. However, if the sample sizes vary a great deal, several complications can result. Most computer packages offer several options for the computation of the ANOVA table in this case. When the sample sizes are approximately equal, all methods give essentially the same results.

6 The data in Examples 11.18 and 11.19 were taken from an article on the results of a National Science Foundation study, M. Scott, "Women scientists earn 79% of men's salaries," *The Scientist*, November 16, 1987.

7 This example is based on a study described by P. Payne, "Nutrition adaptation in man: social adjustments and their nutritional implications," in K. Blaxter and J. C. Waterlow (eds.), *Nutrition Adaptation in Man*, Libbey, London, 1985.

8 Example 11.21 is based on a study described in S. S. Zentall and J. H. Shaw, "Effects of classroom noise on performance and activity of second-grade hyperactive and control children," *Journal of Education Psychology*, 72 (1980), pp. 830–840.

9 This example is based on a study described in P. D. Wood et al., "Plasma lipoprotein distributions in male and female runners," in P. Milvey (ed.), *The Marathon: Physiological, Medical, Epidemiological, and Psychological Studies*, New York Academy of Sciences, 1977.

CHAPTER 11 EXERCISES

11.41 Example 11.6 describes an experiment to compare three methods of teaching reading. One of the pretest measures is analyzed in that example and one of the posttest measures is analyzed in Example 11.10. In the actual study there were two pretest and three posttest measures. A computer listing of the data is given in Appendix A of this chapter. The pretest variable analyzed in Example 11.6 is PRE1 and the posttest variable analyzed in Example 11.10 is POST3. For this exercise, use the data for the pretest variable PRE2.

   (a) Summarize the data with a table giving the sample sizes, means, and standard deviations. Plot the means.

(b) Examine the distribution of the scores in the three groups by constructing normal quantile plots. Do the data look normal?

(c) What is the ratio of the largest to the smallest standard deviation? Is it reasonable to proceed with the ANOVA?

(d) Run a one-way ANOVA on these data. State H_0 and H_a for the ANOVA significance test. Report the value of the F statistic and its P-value. Do you have strong evidence against H_0?

(e) Summarize your conclusions from this analysis.

11.42 Refer to the previous exercise. To examine the effect on the ANOVA of changing the means, you will do an analysis similar to that described in Example 11.9. Use the pretest variable PRE2 for this exercise. For each of the observations in the Basal group add one point to the PRE2. For each of the observations in the Strat group, subtract one point. Leave the observations in the DRTA group unchanged. Call the new variable PRE2X.

(a) Give the sample means and standard deviations for PRE2X in each group. (Find these from the means and standard deviations for PRE2.)

(b) Run the ANOVA on the variable PRE2X.

(c) Compare the results of this analysis with the ANOVA for PRE2.

(d) Summarize your conclusions.

11.43 Refer to Example 11.10, where the scores for the third posttest variable in the reading comprehension study are analyzed. For this exercise you will analyze the first posttest variable, given as POST1 in Appendix A of this chapter.

(a) Summarize the data with a table giving the sample sizes, means, and standard deviations. Plot the means.

(b) Examine the distribution of the scores in the three groups by constructing normal quantile plots. Do the data look normal?

(c) What is the ratio of the largest to the smallest standard deviation? Is it reasonable to proceed with the ANOVA?

(d) Run the ANOVA and summarize the results.

(e) Following the procedure given in the text in the section on contrasts, analyze the contrast for comparing the Basal group with the average of the other two groups. Test the one-sided H_a and give a 95% confidence interval.

(f) Analyze the contrast that compares the DRTA and the Strat groups. Use a two-sided test and give a 95% confidence interval.

(g) Based on your work in (a), (d), (e), and (f), write a brief nontechnical summary of your conclusions.

11.44 Refer to Example 11.10, where the scores for the third posttest variable in the reading comprehension study are analyzed. For this exercise you will analyze the second posttest variable, given as POST2 in Appendix A of this chapter.

(a) Summarize the data with a table giving the sample sizes, means, and standard deviations. Plot the means.

(b) Examine the distribution of the scores in the three groups by constructing normal quantile plots. Do the data look normal?

(c) What is the ratio of the largest to the smallest standard deviation? Is it reasonable to proceed with the ANOVA?

(d) Run the ANOVA and summarize the results.

(e) Following the procedure given in the text in the section on contrasts, analyze the contrast for comparing the Basal group with the average of the other two groups. Test the one-sided H_a and give a 95% confidence interval.

(f) Analyze the contrast that compares the DRTA and the Strat groups. Use a two-sided test and give a 95% confidence interval.

(g) Based on your work in (a), (d), (e), and (f), write a brief nontechnical summary of your conclusions.

11.45 Refer to the nematode experiment described in Exercise 11.25. Suppose that when entering the data into the computer, you accidentally entered the first obervation as 108 rather than 10.8.

(a) Run the ANOVA with the incorrect observation. Summarize the results.

(b) Compare this run with the results obtained with the correct data set. What does this illustrate about the effect of outliers in an ANOVA?

(c) Compute a table of means and standard deviations for each of the four treatments using the incorrect data. How would this table have helped you to detect the incorrect observation?

11.46 Refer to the color attractiveness experiment described in Exercise 11.26. Suppose that when entering the data into the computer, you accidentally entered the first observation as 450 rather than 45.

(a) Run the ANOVA with the incorrect observation. Summarize the results.

(b) Compare this run with the results obtained with the correct data set. What does this illustrate about the effect of outliers in an ANOVA?

(c) Compute a table of means and standard deviations for each of the four treatments using the incorrect data. How would this table have helped you to detect the incorrect observation?

11.47 Refer to the nematode experiment described in Exercise 11.25. With small numbers of observations in each group, it is often very difficult to detect deviations from normality and violations of the equal standard deviation assumption. The log transformation is often used for variables such as the growth of plants. In many cases this will tend to make the standard deviations more similar across groups and to make the data within each group look more normal. Rerun the

ANOVA using the logarithms of the recorded values. Answer the questions given in Exercise 11.25. Compare these results with those obtained by analyzing the raw data.

11.48 Refer to the color attractiveness experiment described in Exercise 11.26. The square root transformation is often used for variables that are counts, such as the number of insects trapped in this example. In many cases data transformed in this way will conform more closely to the assumptions of normality and equal standard deviations. Rerun the ANOVA using the square roots of the original counts of insects. Answer the questions given in Exericse 11.26. Compare these results with those obtained by analyzing the raw data.

11.49 You are planning a study of the SAT mathematics scores of four groups of students. From Example 11.3, we know that the standard deviations of the three groups considered in that study were 86, 67, and 83. In Example 11.5, we found the pooled standard deviation to be 82.5. Since the power of the F test decreases as the standard deviation increases, use $\sigma = 90$ for the calculations in this exercise. This choice will lead to sample sizes that are perhaps a little larger than we need but will prevent us from choosing sample sizes that are too small to detect the effects of interest. You would like to conclude that the population means are different when $\mu_1 = 620$, $\mu_2 = 600$, $\mu_3 = 580$, and $\mu_4 = 560$.

(a) Pick several values for n (the number of students that you will select from each group) and calculate the power of the ANOVA F test for each of your choices.

(b) Plot the power versus the sample size. Describe the general shape of the plot.

(c) Which n would you choose for your study? Give reasons for your answer.

11.50 Refer to the previous exercise. Repeat all parts for the alternative $\mu_1 = 610$, $\mu_2 = 600$, $\mu_3 = 590$, and $\mu_4 = 580$.

11.51 A large research project was undertaken to study the physical properties of wood materials that were constructed by bonding together small flakes of wood. Different species of trees were used, and the flakes were made of different sizes. One of the physical properties measured was the tension modulus of elasticity in the direction perpendicular to the alignment of the flakes. The tension modulus of elasticity is measured in pounds per square inch (psi) and some of the data are given in the following table. The sizes of the flakes are S1 = 0.015 by 2 inches and S2 = 0.025 by 2 inches. (Data provided by Mike Hunt and Bob Lattanzi of the Purdue University Forestry Department.)

	Size of flakes	
Species	S1	S2
Aspen	308	278
	428	398
	426	331
Birch	214	534
	433	512
	231	320
Maple	272	158
	376	503
	322	220

(a) Compute means and standard deviations for the three observations in each species-size group. For each species compute the marginal means. Do the same for the two sizes of flakes. Display the means and marginal means in a table.

(b) Plot the means of the six groups. Put species on the x axis and modulus of elasticity on the y axis. For each size connect the three points corresponding to the different species. Describe the patterns you see. Do the species appear to be different? What about the sizes? Does there appear to be an interaction?

(c) Run a two-way ANOVA on these data. Summarize the results of the significance tests. What do these results say about the impressions that you described in (b) of this exercise?

11.52 Refer to the previous exercise. Another of the physical properties measured was the strength, measured in kilo-pounds per square inch (ksi), in the direction perpendicular to the alignment of the flakes. Some of the data are given in the following table. The sizes of the flakes are S1 = 0.015 by 2 inches and S2 = 0.025 by 2 inches.

	Size of flakes	
Species	S1	S2
Aspen	1296	1472
	1997	1441
	1686	1051
Birch	903	1422
	1246	1376
	1355	1238
Maple	1211	1440
	1827	1238
	1541	748

(a) Compute means and standard deviations for the three observations in each species-size group. For each species compute the marginal means. Do the same for the two sizes of flakes. Display the means and marginal means in a table.

(b) Plot the means of the six groups. Put species on the x axis and strength on the y axis. For each size connect the three points corresponding to the different species. Describe the patterns you see. Do the species appear to be different? What about the sizes? Does there appear to be an interaction?

(c) Run a two-way ANOVA on these data. Summarize the results of the significance tests. What do these results say about the impressions that you described in part (b) of this exercise?

11.53 Refer to the data given for the computer science study presented in Appendix B of this chapter.* Analyze the data for SAT-V, the SAT verbal score. Your analysis should include a table of sample sizes, means, and standard deviations; normal quantile plots; a plot of the means; and a two-way ANOVA using sex and major as the factors. Write a short summary of your conclusions.

11.54 Refer to the data given for the computer science study presented in Appendix B of this chapter. Analyze the data for HSS, the high school science grades. Your analysis should include a table of sample sizes, means, and standard deviations; normal quantile plots; a plot of the means; and a two-way ANOVA using sex and major as the factors. Write a short summary of your conclusions.

11.55 Refer to the data given for the computer science study presented in Appendix B of this chapter. Analyze the data for HSE, the high school English grades. Your analysis should include a table of sample sizes, means, and standard deviations; normal quantile plots; a plot of the means; and a two-way ANOVA using sex and major as the factors. Write a short summary of your conclusions.

11.56 Refer to the data given for the computer science study presented in Appendix B of this chapter. Analyze the data for GPA, the college grade point average. Your analysis should include a table of sample

---

* The computer science study examined in several of the examples in this chapter was undertaken to answer some important questions that arose when it was noticed that large numbers of freshman computer science majors did not continue in this field of study. All available data were used in the analyses performed, which resulted in sample sizes that were unequal for the groups considered. For a one-way ANOVA, this causes no particular problems. However, for the two-way ANOVA, several complications arise when the sample sizes are unequal. Unfortunately, a detailed discussion of these complications is beyond the scope of this text. To avoid these difficulties and still use these interesting data for illustrative purposes, simulated data based on the results of this study are given in Appendix B of this chapter. ANOVAs based on these simulated data will result in roughly the same qualitative conclusions as those obtained with the original data. The following exercises use the simulated data.

sizes, means, and standard deviations; normal quantile plots; a plot of the means; and a two-way ANOVA using sex and major as the factors. Write a short summary of your conclusions.

Appendix A   Listing of Data for a Reading Comprehension Study

Subject	Group	PRE1	PRE2	POST1	POST2	POST3
1	Basal	4	3	5	4	41
2	Basal	6	5	9	5	41
3	Basal	9	4	5	3	43
4	Basal	12	6	8	5	46
5	Basal	16	5	10	9	46
6	Basal	15	13	9	8	45
7	Basal	14	8	12	5	45
8	Basal	12	7	5	5	32
9	Basal	12	3	8	7	33
10	Basal	8	8	7	7	39
11	Basal	13	7	12	4	42
12	Basal	9	2	4	4	45
13	Basal	12	5	4	6	39
14	Basal	12	2	8	8	44
15	Basal	12	2	6	4	36
16	Basal	10	10	9	10	49
17	Basal	8	5	3	3	40
18	Basal	12	5	5	5	35
19	Basal	11	3	4	5	36
20	Basal	8	4	2	3	40
21	Basal	7	3	5	4	54
22	Basal	9	6	7	8	32
23	DRTA	7	2	7	6	31
24	DRTA	7	6	5	6	40
25	DRTA	12	4	13	3	48
26	DRTA	10	1	5	7	30
27	DRTA	16	8	14	7	42
28	DRTA	15	7	14	6	48
29	DRTA	9	6	10	9	49
30	DRTA	8	7	13	5	53
31	DRTA	13	7	12	7	48
32	DRTA	12	8	11	6	43
33	DRTA	7	6	8	5	55
34	DRTA	6	2	7	0	55
35	DRTA	8	4	10	6	57

(*continued*)

Appendix A (Continued)

Subject	Group	PRE1	PRE2	POST1	POST2	POST3
36	DRTA	9	6	8	6	53
37	DRTA	9	4	8	7	37
38	DRTA	8	4	10	11	50
39	DRTA	9	5	12	6	54
40	DRTA	13	6	10	6	41
41	DRTA	10	2	11	6	49
42	DRTA	8	6	7	8	47
43	DRTA	8	5	8	8	49
44	DRTA	10	6	12	6	49
45	Strat	11	7	11	12	53
46	Strat	7	6	4	8	47
47	Strat	4	6	4	10	41
48	Strat	7	2	4	4	49
49	Strat	7	6	3	9	43
50	Strat	6	5	8	5	45
51	Strat	11	5	12	8	50
52	Strat	14	6	14	12	48
53	Strat	13	6	12	11	49
54	Strat	9	5	7	11	42
55	Strat	12	3	5	10	38
56	Strat	13	9	9	9	42
57	Strat	4	6	1	10	34
58	Strat	13	8	13	1	48
59	Strat	6	4	7	9	51
60	Strat	12	3	5	13	33
61	Strat	6	6	7	9	44
62	Strat	11	4	11	7	48
63	Strat	14	4	15	7	49
64	Strat	8	2	9	5	33
65	Strat	5	3	6	8	45
66	Strat	8	3	4	6	42

Appendix B Listing of Data for a Computer Science Study

OBS	SEX	MAJ	SAT-M	SAT-V	HSM	HSE	HSS	GPA
1	1	1	640	530	8	8	6	4.35
2	1	1	670	600	9	7	10	4.08
3	1	1	600	400	8	7	8	5.21
4	1	1	570	480	7	6	7	4.34
5	1	1	510	530	6	8	8	3.40
6	1	1	750	610	10	9	9	3.43
7	1	1	650	460	8	6	9	4.48
8	1	1	720	630	10	9	10	5.73
9	1	1	760	500	10	9	10	5.80
10	1	1	640	670	9	5	6	4.00
11	1	1	640	490	10	8	9	5.16
12	1	1	520	360	9	7	8	4.73
13	1	1	700	520	7	6	8	3.07
14	1	1	490	550	6	6	8	3.82
15	1	1	640	520	10	7	10	5.12
16	1	1	550	290	9	4	7	4.25
17	1	1	600	520	10	10	10	4.93
18	1	1	710	530	10	9	9	4.83
19	1	1	750	670	9	9	10	5.10
20	1	1	620	480	9	9	9	4.87
21	1	1	630	440	10	9	10	5.61
22	1	1	770	720	10	5	7	4.75
23	1	1	610	560	10	9	10	5.26
24	1	1	640	570	10	10	10	5.67
25	1	1	650	480	10	9	10	5.30
26	1	1	660	630	10	8	10	5.62
27	1	1	570	480	7	8	8	4.55
28	1	1	690	550	9	8	7	5.25
29	1	1	670	500	7	8	7	4.21
30	1	1	660	460	10	9	9	4.50
31	1	1	600	630	8	7	8	5.03
32	1	1	447	320	9	8	10	3.92
33	1	1	580	470	6	6	8	4.70
34	1	1	630	630	9	8	7	4.96
35	1	1	600	560	10	10	10	4.76
36	1	1	550	560	9	9	10	5.40
37	1	1	630	500	8	7	8	4.48
38	1	1	750	760	10	10	10	5.86
39	1	1	491	391	9	7	8	4.62
40	1	2	550	500	7	7	8	5.72
41	1	2	550	500	7	7	8	5.72

(continued)

Appendix B (Continued)

OBS	SEX	MAJ	SAT-M	SAT-V	HSM	HSE	HSS	GPA
42	1	2	630	500	8	8	7	5.50
43	1	2	630	500	8	8	7	5.50
44	1	2	630	500	8	8	7	5.50
45	1	2	690	440	6	7	7	4.83
46	1	2	690	440	6	7	7	4.83
47	1	2	540	400	8	5	6	5.06
48	1	2	540	400	8	5	6	5.06
49	1	2	640	480	9	10	10	6.00
50	1	2	640	480	9	10	10	6.00
51	1	2	520	410	8	8	10	5.70
52	1	2	520	410	8	8	10	5.70
53	1	2	559	488	9	4	7	4.81
54	1	2	559	488	9	4	7	4.81
55	1	2	590	510	8	8	6	3.93
56	1	2	590	510	8	8	6	3.93
57	1	2	590	510	8	8	6	3.93
58	1	2	580	580	10	10	10	5.70
59	1	2	580	580	10	10	10	5.70
60	1	2	670	440	9	6	7	4.96
61	1	2	670	440	9	6	7	4.96
62	1	2	620	590	9	8	9	4.64
63	1	2	620	590	9	8	9	4.64
64	1	2	620	590	9	8	9	4.64
65	1	2	540	470	10	8	10	5.09
66	1	2	540	470	10	8	10	5.09
67	1	2	620	560	4	4	3	5.00
68	1	2	620	560	4	4	3	5.00
69	1	2	620	560	4	4	3	5.00
70	1	2	770	540	10	10	10	4.97
71	1	2	770	540	10	10	10	4.97
72	1	2	620	570	10	10	10	4.81
73	1	2	620	570	10	10	10	4.81
74	1	2	620	570	10	10	10	4.81
75	1	2	660	560	10	10	9	5.32
76	1	2	660	560	10	10	9	5.32
77	1	2	710	500	7	4	8	5.40
78	1	2	710	500	7	4	8	5.40
79	1	3	610	390	9	6	6	3.84
80	1	3	560	690	6	7	6	2.40
81	1	3	690	460	9	6	7	4.88
82	1	3	590	440	6	9	5	4.77
83	1	3	530	560	7	10	10	4.43

Appendix B (Continued)

OBS	SEX	MAJ	SAT-M	SAT-V	HSM	HSE	HSS	GPA
84	1	3	640	500	10	6	10	4.63
85	1	3	590	470	8	3	4	3.66
86	1	3	559	545	7	8	7	4.12
87	1	3	490	370	6	7	7	4.46
88	1	3	670	480	8	7	9	3.59
89	1	3	720	610	10	7	10	3.14
90	1	3	640	520	9	7	7	2.65
91	1	3	520	380	7	7	6	4.12
92	1	3	480	410	8	7	9	4.34
93	1	3	480	390	6	9	9	4.11
94	1	3	530	470	6	8	7	3.34
95	1	3	640	630	9	9	8	5.14
96	1	3	690	580	10	10	10	4.25
97	1	3	470	330	6	6	5	4.39
98	1	3	480	460	6	6	6	4.15
99	1	3	540	590	7	6	6	2.75
100	1	3	510	380	5	9	9	5.06
101	1	3	690	490	8	8	8	4.44
102	1	3	590	480	7	7	7	3.11
103	1	3	580	340	10	10	10	5.12
104	1	3	650	500	8	8	7	4.17
105	1	3	530	320	6	6	5	4.00
106	1	3	620	430	10	6	6	3.88
107	1	3	720	740	10	9	9	4.58
108	1	3	590	440	6	6	6	4.16
109	1	3	630	500	7	10	10	4.50
110	1	3	700	480	10	7	8	3.85
111	1	3	620	400	9	8	9	4.95
112	1	3	586	697	6	7	5	2.91
113	1	3	586	670	9	10	9	4.67
114	1	3	700	500	9	7	8	4.51
115	1	3	550	570	7	5	7	3.79
116	1	3	505	518	6	8	6	4.42
117	1	3	515	285	5	7	7	2.58
118	2	1	570	570	10	10	10	4.76
119	2	1	570	570	10	10	10	4.76
120	2	1	580	540	9	9	9	5.35
121	2	1	580	540	9	9	9	5.35
122	2	1	580	540	9	9	9	5.35
123	2	1	560	530	10	8	9	5.80
124	2	1	560	530	10	8	9	5.80

(continued)

Appendix B (Continued)

OBS	SEX	MAJ	SAT-M	SAT-V	HSM	HSE	HSS	GPA
125	2	1	650	570	9	10	9	4.38
126	2	1	650	570	9	10	9	4.38
127	2	1	440	430	10	9	10	4.58
128	2	1	440	430	10	9	10	4.58
129	2	1	570	750	10	10	10	5.18
130	2	1	570	750	10	10	10	5.18
131	2	1	476	576	8	7	8	4.87
132	2	1	476	576	8	7	8	4.87
133	2	1	680	700	8	8	9	5.16
134	2	1	680	700	8	8	9	5.16
135	2	1	490	480	9	9	8	5.07
136	2	1	490	480	9	9	8	5.07
137	2	1	590	490	10	9	8	5.68
138	2	1	590	490	10	9	8	5.68
139	2	1	590	580	10	10	9	5.34
140	2	1	590	580	10	10	9	5.34
141	2	1	650	490	10	8	8	3.93
142	2	1	650	490	10	8	8	3.93
143	2	1	480	520	9	9	5	4.43
144	2	1	480	520	9	9	5	4.43
145	2	1	670	490	10	10	10	5.28
146	2	1	670	490	10	10	10	5.28
147	2	1	710	600	10	10	10	5.66
148	2	1	710	600	10	10	10	5.66
149	2	1	570	570	7	8	6	4.29
150	2	1	570	570	7	8	6	4.29
151	2	1	540	460	6	6	5	4.19
152	2	1	540	460	6	6	5	4.19
153	2	1	620	510	10	10	10	5.06
154	2	1	620	510	10	10	10	5.06
155	2	1	630	470	8	8	6	5.41
156	2	1	630	470	8	8	6	5.41
157	2	2	630	700	9	10	9	5.14
158	2	2	630	700	9	10	9	5.14
159	2	2	630	700	9	10	9	5.14
160	2	2	630	700	9	10	9	5.14
161	2	2	610	460	10	8	8	4.85
162	2	2	610	460	10	8	8	4.85
163	2	2	610	460	10	8	8	4.85
164	2	2	720	680	10	9	10	5.47
165	2	2	720	680	10	9	10	5.47
166	2	2	720	680	10	9	10	5.47

Appendix B (Continued)

OBS	SEX	MAJ	SAT-M	SAT-V	HSM	HSE	HSS	GPA
167	2	2	720	680	10	9	10	5.47
168	2	2	640	590	10	9	10	5.44
169	2	2	640	590	10	9	10	5.44
170	2	2	640	590	10	9	10	5.44
171	2	2	640	590	10	9	10	5.44
172	2	2	650	500	10	10	10	5.90
173	2	2	650	500	10	10	10	5.90
174	2	2	650	500	10	10	10	5.90
175	2	2	650	500	10	10	10	5.90
176	2	2	640	430	9	9	9	5.65
177	2	2	640	430	9	9	9	5.65
178	2	2	640	430	9	9	9	5.65
179	2	2	640	430	9	9	9	5.65
180	2	2	740	460	9	9	8	3.32
181	2	2	740	460	9	9	8	3.32
182	2	2	740	460	9	9	8	3.32
183	2	2	600	660	10	10	10	5.23
184	2	2	600	660	10	10	10	5.23
185	2	2	600	660	10	10	10	5.23
186	2	2	650	430	10	10	9	4.86
187	2	2	650	430	10	10	9	4.86
188	2	2	650	430	10	10	9	4.86
189	2	2	510	440	8	10	9	4.51
190	2	2	510	440	8	10	9	4.51
191	2	2	510	440	8	10	9	4.51
192	2	2	550	510	8	8	9	4.86
193	2	2	550	510	8	8	9	4.86
194	2	2	550	510	8	8	9	4.86
195	2	2	550	510	8	8	9	4.86
196	2	3	630	560	9	9	7	5.33
197	2	3	710	470	10	8	10	5.69
198	2	3	417	518	9	10	10	4.57
199	2	3	600	600	8	10	10	4.28
200	2	3	460	460	4	7	7	3.60
201	2	3	300	290	2	6	4	4.00
202	2	3	560	480	7	7	6	3.69
203	2	3	590	420	9	9	10	5.06
204	2	3	550	440	9	8	10	4.62
205	2	3	550	660	7	9	10	2.39
206	2	3	650	350	10	9	9	4.44
207	2	3	610	520	9	8	9	5.46

(continued)

Appendix B (Continued)

OBS	SEX	MAJ	SAT-M	SAT-V	HSM	HSE	HSS	GPA
208	2	3	530	480	8	9	7	4.37
209	2	3	550	450	10	9	9	4.80
210	2	3	560	420	5	8	4	4.14
211	2	3	430	330	7	8	7	4.45
212	2	3	490	400	9	10	7	4.71
213	2	3	690	510	9	10	9	4.93
214	2	3	570	480	7	9	6	4.53
215	2	3	470	420	8	9	6	4.69
216	2	3	470	330	8	8	8	3.94
217	2	3	510	360	10	9	8	5.00
218	2	3	450	460	9	8	7	4.09
219	2	3	490	410	10	10	9	5.34
220	2	3	530	510	6	10	9	4.25
221	2	3	580	490	10	10	10	6.00
222	2	3	580	490	6	7	5	4.72
223	2	3	620	420	9	8	7	4.61
224	2	3	500	390	10	10	10	5.39
225	2	3	590	580	8	8	6	5.64
226	2	3	620	600	8	9	7	3.80
227	2	3	520	570	9	10	9	3.52
228	2	3	480	480	6	9	9	5.40
229	2	3	640	560	10	10	9	5.32
230	2	3	600	440	9	6	7	4.07
231	2	3	510	480	7	9	7	2.85
232	2	3	356	350	7	7	9	3.86
233	2	3	630	470	5	7	4	4.59
234	2	3	559	488	9	9	8	4.28

Tables

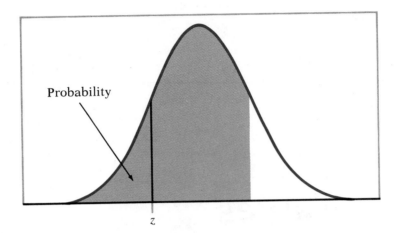

Figure for Table A
Table entry is probability at or below z.

Table A   Standard normal probabilities

z	.00	.01	.02	.03	.04	.05	.06	.07	.08	.09
−3.4	.0003	.0003	.0003	.0003	.0003	.0003	.0003	.0003	.0003	.0002
−3.3	.0005	.0005	.0005	.0004	.0004	.0004	.0004	.0004	.0004	.0003
−3.2	.0007	.0007	.0006	.0006	.0006	.0006	.0006	.0005	.0005	.0005
−3.1	.0010	.0009	.0009	.0009	.0008	.0008	.0008	.0008	.0007	.0007
−3.0	.0013	.0013	.0013	.0012	.0012	.0011	.0011	.0011	.0010	.0010
−2.9	.0019	.0018	.0018	.0017	.0016	.0016	.0015	.0015	.0014	.0014
−2.8	.0026	.0025	.0024	.0023	.0023	.0030	.0029	.0028	.0027	.0026
−2.7	.0035	.0034	.0033	.0032	.0031	.0040	.0039	.0038	.0037	.0036
−2.6	.0047	.0045	.0044	.0043	.0041	.0054	.0052	.0051	.0049	.0048
−2.5	.0062	.0060	.0059	.0057	.0055	.0071	.0069	.0068	.0066	.0064
−2.4	.0082	.0080	.0078	.0075	.0073	.0094	.0091	.0089	.0087	.0084
−2.3	.0107	.0104	.0102	.0099	.0096	.0122	.0119	.0116	.0113	.0110
−2.2	.0139	.0136	.0132	.0129	.0125	.0158	.0154	.0150	.0146	.0143
−2.1	.0179	.0174	.0170	.0166	.0162	.0202	.0197	.0192	.0188	.0183
−2.0	.0228	.0222	.0217	.0212	.0207	.0256	.0250	.0244	.0239	.0233
−1.9	.0287	.0281	.0274	.0268	.0262	.0322	.0314	.0307	.0301	.0294
−1.8	.0359	.0351	.0344	.0336	.0329	.0401	.0392	.0384	.0375	.0367
−1.7	.0446	.0436	.0427	.0418	.0409	.0495	.0485	.0475	.0465	.0455
−1.6	.0548	.0537	.0526	.0516	.0505	.0606	.0594	.0582	.0571	.0559
−1.5	.0668	.0655	.0643	.0630	.0618	.0735	.0721	.0708	.0694	.0681
−1.4	.0808	.0793	.0778	.0764	.0749	.0885	.0869	.0853	.0838	.0823
−1.3	.0968	.0951	.0934	.0918	.0901	.1056	.1038	.1020	.1003	.0985
−1.2	.1151	.1131	.1112	.1093	.1075	.1251	.1230	.1210	.1190	.1170
−1.1	.1357	.1335	.1314	.1292	.1271	.1469	.1446	.1423	.1401	.1379
−1.0	.1587	.1562	.1539	.1515	.1492	.1711	.1685	.1660	.1635	.1611
−0.9	.1841	.1814	.1788	.1762	.1736	.1977	.1949	.1922	.1894	.1867
−0.8	.2119	.2090	.2061	.2033	.2005	.2266	.2236	.2206	.2177	.2148
−0.7	.2420	.2389	.2358	.2327	.2296	.2578	.2546	.2514	.2483	.2451
−0.6	.2743	.2709	.2676	.2643	.2611	.2912	.2877	.2843	.2810	.2776
−0.5	.3085	.3050	.3015	.2981	.2946	.3264	.3228	.3192	.3156	.3121
−0.4	.3446	.3409	.3372	.3336	.3300	.3632	.3594	.3557	.3520	.3483
−0.3	.3821	.3783	.3745	.3707	.3669	.4013	.3974	.3936	.3897	.3859
−0.2	.4207	.4168	.4129	.4090	.4052	.4404	.4364	.4325	.4286	.4247
−0.1	.4602	.4562	.4522	.4483	.4443	.4801	.4761	.4721	.4681	.4641
−0.0	.5000	.4960	.4920	.4880	.4840					

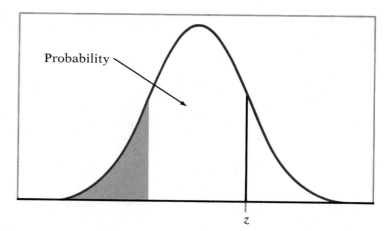

Figure for Table A (continued) Table entry is probability at or below z.

Probability

z

Table A (continued)

z	.00	.01	.02	.03	.04	.05	.06	.07	.08	.09
0.0	.5000	.5040	.5080	.5120	.5160	.5199	.5239	.5279	.5319	.5359
0.1	.5398	.5438	.5478	.5517	.5557	.5596	.5636	.5675	.5714	.5753
0.2	.5793	.5832	.5871	.5910	.5948	.5987	.6026	.6064	.6103	.6141
0.3	.6179	.6217	.6255	.6293	.6331	.6368	.6406	.6443	.6480	.6517
0.4	.6554	.6591	.6628	.6664	.6700	.6736	.6772	.6808	.6844	.6879
0.5	.6915	.6950	.6985	.7019	.7054	.7088	.7123	.7157	.7190	.7224
0.6	.7257	.7291	.7324	.7357	.7389	.7422	.7454	.7486	.7517	.7549
0.7	.7580	.7611	.7642	.7673	.7704	.7734	.7764	.7794	.7823	.7852
0.8	.7881	.7910	.7939	.7967	.7995	.8023	.8051	.8078	.8106	.8133
0.9	.8159	.8186	.8212	.8238	.8264	.8289	.8315	.8340	.8365	.8389
1.0	.8413	.8438	.8461	.8485	.8508	.8531	.8554	.8577	.8599	.8621
1.1	.8643	.8665	.8686	.8708	.8729	.8749	.8770	.8790	.8810	.8830
1.2	.8849	.8869	.8888	.8907	.8925	.8944	.8962	.8980	.8997	.9015
1.3	.9032	.9049	.9066	.9082	.9099	.9115	.9131	.9147	.9162	.9177
1.4	.9192	.9207	.9222	.9236	.9251	.9265	.9279	.9292	.9306	.9319
1.5	.9332	.9345	.9357	.9370	.9382	.9394	.9406	.9418	.9429	.9441
1.6	.9452	.9463	.9474	.9484	.9495	.9505	.9515	.9525	.9535	.9545
1.7	.9554	.9564	.9573	.9582	.9591	.9599	.9608	.9616	.9625	.9633
1.8	.9641	.9649	.9656	.9664	.9671	.9678	.9686	.9693	.9699	.9706
1.9	.9713	.9719	.9726	.9732	.9738	.9744	.9750	.9756	.9761	.9767
2.0	.9772	.9778	.9783	.9788	.9793	.9798	.9803	.9808	.9812	.9817
2.1	.9821	.9826	.9830	.9834	.9838	.9842	.9846	.9850	.9854	.9857
2.2	.9861	.9864	.9868	.9871	.9875	.9878	.9881	.9884	.9887	.9890
2.3	.9893	.9896	.9898	.9901	.9904	.9906	.9909	.9911	.9913	.9916
2.4	.9918	.9920	.9922	.9925	.9927	.9929	.9931	.9932	.9934	.9936
2.5	.9938	.9940	.9941	.9943	.9945	.9946	.9948	.9949	.9951	.9952
2.6	.9953	.9955	.9956	.9957	.9959	.9960	.9961	.9962	.9963	.9964
2.7	.9965	.9966	.9967	.9968	.9969	.9970	.9971	.9972	.9973	.9974
2.8	.9974	.9975	.9976	.9977	.9977	.9978	.9979	.9979	.9980	.9981
2.9	.9981	.9982	.9982	.9983	.9984	.9984	.9985	.9985	.9986	.9986
3.0	.9987	.9987	.9987	.9988	.9988	.9989	.9989	.9989	.9990	.9990
3.1	.9990	.9991	.9991	.9991	.9992	.9992	.9992	.9992	.9993	.9993
3.2	.9993	.9993	.9994	.9994	.9994	.9994	.9994	.9995	.9995	.9995
3.3	.9995	.9995	.9995	.9996	.9996	.9996	.9996	.9996	.9996	.9997
3.4	.9997	.9997	.9997	.9997	.9997	.9997	.9997	.9997	.9997	.9998

Table B Random digits

Line								
101	19223	95034	05756	28713	96409	12531	42544	82853
102	73676	47150	99400	01927	27754	42648	82425	36290
103	45467	71709	77558	00095	32863	29485	82226	90056
104	52711	38889	93074	60227	40011	85848	48767	52573
105	95592	94007	69971	91481	60779	53791	17297	59335
106	68417	35013	15529	72765	85089	57067	50211	47487
107	82739	57890	20807	47511	81676	55300	94383	14893
108	60940	72024	17868	24943	61790	90656	87964	18883
109	36009	19365	15412	39638	85453	46816	83485	41979
110	38448	48789	18338	24697	39364	42006	76688	08708
111	81486	69487	60513	09297	00412	71238	27649	39950
112	59636	88804	04634	71197	19352	73089	84898	45785
113	62568	70206	40325	03699	71080	22553	11486	11776
114	45149	32992	75730	66280	03819	56202	02938	70915
115	61041	77684	94322	24709	73698	14526	31893	32592
116	14459	26056	31424	80371	65103	62253	50490	61181
117	38167	98532	62183	70632	23417	26185	41448	75532
118	73190	32533	04470	29669	84407	90785	65956	86382
119	95857	07118	87664	92099	58806	66979	98624	84826
120	35476	55972	39421	65850	04266	35435	43742	11937
121	71487	09984	29077	14863	61683	47052	62224	51025
122	13873	81598	95052	90908	73592	75186	87136	95761
123	54580	81507	27102	56027	55892	33063	41842	81868
124	71035	09001	43367	49497	72719	96758	27611	91596
125	96746	12149	37823	71868	18442	35119	62103	39244
126	96927	19931	36089	74192	77567	88741	48409	41903
127	43909	99477	25330	64359	40085	16925	85117	36071
128	15689	14227	06565	14374	13352	49367	81982	87209
129	36759	58984	68288	22913	18638	54303	00795	08727
130	69051	64817	87174	09517	84534	06489	87201	97245
131	05007	16632	81194	14873	04197	85576	45195	96565
132	68732	55259	84292	08796	43165	93739	31685	97150
133	45740	41807	65561	33302	07051	93623	18132	09547
134	27816	78416	18329	21337	35213	37741	04312	68508
135	66925	55658	39100	78458	11206	19876	87151	31260
136	08421	44753	77377	28744	75592	08563	79140	92454
137	53645	66812	61421	47836	12609	15373	98481	14592
138	66831	68908	40772	21558	47781	33586	79177	06928
139	55588	99404	70708	41098	43563	56934	48394	51719
140	12975	13258	13048	45144	72321	81940	00360	02428
141	96767	35964	23822	96012	94591	65194	50842	53372
142	72829	50232	97892	63408	77919	44575	24870	04178
143	88565	42628	17797	49376	61762	16953	88604	12724
144	62964	88145	83083	69453	46109	59505	69680	00900
145	19687	12633	57857	95806	09931	02150	43163	58636
146	37609	59057	66967	83401	60705	02384	90597	93600
147	54973	86278	88737	74351	47500	84552	19909	67181
148	00694	05977	19664	65441	20903	62371	22725	53340
149	71546	05233	53946	68743	72460	27601	45403	88692
150	07511	88915	41267	16853	84569	79367	32337	03316

Table B (continued)

Line								
151	03802	29341	29264	80198	12371	13121	54969	43912
152	77320	35030	77519	41109	98296	18984	60869	12349
153	07886	56866	39648	69290	03600	05376	58958	22720
154	87065	74133	21117	70595	22791	67306	28420	52067
155	42090	09628	54035	93879	98441	04606	27381	82637
156	55494	67690	88131	81800	11188	28552	25752	21953
157	16698	30406	96587	65985	07165	50148	16201	86792
158	16297	07626	68683	45335	34377	72941	41764	77038
159	22897	17467	17638	70043	36243	13008	83993	22869
160	98163	45944	34210	64158	76971	27689	82926	75957
161	43400	25831	06283	22138	16043	15706	73345	26238
162	97341	46254	88153	62336	21112	35574	99271	45297
163	64578	67197	28310	90341	37531	63890	52630	76315
164	11022	79124	49525	63078	17229	32165	01343	21394
165	81232	43939	23840	05995	84589	06788	76358	26622
166	36843	84798	51167	44728	20554	55538	27647	32708
167	84329	80081	69516	78934	14293	92478	16479	26974
168	27788	85789	41592	74472	96773	27090	24954	41474
169	99224	00850	43737	75202	44753	63236	14260	73686
170	38075	73239	52555	46342	13365	02182	30443	53229
171	87368	49451	55771	48343	51236	18522	73670	23212
172	40512	00681	44282	47178	08139	78693	34715	75606
173	81636	57578	54286	27216	58758	80358	84115	84568
174	26411	94292	06340	97762	37033	85968	94165	46514
175	80011	09937	57195	33906	94831	10056	42211	65491
176	92813	87503	63494	71379	76550	45984	05481	50830
177	70348	72871	63419	57363	29685	43090	18763	31714
178	24005	52114	26224	39078	80798	15220	43186	00976
179	85063	55810	10470	08029	30025	29734	61181	72090
180	11532	73186	92541	06915	72954	10167	12142	26492
181	59618	03914	05208	84088	20426	39004	84582	87317
182	92965	50837	39921	84661	82514	81899	24565	60874
183	85116	27684	14597	85747	01596	25889	41998	15635
184	15106	10411	90221	49377	44369	28185	80959	76355
185	03638	31589	07871	25792	85823	55400	56026	12193
186	97971	48932	45792	63993	95635	28753	46069	84635
187	49345	18305	76213	82390	77412	97401	50650	71755
188	87370	88099	89695	87633	76987	85503	26257	51736
189	88296	95670	74932	65317	93848	43988	47597	83044
190	79485	92200	99401	54473	34336	82786	05457	60343
191	40830	24979	23333	37619	56227	95941	59494	86539
192	32006	76302	81221	00693	95197	75044	46596	11628
193	37569	85187	44692	50706	53161	69027	88389	60313
194	56680	79003	23361	67094	15019	63261	24543	52884
195	05172	08100	22316	54495	60005	29532	18433	18057
196	74782	27005	03894	98038	20627	40307	47317	92759
197	85228	93264	61409	03404	09649	55937	60843	66167
198	68309	12060	14762	58002	03716	81968	57934	32624
199	26461	88346	52430	60906	74216	96263	69296	90107
200	42672	67680	42376	95023	82744	03971	96560	55148

Table C Binomial probabilities

$$\text{Entry is } P(X = k) = \binom{n}{k} p^k (1 - p)^{n-k}$$

n	k	.01	.02	.03	.04	.05	.06	.07	.08	.09
2	0	.9801	.9604	.9409	.9216	.9025	.8836	.8649	.8464	.8281
	1	.0198	.0392	.0582	.0768	.0950	.1128	.1302	.1472	.1638
	2	.0001	.0004	.0009	.0016	.0025	.0036	.0049	.0064	.0081
3	0	.9703	.9412	.9127	.8847	.8574	.8306	.8044	.7787	.7536
	1	.0294	.0576	.0847	.1106	.1354	.1590	.1816	.2031	.2236
	2	.0003	.0012	.0026	.0046	.0071	.0102	.0137	.0177	.0221
	3	.0000	.0000	.0000	.0001	.0001	.0002	.0003	.0005	.0007
4	0	.9606	.9224	.8853	.8493	.8145	.7807	.7481	.7164	.6857
	1	.0388	.0753	.1095	.1416	.1715	.1993	.2252	.2492	.2713
	2	.0006	.0023	.0051	.0088	.0135	.0191	.0254	.0325	.0402
	3	.0000	.0000	.0001	.0002	.0005	.0008	.0013	.0019	.0027
	4	.0000	.0000	.0000	.0000	.0000	.0000	.0000	.0000	.0001
5	0	.9510	.9039	.8587	.8154	.7738	.7339	.6957	.6591	.6240
	1	.0480	.0922	.1328	.1699	.2036	.2342	.2618	.2866	.3086
	2	.0010	.0038	.0082	.0142	.0214	.0299	.0394	.0498	.0610
	3	.0000	.0001	.0003	.0006	.0011	.0019	.0030	.0043	.0060
	4	.0000	.0000	.0000	.0000	.0000	.0001	.0001	.0002	.0003
	5	.0000	.0000	.0000	.0000	.0000	.0000	.0000	.0000	.0000
6	0	.9415	.8858	.8330	.7828	.7351	.6899	.6470	.6064	.5679
	1	.0571	.1085	.1546	.1957	.2321	.2642	.2922	.3164	.3370
	2	.0014	.0055	.0120	.0204	.0305	.0422	.0550	.0688	.0833
	3	.0000	.0002	.0005	.0011	.0021	.0036	.0055	.0080	.0110
	4	.0000	.0000	.0000	.0000	.0001	.0002	.0003	.0005	.0008
	5	.0000	.0000	.0000	.0000	.0000	.0000	.0000	.0000	.0000
	6	.0000	.0000	.0000	.0000	.0000	.0000	.0000	.0000	.0000
7	0	.9321	.8681	.8080	.7514	.6983	.6485	.6017	.5578	.5168
	1	.0659	.1240	.1749	.2192	.2573	.2897	.3170	.3396	.3578
	2	.0020	.0076	.0162	.0274	.0406	.0555	.0716	.0886	.1061
	3	.0000	.0003	.0008	.0019	.0036	.0059	.0090	.0128	.0175
	4	.0000	.0000	.0000	.0001	.0002	.0004	.0007	.0011	.0017
	5	.0000	.0000	.0000	.0000	.0000	.0000	.0000	.0001	.0001
	6	.0000	.0000	.0000	.0000	.0000	.0000	.0000	.0000	.0000
	7	.0000	.0000	.0000	.0000	.0000	.0000	.0000	.0000	.0000
8	0	.9227	.8508	.7837	.7214	.6634	.6096	.5596	.5132	.4703
	1	.0746	.1389	.1939	.2405	.2793	.3113	.3370	.3570	.3721
	2	.0026	.0099	.0210	.0351	.0515	.0695	.0888	.1087	.1288
	3	.0001	.0004	.0013	.0029	.0054	.0089	.0134	.0189	.0255
	4	.0000	.0000	.0001	.0002	.0004	.0007	.0013	.0021	.0031
	5	.0000	.0000	.0000	.0000	.0000	.0000	.0001	.0001	.0002
	6	.0000	.0000	.0000	.0000	.0000	.0000	.0000	.0000	.0000
	7	.0000	.0000	.0000	.0000	.0000	.0000	.0000	.0000	.0000
	8	.0000	.0000	.0000	.0000	.0000	.0000	.0000	.0000	.0000

Table C (continued)

$$\text{Entry is } P(X = k) = \binom{n}{k} p^k (1 - p)^{n-k}$$

						p				
n	k	.10	.15	.20	.25	.30	.35	.40	.45	.50
2	0	.8100	.7225	.6400	.5625	.4900	.4225	.3600	.3025	.2500
	1	.1800	.2550	.3200	.3750	.4200	.4550	.4800	.4950	.5000
	2	.0100	.0225	.0400	.0625	.0900	.1225	.1600	.2025	.2500
3	0	.7290	.6141	.5120	.4219	.3430	.2746	.2160	.1664	.1250
	1	.2430	.3251	.3840	.4219	.4410	.4436	.4320	.4084	.3750
	2	.0270	.0574	.0960	.1406	.1890	.2389	.2880	.3341	.3750
	3	.0010	.0034	.0080	.0156	.0270	.0429	.0640	.0911	.1250
4	0	.6561	.5220	.4096	.3164	.2401	.1785	.1296	.0915	.0625
	1	.2916	.3685	.4096	.4219	.4116	.3845	.3456	.2995	.2500
	2	.0486	.0975	.1536	.2109	.2646	.3105	.3456	.3675	.3750
	3	.0036	.0115	.0256	.0469	.0756	.1115	.1536	.2005	.2500
	4	.0001	.0005	.0016	.0039	.0081	.0150	.0256	.0410	.0625
5	0	.5905	.4437	.3277	.2373	.1681	.1160	.0778	.0503	.0313
	1	.3280	.3915	.4096	.3955	.3602	.3124	.2592	.2059	.1563
	2	.0729	.1382	.2048	.2637	.3087	.3364	.3456	.3369	.3125
	3	.0081	.0244	.0512	.0879	.1323	.1811	.2304	.2757	.3125
	4	.0004	.0022	.0064	.0146	.0284	.0488	.0768	.1128	.1562
	5	.0000	.0001	.0003	.0010	.0024	.0053	.0102	.0185	.0312
6	0	.5314	.3771	.2621	.1780	.1176	.0754	.0467	.0277	.0156
	1	.3543	.3993	.3932	.3560	.3025	.2437	.1866	.1359	.0938
	2	.0984	.1762	.2458	.2966	.3241	.3280	.3110	.2780	.2344
	3	.0146	.0415	.0819	.1318	.1852	.2355	.2765	.3032	.3125
	4	.0012	.0055	.0154	.0330	.0595	.0951	.1382	.1861	.2344
	5	.0001	.0004	.0015	.0044	.0102	.0205	.0369	.0609	.0937
	6	.0000	.0000	.0001	.0002	.0007	.0018	.0041	.0083	.0156
7	0	.4783	.3206	.2097	.1335	.0824	.0490	.0280	.0152	.0078
	1	.3720	.3960	.3670	.3115	.2471	.1848	.1306	.0872	.0547
	2	.1240	.2097	.2753	.3115	.3177	.2985	.2613	.2140	.1641
	3	.0230	.0617	.1147	.1730	.2269	.2679	.2903	.2918	.2734
	4	.0026	.0109	.0287	.0577	.0972	.1442	.1935	.2388	.2734
	5	.0002	.0012	.0043	.0115	.0250	.0466	.0774	.1172	.1641
	6	.0000	.0001	.0004	.0013	.0036	.0084	.0172	.0320	.0547
	7	.0000	.0000	.0000	.0001	.0002	.0006	.0016	.0037	.0078
8	0	.4305	.2725	.1678	.1001	.0576	.0319	.0168	.0084	.0039
	1	.3826	.3847	.3355	.2670	.1977	.1373	.0896	.0548	.0313
	2	.1488	.2376	.2936	.3115	.2965	.2587	.2090	.1569	.1094
	3	.0331	.0839	.1468	.2076	.2541	.2786	.2787	.2568	.2188
	4	.0046	.0185	.0459	.0865	.1361	.1875	.2322	.2627	.2734
	5	.0004	.0026	.0092	.0231	.0467	.0808	.1239	.1719	.2188
	6	.0000	.0002	.0011	.0038	.0100	.0217	.0413	.0703	.1094
	7	.0000	.0000	.0001	.0004	.0012	.0033	.0079	.0164	.0312
	8	.0000	.0000	.0000	.0000	.0001	.0002	.0007	.0017	.0039

Table C (continued)

						p				
n	k	.01	.02	.03	.04	.05	.06	.07	.08	.09
9	0	.9135	.8337	.7602	.6925	.6302	.5730	.5204	.4722	.4279
	1	.0830	.1531	.2116	.2597	.2985	.3292	.3525	.3695	.3809
	2	.0034	.0125	.0262	.0433	.0629	.0840	.1061	.1285	.1507
	3	.0001	.0006	.0019	.0042	.0077	.0125	.0186	.0261	.0348
	4	.0000	.0000	.0001	.0003	.0006	.0012	.0021	.0034	.0052
	5	.0000	.0000	.0000	.0000	.0000	.0001	.0002	.0003	.0005
	6	.0000	.0000	.0000	.0000	.0000	.0000	.0000	.0000	.0000
	7	.0000	.0000	.0000	.0000	.0000	.0000	.0000	.0000	.0000
	8	.0000	.0000	.0000	.0000	.0000	.0000	.0000	.0000	.0000
	9	.0000	.0000	.0000	.0000	.0000	.0000	.0000	.0000	.0000
10	0	.9044	.8171	.7374	.6648	.5987	.5386	.4840	.4344	.3894
	1	.0914	.1667	.2281	.2770	.3151	.3438	.3643	.3777	.3851
	2	.0042	.0153	.0317	.0519	.0746	.0988	.1234	.1478	.1714
	3	.0001	.0008	.0026	.0058	.0105	.0168	.0248	.0343	.0452
	4	.0000	.0000	.0001	.0004	.0010	.0019	.0033	.0052	.0078
	5	.0000	.0000	.0000	.0000	.0001	.0001	.0003	.0005	.0009
	6	.0000	.0000	.0000	.0000	.0000	.0000	.0000	.0000	.0001
	7	.0000	.0000	.0000	.0000	.0000	.0000	.0000	.0000	.0000
	8	.0000	.0000	.0000	.0000	.0000	.0000	.0000	.0000	.0000
	9	.0000	.0000	.0000	.0000	.0000	.0000	.0000	.0000	.0000
	10	.0000	.0000	.0000	.0000	.0000	.0000	.0000	.0000	.0000
12	0	.8864	.7847	.6938	.6127	.5404	.4759	.4186	.3677	.3225
	1	.1074	.1922	.2575	.3064	.3413	.3645	.3781	.3837	.3827
	2	.0060	.0216	.0438	.0702	.0988	.1280	.1565	.1835	.2082
	3	.0002	.0015	.0045	.0098	.0173	.0272	.0393	.0532	.0686
	4	.0000	.0001	.0003	.0009	.0021	.0039	.0067	.0104	.0153
	5	.0000	.0000	.0000	.0001	.0002	.0004	.0008	.0014	.0024
	6	.0000	.0000	.0000	.0000	.0000	.0000	.0001	.0001	.0003
	7	.0000	.0000	.0000	.0000	.0000	.0000	.0000	.0000	.0000
	8	.0000	.0000	.0000	.0000	.0000	.0000	.0000	.0000	.0000
	9	.0000	.0000	.0000	.0000	.0000	.0000	.0000	.0000	.0000
	10	.0000	.0000	.0000	.0000	.0000	.0000	.0000	.0000	.0000
	11	.0000	.0000	.0000	.0000	.0000	.0000	.0000	.0000	.0000
	12	.0000	.0000	.0000	.0000	.0000	.0000	.0000	.0000	.0000
15	0	.8601	.7386	.6333	.5421	.4633	.3953	.3367	.2863	.2430
	1	.1303	.2261	.2938	.3388	.3658	.3785	.3801	.3734	.3605
	2	.0092	.0323	.0636	.0988	.1348	.1691	.2003	.2273	.2496
	3	.0004	.0029	.0085	.0178	.0307	.0468	.0653	.0857	.1070
	4	.0000	.0002	.0008	.0022	.0049	.0090	.0148	.0223	.0317
	5	.0000	.0000	.0001	.0002	.0006	.0013	.0024	.0043	.0069
	6	.0000	.0000	.0000	.0000	.0000	.0001	.0003	.0006	.0011
	7	.0000	.0000	.0000	.0000	.0000	.0000	.0000	.0001	.0001
	8	.0000	.0000	.0000	.0000	.0000	.0000	.0000	.0000	.0000
	9	.0000	.0000	.0000	.0000	.0000	.0000	.0000	.0000	.0000
	10	.0000	.0000	.0000	.0000	.0000	.0000	.0000	.0000	.0000
	11	.0000	.0000	.0000	.0000	.0000	.0000	.0000	.0000	.0000
	12	.0000	.0000	.0000	.0000	.0000	.0000	.0000	.0000	.0000
	13	.0000	.0000	.0000	.0000	.0000	.0000	.0000	.0000	.0000
	14	.0000	.0000	.0000	.0000	.0000	.0000	.0000	.0000	.0000
	15	.0000	.0000	.0000	.0000	.0000	.0000	.0000	.0000	.0000

Table C (continued)

						p				
n	k	.10 / .01	.15 / .02	.20 / .03	.25 / .04	.30 / .05	.35 / .06	.40 / .07	.45 / .08	.50 / .09
9	0	.3874	.2316	.1342	.0751	.0404	.0207	.0101	.0046	.0020
	1	.3874	.3679	.3020	.2253	.1556	.1004	.0605	.0339	.0176
	2	.1722	.2597	.3020	.3003	.2668	.2162	.1612	.1110	.0703
	3	.0446	.1069	.1762	.2336	.2668	.2716	.2508	.2119	.1641
	4	.0074	.0283	.0661	.1168	.1715	.2194	.2508	.2600	.2461
	5	.0008	.0050	.0165	.0389	.0735	.1181	.1672	.2128	.2461
	6	.0001	.0006	.0028	.0087	.0210	.0424	.0743	.1160	.1641
	7	.0000	.0000	.0003	.0012	.0039	.0098	.0212	.0407	.0703
	8	.0000	.0000	.0000	.0001	.0004	.0013	.0035	.0083	.0176
	9	.0000	.0000	.0000	.0000	.0000	.0001	.0003	.0008	.0020
10	0	.3487	.1969	.1074	.0563	.0282	.0135	.0060	.0025	.0010
	1	.3874	.3474	.2684	.1877	.1211	.0725	.0403	.0207	.0098
	2	.1937	.2759	.3020	.2816	.2335	.1757	.1209	.0763	.0439
	3	.0574	.1298	.2013	.2503	.2668	.2522	.2150	.1665	.1172
	4	.0112	.0401	.0881	.1460	.2001	.2377	.2508	.2384	.2051
	5	.0015	.0085	.0264	.0584	.1029	.1536	.2007	.2340	.2461
	6	.0001	.0012	.0055	.0162	.0368	.0689	.1115	.1596	.2051
	7	.0000	.0001	.0008	.0031	.0090	.0212	.0425	.0746	.1172
	8	.0000	.0000	.0001	.0004	.0014	.0043	.0106	.0229	.0439
	9	.0000	.0000	.0000	.0000	.0001	.0005	.0016	.0042	.0098
	10	.0000	.0000	.0000	.0000	.0000	.0000	.0001	.0003	.0010
12	0	.2824	.1422	.0687	.0317	.0138	.0057	.0022	.0008	.0002
	1	.3766	.3012	.2062	.1267	.0712	.0368	.0174	.0075	.0029
	2	.2301	.2924	.2835	.2323	.1678	.1088	.0639	.0339	.0161
	3	.0852	.1720	.2362	.2581	.2397	.1954	.1419	.0923	.0537
	4	.0213	.0683	.1329	.1936	.2311	.2367	.2128	.1700	.1208
	5	.0038	.0193	.0532	.1032	.1585	.2039	.2270	.2225	.1934
	6	.0005	.0040	.0155	.0401	.0792	.1281	.1766	.2124	.2256
	7	.0000	.0006	.0033	.0115	.0291	.0591	.1009	.1489	.1934
	8	.0000	.0001	.0005	.0024	.0078	.0199	.0420	.0762	.1208
	9	.0000	.0000	.0001	.0004	.0015	.0048	.0125	.0277	.0537
	10	.0000	.0000	.0000	.0000	.0002	.0008	.0025	.0068	.0161
	11	.0000	.0000	.0000	.0000	.0000	.0001	.0003	.0010	.0029
	12	.0000	.0000	.0000	.0000	.0000	.0000	.0000	.0001	.0002
15	0	.2059	.0874	.0352	.0134	.0047	.0016	.0005	.0001	.0000
	1	.3432	.2312	.1319	.0668	.0305	.0126	.0047	.0016	.0005
	2	.2669	.2856	.2309	.1559	.0916	.0476	.0219	.0090	.0032
	3	.1285	.2184	.2501	.2252	.1700	.1110	.0634	.0318	.0139
	4	.0428	.1156	.1876	.2252	.2186	.1792	.1268	.0780	.0417
	5	.0105	.0449	.1032	.1651	.2061	.2123	.1859	.1404	.0916
	6	.0019	.0132	.0430	.0917	.1472	.1906	.2066	.1914	.1527
	7	.0003	.0030	.0138	.0393	.0811	.1319	.1771	.2013	.1964
	8	.0000	.0005	.0035	.0131	.0348	.0710	.1181	.1647	.1964
	9	.0000	.0001	.0007	.0034	.0116	.0298	.0612	.1048	.1527
	10	.0000	.0000	.0001	.0007	.0030	.0096	.0245	.0515	.0916
	11	.0000	.0000	.0000	.0001	.0006	.0024	.0074	.0191	.0417
	12	.0000	.0000	.0000	.0000	.0001	.0004	.0016	.0052	.0139
	13	.0000	.0000	.0000	.0000	.0000	.0001	.0003	.0010	.0032
	14	.0000	.0000	.0000	.0000	.0000	.0000	.0000	.0001	.0005
	15	.0000	.0000	.0000	.0000	.0000	.0000	.0000	.0000	.0000

Table C (continued)

n	k	.01	.02	.03	.04	.05	.06	.07	.08	.09
						p				
20	0	.8179	.6676	.5438	.4420	.3585	.2901	.2342	.1887	.1516
	1	.1652	.2725	.3364	.3683	.3774	.3703	.3526	.3282	.3000
	2	.0159	.0528	.0988	.1458	.1887	.2246	.2521	.2711	.2818
	3	.0010	.0065	.0183	.0364	.0596	.0860	.1139	.1414	.1672
	4	.0000	.0006	.0024	.0065	.0133	.0233	.0364	.0523	.0703
	5	.0000	.0000	.0002	.0009	.0022	.0048	.0088	.0145	.0222
	6	.0000	.0000	.0000	.0001	.0003	.0008	.0017	.0032	.0055
	7	.0000	.0000	.0000	.0000	.0000	.0001	.0002	.0005	.0011
	8	.0000	.0000	.0000	.0000	.0000	.0000	.0000	.0001	.0002
	9	.0000	.0000	.0000	.0000	.0000	.0000	.0000	.0000	.0000
	10	.0000	.0000	.0000	.0000	.0000	.0000	.0000	.0000	.0000
	11	.0000	.0000	.0000	.0000	.0000	.0000	.0000	.0000	.0000
	12	.0000	.0000	.0000	.0000	.0000	.0000	.0000	.0000	.0000
	13	.0000	.0000	.0000	.0000	.0000	.0000	.0000	.0000	.0000
	14	.0000	.0000	.0000	.0000	.0000	.0000	.0000	.0000	.0000
	15	.0000	.0000	.0000	.0000	.0000	.0000	.0000	.0000	.0000
	16	.0000	.0000	.0000	.0000	.0000	.0000	.0000	.0000	.0000
	17	.0000	.0000	.0000	.0000	.0000	.0000	.0000	.0000	.0000
	18	.0000	.0000	.0000	.0000	.0000	.0000	.0000	.0000	.0000
	19	.0000	.0000	.0000	.0000	.0000	.0000	.0000	.0000	.0000
	20	.0000	.0000	.0000	.0000	.0000	.0000	.0000	.0000	.0000

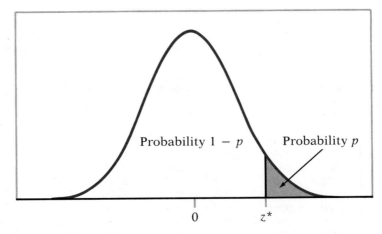

Figure for Table D
Table entry is the point z^* with given probability p lying above it.

n	k					p				
		.01	.02	.03	.04	.05	.06	.07	.08	.09
20	0	.1216	.0388	.0115	.0032	.0008	.0002	.0000	.0000	.0000
	1	.2702	.1368	.0576	.0211	.0068	.0020	.0005	.0001	.0000
	2	.2852	.2293	.1369	.0669	.0278	.0100	.0031	.0008	.0002
	3	.1901	.2428	.2054	.1339	.0716	.0323	.0123	.0040	.0011
	4	.0898	.1821	.2182	.1897	.1304	.0738	.0350	.0139	.0046
	5	.0319	.1028	.1746	.2023	.1789	.1272	.0746	.0365	.0148
	6	.0089	.0454	.1091	.1686	.1916	.1712	.1244	.0746	.0370
	7	.0020	.0160	.0545	.1124	.1643	.1844	.1659	.1221	.0739
	8	.0004	.0046	.0222	.0609	.1144	.1614	.1797	.1623	.1201
	9	.0001	.0011	.0074	.0271	.0654	.1158	.1597	.1771	.1602
	10	.0000	.0002	.0020	.0099	.0308	.0686	.1171	.1593	.1762
	11	.0000	.0000	.0005	.0030	.0120	.0336	.0710	.1185	.1602
	12	.0000	.0000	.0001	.0008	.0039	.0136	.0355	.0727	.1201
	13	.0000	.0000	.0000	.0002	.0010	.0045	.0146	.0366	.0739
	14	.0000	.0000	.0000	.0000	.0002	.0012	.0049	.0150	.0370
	15	.0000	.0000	.0000	.0000	.0000	.0003	.0013	.0049	.0148
	16	.0000	.0000	.0000	.0000	.0000	.0000	.0003	.0013	.0046
	17	.0000	.0000	.0000	.0000	.0000	.0000	.0000	.0002	.0011
	18	.0000	.0000	.0000	.0000	.0000	.0000	.0000	.0000	.0002
	19	.0000	.0000	.0000	.0000	.0000	.0000	.0000	.0000	.0000
	20	.0000	.0000	.0000	.0000	.0000	.0000	.0000	.0000	.0000

Table D Normal critical values

p	z*	p	z*
.25	.674	.02	2.054
.20	.841	.01	2.326
.15	1.036	.005	2.576
.10	1.282	.0025	2.807
.05	1.645	.001	3.091
.025	1.960	.0005	3.291

Table E *t* distribution critical values

df	.25	.20	.15	.10	.05	.025	.02	.01	.005	.0025	.001	.0005
							p					
1	1.000	1.376	1.963	3.078	6.314	12.71	15.89	31.82	63.66	127.3	318.3	636.6
2	.816	1.061	1.386	1.886	2.920	4.303	4.849	6.965	9.925	14.09	22.33	31.60
3	.765	.978	1.250	1.638	2.353	3.182	3.482	4.541	5.841	7.453	10.21	12.92
4	.741	.941	1.190	1.533	2.132	2.776	2.999	3.747	4.604	5.598	7.173	8.610
5	.727	.920	1.156	1.476	2.015	2.571	2.757	3.365	4.032	4.773	5.893	6.869
6	.718	.906	1.134	1.440	1.943	2.447	2.612	3.143	3.707	4.317	5.208	5.959
7	.711	.896	1.119	1.415	1.895	2.365	2.517	2.998	3.499	4.029	4.785	5.408
8	.706	.889	1.108	1.397	1.860	2.306	2.449	2.896	3.355	3.833	4.501	5.041
9	.703	.883	1.100	1.383	1.833	2.262	2.398	2.821	3.250	3.690	4.297	4.781
10	.700	.879	1.093	1.372	1.812	2.228	2.359	2.764	3.169	3.581	4.144	4.587
11	.697	.876	1.088	1.363	1.796	2.201	2.328	2.718	3.106	3.497	4.025	4.437
12	.695	.873	1.083	1.356	1.782	2.179	2.303	2.681	3.055	3.428	3.930	4.318
13	.694	.870	1.079	1.350	1.771	2.160	2.282	2.650	3.012	3.372	3.852	4.221
14	.692	.868	1.076	1.345	1.761	2.145	2.264	2.624	2.977	3.326	3.787	4.140
15	.691	.866	1.074	1.341	1.753	2.131	2.249	2.602	2.947	3.286	3.733	4.073
16	.690	.865	1.071	1.337	1.746	2.120	2.235	2.583	2.921	3.252	3.686	4.015
17	.689	.863	1.069	1.333	1.740	2.110	2.224	2.567	2.898	3.222	3.646	3.965
18	.688	.862	1.067	1.330	1.734	2.101	2.214	2.552	2.878	3.197	3.611	3.922
19	.688	.861	1.066	1.328	1.729	2.093	2.205	2.539	2.861	3.174	3.579	3.883
20	.687	.860	1.064	1.325	1.725	2.086	2.197	2.528	2.845	3.153	3.552	3.850
21	.686	.859	1.063	1.323	1.721	2.080	2.189	2.518	2.831	3.135	3.527	3.819
22	.686	.858	1.061	1.321	1.717	2.074	2.183	2.508	2.819	3.119	3.505	3.792
23	.685	.858	1.060	1.319	1.714	2.069	2.177	2.500	2.807	3.104	3.485	3.768
24	.685	.857	1.059	1.318	1.711	2.064	2.172	2.492	2.797	3.091	3.467	3.745
25	.684	.856	1.058	1.316	1.708	2.060	2.167	2.485	2.787	3.078	3.450	3.725
26	.684	.856	1.058	1.315	1.706	2.056	2.162	2.479	2.779	3.067	3.435	3.707
27	.684	.855	1.057	1.314	1.703	2.052	2.158	2.473	2.771	3.057	3.421	3.690
28	.683	.855	1.056	1.313	1.701	2.048	2.154	2.467	2.763	3.047	3.408	3.674
29	.683	.854	1.055	1.311	1.699	2.045	2.150	2.462	2.756	3.038	3.396	3.659
30	.683	.854	1.055	1.310	1.697	2.042	2.147	2.457	2.750	3.030	3.385	3.646
40	.681	.851	1.050	1.303	1.684	2.021	2.123	2.423	2.704	2.971	3.307	3.551
50	.679	.849	1.047	1.299	1.676	2.009	2.109	2.403	2.678	2.937	3.261	3.496
60	.679	.848	1.045	1.296	1.671	2.000	2.099	2.390	2.660	2.915	3.232	3.460
80	.678	.846	1.043	1.292	1.664	1.990	2.088	2.374	2.639	2.887	3.195	3.416
100	.677	.845	1.042	1.290	1.660	1.984	2.081	2.364	2.626	2.871	3.174	3.390
1000	.675	.842	1.037	1.282	1.646	1.962	2.056	2.330	2.581	2.813	3.098	3.300
∞	.674	.841	1.036	1.282	1.645	1.960	2.054	2.326	2.576	2.807	3.091	3.291

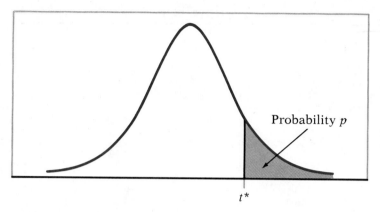

Figure for Table E
Table entry is the point *t** with given probability *p* lying above it.

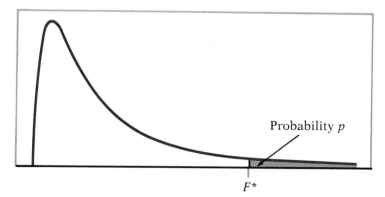

Figure for Table F
Table entry is the point *F** with given probability *p* lying above it.

Table F *F* Critical values

DFD	p	Degrees of freedom in the numerator								
		1	2	3	4	5	6	7	8	9
1	.100	39.86	49.50	53.59	55.83	57.24	58.20	58.91	59.44	59.86
	.050	161.45	199.50	215.71	224.58	230.16	233.99	236.77	238.88	240.54
	.025	647.79	799.50	864.16	899.58	921.85	937.11	948.22	956.66	963.28
	.010	4052.2	4999.5	5403.4	5624.6	5763.6	5859.0	5928.4	5981.1	6022.5
	.001	405284	500000	540379	562500	576405	585937	592873	598144	602284
2	.100	8.53	9.00	9.16	9.24	9.29	9.33	9.35	9.37	9.38
	.050	18.51	19.00	19.16	19.25	19.30	19.33	19.35	19.37	19.38
	.025	38.51	39.00	39.17	39.25	39.30	39.33	39.36	39.37	39.39
	.010	98.50	99.00	99.17	99.25	99.30	99.33	99.36	99.37	99.39
	.001	998.50	999.00	999.17	999.25	999.30	999.33	999.36	999.37	999.39
3	.100	5.54	5.46	5.39	5.34	5.31	5.28	5.27	5.25	5.24
	.050	10.13	9.55	9.28	9.12	9.01	8.94	8.89	8.85	8.81
	.025	17.44	16.04	15.44	15.10	14.88	14.73	14.62	14.54	14.47
	.010	34.12	30.82	29.46	28.71	28.24	27.91	27.67	27.49	27.35
	.001	167.03	148.50	141.11	137.10	134.58	132.85	131.58	130.62	129.86
4	.100	4.54	4.32	4.19	4.11	4.05	4.01	3.98	3.95	3.94
	.050	7.71	6.94	6.59	6.39	6.26	6.16	6.09	6.04	6.00
	.025	12.22	10.65	9.98	9.60	9.36	9.20	9.07	8.98	8.90
	.010	21.20	18.00	16.69	15.98	15.52	15.21	14.98	14.80	14.66
	.001	74.14	61.25	56.18	53.44	51.71	50.53	49.66	49.00	48.47
5	.100	4.06	3.78	3.62	3.52	3.45	3.40	3.37	3.34	3.32
	.050	6.61	5.79	5.41	5.19	5.05	4.95	4.88	4.82	4.77
	.025	10.01	8.43	7.76	7.39	7.15	6.98	6.85	6.76	6.68
	.010	16.26	13.27	12.06	11.39	10.97	10.67	10.46	10.29	10.16
	.001	47.18	37.12	33.20	31.09	29.75	28.83	28.16	27.65	27.24
6	.100	3.78	3.46	3.29	3.18	3.11	3.05	3.01	2.98	2.96
	.050	5.99	5.14	4.76	4.53	4.39	4.28	4.21	4.15	4.10
	.025	8.81	7.26	6.60	6.23	5.99	5.82	5.70	5.60	5.52
	.010	13.75	10.92	9.78	9.15	8.75	8.47	8.26	8.10	7.98
	.001	35.51	27.00	23.70	21.92	20.80	20.03	19.46	19.03	18.69
7	.100	3.59	3.26	3.07	2.96	2.88	2.83	2.78	2.75	2.72
	.050	5.59	4.74	4.35	4.12	3.97	3.87	3.79	3.73	3.68
	.025	8.07	6.54	5.89	5.52	5.29	5.12	4.99	4.90	4.82
	.010	12.25	9.55	8.45	7.85	7.46	7.19	6.99	6.84	6.72
	.001	29.25	21.69	18.77	17.20	16.21	15.52	15.02	14.63	14.33
8	.100	3.46	3.11	2.92	2.81	2.73	2.67	2.62	2.59	2.56
	.050	5.32	4.46	4.07	3.84	3.69	3.58	3.50	3.44	3.39
	.025	7.57	6.06	5.42	5.05	4.82	4.65	4.53	4.43	4.36
	.010	11.26	8.65	7.59	7.01	6.63	6.37	6.18	6.03	5.91
	.001	25.41	18.49	15.83	14.39	13.48	12.86	12.40	12.05	11.77
9	.100	3.36	3.01	2.81	2.69	2.61	2.55	2.51	2.47	2.44
	.050	5.12	4.26	3.86	3.63	3.48	3.37	3.29	3.23	3.18
	.025	7.21	5.71	5.08	4.72	4.48	4.32	4.20	4.10	4.03
	.010	10.56	8.02	6.99	6.42	6.06	5.80	5.61	5.47	5.35
	.001	22.86	16.39	13.90	12.56	11.71	11.13	10.70	10.37	10.11
10	.100	3.29	2.92	2.73	2.61	2.52	2.46	2.41	2.38	2.35
	.050	4.96	4.10	3.71	3.48	3.33	3.22	3.14	3.07	3.02
	.025	6.94	5.46	4.83	4.47	4.24	4.07	3.95	3.85	3.78
	.010	10.04	7.56	6.55	5.99	5.64	5.39	5.20	5.06	4.94
	.001	21.04	14.91	12.55	11.28	10.48	9.93	9.52	9.20	8.96
11	.100	3.23	2.86	2.66	2.54	2.45	2.39	2.34	2.30	2.27
	.050	4.84	3.98	3.59	3.36	3.20	3.09	3.01	2.95	2.90
	.025	6.72	5.26	4.63	4.28	4.04	3.88	3.76	3.66	3.59
	.010	9.65	7.21	6.22	5.67	5.32	5.07	4.89	4.74	4.63
	.001	19.69	13.81	11.56	10.35	9.58	9.05	8.66	8.35	8.12
12	.100	3.18	2.81	2.61	2.48	2.39	2.33	2.28	2.24	2.21
	.050	4.75	3.89	3.49	3.26	3.11	3.00	2.91	2.85	2.80
	.025	6.55	5.10	4.47	4.12	3.89	3.73	3.61	3.51	3.44
	.010	9.33	6.93	5.95	5.41	5.06	4.82	4.64	4.50	4.39
	.001	18.64	12.97	10.80	9.63	8.89	8.38	8.00	7.71	7.48

Table F (continued)

Degrees of freedom in the numerator

10	12	15	20	25	30	40	50	60	120	1000
60.19	60.71	61.22	61.74	62.05	62.26	62.53	62.69	62.79	63.06	63.30
241.88	243.91	245.95	248.01	249.26	250.10	251.14	251.77	252.20	253.25	254.19
968.63	976.71	984.87	993.10	998.08	1001.4	1005.6	1008.1	1009.8	1014.0	1017.7
6055.8	6106.3	6157.3	6208.7	6239.8	6260.6	6286.8	6302.5	6313.0	6339.4	6362.7
605621	610668	615764	620908	624017	626099	628712	630285	631337	633972	636301
9.39	9.41	9.42	9.44	9.45	9.46	9.47	9.47	9.47	9.48	9.49
19.40	19.41	19.43	19.45	19.46	19.46	19.47	19.48	19.48	19.49	19.49
39.40	39.41	39.43	39.45	39.46	39.46	39.47	39.48	39.48	39.49	39.50
99.40	99.42	99.43	99.45	99.46	99.47	99.47	99.48	99.48	99.49	99.50
999.40	999.42	999.43	999.45	999.46	999.47	999.47	999.48	999.48	999.49	999.50
5.23	5.22	5.20	5.18	5.17	5.17	5.16	5.15	5.15	5.14	5.13
8.79	8.74	8.70	8.66	8.63	8.62	8.59	8.58	8.57	8.55	8.53
14.42	14.34	14.25	14.17	14.12	14.08	14.04	14.01	13.99	13.95	13.91
27.23	27.05	26.87	26.69	26.58	26.50	26.41	26.35	26.32	26.22	26.14
129.25	128.32	127.37	126.42	125.84	125.45	124.96	124.66	124.47	123.97	123.53
3.92	3.90	3.87	3.84	3.83	3.82	3.80	3.80	3.79	3.78	3.76
5.96	5.91	5.86	5.80	5.77	5.75	5.72	5.70	5.69	5.66	5.63
8.84	8.75	8.66	8.56	8.50	8.46	8.41	8.38	8.36	8.31	8.26
14.55	14.37	14.20	14.02	13.91	13.84	13.75	13.69	13.65	13.56	13.47
48.05	47.41	46.76	46.10	45.70	45.43	45.09	44.88	44.75	44.40	44.09
3.30	3.27	3.24	3.21	3.19	3.17	3.16	3.15	3.14	3.12	3.11
4.74	4.68	4.62	4.56	4.52	4.50	4.46	4.44	4.43	4.40	4.37
6.62	6.52	6.43	6.33	6.27	6.23	6.18	6.14	6.12	6.07	6.02
10.05	9.89	9.72	9.55	9.45	9.38	9.29	9.24	9.20	9.11	9.03
26.92	26.42	25.91	25.39	25.08	24.87	24.60	24.44	24.33	24.06	23.82
2.94	2.90	2.87	2.84	2.81	2.80	2.78	2.77	2.76	2.74	2.72
4.06	4.00	3.94	3.87	3.83	3.81	3.77	3.75	3.74	3.70	3.67
5.46	5.37	5.27	5.17	5.11	5.07	5.01	4.98	4.96	4.90	4.86
7.87	7.72	7.56	7.40	7.30	7.23	7.14	7.09	7.06	6.97	6.89
18.41	17.99	17.56	17.12	16.85	16.67	16.44	16.31	16.21	15.98	15.77
2.70	2.67	2.63	2.59	2.57	2.56	2.54	2.52	2.51	2.49	2.47
3.64	3.57	3.51	3.44	3.40	3.38	3.34	3.32	3.30	3.27	3.23
4.76	4.67	4.57	4.47	4.40	4.36	4.31	4.28	4.25	4.20	4.15
6.62	6.47	6.31	6.16	6.06	5.99	5.91	5.86	5.82	5.74	5.66
14.08	13.71	13.32	12.93	12.69	12.53	12.33	12.20	12.12	11.91	11.72
2.54	2.50	2.46	2.42	2.40	2.38	2.36	2.35	2.34	2.32	2.30
3.35	3.28	3.22	3.15	3.11	3.08	3.04	3.02	3.01	2.97	2.93
4.30	4.20	4.10	4.00	3.94	3.89	3.84	3.81	3.78	3.73	3.68
5.81	5.67	5.52	5.36	5.26	5.20	5.12	5.07	5.03	4.95	4.87
11.54	11.19	10.84	10.48	10.26	10.11	9.92	9.80	9.73	9.53	9.36
2.42	2.38	2.34	2.30	2.27	2.25	2.23	2.22	2.21	2.18	2.16
3.14	3.07	3.01	2.94	2.89	2.86	2.83	2.80	2.79	2.75	2.71
3.96	3.87	3.77	3.67	3.60	3.56	3.51	3.47	3.45	3.39	3.34
5.26	5.11	4.96	4.81	4.71	4.65	4.57	4.52	4.48	4.40	4.32
9.89	9.57	9.24	8.90	8.69	8.55	8.37	8.26	8.19	8.00	7.84
2.32	2.28	2.24	2.20	2.17	2.16	2.13	2.12	2.11	2.08	2.06
2.98	2.91	2.85	2.77	2.73	2.70	2.66	2.64	2.62	2.58	2.54
3.72	3.62	3.52	3.42	3.35	3.31	3.26	3.22	3.20	3.14	3.09
4.85	4.71	4.56	4.41	4.31	4.25	4.17	4.12	4.08	4.00	3.92
8.75	8.45	8.13	7.80	7.60	7.47	7.30	7.19	7.12	6.94	6.78
2.25	2.21	2.17	2.12	2.10	2.08	2.05	2.04	2.03	2.00	1.98
2.85	2.79	2.72	2.65	2.60	2.57	2.53	2.51	2.49	2.45	2.41
3.53	3.43	3.33	3.23	3.16	3.12	3.06	3.03	3.00	2.94	2.89
4.54	4.40	4.25	4.10	4.01	3.94	3.86	3.81	3.78	3.69	3.61
7.92	7.63	7.32	7.01	6.81	6.68	6.52	6.42	6.35	6.18	6.02
2.19	2.15	2.10	2.06	2.03	2.01	1.99	1.97	1.96	1.93	1.91
2.75	2.69	2.62	2.54	2.50	2.47	2.43	2.40	2.38	2.34	2.30
3.37	3.28	3.18	3.07	3.01	2.96	2.91	2.87	2.85	2.79	2.73
4.30	4.16	4.01	3.86	3.76	3.70	3.62	3.57	3.54	3.45	3.37
7.29	7.00	6.71	6.40	6.22	6.09	5.93	5.83	5.76	5.59	5.44

DFD	p	\multicolumn{9}{c}{Degrees of Freedom in the Numerator}
		1	2	3	4	5	6	7	8	9
13	0.100	3.14	2.76	2.56	2.43	2.35	2.28	2.23	2.20	2.16
	0.050	4.67	3.81	3.41	3.18	3.03	2.92	2.83	2.77	2.71
	0.025	6.41	4.97	4.35	4.00	3.77	3.60	3.48	3.39	3.31
	0.010	9.07	6.70	5.74	5.21	4.86	4.62	4.44	4.30	4.19
	0.001	17.82	12.31	10.21	9.07	8.35	7.86	7.49	7.21	6.98
14	0.100	3.10	2.73	2.52	2.39	2.31	2.24	2.19	2.15	2.12
	0.050	4.60	3.74	3.34	3.11	2.96	2.85	2.76	2.70	2.65
	0.025	6.30	4.86	4.24	3.89	3.66	3.50	3.38	3.29	3.21
	0.010	8.86	6.51	5.56	5.04	4.69	4.46	4.28	4.14	4.03
	0.001	17.14	11.78	9.73	8.62	7.92	7.44	7.08	6.80	6.58
15	0.100	3.07	2.70	2.49	2.36	2.27	2.21	2.16	2.12	2.09
	0.050	4.54	3.68	3.29	3.06	2.90	2.79	2.71	2.64	2.59
	0.025	6.20	4.77	4.15	3.80	3.58	3.41	3.29	3.20	3.12
	0.010	8.68	6.36	5.42	4.89	4.56	4.32	4.14	4.00	3.89
	0.001	16.59	11.34	9.34	8.25	7.57	7.09	6.74	6.47	6.26
16	0.100	3.05	2.67	2.46	2.33	2.24	2.18	2.13	2.09	2.06
	0.050	4.49	3.63	3.24	3.01	2.85	2.74	2.66	2.59	2.54
	0.025	6.12	4.69	4.08	3.73	3.50	3.34	3.22	3.12	3.05
	0.010	8.53	6.23	5.29	4.77	4.44	4.20	4.03	3.89	3.78
	0.001	16.12	10.97	9.01	7.94	7.27	6.80	6.46	6.19	5.98
17	0.100	3.03	2.64	2.44	2.31	2.22	2.15	2.10	2.06	2.03
	0.050	4.45	3.59	3.20	2.96	2.81	2.70	2.61	2.55	2.49
	0.025	6.04	4.62	4.01	3.66	3.44	3.28	3.16	3.06	2.98
	0.010	8.40	6.11	5.19	4.67	4.34	4.10	3.93	3.79	3.68
	0.001	15.72	10.66	8.73	7.68	7.02	6.56	6.22	5.96	5.75
18	0.100	3.01	2.62	2.42	2.29	2.20	2.13	2.08	2.04	2.00
	0.050	4.41	3.55	3.16	2.93	2.77	2.66	2.58	2.51	2.46
	0.025	5.98	4.56	3.95	3.61	3.38	3.22	3.10	3.01	2.93
	0.010	8.29	6.01	5.09	4.58	4.25	4.01	3.84	3.71	3.60
	0.001	15.38	10.39	8.49	7.46	6.81	6.35	6.02	5.76	5.56
19	0.100	2.99	2.61	2.40	2.27	2.18	2.11	2.06	2.02	1.98
	0.050	4.38	3.52	3.13	2.90	2.74	2.63	2.54	2.48	2.42
	0.025	5.92	4.51	3.90	3.56	3.33	3.17	3.05	2.96	2.88
	0.010	8.18	5.93	5.01	4.50	4.17	3.94	3.77	3.63	3.52
	0.001	15.08	10.16	8.28	7.27	6.62	6.18	5.85	5.59	5.39
20	0.100	2.97	2.59	2.38	2.25	2.16	2.09	2.04	2.00	1.96
	0.050	4.35	3.49	3.10	2.87	2.71	2.60	2.51	2.45	2.39
	0.025	5.87	4.46	3.86	3.51	3.29	3.13	3.01	2.91	2.84
	0.010	8.10	5.85	4.94	4.43	4.10	3.87	3.70	3.56	3.46
	0.001	14.82	9.95	8.10	7.10	6.46	6.02	5.69	5.44	5.24
21	0.100	2.96	2.57	2.36	2.23	2.14	2.08	2.02	1.98	1.95
	0.050	4.32	3.47	3.07	2.84	2.68	2.57	2.49	2.42	2.37
	0.025	5.83	4.42	3.82	3.48	3.25	3.09	2.97	2.87	2.80
	0.010	8.02	5.78	4.87	4.37	4.04	3.81	3.64	3.51	3.40
	0.001	14.59	9.77	7.94	6.95	6.32	5.88	5.56	5.31	5.11
22	0.100	2.95	2.56	2.35	2.22	2.13	2.06	2.01	1.97	1.93
	0.050	4.30	3.44	3.05	2.82	2.66	2.55	2.46	2.40	2.34
	0.025	5.79	4.38	3.78	3.44	3.22	3.05	2.93	2.84	2.76
	0.010	7.95	5.72	4.82	4.31	3.99	3.76	3.59	3.45	3.35
	0.001	14.38	9.61	7.80	6.81	6.19	5.76	5.44	5.19	4.99
23	0.100	2.94	2.55	2.34	2.21	2.11	2.05	1.99	1.95	1.92
	0.050	4.28	3.42	3.03	2.80	2.64	2.53	2.44	2.37	2.32
	0.025	5.75	4.35	3.75	3.41	3.18	3.02	2.90	2.81	2.73
	0.010	7.88	5.66	4.76	4.26	3.94	3.71	3.54	3.41	3.30
	0.001	14.20	9.47	7.67	6.70	6.08	5.65	5.33	5.09	4.89
24	0.100	2.93	2.54	2.33	2.19	2.10	2.04	1.98	1.94	1.91
	0.050	4.26	3.40	3.01	2.78	2.62	2.51	2.42	2.36	2.30
	0.025	5.72	4.32	3.72	3.38	3.15	2.99	2.87	2.78	2.70
	0.010	7.82	5.61	4.72	4.22	3.90	3.67	3.50	3.36	3.26
	0.001	14.03	9.34	7.55	6.59	5.98	5.55	5.23	4.99	4.80

Table F (continued)

Degrees of Freedom in the Numerator

10	12	15	20	25	30	40	50	60	120	1000
2.14	2.10	2.05	2.01	1.98	1.96	1.93	1.92	1.90	1.88	1.85
2.67	2.60	2.53	2.46	2.41	2.38	2.34	2.31	2.30	2.25	2.21
3.25	3.15	3.05	2.95	2.88	2.84	2.78	2.74	2.72	2.66	2.60
4.10	3.96	3.82	3.66	3.57	3.51	3.43	3.38	3.34	3.25	3.18
6.80	6.52	6.23	5.93	5.75	5.63	5.47	5.37	5.30	5.14	4.99
2.10	2.05	2.01	1.96	1.93	1.91	1.89	1.87	1.86	1.83	1.80
2.60	2.53	2.46	2.39	2.34	2.31	2.27	2.24	2.22	2.18	2.14
3.15	3.05	2.95	2.84	2.78	2.73	2.67	2.64	2.61	2.55	2.50
3.94	3.80	3.66	3.51	3.41	3.35	3.27	3.22	3.18'	3.09	3.02
6.40	6.13	5.85	5.56	5.38	5.25	5.10	5.00	4.94	4.77	4.62
2.06	2.02	1.97	1.92	1.89	1.87	1.85	1.83	1.82	1.79	1.76
2.54	2.48	2.40	2.33	2.28	2.25	2.20	2.18	2.16	2.11	2.07
3.06	2.96	2.86	2.76	2.69	2.64	2.59	2.55	2.52	2.46	2.40
3.80	3.67	3.52	3.37	3.28	3.21	3.13	3.08	3.05	2.96	2.88
6.08	5.81	5.54	5.25	5.07	4.95	4.80	4.70	4.64	4.47	4.33
2.03	1.99	1.94	1.89	1.86	1.84	1.81	1.79	1.78	1.75	1.72
2.49	2.42	2.35	2.28	2.23	2.19	2.15	2.12	2.11	2.06	2.02
2.99	2.89	2.79	2.68	2.61	2.57	2.51	2.47	2.45	2.38	2.32
3.69	3.55	3.41	3.26	3.16	3.10	3.02	2.97	2.93	2.84	2.76
5.81	5.55	5.27	4.99	4.82	4.70	4.54	4.45	4.39	4.23	4.08
2.00	1.96	1.91	1.86	1.83	1.81	1.78	1.76	1.75	1.72	1.69
2.45	2.38	2.31	2.23	2.18	2.15	2.10	2.08	2.06	2.01	1.97
2.92	2.82	2.72	2.62	2.55	2.50	2.44	2.41	2.38	2.32	2.26
3.59	3.46	3.31	3.16	3.07	3.00	2.92	2.87	2.83	2.75	2.66
5.58	5.32	5.05	4.78	4.60	4.48	4.33	4.24	4.18	4.02	3.87
1.98	1.93	1.89	1.84	1.80	1.78	1.75	1.74	1.72	1.69	1.66
2.41	2.34	2.27	2.19	2.14	2.11	2.06	2.04	2.02	1.97	1.92
2.87	2.77	2.67	2.56	2.49	2.44	2.38	2.35	2.32	2.26	2.20
3.51	3.37	3.23	3.08	2.98	2.92	2.84	2.78	2.75	2.66	2.58
5.39	5.13	4.87	4.59	4.42	4.30	4.15	4.06	4.00	3.84	3.69
1.96	1.91	1.86	1.81	1.78	1.76	1.73	1.71	1.70	1.67	1.64
2.38	2.31	2.23	2.16	2.11	2.07	2.03	2.00	1.98	1.93	1.88
2.82	2.72	2.62	2.51	2.44	2.39	2.33	2.30	2.27	2.20	2.14
3.43	3.30	3.15	3.00	2.91	2.84	2.76	2.71	2.67	2.58	2.50
5.22	4.97	4.70	4.43	4.26	4.14	3.99	3.90	3.84	3.68	3.53
1.94	1.89	1.84	1.79	1.76	1.74	1.71	1.69	1.68	1.64	1.61
2.35	2.28	2.20	2.12	2.07	2.04	1.99	1.97	1.95	1.90	1.85
2.77	2.68	2.57	2.46	2.40	2.35	2.29	2.25	2.22	2.16	2.09
3.37	3.23	3.09	2.94	2.84	2.78	2.69	2.64	2.61	2.52	2.43
5.08	4.82	4.56	4.29	4.12	4.00	3.86	3.77	3.70	3.54	3.40
1.92	1.87	1.83	1.78	1.74	1.72	1.69	1.67	1.66	1.62	1.59
2.32	2.25	2.18	2.10	2.05	2.01	1.96	1.94	1.92	1.87	1.82
2.73	2.64	2.53	2.42	2.36	2.31	2.25	2.21	2.18	2.11	2.05
3.31	3.17	3.03	2.88	2.79	2.72	2.64	2.58	2.55	2.46	2.37
4.95	4.70	4.44	4.17	4.00	3.88	3.74	3.64	3.58	3.42	3.28
1.90	1.86	1.81	1.76	1.73	1.70	1.67	1.65	1.64	1.60	1.57
2.30	2.23	2.15	2.07	2.02	1.98	1.94	1.91	1.89	1.84	1.79
2.70	2.60	2.50	2.39	2.32	2.27	2.21	2.17	2.14	2.08	2.01
3.26	3.12	2.98	2.83	2.73	2.67	2.58	2.53	2.50	2.40	2.32
4.83	4.58	4.33	4.06	3.89	3.78	3.63	3.54	3.48	3.32	3.17
1.89	1.84	1.80	1.74	1.71	1.69	1.66	1.64	1.62	1.59	1.55
2.27	2.20	2.13	2.05	2.00	1.96	1.91	1.88	1.86	1.81	1.76
2.67	2.57	2.47	2.36	2.29	2.24	2.18	2.14	2.11	2.04	1.98
3.21	3.07	2.93	2.78	2.69	2.62	2.54	2.48	2.45	2.35	2.27
4.73	4.48	4.23	3.96	3.79	3.68	3.53	3.44	3.38	3.22	3.08
1.88	1.83	1.78	1.73	1.70	1.67	1.64	1.62	1.61	1.57	1.54
2.25	2.18	2.11	2.03	1.97	1.94	1.89	1.86	1.84	1.79	1.74
2.64	2.54	2.44	2.33	2.26	2.21	2.15	2.11	2.08	2.01	1.94
3.17	3.03	2.89	2.74	2.64	2.58	2.49	2.44	2.40	2.31	2.22
4.64	4.39	4.14	3.87	3.71	3.59	3.45	3.36	3.29	3.14	2.99

		Degrees of Freedom in the Numerator								
DFD	p	1	2	3	4	5	6	7	8	9
25	0.100	2.92	2.53	2.32	2.18	2.09	2.02	1.97	1.93	1.89
	0.050	4.24	3.39	2.99	2.76	2.60	2.49	2.40	2.34	2.28
	0.025	5.69	4.29	3.69	3.35	3.13	2.97	2.85	2.75	2.68
	0.010	7.77	5.57	4.68	4.18	3.85	3.63	3.46	3.32	3.22
	0.001	13.88	9.22	7.45	6.49	5.89	5.46	5.15	4.91	4.71
26	0.100	2.91	2.52	2.31	2.17	2.08	2.01	1.96	1.92	1.88
	0.050	4.23	3.37	2.98	2.74	2.59	2.47	2.39	2.32	2.27
	0.025	5.66	4.27	3.67	3.33	3.10	2.94	2.82	2.73	2.65
	0.010	7.72	5.53	4.64	4.14	3.82	3.59	3.42	3.29	3.18
	0.001	13.74	9.12	7.36	6.41	5.80	5.38	5.07	4.83	4.64
27	0.100	2.90	2.51	2.30	2.17	2.07	2.00	1.95	1.91	1.87
	0.050	4.21	3.35	2.96	2.73	2.57	2.46	2.37	2.31	2.25
	0.025	5.63	4.24	3.65	3.31	3.08	2.92	2.80	2.71	2.63
	0.010	7.68	5.49	4.60	4.11	3.78	3.56	3.39	3.26	3.15
	0.001	13.61	9.02	7.27	6.33	5.73	5.31	5.00	4.76	4.57
28	0.100	2.89	2.50	2.29	2.16	2.06	2.00	1.94	1.90	1.87
	0.050	4.20	3.34	2.95	2.71	2.56	2.45	2.36	2.29	2.24
	0.025	5.61	4.22	3.63	3.29	3.06	2.90	2.78	2.69	2.61
	0.010	7.64	5.45	4.57	4.07	3.75	3.53	3.36	3.23	3.12
	0.001	13.50	8.93	7.19	6.25	5.66	5.24	4.93	4.69	4.50
29	0.100	2.89	2.50	2.28	2.15	2.06	1.99	1.93	1.89	1.86
	0.050	4.18	3.33	2.93	2.70	2.55	2.43	2.35	2.28	2.22
	0.025	5.59	4.20	3.61	3.27	3.04	2.88	2.76	2.67	2.59
	0.010	7.60	5.42	4.54	4.04	3.73	3.50	3.33	3.20	3.09
	0.001	13.39	8.85	7.12	6.19	5.59	5.18	4.87	4.64	4.45
30	0.100	2.88	2.49	2.28	2.14	2.05	1.98	1.93	1.88	1.85
	0.050	4.17	3.32	2.92	2.69	2.53	2.42	2.33	2.27	2.21
	0.025	5.57	4.18	3.59	3.25	3.03	2.87	2.75	2.65	2.57
	0.010	7.56	5.39	4.51	4.02	3.70	3.47	3.30	3.17	3.07
	0.001	13.29	8.77	7.05	6.12	5.53	5.12	4.82	4.58	4.39
40	0.100	2.84	2.44	2.23	2.09	2.00	1.93	1.87	1.83	1.79
	0.050	4.08	3.23	2.84	2.61	2.45	2.34	2.25	2.18	2.12
	0.025	5.42	4.05	3.46	3.13	2.90	2.74	2.62	2.53	2.45
	0.010	7.31	5.18	4.31	3.83	3.51	3.29	3.12	2.99	2.89
	0.001	12.61	8.25	6.59	5.70	5.13	4.73	4.44	4.21	4.02
50	0.100	2.81	2.41	2.20	2.06	1.97	1.90	1.84	1.80	1.76
	0.050	4.03	3.18	2.79	2.56	2.40	2.29	2.20	2.13	2.07
	0.025	5.34	3.97	3.39	3.05	2.83	2.67	2.55	2.46	2.38
	0.010	7.17	5.06	4.20	3.72	3.41	3.19	3.02	2.89	2.78
	0.001	12.22	7.96	6.34	5.46	4.90	4.51	4.22	4.00	3.82
60	0.100	2.79	2.39	2.18	2.04	1.95	1.87	1.82	1.77	1.74
	0.050	4.00	3.15	2.76	2.53	2.37	2.25	2.17	2.10	2.04
	0.025	5.29	3.93	3.34	3.01	2.79	2.63	2.51	2.41	2.33
	0.010	7.08	4.98	4.13	3.65	3.34	3.12	2.95	2.82	2.72
	0.001	11.97	7.77	6.17	5.31	4.76	4.37	4.09	3.86	3.69
100	0.100	2.76	2.36	2.14	2.00	1.91	1.83	1.78	1.73	1.69
	0.050	3.94	3.09	2.70	2.46	2.31	2.19	2.10	2.03	1.97
	0.025	5.18	3.83	3.25	2.92	2.70	2.54	2.42	2.32	2.24
	0.010	6.90	4.82	3.98	3.51	3.21	2.99	2.82	2.69	2.59
	0.001	11.50	7.41	5.86	5.02	4.48	4.11	3.83	3.61	3.44
200	0.100	2.73	2.33	2.11	1.97	1.88	1.80	1.75	1.70	1.66
	0.050	3.89	3.04	2.65	2.42	2.26	2.14	2.06	1.98	1.93
	0.025	5.10	3.76	3.18	2.85	2.63	2.47	2.35	2.26	2.18
	0.010	6.76	4.71	3.88	3.41	3.11	2.89	2.73	2.60	2.50
	0.001	11.15	7.15	5.63	4.81	4.29	3.92	3.65	3.43	3.26
1000	0.100	2.71	2.31	2.09	1.95	1.85	1.78	1.72	1.68	1.64
	0.050	3.85	3.00	2.61	2.38	2.22	2.11	2.02	1.95	1.89
	0.025	5.04	3.70	3.13	2.80	2.58	2.42	2.30	2.20	2.13
	0.010	6.66	4.63	3.80	3.34	3.04	2.82	2.66	2.53	2.43
	0.001	10.89	6.96	5.46	4.65	4.14	3.78	3.51	3.30	3.13

Table F (continued)

10	12	15	20	25	30	40	50	60	120	1000
1.87	1.82	1.77	1.72	1.68	1.66	1.63	1.61	1.59	1.56	1.52
2.24	2.16	2.09	2.01	1.96	1.92	1.87	1.84	1.82	1.77	1.72
2.61	2.51	2.41	2.30	2.23	2.18	2.12	2.08	2.05	1.98	1.91
3.13	2.99	2.85	2.70	2.60	2.54	2.45	2.40	2.36	2.27	2.18
4.56	4.31	4.06	3.79	3.63	3.52	3.37	3.28	3.22	3.06	2.91
1.86	1.81	1.76	1.71	1.67	1.65	1.61	1.59	1.58	1.54	1.51
2.22	2.15	2.07	1.99	1.94	1.90	1.85	1.82	1.80	1.75	1.70
2.59	2.49	2.39	2.28	2.21	2.16	2.09	2.05	2.03	1.95	1.89
3.09	2.96	2.81	2.66	2.57	2.50	2.42	2.36	2.33	2.23	2.14
4.48	4.24	3.99	3.72	3.56	3.44	3.30	3.21	3.15	2.99	2.84
1.85	1.80	1.75	1.70	1.66	1.64	1.60	1.58	1.57	1.53	1.50
2.20	2.13	2.06	1.97	1.92	1.88	1.84	1.81	1.79	1.73	1.68
2.57	2.47	2.36	2.25	2.18	2.13	2.07	2.03	2.00	1.93	1.86
3.06	2.93	2.78	2.63	2.54	2.47	2.38	2.33	2.29	2.20	2.11
4.41	4.17	3.92	3.66	3.49	3.38	3.23	3.14	3.08	2.92	2.78
1.84	1.79	1.74	1.69	1.65	1.63	1.59	1.57	1.56	1.52	1.48
2.19	2.12	2.04	1.96	1.91	1.87	1.82	1.79	1.77	1.71	1.66
2.55	2.45	2.34	2.23	2.16	2.11	2.05	2.01	1.98	1.91	1.84
3.03	2.90	2.75	2.60	2.51	2.44	2.35	2.30	2.26	2.17	2.08
4.35	4.11	3.86	3.60	3.43	3.32	3.18	3.09	3.02	2.86	2.72
1.83	1.78	1.73	1.68	1.64	1.62	1.58	1.56	1.55	1.51	1.47
2.18	2.10	2.03	1.94	1.89	1.85	1.81	1.77	1.75	1.70	1.65
2.53	2.43	2.32	2.21	2.14	2.09	2.03	1.99	1.96	1.89	1.82
3.00	2.87	2.73	2.57	2.48	2.41	2.33	2.27	2.23	2.14	2.05
4.29	4.05	3.80	3.54	3.38	3.27	3.12	3.03	2.97	2.81	2.66
1.82	1.77	1.72	1.67	1.63	1.61	1.57	1.55	1.54	1.50	1.46
2.16	2.09	2.01	1.93	1.88	1.84	1.79	1.76	1.74	1.68	1.63
2.51	2.41	2.31	2.20	2.12	2.07	2.01	1.97	1.94	1.87	1.80
2.98	2.84	2.70	2.55	2.45	2.39	2.30	2.25	2.21	2.11	2.02
4.24	4.00	3.75	3.49	3,33	3.22	3.07	2.98	2.92	2.76	2.61
1.76	1.71	1.66	1.61	1.57	1.54	1.51	1.48	1.47	1.42	1.38
2.08	2.00	1.92	1.84	1.78	1.74	1.69	1.66	1.64	1.58	1.52
2.39	2.29	2.18	2.07	1.99	1.94	1.88	1.83	1.80	1.72	1.65
2.80	2.66	2.52	2.37	2.27	2.20	2.11	2.06	2.02	1.92	1.82
3.87	3.64	3.40	3.14	2.98	2.87	2.73	2.64	2.57	2.41	2.25
1.73	1.68	1.63	1.57	1.53	1.50	1.46	1.44	1.42	1.38	1.33
2.03	1.95	1.87	1.78	1.73	1.69	1.63	1.60	1.58	1.51	1.45
2.32	2.22	2.11	1.99	1.92	1.87	1.80	1.75	1.72	1.64	1.56
2.70	2.56	2.42	2.27	2.17	2.10	2.01	1.95	1.91	1.80	1.70
3.67	3.44	3.20	2.95	2.79	2.68	2.53	2.44	2.38	2.21	2.05
1.71	1.66	1.60	1.54	1.50	1.48	1.44	1.41	1.40	1.35	1.30
1.99	1.92	1.84	1.75	1.69	1.65	1.59	1.56	1.53	1.47	1.40
2.27	2.17	2.06	1.94	1.87	1.82	1.74	1.70	1.67	1.58	1.49
2.63	2.50	2.35	2.20	2.10	2.03	1.94	1.88	1.84	1.73	1.62
3.54	3.32	3.08	2.83	2.67	2.55	2.41	2.32	2.25	2.08	1.92
1.66	1.61	1.56	1.49	1.45	1.42	1.38	1.35	1.34	1.28	1.22
1.93	1.85	1.77	1.68	1.62	1.57	1.52	1.48	1.45	1.38	1.30
2.18	2.08	1.97	1.85	1.77	1.71	1.64	1.59	1.56	1.46	1.36
2.50	2.37	2.22	2.07	1.97	1.89	1.80	1.74	1.69	1.57	1.45
3.30	3.07	2.84	2.59	2.43	2.32	2.17	2.08	2.01	1.83	1.64
1.63	1.58	1.52	1.46	1.41	1.38	1.34	1.31	1.29	1.23	1.16
1.88	1.80	1.72	1.62	1.56	1.52	1.46	1.41	1.39	1.30	1.21
2.11	2.01	1.90	1.78	1.70	1.64	1.56	1.51	1.47	1.37	1.25
2.41	2.27	2.13	1.97	1.87	1.79	1.69	1.63	1.58	1.45	1.30
3.12	2.90	2.67	2.42	2.26	2.15	2.00	1.90	1.83	1.64	1.43
1.61	1.55	1.49	1.43	1.38	1.35	1.30	1.27	1.25	1.18	1.08
1.84	1.76	1.68	1.58	1.52	1.47	1.41	1.36	1.33	1.24	1.11
2.06	1.96	1.85	1.72	1.64	1.58	1.50	1.45	1.41	1.29	1.13
2.34	2.20	2.06	1.90	1.79	1.72	1.61	1.54	1.50	1.35	1.16
2.99	2.77	2.54	2.30	2.14	2.02	1.87	1.77	1.69	1.49	1.22

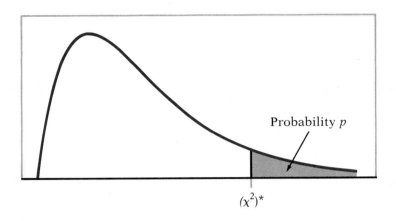

Figure for Table G
Table entry is the point $(x^2)^*$ with given probability p lying above it.

Probability p

$(\chi^2)^*$

Table G χ^2 Critical values

df	.25	.20	.15	.10	.05	.025	.02	.01	.005	.0025	.001	.0005
1	1.32	1.64	2.07	2.71	3.84	5.02	5.41	6.63	7.88	9.14	10.83	12.12
2	2.77	3.22	3.79	4.61	5.99	7.38	7.82	9.21	10.60	11.98	13.82	15.20
3	4.11	4.64	5.32	6.25	7.81	9.35	9.84	11.34	12.84	14.32	16.27	17.73
4	5.39	5.99	6.74	7.78	9.49	11.14	11.67	13.28	14.86	16.42	18.47	20.00
5	6.63	7.29	8.12	9.24	11.07	12.83	13.39	15.09	16.75	18.39	20.51	22.11
6	7.84	8.56	9.45	10.64	12.59	14.45	15.03	16.81	18.55	20.25	22.46	24.10
7	9.04	9.80	10.75	12.02	14.07	16.01	16.62	18.48	20.28	22.04	24.32	26.02
8	10.22	11.03	12.03	13.36	15.51	17.53	18.17	20.09	21.95	23.77	26.12	27.87
9	11.39	12.24	13.29	14.68	16.92	19.02	19.68	21.67	23.59	25.46	27.88	29.67
10	12.55	13.44	14.53	15.99	18.31	20.48	21.16	23.21	25.19	27.11	29.59	31.42
11	13.70	14.63	15.77	17.28	19.68	21.92	22.62	24.72	26.76	28.73	31.26	33.14
12	14.85	15.81	16.99	18.55	21.03	23.34	24.05	26.22	28.30	30.32	32.91	34.82
13	15.98	16.98	18.20	19.81	22.36	24.74	25.47	27.69	29.82	31.88	34.53	36.48
14	17.12	18.15	19.41	21.06	23.68	26.12	26.87	29.14	31.32	33.43	36.12	38.11
15	18.25	19.31	20.60	22.31	25.00	27.49	28.26	30.58	32.80	34.95	37.70	39.72
16	19.37	20.47	21.79	23.54	26.30	28.85	29.63	32.00	34.27	36.46	39.25	41.31
17	20.49	21.61	22.98	24.77	27.59	30.19	31.00	33.41	35.72	37.95	40.79	42.88
18	21.60	22.76	24.16	25.99	28.87	31.53	32.35	34.81	37.16	39.42	42.31	44.43
19	22.72	23.90	25.33	27.20	30.14	32.85	33.69	36.19	38.58	40.88	43.82	45.97
20	23.83	25.04	26.50	28.41	31.41	34.17	35.02	37.57	40.00	42.34	45.31	47.50
21	24.93	26.17	27.66	29.62	32.67	35.48	36.34	38.93	41.40	43.78	46.80	49.01
22	26.04	27.30	28.82	30.81	33.92	36.78	37.66	40.29	42.80	45.20	48.27	50.51
23	27.14	28.43	29.98	32.01	35.17	38.08	38.97	41.64	44.18	46.62	49.73	52.00
24	28.24	29.55	31.13	33.20	36.42	39.36	40.27	42.98	45.56	48.03	51.18	53.48
25	29.34	30.68	32.28	34.38	37.65	40.65	41.57	44.31	46.93	49.44	52.62	54.95
26	30.43	31.79	33.43	35.56	38.89	41.92	42.86	45.64	48.29	50.83	54.05	56.41
27	31.53	32.91	34.57	36.74	40.11	43.19	44.14	46.96	49.64	52.22	55.48	57.86
28	32.62	34.03	35.71	37.92	41.34	44.46	45.42	48.28	50.99	53.59	56.89	59.30
29	33.71	35.14	36.85	39.09	42.56	45.72	46.69	49.59	52.34	54.97	58.30	60.73
30	34.80	36.25	37.99	40.26	43.77	46.98	47.96	50.89	53.67	56.33	59.70	62.16
40	45.62	47.27	49.24	51.81	55.76	59.34	60.44	63.69	66.77	69.70	73.40	76.09
50	56.33	58.16	60.35	63.17	67.50	71.42	72.61	76.15	79.49	82.66	86.66	89.56
60	66.98	68.97	71.34	74.40	79.08	83.30	84.58	88.38	91.95	95.34	99.61	102.7
80	88.13	90.41	93.11	96.58	101.9	106.6	108.1	112.3	116.3	120.1	124.8	128.3
100	109.1	111.7	114.7	118.5	124.3	129.6	131.1	135.8	140.2	144.3	149.4	153.2

Solutions to Selected Exercises

Chapter 1

1.11 1120, 1001, 1017, 982, 989, 961, 960, 1089, 987, 976, 902, 980, 1098, 1057, 913, 999.

1.13 Figure 1.4(a) is strongly skewed to the right with a peak at 0; Figure 1.4(b) is somewhat symmetric with a central peak at 4. The peak is at the lowest value in Figure 1.4(a) and is at a central value in Figure 1.4(b).

1.15 There are two peaks. The ACT states are located in the upper portion of the distribution.

1.17 200 appears to be an outlier. The median is approximately 140.

1.19 The distribution is somewhat skewed to the right. There is one peak. The center is 523 eggs. 915 and 945 appear to be unusual values.

1.21 Yes. Slightly skewed to the right. One peak.

1.23 Stemplot and histogram give similar impression. Slightly skewed to the left.

1.25 There are two very low values. Between 5.4 and 5.5.

1.27 No. The centers are close together.

1.29 Both distributions are skewed to the right. Men tend to have higher salaries than women.

1.31 Shape is roughly symmetric. Center is around 68 inches.

1.33 Mean = 141. Median = 138.5. Distribution has a single outlier that causes the mean to be larger than the median.

1.35 Mean = 224.002. Median = 223.988. Distribution is slightly skewed to the right and the mean is a little larger than the median.

1.37 $x^* = -223{,}000 + 1000x$. Mean = 1002. Median = 988.

1.41 **(a)** 1.12, 1.88, 2.23, 2.86, 4.69. **(b)** IQR = .98. **(c)** Mean should be greater than the median since the distribution is skewed to the right.

1.43 **(a)** Skewed to the right **(b)** 89.67, 61.33, 7.83. **(c)** 88.5, 84.5, 93, 8.5. No outliers. **(d)** Quartiles, since distribution is skewed to the right.

1.45 60, 4.97.

1.49 61.8, .0618 mm. No or maybe.

1.51 Prefer median and quartiles. 435, 523, 590.

1.53 **(a)** 60. Yes.

1.55 **(a)** $x^* = 0 + (1/7)x$. **(b)** 29.09, 7.41, 27.43.

1.57 160, 181, 203, 248.

1.59 Calcium: 102, 107, 111.5, 123, 136. Placebo: 98, 109, 112, 119, 130. No.

1.61 **(a)** $x^* = 0 + (1/28.35)x$. **(b)** Control: 9.59, 11.89, 12.63, 14.13, 16.30. Experimental: 11.22, 13.53, 14.34, 15.11, 16.83.

1.63 10% trimmed mean = 2.32. 20% trimmed mean = 2.32. Median = 2.23. Mean = 2.37.

1.65 **(a)** 20%. **(b)** 40%. **(c)** 50%.

1.69 .65 to 2.25.

1.71 Eleanor: 1.8; Gerald: 1.5. Eleanor.

1.73 **(a)** .9978. **(b)** .0022. **(c)** .9515. **(d)** .9493.

1.75 **(a)** $-.68$. **(b)** .25

1.77 **(a)** .7257. **(b)** .4514.

1.79 .0668.

1.81 69.24.

1.83 .0301, .1995, .4295, .2815, .0594.

1.85 124.6, 134.9.

1.87 44%, 56%, 81%. Yes.

1.89 Approximately normal but the tails are less extreme than expected.

1.91 Approximately normal with one large outlier.

1.93 83%, 97%, 97%. Yes, there are outliers.

1.95 Distribution is not normal. Both tails have some extreme observations.

1.97 The 1940 distribution has a marked skew to the left while the 1980 distribution is less spread out and is roughly symmetric. For 1980 few states have low percentages, which reflects the increased voting by blacks.

1.99 **(a)** Distribution is skewed to the right. **(b)** 59.66, 101.29. **(c)** 5, 20, 30, 56.5, 765, 36.5. **(d)** Yes, because the distribution is skewed. **(e)** 95%. It does not apply. **(f)** No. Yes. 5489.

1.101 **(b)** It is not symmetric, not normal. There are three peaks and no outliers. **(c)** Five-number summary = 145, 183, 210, 235, 295. Peaks = approximately 185, 220, 260. **(d)** Five-number summary = 65.9, 83.2, 95.5, 106.8, 134.1. Peaks = 84.1, 100.0, 118.2. **(e)** Many numbers end in 0 or 5.

1.103 Skewed to the right. Five-number summary = 43, 82.5, 102.5, 151.5, 598.

1.105 $x^* = 68.6393 + .8937x$.

1.107 **(a)** .1587. **(b)** .0013. **(c)** 28.38.

Chapter 2

2.1   Six points outside limits. Run of nine points above center line. Decreasing.

2.3   No values outside control limits. Run of ten points below center line. Nancy's blood pressure is decreasing.

2.5   (a) Scores are decreasing from 1976 to 1980 and are increasing from 1981 to 1987.

2.7   There are seven points above the upper control limit. There is a run of nine points above the mean. The process appears to be shifting to higher values.

2.9 $\bar{x} = 7.75$, $s = 1.88$. Crime was increasing in the 1960s and steady in the 1970s and early 1980s.

2.13   (b) $2500.00 (c) $y = 500 + 200t$.

2.15   (a) $w = 100 + 40t$, 40. (c) No, predicted weight would be 9.4 pounds.

2.17   (b) 5.43, 4.64. (c) $-.0053$, pH is decreasing.

2.19   (b) 10. (c) 20. Extrapolation could be done with caution.

2.21   (c) 163.6, yes. (d) 199.7, rate of growth from 14 to 20 is less.

2.23   (b) Yes, month 21 looks low. (d) 10.5. (e) $-.2$.

2.25   Residuals are increasing from 1 to 5 months and decreasing from 6 to 10 months.

2.27   (a) 1, 2, 4, 8, 16, 32, 64, 128, 256, 512.
(c) Approximately 9,000,000,000,000,000,000.
(d) .00, .30, .60, .90, 1.20, 1.51, 1.81, 2.11, 2.41, 2.71.
(e) .3, $-.3$, 18.9.

2.29   Alice has $3049.17; Fred has $3000.00.

2.31   (a) Exponential. (b) Yes. (d) Predicted log for 1985 is 5.57; predicted amount spent in 1985 is $371,535. (e) Log is 5.50, yes, no.

2.33   (b) The log plot looks linear from 1790 to 1880 and also from 1880 to 1980. The slope is less in the second period. (c) 1945. This was the end of World War II. (d) Log is 2.4; predicted population is 251.2 million.

2.35   (a) Predicted logs are 1.526, 1.609, 2.271. Predicted motor vehicles are 33.6, 40.6, 186.6. (b) No.

2.37   (b) The trend is decreasing. The highs appear from Apr. to June and the lows appear from Oct. to Dec. The warmer weather brings more people outside.

2.39   (b) There was a steady increase from 1963 to 1973, a decrease from 1973 to 1979, and then an increase from 1979 to 1983.

2.41   (b) Although Jan. and Dec. are at opposite ends of the plot, they are close in time. (c) 9:13, difference is 7 minutes. (d) 13:20, difference is 2 minutes. (e) 7 minutes.

2.43   The smoothed plot gives the same picture as the original plot.

2.45   (c) Times are decreasing.

2.47   The highest weights appear in the winter months, and the lowest weights are in the summer months.

2.49 **(a)** Apr., May, June, July are high; Sept. and Dec. are low. **(b)** Yes, the secondary peak is between Oct. and Nov. **(c)** Months with high incidence of diarrhea are followed by months with low weight. Height is not as sensitive to the short-term effects of illness.

2.51 **(a)** Phase One (0 to 6 hours): no increase in growth. Phase Two (9 to 24 hours): exponential growth. Phase Three (36 hours): growth is slower than exponential. **(b)** Predicted log is .4230. Predicted colony size is 2.65. **(c)** 1.16.

2.53 **(b)** This control chart is very similar to the control chart found in Exercise 2.4. **(c)** They are not sensitive to outliers.

2.55 **(b)** Exponential, exponential, linear.

2.57 **(c)** Day 5 is above; day 24 is below. **(d)** No increase or decrease is evident. **(e)** Thomas' flea eggs show periods of increase and decrease.

2.59 Overall the pattern is neither linear nor exponential. Over short periods the pattern is approximately linear.

Chapter 3

3.1 **(a)** Categorical. **(b)** Quantitative. **(c)** Quantitative. **(d)** Categorical. **(e)** Quantitative. **(f)** Quantitative.

3.3 **(a)** Negative. Clearly curved. One observation is high on nitrogen oxides. **(b)** No. Low nitrogen oxide is associated with high carbon monoxide.

3.5 **(b)** As the flow rate increases, the amount of eroded soil increases. Yes. Positive.

3.7 **(a)** Heavier cars cost more. The association is weak and positive. **(b)** The heaviest cars, greater than 3500 pounds, are all domestic cars. In the range of 2500 to 3500 pounds, the foreign cars generally cost more than domestic cars of similar weight.

3.9 **(b)** No clear relationship is evident. **(c)** Those who survive appear to be older and have longer incubation periods. **(d)** The two survivors with incubation periods greater than 70.

3.11 **(a)** The number of fleas increases and then decreases. **(b)** The same pattern is evident. The 3-day median trace gives a slightly better picture of the pattern.

3.13 **(a)** Means = 10.65, 10.43, 5.60, 5.45. **(b)** The introduction of 1000 nematodes per pot has no effect on seedling growth. With 5000 nematodes there is a substantial reduction in seedling growth. Introduction of 10,000 nematodes causes essentially the same growth reduction as 5000.

3.15 **(a)** Means = 1520, 1707, 1540, 1816. **(b)** Against. Pecking order 1 has the lowest mean weight, and pecking order 4 has the highest mean weight.

3.17 **(a)** .38, 71.95, .38 cm. **(c)** 87.91 cm.

3.19 Predicted heights = 85.75, 90.35, 91.50, 92.65, 93.80, 94.95. Residuals = .25, −.35, −.50, .35, .20, .05. There is no clear pattern in the residuals.

3.21 **(a)** Yes. **(b)** b = .080, a = 1.766. **(c)** 3.039, 3.121, 3.171, 3.261, 3.369, 3.457, 3.541. **(d)** .011, −.001, −.001, −.011, −.009, .003, .009. **(e)** The residuals are all very small, indicating that the line fits the data well. Positive residuals are associated with high and low speeds, and negative residuals are associated with intermediate speeds.

3.23 **(b)** $\hat{y} = 1.03 + .90x$. **(c)** Predicted = 5.52, 10.03, 14.53, 19.03. Yes.

3.25 **(a)** Rural = 108, city = 123. No. **(b)** 7.4820. **(c)** Yes. **(d)** 93.65.

3.27 **(a)** The relationship is linear with two clusters and one observation that is low in calories and sodium. **(b)** Black. It is closer to the influential observation. **(c)** 407.05.

3.29 **(a)** Yes. The round one scores of 102 and 105 are influential, and the round one score of 105 appears to be an outlier. **(b)** $\hat{y} = 26.332 + .688x$. **(c)** Predicted = 87.54, 88.23, 86.17, 91.67, 85.48, 82.04, 96.48, 98.55, 83.42, 86.85, 88.92, 80.66. **(d)** Residuals = 6.46, −3.23, 2.83, −2.67, −4.48, −6.04, 10.52, −9.55, 3.58, 4.15, −.92, −.66. **(e)** There is a random pattern with large residuals for the high round one scores.

3.31 **(a)** Yes. **(b)** $\hat{y} = -.033 + 1.018x$. **(d)** No clear pattern. The linear fit is adequate. **(e)** The last 15 residuals are all positive. **(f)** The distribution is approximately symmetric, approximately normal.

3.33 **(b)** There is one large outlier. **(c)** No. No.

3.35 **(a)** Positive. Not. **(b)** $r = .56533$. **(c)** It does not change. No. **(d)** $r = 1$.

3.37 Means = 69, 66, Standard deviations = 2.52982, 2.09762. Slopes = .6818, .4687. At the means.

3.39 **(a)** $r = .77440$. **(b)** $r = .99824$. The additional point is very extreme in both x and y.

3.41 The paper suggests a negative relationship. The psychologist is saying there is no linear relationship.

3.43 $r = .4$

3.45 **(b)** $r = .25$. **(c)** No. It will be 100 times as large.

3.47 It is the percentage of variation in the number of students enrolled in 100 level mathematics courses explained by the linear relationship with the number of students in the freshman class.

3.49 93.5%.

3.51 **(a)** .16. **(b)** 78.2.

3.53 $r = .9519$.

3.55 $r = -.52$, $r^2 = .27$, $\hat{y} = 107.585 - 1.050x$. There is a substantial decrease in r^2.

3.57 **(a)** .99899. **(b)** .9980. **(c)** $r = .99967$, $r^2 = .9993$. No appreciable change in the correlation or the line.

3.59 **(b)** There is a positive linear relationship. No outliers. Sea scallops is an influential observation. **(c)** .96704. **(d)** .9352. **(e)** .93996. **(f)** The correlation indicates that the linear relationship is strong.

3.61 **(a)** 45.90%, 45.08%, 9.02%. **(b)** 60.66%, 39.34%. **(d)** Yes. The percentages for mild, moderate, and severe are similar for each type of operation.

3.63 **(a)** 23,403. Round-off error. **(b)** 56.19%, 32.32%, 7.75%, 3.73%. **(c)** 4 years of high school: 30.27%, 43.05%, 16.68%, 10.00%. 1 to 3 years of college: 22.38%, 42.36%, 19.48%, 15.78%. 4 or more years of college: 10.22%,

32.47%, 21.34%, 35.97%. **(e)** Higher income is associated with more education.

3.65 Education: 26.46%, 35.36%, 17.31%, 20.86%. Income: 31.58%, 37.88%, 15.78%, 14.76%. $15,000. to $34,000.

3.67 38% of the women and 14% of the men use poison. 41% of the women and 64% of the men use firearms.

3.69 **(a)** .263. **(b)** .262, .357. Yes.

3.71 **(a)** Joe: 120, 380. Moe: 130, 370. **(b)** .240, .260, Moe. **(d)** Right: .400, .300, Joe. Left: .200, .100, Joe. **(e)** Joe hits better against both right- and left-handed pitchers. Moe's batting average is higher because 80% of his at bats were against right-handed pitchers, and right-handed pitchers gave up more hits. Only 20% of Joe's at bats were against these pitchers.

3.73 Older children have bigger shoes and better reading comprehension.

3.75 Yes. Anesthetic C may be used for patients who have a higher risk of death.

3.77 Not necessarily. The larger hospitals might admit the more seriously ill patients, who require longer hospital stays.

3.79 Explanatory variable is foreign language study. Response variable is English achievement score. Students who are more skilled in languages may study foreign languages and score higher on English achievement tests.

3.81 Type of operation and patient characteristics that may be related to risk of death.

3.83 **(a)** Yes. **(b)** Meat and beef: $\hat{y} = -162.16 + 3.62x$. Meat and beef without veal outlier: $\hat{y} = -120.65 + 3.34x$. Yes. **(c)** Slope for poultry is less than slope for meat and beef. Intercept for poultry is greater than the intercept for meat and beef. **(d)** $\hat{y} = 169.23 + 2.37x$. Meat and beef: $r^2 = .76$. Poultry: $r^2 = .51$. No.

3.85 **(a)** SAT-V increases with family income. The pattern is curved. **(b)** Both ethnic group and SAT scores are associated with income.

3.87 The plot shows no clear association. The correlation is -.08. Three teams have high batting averages and low home runs. Without these three there appears to be a positive relationship.

3.89 **(b)** HC versus CO: positive linear. HC versus NOX: negative curved. NOX versus CO: negative curved. **(c)** Engines 22 and 24 are unusual in all three plots. Engine 32 has a very high value of NOX. **(d)** All data—CO, HC: $r = .90$; CO, NOX: $r = -.69$; HC, NOX: $r = -.56$. Without 22 and 24—CO, HC: $r = .84$; CO, NOX: $r = -.69$; HC, NOX: $r = -.52$. Without 22, 24, and 32—CO, HC: $r = .84$; CO, NOX: $r = -.74$; HC, NOX: $r = -.60$. **(e)** All data—$\widehat{HC} = .32 + .029CO$; $\widehat{NOX} = 1.83 - .06CO$; $\widehat{HC} = .81 - .19NOX$. **(f)** Without 22 and 24—$\widehat{HC} = .34 + .026CO$; $\widehat{NOX} = 1.92 - .08CO$; $\widehat{HC} = .72 - .14NOX$. Without 22, 24, and 32—$\widehat{HC} = .34 + .026CO$; $\widehat{NOX} = 1.85 - .07CO$; $\widehat{HC} = .78 - .19NOX$. **(g)** The unusual points do not change the regression lines very much. The strongest relationship is between CO and HC.

3.91 **(a)** If there are not enough states in a region, plot the individual points. **(c)** There appear to be differences by region in both spending and pay. The differences are somewhat consistent for the two variables. **(e)** $\widehat{PAY} = 16.37 + 2.05 SPENDING$. **(f)** The slope is lower and the intercept is higher. The original line is much more influenced by Alaska. $r^2 = .55$ and is lower.

3.93 **(b)** As length increases, strength decreases. From 9 to 14 inches the change in strength is fairly small. No. **(c)** One line would fit the data from 5 to 9 inches and another from 9 to 14 inches. **(d)** $\hat{y} = 488.38 - 20.75x$. **(e)** The residuals show a V-shaped pattern. The line is not an adequate summary of these data. **(f)** 5 to 9 inches: $\hat{y} = 667.50 - 6.90x$. 9 to 14 inches: $\hat{y} = 283.10 - 3.37x$. **(g)** The two lines fit the data much better than one.

3.95 **(a)** The percentage of people voting decreases over time. **(b)** The proportion of younger people eligible to vote has increased from 1960 to 1984, and younger people are less likely to vote than older people.

Chapter 4

4.1 This was not an experiment. The okra may have been planted in an area of the garden that was not attractive to stinkbugs.

4.3 **(a)** Treatments were not actively imposed. **(b)** This was a survey and no treatment was imposed.

4.5 No. No treatment was imposed. The explanatory variable was whether or not they live in public housing. The response variable was family stability and other variables.

4.7 Yes. The treatment was walking briskly on the treadmill. The conclusions that can be drawn are limited because eating was not recorded. The explanatory variable was time after exercise, and the response variable was the metabolic rate.

4.9 Experimental units: pairs of pieces of package liner. Explanatory variable: temperature of jaws. Response variable: peel strength of the seal.

4.11 Experimental units: 1-day-old male chicks. Explanatory variables: corn variety and protein level. Response variable: weight gain.

4.13 250°, 275°, 300°, 325°.

4.15 **(a)** Patients could be randomized to the two treatments. **(b)** No, there is no treatment that can be applied.

4.17 Assume the field has four rows and five plots per row. First 10 numbers between 1 and 20 correspond to plots receiving treatment A. Row 1: AABBB. Row 2: AAAAA. Row 3: BBBBB. Row 4: ABBAA.

4.19 250°: units 4, 7, 10, 16, 19. 275°: units 5, 8, 9, 13, 15. 300°: units 1, 3, 6, 11, 18. 325°: units 2, 12, 14, 17, 20.

4.21 First two numbers between 1 and 20 correspond to plots receiving treatment A and genetic line one; next two numbers correspond to plots receiving treatment A and genetic line two; etc. Row 1: A3, A5, B2, B4, B3. Row 2: A1, A5, A4, A2, A2. Row 3: B3, B2, B4, B5, B1. Row 4: A3, B5, B1, A1, A4.

4.23 The experimenter knew which subjects were assigned to the meditation group and might rate these subjects as having lower anxiety.

4.25 For each person, flip the coin. If heads, measure the right hand and then the left hand. If tails, measure in reverse order.

4.27 First, assign the numbers 1 to 6 to the schools in district one. From line 125 of Table B we use the following digits for the randomization: 6, 4, 1, 2,

3, 5. Assign school number 6 to Fact-Before, 4 to Compute-Before, 1 to Word-Before, 2 to Fact-After, 3 to Compute-After, and 5 to Word-After. For district two the randomization is 1, 6, 2, 3, 4, 5.

4.29 **(a)** The basic unit is a person. The population is all adult U.S. residents. **(b)** The basic unit is a household. The population is all U.S. households. **(c)** The basic unit is a voltage regulator. The population is all voltage regulators from the last shipment.

4.31 The sample was not representative of women in general because the questionnaires were distributed through women's groups and the response was voluntary. It is known that voluntary respondents are more likely to have stronger feelings about their opinions than nonrespondents. It is likely that these facts would cause the percents to be higher.

4.33 Beginning with A1096 and going across rows, label the control numbers with the numbers from 1 to 25. From line 111 of Table B we select the following: 12-B0986, 04-A1101, 11-A2220.

4.35 Beginning with Agarwal and going down the columns, label the people with the numbers 1 to 28. From line 139 of Table B we select the following: 04-Bowman, 10-Frank, 17-Mihalko, 19-Naber, 12-Goel, 13-Gupta.

4.37 We will choose one of the first 40 at random and then the addresses 40, 80, 120, and 160 places down the list from it. The addresses selected are 35, 75, 115, 155, 195.

4.39 Give each name on the alphabetized lists a number; 001 to 500 for females and 0001 to 2000 for males. From line 122 of Table B, the first five females selected are 138, 159, 052, 087, and 359. If we start at the beginning of line 122 of Table B, the first five males selected are 1387, 0529, 0908, 1369, 0815.

4.41 **(a)** Households without telephones and those with unlisted numbers are omitted. These households would include people who cannot afford a telephone, people who choose not to have a telephone, and people who prefer not to list their telephone number. **(b)** The random digit method includes unlisted numbers in the sampling frame.

4.43 **(a)** Many people think that "food stamps" refer to the prize stamps that some grocery stores give away. So the question is not clear. **(b)** The question is clear but it is slanted because it asks to choose between an extreme position (confiscate) and a general statement taken from the Constitution. **(c)** The question is clear but it is slanted toward agreement because it gives reasons to support a freeze. **(d)** The question is unclear. It uses fancy words and technical phrases that many people will not understand. It is slanted toward a positive response because it gives reasons to favor recycling.

4.45 Parameter, statistic.

4.47 Statistic, statistic.

4.49 **(a)** Use digits 0, 1, 2, 3 for "Yes" and 4 to 9 for "No." Starting at the beginning of line 110, we obtain $\hat{p} = .25$. **(b)** For lines 111 to 119, the values of $\hat{p}$ are .30, .30, .45, .40, .40, .45, .50, .50, .25. The median is .40. Yes.

4.51 **(a)** Large bias, large variability. **(b)** Small bias, small variability. **(c)** Small bias, large variability. **(d)** Large bias, small variability.

4.53 For this exercise we assume the population proportions in all states are about the same. The effect of the population proportion on the variability will be studied further in Chapter 6. **(a)** No. The variability is controlled by the size of the sample. **(b)** Yes. The sample sizes will vary from 2100 in Wyoming to 120,000 in California, and larger samples are less variable.

4.55 Both the center and the spread change with p. The variability increases as p goes from .1 to .3 to .5. The distributions are approximately normal.

4.57 For each taster, flip a coin. If heads, taste Pepsi first and then Coke. If tails, taste Coke first and then Pepsi.

4.61 (a) Fifteen patients are randomly assigned to receive β-blockers. The other 15 patients will receive a placebo. **(b)** The patients are numbered from 01 to 30. Those receiving the β-blockers are 21, 18, 23, 19, 10, 08, 03, 25, 06, 11, 15, 13, 24, 09, and 28.

4.63 (a) Again the numbers 0001 to 3478 to the alphabetized list of students. **(b)** 2940, 0769, 1481, 2975, 1315.

4.65 (a) The population is all students classified as full-time undergraduates at the beginning of the fall semester on a list provided by the registrar. **(b)** Stratify on class and randomly select 125 students in each class. **(c)** If questionnaires are mailed, some students will not respond. If telephones are used, the students without a telephone will not be contacted.

4.67 (a) There are two factors and six treatments. The treatments are 50°-60 rpm, 50°-90 rpm, 50°-120 rpm, 60°-60 rpm, 60°-90 rpm, and 60°-120 rpm. Twelve experimental units. **(c)** Label the batches with the numbers 01 to 12. The first two numbers that appear in the table will be given treatment 50°-60 rpm; the next two will be given treatment 50°-90 rpm; etc. The assignment is 06 and 09 to 50°-60 rpm, 03 and 05 to 50°-90 rpm, 04 and 07 to 50°-120 rpm, 02 and 08 to 60°-60 rpm, 10 and 11 to 60°-90 rpm, 12 and 01 to 60°-120 rpm.

4.69 The results have no statistical validity because the voluntary letter writers are not necessarily representative of voters in general.

4.71 In Example 4.8, 4 rats with genetic defects are in the experimental group and 6 are in the control group.

Chapter 5

5.3 (a) $S = \{$germinates, does not germinate$\}$. **(b)** If measured in weeks, for example, $S = \{0, 1, 2, \ldots\}$. **(c)** $S = \{A, B, C, D, F\}$. **(d)** $S = \{$makes the shot, misses the shot$\}$. **(e)** $S = \{1, 2, 3, 4, 5, 6, 7\}$.

5.5 $S = \{$all numbers between 0.00 and 5.00$\}$.

5.7 .04.

5.9 Model 1: Legitimate. Model 2: Legitimate. Model 3: Sum is less than 1. Model 4: Probabilities cannot be negative.

5.11 No. The sum of the probabilities is greater than 1.

5.13 .54.

5.15 .70, .30.

5.17 (a) .64, .07. **(b)** The event that the student selected ranked in the lower 60% in high school. 36. **(c)** .71.

5.19 (b) .435. **(c)** .97. **(d)** .29. **(e)** .71.

5.21 .886.

5.23 .668.

5.25 .25, .25, .50.

5.27 Democrat: .060, .324, .139, .043, .016. Republican: .043, .232, .100, .031, .012.

5.29 (b) .111. (c) .041. (d) .335. (e) .763. (f) $P(X > 2) = .446$.

5.31 (a) $X = 0$: BBBB. $X = 1$: GBBB, BGBB, BBGB, BBBG. $X = 2$: GGBB, GBGB, GBBG, BGGB, BGBG, BBGG, $X = 3$: GGGB, GGBG, GBGG, BGGG. $X = 4$: GGGG. (b) 5/16 = .9375. (c) 6/16 = .3750. (d) 8/16 = .50.

5.33 (a) (1, 1), (1, 2), (1, 3), (1, 4), (1, 5), (1, 6), (2, 1), (2, 2), (2, 3), (2 ,4), (2, 5), (2, 6), (3, 1), (3, 2), (3, 3), (3, 4), (3, 5), (3, 6), (4, 1), (4, 2), (4, 3), (4, 4), (4, 5),(4, 6), (5, 1), (5, 2), (5, 3), (5, 4), (5, 5), (5, 6), (6, 1), (6, 2), (6, 3), (6, 4), (6, 5), (6, 6). (b) 1/36. (c) (x, prob.): (2, 1/36), (3, 2/36), (4, 3/36),(5, 4/36), (6, 5/36), (7, 6/36), (8, 5/36), (9, 4/36), (10, 3/36), (11, 2/36), (12, 1/36). (d) 2/9. (e) 5/6.

5.35 (a) .4. (b) .6. (c) .2. (d) .2. (e) .487.

5.37 (a) .5. (b) .5. (c) .4 (d) .6. (e) .3. (f) 0.

5.39 (a) 0.00. (b) .015. (c) .970. (d) 0.

5.41 2.644.

5.43 2.25.

5.45 The company will earn $309.74 per person in the long run because of the law of large numbers.

5.47 (a) Independent. (b) Not independent. (c) Not independent.

5.49 (a) The spins are independent. (b) Wrong. The probabilities are not equally likely, because the deck now contains more black cards than red cards.

5.51 70.

5.53 2.68, 1.3106.

5.55 2.2361.

5.57 (a) 7,800,000. 2793. (b) 2796. (c) 5,606,738.

5.59 1951.

5.61 .282.

5.63 (a) .075, the household is prosperous and educated. (b) .073, the household is prosperous and not educated. (c) .134, the household is not prosperous and is educated. (d) .718, the household is not prosperous and not educated.

5.65 (a) .53. (b) .6923. (c) .57.

5.67 (a) .1476. (b) .3597. (c) .5084. (d) Not independent. (e) .4317.

5.69 .1032.

5.71 (a) $P(A) = .865$, $P(B|A) = .938$, $P(B|A^c) = .861$. (c) .8114, .1162, .9276.

5.73 .8747.

5.75 .58.

5.77 (a) A, A, A, B. (b) .1024. (c) $P(X = x) = (.8)^{x-1}.2$.

5.79 Contracting the production to Hong Kong is more profitable.

5.81 (a) 3.00, 2.52. (b) 2.50, 2.27.

5.83 (a) .1587. (b) 36.63 seconds.

5.85 (a) $S = 3, 4, 5, 6, \ldots, 18$. (b) $6/216 = .0278$. (c) 10.5, 2.96.

5.87 .0384.

5.89 (a) .067. (b) .1493.

5.91 (a) .43. (b) .0481. (c) .0962 (d) .3168. (e) .3510.

5.93 .35, .40, .22.

Chapter 6

6.1 (a) Yes. (b) No. n is random. (c) No. Trials are not independent.

6.3 (a) Yes. (b) No. Trials are not independent. (c) No. Each trial is performed under different conditions.

6.5 (a) .0020. (b) .0026.

6.7 $P(X = 0) = .0018$, $P(X = 1) = .0205$, $P(X = 2) = .0951$, $P(X = 3) = .2355$, $P(X = 4) = .3280$, $P(X = 5) = .2437$, $P(X = 6) = .0754$.

6.9 (a) There is only one way in which n successes can be chosen from among n trials. (b) There are n ways in which $n - 1$ successes can be chosen from among n trials. (c) The number of ways in which k successes can be chosen from among n trials is the same as the number of ways in which $n - k$ successes can be chosen from among n trials.

6.11 3.9, 1.17, .8072.

6.13 (a) .3174. (b) .2131.

6.15 (a) 66.00, 7.17. (b) .0256 (with continuity correction, .0301).

6.17 (a) 1050, 17.75. (b) .9976 (with continuity correction, .9978). (c) .0000. (d) .2981 (with continuity correction, .2877).

6.19 (a) .1251 (with continuity correction, .1492). (b) .0336 (with continuity correction, .0401). (c) 400. (d) Yes.

6.21 (a) .6676. (b) .0071. (c) $\mu = 19.6$ $\sigma = .6261$.

6.23 $\sigma = .0462$.

6.25 $\mu = 18.6$, $\sigma = .6768$.

6.27 (a) .3409. (b) .0039.

6.29 (a) .0668. (b) .0013.

6.31 (a) $N(55,000, 1591)$. (b) .0222.

6.33 (a) $N(.001, .00000025)$. (b) .0228.

6.35 (a) $N(34, 2.3534)$. (b) $N(37, 2.2454)$. (c) $N(3, 3.2527)$. (d) .6217.

6.37 (a) .004, .002. (b) .0024, .0048, no.

6.39 .0768.

6.41 (a) Yes, by the rules for means. (b) No, by the rules for variances. Variances add only if the two variables are uncorrelated.

6.43 (a) $N(2.2, .1941)$. (b) .1515. (c) .0764.

6.45 1.4623.

6.47 Center line = 75°. Control limits are 75.75 and 74.25.

6.49 **(a)** Center line = 11.5. Control limits are 11.80 and 11.20. **(c)** B.C. at time 11 or 12. A.

6.51 **(a)** Center line = 10. Control limits are 12.08 and 7.92. **(b)** Process is out of control.

6.53 .0058.

6.55 From the table the following would be correct: 3.08, 3.09, and 3.10.

6.57 Center line = 325. Control limits are 463 and 187. All of the June values are low. The family probably takes a vacation at this time.

6.59 Center line = p. Control limits are $p \pm 3\sqrt{p(1-p)/n}$.

6.61 .0704, yes.

6.63 **(a)** 3525, 55. **(b)** .6736.

6.65 **(a)** .3115. **(b)** 60. **(c)** .9951 (with continuity correction, .9966).

6.67 **(a)** $B(500, .464)$. **(b)** .0537 (with continuity correction, .0582).

6.69 .0122 (with continuity correction, .0139).

6.71 Center line = 10. Control limits are 11.47 and 8.53.

6.73 **(a)** $N(32, 1.25)$, $N(29, 1.04)$. **(b)** $N(-3, 1.63)$. **(c)** .0329.

Chapter 7

7.1 **(a)** 1.02. **(b)** (58.00, 62.00).

7.3 (62.63, 57.37), wider, more confidence requires a larger interval.

7.5 **(a)** (2.87, 3.53). **(b)** (3.01, 3.39).

7.7 2.326, 1.282.

7.9 (224.031, 223.973).

7.11 **(a)** 132. **(b)** 11. **(c)** 2.2454, $\sigma_{\bar{y}} = 2.2\sigma_{\bar{x}}$. **(d)** (127.60, 136.40). **(e)** Yes.

7.13 35.

7.15 **(a)** (10.0021, 10.0025). **(b)** 22.

7.17 **(a)** No. **(b)** (27, 33). **(c)** 1.5306. **(d)** No.

7.19 **(a)** .6983. **(b)** .9556.

7.21 **(a)** $\mu = 18$, $\mu < 18$. **(b)** $\mu = 50$, $\mu > 50$. **(c)** $\mu = 24$, $\mu \neq 24$.

7.23 If the average blood pressures in the two groups are equal, the probability of observing a difference as extreme as that actually observed is 0.008. Since the P-value is small, we conclude that the means are different.

7.25 If the two sexes had the same average earnings, then the chance of observing a difference as extreme as that actually observed is .038. For race the corresponding probability is .476. Since the probability is small for sex and large for race, we conclude that there is evidence for a sex difference but no clear evidence for a race difference.

7.27 **(a)** $z = 1.7$, $P = .0446$. **(b)** An SRS selected from a population with a normal distribution. SRS.

7.29 (a) $\mu = 20$, $\mu > 20$. $P = .1151$. The data do not provide sufficient evidence to conclude that the new course has improved the students' ACT scores. (b) Randomly assign students to two groups; one group will take the course and the other will not.

7.31 (a) $\mu = 9.5$, $\mu \neq 9.5$. (b) $z = 2.3479$, $P = .0188$. There is evidence to conclude that the mean differs from 9.5. (c) (9.51, 9.63).

7.33 (a) $z = -2.20$. (b) Yes. (c) No.

7.35 (a) No. (b) No.

7.37 .01, .02.

7.39 .05, .10, $z(.0853) = .1706$.

7.41 (a) Yes. $\mu = 7$, $\mu \neq 7$. (b) No.

7.43 b.

7.45 (a) .3821. (b) .1711. (c) .0013.

7.47 (a) $z = 1.64$, no. (b) $z = 1.65$, yes. (c) No.

7.49 (a) No, you would expect five results to be significant at the .01 level by chance alone. (b) They should be further tested.

7.51 .4641. The test is not sensitive enough to detect an increase of 10 points.

7.53 (a) .5080. (b) .9545.

7.55 (a) .50. (b) .1841. (c) .0013.

7.57 .01, .5359.

7.59 (a) Patient does not need medical attention; patient needs medical attention. Referring a patient who does not need medical attention; clearing a patient who needs medical attention. (b) The second error probability should be made small. Failing to provide medical attention to a patient needing it could have very serious consequences.

7.61 (4.76, 5.98).

7.63 (a) Narrower. If the probability that the interval covers the true value is less, then the interval is smaller. (b) No. (c) Yes. From the interval given we calculate $\sigma_x = 138.587$. Therefore, $z = -1.91$, and we do not reject H_0 at the 5% level.

7.65 (12.285, 13.515). An SRS from a normal distribution.

7.67 $p = 18/38$, $p \neq 18/38$.

7.69 Yes. The chance variation in the statement is calculated under the assumption that H_0 is true.

Chapter 8

8.1 (a) 2.262. (b) 2.861. (c) 1.440.

8.3 (a) 24. (b) .10, .15. (c) .20, .30. (d) No, no.

8.5 18.5.

8.7 (a) Yes. $\mu = 0$, $\mu > 0$. $t = 43.47$. (b) (312.139, 351.861). (c) When the sample size is large, the t procedure can be used for skewed distributions. (d) Randomly assign the customers to two groups. One group will receive the no-fee offer, and the other will not receive the offer.

8.9 (a) 1.75, .065. (b) (1.60, 1.90).

8.11 Yes. $\mu = 1.3$, $\mu > 1.3$. $t = 6.971$. .0025 and .005. The data give evidence that the mean is larger in rats poisoned with DDT.

8.13 (a) (1.54, 1.80). (b) An SRS is selected from a population with a normal distribution. SRS.

8.15 (a) (1.137, 1.521). (b) Yes.

8.17 (a) Standard error of the mean. (b) .0173. (c) (.811, .869).

8.19 $(-21.17, -5.47)$. No. .8866.

8.21 (a) Toss a coin to decide which test is given first for each person. (b) $t = 4.27$. $P = .0003$. The two tests do not have the same means. (c) (.129, .375).

8.23 No. $\mu_A = \mu_B$, $\mu_A > \mu_B$. $P = .133$.

8.25 The results are based on a census and there is no sample.

8.27 (a) .5279. (b) .9032.

8.29 (a) $\eta = 0$, $\eta > 0$, $p = .5$, $p > .5$. (b) .0033. The right-hand thread can be completed faster than the left-hand thread.

8.31 The test cannot be done because the number of people with positive differences is not given.

8.33 (a) (116.18, 167.51). (c) (2.015, 2.129).

8.35 $\mu_1 = \mu_2$, $\mu_1 > \mu_2$. $t = 5.81$. Yes, reject H_0.

8.37 (a) $(-1.274, 7.274)$. (b) We are 95% confident that the true change in sales is covered by our interval. Since the interval includes 0 and negative values, we cannot be certain that the sales have increased.

8.39 (a) $\mu_1 = \mu_2$, $\mu_1 > \mu_2$. $t = 1.65$, df $= 18$, $P = .057$ (from Table E, the P-value is between .05 and .10). The difference between the hemoglobin levels is not significant at the .05 level. (b) $(-.24, 2.04)$. (c) The two means are calculated from independent SRSs from normal populations. The standard deviations are not assumed to be equal.

8.41 (a) $\mu_1 = \mu_2$, $\mu_1 > \mu_2$. (b) $t = 3.15$, df $= 7$, $P = .008$ (from Table E, the P-value is between .01 and .005). The skilled rowers have a higher angular velocity. (c) (.47, 1.87).

8.43 (27.91, 32.09).

8.45 (a) No. (b) $\mu_1 = \mu_2$, $\mu_1 < \mu_2$. $t = 2.46$, df $= 19$, $P = .012$, yes, yes, no. (c) (5.58, 67.72).

8.47 (a) (17.38, 51.18). (b) Yes. The confidence interval does not include 0. (c) The two means are calculated from independent SRSs from normal populations. The standard deviations are not assumed to be equal. No clear deviation from the normal assumption is evident.

8.49 If there are no differences between the children and adults, approximately 5% of the tests would be significant.

8.53 (a) $\mu_1 = \mu_2$, $\mu_1 > \mu_2$. $t = 1.66$, df $= 40$, $P = .052$ (from Table E, the P-value is between .05 and .10). The difference between the hemoglobin levels is not significant at the .05 level. **(b)** $(-.19, 1.99)$. The results are essentially the same.

8.55 (a) $t = 1.09$, df $= 28$, $P = .28$. **(b)** $(-17.00, 56.00)$. **(c)** The results are very similar.

8.57 (a) $\mu_1 = \mu_2$, $\mu_1 \neq \mu_2$. $t = (\bar{x}_1 - \bar{x}_2)/\sqrt{(s_1^2/n_1) + (s_2^2/n_2)}$. **(b)** 1.984, -1.984. **(c)** .5675.

8.59 (a) 3.68. **(b)** No, no.

8.61 10.58. From Table F, the P-value is between .02 and .05. Yes. No.

8.63 (a) $\sigma_1 = \sigma_2$, $\sigma_1 \neq \sigma_2$. **(b)** $F = 2.20$. The P-value is greater than .2.

8.65 (a) $\sigma_1 = \sigma_2$, $\sigma_1 < \sigma_2$, where group 1 is the females. **(b)** $F = 1.54$. **(c)** Since there is no entry for 19 df in the numerator, we use the entries for $F(20, 17)$. $P > .1$. There is no clear evidence that the men are more variable than the women.

8.67 (a) .045 **(b)** .075.

8.69 (a) $\mu = 0$, $\mu < 0$. $t = -2.46$, df $= 47$, $P = .009$. There is evidence that the nutrition program was effective. **(b)** The population is all employees with high blood pressure who will volunteer for attendance at a nutrition education program. **(c)** Since the sample size is large, the t procedure can be used even for clearly skewed distributions.

8.71 $\mu_1 = \mu_2$, $\mu_1 < \mu_2$. $t = -8.95$, df $= 411$, $P < .0005$. **(b)** With large samples the t procedure can be used. **(c)** $(29.81, 44.83)$. **(d)** The distribution for the fifteenth minute is more variable and more symmetric.

8.73 (a) Standard error of the mean. Drivers: 2821, 436, .24, .59. Conductors: 2844, 437, .39, 1.00. **(b)** $t = .3532$, df $= 82$, not significant at the 5% level. **(c)** $t = 1.20$, df $= 82$, $P = .12$. **(d)** $(.21, .57)$. **(e)** $(-.012, .312)$.

8.75 (a) No. With large samples skewness does not invalidate the test. **(b)** Yes. This test is very sensitive to the normality assumption.

8.77 No. This is a census. The mean population can be computed with no error.

8.79 (a) $t = -1.90$, df $= 39$, not significant at the 1% level. **(b)** The two methods should be tested in a comparative experiment.

8.81 (a) $\mu_1 = \mu_2$, $\mu_1 > \mu_2$. **(b)** Paired comparison. **(d)** $t = 2.38$, df $= 25$, $P = .012$, there is evidence that the city particulate level exceeds the rural level. **(e)** $(.619, 3.761)$.

8.83 552.63. For 95% confidence the margin of error is 60.53.

8.85 95% confidence intervals: abdomen $(11.51, 15.49)$, thigh $(9.65, 12.95)$.

Chapter 9

9.1 (a) No. **(b)** Yes. **(c)** No. **(d)** Yes.

9.3 $(.80, .88)$.

9.5 $(.595, .725)$.

9.7 1052.

9.9 $p = .64$, $p \neq .64$. $z = -9.78$, $P < .0002$. The telephone survey technique has a probability of selecting a household with an income of $25,000 or less that is less than the proportion reported in the 1980 census.

9.11 (a) $z = .155$, H_0 is not rejected at $\alpha = .05$, $P = .88$. (b) $(.492, .509)$.

9.13 (a) $z = -1.70$, reject H_0 for $\alpha = .05$, $P = .045$. (b) $(.267, .493)$.

9.15 323.

9.17 404. .082.

9.19 (a) .118, .157, .180, .192, .196, .192, .180, .157, .118. (b) No np is too small.

9.21 (a) .703, .575. (b) .108. (c) $(-.051, .306)$.

9.23 (a) .636. (b) .110. (c) $p_1 = p_2$, $p_1 > p_2$. (d) $z = 1.164$, $P = .122$. There is not sufficient evidence to conclude that it is easier to win at home.

9.25 (a) .297, .425. (b) .108. (c) $(-.306, .051)$. The results give the same information in a different form. The standard error and the width of the confidence interval are the same.

9.27 Let p be the proportion of losses. (a) .364. (b) .110. (c) $p_1 = p_2$, $p_1 < p_2$. (d) $z = -1.164$, $P = .122$. There is not sufficient evidence to conclude that the chance of losing is less at home. The results and conclusions are essentially the same. The size of z has changed.

9.29 $\hat{p}_1 = .603$, $\hat{p}_2 = .591$, $z = .265$, $P = .79$. There is no evidence that the Protestants and Catholics differ on this issue.

9.31 (a) $p_1 = p_2$, $p_1 \neq p_2$. $\hat{p}_1 = .901$, $\hat{p}_2 = .810$, $z = 5.07$, $P < .0004$. (b) $(.045, .137)$.

9.33 (a) .808, .558. (b) $(.108, .391)$. (c) $p_1 = p_2$, $p_1 > p_2$. (d) $z = 3.34$, $P < .0004$. Use of aspirin increases the probability of a favorable outcome for patients with cerebral ischemia.

9.35 (a) $z = 1.41$, $P = .080$. There is not sufficient evidence to conclude that aspirin is effective. (b) In Exercise 9.21 H_0 was rejected. With the smaller sample sizes in this exercise it could not be rejected even though the sample proportions are similar. Larger sample sizes give more power.

9.37 (a) 71.74, 28.26, 50.00, 50.00. For the years in which the market is up in January, 71.74% of the time the market is up during the rest of the year, etc. (b) 71.74, 28.26, 50.00, 50.00. For the years in which the market is up from February to December, 71.74% of the time the market is up in January, etc. (c) There is no relation between the behavior of the market in January and the behavior during the rest of the year. The January behavior predicts the rest of the year. (d) Reading down columns: 29.4, 16.6, 16.6, 9.4. The expected counts exceed sample counts in the 1, 2 and 2, 1 cells; they are smaller than the sample counts in the other cells. (e) $X^2 = 3.403$, df = 1, $P = .065$. Using $\alpha = .05$, we do not have evidence for rejecting H_0. If we had used a one-sided test to compare two proportions, we would obtain a P-value one-half as large, and H_0 would be rejected. (f) The January performance of the market can be used to predict the performance during the rest of the year.

9.39 (a) 45.95, 44.59, 9.46; 45.83, 45.83, 8.33; 45.90, 45.08, 9.02. The column percentages are very similar. (b) The probability of selecting an egg operation with a mild rodent problem is p_{11}. The probability of selecting a turkey operation with a severe rodent problem is p_{32}. (c) $p_{ij} = r_i c_j$ for all i and j; $p_{ij} \neq r_i c_j$ for some i and j. (d) Reading down columns: 34.0, 33.4, 6.7; 22.0, 21.6, 4.3. (e) $X^2 = .051$, df = 2, $P = .975$. There is no evidence to conclude that the severity of the rodent problem is related to the type of operation.

9.41 **(a)** 327, 443; 214, 226, 330. **(b)** Urban: 30.37, 69.63. Intermediate: 39.82, 60.18. Rural: 52.12, 47.88. Combined sample: 42.47, 57.53. **(c)** $H_0: p_{ij} = r_i c_j$ for all i and j. $H_a: p_{ij} \neq r_i c_j$ for some i and j. **(d)** Reading down columns: 90.9, 123.1; 96.0, 130.0; 140.1, 189.9. **(e)** $X^2 = 26.04$, df = 2, $P < .001$. **(f)** There is evidence to conclude that the county of the practice is related to whether tetracycline is prescribed.

9.43 **(a)** 1202, 1617; 366, 498, 442, 602, 911. **(b)** 1961: 52.46, 47.54. 1966: 48.19, 51.81. 1971: 41.18, 58.82. 1976: 43.69, 56.31. 1981: 35.68, 64.32. Combined: 42.64, 57.36. **(c)** $H_0: p_{1(1)} = p_{1(2)} = \cdots = p_{1(5)}, p_{2(1)} = p_{2(2)} = \cdots = p_{2(5)}$. H_a: at least one of the equalities in H_0 does not hold. **(d)** Reading down columns: 156.1, 209.2; 212.3, 285.7; 188.5, 253.5; 256.7, 345.3; 388.4, 522.6. **(e)** $X^2 = 39.43$, df = 4, $P < .001$. There is evidence to conclude that the proportion of alumni contributing depends upon class year. In 1961 and 1966, a larger proportion contributed than would be expected, whereas in 1981 a smaller proportion contributed.

9.45 **(a)** 34.54, 19.08, 27.63, 18.75; 10.77, 7.69, 56.92, 24.62; 13.25, 15.66, 50.50, 20.48. For women with higher nicotine consumption there is a tendency for greater alcohol consumption. **(b)** 85.37, 5.69, 8.94; 76.32, 6.58, 17.11; 51.53, 22.70, 25.77; 63.33, 17.78, 18.89. For women with a higher alcohol consumption there is a higher nicotine consumption. **(c)** $H_0: p_{ij} = r_i c_j$, H_a: $p_{ij} \neq r_i c_j$ for some i and j. **(d)** Reading down columns: 82.7, 51.1, 109.6, 60.5; 17.7, 10.9, 23.4, 12.9; 22.6, 14.0, 29.9, 16.5. **(e)** $X^2 = 42.25$, df = 6, $P < .001$. **(f)** There is evidence to conclude that there is a relationship between alcohol consumption and nicotine consumption.

9.47 **(a)** .2515. **(b)** .5. **(c)** $p = .2515$, $p \neq .2515$. **(d)** $z = 16.71$, $P < .0004$, H_0 is rejected. More women were among the top 30 students than would be expected.

9.49 47% of the blacks are vegetarians, and 61% of the whites are vegetarians. Testing H_0 that the proportions are the same versus the two-sided alternative gives $z = -2.15$ and $P = .0317$. There is evidence to conclude that the proportion of vegetarians among blacks is higher for the people attending this meeting. Inferences to Seventh Day Adventists in general and blacks and whites in general are beyond the realm of this particular study.

9.51 **(a)** $p_1 = p_2$, $p_1 < p_2$. $z = -9.81$, $P < .0002$. The program has increased the proportion of mothers who feed children suffering from diarrhea. **(b)** $(-.297, -.202)$.

9.53 For men, the column percentages are 53.21, 16.86, 29.94. For women, 42.79, 14.41, 42.79. The column proportions for men and women are the same versus the alternative that they are not. $H_0: p_{i(m)} = p_{i(f)}$ for $i = 1, 2, 3$. H_a: at least one of the equalities in H_0 is not true. $X^2 = 13.40$, df = 2, $P < .0025$. The distributions of the male and female students are not the same. The females appear to have a higher dropout rate. The results may be very distorted because area of study of the students is not taken into account.

9.55 The column percentages are: Hawaiians: 40.75, 53.32, 3.81, 2.12. Hawaiian-white: 44.77, 46.79, 6.07, 2.36. Hawaiian-Chinese: 40.97, 43.97, 10.55, 4.51. White: 43.00, 44.00, 13.00, 4.00. $X^2 = 1078.60$, df = 9, $P < .0005$. There is evidence to conclude that the proportions of the four blood types are not the same in all four ethnic groups.

9.57 (a) $p_1 = p_2$, $p_1 \neq p_2$. $z = -.567$, $P = .57$. **(b)** 28, 30; 54, 48. $p_{1(1)} = p_{1(2)}$ and $p_{2(1)} = p_{2(2)}$. The alternative is that one of the equalities in the null hypothesis is false. $X^2 = .322$, $P = .57$. $z^2 = (-.567)^2 = .321$. The difference is due to round-off. **(c)** Gastric freezing is not an effective treatment for ulcers.

9.59 $\hat{p} = .5$. $(n, z, P) = (15, 1.10, .27), (25, 1.41, .16), (50, 2.00, .05), (75, 2.45, .014), (100, 2.83, .005), (500, 6.32, .0000)$.

9.61 $(n, \text{width}) = (10, 1.15), (30, .67), (50, .52), (100, .36), (200, .26), (500, .16)$.

9.63 It is not possible to find the value of n_2 to guarantee the desired result. Even if n_2 is so large that the term $.25/n_2$ can be neglected, then $s(D) = \sqrt{.0125}$ and $w = .3678$ which is greater than .2.

9.67 The small sample sizes will probably not allow you to see any meaningful effect.

Chapter 10

10.1 (a) (.252, .286). **(b)** The width of this interval is smaller.

10.5 (a) (6.69, 791). **(b)** (15.01, 19.81). **(c)** The interval for 80-degree days is wider because 80 is farther from $\bar{x}$ than 30. **(d)** Yes. 80-degree days constitute a value much higher than any value in the data set.

10.7 (a) (5.66, 8.94). **(b)** (14.57, 20.25). **(c)** The prediction intervals are wider than the confidence intervals.

10.9 (a) 13.07, 443.20. **(b)** $t = 10.81$, df = 8, $P < .0005$. **(c)** (.622, 1.182). **(d)** 11.85, (10.56, 13.15). **(e)** (11.12, 18.00).

10.11 (a) No. **(b)** 1.184. **(c)** (.937, 1.432). **(d)** $t = -.568$, df = 3, $P = .61$. The null hypothesis that the intercept is 0 is not rejected. **(e)** .844, (.698, 1.068).

10.13 .875, (.8218, .9347).

10.15 (a) No. **(b)** $\hat{y} = -2.804 + .039x$. **(c)** $t = 16.10$, df = 17, $P < .0005$, for the one-sided alternative. **(d)** (.600, 1.135), (1.186, 1.709). **(e)** Yes. The line fits the data very well and can be used in subsequent experiments.

10.17 The first two observations have large negative residuals. In this experiment, the apparatus was connected and no measurements were taken during a warmup period. The data suggest that these first two observations should be considered part of the warmup period. The regression should be rerun without these two points.

10.19 (a) $t = 12.77$, $\beta_1 > 0$, df = 79, $P < .0005$. **(b)** (.658, 1.002). **(c)** Although the significance test clearly rejects H_0, the clinicians' judgement that the prediction intervals are too large implies that the easy method is not useful.

10.21 (a)

Source	df	SS	MS	F	P
Model	1	2.093	2.093	231.10	.0006
Error	3	.027	.009		
Total	4	2.120			

(b) $\beta_1 = 0$. **(c)** $F(1,3)$, $P < .001$.

10.23 (a)

Source	df	SS	MS	F	P
Model	1	3.762	3.762	259.27	.0001
Error	17	.247	.015		
Total	18	4.009			

(b) $\beta_1 = 0$. **(c)** $F(1,17)$, $P < .001$. **(d)** $16.102^2 = 259.27$. **(e)** .94.

10.25 (a) 2.61. **(b)** $\rho > 0$. **(c)** .006.

10.27 (a) $\mu_{GPA(9,8,7)} = \beta_0 + 9\beta_1 + 8\beta_2 + 7\beta_3$. **(b)** 4.697.

10.29 (a) (.099, .239). **(b)** (−.031, .121).

10.31 (a) $\hat{y} = 2.590 + .169x_1 + .034x_2 + .045x_3$. **(b)** .6998. **(c)** $\beta_1 = \beta_2 = \beta_3 = 0$, $\beta_j \neq 0$ for at least one $j = 1, 2, 3$. **(d)** $F(3,220)$. **(e)** 30.46%.

10.33 For SAT-M and GPA a hint of a positive relationship can be seen. This corresponds to the statistically significant ($P = .0001$) correlation of .251 reported in Figure 10.20. No clear relationship is evident between SAT-V and GPA. There are no outliers or unusual points.

10.35 No unusual points or patterns are evident in the residual plots.

10.39 (a) $\hat{y} = 2.666 + .193$HSM $+ .0006$ATM. **(b)** $\beta_1 = \beta_2 = 0$, $\beta_1 \neq 0$ for at least one $j = 1, 2$. $F = 26.63$. $P < .0001$. Yes. **(c)** (.1295, .2565), (−.0006, .0018). The second interval includes 0. **(d)** $t = 5.99$, $P < .001$. $t = 1.00$, $P = .319$. For this model we reject H_0: $\beta_1 = 0$ and we do not reject H_0: $\beta_2 = 0$. Given that HSM is in the model, SAT-M does add statistically significant information for predicting GPA. **(e)** .7028. **(f)** 19.42%.

10.41 (a) There appears to be a strong positive linear relationship between the two measurements. There are no outliers or unusual points. **(b)** $\widehat{T2} = .0048 + .9920T1, .0265$. **(c)** .988, .977. **(d)** $t = 45.22$. $\beta_1 > 0$, $P < .0001$. **(e)** $45.22^2 = 2044.85$, $F = 2044.66$. Difference is due to round-off.

10.43 (a) $\hat{y} = 109.87 − 1.127x$. **(b)** $t = −3.63$, $t(19)$, $P = .0009$. **(c)** (−1.776, −.478). **(d)** 41.00%. **(e)** 11.023.

10.45 Plots of GPA versus each predictor do not indicate any very strong relationships. The F statistic for the model is 10.63 with 3 and 141 degrees of freedom and $P < .0001$. The proportion of variation explained is .1844. The model with the three high school variables as predictors is statistically significant. For the tests on the individual regression coefficients, only HSM is significant ($t = 3.46$, df = 141, $P = .0007$). No unusual patterns or observations are evident in the residual plots. The model should be rerun deleting either HSS or HSE.

10.47 (a) $\hat{y} = −9.102 + 1.086x$. **(b)** .6529. **(c)** $t = 1.66$, $P = .14$. All of the results are quite different. The intercept is no negative. The value of t has changed from 17.64 (see Example 10.5) to 1.66 and is no longer statistically significant. The incorrect observation has caused the results of the regression analysis to be meaningless.

10.49 The relationships are weak and positive. The correlations and their *P*-values are: taste with acetic .55, .0017; taste with H2S .76, .0001; taste with lactic .70, .0001; acetic with H2S .62, .0003; acetic with lactic .60, .0004; H2S with lactic .64, .0001.

10.51 $\hat{y} = -9.77 + 5.78x$, $F = 37.29$, $P = .0001$, $R^2 = .57$. The residuals are smaller for low values of H2S. The residual plots versus acetic and lactic show no clear patterns. There is some tendency for smaller residuals to be associated with smaller values of these variables. The normal quantile plot of the residuals indicates that they are approximately normal with a small amount of skewness to the right.

10.53 F, P, R^2, $\sqrt{MSE}$, b_0, b_1: Acetic 12.114, .0017, .30, 13.82, -61.50, 15.65. H2S 37.293, .0001, .57, 10.83, -9.79, 5.78. Lactic 27.55, .0001, .50, 11.75, -29.86, 37.72. The intercepts are not the same because the models are different.

10.55 For the multiple regression $\hat{y} = -27.59 + 3.95H2S + 19.89Lactic$. The coefficient for H2S is significant ($t = 3.475$, $P = .0018$). The same is true for the coefficient for Lactic ($t = 2.499$, $P = .0188$). For this model $F = 25.26$, $P = .0001$, $R^2 = .652$. The R^2s for the simple regressions are .5712 for H2S and .4759 for Lactic. We prefer the multiple regression because both coefficients are significant and the R^2 is considerably higher.

Chapter 11

11.1 (a) Yield of tomatoes in pounds. Variety with four levels. (b) Attractiveness rating. Type of packaging with six levels. (c) Weight loss. Type of weight loss program with three levels.

11.3 Yes. 69.42, 8.33.

11.5 (a) Source, df: Variety, 3; Error, 36; Total, 39. (b) Source, df: Type, 5; Error, 714; Total, 719. (c) Source, df: Type, 2; Error, 27; Total, 29.

11.7 (a) $\mu_1 = \mu_2 = \mu_3$. Not all of the μ_i are equal. (b) Source, df: Major, 2; Error, 253; Total, 255. (c) $F(2, 253)$. (d) 3.04 for $F(2, 200)$; 3.00 for $F(2, 1000)$. From computer, 3.03 for $F(2, 253)$.

11.9 (a) SST $= 175,356.46$; DFT $= 35$; MSG $= 34,951.96$; MSE $= 2203.14$; $F = 15.86$. (b) $\mu_1 = \mu_2 = \mu_3 = \mu_4$. Not all of the μ_i are equal. (c) $F(3, 32)$. $P < .001$. The mean fitness scores for the four groups are not all the same. (d) 2203.14, 46.94.

11.11 (a) 3.90, MSE. (b) Source, SS, df, MS: Country, 17.22, 2, 8.61; Error, 802.89, 206, 3.90; Total, 820.11, 208. $F = 2.21$. (c) $\mu_1 = \mu_2 = \mu_3$. Not all of the μ_i are equal. (d) $F(2, 206)$. $P > .1$. From computer, $P = .112$. The data do not provide evidence to conclude that the mean birth weights are different in the three countries. (e) .0210.

11.13 (a) $\psi_1 = (1/2)(\mu_1 + \mu_2) - \mu_3$. (b) $\psi_2 = \mu_1 - \mu_2$.

11.15 (a) H_{01}: $(1/2)(\mu_1 + \mu_2) = \mu_3$, H_{a1}: $(1/2)(\mu_1 + \mu_2) > \mu_3$. H_{02}: $\mu_1 = \mu_2$, H_{a2}: $\mu_1 \neq \mu_2$. (b) 49, -10. (c) 11.28, 16.90. (d) $t_1 = 4.34$, df $= 253$, $P < .0005$. $t_2 = -.59$, df $= 253$, $P > .50$. From computer, $P = .555$. The average of the means for the first two groups is greater than the mean for the third group. There is no evidence to conclude that the first two groups have different means. (e) (26.78, 71.22), (-43.29, 23.29).

11.17 **(a)** $\psi_1 = \mu_1 - \mu_2$, $H_{01}: \mu_1 = \mu_2$, $H_{a1}: \mu_1 > \mu_2$. $\psi_2 = \mu_1 - (1/2)(\mu_2 + \mu_4)$, $H_{02}: \mu_1 = (1/2)(\mu_2 + \mu_4)$, $H_{a2}: \mu_1 > (1/2)(\mu_2 + \mu_4)$. $\psi_3 = \mu_3 - (1/3)$ $(\mu_1 + \mu_2 + \mu_4)$, $H_{03}: \mu_3 = (1/3)(\mu_1 + \mu_2 + \mu_4)$, $H_{a3}: \mu_3 > (1/3)(\mu_1 + \mu_2 + \mu_4)$. **(b)** $C_1 = -17.06$, $s(C_1) = 25.71$. $C_2 = 24.39$, $s(C_2) = 19.64$. $C_3 = 91.22$, $s(C_3) = 17.27$. **(c)** $t_1 = -.66$, df $= 32$, $P > .25$. From computer, $P = .256$. $t_2 = 1.24$, df $= 32$, $P < .15$. From computer, $P = .112$. $t_3 = 5.28$, df $= 32$, $P < .0005$. **(d)** There is no evidence to conclude that the treatment and control groups have different means. There is no evidence to conclude that the mean of the treatment group is greater than the average of the means of the control and sedentary groups. The mean for the joggers is greater than the average of the means for the other three groups. **(e)** No conclusion about causation can be drawn.

11.19 The Egypt and Kenya means are not distinguishable ($t = 1.73$). The Egypt and Mexico means are not distinguishable ($t = 2.00$). The Kenya and Mexico means are not distinguishable ($t = .60$).

11.21 $t_{12} = -.66$, not distinguishable. $t_{13} = -3.65$, different. $t_{14} = 3.14$, different. $t_{23} = -2.29$, not distinguishable. $t_{24} = 3.22$, different. $t_{34} = 6.86$, different.

11.23

n	DFG	DFE	F^*	λ	Power
25	2	72	3.12	2.36	.25
50	2	147	3.06	4.73	.47
75	2	222	3.04	7.09	.66
100	2	297	3.03	9.45	.79
125	2	372	3.02	11.81	.88
150	2	447	3.02	14.18	.93
175	2	522	3.01	16.54	.96
200	2	597	3.01	18.90	.98

A sample size of 175 appears to be adequate.

11.25 **(a)** Mean, standard deviation: 10.65, 2.05; 10.43, 1.49; 5.60, 1.24; 5.45, 1.77. **(b)** $\mu_1 = \mu_2 = \mu_3 = \mu_4$. Not all of the μ_i are equal. **(c)** $F = 12.08$, $P = .0006$. There is evidence to conclude that not all of the means are equal. $s_p = 1.67$, $R^2 = .75$.

11.27 **(a)** $\psi = \mu_1 - (1/3)(\mu_2 + \mu_3 + \mu_4)$. **(b)** $H_0: \mu_1 = (1/3)(\mu_2 + \mu_3 + \mu_4)$. $H_a: \mu_1 > (1/3)(\mu_2 + \mu_3 + \mu_4)$. **(c)** $t = 3.63$, df $= 12$, $P < .0025$. From computer, $P = .002$. Yes. **(d)** $\psi = \mu_1 - \mu_4$. (2.63, 7.77).

11.29 **(a)** Yield of tomatoes in pounds. Variety with four levels, and fertilizer with two levels. **(b)** Attractiveness rating. Type of packaging with six levels, and parts of country with four levels. **(c)** Weight loss. Type of weight loss program with three levels, and sex with two levels.

11.31 **(a)** Source, df: Variety, 3; Fertilizer, 1; Variety x Fertilizer, 3; Error, 32; Total, 39. **(b)** Source, df: Packaging, 5; Part, 3; Packaging x Part, 15; Error, 696, Total, 719. **(c)** Source, df: Program, 2; Sex, 1; Program x Sex, 2; Error, 24, Total, 29.

11.33 (b) Blood pressure for nonwhites is higher than for whites in all age groups. Yes, blood pressure increases with age. There does not appear to be an interaction. (c) Race: 135.98, 137.82. Age group: 131.65, 133.25, 136.20, 140.35, 143.05. Differences: 1.3, 1.9, 2.0, 1.9, 2.1.

11.35 (b) The mean amount of GITH for the low chromium diet is about the same as the amount for the normal chromium diet. The GITH for the R diet is greater than for the M diet. Yes, there is an increase in GITH for the R diet as the level of chromium is increased; but for the M diet there is a decrease. (c) Diet: 4.48, 5.25. Chromium: 4.86, 4.87. Differences: .63, .89. There is a larger difference for the normal chromium level.

11.37 (b) Yes, the average score for the males is higher than for the females. Yes, the scores in engineering are higher than the scores in computer science. The other group has the lowest scores. Yes. (c) Sex: 611.7, 585.3. Majors: 605.0, 624.5, 566.0. Differences: 46, -13, 46.

11.39 (a) SST = 7.04487, DFB = 1, DFAB = 1, DFE = 36, DFT = 39, MSA = .00121, MSB = 5.79121, MSAB = .17161, MSE = .03002, $F_A = .04$, $F_B = 192.89$, $F_{AB} = 5.72$. (b) 5.72, $F(1, 36)$, $P < .025$. (c) .04, $F(1, 36)$, $P > .10$; 192.89, $F(1, 36)$, $P < .001$. (d) .03, .17. (e) The interpretation given in Exercise 11.35 is supported by the significance tests.

11.41 (a) n, $\bar{x}$, s: 22, 5.27, 2.76; 22, 5.09, 2.00; 22, 4.95, 1.86. (b) Yes. (c) 1.48, yes. (d) $\mu_1 = \mu_2 = \mu_3$, not all of the μ_i are equal. $F = .11$, $P = .8948$. H_0 is not rejected. (e) There is no evidence to conclude that the three means are different.

11.43 (a) n, $\bar{x}$, s: 22, 6.68, 2.77; 22, 9.77, 2.72; 22, 7.77, 3.93. (b) Yes. (c) 1.44, yes. (d) $F = 5.32$, $P = .0073$. H_0 is rejected. (e) $\psi = -\mu_1 + (1/2)(\mu_2 + \mu_3)$. $H_0: \mu_1 = (1/2)(\mu_2 + \mu_3)$, $H_a: \mu_1 < (1/2)(\mu_2 + \mu_3)$. $C = 2.09$, $s(C) = .833$, $t = 2.51$, df = 63, $P < .01$. From computer, $P = .007$. (.43, 3.75). (f) $\psi = \mu_2 - \mu_3$. $H_0: \mu_2 = \mu_3$, $H_a: \mu_2 \neq \mu_3$. $C = 2.00$, $s(C) = .96$, $t = 2.08$, df = 63, $P < .05$. From computer, $P = .0415$. (.079, 3.921). (g) Not all of the means are equal. The Basal group mean is smaller than the average of the DRTA and Strat group means, and the DRTA group mean is larger than the Strat group mean.

11.45 (a) $F = 1.33$, $P = .31$. H_0 is not rejected. (b) $F = 12.08$, $P = .0006$, from Exercise 11.25. The outlier has caused a statistically significant result to disappear. (c) $\bar{x}$, s: 34.95, 48.74; 10.43, 1.49; 5.60, 1.24; 5.45, 1.77. The large standard deviation and mean for the first group relative to the other groups should lead us to check the data in this group.

11.47 (a) $\bar{x}$, s: 1.02, .08; 1.01, .07; .74, .09; .72, .15. (c) $F = 10.39$, $P = .0012$. The number of nematodes affects the growth of the seedlings. $s_p = .10$, $R^2 = .72$. The results are qualitatively the same.

11.49 (a) n, power: 20, .42; 40, .74; 60, .91; 80, .97; 100, .99. (b) The power increases rapidly from 20 to 60 and levels off afterward. (c) A sample size between 80 and 100 will give good power.

11.51 (a) $\bar{x}$, s: Aspen 387, 69; 336, 60. Birch 293, 122; 455, 118. Maple 323, 52; 294, 184. Species 362, 374, 308. Size 334, 362. (b) Yes, yes. The interaction is so strong that statements about the marginal means are meaningless. (c) No statistically significant effects are present. The impressions obtained from the plot are not substantiated by the analysis and are probably due to chance variation. Note that there is a large range of standard deviations and the sample sizes are small.

11.53 n, $\bar{x}$, s: Males 39, 526.95, 100.94; 39, 507.85, 57.21; 39, 487.56, 108.78. Females 39, 543.38, 77.65; 39, 538.21, 102.21; 39, 465.03, 82.18. The interaction is not significant ($F = 1.81$, $P = .17$). The sex effect is not significant ($F = .47$, $P = .49$). The majors differ significantly ($F = 9.32$, $P = .0001$).

11.55 n, $\bar{x}$, s: Males 39, 7.79, 1.51; 39, 7.49, 2.15; 39, 7.41, 1.57. Females 39, 8.85, 1.14; 39, 9.26, .75; 39, 8.62, 1.16. The high school grade variables are clearly not normal. We rely upon the robustness of the ANOVA procedures for nonnormal data for this analysis. The interaction is not significant ($F = 1.33$, $P = .27$). There is a significant sex difference ($F = 50.32$, $P = .0001$). The majors are not significantly different ($F = 1.33$, $P = .27$).

Index